T0134757

Lecture Notes in Computer Science 762

Edited by G. Goos and J. Hartmanis

Advisory Board: W. Brauer D. Gries J. Stoer

K. W. Ng   P. Raghavan
N. V. Balasubramanian   T. Y. L. Chin (eds.)

Algorithms and Computation

4th International Symposium, ISAAC '93
Hong Kong, December 15-17, 1993
Proceedings

Springer-Verlag
Berlin Heidelberg New York
London Paris Tokyo
Hong Kong Barcelona
Budapest

K. W. Ng   P. Raghavan
N. V. Balasubramanian   F. Y. L. Chin (Eds.)

Algorithms and Computation

4th International Symposium, ISAAC '93
Hong Kong, December 15-17, 1993
Proceedings

Springer-Verlag

Berlin Heidelberg New York
London Paris Tokyo
Hong Kong Barcelona
Budapest

Series Editors

Gerhard Goos
Universität Karlsruhe
Postfach 69 80
Vincenz-Priessnitz-Straße 1
D-76131 Karlsruhe, Germany

Juris Hartmanis
Cornell University
Department of Computer Science
4130 Upson Hall
Ithaca, NY 14853, USA

Volume Editors

K. W. Ng
Department of Computer Science, The Chinese University of Hong Kong
Hong Kong

P. Raghavan
IBM T. J. Watson Research Center
Box 218, Yorktown Heights, NY 10598, USA

N. V. Balasubramanian
Department of Computer Science, City Polytechnic of Hong Kong
Hong Kong

F. Y. L. Chin
Department of Computer Science, The University of Hong Kong
Hong Kong

CR Subject Classification (1991): F.1-2, G.2-3, H.3, I.3.5

ISBN 3-540-57568-5 Springer-Verlag Berlin Heidelberg New York
ISBN 0-387-57568-5 Springer-Verlag New York Berlin Heidelberg

Typesetting: Camera-ready by author
Printing and binding: Druckhaus Beltz, Hemsbach/Bergstr.
45/3140-543210 - Printed on acid-free paper

Preface

The first International Symposium on Algorithms was held in Tokyo in 1990 and the second one in Taipei in 1991. The title was changed to International Symposium on Algorithms and Computation in 1992 when it was held in Nagoya. The papers in this volume were presented at the fourth symposium in Hong Kong during December 15-17, 1993. The job of the Program Committee was quite challenging in that it had to sieve nearly 200 papers! The international nature of the symposium is apparent not only from the composition of the Program Committee but also in the number of countries represented in the papers received (some 26). We are grateful to all those who submitted papers for consideration and to the Program Committee members who helped in the evaluation of these many papers. Three eminent invited speakers, Prof. Allan Borodin of the University of Toronto, Prof. Tom Leighton of MIT, and Dr. Mihalis Yannakakis of AT&T, have contributed to these proceedings. We are grateful to them. We hope that the international research community will benefit significantly from these proceedings and we are grateful to Springer-Verlag for undertaking to publish them. We are grateful to several sponsors who made this symposium financially viable, and to the IEEE Computer Society for cooperation.

We, the Organizing Committee, the Program Committee Co-chairs, and the Symposium Co-chairs, have worked hard to provide for a quality symposium and quality proceedings. The selected papers were chosen on the basis of originality and relevance to the field of algorithms and computation. The selections were based on extended abstracts, which represent a preliminary report of ongoing research. It is anticipated that many of these reports will appear in more polished and complete form in recognized scientific journals.

N V Balasubramanian   Francis Y.L. Chin
Kam Wing Ng   Prabhakar Raghavan

December 1993

Symposium Chairs:
N.V. Balasubramanian (City Polytechnic of Hong Kong, HK)
Francis Y.L. Chin (The University of Hong Kong, HK)

Program Committee Chairs:
Kam Wing Ng (Chinese University of HK, HK)
Prabhakar Raghavan (IBM, USA)

Organising Committee Members:
Siu Wing Cheng (Hong Kong University of Science and Technology, HK)
Horace Ip (City Polytechnic of Hong Kong, HK)
Tak Wah Lam (The University of Hong Kong, HK)
Karl Leung (Hong Kong Polytechnic, HK)
K.F. Suen (Hong Kong Baptist College, HK)

Advisory Committee Members:
Francis Y.L. Chin (The University of Hong Kong, HK)
W.L. Hsu (Academia Sinica, R.O.C.)
T. Ibaraki (Kyoto University, Japan)
D.T. Lee (Northwestern University, USA)
R.C.T. Lee (Tsing Hua University, R.O.C.)
T. Nishizeki (Chair, Tohoku University, Japan)

Program Committee Members:
Pranay Chaudhury, UNSW, Australia
Ding-Zhu Du, U. of Minnesota, USA
Wen-Lian Hsu, Academia Sinica, Taiwan
Hiroshi Imai, U. of Tokyo, Japan
Maria Klawe, U. of British Columbia, Canada
Rao Kosaraju, Johns Hopkins University, USA
R.C.T. Lee, Tsing Hua University, Taiwan
J. van Leeuwen, U. of Utrecht, The Netherlands
Christos Papadimitriou, UC San Diego, USA
Martin Tompa, U. of Washington, USA
Eli Upfal, IBM, USA
Osamu Watanabe, TITech, Japan

In Cooperation with
IEEE Computer Society

Acknowledgements

The ISAAC '93 Program Committee thanks the following people for their help in evaluating the submissions:

Pankaj Agarwal, Alok Aggarwal, Paul Callahan, S.W. Cheng, Ron Fagin, Mike Goodrich, Joe Halpern, Susumu Hayashi, Keiko Imai, Russell Impagliazzo, Amos Israeli, Anna Karlin, Hoong-Chuin Lau, Rajeev Motwani, Ketan Mulmuley, Nick Pippenger, Balaji Raghavachari, V.T. Rajan, Baruch Schieber, Madhu Sudan, H.F. Ting, Seinosuke Toda, Takeshi Tokuyama and Gilbert Young.

The Organizing Committee thanks:

(i) Miss Monica Lau, Secretary to Dr N V Balasubramanian, for all the assistance rendered to him in his correspondence with the donors, Hong Kong Tourist Association, IEEE Computer Society, etc. Monica was also secretary to the Organizing Committee and handled its correspondence, scheduling of meetings, taking minutes, follow-up, sending and receiving faxes, answering telephone calls etc. Her patience, promptness and courtesy has made the organizing work effective and pleasant.

(ii) Miss Maria Lam, Secretary to Professor Francis Chin, for typing the list of accepted papers, helping with the accommodation of PC members, helping with the camera-ready papers, registration, etc.

(iii) Ms Nancy Donovan of the IBM, T J Watson Research Centre, Yorktown Heights, New York for her help in receiving, sorting and distributing submissions to the conference.

(iv) Miss Doris Lam of the Hong Kong Tourist Association for all the assistance, information and the colourful stationery provided by her with keen interest and enthusiasm.

ISAAC '93 was supported by:

The Croucher Foundation
Epson Foundation
Hong Kong Computer Society
Lee Hysan Foundation
IEEE Society HK Section Computer Chapter
The Hong Kong Chapter of ACM
SIGAL, IPSJ (Japan)
TGCOMP, IEICE (Japan)
The University of Hong Kong
The Chinese University of Hong Kong
Hong Kong University of Science & Technology
Hong Kong Polytechnic
City Polytechnic of Hong Kong
Hong Kong Baptist College

Best Paper Award

The Best Paper Award this year is awarded in recognition of the most outstanding paper written solely by a student, as judged by the program committee, and is presented for the paper, "Multicommodity Flows in Even, Planar Networks," by Karsten Weihe.

Contents

SESSION 6B

SESSION 7   Invited Paper

SESSION 8A

SESSION 8B

SESSION 9   Invited Paper

SESSION 10A

SESSION 12B

Reaching a Goal with Directional Uncertainty*

Mark de Berg[1]   Leonidas Guibas[2]   Dan Halperin[3]

Mark Overmars[1] Otfried Schwarzkopf[1] Micha Sharir[4]

Monique Teillaud[5]

Abstract. We study two problems related to planar motion planning for robots with imperfect control, where, if the robot starts a linear movement in a certain commanded direction, we only know that its actual movement will be confined in a cone of angle α centered around the specified direction.

1 Introduction and Statement of Result

In this paper we look at regions of the plane defined as the locus of all points for which a cone of angle α can be placed at the point so that certain regions of the plane (the obstacles) are completely avoided, while other regions (the goal regions) are intersected by all rays in the cone. As we explain below, such "visibility" questions arise primarily in robotics, but also in computer graphics and other areas. We give combinatorial bounds and algorithms for the computation of such regions in two special cases.

Many motion planning algorithms in the literature assume that we know the precise geometry of the workspace, and that the robot has precise control over its movements. In practice, however, this will rarely be the case. In most cases, our knowledge of the workspace will be incomplete or erroneous, and the robot can only control its movement imperfectly. As the robot executes a prepared plan to move around the workspace, it will have to deal with uncertainty in the execution of its commanded motions. In many cases it may need to recalibrate its position by sensing the environment or taking equivalent steps.

---

*This research was supported by the Netherlands' Organization for Scientific Research (NWO) and partially by ESPRIT Basic Research Actions No. 6546 (project *PROMotion*) and No. 7141 (project ALCOM II: *Algorithms and Complexity*). Part of the research was done during the Second Utrecht Workshop on Computational Geometry and its Application, supported by NWO. L.G. acknowledges support by NSF grant CCR-9215219, by a grant from the Stanford SIMA Consortium, and by grants from the Digital Equipment, Mitsubishi, and Toshiba Corporations. D.H. was supported by a Rothschild Postdoctoral Fellowship, by a grant from the Stanford Integrated Manufacturing Association (SIMA), and by NSF/ARPA Research Grant IRI-9306544. Work on this paper by M.S. has been supported by National Science Foundation Grant CCR-91-22103, and by grants from the U.S.-Israeli Binational Science Foundation, the Fund for Basic Research administered by the Israeli Academy of Sciences, and the G.I.F., the German-Israeli Foundation for Scientific Research and Development.

[1]Vakgroep Informatica, Universiteit Utrecht, Postbus 80.089, 3508 TB Utrecht, the Netherlands
[2]Dept. of Computer Science, Stanford University, and DEC Systems Research Center, Palo Alto
[3]Robotics Laboratory, Dept. of Computer Science, Stanford University, Stanford, CA 94305
[4]School of Mathematical Sciences, Tel Aviv University, and Courant Institute of Mathematical Sciences, New York University
[5]INRIA, B.P. 93, 06902 Sophia-Antipolis Cedex, France

A motivation for this paper is to understand the effect of uncertainty within a single commanded motion. We have a goal region that the robot wants to reach in one step, while avoiding a certain set of obstacles. We treat the robot as a point—the usual Minkowski sum techniques can be used to reduce to the point case if the robot has finite extent. While we assume perfect knowledge about the scene, our robot does not have full control of its movement: if it starts a linear movement in a certain commanded direction, we only know that its actual movement will be confined in a cone of angle α centered around the specified direction. We are interested in the region from which a certain goal can be reached under these circumstances, and in its complexity and computation.

Such a model was first proposed by Lozano-Pérez, Mason, and Taylor [12] and was further developed in Erdmann's thesis [5] at MIT. For a detailed discussion see the recent book by Latombe [10]. In computational geometry such a model of uncertainty was used for planning compliant motions within a polygonal environment in the works of Briggs [2], Donald [3], and Friedman and others [6, 7]. Most recently Latombe and Lazanas [11] used this model to develop a complete planner for an environment consisting of circular initial, goal, and obstacle regions, as well as circular landmark regions in which the robot has perfect sensing and control.

Similar geometric issues arise in "graphics in flatland", where the goal is to compute global illumination in a two-dimensional scene. Here the goal regions play the role of light sources, and the obstacles are just opaque objects in the environment. In order to obtain a radiosity solution, the environment needs to be meshed, and this meshing needs to be done in accordance with discontinuities in the illumination function. See Heckbert and Winget [8].

In the present paper, we consider two special cases of the general problem presented above. In the first situation, we consider a single goal region, namely the "region at infinity", and a set of polygonal obstacles, modeled as a set S of n disjoint line segments. We are interested in the region $\mathcal{R}_\alpha(S)$ from where we can reach infinity with a directional uncertainty of α. We first observe that if the uncertainty angle α is not bounded from below, the complexity of $\mathcal{R}_\alpha(S)$ can be $\Theta(n^4)$. In practice, however, we can assume that α is bounded from below by some constant. Under this condition, we obtain a much better complexity of $O(n/\alpha^5)$. Our proof techniques for this case use recent geometric results of Matoušek et al. [13] and van Kreveld [14] about the arrangements of fat geometric objects. Our result generalizes the case considered by Bhattacharya et al. [1], where the obstacles form a single simple polygon.

In the second situation, we consider a collection of k polygonal goal regions of total complexity m, but without any obstacles. We are again interested in the region from where we can reach some goal region (we do not care which one) within the specified uncertainty. Surprisingly, it turns out that in this case it doesn't help to assume that α is bounded from below, since we can construct an example where the complexity of the region is $\Omega(k^4 + k^2 m)$ even for constant α. For this case we prove an upper bound of $O(mk^3)$.

We can also show corresponding computational results. We obtain an algorithm of running time $O((n/\alpha^5) \log n)$ for the first problem, and a rather naive algorithm running in time $O(k^5 m + k^3 m \log m)$ for the second problem. For reasons of space, we omit the description of these algorithms from this extended abstract.

2 Moving to infinity

In this section we assume that we are given a set S of n line segments with disjoint interiors—we will just call them "disjoint segments" in the sequel—as well as an angle $\alpha > 0$, and we want to find the region $\mathcal{R}_\alpha(S)$ of all points from which we can reach infinity with directional uncertainty α without hitting any obstacle segment in S. Observe that, since the segments are allowed to touch, our setting subsumes that of disjoint simple polygons as obstacles. More formally, let us define an α-cone to be a cone with apex angle α. We assume α to be less than π, and consider α-cones as oriented, so an α-cone has a left ray and a right ray that form an angle of α. We call an α-cone *safe (with respect to S)* if its interior does not intersect any segment in S. A point $x \in \mathbb{E}^2$ is safe if and only if there is a safe α-cone with apex x. Finally, the region $\mathcal{R}_\alpha(S)$ is defined as the locus of all safe points in $\mathbb{E}^2$.

We will prove bounds on the maximum complexity of the safe region $\mathcal{R}_\alpha(S)$. We will consider bounds depending on both n and α, because—due to practical considerations—we are mostly interested in the case where α is a fixed constant. Indeed, for constant α the safe region will be shown to have linear complexity, whereas the best bound that is independent of the value of α is $\Theta(n^4)$, as we show first.

We start with a few general observations. A point on the boundary of $\mathcal{R}_\alpha(S)$ is either on a segment of S or is the apex of an α-cone w that has endpoints p and q of some segments of S on its left and right rays. We say that such an α-cone is *determined* by p and q. The apices of all α-cones determined by two fixed endpoints p and q form two circular arcs, see Figure 1. This implies that the boundary of $\mathcal{R}_\alpha(S)$ is bounded by circular arcs and straight line segments that are pieces of the original segments in S. We only state the following theorem—see the full paper for details.

Theorem 1 *Given a set S of n disjoint line segments, and an angle $\alpha < \pi$. Then the complexity of the region $\mathcal{R}_\alpha(S)$ is bounded by $O(n^4)$. Furthermore, for every n there is a set S of n line segments and an angle $\alpha > 0$ (which decreases with n) such that $\mathcal{R}_\alpha(S)$ has complexity $\Omega(n^4)$.*

Note that to realize the lower bound, we have to use a value of α that decreases quite fast when n grows. Therefore, we turn our attention to more useful bounds in terms of α. Especially for the case where safe cones must have an angle that is bounded from below by a constant, we will be able to show a much stronger result.

It turns out to be useful to consider the following directed version of the problem. Let $\vec{u}$ be a direction vector, and let $\mathcal{R}_{\alpha,\vec{u}}(S)$ be the region of all points $x \in \mathbb{E}^2$ such that there is a safe α-cone w with apex x such that the ray with origin x and direction $\vec{u}$ lies in the closure of w. We proceed to analyze the complexity of $\mathcal{R}_{\alpha,\vec{u}}(S)$. We assume without loss of generality that the preferred direction $\vec{u}$ is the upward vertical direction, i.e. the positive y-direction.

Notice that the boundary γ of $\mathcal{R}_{\alpha,\vec{u}}(S)$ is a chain with the property that its intersection with any line with direction $\vec{u}$ is a point or a segment. We will call such a chain *semi-monotone (in direction $\vec{u}$)*. Furthermore, γ consists of circular arcs (determined by two endpoints of S), line segments (pieces of the segments of S), and vertical segments (below an endpoint of a segment of S).

4

Let P be the set of endpoints of S. We define $\mathcal{R}_{\alpha,\vec{u}}(P)$ analogously to $\mathcal{R}_{\alpha,\vec{u}}(S)$, i.e. $x \in \mathcal{R}_{\alpha,\vec{u}}(P)$ if there is an α-cone w with apex x whose interior does not contain a point of P and such that the ray from x with direction $\vec{u}$ is contained in w. The following lemma follows then quite easily.

Lemma 2 $\mathcal{R}_{\alpha,\vec{u}}(S)$ *is the intersection of* $\mathcal{R}_{\alpha,\vec{u}}(P)$ *with the region above the upper envelope of* S.

Lemma 3 *The complexity of* $\mathcal{R}_{\alpha,\vec{u}}(S)$ *is* $O(n)$.

Proof: Since the boundary of $\mathcal{R}_{\alpha,\vec{u}}(P)$ and the upper envelope of S are both semi-monotone chains, and the latter has complexity $O(n)$, the result follows from Lemma 2 if we can prove that the complexity of $\mathcal{R}_{\alpha,\vec{u}}(P)$ is $O(n)$.

The boundary of $\mathcal{R}_{\alpha,\vec{u}}(P)$ consists of circular arcs and vertical segments. A vertex x of $\mathcal{R}_{\alpha,\vec{u}}(P)$ either lies below a point of P—there are at most $4n$ such vertices, namely two for each of the $2n$ points of P—or is the apex of an α-cone with at least three points of P on its bounding rays. We first count the vertices where there are at least two points of P on the left ray of this α-cone. To this end we observe that the rightmost of these points cannot play this role for more than one vertex: Suppose that there are two α-cones, both having at least two points on their left ray, which share the rightmost point on their left ray. Then one of the two cones must contain the leftmost point on the left ray of the other cone, as illustrated in Figure 2. Thus, there are at most $2n$ such vertices. Vertices with at least two points on the right ray of the corresponding α-cone are counted in the same way. This proves Lemma 3. ⊟

We will exploit this lemma to bound the complexity of $\mathcal{R}_\alpha(S)$. We first observe that there is a collection U of $O(1/\alpha)$ different orientations such that

$$\mathcal{R}_\alpha(S) = \bigcup_{\vec{u} \in U} \mathcal{R}_{\alpha,\vec{u}}(S).$$

Next we note that any vertex of $\mathcal{R}_\alpha(S)$ is a vertex of $\mathcal{R}_{\alpha,\vec{u}}(S) \cup \mathcal{R}_{\alpha,\vec{v}}(S)$ for some pair of $\vec{u}$, $\vec{v}$ in U. We will show that the complexity of such a union $\mathcal{R}_{\alpha,\vec{u}}(S) \cup \mathcal{R}_{\alpha,\vec{v}}(S)$ is $O(n/\alpha^3)$. Since there are $O(1/\alpha^2)$ possible pairs of $\vec{u}$ and $\vec{v}$, this will prove an upper bound of $O(n/\alpha^5)$ on the complexity of $\mathcal{R}_\alpha(S)$.

So let us fix two directions $\vec{u}$ and $\vec{v}$, and consider the regions $\mathcal{R}_{\alpha,\vec{u}}(S)$ and $\mathcal{R}_{\alpha,\vec{v}}(S)$. These regions cannot have any long and skinny parts—after all, they

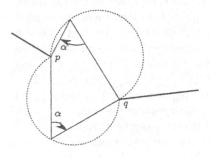

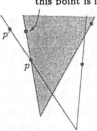

Figure 1:   Figure 2:

are unions of (infinitely many) α-cones, so the value of α gives a lower bound on the "skinniness" of $\mathcal{R}_{\alpha,\vec{u}}(S)$. In fact, this is the concept of *fatness* employed by Matoušek et al. [13] and van Kreveld [14]. They have proven results on the number of holes in the union of fat regions, which can in turn be used to bound the complexity of their union. Unfortunately, these results are only proven for polygonal regions, and our regions are bounded by circular segments. We will circumvent this problem by approximating the circular arcs by line segments, and proving that this does not increase the complexity of the union too much.

Let γ be the boundary of $\mathcal{R}_{\alpha,\vec{u}}(S)$, and μ be the boundary of $\mathcal{R}_{\alpha,\vec{v}}(S)$. γ and μ are semi-monotone with respect to the directions $\vec{u}$ and $\vec{v}$, resp. We partition γ and μ into pieces at their break points. We denote the resulting set of pieces by $\gamma_1, \gamma_2, \ldots$ and $\mu_1, \mu_2, \ldots$, respectively. Note that each γ_i (or μ_i) is a line segment or a circular arc. We will treat all these pieces separately. For a piece γ_i, let γ_i' be the segment connecting the two endpoints of γ_i, and let γ_i'' be the polygonal chain obtained by replacing γ_i by two vertical segments and a horizontal segment through its lowest point, as in Figure 3. (Here, $\vec{u}$ is assumed to be vertical). We define γ' and γ'' to be the union of the pieces γ_i' and the γ_i'', respectively. Let Δ_i^γ be the possibly degenerated trapezoid enclosed between γ_i' and γ_i''. Define μ_j', μ_j'', μ', μ'' and Δ_j^μ in the same way.

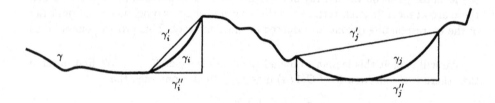

Figure 3:

Consider now a pair γ_i and μ_j. Because of their simple shape, those two pieces can have at most a constant number of intersections, or, equivalently, can contribute at most a constant number of vertices to the union of $\mathcal{R}_{\alpha,\vec{u}}(S)$ and $\mathcal{R}_{\alpha,\vec{v}}(S)$. To estimate the complexity of $\mathcal{R}_{\alpha,\vec{u}}(S) \cup \mathcal{R}_{\alpha,\vec{v}}(S)$ it is therefore sufficient to bound the number of pairs γ_i, μ_j that intersect.

Consider now a pair γ_i, μ_j that intersect. If μ_j lies completely within the trapezoid Δ_i^γ, then μ_j cannot intersect any other $\gamma_{i'}$. It follows that there are at most $O(n)$ such intersections, and the same reasoning holds for the case that γ_i lies in Δ_j^μ. For all remaining intersecting pairs γ_i, μ_j, there must also be an intersection between two of the curves γ_i', γ_i'', μ_j', and μ_j''. Or, equivalently, for every such pair there is a vertex in $\mathcal{R}_{\alpha,\vec{u}}'(S) \cup \mathcal{R}_{\alpha,\vec{v}}'(S)$, $\mathcal{R}_{\alpha,\vec{u}}'(S) \cup \mathcal{R}_{\alpha,\vec{v}}''(S)$, $\mathcal{R}_{\alpha,\vec{u}}''(S) \cup \mathcal{R}_{\alpha,\vec{v}}'(S)$, or $\mathcal{R}_{\alpha,\vec{u}}''(S) \cup \mathcal{R}_{\alpha,\vec{v}}''(S)$, where $\mathcal{R}_{\alpha,\vec{u}}'(S)$ $(\mathcal{R}_{\alpha,\vec{v}}'(S))$ is the region above γ' (μ') and $\mathcal{R}_{\alpha,\vec{u}}''(S)$ $(\mathcal{R}_{\alpha,\vec{v}}''(S))$ is the region above γ'' (μ''). It is important here that the regions $\mathcal{R}_{\alpha,\vec{u}}'(S)$, etc., are defined to be open; otherwise some intersections can be missed. So, it will be sufficient to prove that the complexity of all these pairwise unions is $O(n/\alpha^3)$.

We now use the *combination lemma* by Edelsbrunner et al. [4]. It states that the complexity of the union of two polygonal regions R and R' is bounded by the complexities of the two components plus the number of holes in $R \cup R'$. It remains to show that the number of holes in the above unions is in $O(n/\alpha^3)$.

To this end we first show that both $\mathcal{R}'_{\alpha,\vec{u}}(S)$ and $\mathcal{R}''_{\alpha,\vec{u}}(S)$ can be covered by $O(n)$ β-fat triangles—triangles whose smallest angle is bounded from below by β—with $\beta \geqslant c\alpha$ for some constant $c > 0$. We use the technique by van Kreveld [14]. He defines a polygon $\mathcal{P}$ to be δ-wide if it does not contain a γ-corridor for $\gamma < \delta$; here a γ-corridor is defined as a quadrilateral with vertices v_1, v_2, v_3, v_4 such that v_1, v_2 lie on some edge e of $\mathcal{P}$, v_3, v_4 lie on some edge e', $\angle v_1 v_2 v_3 = \angle v_2 v_3 v_4$, $\angle v_3 v_4 v_1 = \angle v_4 v_1 v_2$ and $|\overline{v_1 v_2}| = |\overline{v_3 v_4}| = \frac{1}{\gamma} \max\{|\overline{v_2 v_3}|, |\overline{v_4 v_1}|\}$. Informally speaking, a γ-corridor is a symmetric trapezoid with vertices on two edges of $\mathcal{P}$ whose width-length ratio is γ. Van Kreveld has proven that any δ-wide polygon can be covered with $O(n)$ $(c'\delta)$-fat triangles for some constant $c' > 0$. Since any corridor in $\mathcal{R}'_{\alpha,\vec{u}}(S)$ must contain an α-cone, its width-length ratio cannot be worse than $\sin \alpha$. It follows that we can cover $\mathcal{R}'_{\alpha,\vec{u}}(S)$ and $\mathcal{R}''_{\alpha,\vec{u}}(S)$ with $O(n)$ β-fat triangles with $\beta \geqslant c\alpha$ for some constant $c > 0$. Now we can apply a result by Matoušek et al. [13] which states that the union of n β-fat triangles has at most $O(n/\beta^3)$ holes. Applied to our case, this gives us the $O(n/\alpha^3)$ bound on the union of $\mathcal{R}'_{\alpha,\vec{u}}(S)$ and $\mathcal{R}''_{\alpha,\vec{u}}(S)$ we are looking for.

Theorem 4 *Let S be a set of n disjoint line segments in the plane, and let $\alpha < \pi$ be given. The complexity of $\mathcal{R}_\alpha(S)$ is $O(n/\alpha^5)$. Moreover, there is an example of n line segments where the complexity of $\mathcal{R}_\alpha(S)$ is $\Omega(n/\alpha)$. $\mathcal{R}_\alpha(S)$ can be computed in time $O((n/\alpha^5) \log n)$.*

The lower bound example can be found in the full paper. There, we also give the algorithm to actually construct $\mathcal{R}_\alpha(S)$, essentially following the ideas used in the combinatorial proof. We first show how to compute $\mathcal{R}_{\alpha,\vec{u}}(S)$, for a fixed direction $\vec{u}$. This can be done by constructing the chain γ from left to right, and takes time $O(n \log n)$. We then use a divide and conquer algorithm on the set U of $O(1/\alpha)$ directions, merging the different $\mathcal{R}_{\alpha,\vec{u}}(S)$ to compute $\mathcal{R}_\alpha(S)$.

3 Multiple goal regions

In this second part of the paper we study the following problem. We are given a family $\mathcal{B}$ of k pairwise disjoint polygonal goal regions with a total complexity m, and we are interested in the region $\mathcal{R}_\alpha(\mathcal{B})$ from where some goal in $\mathcal{B}$ can be reached with directional uncertainty $\alpha > 0$. More formally, we will say that an α-cone w with apex x is *safe* if and only if every ray with origin x that lies in w intersects an element of $\mathcal{B}$. We call a point $x \in \mathbb{E}^2$ *safe* if there is a safe α-cone w with apex x, and define $\mathcal{R}_\alpha(\mathcal{B})$ as the region of all safe points. Again, we want to prove bounds on the maximum complexity of the region $\mathcal{R}_\alpha(\mathcal{B})$. We first observe that we can assume that the polygons in $\mathcal{B}$ are convex if $\alpha < \pi$. This is true because for $\alpha < \pi$, we can always reach a polygon B from any point within its convex hull, and a ray with origin outside the convex hull of B intersects B exactly if it intersects its convex hull, see Figure 4. Notice that the convex hulls of a set of disjoint polygons are not necessarily disjoint. However, if two or more of the convex hulls intersect then we can repeat the above argument, and replace them by the convex hull of their union. This process continues until we are left with a set of disjoint convex polygons. Notice that every vertex of the remaining polygons must be a vertex of one of the original polygons, so the total complexity of the polygons has not increased.

Let us start by considering a single convex goal polygon B with m vertices. The region $\mathcal{R}_\alpha(\{B\})$ is a flower-shaped region, bounded by circular arcs. Let γ be the closed boundary curve of $\mathcal{R}_\alpha(\{B\})$. For a point x on γ, there is an α-cone w whose boundary rays are tangent to B. For vertices of γ, one of the boundary rays is flush with an edge of B. To bound the number of vertices of γ we thus have to bound the number of edge-vertex pairs such that there is an α-cone with one of its rays containing the edge, and the other ray being tangent to B at the vertex. Observe that each edge of B defines at most two such pairs: an edge with orientation θ forms a pair exactly with the two extreme vertices of B in the directions orthogonal to $\theta - \alpha$ and $\theta + \alpha$. Consequently, the complexity of $\mathcal{R}_\alpha(\{B\})$ is in $O(m)$. The example of a regular convex m-gon shows that this bound can actually be achieved. It is not difficult to compute $\mathcal{R}_\alpha(\{B\})$ in linear time: the relevant edge-vertex pairs can easily be computed after merging the ordered list of all orientations of edges of B with the same list with α added to the orientations.

The above discussion is summarized in the following lemma.

Lemma 5 *The maximum complexity of the region $\mathcal{R}_\alpha(\{B\})$ of a convex polygon B with m vertices is $\Theta(m)$. Moreover, $\mathcal{R}_\alpha(\{B\})$ can be computed in $O(m)$ time.*

We now turn our attention to the case where we have a family $\mathcal{B} = \{B_1, \ldots, B_k\}$ of k disjoint convex goal regions. Let m_i denote the number of vertices of B_i and let $m = \sum_{i=1}^{k} m_i$ be the total number of vertices. Notice that is is not sufficient to simply take the union of the regions $\mathcal{R}_\alpha(\{B_i\})$, because some points may not have an α-cone that is safe by any single goal region but only an α-cone safe due to several goal regions.

Consider a (circular) piece of the boundary of $\mathcal{R}_\alpha(\mathcal{B})$ which is defined by more than one goal region. There can be more than two goal regions which are needed to make sure that points on this boundary piece have a safe α-cone. However, for points on the boundary of $\mathcal{R}_\alpha(\mathcal{B})$ there is an α-cone that touches only two of them, each in a vertex. So the question becomes: how many pairs of vertices, one from B_i and one from B_j, can there be such that there is an α-cone touching B_i at one vertex and touching B_j at the other vertex? Now we note that such a pair of vertices also defines an α-cone which touches the convex hull of B_i and B_j in two points (namely, in the two vertices). This convex hull has at most $m_i + m_j$ vertices, so by Lemma 5 there are only $O(m_i + m_j)$ such pairs. Summing over all pairs of polygons, we obtain

$$O\left(\sum_{1 \leqslant i \leqslant j \leqslant k} (m_i + m_j) \right) = O\left(\sum_{i=1}^{k} \sum_{j=1}^{k} m_j \right) = O(km).$$

It follows that there are only $O(km)$ possible pairs of vertices that can determine an arc of the boundary of $\mathcal{R}_\alpha(\mathcal{B})$.

However, the complexity of $\mathcal{R}_\alpha(\mathcal{B})$ can be a lot higher, because the circular arc defined by a pair of vertices can appear in several pieces on the boundary of $\mathcal{R}_\alpha(\mathcal{B})$. To see what happens it is useful to go back to the case of one goal region B, and to take a somewhat different view on $\mathcal{R}_\alpha(\{B\})$. For every pair p, q of vertices of B

8

let $C(p,q)$ be the region $\mathcal{R}_\alpha(\{\overline{pq}\})$. $C(p,q)$ is the union of two discs as in Figure 5. Clearly, $\mathcal{R}_\alpha(\{B\})$ is just the union of all $C(p,q)$, for all pairs of p and q. Lemma 5 tells us that only a linear number of pairs is relevant.

Now we return to the case of multiple goal regions. Here we have $O(km)$ pairs (p,q) that define a region $C(p,q)$ which is relevant. The complication is that for vertices p, q of different polygons the whole region $C(p,q)$ is not necessarily contained in $\mathcal{R}_\alpha(\mathcal{B})$: we know that for points in $C(p,q)$ there is an α-cone whose bounding rays intersect two of the goal regions but this α-cone need not be safe.

To obtain this extra information we consider the arrangement $\mathcal{A}(L)$ formed by the set L of lines tangent to two polygons in $\mathcal{B}$. Since there are $O(k^2)$ such lines, the arrangement $\mathcal{A}(L)$ consists of $O(k^4)$ cells. Consider a cell c of $\mathcal{A}(L)$. With each cell c of $\mathcal{A}(L)$ we associate a *visibility cycle* $\mathcal{V}_c$, defined as the circularly ordered list of visible polygons intersected by a ray rotating clockwise around any given point in the cell. Whenever a ray does not intersect any polygon, the corresponding element in the cycle is denoted as ∞. Each visibility cycle contains $O(k)$ elements, and it consists of several connected components, separated by ∞. Observe that the visibility cycle of a cell c of $\mathcal{A}(L)$ is well defined, that is, $\mathcal{V}_c$ does not depend on which point in c is chosen. But then it readily follows that within every cell c of $\mathcal{A}(L)$, the region $\mathcal{R}_\alpha(\mathcal{B})$ is equal to the union of the regions $C(p,q)$, where the union is taken over all pairs of p and q that come from polygons in the same connected component in $\mathcal{V}_c$. Thus within every cell the region is equal to the union of $O(km)$ discs, which has $O(km)$ complexity [9]. Since $\mathcal{A}(L)$ has $O(k^4)$ cells the total complexity of $\mathcal{R}_\alpha(\mathcal{B})$ is $O(k^5m)$. However, it is possible to do better if we observe that a disc is interesting in a certain cell of $\mathcal{A}(L)$ only if its boundary intersects the cell—otherwise the disc either makes the whole cell part of $\mathcal{R}_\alpha(\mathcal{B})$, or it cannot participate in the complexity within this cell at all. A circle can intersect a line at most twice, and hence can intersect at most $O(k^2)$ cells of our arrangement. Since we have $O(km)$ discs, we find that the number of interesting cell-disc pairs is only $O(k^3m)$. It follows that the total complexity of $\mathcal{R}_\alpha(\mathcal{B})$ is at most $O(k^3m)$.

Theorem 6 *Given a family $\mathcal{B}$ of k polygons of total complexity m and an angle $0 < \alpha < \pi$, the total complexity of $\mathcal{R}_\alpha(\mathcal{B})$ is at most $O(k^3m)$. There is an example*

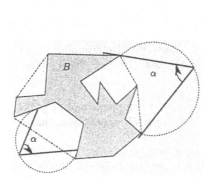

Figure 4:

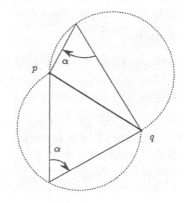

Figure 5:

of k goal polygons with total complexity m such that the complexity of $\mathcal{R}_\alpha(\mathcal{B})$ is $\Omega(k^4 + k^2 m)$. The region $\mathcal{R}_\alpha(\mathcal{B})$ can be computed in $O(k^5 m + k^3 m \log m)$ randomized time.

Again, we omit the lower bound examples as well as the algorithm from this extended abstract.

4 Conclusion and Extensions

We have studied combinatorial and algorithmic aspects of motion planning for robots with imperfect control, where, if the robot starts a linear movement in a certain commanded direction, we only know that its actual movement will be confined to a cone of angle α around the specified direction. We have studied the case where we have a set of obstacle line segments and we are interested in the locus of all points from where infinity can be reached with directional uncertainty α. We also studied the case where there are no obstacles, but we have a set of polygonal goal regions that we want to reach.

A number of questions is left open. First of all, it would be nice to tighten the gaps in our combinatorial bounds. A related question is the following: We have proven the upper bound in Theorem 4 by first approximating the region by a polygonal region, and then employing bounds from the literature for the union of fat triangles. It would be much nicer if we had tools that allowed us to directly bound the complexity of the union of fat objects with curved boundaries. In that case we could argue that the final region is the union of $O(1/\alpha)$ α-wide "curved polygons", and could probably obtain better bounds than in the present paper.

On the algorithmic side there are also some open problems. For example, the algorithm that we gave for the second problem is not as close to the combinatorial upper bound as we would like it to be. Another challenge is to design an output-sensitive algorithm for computing the region, as the region will not have the worst-case complexity in many practical situations.

Finally, it would be interesting to try to combine our results to obtain results for the general setting, where we have both goal regions and obstacles.

References

1. B. Bhattacharya, D. G. Kirkpatrick, and G. T. Toussaint. Determining sector visibility of a polygon. In *Proc. 5th Annu. ACM Sympos. Comput. Geom.*, pages 247–254, 1989.

2. A. J. Briggs. An efficient algorithm for one-step planar compliant motion planning with uncertainty. In *Proc. 5th Annu. ACM Sympos. Comput. Geom.*, pages 187–196, 1989.

3. B. R. Donald. The complexity of planar compliant motion planning under uncertainty. *Algorithmica*, 5:353–382, 1990.

4. H. Edelsbrunner, L. J. Guibas, and M. Sharir. The complexity and construction of many faces in arrangements of lines and of segments. *Discrete Comput. Geom.*, 5:161–196, 1990.

5. M. Erdmann. On motion planning with uncertainty. Technical Report 810, AI Laboratory, MIT, 1984.

6. J. Friedman, J. Hershberger, and J. Snoeyink. Compliant motion in a simple polygon. In *Proc. 5th Annu. ACM Sympos. Comput. Geom.*, pages 175–186, 1989.

7. J. Friedman, J. Hershberger, and J. Snoeyink. Input-sensitive compliant motion in the plane. In *Proc. 2nd Scand. Workshop Algorithm Theory*, volume 447 of *Lecture Notes in Computer Science*, pages 225–237. Springer-Verlag, 1990.

8. P. Heckbert and J. Winget. Finite-element methods for global illumination. To appear.

9. K. Kedem, R. Livne, J. Pach, and M. Sharir. On the union of Jordan regions and collision-free translational motion amidst polygonal obstacles. *Discrete Comput. Geom.*, 1:59–71, 1986.

10. J.-C. Latombe. *Robot Motion Planning*. Kluwer Academic Publishers, Boston, 1991.

11. A. Lazanas and J.-C. Latombe. Landmark-based robot navigation. Submitted to Algorithmica.

12. T. Lozano-Pérez, M. T. Mason, and R. H. Taylor. Automatic synthesis of fine-motion strategies for robots. *Internat. J. Robotics Research*, 3:3–24, 1984.

13. J. Matoušek, N. Miller, J. Pach, M. Sharir, S. Sifrony, and E. Welzl. Fat triangles determine linearly many holes. In *Proc. 32nd Annu. IEEE Sympos. Found. Comput. Sci.*, pages 49–58, 1991.

14. M. van Kreveld. On fat partitioning, fat covering, and the union size of polygons. In *Proc. 3rd Workshop Algorithms Data Struct.*, Lecture Notes in Computer Science, 1993.

Constructing Degree-3 Spanners with Other Sparseness Properties

Gautam Das*
Paul J. Heffernan

Department of Mathematical Sciences, Memphis State University
Memphis, TN 38152

Abstract

Let V be any set of n points in k-dimensional Euclidean space. A subgraph of the complete Euclidean graph is a *t-spanner* if for any u and v in V, the length of the shortest path from u to v in the spanner is at most t times $d(u,v)$. We show that for any $\delta > 1$, there exists a polynomial-time constructible t-spanner (where t is a constant that depends only on δ and k) with the following properties. Its maximum degree is 3, it has at most $n \cdot \delta$ edges, and its total edge weight is comparable to the minimum spanning tree of V (for $k \leq 3$ its weight is $O(1) \cdot wt(MST)$, and for $k > 3$ its weight is $O(\log n) \cdot wt(MST)$).

1 Introduction

Let V be any set of n points in k-dimensional Euclidean space. A subgraph of the complete Euclidean graph is a *t-spanner* if for any u and v in V, the length of the shortest path from u to v in the spanner is at most t times $d(u,v)$. The value of t is called the *stretch factor* of G. We consider the problem of constructing sparse spanners, with the objective of keeping the stretch factor a constant (that is, independent of n). The motivation is, complete graphs represent ideal communication networks but are expensive to build in practice, and sparse spanners represent low cost alternatives. Sparseness of spanners is usually measured by various criteria such as small degree, few edges, and/or small edge weight. This paper is significant because we construct a single spanner which is extremely sparse by *all the above criteria*. Some of the sparseness results are optimal in a very strong sense.

Spanners for complete Euclidean graphs as well as for general graphs find applications in robotics, network topology design, distributed systems, design of parallel machines, and have also been investigated by graph theorists. Recent surveys on spanners may be found in [2, 3, 12].

The minimum spanning tree of V, MST, is obviously the sparsest possible connected spanner both in number of edges and weight (although not in degree), but its

---

*Supported in part by NSF grant CCR-9306822

stretch factor can be as bad as $n - 1$ [1]. The sparseness of a spanner is frequently judged by comparing it to the number of edges and weight of the MST. In this paper we prove the following theorem.

Theorem 1.1 *Let V be any set of n points in k-dimensional Euclidean space. For any $\delta > 1$, there exists a polynomial-time constructible t-spanner (where t depends only on δ and k) with the following properties:*

1. *Its maximum degree is 3.*

2. *It has at most $n \cdot \delta$ edges.*

3. *For $k \leq 3$ its weight is $O(1) \cdot wt(MST)$, and for $k > 3$ its weight is $O(\log n) \cdot wt(MST)$. The constants implicit in the O-notation depend only on δ and k.*

Let us first consider the problem of constructing spanners with a small degree. In [2] an algorithm is described which constructs a spanner in k-dimensions with a constant maximum degree (that is, not dependent on n), however the degree is exponentially dependent on k and hence quite large for higher dimensions. In 2 dimensions, a degree-7 spanner is reported in [6], and a degree-5 spanner in [13]. It is known that the lower bound for the degree is 3 because, given a set of points arranged in a grid, a Hamiltonian path or circuit cannot be a spanner with a constant stretch factor. A breakthrough in this problem appeared in [11] where it is shown that there exists a degree-4 spanner with a constant stretch factor in k-dimensional space. In that paper the author reports that it may be difficult to extend the technique to construct degree-3 spanners, but conjectures that such spanners exist.

In this paper we prove the existence of a degree-3 spanner in k-dimensional space. With this result, the problem of constructing spanners with small degree is satisfactorily solved. The proof employs some of the techniques in [11], but several new and non-trivial ideas are used.

Now consider the problem of generating spanners with few edges. A lower bound on the number of edges is $n - 1$ because a spanning tree is a connected graph with the fewest edges. Spanners with $O(n)$ edges and constant stretch factors have been designed in several papers, but in most of these spanners the constant implicit in the O-notation depends on the dimension [1, 2, 3, 4, 7, 9, 10, 14]. The degree-4 spanner constructed in [11] is an improvement, because it has at most $2 \cdot n$ edges. The next question is, *can the proportionality constant be reduced even further?* We are thus seeking edge sparseness in a very strong sense. This has been answered in the affirmative in 2-dimensions. In [1] it is shown that, given any $\delta > 1$, a spanner with a constant stretch factor (which depends on δ) can be constructed with at most $n \cdot \delta$ edges. However, the techniques employ planarity properties and do not extend to higher dimensions.

In this paper we show that, given any $\delta > 1$, a degree-3 spanner in k-dimensional space can be constructed with a constant stretch factor (which depends on k and δ), and with at most $n \cdot \delta$ edges. Thus the number of edges can be made *arbitrarily small*, with a corresponding increase in the stretch factor. This result satisfactorily solves the general problem of constructing spanners with few edges.

We now turn our attention to the third criterion, that is the weight of the spanner. Let $wt(G)$ be the total edge weight of a graph G. In [5, 8] it is shown that a

2-dimensional spanner exists with weight at most $O(1) \cdot wt(MST)$, which is an asymptotically optimal result. Both papers exploit planarity and the techniques do not extend to higher dimensions. In [2] it is shown that a k-dimensional spanner exists with weight at most $O(\log n) \cdot wt(MST)$. In [4] it is shown that for $k \leq 3$, a spanner exists with weight at most $O(1) \cdot wt(MST)$. Thus asymptotic optimality has been achieved up to the third dimension. But the spanners in [2, 4] do not have the degree and edges bounds presented here.

In addition to having a degree of 3 and the number of edges bounded by $n \cdot \delta$, the spanner in this paper has a weight bound that match the spanners in [2, 4]. However, the general problem of constructing spanners with low weight has not yet been satisfactorily solved. The $O(\log n)$ factor in dimensions higher than 3 is probably suboptimal, and one would like to go beyond asymptotic optimality even for $k \leq 3$. For example, given any $\gamma > 1$, are there spanners with weight at most $wt(MST) \cdot \gamma$?

The paper is organized as follows. In Section 2 we introduce some notations and summarize previous research relevant to our proofs. Section 3 describes the algorithm and shows that the resulting spanner is degree-3. Sections 4 and 5 are devoted to proving upper bounds on the number of edges and the total weight respectively. We conclude with some open problems.

Remark: Throughout the paper various constants are used, but lack of space prohibits us from explicitly presenting their values or their functional dependencies on other parameters such as k and δ. They are available in the complete version of this paper.

2 Preliminaries

In this section we introduce a few basic definitions and also summarize some previous relevant results. The Euclidean distance between two points u and v is $d(u, v)$. The length of the shortest path between two vertices u and v of a graph G is $d_G(u, v)$.

In [2, 4] a greedy algorithm is developed for constructing spanners of any set V of n points in k-dimensional space. Given any $t' > 1$, this algorithm constructs a degree-s t'-spanner, where s depends on k and t'. Such spanners are called *greedy spanners* because the algorithm is greedy. Select a fixed value of t' (such as $t' = 2$). In the rest of this paper, assume that whenever the greedy algorithm is run, it uses this value of t'. Thus the degree of the resulting spanner, s, depends only on k. It is known that for $k \leq 3$ the weight of the greedy spanner is $O(1) \cdot wt(MST)$, and for $k > 3$ the weight is $O(\log n) \cdot wt(MST)$. The constants implicit in the O-notation depend only on k.

The technique in [11] generates a degree-4 spanner whose stretch factor depends only on k. This result uses the concept of *nearest-neighbor graphs*. Since our paper also relies on the same concept, we present it in detail here.

Assume no two interdistances among the points in V are identical. The *nearest-neighbor graph* is a digraph over the vertex set V, such that from each vertex u, there is a directed edge to its nearest neighboring vertex, $n_V(u)$. The following are true. Every simple path in this digraph follows edges of decreasing length. The only cycles are *short cycles*, containing exactly two vertices. In every simply-connected component of this digraph, exactly one short cycle appears. Thus, each simply-connected component has at least two vertices.

This digraph can be transformed into an undirected forest as follows. Replace directed edges by undirected edges. Since short cycles will be transformed into multi-edges, replace multi-edges by single edges. The result is the undirected *nearest-neighbor forest*. In this forest, each tree contains a special edge created from the unique short cycle of the corresponding simply-connected component of the digraph. Designate one of the two adjacent vertices of this edge the root of the tree. We use the notation $r(T)$ to represent the root of a tree T. Thus the nearest-neighbor forest is a forest of rooted trees. It is noted in [11] that the degree of a nearest-neighbor forest is at most 3^k.

3   Constructing a Degree-3 Spanner

In this section we show how to construct a degree-3 spanner in k-dimensional space, whose stretch factor depends only on k. In [11] an algorithm is described for constructing degree-4 spanners. We borrow two techniques from that paper, namely the use of nearest-neighbor forests, and a method of transforming a fixed degree tree into a degree-3 tree. However, to construct a degree-3 spanner with other sparseness properties requires several new and non-trivial ideas.

Recall that s is the maximum degree of greedy spanners, where s depends only on k. The intuitive idea is to partition V into clusters such that each cluster has a representative vertex u (called the *root*) and several other vertices, of which at least $s-2$ vertices (say $w_1, w_2, \cdots, w_{s-2}$) are quite close to u as compared to the distance between u and roots of other clusters. Then an individual spanner for each cluster is constructed, such that each root is degree-1, each of $w_1, w_2, \cdots, w_{s-2}$ is degree-2, and the remaining vertices in the cluster are at most degree-3. A spanner G for V is created by taking the union of all individual spanners with a greedy spanner constructed for the set of roots of all clusters. The degree of G is reduced to 3 by shifting $s-2$ of the greedy spanner edges incident to a root u to its nearby vertices $w_1, w_2, \cdots, w_{s-2}$.

The result in [11] also uses a clustering concept. There each cluster may have several vertices, but only one other vertex is guaranteed to be close to the representative vertex. The drawback is that a degree-3 spanner cannot be constructed. The advantage of our clustering technique is two-fold. Apart from constructing degree-3 spanners, we also reduce the total number of edges in the spanner. Our clusters are constructed from a hierarchy of nearest-neighbor forests, whereas the clusters in [11] are constructed from a single level nearest-neighbor forest.

The construction will consist of a series of steps.

Step 1: Let $V^1 = V$. Construct the nearest-neighbor forest of V^1, which is called the *first-level nearest-neighbor forest* and denoted NNF^1. Then construct the *fragmented first-level nearest-neighbor forest* (denoted $FNNF^1$) as follows. Select any $c > 1$. Label each edge of NNF^1 as either *short* or *long*. An edge in a tree is labeled short if it is adjacent to the root or if its length is less than

c times the length of its parent edge. Otherwise the edge is labeled long. After the labeling is over, repeat the following step until no tree in the forest has three consecutive short edges on a root to leaf path. Locate a tree which has three consecutive short edges, with the middle short edge being (u, w), with u the parent of w. Fragment the tree into two by removing the middle short edge, and make w the root of the new tree.

Step 2: Let $l = s - 2$. (A larger l will also work, and in fact, in the next section the algorithm is run with a larger l). Thus l depends only on k. Repeat the following procedure for $i = 2, 3, \cdots, l$. Set V^i to be the set of all the roots in $FNNF^{i-1}$. Construct the nearest-neighbor forest of V^i, denoted NNF^i. Then construct $FNNF^i$ by fragmenting NNF^i exactly as in Step 1, using the same value of c.

At this stage we have constructed a hierarchy of l fragmented forests. Several properties hold. The forests are pairwise edge disjoint. Each tree in $FNNF^i$ has at least two vertices. The set V^i is a proper subset of V^{i-1}. The following lemmas describe additional properties.

Lemma 3.1 *Let u be a vertex of tree T^i belonging to $FNNF^i$. Let v be any other vertex of V^i. Then $d_{T^i}(u, r(T^i))$ is at most a constant times $d(u, v)$.*

Proof : Clearly $d(u, n_{V^i}(u)) \leq d(u, v)$. But $(u, n_{V^i}(u))$ is the first edge along the path in T^i from u to the root. The way $FNNF^i$ is constructed, no three consecutive edges on this path are short edges. Thus the length of this path converges to at most a constant times the length of the first edge, $d(u, n_{V^i}(u))$. ∎

We next show that each root has a close child compared to the distance between the root and other roots in the same level.

Lemma 3.2 *Let u and v be two roots of $FNNF^i$. Let w be the closest child of u. Then $d(u, v)$ is at least a constant times $d(w, u)$.*

Proof : The vertex u could either be a root of NNF^i, or a root of a tree created by the fragmentation process. In the first case, $w = n_{V^i}(u)$ so the lemma trivially holds. In the second case, u became a root because, both (w, u) and $(u, n_{V^i}(u))$ were short edges in NNF^i and $(u, n_{V^i}(u))$ was removed. Thus $d(w, u) < c \cdot d(u, n_{V^i}(u)) \leq c \cdot d(u, v)$. ∎

Step 3: Construct the *composite forest* by taking the union of all $FNNF^i$, for $i = 1, 2, \cdots, l$.

Each connected component R in the composite forest is a tree, and is basically the union of a single l^{th} level tree T^l, with all $(l - 1)^{th}$ level trees whose roots are vertices of T^l, with all $(l - 2)^{th}$ level trees whose roots are vertices of these $(l - 1)^{th}$ level trees, and so on. Define the root of R as $r(R) = r(T^l)$. The maximum degree of R depends only on k (it is at most l times the maximum degree of any nearest-neighbor forest). Each R has at least 2^l vertices. Furthermore, $r(R)$ has at least l children because it is the common root to l different level trees.

Lemma 3.3 *Each composite tree R is a spanner over its own vertices, with a stretch factor dependent only on k.*

Proof : Let u and v be two vertices in R. We shall show that $d_R(u,v)$ is at most a constant times $d(u,v)$. Let w be the nearest common ancestor of u and v in R. Let $T^1, T^2, \cdots, T^l$ be trees such that u is in T^1, $r(T^1)$ is in T^2, $r(T^2)$ is in T^3, and so on. Let $H^1, H^2, \cdots, H^l$ be trees such that v is in H^1, $r(H^1)$ is in H^2, $r(H^2)$ is in H^3, and so on. Let j be the smallest integer such that $T^j = H^j$. Thus w belongs to T^j. The path from u to w in R goes via $r(T^1), r(T^2), \cdots, r(T^{j-1})$. The path from v to w in R goes via $r(H^1), r(H^2), \cdots, r(H^{j-1})$.

By applying Lemma 3.1, both $d_{T^1}(u, r(T^1))$ and $d_{H^1}(v, r(H^1))$ are at most a constant times $d(u,v)$. This also implies that $d(r(T^1), r(H^1))$ is at most a constant times $d(u,v)$. If we carry this argument to higher levels we can conclude that both $d_{T^i}(r(T^{i-1}), r(T^i))$ and $d_{H^i}(r(H^{i-1}), r(H^i))$ are at most a constant times $d(u,v)$. This also implies that $d(r(T^i), r(H^i))$ is at most a constant times $d(u,v)$. At level j we can conclude that both $d_{T^j}(r(T^{j-1}), w)$ and $d_{H^j}(r(H^{j-1}), w)$ are at most a constant times $d(u,v)$. However, these constants now depend on the level.

Since we have shown that each fragment of the path from u to v via w is at most a constant times $d(u,v)$, and there are $2 \cdot j$ such fragments, the lemma is proved. Of course, the constant depends on l, which in turn depends on k. ∎

We next show that in the composite forest each root has l distinct close children compared to the distance between the root and other roots. This fact is crucial in the eventual construction. This is different from Lemma 3.2 where trees at any given level have only one child guaranteed to be close to the root.

Lemma 3.4 *Let u and v be two roots of the composite forest. Let R be the composite tree such that $u = r(R)$. For $i = 1, 2, \cdots, l$, let T^i be the ith level tree such that $u = r(T^i)$, and let w_i be the closest child of u in T^i. Then each w_i is a distinct child of u in R, and $d(u,v)$ is at least a constant times $d(w_i, u)$.*

Proof : By applying Lemma 3.2 at all levels. ∎

We now introduce a *tree transformation* called f, which is similar to one used in [11]. Let R be any rooted tree with a constant maximum degree. $f(R)$ is another tree defined as follows. The root and the vertex set remain unchanged. For every vertex u in R, order its children in terms of increasing distance from u, say $x_1, x_2, \cdots, x_m$. Remove all edges between u and its children, and reconnect by creating the chain $(u, x_1), (x_1, x_2), \cdots (x_{m-1}, x_m)$. We see that the root of $f(R)$ is degree-1, and the maximum degree of $f(R)$ is 3. If we apply the transformation again and create $f(f(R))$, the degree of the root will remain 1, the maximum degree will remain 3, and additionally the child of the root will have degree at most 2. In general, for the tree $f^i(R)$, the root will form one end of a chain of $i + 1$ vertices, with the rest of the tree connected to the other end of the chain. The maximum degree will remain 3.

Lemma 3.5 *Let R be a tree with a constant degree, such that it is a spanner over its own vertices, with a constant stretch factor. Then $f^i(R)$ is also a spanner with a stretch factor dependent upon i.*

Proof : Let (u, x) be an edge removed from R where u was the parent of x. An alternate path is created from u to x in $f(R)$ consisting of a constant number of edges, the length of each being no more than a constant times $d(u, x)$. Thus $f(R)$ is a spanner. Applying this argument i times proves the lemma. ∎

Step 4: Construct a *transformed forest* from the composite forest as follows. Replace each tree R in the composite forest by the transformed tree $F = f^{l+1}(R)$.

Each F is a spanner over its vertices with a stretch factor dependent only on k, due to Lemmas 3.3 and 3.5. In addition, the root is connected to $l+1$ other vertices by a chain. In this chain the root is degree-1 and the l intermediate vertices are each degree-2. The last vertex on this chain is attached to the rest of the tree, and may be degree-3.

We next show that in the transformed forest the chain attached to each root is quite short compared to the distance between the root and other roots.

Lemma 3.6 *Let u and v be two roots of the transformed forest. Let F be the transformed tree such that $u = r(F)$. Then $d(u, v)$ is at least a constant times the path length in F from u to its lth descendent, where this constant depends on k.*

Proof : Let R be the composite tree such that $u = r(R)$. In R, u has l distinct children $w_1, w_2, \cdots, w_l$ satisfying Lemma 3.4. The tree F is transformed from R, and thus has a chain whose edges are say, $(u, x_1), (x_1, x_2), \cdots (x_l, x_{l+1})$. The final vertex x_{l+1} is connected to the rest of the tree. Since the vertex sets of R and F are the same, one of $w_1, w_2, \cdots, w_l$ has to belong to the subtree of F with root x_l. Suppose this is vertex w_i. We know F is a spanner, so $d_F(w_i, u)$ is at most a constant (which depends on k) times $d(w_i, u)$. But $d_F(w_i, u) \geq d_F(x_l, u)$ because the path from w_i to u has to traverse x_l. So $d_F(x_l, u)$ is at most a constant times $d(w_i, x)$, which in turn is at most a constant times $d(u, v)$ due to Lemma 3.4. ∎

Step 5: Let V^{l+1} be the set of roots of the transformed forest. Construct a degree-s greedy spanner of V^{l+1} using the greedy algorithm in [2, 4]. Let G be the graph formed by the union of the greedy spanner and the transformed forest.

Lemma 3.7 *G is a spanner of V with a stretch factor dependent only on k.*

Proof : Let u and v be two vertices in V. We have to construct a short path between them within G. If both belong to the same transformed tree F, there exists a short path between them within F since F is a spanner over its vertices. Otherwise let u belong to F_1 and v belong to F_2. Let F_1 and F_2 be transformed from the composite trees R_1 and R_2. Arguments similar to those in Lemma 3.3 can be used to show that both $d_{R_1}(u, r(R_1))$ and $d_{R_2}(v, r(R_2))$ are at most a constant (which depends only on k) times $d(u, v)$. Since F_1 and F_2 are themselves spanners, this implies that both $d_{F_1}(u, r(F_1))$ and $d_{F_2}(v, r(F_2))$ are at most a constant (which depends only on k) times $d(u, v)$. This in turn implies that $d(r(F_1), r(F_2))$ is also at most a constant (which depends only on k) times $d(u, v)$. Since the greedy spanner has a short path between $r(F_1)$ and $r(F_2)$, we conclude that the path from u to $r(F_1)$ within F_1, then from $r(F_1)$ to $r(F_2)$ within the greedy spanner, then from $r(F_2)$ to v within F_2 is at most a constant times $d(u, v)$, where the constant depends only on k. ∎

The vertices of G are at most degree-3, except for the vertices in V^{l+1}. Such vertices could potentially be degree-$(s + 1)$, one edge belonging to the transformed tree, and s edges belonging to the greedy spanner. We now show how to reduce these to degree-3.

Step 6: Perform the following at each vertex u in V^{l+1}. Let F be the transformed tree whose root is u. Suppose the chain in F originating from u is $(u, x_1), (x_1, x_2), \cdots, (x_l, x_{l+1})$. Suppose the greedy spanner edges incident at u are $e_1, e_2, \cdots, e_s$. Recall that $l = s - 2$. Shift $e_1, e_2, \cdots, e_{s-2}$ away from u so that they are incident to $x_1, x_2, \cdots, x_{s-2}$ respectively.

Lemma 3.8 *G is a degree-3 spanner of V with a stretch factor dependent only on k.*

Proof : It is easy to see that G is a degree-3 graph. Now consider an edge (u, v) that was shifted in Step 6. Let the new endpoints be (u_1, v_1). The vertex u_1 can be no further away from u than the lth descendent of u in the transformed forest. Similarly the vertex v_1 can be no further away from v than the lth descendent of v in the transformed forest. By applying Lemmas 3.6 and 3.7, there is a path in G from u to v which is at most a constant times $d(u, v)$, where this constant depends only on k. ∎

4   Reducing the Number of Edges

In the previous section we showed how to construct a degree-3 spanner G with a stretch factor dependent only on k. However, graphs with maximum degree of 3 may still have as many as $n \cdot (3/2)$ edges. It is of interest to see whether the proportionality factor can be reduced even further. In this section we show that given any $\delta > 1$, there exists a degree-3 spanner with at most $n \cdot \delta$ edges, and with a stretch factor dependent only on k and δ.

Earlier we had chosen $l = s - 2$. A larger l will work, and the resulting degree-3 spanner will have fewer edges. Select the smallest integer l such that $l \geq s - 2$ and $\delta > 1 + \frac{s}{2^{l+1}}$, and then construct the spanner G as in the previous section. The stretch factor now depends on k and δ.

Lemma 4.1 *G has at most $n \cdot \delta$ edges.*

Proof : There are two types of edges in G: the *forest edges* that belong to the transformed forest, and the *shifted greedy spanner edges* that belonged to the greedy spanner over V^{l+1} but were eventually shifted in Step 6 (see previous section). There are at most $n - 1$ forest edges. Since each transformed tree has at least 2^l vertices, the number of roots in the transformed forest is at most $n/2^l$. Thus the number of shifted greedy spanner edges is at most $n \cdot (\frac{s}{2^{l+1}})$, which implies that the number of edges in G is at most $n \cdot (1 + \frac{s}{2^{l+1}}) - 1$. ∎

5   Estimating the Weight of the Spanner

In this section we provide an upper bound on the total edge weight of the spanner G constructed in Section 3. Let $MST(V)$ denote a minimum spanning tree of the point set V. We show that for $k \leq 3$, $wt(G) = O(1) \cdot wt(MST(V))$. This is an asymptotically optimal result. For $k > 3$ we show that $wt(G) = O(\log n) \cdot wt(MST(V))$. The constants implicit in the O-notation depend on k and δ.

Lemma 5.1 *Let U be a subset of V. The following are true.*

1. *The nearest-neighbor forest of V is contained in $MST(V)$.*

2. *$wt(MST(U)) = O(wt(MST(V)))$.*

Proof : Straightforward.   ∎

The edges of G consist of forest edges and shifted greedy spanner edges. Consider the forest edges first. Let F be a transformed tree which is constructed from a composite tree R. The nature of the transformation makes it easy to see that the weight of F is at most a constant (which depends on k and δ) times the weight of R. Thus the weight of the transformed forest is at most a constant times the weight of the composite forest. But the weight of the composite forest is the sum of the weight of each $FNNF^i$, where $FNNF^i$ is constructed over V^i. But V^i is a subset of V. By applying Lemma 5.1, we can conclude that the transformed forest weighs at most a constant times $MST(V)$, and this constant depends on k and δ.

Now consider the shifted greedy spanner edges. Each edge (u_1, v_1) corresponds to an edge (u, v) of the greedy spanner over V^{l+1}. But $d(u_1, v_1)$ is at most a constant (which depends on k and δ) times $d(u, v)$. Using the known weight bounds on greedy spanners (see Section 2), the fact that V^{l+1} is a subset of V, and Lemma 5.1, we can conclude that the total weight of these edges is $O(1) \cdot wt(MST(V))$ for $k \leq 3$, and $O(\log n) \cdot wt(MST(V))$ for $k > 3$. Thus the weight of the shifted greedy spanner edges dominates the weight of forest edges in an asymptotic sense, and we have proven the following lemma.

Lemma 5.2 *For $k \leq 3$, $wt(G) = O(1) \cdot wt(MST(V))$, and for $k > 3$, $wt(G) = O(\log n) \cdot wt(MST(V))$. The constants implicit in the O-notation depend only on δ and k.*

6   Open Problems

We conclude with some open problems. Most of the problems considered in this paper have their optimization counterparts. For example, given V, design a degree-3 spanner with the minimum stretch factor. Another version is, given a stretch factor, design a spanner with minimum number of edges, or minimum weight. It is likely that most of these are intractable. If so, then good approximation algorithms are necessary.

The weight result has room for improvement. For $k > 3$ the $O(\log n)$ factor needs to be eliminated. Even if that is achieved, we can still ask stronger questions such as, given $\gamma > 1$, are there spanners with weight $wt(MST(V)) \cdot \gamma$?

References

[1] I. Althöfer, G. Das, D.P. Dobkin, D. Joseph, J. Soares: On Sparse Spanners of Weighted Graphs. Discrete and Computational Geometry, 9, 1993, pp. 81-100

[2] B. Chandra, G. Das, G. Narasimhan, J. Soares: New Sparseness Results on Graph Spanners. ACM Symposium on Computational Geometry, 1992, pp. 192-201

[3] G. Das: Approximation Schemes in Computational Geometry. PhD Thesis, CS Dept, Univ of Wisconsin-Madison, 1990

[4] G. Das, P. Heffernan, G. Narasimhan: Optimally Sparse Spanners in 3-Dimensional Euclidean Space. ACM Symposium on Computational Geometry, 1993, pp. 53-62

[5] G. Das, D. Joseph: Which Triangulations Approximate the Complete Graph? International Symposium on Optimal Algorithms, LNCS, Springer-Verlag, 1989

[6] D.P. Dobkin, S.J. Friedman, K.J. Supowit: Delaunay Graphs are Almost as Good as Complete Graphs. Discrete and Computational Geometry, 5, 1990, pp. 399-407

[7] J.M. Keil: Approximating the Complete Euclidean Graph. SWAT, LNCS, Springer-Verlag, 1989

[8] C. Levcopoulos, A. Lingas: There are Planar Graphs Almost as Good as the Complete Graphs and as Short as the Minimum Spanning Trees. Symposium on Optimal Algorithms, LNCS, Springer-Verlag, 1989, pp. 9-13

[9] D. Rupert, R. Seidel: Approximating the d-Dimensional Complete Euclidean Graph. Canadian Conference on Computational Geometry, 1991, pp. 207-210

[10] J.S. Salowe: Construction of Multidimensional Spanner Graphs with Applications to Minimum Spanning Trees. ACM Symposium on Computational Geometry, 1991, pp. 256-261

[11] J.S. Salowe: On Euclidean Spanner Graphs with Small Degree. ACM Symposium on Computational Geometry, 1992, pp. 186-191

[12] J. Soares: Graph Spanners. Ph.D Thesis, Univ. of Chicago Technical Report CS 92-14, 1992

[13] J. Soares: Approximating Complete Euclidean Graphs by Bounded Degree Graphs. Manuscript, 1991

[14] P.M. Vaidya: A Sparse Graph Almost as Good as the Complete Graph on Points in K Dimensions. Discrete and Computational Geometry, 6, 1991, pp. 369-381

Remembering Conflicts in History Yields Dynamic Algorithms*

Katrin Dobrindt[1] and Mariette Yvinec[2]

[1] INRIA, B.P.93, 06902 Sophia-Antipolis cedex, France.
e-mail : dobrindt@sophia.inria.fr
[2] INRIA and Laboratoire I3S, CNRS-URA 1376, 06902 Sophia-Antipolis, France.
e-mail : yvinec@sophia.inria.fr

Abstract. A dynamic algorithm can maintain the solution of a given problem under insertions and deletions of input objects. In this paper we propose a general scheme to obtain dynamic algorithms which is based on the abstract setting introduced by Clarkson and Shor. This scheme uses a novel data structure that combines the conflict graph and the history structure used by incremental algorithms. The randomized analysis of the dynamic algorithms assumes a probabilistic model of the update sequence, in which each currently present input object is equally likely to have been added by the previous insertion or to be deleted by the next deletion. We apply our general technique to obtain new and efficient algorithms for dynamically maintaining arrangements of line segments, lower envelopes of triangles, convex hulls and Voronoi diagrams of points in any dimension, and Voronoi diagrams of line segments in a plane.

1 Introduction

Let us consider an algorithm for solving some problem, like for example the construction of some geometrical structure, for the input objects. Such an algorithm is said to be *static* if it has to know the whole set of input objects in advance. It is *semidynamic* if the solution of the problem can be maintained while new objects are inserted, and *dynamic* when it can deal with deletions of objects as well as with insertions. Designing a dynamic algorithm is an important problem in computational geometry and a much more difficult challenge than solving its static (or semidynamic) counterpart.

Recently, the randomized incremental paradigm introduced by Clarkson and Shor [5] has become very popular in the field of computational geometry, since it leads to simple and efficient algorithms for a lot of problems. In the incremental paradigm, the input objects are added one by one while maintaining the current structure during the course of the computation. The early randomized incremental algorithms [5, 13] are static as they had to maintain a *conflict graph*, which is a bipartite graph between the already constructed structure and the not

* This work has been partly supported by the ESPRIT Basic Research Action Program, under contract No. 7141 (project ALCOM II).

yet inserted objects. Later, semidynamic algorithms [1, 10] appeared where the conflict graph is replaced by a data structure, called *history*, which remembers the successive stages of the construction and which allows to insert a new object without prior knowledge of the whole set of objects. Furthermore, this structure can be used for answering queries.

In this paper we show how to obtain a generic dynamic algorithm by combining conflict graph and history. These algorithms maintain a data structure, called *augmented history*, which can be roughly described as a history structure augmented by a conflict graph. The main principle of the method is that, when handling a deletion, the augmented history is restored in that state where it would have been, if the deleted object had never been inserted. The augmented history is thus a data structure which forgets everything about objects which have been deleted. This is a significant advantage when this structure is to be used additionally to perform some type of queries, like point location queries in arrangements of segments or Voronoi diagrams. The randomized analysis estimates the expected performances of those dynamic algorithms under the following probabilistic model:
– each insertion concerns any object present in the structure after the insertion with the same probability,
– each deletion concerns any object currently present in the structure with the same probability.
This model is a simple extension of the probabilistic model underlying the randomized analysis of static and semidynamic incremental algorithm which assume that the input order is a random permutation of the input.
Our generic algorithm allows the dynamic maintenance of the following geometric structures

1. the arrangement of n line segments in the plane with expected time $O(\log n + \frac{A}{n})$ for an insertion and $O(\log n + \frac{A}{n} \log \log n)$ for a deletion, where A is the current complexity of the arrangement.
2. the upper envelope of a set of n triangles in $\mathbb{R}^3$ with $O(n\alpha(n) \log n)$ expected update time[3],
3. the Voronoi diagram of n points in the plane with $O(\log n)$ expected update time,
4. the Voronoi diagram of n line segments in the plane with $O(\log n)$ expected update time,
5. the convex hull of a set of n points in $\mathbb{R}^d$ with an expected update time of $O\left(\log n + n^{\lfloor \frac{d}{2} \rfloor - 1}\right)$,
6. the Voronoi diagram of a set of points in higher dimensions (which follows by the usual lifting map from 5).

Furthermore in the Case 1 (resp. Cases 2 and 3) point location queries can be answered in $O(\log n)$ (resp. $O(\log^2 n)$) expected time. Note that n is the number of objects currently in the structure. For the Cases 2 and 4 we present the – to our knowledge – first dynamic algorithms.

---

[3] $\alpha(n)$ is the extremely slow growing functional inverse of the Ackermann function.

Schwarzkopf [16] and Mulmuley [15] suggest to make dynamic algorithm by a lazy maintenance of the history structure. In this scheme, the history structure keeps track of every insertion or deletion. Whenever the structure gets too large, it has to be rebuilt by only inserting the present objects. In [15], Mulmuley presents also a dynamisation technique based on the history of a sequence of actual or imaginary updates. Our method can be viewed as an alternative to this method. Apart from these works, dynamisation has been achieved for specific problems, namely in the case of planar points Voronoi diagrams [6] and also in the case of convex hulls in any dimension [4]. In [14], Mulmuley proposes a different approach which is not based on the notion of history, but uses the paradigm of dynamic random sampling. However, this method does not yield efficient algorithms for the dynamic maintenance of convex hulls or upper envelopes.

The paper is organized as follows. Section 2 presents the method of the augmented history in an abstract setting, and Section 3 describes exemplarily the dynamic maintenance of the convex hull and summarizes the expected performances for the other problems listed above.

2 General Framework for Dynamic Algorithms

Geometric problems are formulated according to the abstract setting introduced by Clarkson and Shor [5]. The input of a problem is a finite subset S of a universe of *objects* which typically are points in $\mathbb{R}^d$, segments in the plane, or halfspaces in $\mathbb{R}^d$ etc. The *regions* are defined by subsets of objects of cardinality less than a constant b. Moreover, there is a notion of *conflicts* between objects and regions: each region has a *conflict set* including all the objects of the universe in conflict with this region. Objects, regions and conflicts are defined in such a way that the goal of the algorithm is to construct the regions which are defined by objects in S and which do not conflict with any object in S. Such a region will be called *empty* with respect to S. The set of all empty regions with respect to S is denoted by $\mathcal{F}^0(S)$. As an illustration the reader may think of the convex hull problem. In this case the objects are points in the d-dimensional space, and regions are halfspaces. Each subset of d independent points define two regions: the two halfspaces bounded by the hyperplane passing through these points. A point is in conflict with a halfspace if it is contained in this halfspace.

The current set S of objects present in the structure is the result of a sequence of insertions and deletions. According to the strategy we adopted for dealing with deletions, the history does not keep track of the deleted objects. Indeed, the same object may have been inserted and deleted several times, but the history only keeps track of the last insertion of each object. Thus the structure depends only on the *chronological sequence* $\pi = (x_1, \ldots, x_n)$ which is the sequence of the objects of S sorted by increasing order of the last insertion time. The randomized analysis of the dynamic algorithms assumes that
– each insertion concerns any object present in the structure after the insertion with the same probability,
– each deletion concerns any object present in the structure with the same probability.

Consequently, the sequence π is a random permutation of $\mathcal{S}$. Let $\mathcal{S}_i$ denote the subset of $\mathcal{S}$ consisting of the first i objects $\{x_1, \ldots, x_i\}$ of the chronological sequence.

The *augmented history structure* is obtained for the chronological sequence π of the set $\mathcal{S}$ of objects and denoted by $\mathrm{H}(\pi)$. It is a rooted, directed and acyclic graph, which comprises a node for each region belonging to $\bigcup_{i=1}^{n} \mathcal{F}^0(\mathcal{S}_i)$. In the following, we do not distinguish between a node and its associated region. The augmented history structure is characterized by the two following fundamental properties. Firstly, at each stage in the construction, the empty regions are leaves of the augmented history. Secondly, the conflict set of a region is included in the union of the conflict set of its parents. The root of augmented history is associated to a dummy region whose conflict set is the whole universe of objects. For the deletions, the structure includes a conflict graph between the objects of $\mathcal{S}$ and the nodes of the history: each node has a conflict list containing the objects of $\mathcal{S}$ which conflicts with that node, and each object of $\mathcal{S}$ has a conflict list containing the set of nodes conflicting with this object. Furthermore, the entry for an object x in the conflict list of a region F is linked by a bidirectional pointer to the entry for F in the conflict list of x.

2.1 The Insertion Algorithm

Let x_n be the new object to be added to the structure constructed for $\mathcal{S}_{n-1}$. The regions of $\mathcal{F}^0(\mathcal{S}_{n-1})$ which conflict with x_n are said to be *killed* by x_n, and x_n is called their *killer*. The regions of $\mathcal{F}^0(\mathcal{S}_n)$ defined by subsets including x_n are said to be *created* by x_n, and x_n is called their *creator*. The insertion of the new object x_n is done in two steps. First, in a *location step* we find the regions killed by x_n. This is done by a graph traversal which starts at the root of the history and visits recursively all the nodes which are in conflict with x_n and have not yet been visited. Doing this x_n is added in the conflict list of the nodes it conflicts with and its own conflict list is created. Then, in the *update step* we add a new node to the history structure for each region created by x_n. This node is linked to parents nodes such that its conflict set is included in the union of the conflict set of its parents. Typically, the parents of a node created by x_n are found among nodes which are killed by x_n. In the following we assume that this is actually the case. The details of the update step depend obviously on the application.

2.2 The Deletion Algorithm

Assume that the object x_k with $1 \leq k \leq n$ is deleted from the chronological sequence π. Let $\pi \backslash k = (x_1, \ldots, x_{k-1}, x_{k+1}, \ldots, x_n)$. The algorithm must restore the augmented history structure as it would have been, if this object had never been inserted, i.e. the graph $\mathrm{H}(\pi \backslash k)$.

The differences between $\mathrm{H}(\pi)$ and $\mathrm{H}(\pi \backslash k)$ can be characterized in the following way. A region of $\mathrm{H}(\pi)$ does not belong to $\mathrm{H}(\pi \backslash k)$ if and only if it is defined by x_k. Such a region is said to be *removed*. A region of $\mathrm{H}(\pi \backslash k)$ does not belong to $H(\pi)$ if and only if it is in conflict with x_k and created by an object x_i with $i > k$. Such a region is said to be *new*. A region which belongs to $H(\pi)$ and $H(\pi\backslash k)$ may change some of its parents. This happens for regions which are not defined by x_k, but have parents which are defined by x_k and hence are removed. These regions are called *unhooked* and must be linked to new nodes of $H(\pi\backslash k)$. Since the conflict set of a region is included in the union of the conflict sets of its parents, a new region of $H(\pi\backslash k)$ has at least one of its parents which is in conflict with x_k and thus is either a new region or a region of $H(\pi)$ killed by x_k. However, the other parents of a new node are not necessarily in conflict with x_k.

According to above observations the regions of $\mathcal{F}^0(\mathcal{S}_i\backslash\{x_k\})$ in conflict with x_k play an important role during the reconstruction. In the following, we call *critical* and denote it by C_i the set of regions of $\mathcal{F}^0(\mathcal{S}_i\backslash\{x_k\})$ in conflict with x_k. The critical regions of C_k are the regions of $H(\pi)$ killed by the insertion of x_k. Note, that possibly $C_j = C_{j+1} = \ldots = C_{j+l}$ for certain values of l and j.

We next describe the parts of the deletion algorithm which are independent of the special application. When an object is deleted, we perform a location and an update step.

Location step: In this step we find the regions of C_k, and the removed and unhooked nodes and delete x_k from the conflict lists. This can be done by a graph traversal that starts at the root of the history and visits the nodes in conflict with x_k and continues visiting their descendants defined by x_k. The removed and unhooked nodes are inserted into a dictionary which is used during the update step.

Update step: In this step we create the new nodes of $H(\pi\backslash k)\backslash H(\pi)$ and find new parents for the unhooked nodes of $H(\pi)$. The details of the update depend obviously on the specific problem. However, the general scheme is always the same: we reinsert in chronological order the objects which were inserted after x_k and which are the creator of at least one new or unhooked region. Reinsertion of an object x_i with $i > k$ means that the new nodes created by x_i are added to the history and linked to their parents, and their conflict lists are established, and, furthermore, the unhooked nodes created by x_i are linked to new parents.

Because the parents of a node created by a given object are nodes killed by that object, an object x_i with $i > k$ is reinserted if and only if it is in conflict with at least one critical region of C_{i-1}. Thus, the next object to be reinserted is the object with smallest chronological rank among the objects in conflict with at least one critical region. Let the *new killer* of a region be the first object of $\pi\backslash k$ in conflict with that region. To determine the next object to be reinserted and its conflicting critical regions, the algorithm maintains a priority queue whose entries are the new killers of critical regions and the keys are the chronological rank. For each entry we also keep the set of critical regions it conflicts with.

Let x_i be the next object to be reinserted (at the same time we know the set of regions in C_{i-1} in conflict with x_i). The parents of new nodes created by x_i which are not critical regions of C_{i-1} and the nodes associated to unhooked regions created by x_i are found using the dictionary of removed and unhooked nodes. The conflict lists of new nodes are obtained by merging conflict lists of its critical parents and conflict lists of removed regions. The priority queue of new killers and the associated critical conflict lists are updated accordingly.

2.3 Analysis of the Dynamic Algorithm

The details of the update step for an insertion or a deletion depend on the special application, therefore for a general analysis, we assume that the following update conditions are fulfilled.

Update Conditions.

- we can decide in constant time whether a given object is in conflict with a given region,
- the outdegree of each node of the history is bounded,
- the complexity of the update step of an insertion is proportional to the number of killed or created nodes,
- except for the dictionary and priority queue operations the complexity of the update step of a deletion is proportional to the number of structural changes which include the new and removed nodes, the edges of the history incident to new or removed nodes, the conflict lists of new or removed nodes.

The analysis of the algorithm relies on the following results which are expressed using the function $f^0(r)$. This function describes the expected number of empty regions defined by r-random sample S_r of S.

Lemma 1. a) *The expected number of nodes created by the object x_k is $O(\frac{f^0(k)}{k})$
and the expected number of nodes killed by the object x_k is $O(\frac{f^0(\lfloor k/2 \rfloor)}{k})$.*
b) *The expected total number of nodes in the history structure in conflict with the object x_k is $O(\sum_{i=1}^{k-1} \frac{f^0(i)}{i^2})$.*
c) *The expected number of removed, new and unhooked nodes during a deletion is $O(\frac{1}{n}\sum_{i=1}^{n} \frac{f^0(i)}{i})$.*
d) *The expected number of edges incident to removed or new nodes during a deletion is $O(\frac{1}{n}\sum_{i=1}^{n} \frac{f^0(i)}{i})$.*
e) *The expected total length of conflict lists of removed and new nodes during a deletion is $O(\sum_{i=1}^{n} \frac{f^0(i)}{i^2})$.*

Proof. This lemma is proved in [8, 3].

Theorem 2. *Under the update conditions, the augmented history structure for a set of n objects can be maintained dynamically with*

(i) $O\left(\sum_{i=1}^{n} \frac{f^0(i)}{i^2}\right)$ *expected time per insertion,*

(ii) $O\left(\min\left\{n, \frac{\log\log n}{n}\sum_{i=1}^{n} \frac{f^0(i)}{i}\right\} + \sum_{i=1}^{n} \frac{f^0(i)}{i^2}\right)$ *expected time per deletion,*

(iii) *and* $O\left(n\sum_{i=1}^{n} \frac{f^0(i)}{i^2}\right)$ *expected space.*

Proof. (i) When inserting an objet x_n, it can be located in time $O(\sum_{i=1}^{n} \frac{f^0(i)}{i^2})$ (Lemma 1(b)), and the structure can be updated in time $O(\frac{f^0(|k|)}{k})$ (Lemma 1(a)).

(ii) When performing a deletion, the location step takes time $O(\sum_{i=1}^{n} \frac{f^0(i)}{i^2})$ (Lemma 1(b) and (c)). Except for the dictionary and priority queue operations, the update step can be done in time $O(\sum_{i=1}^{n} \frac{f^0(i)}{i^2})$ (Lemma 1(c), (d) and (e)). Since a region corresponds to a tuple of a finite universe, we can realize the dictionary of removed and unhooked regions as a trie [11] whose nodes are realized by perfect dynamic hashing [7]. In this way all dictionary operations take randomized time $O(1)$. Since the number of dictionary operations is at most proportional to the number of removed and unhooked nodes, their total complexity is $O(\frac{1}{n}\sum_{i=1}^{n} \frac{f^0(i)}{i})$ (Lemma 1(c)). The priority queue contains at most $min\{n, O(\frac{1}{n}\sum_{i=1}^{n} \frac{f^0(i)}{i})\}$ entries (Lemma 1(c))[4]. If $f^0(r)$ is linear or sublinear, it is realized as a bounded priority queue with $O(\log\log n)$ access time ([12]), and otherwise as an array of size n.

(iii) The expected total number of nodes in the history is $O(\sum_{i=1}^{n} \frac{f^0(i)}{i})$ (Lemma 1(a)), and the expected total length of the conflict lists is $O(n\sum_{i=1}^{n} \frac{f^0(i)}{i^2})$, (Lemma 1(b)).

3 Applications

Due to space limitations, we cannot describe in detail how the general scheme applies to each of the problem listed in the introduction. Therefore, we describe exemplarily the dynamic convex hull algorithm, and we only state the performances of the other algorithms. Although the convex hull algorithm is not the most original achievement in this paper (which perhaps is the dynamic maintenance of Voronoi diagram of line segments and of the upper envelope of triangles), it is the most simple application of our method.

3.1 Dynamic Convex Hull

The general framework described in the preceding section can be applied for dynamically maintaining the convex hull of a set of points in $\mathbb{R}^d$. The description given below assumes some familiarity with the incremental construction of a convex hull (see for example [9]).

To employ the abstract setting we redefine the problem in terms of objects, regions, and conflicts. Here the object are points in $\mathbb{R}^d$ and each region is the union of two halfspaces[5]. A region is defined by $d+1$ points. The region defined by the $(d+1)$-tuple of independent points $\{x_0, x_1, \ldots, x_{d-1}, x_d\}$ is the union of the half-space H_d^+ bounded by the hyperplane through the points $\{x_0, x_1, \ldots, x_{d-1}\}$

---

[4] Note, that the sequence of keys returned by a delete-min operation is increasing.
[5] For technical reasons this definition of a region differs slightly from that we suggested in Section 2.

and not containing x_d and of the half-space H_0^+ bounded by the hyperplane through the points $\{x_1, \ldots, x_d\}$ and not containing x_0. A point is in conflict with a region if and only if it is contained in the region. If the set S of points is in general position, each $(d-1)$-face of the convex hull $\mathrm{conv}(S)$ is a simplex, and the regions empty with respect to S may be thought of as defined by pairs of adjacent $(d-1)$-faces of $\mathrm{conv}(S)$. These regions are in bijection with the $(d-2)$-faces of $\mathrm{conv}(S)$.

When a new point x_n is added to the history constructed for the points in S_{n-1}, the location step is standard. If there is no region empty with respect to S_{n-1} in conflict with x_n, the point x_n lies in the interior of the convex hull and we are done. So in the following we suppose that x_n is in conflict with at least one region. A *horizon ridge* is a $(d-2)$-face of $\mathrm{conv}(S_{n-1})$ with the property that exactly one of the halfspaces of the corresponding region contains x_n[6]. For each horizon ridge f of $\mathrm{conv}(S_{n-1})$ we add a node corresponding to the face f of $\mathrm{conv}(S_n)$ to the history. This node is made child of the node corresponding to the face f of $\mathrm{conv}(S_{n-1})$. For each $(d-3)$-face g of $\mathrm{conv}(S_{n-1})$ incident to two horizon ridges f_1 and f_2, a node corresponding to the $(d-2)$-face $\mathrm{conv}(g \cup \{x_n\})$ of $\mathrm{conv}(S_n)$ is added to the history and linked to the the nodes corresponding to the faces f_1 and f_2 of $\mathrm{conv}(S_{n-1})$. It follows that each region has at most d children and at most two parents.

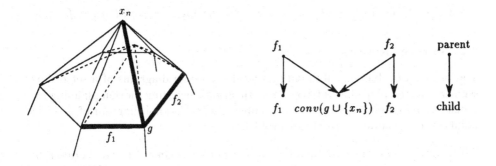

Fig. 1. Insertion of the point x_n in the history in three dimensions

We next show how to delete a point x_k from S_n. The location step and the management of the priority queue of critical killers are standard, so we focus below on the reinsertion of a point x_i. We call *critical horizon ridge* a $(d-2)$-face of $\mathrm{conv}(S_{i-1} \backslash \{x_k\})$ which is horizon ridges with respect to x_i and corresponds to a critical region of C_{i-1}. In the following, we denote the $(d-2)$-faces of $\mathrm{conv}(S_{i-1} \backslash \{x_k\})$ by a lower case letter and their corresponding regions (or nodes) by the corresponding capital letter.

---

[6] The set of horizon ridges of $\mathrm{conv}(S_{n-1})$ with respect to x_n is isomorphic to the set of $(d-2)$-faces of a $(d-1)$-polytope.

Each critical horizon ridge f is a face of $\text{conv}(\mathcal{S}_i\backslash\{x_k\})$ corresponding to a region created by x_i. This region is new, if it conflicts with x_k, and unhooked otherwise. In the former case, a new node is added to the history and linked to F. In the latter case, the unhooked node is searched in the dictionary of unhooked or removed nodes and linked to F.

Let g be a $(d-3)$-face which is incident to two horizon ridges f_1 and f_2 of $\text{conv}(\mathcal{S}_{i-1}\backslash\{x_k\})$ (with respect to x_i), where at least one of them is critical. For each such a $(d-3)$-face g we consider the region Gx_i corresponding to the $(d-2)$-face $\text{conv}(g \cup \{x_i\})$ of $\text{conv}(\mathcal{S}_i\backslash\{x_k\})$. This region Gx_i is either new or unhooked and has the nodes F_1 and F_2 as parents. If Gx_i is new, its associated node is added to the history. If f_1 (resp. f_2) is a critical ridge, the region F_1 (resp. F_2) belongs to C_{i-1} and the node Gx_i can be directly linked to it. Otherwise, the region F_1 (resp. F_2) has necessarily lost a child which was created by x_i and defined by x_k. This child can be found using the dictionary, and F_1 (resp. F_2) is the unremoved parent of this node. If Gx_i is a unhooked node, it is searched in the dictionary and linked to F_1 (resp. F_2) if f_1 (resp. f_2) is a critical ridge.

The main result of this section follows from the preceding paragraph with Theorem 2 knowing that the expected size $f^0(r)$ of the convex hull of a r-random sample of $\mathcal{S}_n$ is $O(r^{\lfloor\frac{d}{2}\rfloor})$.

Theorem 3. *The convex hull of n points in d-space can be dynamically maintained with $O\left(n^{\lfloor\frac{d}{2}\rfloor-1}\right)$ expected update time and $O\left(n^{\lfloor\frac{d}{2}\rfloor}\right)$ expected space if $d > 3$ and $O(\log n)$ expected update time and $O(n\log n)$ expected space if $d \leq 3$.*

3.2 Further Results

The problem of dynamically maintaining the Voronoi diagrams of a set of line segments in the plane has interesting applications including motion planning for a disk in a changing polygonal environment. The Voronoi diagram of a set of points is a special instance of this problem.

Theorem 4. *The expected time to insert a random segment to the Voronoi diagram of n segments or to delete a random segment from it is $O(\log n)$ with $O(n\log n)$ expected space.*

Our framework can be applied to maintain dynamically arrangements of line segments in the plane.

Theorem 5. *The arrangement of n line segments in the plane can be maintained dynamically with $O(\log n + \frac{A}{n})$ expected insertion time and $O(\log n + \min\{n, \frac{A}{n}\log\log n\})$ expected deletion time, where A is the current complexity of the arrangement. The expected space complexity is $O(n\log n + A)$. The structure allows to answer point location queries in $O(\log n)$ time in average[7].*

---

[7] Using Chernoff bounds we can obtain the same bounds with high probability (see [16])

As another interesting application of our framework, we consider the maintenance the upper envelope of a set of triangles in three dimensions under insertions and deletions. The case of insertion was treated in [2].

Theorem 6. *The expected time to insert a random triangle in the upper envelope of n triangles or to delete a random triangle from it is $O(n\alpha(n)\log n)$ with $O(n^2\alpha(n))$ expected space, where $\alpha(n)$ is the extremely slow growing functional inverse of the Ackermann function.*

References

1. J.-D. Boissonnat, O. Devillers, R. Schott, M. Teillaud, and M. Yvinec. Applications of random sampling to on-line algorithms in computational geometry. *Discrete Comput. Geom.*, 8:51–71, 1992.
2. J. D. Boissonnat and K. Dobrindt. Randomized construction of the upper envelope of triangles in $\mathbb{R}^3$. In *Proc. 4th Canad. Conf. Comput. Geom.*, pages 311–315, 1992.
3. J-D. Boissonnat and M. Yvinec. *Structures et algorithmes géométriques*. To appear, 1994.
4. K. L. Clarkson, K. Mehlhorn, and R. Seidel. Four results on randomized incremental constructions. In *Proc. 9th Sympos. Theoret. Aspects Comput. Sci.*, volume 577 of *Lecture Notes in Computer Science*, pages 463–474. Springer-Verlag, 1992.
5. K. L. Clarkson and P. W. Shor. Applications of random sampling in computational geometry, II. *Discrete Comput. Geom.*, 4:387–421, 1989.
6. O. Devillers, S. Meiser, and M. Teillaud. Fully dynamic Delaunay triangulation in logarithmic expected time per operation. *Comput. Geom. Theory Appl.*, 2(2):55–80, 1992.
7. M. Dietzfelbinger, A. Karlin, K. Mehlhorn, F. Meyer auf der Heide, H. Rohnert, and R. E. Tarjan. Dynamic perfect hashing — upper and lower bounds. In *Proc. 29th Annu. IEEE Sympos. Found. Comput. Sci.*, pages 524–531, 1988.
8. K. Dobrindt. Ph.D. thesis, Ecole de Mines de Paris, Paris, France, 1994.
9. H. Edelsbrunner. *Algorithms in Combinatorial Geometry*. Springer-Verlag, Heidelberg, West Germany, 1987.
10. L. J. Guibas, D. E. Knuth, and M. Sharir. Randomized incremental construction of Delaunay and Voronoi diagrams. *Algorithmica*, 7:381–413, 1992.
11. K. Mehlhorn. *Sorting and Searching*, volume 1 of *Data Structures and Algorithms*. Springer-Verlag, Heidelberg, West Germany, 1984.
12. K. Mehlhorn and S. Näher. Bounded ordered dictionaries in $O(\log \log n)$ time and $O(n)$ space. *Inform. Process. Lett.*, 35:183–189, 1990.
13. K. Mulmuley. A fast planar partition algorithm, I. In *Proc. 29th Annu. IEEE Sympos. Found. Comput. Sci.*, pages 580–589, 1988.
14. K. Mulmuley. Randomized multidimensional search trees: dynamic sampling. In *Proc. 7th Annu. ACM Sympos. Comput. Geom.*, pages 121–131, 1991.
15. K. Mulmuley. Randomized multidimensional search trees: lazy balancing and dynamic shuffling. In *Proc. 32nd Annu. IEEE Sympos. Found. Comput. Sci.*, pages 180–196, 1991.
16. O. Schwarzkopf. Dynamic maintenance of geometric structures made easy. In *Proc. 32nd Annu. IEEE Sympos. Found. Comput. Sci.*, pages 197–206, 1991.

Coloring Random Graphs in Polynomial Expected Time

Martin Furer [*]   C.R. Subramanian [†]   C.E. Veni Madhavan [‡]

Abstract

We consider the problem of vertex coloring random k-colorable graphs using k colors. We consider two different models for generating random graphs. We give algorithms for coloring random graphs in these models, with running times polynomial on the average. The first model is discussed in Turner [6] and the second model is discussed in Dyer and Frieze [3]. Our results improve the these current results for this problem by removing the assumption of *constant* edge probability used in these models.

1   Introduction

The problem of 3-coloring 3-colorable graphs is not only NP-complete, it is also quite difficult to approximate. The best known approximation algorithm of Blum [1] is the result of a sequence of impressive improvements but still needs $\Omega(n^{3/8})$ colors. On the other hand, Lund and Yannakakis [5] have shown recently that if $P \neq NP$, then it is not possible to approximate chromatic number within a ratio of $O(n^\epsilon)$ for some constant $\epsilon > 0$. It is still an open question whether similar non-approximability results exist for bounded chromatic graphs.

These difficulties in handling the worst cases, sharply contrast the average case complexity. It is possible to k-color random k-colorable graphs with high probability under suitable probability distributions, provided that the distributions favor graphs with reasonably high edge density. A simple random graph model is $G(n, p, k)$, where the n vertices are partitioned into k equal sized color classes, and then for each pair u, v of vertices belonging to different color classes, the edge u, v is placed in the graph with probability p. We consider two different variations of this basic model.

Model 1: Let $0 < \epsilon < 1$ be fixed, and $p = p(n)$ and $k = k(n)$ be such that $k(n) * p(n) * (1/p(n))^{k(n)} = o(n^{1-\epsilon})$ and $(1/p(n)) = o((logn)^N)$ for some constant N

[*]Department of Computer Science, The Pennsylvania State University, State College, PA - 16801, USA, *email*: furer@cs.psu.edu

[†]Dept. of Computer Science and Automation, Indian Institute of Science, Bangalore 560 012, India, *email*: crs@csa.iisc.ernet.in

[‡]Dept. of Computer Science and Automation, Indian Institute of Science, Bangalore 560 012, India, *email*: cevm@csa.iisc.ernet.in

and $p(n)$ is bounded above by a constant less than one. Then the graphs of model $G1(n, p, k)$ are generated as follows: Let $V = \{1, \ldots, n\}$. Randomly partition V into $V_1, \ldots, V_k$ such that each V_i has at least $n/2k$ vertices. For all i, for all $u \in V_i$, let $c(u) = i$. For each pair $u, v \in V$ such that $c(u) \neq c(v)$, include the edge u, v in E with probability at least $p(n)$.

In the above model, neither the bound on the chromatic number k, nor the edge probability p is fixed; but both are allowed to vary with the size of the graph. It follows from the definition above that $k(n) = O(\log n)$ as long as $p(n)$ is bounded above by a constant less than one.

Model 2: Let $0 < \epsilon < 1$ be fixed, and let k be a fixed positive integer. Let $V = \{1, \ldots, n\}$. Partition V into k subsets $V_1, \ldots, V_k$, where $|V_i| = \Theta(n)$ for each i. Then the graphs of model $G2(n, p, k)$ are generated as follows: For each pair u, v of vertices belonging to different subsets in the collection $\{V_1, \ldots, V_k\}$, include the edge u, v in E with probability at least $p(n)$, where $p(n)$ is the probability function. In both models, the edges are chosen independently.

Turner [6] considers a restricted version of Model 1, in which the edge probability $p(n)$ is required to be a fixed constant. For graphs given by this model, his algorithm colors them with high probability, i.e. with probability $1 - o(1)$. However, as per the analysis given in Turner [6], the probability of failure of his algorithm is *not* exponentially low. In this paper, we show that Turner's algorithm works even on the more general Model 1. We also give an improved analysis of his algorithm, which enables us to prove that the probability of failure is exponentially low.

In the case of Model 2, Furer and Subramanian [4] have shown that a conceptually very simple algorithm succeeds with very high probability, i.e. exponentially low failure probability. Their analysis shows that the algorithm of [4] succeeds even when the edge probability is as low as $p \geq n^{(-2k)/((k-1)(k+2))+\epsilon}$. Actually, their analysis carries over to the more general semi-random balanced model $G_{SB}(n, p, k)$ also. Independently, Blum and Spencer [2] have also obtained the same result.

The above mentioned algorithms can always be translated into an algorithm whose running time is polynomial on the average, provided the failure probability is very very low. This is the approach used by Dyer and Frieze [3] in their paper. However, their approach works only for model $G2$ and when both the bound on the chromatic number k, and the edge probability p are constant. In this paper, we give a different algorithm, which can be combined with the improved analyses that we have mentioned before, to give an algorithm with polynomial expected running time and which works for both models $G1$ and $G2$. In particular, in the case of model $G2$, the edge probability need not be constant and can go to zero asymptotically. We derive our results for both models only when all edge probabilities are equal to $p(n)$. It is easy to extend the results to the case when all edge probabilities are at least $p(n)$.

2 Algorithms for Model 1

We describe the algorithm given in Turner [6] for coloring random graphs given by $G1(n, p, k)$. In the following the parameter k is assumed to be an input parameter.

Color1(G,n,k)
1. Find a $k - clique$ of G.
2. Color the k-clique found in *step1* with k different colors.
3. *while* there are uncolored vertices *do*

> Select an uncolored vertex u such that all but one of the k
> colors have already been used to color some or all neighbors of u.
> (In this case the vertex u is forced to take the
> remaining unused color).
> If there is no such vertex u, exit. /* failure */

*end (* Color 1 *)*
*Note*1 : Once the clique is chosen in *Step*1, the success or failure of the algorithm in coloring all the vertices does not depend on the choice of vertex u in *Step*3.

We give below an analysis of the above algorithm which gives a better estimate on the probability of success and which is also simpler than the analysis presented in Turner [6].

Let $G \in G1(n, p, k)$ where p and k satisfy the condition mentioned in the definition of Model 1 for some constant $\epsilon > 0$. We say G has the *clique property* if for every $r < k$, all cliques on r vertices can be extended to $r + 1$-cliques. We have the following fact [6].

Lemma 1 *Let $G \in G1(n, p, k)$. Then $Prob\{$clique property does not hold for $G\} \leq e^{-n^\delta}$ for some constant $0 < \delta < \epsilon$.*

As a corollary of *Lemma 1*, we deduce that *Step*1 of the algorithm can be done in polynomial time ; just the greedy algorithm will give us a k-clique.
We have the following simpler and refined analysis.

Lemma 2 *Let $G \in G1(n, p, k)$. Let $V_1, \ldots, V_k$ be the k color classes of G. Let $x_i \in V_i, 1 \leq i \leq k$, be fixed vertices of G. Assume $\{x_1, \ldots, x_k\}$ is the clique found in Step1 of Color1. Suppose for all i, vertex x_i is given color i. Then, with probability $\geq 1 - e^{-n^\alpha}$, for some constant $\alpha > 0$, the algorithm succeeds in coloring all vertices of V_i with color i, for all i.*

Proof : Let $S = \{x_1, \ldots, x_k\}$. Fix some i, $1 \leq i \leq k$. Then, any vertex $u \in V_i$ is adjacent to all vertices of $S - \{x_i\}$ with probability p^{k-1}. Since $n_i \geq n/2k$, with probability greater than $1 - e^{-n^\epsilon}$, at least n^ϵ vertices of V_i will be adjacent to all vertices of S except x_i. Hence, by *Note*1 mentioned above, they will all be given color i. Let S_i denote the set of all such colored vertices in V_i. Similar statements hold for other values of i also. Now for each i, $1 \leq i \leq k$, at least n^ϵ vertices of V_i have been given color i. Now take any uncolored vertex u in any V_i. Then with probability greater than $1 - ke^{-p*n^\epsilon}$, u will be adjacent to at least one vertex in each S_j, for $j \neq i$. Hence with that much probability, u will be given color i. Hence for some constant $\alpha > 0$, with probability greater than $1 - e^{-n^\alpha}$, all vertices of V_i will be given color i, for all $1 \leq i \leq k$. ∎

Lemma 2 holds true for a fixed choice of vertices $x_1, \ldots, x_k$. But since $k(n) = O(logn)$, Lemma 2 can be extended to hold true for all choices of $x_1, \ldots, x_k$. Hence we have the following

Theorem 1 *Let $G \in G1(n, p, k)$. Then, Color1 will succeed on G with probability $\geq 1 - e^{-n^\alpha}$, for some constant $\alpha > 0$.*

The failure probability of $Color1$ is exponentially low, but still not sufficiently low for brute-force coloring. We describe below a randomized algorithm which is applied when $Color1$ fails, whose failure probability is $o(e^{-n*logk(n)})$.
Color2(G,n,k)

if ($k = 2$) *then* obtain a 2-coloring of G.
$m = (logn)^M$; $r = n^2 * n^m$; $l = 4 * (k(n))^2/p(n)$;
(* Here $m(n)$ is such that $(1/p(n)) * logn = o(m(n))$. *)
repeat r times
begin
    Choose $W \subseteq V(G)$ at random with $|W| = m$;
    If W is not independent go to the next random choice of W;
    $U = \{u \in V(G)|u$ is not adjacent to all vertices in $W\}$;
    $Y = V(G) - U$;
    for all subsets $X \subseteq U$ of size at most l *do*
    begin
        if $U - X$ is *not* independent go to the next subset;
        $G1 =$ subgraph induced by vertices in Y and X.
        Color2(G1, n, k-1) ;
        if *Color2(G1)* succeeds and gives a valid $(k - 1)$ coloring *then*
        begin output the k-coloring of G; /* success */
        end;
    end;
end;
end (Color2 *)*

Note that in the recursive call $Color2(G1, n, k - 1)$, we are using n instead of $|X| + |Y|$. This is to make sure that the values of m, r, l remain the same in all recursive calls. This fact is used in the analysis of the algorithm.

One can verify that the running time T2(n) of Color2 is $O(n^{(logn)^M * k(n) + 4*(k(n))^3/p(n)})$.

Theorem 2 *Prob$\{Color2$ fails on $G1(n, p, k)\} = o(e^{-n*k})$.*

Proof: For $1 \leq i \leq k$, $1 \leq j \leq \lfloor n_i/m \rfloor$, let $V_{i,j}$ be fixed partitions of the color classes V_i, such that size of each $V_{i,j}$ is m. In what follows we merely use n_i/m for $\lfloor n_i/m \rfloor$. In each V_i, at most m-1 vertices remain uncovered by the subsets $V_{i,j}$. We already have $|V_i| = n_i \geq n/2k$.

For $1 \leq i \leq k - 2$, let $W_i = \{u | u \in V_j, \ for \ some \ j > i\}$. For $W \subseteq V_i$, denote by $f(W)$ the set $\{u \in W_i | \ u \ is \ not \ adjacent \ to \ all \ vertices \ in \ W\}$. Then we have

Claim1: For any i, $1 \leq i \leq k - 2$, with probability at least $(1 - e^{-n*k*3/2})$, there exists some subset $V_{i,j}, 1 \leq j \leq n_i/m$, such that the size of the set $f(V_{i,j})$ is at most $4 * (k(n))^2/p(n)$.

Proof of Claim1: Let $U_i = \{V_{i,1}, \ldots, V_{i,n_i/m}\}$. Let $l = 4 * (k(n))^2/p(n).P(W_i, l) = \{X \subseteq W_i | |X| = l\}$. Let $G(U_i, l) = \{h : U_i \rightarrow P(W_i, l)\}$ = set of all possible assignments of elements from $P(W_i, l)$ to elements in U_i. We have $|G(U_i, l)| = O(n^{l*n_i/m})$.

For any set A of vertices, define $g(A) \subseteq A$ to be the set of l lowest numbered vertices in A. Now consider all possible assignments of values to $f(V_{i,j})$. $F(U_i, l) =$ set of all possible assignments of values of to each $f(V_{i,j})$ such that each $f(V_{i,j})$ has at least l elements. Now we can use $G(U_i, l)$ to partition $F(U_i, l)$ as follows: For each $h \in G(U_i, l)$, we have $F_h \subseteq F(U_i, l)$ defined by $F_h = \{h1 \in F(U_i, l) | \ for \ all \ j, \ g(h1(V_{i,j})) = h(V_{i,j})\}$. In other words, all those assignments of values to $f(V_{i,j})$ which have the same *g-value* are grouped into a single subset. Let us estimate the probability that a given random graph $G \in G1(n, p, k)$ has assignment of values for $f(V_{i,j})$ belonging to F_h, for a fixed $h \in G(U_i, l)$.

$Prob\{G \in G1(n, p, k) \ has \ (f(V_{i,1}), \ldots, f(V_{i,n_i/m})) \in F_h\}$
$\leq (1 - p)^{m*l*n_i/m} * (\sum_{0 \leq m_1, \ldots, m_{n_i/m} \leq |W_i| - l} (W_i - l)^{\sum m_j} * (1 - p)^{m * \sum m_j})$
$\leq (1 - p)^{l*n_i} * (\sum_{0 \leq m_j \leq |W_i| - l, 1 \leq j \leq n_i/m} (n * (1 - p)^m)^{\sum m_{\cdot \cdot}})$
$\leq (1 - p)^{l*n_i} * n^{n_i/m}$ since $(n * (1 - p)^m) \leq 1$.

Thus the required probability is given by $Prob\{|f(V_{i,j})| \geq l$, for all $j, 1 \leq j \leq n_i/m\} = O(n^{l*n_i/m} * n^{n_i/m} * e^{-p*n_i*l}) = O(e^{n_i*l*((logn/m)-p)}) = O(e^{-(3/2)*n*k(n)})$ since $(1/p(n)) * logn = o(m)$ by definition of m. This proves *Claim1*.

Hence with probability at least $(1 - ke^{-n*k*3/2})$, *Claim1* is true for every i, $1 \leq i \leq k - 2$. Call a subset $W \subseteq V_1$ of size m, *good* if the size of $f(W)$ is at most $4 * (k(n))^2/p(n)$. Now with probability at least $q(n, k) = (1 - e^{-n*k*3/2})$, we have a good subset $W \subseteq V_1$. Let $l = 4 * (k(n))^2/p(n)$. Once a good subset W has been chosen by the algorithm in the repeat loop, the set U will contain the whole of V_1 plus at most l vertices from W_1. Now this set of at most l vertices from W_1 will eventually be removed by the for loop from U and $G1$ becomes the subgraph induced by precisely the vertices of W_1. Since there are only $O(n^m)$ subsets of size m, with probability at least $(1 - e^{-n^2})$ the repeat loop picks such a good subset W. Now using the independence of edges, we have $Prob\{ Color2(G, n, k) \ succeeds\} \geq q(n, k) * (1 - e^{-n^2}) * Prob\{ Color2(G1, n, k - 1) \ succeeds\}$.

Now since $|V_i| \geq n/2k$, and *Claim1* is true for all $i, 1 \leq i \leq k - 2$, using induction on k, we have

$$Prob\{Color2\ fails\ on\ G1(n,p,k)\} = o(e^{-n*k})$$

∎

Now when $Color2$ fails, we apply the brute-force coloring algorithm to color G. This will take $O(n^2 * k^n)$ time. If $T(n)$ denotes the total running time of the algorithm, we have $E(T(n)) = O(n^2) + T2(n) * Prob\{Color1\ fails\} + O(n^2 * k^n) * Prob\{both\ Color1\ and\ Color2\ fail\} = O(n^2)$.

Hence graphs from Model $G1(n,p,k)$ can be colored with k colors in polynomial expected time.

3 Algorithms for Model 2

Now consider graphs given by $Model2$. We give polynomial expected time algorithms for coloring graphs given by this model. Dyer and Frieze [3] have given such algorithms for graphs from this model. However, their algorithms work only when both k, the chromatic number, and $p(n)$, the edge probability are constant. In this section, we describe coloring algorithms with polynomial expected running time which work even for graphs with edge probability as low as $p \geq n^{-\beta}$, for some positive constant $\beta > 0$. Now consider the following coloring algorithm.

$Color3(G,n,k)$
$while$ there are vertices u, v adjacent to the same $(k-1)$-clique do
    identify u and v.
end (* $Color3$ *)

If the result of $Color3$ is a graph on k vertices we have obtained a k-coloring of the input graph G.

Let k be a fixed positive integer and $G \in G2(n,p,k)$ be a random graph where $p(n) \geq n^{-\alpha+\epsilon}$ with $\alpha = (2k)/((k-1)(k+2))$ and $\epsilon > 0$ is any constant. Furer and Subramanian [4] have shown that $Color3$ succeeds on G with probability $\geq 1 - e^{-n^{A\epsilon}}$ where A is a positive real constant depending on k. We can take A to be 1 for all values of k. We hope that a more careful analysis yields larger values of A. Let $\beta < \alpha$ be any positive constant such that $\beta < A(\alpha - \beta)$. Let $G \in G2(n,p,k)$ be a random graph where k is fixed and $p \geq n^{-\beta}$. We first apply $Color3$ over G. If it succeeds in k-coloring G, we stop. Otherwise we apply the algorithm $Color2$ described in the previous section with the following values for the parameters used in $Color2$: $m(n) = (1/p(n)) * (logn)^2$; $l(n) = (2*k*C)/(p(n))$, where C is the constant hidden in the $\Theta(n)$ notation used in the definition of $Model2$. That is, each color class has size atleast n/C.

It can be verified that the running time of Color2 is $O(n^{m(n)*k(n)+l(n)*k(n)})$ with the values of $m(n)$ and $l(n)$ defined as above. Also, using the same analysis given in the proof of Theorem 2, we can derive

Theorem 3 $Prob\{Color2\ fails\ on\ G2(n,p,k)\} = o(e^{-n*k})$.

When *Color2* fails, we apply brute-force coloring over G. This will take $O(n^2 * k^n)$ time. Now if $T(n)$ denotes the running time of coloring random graph $G \in G2(n, p, k)$, then we have

$E(T(n)) = O(n^k) + (Running\ time\ of\ Color2) * Prob\{Color3\ fails\} +$
$O(n^2 * k^n) * Prob\{both\ Color3\ and\ Color2\ fail\} = O(n^k).$

Hence as before we conclude that graphs from the model $G2(n, p, k)$ can be colored with k colors in polynomial expected time.

4 Conclusions

We have given randomized algorithms, with polynomial expected running time, for coloring random graphs from two different models. In both models, our results improve the previous bounds on the probability of choosing edges. We believe that our algorithms can still be used, even when the edge probabilities are further reduced. We have made some significant progress in this direction. One disadvantage of our algorithms is that they require the knowledge of k and $p(n)$ beforehand. It would be interesting to verify if we can overcome this disadvantage. In case of *Model2*, k is part of the definition and only $p(n)$ is the unknown. However, the knowledge of $p(n)$ is not essential, and we can modify the algorithm so that it will work even without knowing $p(n)$. The details of doing this alongwith improvements of the results mentioned here are to appear in a forthcoming paper. An interesting problem is to devise similar algorithms for other models of random graphs.

References

[1] A. Blum, Some Tools for Approximate 3-Coloring, *FOCS*, 1990,554-562.

[2] A. Blum and J. Spencer, Coloring Random and Semi-Random k-Colorable Graphs, *Submitted to Journal of Algorithms*.

[3] M.E. Dyer and A.M. Frieze, The solution of Some Random NP-Hard Problems in Polynomial Expected Time, *J. Alg.*,**10**, 1989, 451-489.

[4] M.Furer and C.R. Subramanian, Coloring Random Graphs, *SWAT*,1992.

[5] C. Lund and M. Yannakakis, On the Hardness of Approximating Minimization Problems, Proc. of Worshop on Approx. Algos. New Delhi, Dec 1992.

[6] J.S. Turner, Almost All k-colorable Graphs are Easy to Color, *J. Alg.*,**9**, 1988, 63-82.

Graphical Degree Sequence Problems
with Connectivity Requirements

Takao Asano *

Department of Information and System Engineering
Chuo University, Bunkyo-ku, Tokyo 112, Japan

Abstract. A sequence of integers $D = (d_1, d_2, ..., d_n)$ is *k-conneted graphical* if there is a k-connected graph with vertices $v_1, v_2, ..., v_n$ such that $deg(v_i) = d_i$ for each $i = 1, 2, ..., n$. The k-connected graphical degree sequence problem is: Given a sequence D of integers, determine whether it is k-connected graphical or not, and, if so, construct a graph with D as its degree sequence. In this paper, we consider the k-connected graphical degree sequence problem and present an $O(n \log \log n)$ time algorithm.

1 Introduction

A sequence of integers $D = (d_1, d_2, ..., d_n)$ is a *(k-connected) graphical degree sequence* (for short, *(k-connected) graphical*) if there is a *(k-connected)* graph with vertices $v_1, v_2, ..., v_n$ such that $deg(v_i) = d_i$ for each $i = 1, 2, ..., n$. Here, $deg(v_i)$ is the degree of v_i. The *(k-connected)* graphical degree sequence problem is: Given a sequence D of integers, determine whether it is *(k-connected)* graphical or not, and, if so, construct a graph with D as its degree sequence. The graphical degree sequence problem is one of the most fundamental problems in graph theory [1, 5, 9] and was first considered by Havel [6] and then considered by Erdös and Gallai [3] and Hakimi [4]. They gave simple characterizations which lead to efficient algorithms. The k-connected graphical degree sequence problems were considered by Berge [1]$(k = 1, 2)$, by Rao and Ramachandra Rao [7] $(k = 3)$ and by Wang and Kleitman [12] $(k \geq 2)$. They gave characterizations for a graphical degree sequence to be k-connected.

In this paper, we consider the k-connected graphical degree sequence problem for any positive integer k and present efficient algorithms including an $O(n)$ time algorithm to determine whether a sequence of integers $D = (d_1, d_2, ..., d_n)$ is k-connected graphical or not and an $O(m)$ time algorithm to construct a k-connected graph with D as its degree sequence if D is k-connected graphical $(m = \sum_{i=1}^{n} s_i/2)$.

This $O(m)$ time algorithm represents graphs explicitly. In some applications, however, implicitly represented graphs are satisfactory and faster algorithms may be required. In this paper, we also consider an implicit representation of graphs and present a faster $O(n \log \log n)$ time algorithm for constructing a k-connected graph with D as its degree sequence for a given k-connected graphical degree sequence D.

---

*Supported in part by Grant in Aid for Scientific Research of the Ministry of Education, Science and Culture of Japan and by the Alexander von Humboldt Foundation.

2 k-Connected Graphical Degree Sequences

In this section, we consider the k-connected graphical degree sequence problem. We first recall the previous results. Havel [6] and Hakimi [4] gave Proposition 1 independently and Erdös and Gallai [3] gave Proposition 2 below. Their proofs can be found in standard books of graph theory [1, 5, 9].

Proposition 1. Let $D = (d_1, d_2, ..., d_n)$ be a sequence of integers with $n > d_1 \geq d_2 \geq ... \geq d_n \geq 0$ and let $D' = (d'_1, d'_2, ..., d'_{n-1})$ be a sequence of integers obtained from D by setting $d'_i = d_{i+1} - 1$ $(i = 1, 2, ..., d_1)$ and $d'_j = d_{j+1}$ $(j = d_1 + 1, ..., n - 1)$. Then D is graphical if and only if D' is graphical.

Proposition 2. Let $D = (d_1, d_2, ..., d_n)$ be a sequence of integers with $n > d_1 \geq d_2 \geq ... \geq d_n \geq 0$. Then D is graphical if and only if $\sum_{j=1}^{i} d_j \leq i(i - 1) + \sum_{j=i+1}^{n} \min\{i, d_j\}$ for each $i = 1, 2, ..., n$.

For a sequence of integers $D = (d_1, d_2, ..., d_n)$ with $n > d_1 \geq d_2 \geq ... \geq d_n \geq 0$, if $\sum_{i=1}^{n} d_i$ is odd then D is not graphical. Thus, we assume throughout this paper that $\sum_{i=1}^{n} d_i$ is even and $m = \sum_{i=1}^{n} d_i / 2$.

For a graphical degree sequence D, it is trivial to obtain a graph G with D as its degree sequence in $O(n^2)$ time based on Proposition 1. A drawback to use Proposition 1 is that $d'_1 \geq d'_2 \geq ... \geq d'_{n-1}$ does not always hold. Thus we have to sort again to use Proposition 1 recursively. If the proposition is modified to avoid sorting in the following way, then the time complexity can be reduced to $O(m)$ [8].

Proposition 3. Let $D = (d_1, d_2, ..., d_n)$ be a sequence of integers with $n > d_1 \geq d_2 \geq ... \geq d_n > 0$ and let $C = (c_1, c_2, ..., c_{n-1})$ be defined by using $h = d_n \geq 0$, $x = \min\{j | d_j = d_h\}$ and $y = \max\{j | j \leq n - 1, d_j = d_h\}$ as follows.

$$c_i = \begin{cases} d_i - 1 & \text{if } 1 \leq i \leq x - 1 \text{ or } y - h + x \leq i \leq y, \\ d_i & \text{if } x \leq i \leq y - h + x - 1 \text{ or } y + 1 \leq i \leq n - 1. \end{cases}$$

Then $c_1 \geq c_2 \geq ... \geq c_{n-1} \geq 0$ and D is graphical if and only if C is graphical.

Proposition 3 can be shown in almost the same way as Proposition 1. The above results are all without connectivity requirement. For 1-connectivity requirement, the following Proportion 4 is in Berge [1] and Proposition 5 is obtained similalry.

Proposition 4. Let $D = (d_1, d_2, ..., d_n)$ be a sequence of integers with $n > d_1 \geq d_2 \geq ... \geq d_n \geq 0$ $(n \geq 2)$. Then D is 1-connected graphical if and only if (i) D is graphical, (ii) $d_n \geq 1$, and (iii) $\sum_{i=1}^{n} d_i \geq 2(n - 1)$.

Proposition 5. Let $D = (d_1, d_2, ..., d_n)$ be a sequence of integers with $n > d_1 \geq d_2 \geq ... \geq d_n > 0$ and let $C = (c_1, c_2, ..., c_{n-1})$ be the same as in Proposition 3. Then D is 1-connected graphical if and only if C is 1-connected graphical.

Base on Proposition 4, we can determine whether D is 1-connected graphical or not in $O(n)$ time, since (ii) and (iii) can be trivially checked in $O(n)$ time, and (i) can be checked in $O(n)$ time based on Proposition 2 [8].

On the other hand, based on Proposition 5, one can easily obtain an $O(m)$ time algorithm for constructing a connected graph G with D as its degree sequence if D is 1-connected graphical. Thus the following theorem is obtained.

Theorem 1. For a sequence of integers $D = (d_1, d_2, ..., d_n)$ with $d_1 \geq d_2 \geq ... \geq d_n \geq 0$, it can be determined whether D is 1-connected graphical or not in $O(n)$ time. Furthermore, if D is 1-connected graphical, then a connected graph with D as its degree sequence can be obtained in $O(m)$ time, where $m = \sum_{i=1}^{n} d_i/2$.

Note that a graph constructed based on Proposition 1 is not always connected even if D is 1-connected graphical. For the k-connected graphical degree sequence problem ($k \geq 2$), Proposition 6 below was obtained by Wang and Kleitman [12]. Here we give another proof which leads to an efficient algorithm based on Propositions 1 and 3. The proof is almost similar to the proof given by Wang and Kleitman. In fact, our proof can be considered to be a simplified version of their proof.

Proposition 6. Let $D = (d_1, d_2, ..., d_n)$ be a sequence of integers with $n > d_1 \geq d_2 \geq \cdots \geq d_n \geq 0$ ($n \geq k + 1$). Then D is k-connected graphical if and only if (i) D is graphical, (ii) $d_n \geq k$, and (iii) $\sum_{j=1}^{n} d_j \geq 2(n + \sum_{j=1}^{k-1} d_j - k(k-1)/2 - 1)$.

Proof. (Necessity) Let G be a k-connected graph with D as its degree sequence. Then (i) and (ii) are trivially true and $G - \{v_1, v_2, ..., v_{k-1}\}$ is connected. Thus, the number of edges in $G - \{v_1, v_2, ..., v_{k-1}\}$ is greater than or equal to $-1+$ the number of vertices in $G - \{v_1, v_2, ..., v_{k-1}\}$. Since the number of edges in $G - \{v_1, v_2, ..., v_{k-1}\}$ is at most $\sum_{j=1}^{n} d_j/2 - (\sum_{j=1}^{k-1} d_j - (k-1)(k-2)/2)$, we have (iii).

(Sufficiency) We prove the sufficiency by induction on $n \geq k+1$. For $n = k+1$, if D satisfies (i)-(iii) in Proposition 6, then $d_1 = d_2 = \cdots = d_n = n - 1 = k$ since $k = n - 1 \geq d_1 \geq d_n \geq k = n - 1$ and the complete graph K_{k+1} is k-connected and has D as its degree sequence. Thus, we now assume that it is true for all integers less than $n \geq k + 2$. Suppose that D satisfies (i)-(iii) in Proposition 6. We divide into two cases: $d_1 = n - 1$ and $d_1 \leq n - 2$.

Case I: $d_1 = n - 1$. Let $D' = (d'_1, ..., d'_{n-1})$ be the same as in Proposition 1 ($d'_j = d_{j+1} - 1$ for each $j = 1, 2, ..., n - 1$). Since D' satisfies (i)-(iii) of Proposition 6 for $k - 1$, D' is $(k-1)$-connected graphical and there is a $(k-1)$-connected graph G' with D' as its degree sequence by induction hypothesis. The graph G obtained from G' by adding a new vertex v_1 and the edges joining vertex v_1 and all vertices of G' is clearly k-connected and has D as its degree sequence.

Case II: $d_1 \leq n - 2$. Let $C = (c_1, c_2, ..., c_{n-1})$ be the same as in Proposition 3. We have two cases (II-A): $c_{n-1} = k - 1$ and (II-B): $c_{n-1} \geq k$.

Case II-A: $c_{n-1} = k-1$. In this case, $d_n = d_{n-1} = k$ and $d_k = d_{k+1} = \cdots = d_n = k$ by the definition of C, since if $d_k > d_{n-1}$ then c_{n-1} would be $c_{n-1} = d_{n-1} = k$. Furthermore, we have

$$k(k-1) \leq \sum_{j=1}^{k-1} d_j \leq (k-2)n + 2 \tag{1}$$

since $d_1 \geq d_2 \geq ... \geq d_{k-1} \geq k$ and $\sum_{j=1}^{n} d_j = \sum_{j=1}^{k-1} d_j + k(n - k + 1) \geq 2(n + \sum_{j=1}^{k-1} d_j - k(k-1)/2 - 1)$ by (iii). We will actually construct a graph G with D as its degree sequence by dividing this case into two subcases $n - k =$ even and $n - k =$ odd.

Case II-A-1: $n - k =$ even. By Eq.(1), we have

$$k \leq \sum_{j=1}^{k} d_j - k(k-1) \leq (k-2)(n-k) + 2. \tag{2}$$

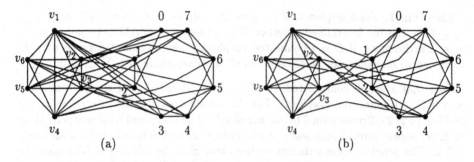

Fig. 1. (a) $G(8,4,(10,9,9,8,7,7,6,6,6,6,6,6,6,6))$ $(n=14,k=6,q=2,r=4,p=4)$ (b) $H(8,5,(8,8,7,7,6,6,6,6,6,6,6,6,6,6))$ $(n=14,k=6,q=1,r=4,p=3)$.

Let $q = \lfloor (\sum_{j=1}^{k} d_j - k(k-1))/(n-k) \rfloor$ and $p = \min\{i | \sum_{j=1}^{i} d_j - i(k-1) \geq q(n-k)\}$. Thus, $0 \leq q \leq k-2$, $\sum_{j=1}^{k} d_j - k(k-1) = q(n-k) + r$ $(0 \leq r < n-k)$ and $p \leq k$. Since $\sum_{j=1}^{n} d_j$ and $n-k$ are both even, $\sum_{j=1}^{k} d_j = \sum_{j=1}^{n} d_j - k(n-k)$ is even and thus r is also even. We have two subcases $k-q$ =even and $k-q$ =odd.

Case II-A-1-a: $k-q$ =even. Let $G(2a,2b)$ $(2a > 2b)$ be the graph with vertex set $\{0,1,...,2a-1\}$ and edge set $\{(i,j)|0 < |(j-i)\bmod 2a| \leq b, 0 \leq i,j \leq 2a-1\}$. Here we assume $(j-i)\bmod 2a$ takes a value in $[-(a-1),(a-1)]$. Note that $G(2a,2b)$ is $2b$-regular and $2b$-connected. We construct a k-connected graph with D as its degree sequence by setting $a := (n-k)/2$; $b := (k-q)/2$; $(k = n-2a, q = k-2b = n-2a-2b)$ and using $G(2a,2b)$ and the complete graph K_{n-2a} with vertices $\{v_1, v_2, ..., v_{n-2a}\}$.

First we add $2aq$ edges between vertices of K_{n-2a} and vertices of $G(2a,2b)$. More specifically, we add $2aq$ edges in a way that

$d_1 - (n-2a-1)$ edges join v_1 with $0,...,d_1 - (n-2a-1) - 1$,

$\cdots$,

$d_{p-1} - (n-2a-1)$ edges join v_{p-1} with

$\quad \sum_{j=1}^{p-2} d_j - (p-2)(n-2a-1),..., \sum_{j=1}^{p-1} d_j - (p-1)(n-2a-1) - 1$

and

the remaining $2aq - (\sum_{j=1}^{p-1} d_j - (p-1)(n-2a-1))$ edges join v_p with

$\quad \sum_{j=1}^{p-1} d_j - (p-1)(n-2a-1),...,2a-1$

by visiting vertices $0,1,...,2a-1$ of $G(2a,2b)$ cyclically q times (thus, $+/-$ is mod $2a$ addition/subtraction). Then we delete $r/2$ edges $(0,1),(2,3),...,(r-2,r-1)$ from $G(2a,2b)$ and add r edges between $\{v_p,...,v_k\}$ and $\{0,1,...,r-1\}$. More specifically,

$\sum_{j=1}^{p} d_j - p(n-2a-1) - 2aq$ edges join v_p with

$\quad 0,..., \sum_{j=1}^{p} d_j - p(n-2a-1) - 2aq - 1,$

$d_{p+1} - (n-2a-1)$ edges join v_{p+1} with

$\quad \sum_{j=1}^{p} d_j - p(n-2a-1) - 2aq,..., \sum_{j=1}^{p+1} d_j - (p+1)(n-2a-1) - 2aq - 1,$

$\cdots$,

$d_{n-2a} - (n-2a-1)$ edges join v_{n-2a} with

$\quad \sum_{j=1}^{n-2a-1} d_j - (n-2a-1)^2 - 2aq,...,r-1.$

Finally we identify each vertex i of $G(2a,2b)$ with vertex $v_{n-2a+i+1}$. It is an easy exercise to show that the graph $G(2a,2b,D)$ obtained in this way (Fig.1(a)) is k-connected and has D as its degree sequence.

Case II-A-1-b: $k - q$ =odd. Let $H(2a, 2b + 1)$ $(2a > 2b + 1)$ be the graph with vertex set $\{0, 1, ..., 2a - 1\}$ and edge set $\{(i, j)|0 < |(j - i)\bmod 2a| \leq b, 0 \leq i, j \leq 2a-1\}\cup\{(i, i+a)|i = 0, 1, ..., a-1\}$ ($(j-i)\bmod 2a$ takes a value in $[-(a-1), (a-1)]$). Note that $H(2a, 2b + 1)$ is obtained from $G(2a, 2b)$ by adding edges $\{(i, i + a)|i = 0, 1, ..., a - 1\}$ and thus $H(2a, 2b + 1)$ is $(2b + 1)$-regular and $(2b + 1)$-connected. We construct a k-connected graph with D as its degree sequence by setting $a :=$ $(n - k)/2$; $b := (k - q - 1)/2$; $(k = n - 2a, q = k - 2b - 1 = n - 2a - 2b - 1)$ and using $H(2a, 2b + 1)$ and the complete graph K_{n-2a} with vertices $\{v_1, v_2, ..., v_{n-2a}\}$.

First we add $2aq$ edges between vertices of K_{n-2a} and vertices of $H(2a, 2b + 1)$ $(G(2a, 2b))$ in the same way as in Case II-A-1-a. Then we delete $r/2$ edges $(0, a), (1, 1+a), ..., (-1+r/2, -1+a+r/2)$ from $H(2a, 2b+1)$ and, by labeling these vertices $0, a, 1, 1 + a, ..., -1 + r/2, -1 + a + r/2$ as $\pi(0), \pi(1), ..., \pi(r - 1)$, we add r edges between $\{v_p, ..., v_k\}$ and $\{\pi(0), \pi(1), ..., \pi(r - 1)\}$. More specifically,

$\sum_{j=1}^{p} d_j - p(n - 2a - 1) - 2aq$ edges join v_p with
$\quad \pi(0), ..., \pi(\sum_{j=1}^{p} d_j - p(n - 2a - 1) - 2aq - 1)$,
$d_{p+1} - (n - 2a - 1)$ edges join v_{p+1} with
$\quad \pi(\sum_{j=1}^{p} d_j - p(n-2a-1) - 2aq), ..., \pi(\sum_{j=1}^{p+1} d_j - (p+1)(n-2a-1) - 2aq - 1)$,
$\cdots$,
$d_{n-2a} - (n - 2a - 1)$ edges join v_{n-2a} with
$\quad \pi(\sum_{j=1}^{n-2a-1} d_j - (n - 2a - 1)^2 - 2aq), ..., \pi(r - 1)$.

Finally we identify each vertex i of $H(2a, 2b + 1)$ with vertex $v_{n-2a+i+1}$. The graph $H(2a, 2b + 1, D)$ obtained in this way (see Fig. 1(b)) is k-connected and has D as its degree sequence.

Case II-A-2: $n - k$ =odd. This case is almost the same as Case II-A-1. By Eq.(1), we have

$$2k - 2 \leq \sum_{j=1}^{k-1} d_j - (k - 1)(k - 2) \leq (k - 2)(n - k + 1) + 2. \tag{3}$$

Let $q = \lfloor(\sum_{j=1}^{k-1} d_j - (k - 1)(k - 2))/(n - k + 1)\rfloor$. Thus, $0 \leq q \leq k - 2$ and $\sum_{j=1}^{k-1} d_j - (k - 1)(k - 2) = q(n - k + 1) + r$ $(0 \leq r < n - k + 1)$. Since $\sum_{j=1}^{k-1} d_j$ and $n - k + 1$ are both even, $\sum_{j=1}^{k-1} d_j = \sum_{j=1}^{n} d_j - k(n - k + 1)$ is even and thus r is also even. Let $p = \min\{i | \sum_{j=1}^{i} d_j - i(k - 2) \geq q(n - k + 1)\}$. Then $p \leq k - 1$. We have two subcases $k - q$ =even and $k - q$ =odd.

Case II-A-2-a: $k - q$ =even. If we set $a := (n - k + 1)/2$; $b = (k - q)/2$; $(k = n - 2a + 1, q = n - 2a + 1 - 2b)$ then the graph $G(2a, 2b, D)$ defined in Case II-A-1-a is k-connected and has D as its degree sequence.

Case II-A-2-b: $k - q$ =odd. If we set $a := (n - k + 1)/2$; $b = (k - q - 1)/2$; $(k = n - 2a + 1, q = n - 2a - 2b)$ then the graph $H(2a, 2b + 1, D)$ defined in Case II-A-1-b is k-connected and has D as its degree sequence.

Case II-B: $c_{n-1} \geq k$. We have the following two cases II-B-1 and II-B-2.

Case II-B-1: $\sum_{j=1}^{n-1} c_j = \sum_{j=1}^{n} d_j - 2d_n \geq 2(n + \sum_{j=1}^{k-1} c_j - k(k-1)/2 - 2)$. In this case, C is k-connected graphical by Proposition 3 and by induction hypothesis, and there is a k-connected graph G' with C as its degree sequence. The graph G obtained by adding d_n edges joining vertex v_n and the vertices v_i of G' with $c_i = d_i - 1$ is k-connected and has D as its degree sequence.

Case II-B-2: $\sum_{j=1}^{n-1} c_j = \sum_{j=1}^{n} d_j - 2d_n < 2(n + \sum_{j=1}^{k-1} c_j - k(k-1)/2 - 2)$. We will prove that this case cannot occur.

Let h be an integer such that $c_{n-1} \geq h \geq k$. Let $\alpha = c_{k-1} - c_k$. Clearly $0 \leq \alpha \leq n-k-2$ since $n-2 \geq d_1 \geq d_n \geq k$. Thus, we have $(k-2)(n-2) \geq \sum_{j=1}^{k-2} c_j \geq -\alpha + \sum_{j=k+1}^{n-1} c_j - 2n + k(k-1) + 6 \geq h(n-k-1) - 2n + k(k-1) + 6 - \alpha = (k-2)(n-2) + (h-k)(n-k-1) + 2 - \alpha$ since $\sum_{j=1}^{n-1} c_j \leq 2(n + \sum_{j=1}^{k-1} c_j - k(k-1)/2 - 3)$. This implies that $c_{n-1} = h = k$ and $\alpha \geq 2$. Furthermore, $c_{n-1} = k$ implies $d_n \leq d_{n-1} \leq k+1$ and $\alpha \geq 2$ implies $d_{k-1} \geq d_k + 2$ and that $c_i = d_i - 1$ for all $i \leq k-1$. Thus, we have $(k-1)(n-2) \geq \sum_{j=1}^{k-1} d_j = \sum_{j=1}^{k-1} c_j + k - 1 \geq \sum_{j=k}^{n-1} c_j - 2n + k(k-1) + 6 + k - 1 = \sum_{j=k}^{n-1} d_j - (d_n - (k-1)) - 2n + k(k-1) + k + 5$. This implies that if $d_n = k+1$ then we have $(k-1)(n-2) \geq \sum_{j=k}^{n-1} d_j - (d_n - (k-1)) - 2n + k(k-1) + k + 5 \geq (k+1)(n-k) - 2n + k(k-1) + k + 3 = (k-1)(n-2) + k + 1$ and that if $d_n = k$ then we have by (iii) $\sum_{j=1}^{k-1} d_j \geq \sum_{j=k}^{n-1} d_j - (d_n - (k-1)) - 2n + k(k-1) + k + 5 \geq \sum_{j=1}^{k-1} d_j - 2 - d_n - (d_n - (k-1)) + k + 5 = \sum_{j=1}^{k-1} d_j + 2$. Thus, we have a contradiction in either case.

This completes the proof of Proposition 6.

Based on Proposition 6, it can be determined whether a given sequence of positive integers $D = (d_1, d_2, ..., d_n)$ is k-connected graphical or not in $O(n)$ time. Since the proof of Proposition 6 is constructive, we can obtain an $O(m)$ time algorithm for constructing a k-connected graph with D as its degree sequence if D is k-connected graphical. Thus, we have the following theorem.

Theorem 2. For a sequence of integers $D = (d_1, d_2, ..., d_n)$ with $d_1 \geq d_2 \geq \cdots \geq d_n$, it can be determined whether D is k-connected graphical or not in $O(n)$ time. Furthermore, if D is k-connected graphical, then a k-connected graph with D as its degree sequence can be constructed in $O(m)$ time ($m = \sum_{i=1}^{n} d_i/2$).

3 An $O(n \log \log n)$ Time Algorithm

In this section we present an $O(n \log \log n)$ time algorithm for constructing a k-connected graph represented implicitly based on Proposition 6. Before giving the algorithm, we consider an implicit representation of a graph by using an example and present an $O(n \log \log n)$ time algorithm for constructing an implicitly represented connected graph based on Proposition 5.

Let $D = (6, 5, 5, 4, 3, 3, 3, 3)$. Then D is 1-connected graphical. Based on Proposition 3, we have the following sequence ($D_0 = D, D_1 = C$):

$$D_0 = (6, 5, 5, 4, 3, 3, 3, 3), \ D_1 = (5, 4, 4, 4, 3, 3, 3, 3), \ D_2 = (4, 4, 3, 3, 3, 3),$$
$$D_3 = (3, 3, 3, 3, 2), \ D_4 = (3, 3, 2, 2), \ D_5 = (2, 2, 2), \ D_6 = (1, 1).$$

The graph obtained in this way is shown in Fig.2(a). Note that we can represent vertices adjacent to vertex v_8 by an interval $I_1(v_8) = [1, 3]$ when we obtain D_1 from D_0. Similarly, vertices adjacent to vertex v_7 can be represented by intervals $I_1(v_7) = [1, 1]$ and $I_2(v_7) = [3, 4]$ when we obtain D_2 from D_1. Repeating this, we have a set of intervals shown in Fig.2(b). Thus,

$$E(v_i) = \{(v_j, v_i) | j < i, j \in I_1(v_i) \cup I_2(v_i)\} \cup \{(v_j, v_i) | j > i, i \in I_1(v_j) \cup I_2(v_j)\}$$

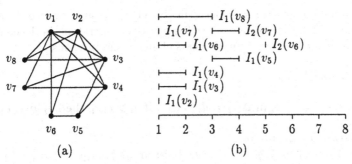

Fig. 2. (a) Graph G with $D = (6, 5, 5, 4, 3, 3, 3, 3)$ as its degree sequence
(b) An implicit representation of G by a set of intervals.

is the set of edges incident on vertex v_i in G ($I_2(v_i)$ and $I_2(v_j)$ may be empty).

A graph G represented by a set of intervals in this way is called *implicitly represented*. Note that, if necessary, the edge set $E(v_i)$ incident on vertex v_i can be obtained efficiently (in $O(|E(v_i)| + \log n)$ time) by solving the 1-dimensional range search and 1-dimensional point enclosure problems in computational geometry [2].

Degree sequence D can also be represented by a set of intervals with weights as follows. Let $\mathcal{J}(D)$ be the set of intervals obtained by partitioning the underlying set $[1, n]$ using the elements i with $d_i > d_{i+1}$. Thus, for each interval $J \in \mathcal{J}(D)$, $d_i = d_j$ if $i, j \in J$. We define the weight $w(J)$ of J to be d_i for any $i \in J$. For example, corresponding to $D = (6, 5, 5, 4, 3, 3, 3, 3)$ we have

$$\mathcal{J}(D) = \{[1, 1], [2, 3], [4, 4], [5, 8]\}$$

with $w([1, 1]) = 6, w([2, 3]) = 5, w([4, 4]) = 4, w([5, 8]) = 3$. If $D = D_0$ is modified to $D_1 = (5, 4, 4, 4, 3, 3, 3)$, then $\mathcal{J}(D)$ and the weigths will be updated to

$$\mathcal{J}(D_1) = \{[1, 1], [2, 4], [5, 7]\}, \quad w([1, 1]) = 5, w([2, 4]) = 4, w([5, 7]) = 3.$$

Thus, if we represent a degree sequence in terms of the corresponding set of intervals as above, we need three kinds of operations (*find*, *delete* and *split*) to efficiently manipulate such disjoint intervals. Here, $find(i)$ returns the maximum element in the interval containing i. $delete(i)$ unites the interval J containing i and the interval J' containing $i + 1$ (old J and J' will be destroyed). If i is the maximum element of a current underlying set, then $delete(i)$ sets $J := J - \{i\}$ (and J will be removed if J becomes empty). $split(i)$ splits the interval J containing i into two intervals $J' := \{j \in J | j \leq i\}$ and $J'' := \{j \in J | j > i\}$ (old J will be destroyed and J'' will be removed if $J'' = \emptyset$).

Initially $\mathcal{J}(D)$ can be obtained by $split(i)$ for each i with $d_i > d_{i+1}$. Thus, in the example above, $\mathcal{J}(D)$ is obtained by a sequence of $split(1), split(3), split(4)$. Similarly, $\mathcal{J}(D_1)$ is obtained from $\mathcal{J}(D)$ by a sequence of $delete(8), find(3), delete(3)$.

For each i, we initially set $\Delta(n) := d_n$ and $\Delta(i) := d_i - d_{i+1}$ for each $i \neq n$. Thus $d_i = \sum_{j=i}^{n} \Delta(j)$. In each iteration with the current degree sequence $C = (c_1, c_2, ..., c_g)$, Δ is maintained to satisfy $\Delta(g) = c_g$ (g is the maximum element of a current underlying set $[1, g]$) and $\Delta(i) = c_i - c_{i+1}$ for $i \neq g$. Thus, $c_i = \sum_{j=i}^{g} \Delta(j)$ for all i. Note that $\Delta(i) > 0$ if and only if $find(i) = i$.

Below we present an $O(n \log \log n)$ time algorithm for constructing a connected graph for a given 1-connected graphical degree sequence $D = (d_1, d_2, ..., d_n)$. In the algorithm, we use data structures proposed in [10, 11] to support three operations *find*, *delete* and *split* described above as well as to support operations in the doubly-linked list representing the maximum elements of all current intervals.

Algorithm CGIMPL;
begin
    for $i := 1$ **to** $n - 1$ **do begin** $\Delta(i) := d_i - d_{i+1}$; **if** $d_i > d_{i+1}$ **then** *split(i)* **end**;
    $\Delta(n) := d_n$;
    for $g := n$ **downto** 2 **do begin**
        delete(g); $\Delta(g - 1) := \Delta(g - 1) + \Delta(g)$; {**comment** $find(g - 1) = g - 1$}
        $j := find(\Delta(g))$; {**comment** $g - 1 \geq j \geq \Delta(g) > pre[j] \geq 0$}
        if $j = \Delta(g)$ **then** $I_1(g) := [1, j]$
        else begin
            $j_{new} := j - \Delta(g) + pre[j]$;
            if $pre[j] \geq 1$ **then begin**
                $I_1(g) := [1, pre[j]]$; $I_2(g) := [j_{new} + 1, j]$; $\Delta(pre[j]) := \Delta(pre[j]) - 1$;
                if $\Delta(pre[j]) = 0$ **then** *delete(pre[j])* **end**
            else {**comment** $pre[j] = 0$} $I_1(g) := [j_{new} + 1, j]$;
            $\Delta(j_{new}) := \Delta(j_{new}) + 1$; *split(j_{new})* **end**;
        $\Delta(j) := \Delta(j) - 1$; **if** $(j \neq g - 1)$ **and** $(\Delta(j) = 0)$ **then** *delete(j)* **end**
end;

The correctness of Algorithm CGIMPL can be easily shown. Since there are only $O(n)$ operations *find*, *delete* and *split* in the Algorithm, it takes $O(n \log \log n)$ time and $O(n)$ space by the data structure of van Emde Boas and Zijstra [10, 11]

Theorem 3. For a 1-connected graphical degree sequence $D = (d_1, d_2, ..., d_n)$ with $d_1 \geq d_2 \geq ... \geq d_n$, Algorithm CGIMPL implicitly constructs a connected graph with D as its degree sequence in $O(n \log \log n)$ time and in $O(n)$ space.

Now we will present $O(n \log \log n)$ time algorithm for constructing a k-connected graph represented implicitly. Before that, we consider an implicit representation of graphs $G(2a, 2b, D)$ and $H(2a, 2b + 1, D)$ defined in the proof of Proposition 6.

For the graph $G(8, 4, D)$ with $D = (10, 9, 9, 8, 7, 7, 6, 6, 6, 6, 6, 6, 6, 6)$ as its degree sequence (Fig.1(a)), we can represent it implicitly by using intervals shown in Fig. 3 based on the construction of $G(2a, 2b, D)$ in the proof of Proposition 6. In this way, an implicit representation of $G(2a, 2b, D)$ can be obtained in $O(n)$ time. Note that there are at most three intervals $I_f(v)$ for each vertex v and $I_2(v)$ and $I_3(v)$ may be empty. Thus, the set of edges adjacent vertex v_i is

$$\{(v_i, v_j) | i \in I_f(v_j) \text{ or } j \in I_f(v_i), \ f = 1, 2, 3\}.$$

Similarly, it is an easy exercise to represent $H(2a, 2b + 1, D)$, $G(2a, 2b, C)$ and $H(2a, 2b + 1, C)$ with $C = (c_\ell, ..., c_g)$ implicitly in $O(n)$ time based on the construction in the proof of Proposition 6. Thus, we can obtain the following algorithm.

Algorithm KCGIMPL;
{**comment** this calls procedure construct_graph_implicitly desrcibed below}
begin
    for $i := 1$ **to** $n - 1$ **do begin** $\Delta(i) := d_i - d_{i+1}$; **if** $d_i > d_{i+1}$ **then** *split(i)* **end**;

46

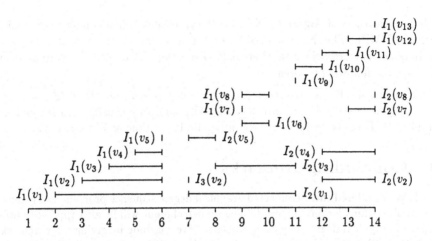

Fig. 3. An implicit representation of $G(8, 4, (10, 9, 9, 8, 7, 7, 6, 6, 6, 6, 6, 6, 6, 6))$ in Fig.1(a).

$\Delta(n) := d_n$; construct_graph_implicitly$(1, n)$
end;
procedure construct_graph_implicitly(ℓ, g);
{comment this constructs an implicitly represented $(k - \ell + 1)$-connected graph
with $C = (c_\ell, c_{\ell+1}, ..., c_g)$ as its degree sequence}
begin
 if $c_\ell = g - \ell$ then begin {comment Case I}
  $I_1(\ell) := [\ell + 1, g]$; delete$(\ell)$;
  if $\ell + 1 < g$ then construct_graph_implicitly$(\ell + 1, g)$ end
 else begin
  delete(g); $\Delta(g - 1) := \Delta(g - 1) + \Delta(g)$; {find$(g - 1) = g - 1$, $\Delta(g) > 0$}
  $j := find(\Delta(g) + \ell - 1)$; {comment $g - 1 \geq j \geq \Delta(g) + \ell - 1 > pre[j]$}
  if $(j = g - 1)$ and $(\Delta(g - 1) = k - \ell + 1)$ then {comment Case II-A }
   construct the implicitly represented $(k - \ell + 1)$-connected graph $G(2a, 2b, C)$
   or $H(2a, 2b + 1, C)$ with $C = (c_\ell, ..., c_g)$ as its degree sequence appropriately
   based on the construction in the proof of Proposition 6
  else begin {comment Case II-B-1 }
   if $j = \Delta(g) + \ell - 1$ then $I_1(g) := [\ell, j]$
   else begin
    $j_{new} := j - (\Delta(g) + \ell - 1) + pre[j]$;
    if $pre[j] \geq 1$ then begin
     $I_1(g) := [\ell, pre[j]]$; $I_2(g) := [j_{new} + 1, j]$; $\Delta(pre[j]) := \Delta(pre[j]) - 1$;
     if $\Delta(pre[j]) = 0$ then delete$(pre[j])$ end
    else {comment $pre[j] = 0$} $I_1(g) := [j_{new} + 1, j]$;
    $\Delta(j_{new}) := \Delta(j_{new}) + 1$; split$(j_{new})$ end;
   $\Delta(j) := \Delta(j) - 1$; if $(j \neq g - 1)$ and $(\Delta(j) = 0)$ then delete(j);
   if $\ell < g - 1$ then construct_graph_implicitly$(\ell, g - 1)$ end
 end
end;

The correctness of Algorithm KCGIMPL can be shown in a similar way as Algorithm CGIMPL. The time complexity $O(n \log \log n)$ and space complexity $O(n)$ can be obtained similarly since there are only $O(n)$ *find*, *delete*, *split* operations. Thus we have the following theorem.

Theorem 4. For a k-connected graphical degree sequence $D = (d_1, d_2, ..., d_n)$ with $d_1 \geq d_2 \geq \cdots \geq d_n$, Algorithm KCGIMPL implicitly constructs a k-connected graph with D as its degree sequence in $O(n \log \log n)$ time and in $O(n)$ space.

4 Concluding Remarks

We have considered the k-connected graphical degree sequence problem and given efficient algorithms for any integer k. The same technique will be also applied to other kinds of graphical degree sequence problems. We conclude by giving a few remarks. In EREW PRAM model, we can easily obtain an $O(\log n)$ time parallel algorithm with $O(n/\log n)$ processors to determine whether a given sequence of nonnegative integers $D = (d_1, d_2, ..., d_n)$ is k-connected graphical or not based on Propositon 6. It will be interesting to improve the time complexity of the algorithm for constructing a k-connected graph presented in this paper and find an $o(n \log \log n)$ time algorithm.

References

[1] C. Berge, *Graphes et Hypergraphes*, Dunod, 1970.

[2] H. Edelsbrunner, *Algorithms in Combinatorial Geometry*, Springer-Verlag, 1987.

[3] P. Erdös and T. Gallai, Graphs with prescribed degrees of vertices (Hungarian), *Mat. Lapok*, 11 (1960), 264-274.

[4] S. Hakimi, On the realizability of a set of integers as degrees of the vertices of a graph, *J. SIAM Appl. Math.*, 10 (1962), 496-506.

[5] F. Harary, *Graph Theory*, Addison-Wesley, 1969.

[6] V. Havel, A remark on the existence of finite graphs (Hungarian), *Časopis Pěst. Mat.*, 80 (1955), 477-480.

[7] S.B. Rao and A. Ramachandra Rao, Existence of 3-connected graphs with prescribed degrees, *Pac. J. Math.*, 33 (1970), 203-207.

[8] M. Takahashi, K. Imai and T. Asano, *Graphical Degree Sequence Problems*, Technical Report TR93-AL-33-15, Information Processing Society of Japan, 1993.

[9] K. Thurlasirman and M.N.S. Swamy, *Graphs: Theory and Algorithms*, John Wilely & Sons, 1992.

[10] P. van Emde Boas, Preserving order in a forest in less than logarithmic time and linear space, *Information Processing Letters*, 6 (1977), 80-82.

[11] P. van Emde Boas and E. Zijstra, Design and implementation of an effective priority queue, *Math. System Theory*, 10 (1977), 99-127.

[12] D.L. Wang and D.J. Kleitman, On the existence of n-Connected Graphs with Prescribed Degrees, *Networks*, 3 (1973), pp.225-239.

How to Treat Delete Requests in Semi-Online Problems

Yang Dai[1], Hiroshi Imai[2], Kazuo Iwano[3], and Naoki Katoh[1]

[1] Dept. of Management Science, Kobe University of Commerce, Kobe 651-21, Japan
[2] Dept. of Information Science, University of Tokyo, Tokyo 113, Japan
[3] Tokyo Research Laboratory, IBM Japan, Kanagawa 242, Japan

1 Introduction

We propose a new approach to obtain an *semi-online* fully dynamic algorithm, given a partially dynamic algorithm, by introducing a new way of handling delete requests. Briefly speaking, we are interested in how online algorithms become more efficient with some partial knowledge of future delete requests. Here, a *dynamic* algorithm solves the problem for the current instance every time when an add/delete request is made to change the instance. If a dynamic algorithm allows only add requests, we call it *partially dynamic*, otherwise we call it *fully dynamic*. An *offline* algorithm gives a set of solutions of a dynamic algorithm when the entire request sequence is known beforehand. A *semi-online* problem is a special case of online problems defined as follows: (1) For each update request σ we are given a superset SS_σ of size $O(k)$ which contains all delete requests in the succeeding k update requests for any positive integer k (Notice that we don't have to know the exact sequence of delete requests.); (2) We are also given the total number of requests, l; (3) No information on future add requests is required; (4) After each update request, a solution of the problem so updated is computed. We call this request for a solution a *query* for an update request. Thus, a semi-online problem has properties both of online and offline problems. As typical semi-online problems, we have offline problems and minimum range problems [13].

Since these online/semi-online/offline dynamic algorithms have practical importance, we can find an extensive list of previous research activities. For example, Frederickson studied online updating of minimum spanning trees [7], Eppstein et al. considered the maintenance of minimum spanning forest in a dynamic planar graph [5], and Buchsbaum et al. studied the path finding problem when only arc insertions are allowed [2]. Eppstein also devised an offline algorithm for dynamic maintenance of the minimum spanning tree problem [4]. In the design of a fully dynamic algorithm, we often face with the following difficulty: that is, since a delete request may drastically change the basic structure of a problem, we have to rebuild necessary data structures from scratch at each delete request. This is a reason why the design of efficient fully dynamic algorithms is hard.

However, if the partial information on the future deletion requests is available, we can overcome the above difficulty for a certain class of problems including the Subset Sum problem, the connectivity problem, the Integer Knapsack problem, and the optimization 0-1 Knapsack problem. We introduce a new mechanism which minimizes the total reconstruction work associated with delete requests by dividing a sequence of requests into several phases. At each phase, we create the base data structure which remains the same during the phase and maintain a data structure for items to be deleted or to be newly added in this phase. In this way, we can limit the amount of work for maintaining data structures at each update request. As an instance, we develop an $O(\sqrt{\#d+1} \cdot lK)$ time algorithm for the semi-online dynamic Subset Sum problem with a target value K, when a series of l requests including $\#d$ deletions is made to the initially empty set. We also devise semi-online dynamic algorithms for the connectivity problem, Integer Knapsack problem, and optimization 0-1 Knapsack problem which runs in $O(\sqrt{l(l+n)}\sqrt{\#d} + l \cdot \alpha(l,n))$ time, $O(l\sqrt{\#d \cdot K} + l \cdot \alpha(lK^{1.5}, K))$ time, and $O((\#d+1)^{1/2}lK)$ time, respectively, where l and $\#d$ are the numbers of requests and delete requests, n is the number of vertices in the connectivity problem, and K is the target value in the Integer Knapsack problem or the Knapsack capacity in the optimization 0-1 Knapsack problem. Notice that $\alpha(\cdot)$ is the functional inverse of the Ackermann function. To the authors' knowledge, these bounds are new and nontrivial, and these algorithms are faster than trivial ones.

We have also devised an $O(mnK)$ time algorithm for the m best solutions for the optimization 0-1 Knapsack problem by applying the mechanism of maintaining two data structures at each phase. Notice that the currently best known solution, the naive method, requires $O(mn^2K)$ time [12].

Finally, as an application of the semi-online Subset Sum problem, we consider the problem of finding a cut with the minimum range among all balanced cuts. Here the range of a cut is the maximum difference of edge weights in the cut. As discussed in [3], a minimum range balanced cut algorithm can be used for finding an approximate solution for the minimum balanced cut problem. Since the minimum balanced cut problem, a NP-complete problem [9], has important applications such as the circuit partitioning problem in the VLSI design and has been studied well [6, 11], an approximate solution by using an efficient minimum range balanced cut algorithm would be of broad interest. We then develop an $O(m + n^{2.5})$ time minimum range balanced cut algorithm, which improves an $O(m + n^3)$ time algorithm based on Martello et al.'s general approach to minimum range problems [13].

This paper is organized as follows: In Section 2, we solve the semi-online dynamic Subset Sum problem in $O(\sqrt{\#d+1} \cdot lK)$ time. In Section 3, we generalize a technique developed in Section 2, and apply for developing semi-online dynamic algorithms for the connectivity problem, the Integer Knapsack problem, and the optimization 0-1 Knapsack problem. We, finally, introduce an $O(m + n^{2.5})$ time minimum range balanced cut algorithm.

2 The semi-online dynamic Subset Sum problem

In this section, we consider the semi-online dynamic Subset Sum problem and introduce a new approach for handling delete requests. Our algorithm takes

$O(\sqrt{\#d+1}\cdot lK)$ time with a target value K, when a series of l requests, including $\#d$ deletions, is made to the initially empty set.

For simplicity, we first consider the offline dynamic Subset Sum problem. In the next section, we will see that this offline algorithm can be easily applied to the semi-online case with an appropriate modification. We assume that at each update request, a query for the feasibility for the Subset Sum problem is issued. Then, the *offline dynamic Subset Sum problem* can be formulated as follows:

Input: A positive integer K and a sequence, $Request = (r_1, r_2, \ldots, r_l)$, of l requests. Each request is either an addition request (add, x) or a deletion request $(delete, x)$ where x indicates an item to be added or deleted.

Output: A sequence $(bit_1, bit_2, \ldots, bit_l)$ where $bit_i = 1$ (*resp.* 0) indicates the feasibility (*resp.* infeasibility) of the Subset Sum problem for S_i.

For the Subset Sum problem, there is an $O(sK)$ time algorithm based on dynamic programming ([15]) for an s-item set, as shown in Figure 1. In the algorithm *Subset Sum*, the set M, which we call the *marking set*, consists of all values which can be expressed as $\sum_{k=1}^{s} x_k \cdot c_k$ where $x_k \in \{0, 1\}$. Notice that we can regard the above algorithm *Subset Sum* as an $O(sK)$ time online algorithm which allows only add requests. However, if there exists a delete request, we have to reconstruct the marking set M, which takes $O(sK)$ time. Therefore, an offline dynamic Subset Sum algorithm using the above algorithm takes $O((\#d+1)lK)$ time for l requests including $\#d$ delete requests. We, thus, develop a new technique which enables us to avoid costly reconstruction of the marking set at each delete request.

Input: A positive integer K and the initial set $S = \{c_1, c_2, \ldots, c_s\}$.
Output: A bit indicating the feasibility.
Procedure *Subset Sum*
(1)   Mark the node 0. $M := \{0\}$;
(2)   **for** $j = 1, 2, \ldots, s$ **do**
            For each marked node v, mark the node u and add it to M
            such that $u = v + c_j$;
        end
(3)   **if** $K \in M$, **then** return (1); **else** return (0);

Figure 1. Algorithm *Subset Sum*

We first divide l requests into disjoint $\lceil l/k \rceil$ stages each of which, possibly except the last, consists of consecutive k requests. We now consider the i-th stage. Suppose that we have a set of items $S(i)$ at the beginning of this stage. Let $S(i) = Remain(i) \cup Update(i)$ where $Remain(i)$ (*resp.* $Update(i)$) consists of items in $S(i)$ which will not (*resp.* will) be deleted in the stage. Note that $|Update(i)| = O(k)$. We first create a marking set R (*resp.* D) with respect to $Remain(i)$ (*resp.* $Update(i)$) in $O(lK)$ time. If $K \in R$, this implies that for all requests in this stage we have already had a feasible solution. If $K \notin R$, we do the following. For each item addition/deletion request, we maintain $Update(i)$ by adding/deleting a requested item, and update a marking set D with respect to this request by using the above naive method. That is, for each add request we incrementally update D, which takes $O(K)$ time, and for each delete request we reconstruct D from scratch with respect to the current $Update(i)$, which takes $O(kK)$ time. Whenever we obtain a new marked node x in D, we check whether $K - x \in R$ or not. If so, we have a feasible solution with the current request. Thus, each stage takes $O(lK + \#a(i) \cdot K + \#d(i) \cdot kK)$ time, where $\#a(i)$ (resp. $\#d(i)$) is the number of add (resp. delete) requests in stage i. Therefore, our algorithm in total takes $O(\lceil l/k \rceil lK + \#a \cdot K + \#d \cdot kK) = O(\lceil l/k \rceil lK + \#d \cdot kK)$, where $\#a$ (resp. $\#d$) is the number of add (resp. delete) requests among l requests. Hence we have the following theorem:

Theorem 1. *The offline dynamic Subset Sum problem with l requests, including $\#d$ delete requests, and a target value K can be correctly computed in $O(\sqrt{\#d+1} \cdot lK)$ time.*

Sketch of proof: The analysis of the time complexity follows from the above argument. Notice that we choose $k = l/\sqrt{\#d+1}$ when $\#d > 1$ and $k = l$ when $\#d \leq 1$. The correctness of the algorithm is confirmed by the following: At the j-th addition request of stage i, all possible values for the feasibility test are enumerated by the combination of the marking sets R and D. That is, every value $K - x$ for $x \in D$ is checked to determine whether $K - x \in R$. $\square$

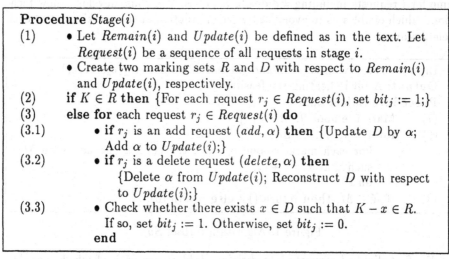

Procedure $Stage(i)$
(1)   • Let $Remain(i)$ and $Update(i)$ be defined as in the text. Let $Request(i)$ be a sequence of all requests in stage i.
      • Create two marking sets R and D with respect to $Remain(i)$ and $Update(i)$, respectively.
(2)   **if** $K \in R$ **then** {For each request $r_j \in Request(i)$, set $bit_j := 1$;}
(3)   **else for** each request $r_j \in Request(i)$ **do**
(3.1)   • **if** r_j is an add request (add, α) **then** {Update D by α; Add α to $Update(i)$;}
(3.2)   • **if** r_j is a delete request $(delete, \alpha)$ **then** {Delete α from $Update(i)$; Reconstruct D with respect to $Update(i)$;}
(3.3)   • Check whether there exists $x \in D$ such that $K - x \in R$. If so, set $bit_j := 1$. Otherwise, set $bit_j := 0$.
      end

Figure 2. $Stage(i)$ of Algorithm *Offline Dynamic Subset Sum*

Since the naive algorithm takes $O((\#d+1)lK)$ time, our algorithm speeds up the running time by a factor of $\sqrt{\#d+1}$. Figure 2 shows one stage of our algorithm *Offline Dynamic Subset Sum* described above.

This algorithm can be directly applied to the semi-online case, noting that $Stage(i)$ in Figure 2 can run correctly as long as it uses a subset $A(i)$ of $Remain(i)$ (resp. superset $B(i) = S(i) - A(i)$ of $Update(i)$) instead of $Remain(i)$ (resp. $Update(i)$), even if it does not know either $Update(i)$ or $Remain(i)$ in advance. This is because, since $A(i)$ is a subset of $Remain(i)$, $A(i)$ will not be destroyed in this stage. Moreover, the time complexity of the algorithm remains the same if $|B(i)| = O(k)$.

3   Further Applications of the Technique

We first assume that a query for the feasibility check is issued after each update request. Let $T_q(l)$ be the running time of each query where l is the total number of requests. We denote a query made right after all requests in S by *the query for S*. We assume that each add (*resp.* delete) request is an addition (*resp.* deletion) of an item to (*resp.* from) a problem instance. Suppose we have a problem which has the following properties:

(1) There exists an offline partially dynamic algorithm which runs in $T_a(\cdot)$ time for each add request.

(2) For a set S of add requests, the result of the query for S does not depend on the order of requests in S, but only depends on S itself. Let $T_f(|S|)$ denote the running time to answer the query for S by an algorithm A_f.

(3) For a set S of add requests, let it be separated into two disjoint subsets S_1 and S_2. Then we assume that the result of the query for S can be obtained in $T_c(|S_1|, |S_2|)$ time from the results of two queries for S_1 and S_2.

For a problem with the above properties, we can construct a fully dynamic semi-online algorithm in the same way as in the previous section. Remark that in a semi-online problem for each update request σ we are given a superset SS_σ of size $O(k)$ which contains a set DS_σ of all items to be deleted in the succeeding k update requests. That is, $DS_\sigma \subset SS_\sigma$ and $|SS_\sigma| = O(k)$. Notice that the technique in the previous section works correctly by handling SS_σ instead of handling DS_σ.

Theorem 2. *For a problem with the above properties, there is an*

$$O(\lceil l/k \rceil \cdot T_f(l) + \#d \cdot T_f(k) + \#a \cdot T_a(\cdot) + l \cdot T_c(l, k) + l \cdot T_q(l))$$

time semi-online dynamic algorithm where l, $\#d$, and $\#a$ are the total number of requests, the number of delete requests, and the number of add requests, respectively, and k is the number of requests in each stage.

Proof: We use the same scheme as in the previous section. That is, we divide a sequence of l requests into $\lceil l/k \rceil$ stages. Let $S(i)$ be a set of items remained at the beginning of the i-th stage. Let $Update(i)$ be a subset of $S(i)$ of size $O(k)$ which is a superset of all items to be deleted in the i-th stage, and let $Remain(i) = S(i) - Delete(i)$. Since at the beginning of each stage, we construct a data structure associated with $Remain(i)$ (a marking set R for $Remain(i)$ in the Subset Sum problem), it takes $T_f(l)$ time for each of $\lceil l/k \rceil$ stages. For each delete/add request, we maintain a data structure $Update(i)$ (another marking set D in the Subset Sum problem) whose size is at most k. For each delete request we reconstruct $Update(i)$ in $T_f(k)$ time, while for each add request we incrementally update $Update(i)$ in $T_a(\cdot)$ time ($O(K)$ time in the Subset Sum problem). Finally, for each add/delete request, we have to combine results of *Remain(i)* and the current *Update(i)*, which takes $O(T_c(l, k))$ time ($O(1)$ time for the Subset Sum problem). For each query time, it takes $T_q(l)$ time ($O(1)$ time for the Subset Sum problem). In summing up the above, we obtain the desired bound. □

3.1 The Connectivity Problem

The semi-online dynamic version of the connectivity problem is as follows: Given an initial undirected graph $G_0 = (V, \emptyset)$ with n vertices and no edges, we have a sequence of the following requests: *insert-edge(v, w)* requests for an insertion of an edge (v, w), *delete-edge(v, w)* requests for a deletion of an edge (v, w), and *check(v, w)* requests for checking whether v and w are in the same component or not.

Notice that we have an partially online algorithm which takes $\alpha(l, n)$ amortized time for each insertion where l is the total number of add requests by using UNION-FIND data structures [17]. Note that $\alpha(\cdot)$ is the functional inverse of the Ackermann function. Notice also that the second term of the time complexity appeared in Theorem 2 (that is, $\#d \cdot T_f(k)$) can be replaced by $\#d \cdot k \cdot T_a(\cdot)$, since instead of using algorithm A_f we may use k incremental updates for each delete requests. Therefore, we have the following theorem:

Theorem 3. *We have an* $O(\sqrt{l(l+n)}\sqrt{\#d} + l \cdot \alpha(l, n))$ *time semi-online dynamic algorithm for the connectivity problem.*

Sketch of proof: We use the same scheme as in the previous section. Suppose we are at the beginning of Stage i. We first run the ordinary linear time connectivity algorithm for *Remain(i)*, and determine an component id $b(v)$ for each vertex $v \in V$. For each add request (say (add, u, v)) in this stage, we first check if $b(u) = b(v)$. If so, u and v are in the same component. Otherwise, we will merge these two components and maintain this information by using an UNION-FIND data structure for component ids, which takes at most $O(\alpha(l, n))$ amortized time. At each delete request, we reconstruct the UNION-FIND data structure for component ids with respect to the current *Update(i)*, which takes $O(k\alpha(l, n))$ amortized time. Since at most k components of *Remain(i)* have been processed during this stage, the reconstruction work associated with a delete request takes $O(k)$ time by regarding each component of *Remain(i)* as a single vertex and running a linear time connectivity algorithm. Thus, we have the following time complexity: $O(\lceil l/k \rceil (l + n) + \#d \cdot k + \#a \cdot \alpha(l, n) + l \cdot \alpha(l, n))$ which becomes $O(\sqrt{l(l+n)}\sqrt{\#d} + l \cdot \alpha(l, n))$ by setting $k = \sqrt{l(l+n)}$. □

Notice that Frederickson's fully dynamic online algorithm [7] takes $O(\sqrt{m_i})$ amortized time for each add/delete request where m_i is the current number of edges. For the offline dynamic version of the connectivity problem, we can implement each update request in $O(\log n)$ amortized time by using Sleator and Tarjan's dynamic trees [16] as follows: We first define an edge weight as ∞ for an edge which will not be deleted and i for an edge which will be deleted by the i-th delete request. Then, the problem becomes the maintenance of maximum spanning tree, which takes $O(\log n)$ amortized time for each request. Since our algorithm above takes $O(\sqrt{(l+n)/l}\sqrt{\#d} + \alpha(l, n))$ amortized time for each request, our algorithm runs faster than Frederickson's algorithm and the above algorithm using dynamic trees when $\#d$ is small.

3.2 The Integer Knapsack Problem

As an application of the semi-online dynamic connectivity problem, we devise an semi-online dynamic Integer Knapsack algorithm in this subsection. Here, the Integer Knapsack problem is defined as follows:

The Integer Knapsack problem [15]: Given integers c_j, $j = 1, \ldots, n$ and K, are there integers $x_j \geq 0$, $j = 1, \ldots, n$ such that $\sum_{j=1}^{n} c_j x_j = K$?

It is well known that the Integer Knapsack problem can be solved by checking whether 0 and K are in the same component or not in the graph $G = (V, E)$ which is defined as follows: $V = \{0, 1, \ldots, K\}$, $E = \{(i, i') \mid i, i' \in V, i' - i = c_j$ for some $j \in \{1, \ldots, n\}\}$. Therefore, we can implement a semi-online dynamic algorithm for the Integer Knapsack problem by making use of the above defined semi-online dynamic algorithm for the connectivity problem. Notice that at each add (resp. delete) request (add, c_j) (resp. $(delete, c_j)$), we have to add (resp. delete) at most K edges to (resp. from) G. Therefore, we have the following theorem:

Theorem 4. *We have an $O(l\sqrt{\#d \cdot K} + l \cdot \alpha(lK^{1.5}, K))$ time semi-online dynamic algorithm for the Integer Knapsack problem.* □

Notice that the above time complexity is faster than the naive bound of $O(l^2 K)$ which creates an associated graph and runs a linear time connectivity algorithm at each request. The above time complexity is also faster than $O((lK)^{1.5})$ time obtained by using Frederickson's fully dynamic online algorithm [7] for the connectivity problem.

3.3 The Optimization 0-1 Knapsack Problem

The optimization 0-1 Knapsack problem is defined as follows:

The optimization 0-1 Knapsack problem [15]: Given the integers $(w_1, \ldots, w_n; c_1, \ldots, c_n; K)$, maximize $\sum_{j=1}^{n} c_j x_j$ subject to $\sum_{j=1}^{n} w_j x_j \leq K$ and $x_j = 0, 1$.

We now have the following theorem:

Theorem 5. *We have an $O((\#d+1)^{1/2} lK)$ time semi-online dynamic algorithm for the optimization 0-1 Knapsack problem where c is the optimal value.*

Proof: We omit the details in this abstract. Our algorithm behaves similarly as the offline dynamic Subset Sum algorithm in Section 2. The time complexity is $O(\lceil l/k \rceil lK + \#d \cdot kK + \#a \cdot K + lK)$, which becomes the desired bound by setting $k = l/(\#d+1)^{1/2}$. Notice that we use $T_f(n) = O(nK)$ time algorithm for the optimization 0-1 Knapsack problem [14], and we can show $T_c(l, k) = O(K)$. □

Notice that the above time complexity is faster than the naive $O(l(\#d+1)K)$ bound by a factor of $(\#d+1)^{1/2}$.

The m Best Solutions for the Optimization 0-1 Knapsack Problem

Using ideas developed here, we can devise an $O(mnK)$ time algorithm for finding the m best solutions for the optimization 0-1 Knapsack problem. The currently best known algorithm to the authors' knowledge takes takes $O(mn^2K)$ time, which is based on the general procedure for computing k best solutions of combinatorial optimization problems [12].

Theorem 6. *We have an $O(mnK)$ time algorithm for finding m best solutions for the optimization 0-1 Knapsack problem.*

Sketch of proof: Suppose we have an optimal solution of the optimization 0-1 Knapsack problem (let the set of indices of an optimal solution be $\{1, 2, \ldots, p\}$). Then, we can obtain the second best solution from solutions of the following p optimization 0-1 Knapsack instances $\{P_j \mid j = 1, 2, \ldots, p\}$ according to the general scheme of Lawler [12]: $P_j = (w_{j+1}, \ldots, w_n; c_{j+1}, \ldots, c_n; K - \sum_{i=1}^{j-1} w_i)$. Let j be the minimum index such that the second best solution uses the item with index j. Then this second best solution can be obtained from a solution of P_j. When we regard that P_j is obtained by deleting w_j from P_{j-1}, we can regard this sequence of instances as an offline dynamic problem. Notice that we can handle delete requests as add requests by reversing the sequence, and we treat the sequence of processes solving P_j for $1 \leq j \leq n$ as a stage. Initially, we construct *Remain* by using $O(nK)$ time algorithm. For each addition, we need $O(K)$ time for maintaining *Update* and merging the result, which we discuss in detail in the full version of this paper. Therefore, the second best solution can be obtained by $O(nK + pK) = O(nK)$ time. In the same way as above we can compute j-th best solution by by $O(nK)$ time. Therefore, our algorithm takes $O(mnK)$ time in total. $\square$

3.4 The minimum range balanced cut problem

In this subsection, we devise an $O(m + n^{2.5})$ time minimum range balanced cut algorithm as an application of the semi-online Subset Sum problem.

Let $G = (V, E)$ be a connected undirected multigraph with n vertices and m edges. A *cut* C associated with a partition $(X, V - X)$ of the vertex set V with $X \neq \emptyset, V$ is defined as $C = \{(u, v) \in E \mid u \in X, v \in V - X\}$, and $|C|$ is called a *cut value* of C. Moreover, C is *balanced* when $|X| = |V - X|$. Given an edge weight function $w(\cdot)$, the *range* of a cut C is defined as the maximum difference of its edge weights: that is, $range(C) = \max_{e \in C} w(e) - \min_{e \in C} w(e)$. Then, the *minimum range balanced cut* problem is the problem of finding a cut with the minimum range among all balanced cuts. As discussed in Introduction, we can make use of a minimum range balanced cut algorithm for finding an approximate solution for the minimum balanced cut problem [3]. Thus, it is important to devise an efficient minimum range balanced cut algorithm. From now on, we assume for simplicity that all edge weights are distinct. We will discuss the case in which some edge weights are not distinct in a full version of this paper. Let $E[\alpha, \beta] = \{e \in E \mid \alpha \leq w(e) \leq \beta\}$. An interval $[\alpha, \beta]$ is said to be *feasible* if $E[\alpha, \beta]$ contains a balanced cut. Otherwise, we say that the interval $[\alpha, \beta]$ is *infeasible*. An interval $[\alpha, \beta]$ is said to be *critical* when it is feasible and any proper sub-interval is infeasible. From now on, let T_{min} (*resp.* T_{max}) be a minimum (*resp.* maximum) spanning tree of G.

The feasibility of an interval $[\alpha, \beta]$ can be tested in $O(m+n^2)$ time as follows. First, we contract all edges in $E - E[\alpha, \beta]$ because any of these edges cannot be a member of any cut in $E[\alpha, \beta]$. When an edge (u, v) is contracted, u and v are merged into one to form a supernode. Let $G' = (V', E[\alpha, \beta])$ be the resulting graph, where V' is the set of supernodes. G' can be constructed in $O(m)$ time. Let $f(v)$ for $v \in V'$ denote the number of vertices of V which are contracted into a single supernode v. The feasibility of $[\alpha, \beta]$ is then reduced to the problem of whether there exists a subset $V'' \subset V'$ such that $\sum_{v \in V''} f(v) = n/2$. This is exactly equivalent to the Subset Sum problem, and can be solved in $O(n^2)$ time [15]. Since for any cut C, an edge e with the maximum (*resp.* minimum) weight among the edges in C belongs to T_{max} (*resp.* T_{min}) on the assumption that the edge weights are distinct [10], we shall assume in this paper that E has been already reduced to $T_{min} \cup T_{max}$, which can be computed in $O(m + n \log n)$ time [8]. Furthermore, applying the general approach proposed by Martello et al. [13], we can show that $O(n)$ feasibility tests are sufficient.

We now briefly explain our algorithm called MRBC based on [13]. Let w_1, w_2, $\cdots, w_p$ be edge weights sorted in ascending order, and let e_i be an edge such that $w(e_i) = w_i$. We start with the feasibility test of $[w_1, w_1]$. In general, the feasibility test of an interval $[w_i, w_j]$ corresponds to the feasibility test of the Subset Sum problem of a set $\{f(v) \mid v \in V'\}$ such that V' is the vertex set obtained by contracting all edges in $E - E[w_i, w_j]$. Let $G_{i,j}$ be the graph obtained by the above contractions, and let $f(v)$ be the size of a supernode v in $G_{i,j}$. When $[w_i, w_j]$ is infeasible, Algorithm MRBC unfolds the edge e_{j+1} and creates $G_{i,j+1}$ from $G_{i,j}$. This unfolding corresponds to the following three requests to the current Subset Sum instance: $(delete, f(p) + f(q))$, $(add, f(p))$, and $(add, f(q))$, where $e_{j+1} = (p, q)$ in $G_{i,j+1}$. On the other hand, when $[w_i, w_j]$ is feasible, Algorithm MRBC contracts the edge e_i and creates $G_{i+1,j}$ from $G_{i,j}$. This contraction corresponds to the following three requests to the current Subset Sum instance: $(delete, f(q))$, $(delete, f(q))$, and $(add, f(p) + f(q))$, where $e_i = (p, q)$ in $G_{i,j}$.

If each instance is solved in $O(n^2)$ time algorithm by [15], the minimum range balanced cut problem can be solved in $O(m + n^3)$ time. However, we can improve this time bound by interpreting this problem as an semi-online problem. First notice that the result of the feasibility test of the current interval $[w_i, w_j]$ is used for determining which of $[w_{i+1}, w_j]$ or $[w_i, w_{j+1}]$ is tested next time. Therefore, we cannot determine the entire sequence of feasibility test in advance. However, we can obtain partial information on future feasibility tests as follows, and thus we can formulate this problem as a semi-online problem.

Suppose that we are given the current interval $[w_i, w_j]$ and the corresponding graph $G_{i,j}$. In this case we have a full knowledge of the k edges to be contracted (*resp.* unfolded) if k consecutive contractions (*resp.* unfoldings) occur in the next

k feasibility tests. This is because contractions and unfoldings of edges occur in the ascending order of edge weights. Therefore, we can obtain a superset of $2k$ edges that are contracted or unfolded in the next k feasibility tests. In terms of the Subset Sum problem, in the succeeding k feasibility tests, the items corresponding to both endpoints of these edges may be deleted, while others may not be deleted. This implies that we can determine $O(k)$ items that may be deleted, but that the other items are not deleted during solving the succeeding k Subset Sum instances. Therefore, the minimum range balanced cut problem can be solved as an instance of the semi-online dynamic problem. In the following theorem, we take $k = \sqrt{n}$.

Theorem 7. *The minimum range balanced cut problem can be solved in $O(m + n^{2.5})$ time.* □

References

1. Ahuja, R.K., T.L. Magnanti, and J.B. Orlin, *Network Flows: Theory, Algorithms, and Applications*, Prentice Hall, Englewood Cliffs, N.J., 1992.
2. Buchsbaum, A.L., P.C. Kanellakis, and J.S. Vitter, A data structure for arc insertion and regular path finding, *ACM/SIAM SODA*, (1990), pp. 22-31.
3. Dai, Y., H. Imai, K. Iwano, N. Katoh, K. Ohtsuka, and N. Yoshimura, A new unified approximate approach to the minimum cut problem and its variants using minimum range cut algorithms, manuscript, 1992.
4. Eppstein, D., Offline algorithms for dynamic minimum spanning tree problems, *WADS*, LNCS 519 Springer-Verlag, (1991), pp. 392-399.
5. Eppstein, D., G.F. Italiano, R. Tamassia, R.E. Tarjan, J. Westbrook, and M. Yung, Maintenance of a minimum spanning forest in a dynamic planar graph. *ACM/SIAM SODA*, (1990), pp. 1-11.
6. Fiduccia, C.M. and R.M. Mattheyses, A linear time heuristic for improving network partitions, ACM/IEEE DAC, (1982), pp. 175-181.
7. Frederickson, G.N., Data structures for on-line updating of minimum spanning trees, *SIAM J. Computing*, 14 (1985), pp. 781-798.
8. Fredman, M.L. and R.E. Tarjan, Fibonacci heaps and their uses in improved network optimization algorithms, *JACM*, Vol. 34, No. 3, (1987) pp. 596-615.
9. Garey, M.R. and D.S. Johnson, *Computers and Intractability - A Guide to the Theory of NP-completeness*, W. H. Freeman and Company, New York, NY, 1979.
10. Katoh, N. and K. Iwano, Efficient algorithms for minimum range cut problems, *WADS*, LNCS 519 Springer-Verlag, (1991), pp. 80-91.
11. Kernighan, B.W. and S. Lin, An effective heuristic procedure for partitioning graphs, *BSTJ*, Vol.49, No.2, (1970), pp. 291- 307.
12. Lawler, E.L., A procedure for computing the k best solutions to discrete optimization problems and its application to the shortest path problem, *Management Science*, Vol. 18, 1972, pp. 401-405.
13. Martello, S., W.R. Pulleyblank, P. Toth, and D. de Werra, Balanced optimization problems. *Operations Research Letters*, Vol. 3, No. 5, 275-278. 1984.
14. Martello, S. and P. Toth, *Knapsack Problems: Algorithms and Computer Implementations*, John Wiley & Sons, West Sussex, England, 1990.
15. Papadimitriou, C.H. and K. Steiglitz, *Combinatorial Optimization: Algorithms and Complexity*, Prentice-Hall, Englewood Cliffs, New Jersey, 1982.
16. Sleator, D.D. and R.E. Tarjan, A data structure for dynamic trees, *Journal of Computer and System Sciences*, 26, pp. 362-391, (1983).
17. Tarjan, R.E., *Data Structures and Network Algorithms*, SIAM, Philadelphia, PA, 1983.

Finding the Shortest Watchman Route in a Simple Polygon

Svante Carlsson[*]   Håkan Jonsson[*]   Bengt J. Nilsson[†]

Abstract

We present the first polynomial-time algorithm that finds the shortest route in a simple polygon such that all points of the polygon is visible from some point on the route. This route is sometimes called the shortest watchman route, and it does not allow any restrictions on the route or on the simple polygon. Our algorithm runs in $O(n^3)$ time.

1   Introduction

It has been known for a long time [1, 6] that the so called art gallery problem is NP-hard. This is the problem of finding the smallest set of guards within a simple polygon such that each point of the polygon is visible from at least one guard. At the same time there are many examples of optimization problems and in particular shortest route problems (for instance the Travelling Salesperson Problem) are NP-hard. The combined problem, to find the shortest closed curve (watchman route) inside a simple polygon such that each point of the polygon is visible to at least one point on the curve seems to be as least as hard as the two above. Therefore, it was quite surprising when Chin and Ntafos showed that it was possible to find, in $O(n^4)$ time, the shortest watchman route that is forced to pass a given point on the boundary of the polygon [4]. This result was improved to $O(n^3)$ by Tan *et al.* [11] using a variant of the same algorithm.

In some practical applications this restriction is of minor importance, for instance if we would like to patrol a building with a robot that has to enter the building at a door. In other cases, as for instance in illumination problems, the restriction of forcing the route through a specific point can be devastating since the route can be arbitrarily longer than the watchman route without any restrictions. Despite the importance of the problem and a number of attempts to solve it the problem has stayed open until now.

In this paper we make three important observations to solve the general problem of finding the shortest watchman route in a simple polygon. The first observation is that the route has to touch at least one out of three line segments inside the polygon. These line segments can be found in linear time. Our aim now is to compute the shortest watchman route forced through each point of these segment. The second observation is that although we are trying to compute an infinite number of routes, only a polynomial (quadratic) number of points (the event points) on the segments need to be considered. The third observation is that we can compute the shortest path forced through an event point in linear time given the shortest path at the previous event point. We can also find the next event point in linear time.

By finding an initial shortest watchman route on the boundary using the algorithm of Tan *et al.* and applying the observations above, we can construct the shortest watchman route in $O(n^3)$ time. This is asymptotically of the same order as the previously best known bound for the restricted case.

---

[*]Department of Computer Science, Luleå University of Technology, 951 87 Luleå, Sweden

[†]Department of Computer Science, Lund University, Box 118, 221 00 Lund, Sweden

59

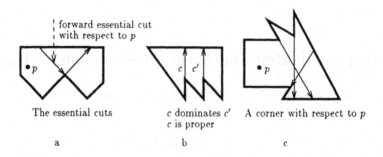

Figure 1: Illustration to the definition of essential cut, domination, and corner.

2 Definitions

Let **P** be a simple polygon having n edges. We assume a representation of **P** as a list of the coordinates of the vertices as they are encountered during a counterclockwise scan of the boundary of **P**. This representation implies an orientation on the edges of **P** and hence, we can say that the interior of the polygon is (locally) to the left of an edge.

A point p in **P** is said to *see* a point q in **P** if the line segment between the two points is contained in **P**. We also say that the two points are *visible*. A *guard set* for **P** is a set of points in **P** such that for each point p in **P** there is a point q in the guard set that sees p.

A *watchman route* is a closed curve W in **P** such that W is a guard set for **P**. If we specify a point d on the boundary of **P** and force the watchman route to pass through this point, we talk about a *fixed watchman route* with the point d being the *door* of the route. If no such point is specified, the route is called a *floating watchman route*. In the following, when we talk about a shortest watchman route we mean a shortest floating watchman route unless otherwise specified.

The shortest watchman route, whether floating or fixed, consists of line segments such that no two consecutive segments are collinear. Similarly as for polygons we represent a watchman route by a list of the vertices as they are encountered during a counterclockwise scan of the route.

We define a *cut* to be a directed line segment in **P** with the following properties. The end points of a cut must coincide with the boundary of **P** and part of the cut must lie in the interior of the polygon. A cut separates **P** into two sub-polygons. If a cut is represented by the segment $[p, q]$ we say that the cut is directed from p to q and we call p the *start point* of the cut. We say that a point lies to the right/left of a cut, if the point lies locally to the right/left in the sub-polygon separated by the cut.

Consider a reflex vertex of a polygon. The two edges connecting at the vertex can each be extended inside **P** until the extensions reach a boundary point. These extended segments are given the same direction as has the edge they are collinear to. We call the cuts thus constructed *essential cuts*. Now, it is easy to see that all guard sets must have a point to the left of (or on) each essential cut, since otherwise the edge collinear to the cut will not be seen by the guard set; see Figure 1a.

Let p be a point of a polygon. We say that an essential cut c is a *forward essential cut with respect to p* if p lies to the right of the cut c; see Figure 1a.

An essential cut c *dominates* an other essential cut c', if all points in **P** to the left of c are also to the left of c'; see Figure 1b.

Algorithm	*Shortest-Watchman-Route*
Input:	A simple polygon **P** of n edges
Output:	A shortest watchman route W
Step 1	Compute the set $\mathcal{C}$ of proper essential cuts
Step 2	Compute a constant sized subset $\mathcal{R}$ of $\mathcal{C}$ such that at least one of the cuts in $\mathcal{R}$ is intersected by some shortest watchman route
Step 3	**for** each cut c in $\mathcal{R}$ **do**
Step 3.1	Let d be one of the end points of c and apply the algorithm by Tan *et al.* [11] on **P** with d as door, giving W_d
Step 3.2	Apply a sliding process on c and compute a representation of the set of shortest watchmen forced to pass through each point on c
	endfor
Step 4	Return the shortest of the computed watchmen
End	*Shortest-Watchman-Route*

We say that an essential cut is *proper* if it is not dominated by any other essential cut; see Figure 1b. We have the following lemma.

LEMMA 2.1 *A closed curve is a watchman route if and only if the curve has at least one point to the left of (or on) each proper essential cut.*

We can view the proper forward essential cuts with respect to some point p as having a cyclic ordering specified by the start points of the cuts as they are encountered during a counterclockwise scan of the polygon boundary. In this way each cut has a predecessor and a successor. A *corner with respect to the point p* is a maximal subset of consecutive proper forward essential cuts with respect to p such that each cut intersects its predecessor. It is clear that any pair of cuts from different corners never intersect; see Figure 1c.

Consider a corner with respect to some point p. The corner consists of intersecting essential cuts. Each cut is intersected by at most $k-1$ other cuts, k being the number of proper forward essential cuts with respect to p, and hence, each cut in a corner is subdivided into at most k segments spanning between the cut intersection points. We call these segments the *fragments* of a cut. As before we can define the dominance relation between a fragment f and a cut c. We say that f dominates c, if f lies to the left of (or on) c. Hence, a fragment of a cut dominates that cut.

We can now formulate the shortest watchman route problem as: "Compute the shortest closed curve that intersects all proper essential cuts."

The following result is known about shortest watchman routes. Let us assume that we know one point of the watchman route and that we have the corners with respect to this point.

LEMMA 2.2 (CHIN, NTAFOS [4]) *The shortest watchman route visits the corners in the order that they appear as the boundary of the polygon is traversed.*

3 Finding a Shortest Watchman Route

The following algorithm computes a shortest watchman route in a simple polygon. To simplify our presentation we assume that the input polygon is not starshaped. In this case the problem of computing the shortest watchman route has a linear time solution; compute the kernel of the polygon [7] and select any point of the kernel as the resulting route.

Let us show how to perform the different steps of the algorithm. In Step 1 we compute the set C of proper essential cuts using the same $O(n^2)$ time algorithm as Chin and Ntafos [4].

To perform Step 2 we apply the linear time kernel algorithm of Lee and Preparata [7] that constructs the half plane intersection of all the half planes associated to the proper essential cuts. Hence, the algorithm, at each step, takes the previous potential kernel and intersects it with the half plane associated to a proper essential cut, yielding a new potential kernel. As soon as the algorithm produces an empty potential kernel we remember c, the proper essential cut associated to the last half plane of the intersection. The last step to compute $\mathcal{R}$ consists of two steps, i) scan through the proper essential cuts associated to the half plane intersections previous to c and check if there is a cut c' that does not intersect c, in which case $\mathcal{R} = \{c, c'\}$, ii) otherwise we follow the boundary of the last non-empty potential kernel in counterclockwise order to find two consecutive cuts c' and c'', where c' is directed towards the left of c and c'' is directed towards the right of c, in which case $\mathcal{R} = \{c, c', c''\}$. One of the cases will always occur.

LEMMA 3.1 *For every non-starshaped polygon* **P**, *there is a set* $\mathcal{R}$ *of two or three proper essential cuts such that any shortest watchman route intersects at least one of the cuts in* $\mathcal{R}$.

PROOF: (Sketch only) In both cases it is impossible for any watchman route to be completely to the left of all the cuts in $\mathcal{R}$, hence, the route must have points to the right of at least one of the cuts in $\mathcal{R}$ implying that this cut must be intersected by the route. □

We give an overview of the main ideas behind the algorithms by Chin and Ntafos [4] and Tan *et al.* [11] to explain Step 3.1 of our algorithm. Both algorithms start by constructing an initial watchman route through d, i.e., a closed curve that intersects all proper forward essential cuts with respect to d, and progress by applying a sequence of *adjustments* to the initial route. The only difference between the two algorithms is in the way they construct the initial watchman route. The way the adjustments are done is the same in both algorithms. In order to explain these adjustments it is important to know what kind of intersections a watchman route can have with the cuts belonging to the corners with respect to d. A watchman route makes a *reflection contact* with a cut c, if the intersection of the route and c is one point and all other points of the route lie to the right of c. A reflection contact is *perfect* if the incoming angle equals the outgoing angle of the reflection. A watchman route makes a *crossing contact* with c, if the intersection is one or two points and the contact is not a reflection contact. Finally, the route makes a *tangential contact* with c, if the intersection is a line segment and all other points of the route lie to the right of c.

Consider the cuts of a corner such that a watchman route makes reflection contact with these cuts. We call these cuts the *active cuts* and the fragments that contain the intersection points are the *active fragments*. Now the following lemma is trivial to show.

LEMMA 3.2 (TAN, HIRATA, INAGAKI [11]) *The following two characteristics hold for the active fragments of a shortest watchman route.*

COMPLETENESS *The set of active fragments dominate all essential cuts of* **P**.

INDEPENDENCE *An active fragment not part of an active cut does not dominate that cut.*

Furthermore, the following lemma provides a way to construct the shortest fixed watchman route given an initial watchman route.

LEMMA 3.3 (CHIN, NTAFOS [4]) *The shortest fixed watchman route through d makes perfect reflections in the interior of the active fragments.*

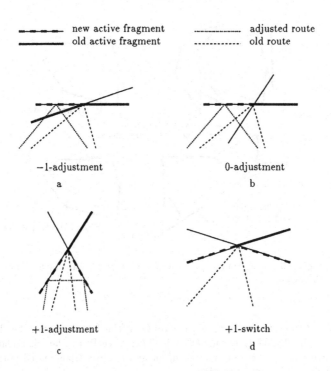

Figure 2: The possible type of adjustments to shorten a watchman route.

The problem thus becomes that of adjusting the initial route until all the reflection contacts are perfect or until no more reflection contacts can be made perfect as shown by the Greek geometer Heron in 100 AD. Each adjustment changes the set of active fragments as depicted in Figure 2. There are essentially three types of adjustments. The −1-adjustment moves the reflection contact past the cut intersection and changes the set of active fragments so that this set contains one less active fragment; see Figure 2a. The 0-adjustment moves the reflection contact towards the point of perfect reflection and keeps the number of active fragments the same; see Figure 2b. Finally, the +1-adjustment and +1-switch take care of the case when the reflection contact may be split into two reflection contacts; see Figure 2c–d.

It can be shown that the maximum number of adjustments that have to be made is bounded above by a low order polynomial in the size of the polygon. Each adjustment takes linear time which gives polynomial time algorithms for the shortest fixed watchman route problem.

Given a set of active fragments, how is the shortest fixed watchman route with reflection contacts at these fragments computed? The approach taken is by *unfolding* the polygon **P** which is a process that produces a polygonal shape that we call an *hourglass* such that the shortest path from d to its image in the hourglass corresponds to the shortest fixed watchman route through d that reflects on the active fragments. The hourglass is constructed from **P** by cutting off the parts of **P** that lie to the left of the active fragments. In this way we get a new polygon **P'** with the active fragments on the boundary. The polygon **P'** is then triangulated using Chazelle's algorithm [3]. Now **P'** is unfolded using the active fragments as mirrors; see Figure 3. The shortest path through the active fragments is computed in the

63

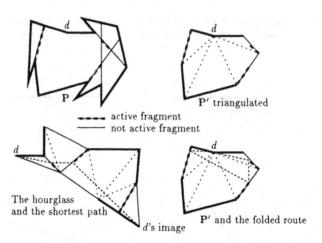

active fragment
not active fragment

The hourglass
and the shortest path

d's image

P′ and the folded route

Figure 3: Computation of the fixed shortest watchman route through d.

hourglass from d to its image point and the route is folded back to give the shortest fixed watchman route in **P**. The time complexity is linear since **P′** can be triangulated in linear time and shortest paths can be computed in linear time in a triangulated polygon [5], the polygon in our case being the hourglass.

4 The Sliding Process

At this point we are ready to present the new technique that together with Lemma 3.1 enables us to compute the shortest watchman route without the assumption of a fixed door the route must pass through.

The sliding process we apply in Step 3.2 is done as follows: We denote by W_p the shortest watchman route passing through the point p of c. The route W_p corresponds to a shortest path S_p in some hourglass that we denote by $\mathbf{H}_p$. Note that there is a direct correspondence between W_p in **P** and S_p in $\mathbf{H}_p$ and therefore, we will refer to W_p and S_p interchangeably. Initially we have the route W_d passing through the end point d of the cut c that corresponds to the shortest path in $\mathbf{H}_d$. Let $p = d$ and suppose we move the point p slightly towards the other end point d' of c. We are interested in the topological changes that occur to W_p and also to $\mathbf{H}_p$ as the sliding proceeds. Our first assumption is that the fragment f of c containing the point p is part of the boundary of $\mathbf{H}_p$ but this assumption can easily be made to hold when we build the hourglass.

We have the following lemma.

LEMMA 4.1 *The shortest path S_p makes turns only at vertices of $\mathbf{H}_p$ that correspond to fragment end points in* **P** *or to vertices of* **P**.

It follows from this lemma that the only change to S_p when p moves occurs with the first and last segments of S_p. Now, when does topological changes occur to the path S_p? As the links intersect the interior of the active fragments in $\mathbf{H}_p$ we have corresponding perfect reflections of W_p. Hence, we may have topological changes to the route only when the following events occur.

64

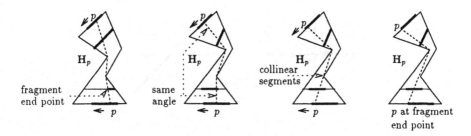

Figure 4: The event points of the sliding process.

1. When the first or last segments of S_p intersects fragment end points,

2. when the first and the last segment of S_p have the same angle to f. This corresponds to perfect reflection or straight crossing depending on whether c is active or not,

3. as the first and second, or the last and penultimate segments of S_p become collinear, and

4. when p reaches the end point of the fragment f on c.

The points p of c where these events occur we call the *event points* of the sliding process and we have four types of event points according to the enumeration above. Refer to Figure 4 for an illustration of the different types of event points.

LEMMA 4.2 *There are $O(n^2)$ event points.*

PROOF: A counting argument yields $O(n^2)$ fragment end points in **P** which also dominates the number of event points, since S_p consists of $O(n)$ segments and c of $O(n)$ fragments. □

Now, the question is what type of changes are to be made when p reaches an event point on f? For the Type 1 event points we claim that the adjustments of Figure 2 are sufficient to maintain the shortest path S_p as the sliding proceeds. The Type 2 event point does not induce any change in the path. However, this point gives a (local) minimum of the route length and is therefore interesting to maintain. For the Type 3 event points it is easy to see that the only operation needed is to merge the two collinear segments into one first or last segment of S_p.

The handling of the Type 4 event point is slightly more involved. This occurs as p reaches the end point of f, i.e., there is a proper essential cut e that intersects c at p. The problem is to decide how to continue the sliding process of p and what changes have to be made to the set of active fragments. We look at the route W_p and show the set of adjustment that are to be made; see Figure 5. Symmetric cases are not shown. The -1-adjustments decreases the number of reflection contacts as the sliding process "crosses over" a reflected cut and makes the contact a crossing contact. The 0-adjustment simply moves the point p past the intersection point of c with e. The $+1$-adjustments splits the reflection contact into two new reflections.

Note that after a Type 4 change it is possible that a Type 1 change may have to be made to adjust the route and shorten the path S_p. Each adjustment can be done in linear time using the unrolling technique of Chin and Ntafos [4].

LEMMA 4.3 *The changes to the path S_p maintains the shortest watchman route W_p as p slides from d to d'.*

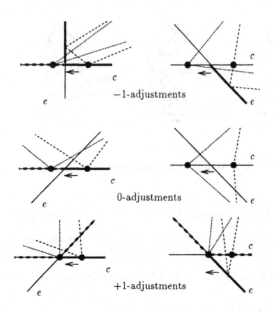

Figure 5: Adjustments to be made as Type 4 changes occur.

PROOF: The Type 2 and Type 3 event points induce no changes to the route, and hence, we need only concern ourselves with the Type 1 and Type 4 event points. We have to show that after the changes the route W_p is the shortest watchman route through p, i.e., reflection contacts in the interior of fragments are all perfect reflections.

Suppose S_p hits a fragment end point during the sliding process. After the Type 1 change, the adjustments of Figure 2, the route W_p still maintains perfect reflections at all points except where the adjustment was made, since adjustments are local transformations. However the adjustment made may introduce a non-perfect reflection at the fragment end point. Further application of the sliding process will yield a Type 3 change at this point making the reflection perfect with respect to the new active fragment.

Suppose p reaches the end point of f during the sliding process. After the Type 4 change, the adjustments of Figure 5, the route W_p does not necessarily maintain perfect reflections at the induced active fragments of the adjustment. Hence, a Type 1 change may be necessary to introduce perfect reflection where possible. We can view this as S_p hitting a fragment end point (the Type 1 event point) immediately after the Type 4 change has been made.

Hence, after a Type 1 or Type 4 change perfect reflections are maintained in the interior of all active fragments, and no non-perfect reflections can be made perfect. Therefore, the route W_p is the shortest watchman route through p. □

After an adjustment it is important to be able to compute the next event point efficiently. We do this by actually maintaining four potential next event points on c, each determined by one of the four event point types. The next event point is always the one of the four potential event points that lies closest to the sliding point p. Once p reaches the next event point, the four potential next event points are updated in the following way: To compute the potential next event point of Type 1 we find the next fragment end point that will be intersected by the first and last segment of S_p as p slides along c. This can, trivially, be done in linear time by looking at each active fragment that is intersected by the first and last segment of S_p. To compute the potential next event points of Types 2 and 3 it is sufficient to look at the path S_p and identify the end points for which the first and last segments become collinear and the end points for which the first and second, and the last and second to last segments become collinear. To compute the potential next event point of Type 4 we keep a reference to the end point of the fragment on c that currently contains the sliding point p. Hence, the time to compute the next event point is in total of linear order in the size of the polygon.

If we maintain the length of the shortest watchman routes passing through each event point, it is easy to find the shortest of these. This follows because the shortest route will pass through one of the event points since the length of a route is either monotonically increasing or decreasing as the sliding process proceeds between event points.

We prove the following theorem.

THEOREM 1 *The Shortest-Watchman-Route algorithm presented above computes a shortest watchman route in a simple polygon* **P** *in $O(n^3)$ time.*

PROOF: The correctness of the algorithm follows from Lemma 4.3 and the discussion above. Hence, it remains to analyze the complexity of the algorithm.

Step 1 can be performed in $O(n^2)$ time [4]. Step 3.1 takes $O(n^3)$ time [11].

To show the complexity of Step 3.2 we first note that by Lemma 4.2 there are $O(n^2)$ event points. Now, for each Type 1, 2, or 3 event point we make one change taking linear time to perform. For the Type 4 event point we make one Type 4 adjustment possibly followed by a Type 1 adjustment, each taking linear time. Hence, this step takes $O(n^3)$ time.

Finally, since we perform Steps 3.1 and 3.2 at most three times, the algorithm uses $O(n^3)$ time. □

5 Conclusion

In this paper we have presented a polynomial time solution to the problem of computing the shortest floating watchman route in a simple polygon. The fact that there is a polynomial time solution for this problem settles an open question in computational geometry.

It turns out that our algorithm works much faster in most cases. It is adaptive in the sense that the complexity depends not only of the size of the polygon but also of the number of intersection points between proper essential cuts. In fact, if the number of intersection points between proper essential cuts is denoted by I, we can show that the time complexity is bounded by $O(In + n \log n)$. The parameter I can have quadratic size but in practice it will be much smaller.

Related problems are that of computing several watchman routes in a polygon using different optimization criteria. Most versions of these problems turn out to be NP-hard but there exists polynomial time algorithms for some of these problems in certain restricted classes of polygons [2, 8, 9, 10]. One important open question in this area is whether the problem of computing the two watchman routes in a polygon such that the sum of the lengths of the two routes is minimized has a polynomial time solution or not.

References

[1] A. AGGARWAL. *The Art Gallery Theorem: Its Variations, Applications and Algorithmic Aspects.* PhD thesis, Johns Hopkins University, 1984.

[2] S. CARLSSON, B.J. NILSSON, S. NTAFOS. Optimum Guard Covers and m-Watchmen Routes for Restricted Polygons. *International Journal of Computational Geometry and Applications*, 3(1):85–105, 1993.

[3] B. CHAZELLE. Triangulating a Simple Polygon in Linear Time. In *Proc. 31st Symposium on Foundations of Computer Science*, pages 220–230, 1990.

[4] W. CHIN, S. NTAFOS. Shortest Watchman Routes in Simple Polygons. *Discrete and Computational Geometry*, 6(1):9–31, 1991.

[5] L. GUIBAS, J. HERSHBERGER, D. LEVEN, M. SHARIR, R. TARJAN. Linear Time Algorithms for Visibility and Shortest Path Problems inside Triangulated Simple Polygons. *Algorithmica*, 2:209–233, 1987.

[6] D.T. LEE, A.K. LIN. Computational Complexity of Art Gallery Problems. *IEEE Transactions on Information Theory*, IT-32:276–282, 1986.

[7] D.T. LEE, F.P. PREPARATA. An Optimal Algorithm for Finding the Kernel of a Polygon. *Journal of the ACM*, 26:415–421, 1979.

[8] J.S.B. MITCHELL, E.L. WYNTERS. Watchman Routes for Multiple Guards. In *Proc. 3rd Canadian Conference on Computational Geometry*, pages 126–129, 1991.

[9] B.J. NILSSON, S. SCHUIERER. Shortest m-Watchmen Routes for Histograms: The MinMax Case. Technical Report "Bericht 43", Institut für Informatik, Universität Freiburg, Germany, January 1992. An extended abstract was presented at the *4th International Conference on Computing and Information*, pages 31–34, 1992.

[10] B.J. NILSSON, D. WOOD. Watchmen Routes in Spiral Polygons. Technical Report LU-CS-TR:90-55, Dept. of Computer Science, Lund University, 1990. An extended abstract of a preliminary version was presented at the *2nd Canadian Conference on Computational Geometry*, pages 269–272, 1990.

[11] X.-H. TAN, T. HIRATA, Y. INAGAKI. An Incremental Algorithm for Constructing Shortest Watchman Routes. In *Proc. ISA '91 Algorithms*, pages 163–175. Springer Verlag, Lecture Notes in Computer Science 557, 1991.

Constructing shortest watchman routes by divide-and-conquer *

Xuehou TAN
School of High-Technology for Human Welfare, Tokai University
317 Nishino, Numazu 410-03, Japan
Tomio HIRATA
Faculty of Engineering, Nagoya University
Chikusa-ku, Nagoya 464, Japan

Abstract

We study the problem of finding shortest watchman routes in simple polygons from which polygons are visible. We develop a divide-and-conquer algorithm that constructs the shortest watchman route in $O(n^2)$ time for a simple polygon with n edges. This improves the previous $O(n^3)$ bound [8] and confirms a conjecture due to Chin and Ntafos [4].

1 Introduction

The **watchman route problem** [3, 4, 8], an interesting variation of the well-known art gallery problem, deals with finding a route (actually a cycle) in a simple polygon with n edges so that every point in the polygon can be seen from at least one point along the route. Two points inside a polygon are said to be mutually visible if the segment between them entirely lies in the polygon. The objective is to minimize the length of the route. The watchman route problem concerns with not only visibility but also metric information.

We consider the watchman route problem for simple polygons which have a starting point s specified on their boundaries, i.e., watchman routes start at s and also end at s. In 1991, Chin and Ntafos [4] presented an $O(n^4)$ algorithm for constructing shortest watchman routes in simple polygons, and then made a conjecture that the watchman route problem for single polygons can be solved in $O(n^2)$ time. Their algorithm first finds an initial watchman route and then computes the shortest watchman route by repeatedly adjusting the current one. (An adjustment always produces a new and shorter watchman route.) The time complexity of their algorithm mainly depends on the number of the required adjustments, which is $O(n^3)$ for their algorithm. (Each adjustment requires linear time.) The watchman route is adjusted with the "adjust at first choice" selection rule. As pointed out in [4], this selection rule is critical in their analysis of time complexity. Later, Tan et al. [8] presented an incremental method which resembles Chin and Ntafos' algorithm in that it empolies the same adjusting technique, but differs in that it starts with a partial watchman route which has a small visibility, then gradually increases its visibility until it becomes the whole watchman route, and has $O(n^2)$ adjustment requirements. It is shown in [8] that a "one-place-adjustable" watchman route can be efficiently adjusted into the shortest one. The selection rule is not crirical for the algorithm in [8].

In this paper, we approach the watchman route problem by divide-and-conquer. Studying the geometric properties of the watchman route problem, we devise an effective method to divide it into subproblems. In the merge step of our algorithm, we first find a "one-place-adjustable" watchman route and then adjust it into the shortest one. Although the divide-and-conquer method requires a total of $O(n \log n)$ adjustments, many of them can be performed together. A careful analysis shows

---

*This work was supported in part by the Hori Information Science Promotion Foundation and the International Information Science Foundation.

that our algorithm runs in $O(n^2)$ time. This improves the previous $O(n^3)$ time bound [8] and confirms Chin and Ntafos' conjecture [4].

Section 2 of this paper reviews some basic properties of shortest watchman routes in simple polygons. Section 3 discusses how to decompose the watchman route into subproblems and then presents the divide-and-conquer algorithm in detail. Concluding remarks are finally given in Section 4.

2 Preliminaries

Let P be an n-sided simple polygon, given by the sequence of its vertices in the clockwise order from s. A vertex is **reflex** if the internal angle is strictly greater than 180^0. P can be partitioned into two pieces by a "cut" that starts at a reflex vertex v and extends either edge incident to v until it first intersects the boundary. We say a cut is a **visibility cut** if it produces a **convex angle** ($< 180^0$) at v in the piece of P containing s. (Some reflex vertices may not contribute to any visibility cut.) In order to see the corner incident to a visibility cut, a watchman route needs to visit only one point on that cut. A cut C is described by a pair of points (l, r), where l (left endpoint) is the endpoint of C that is first visited in the clockwise scan of the boundary of P starting at s and r is the other endpoint. The orientation of a cut C is supposed to be from l to r. Thus, s always falls to the right side of visibility cuts. The piece of P containing s is called the **essential piece** of C.

We say cut C_j **dominates** cut C_i if the essential piece of C_j contains that of C_i. Clearly, if C_j dominates C_i, any route that visits C_j will automatically visit C_i. A cut is called an **essential cut** if it is not dominated by any other cuts. It is important to observe that any watchman route must visit these essential cuts and any route that visits them is a watchman route. The essential cuts can be identified in $O(n)$ time by applying the clockwise scanning scheme. In the rest of this paper, we consider only the essential cuts.

Let m be the number of essential cuts and let $C_1, C_2, \cdots, C_m$ be the sequence of essential cuts indexed in the clockwise order of their left endpoints. The set of essential cuts is then partitioned into cut corners. A **cut corner** is a subset of consecutive cuts $C_i, C_{i+1}, \cdots, C_j$ such that each C_k intersects with C_{k+1} ($i \leq k < j$), and C_i and C_j do not intersect with C_{i-1} and C_{j+1}, respectively. Clearly, any pair of the cuts derived from different cut corners can not intersect each other. This partition of essential cuts into cut corners takes $O(n)$ time.

Let p_{ij} denote the intersection between two essential cuts C_i and C_j. If $i < j$, any route which visits the segment (l_j, p_{ij}) of C_j will visit C_i, and on the other hand, any route which visits the segment (p_{ij}, r_i) of C_i will also visit C_j. That is, a cut intersection corresponds to a switch in visibility dominance between the intersecting cuts. The concept of essential cuts then needs to be refined further. A **fragment** is any line segment along an essential cut C_i that starts at l_i or at a cut intersection p_{ij}, ends at r_i or at a cut intersection p_{ik}, and does not contain any cut intersection in its interior. In a cut corner, a cut can intersect with at most $m - 1$ cuts and is thus divided into at most $m - 1$ **fragments**. We say fragment f (point p) **dominates** cut C if $f(p)$ lies in the non-essential piece of C. That is, any route that visits $f(p)$ also visits the cut C. We also say fragment f **dominates** fragment g if f dominates the cut to which g belongs. (Note that any fragment dominates the cut to which it belongs and two fragments may dominate each other.) Then any watchman route which goes through p will visit C. Since there are cut intersections in a cut corner, the problem faced us is to select a set of the fragments so that visiting them in some appropriate order will visit all the cuts in this cut corner and the distance travelled will be minimum.

Theorem 1 *[4] The shortest watchman route should visit the cut corners in the order in which they appear in the boundary of P.*

Theorem 1 states that the shortest watchman route need not cross itself among cut corners. This property also applies inside a cut corner, i.e., the shortest watchman route does not properly intersect itself (overlapping sections are possible). It is also noted in [4] that the shortest watchman route consists of convex sections within cut corners. If a concave section occurs inside a cut corner, this section can be stretched out to obtain a shorter route. Concave sections can occur only among cut corners.

Corollary 1 *[4] The shortest watchman route should consist of convex chains within each cut corner.*

Depending on whether a shortest watchman route goes over a cut (as viewed from s) or just reflects on a cut, we say that the shortest watchman route makes a **crossing contact** or a **reflection contact** with the cut, respectively. The degenerate cases of reflection contacts and crossing contacts, where the shortest watchman route shares a line segment with the cut, are called **tangential contacts** (see [4, 8] for exact definitions). For a shortest watchman route, the set of the fragments with which the route makes reflection contacts is called the **watchman fragment set**. In other words, if a route visits all the fragments of a watchman fragment set, it will visit all the cuts, and is thus a watchman route. It is easy to see that no fragment is dominated by any other fragments in the fragment set. With respect to a watchman fragment set, we distinguish a fragment as an **active** or **unactive** fragment according to whether it belongs to the fragment set or not. A cut is active if it contains an active fragment. Otherwise, it is unactive.

Conversely, given a watchman fragment set, we can construct the corresponding (optimum) watchman route in P [3]. Specifically, the non-essential pieces of all active cuts are removed (since the optimum watchman route never needs to enter them) and the resulting polygon P' is then triangulated [2]. The active fragments are used as mirrors to "roll-out" the triangulation of P'. Now, the optimum watchman route can be determined by finding the shortest route between s and its image s' in the rolled-out polygon, and by folding back the shortest route along the active fragments. The whole procedure takes $O(n)$ time. (See [3] for details.) We call this method as the rolled-out method. Note that the watchman route found by the rolled-out method is optimum with respect to the given watchman fragment set. It is thus possible to make it shorter by changing the watchman fragment set.

Definition 1 *A watchman route R is **adjustable** on an active cut C if (i) R makes a reflection contact with C, (ii) the incoming angle of R with C is not equal to the outgoing angle and (iii) the contact point of R with C is not the endpoint of C. (That is, we can adjust the contact point on C to get a shorter watchman route.)*

Definition 2 *A watchman route R is **one-place-adjustable** if R is adjustable only on one active cut.*

An adjustment can only occur at the intersection of two essential cuts. As showed in [4, 8], there are three types of adjustments on an active cut C_i (Fig. 1). In Fig. 1, the incoming angle of R with C_i is assumed to be smaller than the outgoing angle. (The symmetric case is omitted). The bold and discontinuous segments in Fig. 1 stand for the active fragments before and after an adjustment, respectively. A possible next route R' is also shown.

In Fig. 1a, R makes reflection contacts with both C_i and C_h at their intersection. The adjustment involves moving the contact point of C_i to the left. The next route, R', will make a reflection with C_i but a crossing contact with C_h. Thus, the current fragment of C_i is replaced by the next fragment and the fragment of C_h is deleted from the fragment set. We call this a **(-1)-adjustment** since the number of active fragments is decreased by 1. Note that a (-1)-adjustment also arises if the contact with C_h is tangential. In Fig. 1b, R makes a reflection contact with C_i and a normal crossing contact with $C_{i'}$. The adjustment involves moving the contact point of C_i to the left, i.e., replacing the current fragment of C_i by the next fragment. The next route R' still makes a crossing contact with $C_{i'}$. This is called a **0-adjustment**. In Fig. 1c, R makes a reflection contact with C_i but a special crossing contact with C_j, i.e., the crossing contact with C_j has degenerated into a reflection or a tangential contact. The adjustment involves substituting the current fragment of C_i with the next fragment and inserting the fragment of C_j next to p_{ij} (the intersection of C_i and C_j) into the fragment set (Fig. 1-c1,c2). For the tangential contact case, the next active cut C_k should be also considered (Fig. 1-c2). In order to shorten the watchman route, the incoming angle of R with C_k must be greater than the outgoing angle. Thus, the current fragment of C_k should be also substituted by the next fragment. We call it a (+1)-adjustment since the number of active fragments is increased by 1. Depending on whether the incoming angle of R with C_j is greater (Fig. 1-c1,c2) or smaller (Fig. 1-c3) than the outgoing angle, the next route R' will be shorter, or the same as R.

Theorem 2 *[4] There is a unique non-adjustable watchman route in a simple polygon P.*

Corollary 2 *[4] A watchman route R is a shortest watchman route if and only if R is non-adjustable.*

Based on Theorem 2 and Corollary 2, Chin and Ntafos gave the following method for solving the watchman route problem.

1. Find an intial watchman fragment set (which is equivalent to the set of extended line segments of [4]) and compute its corresponding watchman route by the rolled-out method. (Note that the initial route should consist of convex chains within cut corners.)

2. Repeatedly adjust the current watchman route with three kinds of adjustments until it becomes the shortest one. (An adjustment always produces a new and shorter route. Clearly, the new route has the convex property within each cut corner.)

In [8], Tan et al. showed that a "one-place-adjustable" watchman route can be efficiently adjusted into the shortest one and presented an $O(n^3)$ time algorithm for solving the watchman route problem. Chin and Ntafos conjectured that the problem can be solved in $O(n^2)$ time.

3 The divide-and-conquer algorithm

We will present a divide-and-conquer algorithm that solves the watchman route problem in $O(n^2)$ time. An effective method to decompose the watchman route problem into subproblems is given in Section 3.1. The key of our algorithm in the merge step is to find an initial watchman route that is one-place-adjustable. The details of the divide-and-conquer algorithm are presented in Section 3.2.

3.1 Decomposing the watchman route problem into subproblems

For geometric problems concerning with simple polygons, it is an usual method to apply Chazelle's polygon-cutting theorem [1] to divide the given problem. But it does not work for the watchman route problem, since the shape of the shortest watchman route is solely determined by the set of essential cuts. The time complexities of all known watchman route algorithms [4, 8] are mainly determined by the number of required adjustments, which is proportional to the number of essential cuts. Therefore, a better way is to divide the watchman route problem on the set of the cuts that a watchman route should visit.

Imagine that we have an optimum watchman route that is obtained by the rolled-out method (and thus consists of convex chains within cut corners) and we want to decompose it into two subroutes, one visits a half of the essential cuts and the other visits another half of the cuts. Since the route reflects on some essential cuts, we divide the route at the reflection point of the cut with the medium index, e.g., $C_{\lfloor m/2 \rfloor}$. Let s' denote the chosen point on $C_{\lfloor m/2 \rfloor}$. Then one subroute starts at s and ends at s', and the other starts at s' and ends at s. Since the original route reflects on $C_{\lfloor m/2 \rfloor}$, both of subroutes lie in the essential piece of $C_{\lfloor m/2 \rfloor}$. When two shortest subroutes are merged into the whole one, this property avoids a concave connection to happen at the chosen point on $C_{\lfloor m/2 \rfloor}$ and thus make the merge algorithm simple and efficient (see Section 3.2). Extending the above idea to the general situation, we have the following definition:

Definition 3 *A route R_{ij} from a point s_i on C_i to a point s_j on C_j $(i < j)$ is a **partial watchman route** if the cuts $C_i, C_{i+1}, \ldots, C_j$ are either visited by R_{ij} or dominated by the starting point s_i, and R_{ij} lies in the polygon $P(C_i, C_j)$, where $P(C_i, C_j)$ denotes the intersection of essential pieces of C_i and C_j.*

Let the edge containing s be the cuts C_0 and C_{m+1} and let the polygon P be the essential pieces of C_0 and C_{m+1}. We set $s = s_0 = s_{m+1}$. Note that we do not insist that all the cuts $C_i, \cdots, C_j$ should be visited by R_{ij}. The cuts dominated by s_i may not be visited by R_{ij}, but they will be visited by the partial watchman routes R_{0i} in our divide-and-conquer algorithm. When $i = 0$ and $j = m + 1$, all the cuts are visited by the watchman routes from s_0 to s_{m+1}.

To recursively divide the watchman route problem, we need to define a set of points s_i on the cuts C_i as described in Definition 3. These points s_i are called "images". For a point p and a segment Q in polygon P, p's **image** on Q is the point of Q that is visibe from p and closest to p. As R_{ij} is defined in the polygon $P(C_i, C_j)$, for any pair of images s_i and s_j, s_i and s_j should lie in the essential pieces of C_j and C_i, respectively. (Note that s_i and s_j lie on the boundary of $P(C_i, C_j)$.) In other words, all the images in a cut corner should give the vertices of a convex polygon.

The images in a cut corner are specified as follows. Let C_i be the first (least indexed) cut in a cut corner. We take the left endpoint l_i as the current starting image s_i and repeat the following procedure:

1. If s_i lies in the essential piece of the next cut C_{i+1}, image s_{i+1} on C_{i+1} is defined as the image of s_i on the portion of C_{i+1} that lies in the essential piece of C_i. (Thus, s_{i+1} also lies in the essential piece of C_i.) If C_{i+1} is the last cut in the cut corner, we are done; otherwise, s_{i+1} is taken as the current image.

2. If s_i lies in the non-essential piece of the next cut C_{i+1}, image s_{i+1} on C_{i+1} is undefined. If C_{i+1} is the last cut in the cut corner, we are done; otherwise, the image s_{i+2} of s_i on C_{i+2} is then considered, i.e., C_{i+2} is taken as the next cut of s_i.

An example of how to specify the images is given in Fig. 2, where a cut corner with five essential cuts is shown. Note that there is no image defined on C_3. Clearly, no image chosen as above can lie in the non-essential piece of any image-defined cuts. Thus, an image-defined cut cannot be dominated by any other images. It may require $O(n)$ time in the worst case to find an image. Thus, the computation of images in polygon P takes $O(nm)$ time.

Lemma 1 *Let C_i and C_j be image-defined cuts. Then the shortest partial watchman route R_{ij} from s_i to s_j is unique.*

Proof. First, we can ignore all the cuts dominated by s_i or s_j, as the cuts dominated by the starting image s_i need not be visited by the partial watchman routes R_{ij} and the cuts which are dominated by s_j but not by s_i must be visited by any route from s_i to s_j. Let the rest cuts, which must be considered in determining the shortest partial watchman route R_{ij}, be $C'_{i'}, \cdots, C'_{j'}$ $(i < i' \leq j' < j)$. (Note that the cuts C'_k $(i' \leq k \leq j)$ need not be consecutive.) We claim that for any C'_k $(i' \leq k \leq j')$, it either entirely lies in $P(C_i, C_j)$ or intersects with the boundary of $P(C_i, C_j)$. It is easy to verify this fact when C_i and C_j lie in the different cut corners. Let C_i and C_j be in the same cut corner. Suppose that there is a cut C'_k which lies outside $P(C_i, C_j)$ (Fig. 3). Then C'_k must intersect with both C_i and C_j; otherwise, either C_i or C_j is dominated by C'_k. If C'_k is an image-defined cut, then the image s_k must lie in the essential piece of C_i but in the non-essential piece of C_j. Thus, s_k dominates C_j, and the image s_j is undefined, a contradiction. If C'_k has no image defined on it, then there exists at least one image which dominates C'_k. This image must lie in the intersection of the essential piece of C_i and the non-essential piece of C'_k. Since this image also dominates C_j, the image on C_j is still undefined, a contradiction again. Thus, the claim is proved.

Since no cut C'_k with $i' \leq k \leq j'$ is dominated by s_i or s_j, then s_i and s_j lie in the same side (essential piece) of any C'_k. Contracting s_j into s_i along the boundary of $P(C_i, C_j)$ reduces the partial watchman route problem to the whole watchman route problem. Thus, the proof is completed. □

As did for the (whole) watchman routes, we also call the set of the fragments with which a partial watchman route R_{ij} makes reflection contacts the watchman fragment set or WFS_{ij} for short.

Corollary 3 *A partial watchman route R_{ij} is the shortest partial watchman route if and only if it is non-adjustable.*

3.2 The details of our algorithm

Before describing the divide-and-conquer algorithm, we give an important lemma. It is actually proved in [8] that $O(j - i)$ adjustments suffice to make a one-place-adjustable route R_{ij}^0 non-adjustable.

Lemma 2 *[8] If a partial watchman route R_{ij}^0 is one-place-adjustable, $O(j - i)$ adjustments suffice to make it non-adjustable.*

Now we present the algorithm in detail. For simplicity, we assume first that all the cuts have the images defined on them. We will show later how this assumption can be removed. The cut set $\{C_0, \cdots, C_{m+1}\}$ is first divided into two subsets $\{C_0, \cdots, C_{\lfloor m+1/2 \rfloor}\}$ and $\{C_{\lfloor m+1/2 \rfloor}, \cdots, C_{m+1}\}$. The watchman route problem is recursively solved for both $R_{0 \lfloor m+1/2 \rfloor}$ and $R_{\lfloor m+1/2 \rfloor m+1}$. When the cut set contains one element C_x, the shortest partial watchman route R_{xx} is degenerated to the point s_x. The WFS_{xx} for R_{xx} is empty. When the cut set contains two elements C_x and C_y ($y = x + 1$), the shortest partial watchman route R_{xy} is the shortest path between s_x and s_y. (The shortest path between two points in a simple polygon can be found by using linear time algorithms [6].) The WFS_{xy} for R_{xy} is still empty.

Suppose that we have obtained the shortest partial watchman routes R_{ik} and R_{kj} ($i \leq k \leq j$). We have to construct the shortest partial watchman route R_{ij} from R_{ik} and R_{kj} (the merge step). We will consider only the cuts with index between i and j. First, we need to compute the intersections among the cuts $C_i, \cdots, C_j$. The intersections on a cut C should be ordered from l to r so that the fragments of C can be easily found. It is trivial to compute these cut intersections in $O((j - i)^2)$ time. Since all the intersections on a cut need to be ordered from left to right, a naive algorithm takes $O((j - i)^2 \log(j - i))$ time. Using Edelsbrunner et al.'s algorithm for arrangements of lines [5], we can do a little better. We first construct, in $O((j - i)^2)$ time, the arrangement of lines $C'_i, \cdots, C'_j$, where C'_k is the supporting line of C_k. Then we walk along each line C'_k to find the (sorted) line intersections on line C'_k. It is easy to check in $O(1)$ time whether a line intersection of C'_k and C'_l is a cut intersection of C_k and C_l or not. This allows us to obtain all cut intersections and the orders of cut intersections in $O((j - i)^2)$ time.

Secondly, we merge R_{ik} and R_{kj} into R_{ij}. If the dominances of s_i and s_j give the cut set $\{C_i, \cdots, C_j\}$, then the shortest path between s_i and s_j is the shortest watchman route R_{ij} and WFS_{ij} is empty. In the following, we assume that the shortest path between s_i and s_j does not form the shortest watchman route R_{ij}. From Lemma 2, the most important thing in the merge step is to find an initial watchman route that is one-place-adjustable. Since both R_{ik} and R_{kj} consist of convex chains within cut corners, the projections of the last segment of R_{ik} and the first segment of R_{kj} are behind and ahead of s_k, respectively. Let R_{ij}^0 denote the route that is obtained by combining R_{ik} and R_{kj} at s_k. We first note that R_{ij}^0 satisfies all the requirements of a partial watchman route from s_i to s_j, i.e., it consists of convex chains within cut corners, lies in polygon $P(C_i, C_j)$ and visits all the cuts which are not dominated by s_i. In most cases, R_{ij}^0 has the one-place-adjustable property. When R_{ij}^0 is one-place-adjustable, we can get a WFS_{ij} from the fragments with which R_{ij}^0 makes reflection contacts. Usually, R_{ij}^0 is not optimal with respect to this WFS_{ij}. The optimum partial watchman route for this WFS_{ij}, denoted by R_{ij}^1, should be computed by the rolled-out method. The shortest partial watchman route R_{ij} can then be obtained by adjusting the current route R_{ij}^l ($l \geq 1$) until it is non-adjustable.

Let the initial WFS_{ij} be the union of the WFS_{ik} (for R_{ik}) and the WFS_{kj} (for R_{kj}). We shall update it into the WFS_{ij} for R_{ij}^1 when R_{ij}^0 satisfies the one-place-adjustable requirement. Note that either R_{ik} or R_{kj} may overlap with C_k. In the former case, the incoming angle of R_{ik} with the cut C_{k-}, which intersects with C_k and is the last cut reflected by R_{ik}, must be greater than the outgoing angle of R_{ik} with C_{k-}. (But we cannot move the contact point to the right since R_{ik} is defined within $P(C_i, C_k)$. Thus, R_{ik} is non-adjustable.) In the latter case, the outgoing angle of R_{kj} with the cut C_k+, which intersects with C_k and is the first cut reflected by R_{kj}, must be smaller than the incoming angle of R_{kj} with C_k+. See Fig. 4. Observe that R_{ij}^0 may not be one-place-adjustable when R_{ik} or R_{kj} overlaps with C_k. We consider the following cases:

Case 1: R_{ik} and R_{kj} do not overlap with C_k (Fig. 5a). This means that R_{ij}^0 makes the reflection contact with C_k at s_k. Then, R_{ij}^0 is adjustable only on C_k and is thus one-place-adjustable. Hence R_{ij}^0 can be adjusted into the shortest partial watchman route R_{ij} by at most $O(j - i)$ adjustments. To obtain the WFS_{ij} for R_{ij}^1, we insert the fragment of C_k containing s_k into the current WFS_{ij}. If s_k happens to be a cut intersection of C_k with $C_{k'}$ and $k' \neq i$ or j, we choose the fragment of C_k left (right) to the intersection when $k > k'$ ($k < k'$). (The chosen fragment dominates the cut $C_{k'}$.) If $k' = i$ (j), we choose the fragment of C_k right (left) to the intersection. (The chosen fragment lies in $P(C_i, C_j)$.)

Case 2: Both R_{ik} and R_{kj} overlap with C_k (Fig. 5b). If one of them consists of only one line segment, then R_{ij}^0 is one-place-adjustable, say, adjustable only on C_{k-}. (Note that from our assumption that the shortest path between s_i and s_j is not R_{ij}, both R_{ik} and R_{kj} can not consist of only one line segment.) To get the WFS_{ij} for R_{ij}^1, we replace the fragment of C_{k-} in the current WFS_{ij} by one right to the intersection $p_{kk}+$. Suppose now that both R_{ik} and R_{kj} consist of more than two line segments. Then R_{ij}^0 is adjustable on both C_{k-} and C_k+. But, these two adjustments are consistent in the sense that both of them require the following routes to move to the left of C_k. All the adjustments on an active cut caused by them have the reflection contact point move in a single direction. We can still say that R_{ij}^0 is one-place-adjustable. Then R_{ij}^0 can be adjusted into R_{ij} by at most $O(j - i)$ adjustments. To get the correct WFS_{ij} for R_{ij}^1, we replace the fragments of C_{k-} and C_k+ in the current WSF_{ij} by those which are right and left to the intersections p_{k-k} and $p_{kk}+$, respectively.

Case 3: R_{kj} overlaps with C_k but R_{ik} does not (Fig. 5c). (Symmetric situation can be analogously handled.)

Case 3.1: R_{kj} is a single line segment. In this case, R_{ij}^0 is adjustable only on C_k and is thus one-place-adjustable. To get the WFS_{ij} for R_{ij}^1, we insert the fragment of C_k containing s_k (or left to s_k when s_k happens to be a cut intersection) into the WFS_{ij}.

Case 3.2: R_{kj} is not a single line segment. Then R_{ij}^0 is adjustable on C_k and C_k+. These two adjustments may not be consistent. The adjustment on C_k makes the following routes move to the left of C_k, but the adjustment on C_k+ requires the following routes to move to the right of C_k. Thus, R_{ij}^0 is not one-place-adjustable. To overcome this difficulty, we first compute a new shortest partial watchman route $R_{ik}+$, which starts at s_i, ends at $p_{kk}+$. Obviously, the segment $(s_k, p_{kk}+)$ of C_k is the shortest partial watchman route $R_{kk}+$. Constructing $R_{ik}+$ from R_{ik} and $R_{kk}+$ becomes Case 3.1. Thus, $R_{ik}+$ can be obtained by doing at most $O(k+ - i)$ adjustments. It is easy to see that $R_{ik}+$ will lie in the essential piece of C_k and thus cannot overlap with C_k+. Let R_k+_j denote the rest part of R_{kj} that starts at $p_{kk}+$ and ends at s_j. Since R_k+_j is non-adjustable, it is the shortest partial watchman route. The new initial route R_{ij}^0, which combines $R_{ik}+$ and R_k+_j at $p_{kk}+$, is adjustable only on C_k+. Constructing R_{ij} from $R_{ik}+$ and R_k+_j becomes Case 1. As a result, the shortest partial watchman route R_{ij} can be computed from R_{ik} and R_{kj} at the cost of $O(j - i)$ adjustments.

By now we have shown that R_{ik} and R_{kj} can be merged into R_{ij} at the cost of $O(j-i)$ adjustments. Each adjustment requires $O(n)$ time, which is determined by the complexity of the polygon the rolled-out method applies to. Thus, R_{ij} can be computed from R_{ik} and R_{kj} in $O((j - i)n)$ time. For completeness, we give below the whole algorithm.

PROCEDURE MERGE(R_{ik}, R_{kj}); (* Merge R_{ik} and R_{kj} into R_{ij}. *)

1. Compute the intersections among the cuts $C_i, \cdots, C_j$.

2. Find a one-place-adjustable watchman route R_{ij}^0 from R_{ik} and R_{kj} as analyzed in Cases 1 to 3, and then get the initial WFS_{ij} from R_{ij}.

3. Compute R_{ij}^1 from the current WFS_{ij} by the rolled out method.

4. $l = 1$;
   WHILE R_{ij}^l is adjustable **DO**
   Take an adjustable cut and update the current WFS_{ij};
   Compute the new route R_{ij}^{l+1} by the rolled-out method;
   $l = l + 1$;
   END { WHILE }

5. Return R_{ij}^l as the shortest partial watchman route R_{ij}.

PROCEDURE SWR(i, j); (* Compute the route R_{ij}. *)
IF the dominances of s_i and s_j form the cut set $\{C_i, \cdots, C_j\}$
THEN Let WFS_{ij} be empty and return the shortest path between s_i and s_j as R_{ij}
ELSE Call MERGE(SWR($i, \lfloor (i+j)/2 \rfloor$), SWR($\lfloor (i+j)/2 \rfloor, j$)).

The Watchman Route Algorithm;

1. Find the essential cuts $C_0, C_1, \cdots, C_{m+1}$. (* C_0 and C_{m+1} are the edge containing s. *)

2. Specify the images in each cut corner. (* For simplicity, we assume that all the images s_0, s_1, $\cdots$, s_{m+1} are defined. *)

3. Call the recursive procedure **SWR($0, m + 1$)**.

4. Report $R_{0(m+1)}$ as the shortest watchman route.

To complete our algorithm, we need to remove the assumption that each cut is an image-defined cut. In our divide-and-conquer algorithm, the images are used temporarily. When computing R_{ij}, we only need to know s_i and s_j. Thus, a cut having no defined image can be handled (e.g., used to compute cut intersections) together with the image-defined cut just after it. In the divide step, the cut with the medium index may have no image. For example, assume that there is no image defined on $C_{\lfloor m/2 \rfloor}$. To effectively divide the problem, we first find two image-defined cuts, one immediately before $C_{\lfloor m/2 \rfloor}$ and the other immediately after $C_{\lfloor m/2 \rfloor}$. Let C_{m1} and C_{m2} denote such two cuts, respectively. If $m1 \geq \lfloor m/4 \rfloor$ or $m2 \leq \lfloor 3m/4 \rfloor$, then the watchman route can be divided into two subroutes, either of them visits no more than $\lfloor 3m/4 \rfloor$ cuts. Otherwise, there are more than $\lfloor m/2 \rfloor$ cuts crossed by the shortest path R_{m1m2} between s_{m1} and s_{m2}. In this case, we first compute the shortest partial watchman routes R_{0m1} and $R_{m2(m+1)}$ by divide-and-conquer, then merge R_{0m1} and R_{m1m2} into R_{0m2}, and R_{0m2} and $R_{m2(m+1)}$ into $R_{0(m+1)}$. Therefore, in the case there exist the cuts having no images, the watchman route problem can be efficiently computed by divide-and-conquer, too.

It is obvious that the divide-and-conquer algorithm needs at most $O(m \log m)$ adjustments, Thus our algorithm requires at first glance $O(nm \log m)$ time. (Note that each adjustment requires $O(n)$ time.) However, many of adjustments can be performed together. A careful analysis shows that our algorithm runs in $O(nm)$ time.

Theorem 3 *The time complexity of the divide-and-conquer algorithm is $O(n^2)$.*

Proof: The time complexity of the divide-and-conquer algorithm can be mainly estimated by the following recursive formula:

$$T(m) = T(\alpha m) + O(mn)$$

where α is a constant and smaller than $3/4$, n is the number of edges of the given polygon and m is the number of essential cuts. The $O(mn)$ time is taken to merge two partial watchman routes because it requires at most $O(m)$ adjustments, each taking $O(n)$ time, to merge two partial watchman routes. Note that the time required to compute two partial watchman routes is $T(\alpha m)$, rather than $2T(\alpha m)$. Since two partial watchman routes lie in the different regions of the rolled-out polygon, the adjustments for these two routes can be performed simultaneously. Therefore, the time required to compute these two routes is the bigger one, rather than the sum of them. (In other words, at most $\lfloor m/2 \rfloor$ adjustments can be performed together at first merge step, $\lfloor m/4 \rfloor$ adjustments at the next step and so on.) Solving the above formula, we obtain that the time required by all the adjustments is $O(mn)$. This dominates the time complexities of the other steps in our algorithm, and thus proves Theorem 3. □

Finally, we note that the divide-and-conquer algorithm uses $O(n^2)$ space to store $O(n^2)$ cut intersections.

4   Concluding Remarks

We have presented an $O(n^2)$ algorithm for constructing shortest watchman routes in simple polygons. This improves the previous $O(n^3)$ time bound and confirms Chin and Ntafos' conjecture. The result obtained in this paper also improves the previous results for other related problems (for instance, the robber route problem [7]), whose solutions mainly depend on the watchman route algorithm.

It is an open problem whether our algorithm is optimal or not. Note that if any watchman algorithm requires to compute all cut intersections, then it takes at least $\Omega(n^2)$ time. Another interesting work is to find a polynomial-time algorithm to construct a shortest watchman route in a simple polygon without specifying any starting point.

References

[1] B. Chazelle, A theorem on polygon cutting with applications, *Proceedings, 23th Annu. IEEE Symp. Found. of Comput. Sci.*, 339-349, 1982.

[2] B. Chazelle, Triangulating a simple polygon in linear time, *Proceedings, 31th Annu. IEEE Symp. Found. of Comput. Sci.*, 220-229, 1990.

[3] W.P.Chin and S.Ntafos, Optimum watchman routes, *Inform. Process. Lett.* **28**, 39-44, 1988.

[4] W.P.Chin and S.Ntafos, Shortest watchman routes in simple polygons, *Discrete Comput. Geometry* **6**, 9-31, 1991.

[5] H.Edelsbrunner, J.O'Rourke and R.Seidel, Constructing arrangements of lines with applications. *SIAM J. Comput.* **15**, 341-363, 1986.

[6] L.Guibas, J.Hershberger, D.Leven, M.Sharir and R.Tarjan, Linear time algorithms for visibility and shortest path problems inside simple triangulated polygons, *Algorithmica* **2**, 209-233, 1987.

[7] S.Ntafos, The robber route problem, *Inform. Process. Lett.* **34**, 59-63, 1990.

[8] X.H.Tan, T.Hirata and Y.Inagaki, An incremental algorithm for constructing shortest watchman routes, To appear in *Internationa Journal of Computational Geometry & Applications* (also in *Lect. Note in Comput. Sci.* **557**, 163-175, 1991).

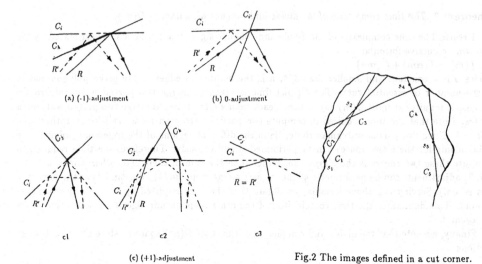

(a) (-1)-adjustment

(b) 0-adjustment

c1

c2

c3

(c) (+1)-adjustment

Fig.1 Three types of adjustments.

Fig.2 The images defined in a cut corner.

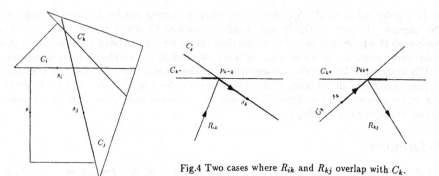

Fig.3 Illustration for the proof of Lemma 1.

Fig.4 Two cases where R_{ik} and R_{kj} overlap with C_k.

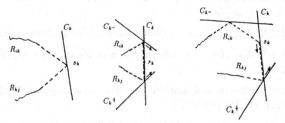

Fig.5 Three cases in merging R_{ik} and R_{kj} into R_{ij}.

A Graph Coloring Result and Its Consequences For Some Guarding Problems

(Extended Abstract)

Frank Hoffmann[1]   Klaus Kriegel[1]

Abstract

We prove the following graph coloring result: Let G be a 2–connected bipartite planar graph. Then one can triangulate G in such a way that the resulting graph is 3–colorable.

This result implies several new upper bounds for guarding problems including the first non–trivial upper bound for the rectilinear Prison Yard Problem:

1. $\lfloor \frac{n}{3} \rfloor$ vertex guards are sufficient to watch the interior of a rectilinear polygon with holes.

2. $\lfloor \frac{5n}{12} \rfloor + 3$ vertex guards resp. $\lfloor \frac{n+4}{3} \rfloor$ point guards are sufficient to watch simultaneously both the interior and exterior of a rectilinear polygon.

Moreover, we show a new lower bound of $\lfloor \frac{5n}{16} \rfloor$ vertex guards for the rectilinear Prison Yard Problem and prove it to be asymptotically tight for the class of ortho-convex polygons.

1   Introduction

The original Art Gallery Problem raised by V. Klee asks how many guards are sufficient to watch the interior of an n-sided simple polygon. In 1975 V. Chvátal [1] gave the answer proving that $\lfloor n/3 \rfloor$ guards are always sufficient and sometimes necessary. Since then many results have been published studying variants of the problem or analyzing algorithmic aspects, see [9], [10], [12] for a detailed discussion.

One of the main open questions in this field is the so called Prison Yard Problem for simple rectilinear polygons (comp. [10]), i.e. one wants to determine the minimal number of vertex guards sufficient to watch simultaneously both the interior and exterior of any n-sided simple rectilinear polygon.

The Prison Yard Problem for general simple polygons has been completely settled by Z. Füredi and D. Kleitman proving that $\lceil \frac{n}{2} \rceil$ vertex guards for convex and $\lfloor \frac{n}{2} \rfloor$ vertex guards for any non–convex simple polygon are sufficient, see [2]. As mentioned in [2] this does not imply new bounds for the rectilinear case. Here, the only upper bound known

---

[1]Institut für Informatik, Freie Universität Berlin, Takustr.9, 14195 Berlin, *hoffmann@tcs.fu-berlin.de*, *kriegel@tcs.fu-berlin.de*
Both authors have been partially supported by the ESPRIT Basic Research Action project ALCOM II and the Wissenschaftler–Integrationsprogramm Berlin.

has been the rather trivial $\lfloor \frac{7n}{16} \rfloor + 5$-bound (see [9]) which can be obtained by combining the $\lfloor \frac{n}{4} \rfloor$-result for the interior (see [3]) with the $\lceil \frac{n}{4} \rceil + 1$ vertex guards for the exterior of an n-sided rectilinear polygon.

Below we are going to derive several new bounds for the original rectilinear Prison Yard Problem as well as for the stronger "Prison Problem" where the guards have to watch not only the inside and outside of the yard but also all cells of the prison. The key tools to prove them are coloring and multicoloring arguments. Especially the new graph coloring result shown in Section 2 is probably also of independent interest. It states that one can triangulate a 2-connected bipartite planar graph in such a way that the resulting graph is 3-colorable.

In Sections 3 and 4 we apply this result to guarding problems by a suitable modeling of the rectilinear polygons. Next in Section 5 we establish lower bounds for the vertex guard number in staircase-like and in orthoconvex rectilinear prison yards. In Section 6 we use a new multicoloring technique to prove these bounds to be tight for the described polygon classes. We close with posing a few related open questions and discussing algorithmic aspects. Because of the space limitation we have to omit some of the proofs. All these missing details can be found in [5].

The following table summarizes the upper bounds on guard numbers shown in this paper for rectilinear polygons, see [9] for previous bounds:

polygon	problem	guard type	previous bound	new bound
simple	prison yard	vertex	$\lfloor \frac{7n}{16} \rfloor + 5$	$\lfloor \frac{5n}{12} \rfloor + 2$
simple	prison yard	point	$\lfloor \frac{7n}{16} \rfloor + 5$	$\lfloor \frac{n+4}{3} \rfloor$
staircase	prison yard	vertex	–	$\lfloor \frac{3n}{10} \rfloor + 2$ (tight)
orthoconvex	prison yard	vertex	–	$\lfloor \frac{5n}{16} \rfloor + 2$ (tight)
h holes	prison	vertex	–	$\lfloor \frac{5n-4h}{12} \rfloor + 2$
$h \geq n/6$ holes	art gallery	vertex	$\lfloor \frac{n+2h}{4} \rfloor$	$\lfloor \frac{n}{3} \rfloor$

2 On a Class of 3–Colorable Planar Graphs

This paragraph is devoted to the following theorem on 3–colorings of planar graphs. Its proof consists of two lemmata.

Theorem 2.1: *Let G be a planar, 2-connected, and bipartite graph. Then there exists a triangulation of G such that the triangulation graph is 3-colorable.*

The first lemma is due to Whitney and can be proved by standard induction arguments, for an elegant proof see [6].

Lemma 2.2: *A planar triangulated graph is 3-colorable iff all vertices have even degree.*

Lemma 2.3: *Let G be a 2-connected, bipartite, and planar graph. Then there exists a triangulation of G such that in the resulting graph H any vertex has even degree.*

Proof: We can assume that the set Q of all faces of G consists of 4–cycles only. For any vertex v let Q_v be the set of all faces having v as a vertex. Consider an auxiliary 2-coloring of G with colors *red* and *blue*. For any face $q \in Q$ we define the main diagonal to be that one connecting the vertices of q colored *red*. Furthermore, we introduce a $\{0,1\}$–valued variable x_q which will be set 1 if we choose the main diagonal in q and 0 if the other diagonal is chosen.

If v is a vertex of q we define $\epsilon_{q,v}$ to be 0 if v is colored *red* and 1 if v is colored *blue*. Obviously, $x_q \oplus \epsilon_{q,v}$ describes the increase of the degree of v by the diagonal of q chosen with respect to x_q. Here and in the following $\oplus$ denotes the addition modulo 2. It is easy to see that the existence of the desired triangulation is equivalent to the condition that the following system of equations has a solution:

$$deg(v) \oplus \bigoplus_{q \in Q_v} (x_q \oplus \epsilon_{q,v}) = 0 \quad (v \in V)$$

or, equivalently,

$$\bigoplus_{q \in Q_v} x_q = deg(v) \oplus \bigoplus_{q \in Q_v} \epsilon_{q,v} \quad (v \in V)$$

The left side of the second system forms the homogeneous part of the system. It is well known that such a system has a solution iff the rank of the homogeneous part is equal to the rank of the full system or, equivalently, any linear dependence of rows in the homogeneous part is also a dependence of rows in the full system. Over $GF(2)$ the only linear combinations of rows are $\oplus - sums$ hence it suffices to prove the following

Claim : *If for some $W \subseteq V$* $\displaystyle\bigoplus_{v \in W} \bigoplus_{q \in Q_v} x_q \equiv 0$ *then* $\displaystyle\bigoplus_{v \in W} (deg(v) \oplus \bigoplus_{q \in Q_v} \epsilon_{q,v}) = 0.$

Suppose that $\bigoplus_{v \in W} \bigoplus_{q \in Q_v} x_q \equiv 0$ for some $W \subseteq V$. Then for any $q \in Q$ the number of vertices of q which are also in W must be even, i.e. it is 0, 2 or 4. We prove the claim by showing that the sums $S_1 = \bigoplus_{v \in W} deg(v)$ and $S_2 = \bigoplus_{v \in W} \bigoplus_{q \in Q_v} \epsilon_{q,v}$ are both zero.

1) Since G is planar and 2–connected for any vertex v the degree equals the cardinality of the set Q_v. Using this fact and changing the order of summation we get:

$$S_1 = \bigoplus_{v \in W} |Q_v| = \bigoplus_{v \in W} \bigoplus_{q \in Q_v} 1 = \bigoplus_{q \in Q} \bigoplus_{v \in q \cap W} 1 = \bigoplus_{q \in Q} |q \cap W|$$

As we have already mentioned all summands are even numbers and consequently $S_1 = 0$.

2) We start as above changing the order of summation.

$$S_2 = \bigoplus_{v \in W} \bigoplus_{q \in Q_v} \epsilon_{q,v} = \bigoplus_{q \in Q} \bigoplus_{v \in q \cap W} \epsilon_{q,v}$$

Since for any $q \in Q$ the number $|q \cap W|$ is even we can subdivide Q according to this cardinality into Q_0, Q_2 and Q_4. Furthermore, we subdivide Q_2 according to the property whether the two vertices in $q \cap W$ lie on a diagonal or on an edge of q. So we get:

$$S_2 = \bigoplus_{q \in Q_0} \bigoplus_{v \in q \cap W} \epsilon_{q,v} \oplus \bigoplus_{q \in Q_2^{diag}} \bigoplus_{v \in q \cap W} \epsilon_{q,v} \oplus \bigoplus_{q \in Q_2^{edge}} \bigoplus_{v \in q \cap W} \epsilon_{q,v} \oplus \bigoplus_{q \in Q_4} \bigoplus_{v \in q \cap W} \epsilon_{q,v}$$

Obviously the first sum is zero and can be deleted. We also delete the sum over Q_2^{diag} since any summand has either the form $1 \oplus 1$ or $0 \oplus 0$. Analogously, the sum over Q_4 is zero, but instead of deleting it, we will add it once more to S_2 and we obtain:

$$S_2 = \bigoplus_{q \in Q_2^{edge}} \bigoplus_{v \in q \cap W} \epsilon_{q,v} \oplus \bigoplus_{q \in Q_4} \bigoplus_{v \in q \cap W} \epsilon_{q,v} \oplus \bigoplus_{q \in Q_4} \bigoplus_{v \in q \cap W} \epsilon_{q,v}$$

Consider the subgraph of G induced by W and denote its edge set by E_W. We will prove that the number of 1's in the sum above is equal to $2 \cdot |E_W|$ (or equivalently to the number of directed edges in E_W) what will finish the proof of the claim. Let us identify each face q with the *directed* cycle obtained by running around the region in counterclockwise order. Now the sum over Q_2^{edge} can be seen as a representation of all directed edges (u, v) in E_W such that their corresponding face q is in Q_2. Note that for any such edge the representing summand $\epsilon_{q,u} \oplus \epsilon_{q,v}$ has the form $1 \oplus 0$ or $0 \oplus 1$. Analogously, the first (resp. second) sum over Q_4 can be seen as a representation of those directed edges in E_W whichs corresponding face is in Q_4 and which are directed from red to blue (resp. from blue to red) vertices. Here any summand represents two directed edges. It is easy to observe that this representation is 1–1 what completes the proof.

3   A Graph Model for the Prison Yard Problem

The idea for our graph model is based on the following nice and simple proof of the classical Art Gallery Theorem due to S. Fisk, see [9]. Consider any triangulation graph of a given simple polygon. One knows that it is 3-colorable. Clearly, any triangle contains each color and choosing guard position corresponding to the smallest color class one can watch the polygon using $\leq \lfloor \frac{n}{3} \rfloor$ guards.

In [3] J. Kahn, M. Klawe, and D. Kleitmann applied a similar idea to rectilinear polygons. They proved that any rectilinear polygon (possibly with holes) has a convex quadrilateralization, i.e. a decomposition into convex 4-gons (called quadrilaterals) using only diagonals (here called chords) of the polygon. Moreover it is easy to see that for simple rectilinear polygons the graph consisting of all polygon edges, all chords, and both inner diagonals of all quadrilaterals is 4-colorable. Hence, they obtained an $\lfloor \frac{n}{4} \rfloor$-upper bound for the rectilinear Art Gallery Problem. However, the 4-colorability of this graph does not hold starting with polygons having holes. But now Theorem 2.1 states that one can select one diagonal per quadrilateral such that the graph formed by all polygon edges, all chords, and the selected diagonals is 3-colorable. So we get at least an $\lfloor \frac{n}{3} \rfloor$-upper bound on the vertex guard number for the Art Gallery Problem in the presence of holes. Below we introduce a graph model which allows us to apply the coloring result also to Prison-type Problems.

Given an n-sided rectilinear polygon P (w.l.o.g. in general position, see [3]) we construct its orthoconvex hull $OConv(P)$, i.e. the smallest point set containing P and such that its intersection with any horizontal or vertical line is convex (see Fig.1a). This partitions the exterior region of P into the exterior region of $OConv(P)$ and those connected components of $OConv(P) \setminus P$ which are different from the interior region of P. They will be called pockets. Since all pockets are bounded by rectilinear polygons there is a quadrilateralization of them as well as of the interior of P. We have to pay for this construction by some additional vertices (u in our example). However, using ideas from [7] one can shift

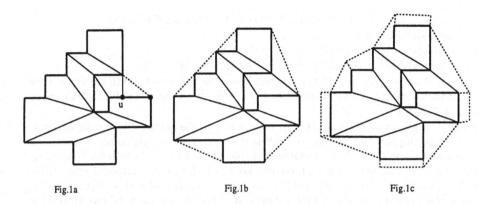

Fig.1a   Fig.1b   Fig.1c

these vertices to neighboring polygon corners on the boundary of $OConv(P)$ in such a way that the resulting polygon $OConv^*(P)$ is also orthoconvex and the quadrilateralizability of the pockets is not destroyed (the dashed line in Fig.1a). Now we have to cover, the exterior region of $OConv^*(P)$ with convex sets. Remark that $OConv^*(P)$ is bounded by four extremal edges (northernmost, westernmost, southernmost, easternmost) which are cyclically connected by monotone staircases. So the exterior of $OConv^*(P)$ is covered by four halfplanes defined by the extremal edges and the cones defined by all concave vertices on the staircases.

Let $G(P)$ be the planar graph (Fig.1b) over the polygon vertices the edge set of which consists of all polygon edges, all quadrilateralization chords, and all pairs of consecutive convex corners on boundary stairs of $OConv^*(P)$. We say that a subset C of the vertex set dominates $G(P)$ if any quadrilateral, any triangle over a staircase, and any of the four extremal edges contains at least one vertex from C. The Prison Yard Problem in this context now reads: Find a small dominating set for $G(P)$.

Since we want to apply Theorem 2.1. it is neccessary to modify $G(P)$ in such a way that it becomes bipartite, i.e. all convex regions (also the exterior cones and halfplanes!) which have to be dominated are represented by convex quadrilaterals. To do this we need some additional vertices. We start as before constructing $OConv^*(P)$. Then we use 8 new vertices to obtain a copy of all extremal edges as indicated in Fig.1c. Finally, for any monotone boundary staircase of $OConv^*(P)$ which contains more than one convex vertex (we do not count the vertices on the extremal edges) we copy every second one of them. Now it is possible to replace any boundary triangle in $G(P)$ by a quadrilateral in the new graph $G^*(P)$, comp. Fig.1c. The number of additional vertices is bounded by $\lfloor \frac{n-12}{4} \rfloor + 8$ and hence the number of vertices of $G^*(P)$ is bounded by $\lfloor \frac{5n}{4} \rfloor + 5$. Obviously, if a new vertex will be chosen as a guard position this guard can be placed onto the original polygon vertex.

In order to demonstrate the power of Theorem 2.1. we introduce the Prison Problem which is a generalization of the Prison Yard Problem as well as of the Art Gallery Problem for polygons with holes. Let a rectilinear polygon P with h rectilinear holes $P_1, \ldots, P_h$ be given, having in total n vertices. We have to select a set of vertices such that any point in the plane can be watched from one of the selected vertices. A graph

$G^*(P, P_1, \ldots, P_h)$ representing this problem can be constructed as follows:
1) quadrilateralize the holes $P_1, \ldots, P_h$,
2) quadrilateralize the interior of P minus the holes,
3) proceed with the exterior of P as in the construction of $G^*(P)$.
Clearly, P has at most $n - 4h$ vertices and hence the number of additional vertices for the construction of $G^*(P)$ is bounded by $8 + \lfloor \frac{n-4h-12}{4} \rfloor$. Thus, $G^*(P, P_1, \ldots, P_h)$ has at most $\lfloor \frac{5n-4h}{4} \rfloor + 5$ vertices.
Finally, we generalize the concept of graph coloring to the notion of labellings and multicolorings. Suppose, we have given k different colors. Then a function which labels any vertex of a graph $G(P)$ with a certain set of colored pebbles will be called a k-labelling. It is a k-multicoloring if adjacent vertices are labelled with disjoint color sets. A labelling is called l-uniform if the pebble sets have cardinality l for all vertices. Clearly, any k-coloring is a 1-uniform k-multicoloring. A multicoloring is called dominating if for any color the set of those vertices labelled with a pebble of this color dominates $G(P)$. Hence, a dominating k-multicoloring of $G(P)$ which uses in total $f(n)$ pebbles implies the existence of an $\lfloor \frac{f(n)}{k} \rfloor$ solution of the Prison Yard Problem for P.

4   General Upper Bounds

We start with a straightforward application of Theorem 2.1 to the classical rectilinear Art Gallery Problem. Consider a rectilinear polygon P with h holes and a total of n vertices. Both the polygon and the holes can be quadrilateralized, so we have the following.
Corollary 4.1: $\lfloor \frac{n}{3} \rfloor$ *vertex guards are sufficient to solve the Art Gallery Problem for rectilinear polygons with holes.*
Observe, that the guards watch the interior of the holes, too. Further let us remark that this improves the previously known $3n/8$-bound. However, there is some evidence that the $2n/7$-lower bound, see [4], for the vertex guard number is tight.
Conjecture 4.2: *For any quadrilateralized rectilinear polygon possibly with holes there is a 2-uniform dominating 7-coloring.*
The second application of Theorem 2.1 deals with the weak version of the rectilinear Prison Yard Problem where point guards are allowed, see [9]. Here the spiral polygon gives an $(\lceil \frac{n}{4} \rceil + 1)$-lower bound. However, the best upper bound up to now has been the same as for the vertex guard version: $\lfloor \frac{7n}{16} \rfloor + 5$.
Corollary 4.3: *For any rectilinear polygon (possibly with holes) P on n vertices $\lfloor \frac{n+4}{3} \rfloor$ point guards are sufficient to solve the Prison Problem.*

Proof: Let R be a rectangle enclosing P. We consider R together with P as a polygon P' having P as a hole. After quadrilateralizing P' as well as the original P the resultig graph fulfills the assumptions of Theorem 2.1 and we can choose guard positions corresponding to the minimal color class of the dominating 3-coloring. Observe, one uses at most 2 point guards, the other guards can be chosen to sit in vertices.

Corollary 4.4: *For any rectilinear polygon on n vertices with h holes $\lfloor \frac{5n-4h}{12} \rfloor + 2$ vertex guards are sufficient to solve the Prison Problem.*
Proof: Apply Theorem 2.1 to the graph $G^*(P)$ defined in Section 3.

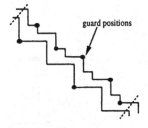

guard positions

Fig. 2a: Dorward's polygon

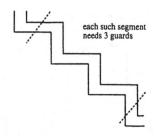

each such segment
needs 3 guards

Fig.2b: 3n/10 guards are necessary

5 Lower Bounds

Any simple convex polygon requires $\lceil \frac{n}{2} \rceil$ vertex guards to solve the Prison Yard Problem. What are candidates for lower bound examples in the rectilinear world? Figure 2a shows an example of a rectilinear polygon due to Dorward who claimed that it required $\lceil \frac{n}{3} \rceil$ guards, see [10]. Continuing, however, periodically the guarding positions indicated in Fig.2a one sees that $7n/24 + 1$ watchmen are sufficient (assume $24|n$). Let P_0 be the simplest staircase shown in Fig.2b. The following can be proved easily.

Proposition 5.1: *The prison yard P_0 requires $\lceil \frac{3n}{10} \rceil$ vertex guards.*

We will show in the next section that $\lfloor \frac{3n}{10} \rfloor + 2$ vertex guards are also sufficient for any strictly monotone rectilinear polygon. Surprisingly, there are other monotone polygons which require even more guards. Let P_1 be the pyramid in Fig.3. Assume that the edge lengths are chosen in such a way that to watch an inner quadrilateral one has to choose one of its vertices as guard position. Again, as indicated $5n/16$ guards are sufficient (up to an additive constant) and we show that we need as many.

Proposition 5.2: *The prison yard P_1 requires $\lceil \frac{5n-10}{16} \rceil$ vertex guards.*
Proof: We distinguish 3 types of guards, comp.Fig.3b. A guard in a concave corner such that it can watch 4 quadrilaterals is called an α-guard. We remark that any such guard must have at least one "partner" on the other side watching the opposite triangle and we choose one of them to form a pair together with the α-guard. An α-guard pair watches together 4 quadrilaterals, 2 type1-triangles, and 1 type2-triangle. β- (resp.γ) guards are sitting in concave (convex) corners and they are not part of α-pairs. They watch each 2 (resp.1) quadrilaterals , 0 (resp.1) type1-triangle, and 1(resp.1) type2-triangle. Since we have a total of $(n-2)/2$ quadrilaterals and $(n-2)/4$ triangles of each type we conclude for any guarding set consisting of $g = 2a+b+c$ guards of α-, β-, and γ-type respectively that: $4a+2b+c \geq (n-2)/2$, $2a+c \geq (n-2)/4$, $a+b+c \geq (n-2)/4$. Adding to the first inequality the second one and then the third multiplied by 2 we get $4g \geq 5(n-2)/4$. But this implies the lower bound.

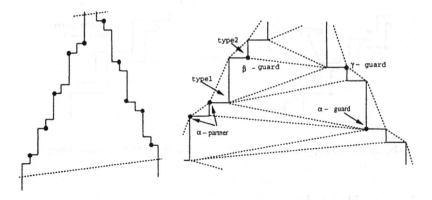

Fig.3: 5n/16 guards are necessary

6   Special Upper Bounds

We are going to show that the lower bounds derived for strictly monotone and for ortho-convex rectilinear polygons are tight up to an additive constant. We use the idea from Section 2 to construct a dominating set of vertices in the graph $G(P)$ by a multicoloring.

Theorem 6.1: $\lfloor \frac{3n}{10} \rfloor + 2$ *guards are sufficient to solve the Prison Yard Problem for any strictly monotone rectilinear polygon P.*

Proof: Let P denote such a polygon, say with north–west orientation (comp. Fig.4). First we remark that the quadrilateralization of P is unique and its dual is a path $W = q_1 q_2 ... q_{(n-2)/2}$; any chord of the quadrilateralization connects a convex with a concave vertex and any quadrilateral has a diagonal connecting two convex vertices (called convex diagonal).

Let $(d_i), i = 1, 2, ...n-1$ be the following sequence of polygon edges, diagonals, and chords obtained by traversing W. We start with d_1 the bottom polygon edge in q_1 followed by the convex diagonal of q_1, d_i is the common edge of $q_{(i-1)/2}$ and $q_{(i+1)/2}$ for an odd $i > 1$, otherwise it is the convex diagonal of $q_{i/2}$. d_{n-1} is the top edge of P. The d_i's induce a canonical numbering of the vertices of P. Starting with d_2 each d_i encounters exactly one new vertex, say v_{i+1}. Let Q_i denote the polygon generated by the first i quadrilaterals.

We show that there is a greedy algorithm which following the path W constructs a dominating 5-multicoloring of $G(P)$ with the following properties:

(1) Both the north–westernmost vertex v_n and the south–easternmost vertex v_2 are labelled by 4 pebbles;

(2) Any other convex (resp. concave) vertex is colored by 2 (resp.1) pebbles.

While building this multicoloring we maintain the following invariant: Each convex diagonal contains exactly 3 colors, i.e. there is one common color on both sides of the diagonal. In total we use $2\frac{n+4}{2} + \frac{n-4}{2} + 4 = \frac{3n+12}{2}$ pebbles. Consequently, there exists a dominating color class of size $\leq \lfloor \frac{3n}{10} \rfloor + 2$.

In a similar way we can prove the following statement.

86

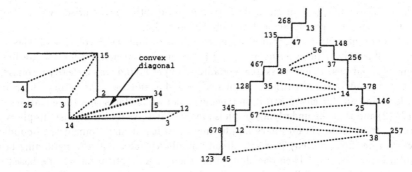

Fig.4: 5–multicoloring a strictly
monotone polygon

Fig.5: 8–multicoloring the lower bound example

Theorem 6.2: $\lfloor \frac{5n}{16} \rfloor + 2$ *guards are always sufficient to solve the Prison Yard Problem for orthoconvex rectilinear polygons.*

Proof: We outline the proof for pyramids. The result follows then for an orthoconvex polygon by decomposing it into at most 2 pyramids and 1 strictly monotone polygon. For a pyramid P we consider again the dual path $W = q_1 q_2 ... q_{(n-2)/2}$ of its unique quadrilateralization. We construct an 8–multicoloring of $G(P)$ such that (comp.Fig.5):
(1) Each convex vertex has 3 pebbles, any concave one gets 2 pebbles;
(2) Each exterior triangle and each quadrilateral has all 8 colors;
(3) Any quadrilateral has exactly 4 colors along the diagonal that contains the convex vertex.
The existence of such an 8–multicoloring can be shown using a greedy algorithm along W. By putting additional pebbles on the extremal edges it can be made dominating.

7 Related Open Problems and Algorithmic Aspects

Let us add two more open problems to those mentioned in Section 4. First, we start with a remark concerning Theorem 2.1. It is essential for the result proved there that all inner faces of the graph G are 4–cycles. The statement is not true if like in the graph $G(P)$ in Section 3 there are also inner triangles. So one needs some new ideas to prove for example an $\lfloor \frac{n}{3} \rfloor$–upper bound for the rectilinear Prison Yard Problem. We think that replacing the multicoloring argument used in Theorem 6.2 by an 8–labelling one can show that the following conjecture is correct.

Conjecture 7.1: *There is an absolute constant c such that any rectilinear prison yard can be watched by $5n/16 + c$ vertex guards.*

Recall, that in a k–labelled graph it is possible for adjacent vertices to be labelled by pebbles of the same color. Further, it would be interesting to find applications of Theorem 2.1 or of some multicoloring or labelling to non–rectilinear art gallery type problems, compare with [2], [11], and Chapter 5.2 in [9].

Although we cannot go into algorithmic details let us mention that all upper bound results here can be converted into efficient algorithms. Since one can quadrilateralize simple rectilinear polygons in linear time we can guard orthoconvex prison yards also in linear time using the greedy algorithm from Theorem 6.2.

A much more interesting algorithmic problem is how to find the 3–coloring of Theorem 2.1 more efficiently than by using a general (superquadratic) method for solving linear systems of equations. Below we sketch a quadratic upper bound which clearly implies the same time bound for the algorithmic problems in 4.1, 4.3, and 4.4 . The main point is, it is possible to find an efficient substitution scheme exploring the facts that the equations are over $GF(2)$ and that the underlying graph is planar. First, we iteratively use simple cycle separators from [8] to obtain a face numbering for which at any moment the boundary (which can be disconnected) between already numbered faces and the remaining faces has total length $O(\sqrt{n})$. Then one shows that there is a substitution scheme based on this numbering with the following properties: (2) Any substitution is applied to at most $O(\sqrt{n})$ equations (which correspond to boundary points only). However, the details of this algorithm are rather involved and we think improvements should be possible.

References

[1] V. Chvátal, *A combinatorial theorem in plane geometry*, J. Combin. Theory Ser. B, 1975, Vol. 18, pp. 39-41.

[2] Z. Füredi and D. Kleitmann, *The Prison Yard Problem*, to appear in Combinatorica

[3] J. Kahn and M. Klawe and D. Kleitman, *Traditional galleries require fewer watchmen*, SIAM Journal of Alg. Disc. Math., 1983, pp. 194-206.

[4] F. Hoffmann, *On the rectilinear art gallery problem*, Proc. ICALP'90, LNCS 443, 1990, pp. 717-728.

[5] F. Hoffmann and K. Kriegel, *A Graph Coloring Result and Its Concequences For Polygon Guarding Problems*, Technical Report B93-08 Institut für Informatik, FU Berlin, June 1993.

[6] L. Lovász, *Combinatorial Problems and Exercises*, North Holland, Amsterdam, 1979.

[7] A. Lubiw, *Decomposing polygonal regions into convex quadrilaterals*, Proc. 1st ACM Symp.Comp.Geometry, 1985, pp. 97-106.

[8] G. Miller, *Finding small simple cycle separators for 2–connected planar graphs*, Journal of Comp. and System Sciences, 1986, pp. 265-279.

[9] J. O'Rourke, *Art gallery theorems and algorithms*, Oxford University Press, 1987

[10] J. O'Rourke, *Computational Geometry Column 15*, SIGACT News 23:2, 1992, pp. 26-28

[11] T. Shermer, *Triangulation graphs that require extra guards*, NYIT, Computer Graphics tech. report No.3D-13, 1984.

[12] T. Shermer, *Recent results in art galleries*, Proc. IEEE, 80(9), 1992, pp. 1384-1399

The maximum k-dependent and f-dependent set problem

Anders Dessmark [*]   Klaus Jansen [†]   Andrzej Lingas [‡]

1 Introduction

Let k be a positive integer. A *k-dependent* set in an undirected graph $G = (V, E)$ is a subset of the set V of vertices such that no vertex in the subset is adjacent to more than k vertices of the subset. This subset induces a subgraph of G of *maximum degree* bounded by k. A 0-dependent set in G is simply an *independent set* of vertices in G. Furthermore, an 1-dependent set is in general a set of independent vertices and edges and a 2-dependent set is a set of independent paths and cycles.

The problem of constructing a maximum k-dependent set and its decision version have been studied in [6]. The NP-completeness of the decision version has been shown for arbitrary graphs and each $k \geq 0$. On the other hand a linear-time algorithm for the construction problem restricted to trees has been presented in [6]. Furthermore, for each constant k the problem of finding a maximum k-dependent set for a graph with constant treewidth has been observed to be solvable in linear time. The problem of finding a maximum k-dependent set for $k = 2$ has several applications, for example in information dissemination in hypercubes with a large number of faulty processors [3].

A generalization of this problem called the maximum *f-dependent* set problem has been given in [7]. Given weights $f(v) \in \mathbb{N}_0$ for $v \in V$, an f-dependent set is a subset A of V such that each vertex $v \in A$ is adjacent to at most $f(v)$ vertices in A. In [7] parallel algorithms for finding maximal k and f-dependent sets have been given.

In this paper we analyze both problems for bipartite graphs, cographs, trees, split graphs and graphs with bounded treewidth. Among others, we show that the decision version of the maximum k-dependent set problem restricted to planar, bipartite graphs is NP-complete for any given $k \geq 1$. This contrast with the well known fact that the maximum 0-dependent (i.e. independent) set in a bipartite graph can be found in polynomial time by reduction to maximum matching (see [10]) via König-Egervary theorem (see [15] and [8]). Next, we give polynomial algorithms for both problems restricted to cographs, trees and graphs with bounded treewidth. On the other hand, we show that the complexity differs for split graphs; we give a polynomial time algorithm for the maximum k-dependent set problem and show the NP-completeness for

---

[*]Department Computer Science, Lund University, Box 118, 221 00 Lund, Sweden.

[†]Fachbereich 11, Informatik, Universität Duisburg, 47048 Duisburg, Germany.

[‡]Department Computer Science, Lund University, Box 118, 221 00 Lund, Sweden.

the maximum f-dependent set problem. Finally, we provide efficient poly-log time PRAM versions of the aforementioned sequential algorithms.

2 Bipartite graphs

Here we show the NP-completeness of the maximum k-dependent set problem, for $k > 0$, restricted to the planar, bipartite graphs.

A graph $G = (V, E)$ is *bipartite*, if its edge set E satisfies $E \subset \{\{v, v'\} | v \in V_1, v' \in V_2\}$ where $V_1 \cup V_2 = V$ and $V_1 \cap V_2 = \emptyset$. Such a partition in two independent sets for a bipartite graph can be obtained in $O(|V| + |E|)$ time [8]. It is known that the maximum 0-dependent problem restricted to bipartite graphs is solvable in polynomial time [8, 10, 15].

Theorem 2.1 *The maximum 1-dependent set for bipartite graphs is NP-complete.*

Proof. Clearly the maximum 1-dependent set for bipartite graphs is in NP. We will reduce 3-SAT to maximum 1-dependent set for bipartite graphs. For each variable in the instance of 3-SAT build the bipartite subgraph with 9 nodes showed in Fig 1. It can be easily seen that the maximum 1-dependent set for this graph contains 6 nodes and any such set contains exactly one of the nodes X_i and $\bar{X}_i$. The maximum 1-dependent set in the subgraph in Fig 1. corresponds to X_i being false. For each clause in the instance of 3-SAT build the bipartite subgraph with 10 nodes showed in Fig 2. It is easily seen that the maximum 1-dependent set for this graph contains 6 nodes. Connect each clause with the appropriate variables as indicated in the figures. The created graph will remain bipartite and the maximum 1-dependent set of the

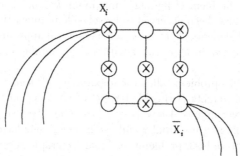

Fig 1. The subgraph corresponding to each variable with a maximum set marked.

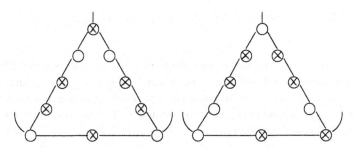

Fig 2. Sufficient (up to symmetry) examples of maximum
1-dependent sets corresponding to clauses.

created graph cannot contain more nodes than the sum of the maximum 1-dependent
set in the subgraphs independently. Only if all three neighbouring nodes of the subgraph corresponding to the clause are included in the 1-dependent set, the maximum 1-dependent set in the clause subgraph will be limited to 5 nodes. Construct the 1-dependent set for the graph. If there exists a satisfying assignment
for the instance of 3-SAT, then there exists a 1-dependent set in the graph contain-
ing $6|clauses| + 6|variables|$. If the instance isn't satisfiable then there must be at
least one clause subgraph with all three neighbouring nodes included in the maximum
1-dependent set for the respective subgraphs corresponding to the variables. And
since not all the subgraphs can have their maximum number of nodes, the maximum
1-dependent set for the complete graph will have less than $6|clauses| + 6|variables|$
nodes. Thus, 3-SAT many-one reduces to maximum 1-dependent set for bipartite
graphs. The construction of the graph can obviously be de done in polynomial time.
□

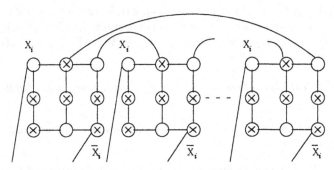

Fig 3. The modified subgraph corresponding to each variable
with a maximum set marked.

By slightly changing the construction and reducing from planar 3-SAT we can
show that the maximum 1-dependent set for planar, bipartite graphs is NP-complete.
An instance to planar 3-SAT is a 3CNF formula F with variables X_i, $1 \leq i \leq n$, and
clauses C_j, $1 \leq j \leq m$, and a planar bipartite graph $G = (\{X_i|1 \leq i \leq n\} \cup \{C_j|1 \leq$
$j \leq m\}, E)$ such that $(X_i, C_j) \in E$ if X_i or $\bar{X}_i$ is a literal of C_j. The construction in
Fig 3. allows X_i and $\bar{X}_i$ to be connected to any number of clauses without crossings.
The required number of nodes in the maximum set will have to be adjusted with the
sizes of the modified variable subgraphs. The reduction is otherwise similar to the
previous case. Hence we have.

Theorem 2.2 *The maximum 1-dependent set for planar, bipartite graphs is NP-complete.*

The result for the maximum 1-dependent set problem can be generalized to the maximum k-dependent set problem. Maximum 1-dependent set in planar bipartite graphs can easily be reduced to maximum k-dependent set in planar bipartite graphs by hanging $k - 1$ pendants on all the original nodes. This will not interfere with the graph being planar and bipartite.

Theorem 2.3 *The maximum k-dependent set problem for planar, bipartite graphs and each $k \geq 1$ is NP-complete.*

FURTHER NOTATION. In the further sections we shall use the following definitions. $\omega_k(G)$ (or $\omega_f(G)$) denotes the maximum cardinality of a k-dependent (or f-dependent) set in G. For a vertex $v \in V$, $\delta(v)$ stands for the number of *neighbours* $|\{u \in V | \{u, v\} \in E\}|$; $\delta(v)$ is called the *degree* of vertex v. For a subset $A \subset V$, the *partial degree* $\delta_A(v)$ is equal to $|\{u \in A | \{u, v\} \in E\}|$. Furthermore, we denote with $\Delta(G)$ the *maximum degree* in G and with $\Delta(G, m)$ the minimum $\Delta(G')$ of a subgraph G' of G induced by m vertices.

3   Cographs

In this section we consider the cographs and give polynomial algorithms for the maximum k and f-dependent set problem.

Definition 3.1 *Let $G_1 = (V_1, E_1)$, $G_2 = (V_2, E_2)$ be two graphs, with V_1 and V_2 disjoint sets. The disjoint union of G_1 and G_2 is the graph $G_1 \cup G_2 = (V_1 \cup V_2, E_1 \cup E_2)$. The product of G_1 and G_2 is the graph $G_1 \times G_2 = (V_1 \cup V_2, E_1 \cup E_2 \cup \{(v, w) \mid v \in V_1, w \in V_2\})$.*

Definition 3.2 *The class of cographs is the smallest set of graphs, fulfilling the following rules:*

1. *Every graph $G = (V, E)$ with one vertex and no edges ($|V| = 1$ and $|E| = 0$) is a cograph.*

2. *If $G_1 = (V_1, E_1)$ is a cograph, $G_2 = (V_2, E_2)$ is a cograph, and V_1 and V_2 are disjoint sets, then $G_1 \cup G_2$ is a cograph.*

3. *If $G_1 = (V_1, E_1)$ is a cograph, $G_2 = (V_2, E_2)$ is a cograph, and V_1 and V_2 are disjoint sets, then $G_1 \times G_2$ is a cograph.*

To each cograph G one can associate a corresponding rooted binary tree T, called the *cotree* of G, in the following way. Each non-leaf node in the tree is labeled with either "∪" (union-nodes) or "×" (product-nodes). Each non-leaf node has exactly two children. Each node x of the cotree corresponds to a cograph $G_x = (V_x, E_x)$. A leaf node corresponds to a cograph with one vertex and no edges. A union-node (product-node) corresponds to the disjoint union (product) of the cographs, associated with the two children of the node. Finally, the cograph that is associated with the root of the cotree is just G, the cograph represented by this cotree.

We remark that the most usual definition of cotrees allows an arbitrary degree of internal nodes. However, it is easy to see that this has the same power, and can easily be transformed in cotrees with two children per internal node. In [4], it is shown that one can decide in $O(|V| + |E|)$ time, whether a graph is a cograph, and build a corresponding cotree.

For each node x of the cotree we compute the minimum degree $\Delta(G_x, m)$ of a subgraph of G_x with m vertices. The value $\Delta(G_x, m) = \infty$ means that a subgraph with m vertices in G_x does not exist.

Lemma 3.3 *1. If x is a leaf, then $\Delta(G_x, 0) = \Delta(G_x, 1) = 0$ and $\Delta(G_x, m) = \infty$ for $m > 1$.*

2. If x is an union node with two children y_1 and y_2, then for each $m \leq |V_x|$:
$$\Delta(G_x, m) = min_{m_i \leq |V_{y_i}|, m_1 + m_2 = m} \; max(\Delta(G_{y_1}, m_1), \Delta(G_{y_2}, m_2)).$$

3. If x is a product node with two children y_1 and y_2, then for each $m \leq |V_x|$:
$$\Delta(G_x, m) = min_{m_i \leq |V_{y_i}|, m_1 + m_2 = m} \; max(\Delta(G_{y_1}, m_1) + m_2, \Delta(G_{y_2}, m_2) + m_1).$$

To compute the size of a maximum k-dependent set in a cograph we generate the values $\Delta(G_x, m)$ in a bottom up order in the cotree for each node x and each $m \leq |V_x|$. Then, given the values $\Delta(G_r, m)$ for the root r we get

$$\omega_k(G) = max\{m | 1 \leq m \leq |V|, \Delta(G_r, m) \leq k\}.$$

To generate a maximum k-dependent set, we store for each inner node x and each $m \leq |V_x|$ the pair (m_1, m_2) which generates the minimum value $\Delta(G_x, m)$. By top-down backtracking in the cotree, we can compute the set of vertices which generates $\omega_k(G)$.

The computing time of the algorithm can be bounded by $O(|V|^3)$, because there are $O(|V|^2)$ comparisons for each node of the cotree. But we improve the calculation and get a better time complexity.

Theorem 3.4 *Given a cograph $G = (V, E)$ and a positive integer k, a maximum k-dependent set can be computed in $O(|V|^2)$ time.*

Proof. For each node x of the cotree we must only compute the values $\Delta(G_x, m)$ for each positive integer m with $m \leq |V_x|$. We compute these values for a product and a union node with two children y_1 and y_2 in $O(|V_{y_1}| \cdot |V_{y_2}|)$ time.

Let $t(n)$ denote the maximum total time to compute all values for cotrees corresponding to cographs with n vertices. Then,

$$t(n) \leq max_{1 \leq i \leq n-1} \; c \cdot i \cdot (n - i) + t(i) + t(n - i)$$

for a constant c. From this formula we can prove by induction that $t(n) = O(n^2)$. □

In the next part of this section we give a linear time algorithm for constant k. The idea is now to compute for each node x the values $\omega_{k'}(G_x)$ for each positive integer k' with $k' \leq k$.

Lemma 3.5   1. *If x is a leaf, then $\omega_k(G_x) = 1$ for each $k \geq 0$.*

   2. *If x is an union node with two children y_1 and y_2, then*
$$\omega_k(G_x) = \omega_k(G_{y_1}) + \omega_k(G_{y_2}).$$

   3. *If x is a product node with two children y_1 and y_2, then*
$$\omega_k(G_x) = max_{k_1,k_2 \leq k}\, min(\omega_{k_2}(G_{y_2}), k - k_1) + min(\omega_{k_1}(G_{y_1}), k - k_2).$$

Proof. We consider only the product case. Let A be a subset of V_x with maximum degree at most k. Then, $A_i = A \cap V_{y_i}$ is a subset of V_{y_i} in G_{y_i} with maximum degree k_i and $A_1 \cup A_2 = A$ and $k_1, k_2 \leq k$. Furthermore, we have $|A_1| + k_2 \leq k$ and $|A_2| + k_1 \leq k$. This implies that $|A_i| \leq min(\omega_{k_i}(G_{y_i}), k - k_{3-i})$ for $i \in \{1, 2\}$. Using $|A| = |A_1| + |A_2|$ we get

$$|A| \leq max_{k_1,k_2 \leq k}\, min(\omega_{k_2}(G_{y_2}), k - k_1) + min(\omega_{k_1}(G_{y_1}), k - k_2).$$

$\square$

Theorem 3.6 *Given a cograph $G = (V, E)$, a corresponding cotree T and a constant $k \in \mathbb{N}_0$, a maximum k-dependent set can be computed in $O(|V|)$ time.*

Proof. Use the recursion in the lemma above. For each node of the cotree we compute k values, and each value k' with $1 \leq k' \leq k$ can be generated in $O(k^2)$ time. In total, we get an algorithm which runs in $O(k^3 \cdot |V|) = O(|V|)$ time. $\square$

In the following, we generalize our algorithm to compute a maximum f-dependent set. For each node of the cotree we have to include the number of vertices in $V \setminus V_x$ which are adjacent to the vertices in G_x. We define a $\{0, 1\}$-variable $\alpha(G_x, m, m')$ where $\alpha(G_x, m, m') = 1$ iff there is a set A of m vertices in the cograph G_x and m' adjacent vertices in $V \setminus V_x$ such that the degree $\delta_A(v)$ for each vertex $v \in A$ is at most $f(v) - m'$. The following lemma shows how we compute these values for each node of the cotree, each $m \leq |V_x|$ and $m' \leq |V \setminus V_x|$.

Lemma 3.7   1. *If x is a leaf with $V_x = \{v\}$, then $\alpha(G_x, 1, m') = 1$ iff $m' \leq f(v)$. Moreover, $\alpha(G_x, 0, m') = 1$ and $\alpha(G_x, m, m') = 0$ for $m > 1$.*

   2. *If x is an union node with two children y_1 and y_2, then $\alpha(G_x, m, m') = 1$ iff there are positive integers $m_1, m_2 \leq m$ with $m_1 + m_2 = m$ and $\alpha(G_{y_i}, m_i, m') = 1$ for $i = 1, 2$.*

   3. *If x is a product node with two children y_1 and y_2, then $\alpha(G_x, m, m') = 1$ iff there are positive integers $m_1, m_2 \leq m$ with $m_1 + m_2 = m$ and $\alpha(G_{y_i}, m_i, m' + m_{3-i}) = 1$ for $i = 1, 2$.*

Theorem 3.8 *The problem to find a maximum f-dependent set in a cograph can be solved in $O(|V|^3)$ time.*

Proof. For each node x of the cotree we compute $O(|V_x| \cdot |V|)$ values. If y_1 and y_2 are the children of x, these values can be generated in $O(|V_{y_1}| \cdot |V_{y_2}| \cdot |V|)$ time. Using the same argument as in theorem 3.4, we get an algorithm which runs in $O(|V|^3)$ time. $\square$

Corollary 3.9 *The problem to find a maximum f-dependent set in a cograph with weights $f(v) \leq k$ and k constant can be solved in $O(|V|^2)$ time.*

4 Split graphs

In this section we analyze the complexity of both problems for split graphs. First, we give a linear algorithm for the k-dependent set problem. Then, we show that the f-dependent set problem for split graphs becomes NP-complete.

A graph $G = (V, E)$ is a split graph, iff there is a partition of the vertex set into a clique C and an independent set U. In other words, if $Q = \{\{v, w\}|v, w \in C, v \neq w\}$, $Q \subset E$ & $E \subset \{\{v, w\}|v \in C, w \in U\} \cup Q$. A representation of a split graph in this form can be computed in linear time $O(|V| + |E|)$ [11]. We show now how we compute for a given number m the minimum maximum degree $\Delta(G, m)$ of a subgraph of G induced by m vertices. Clearly, for $m \leq |U|$ we get $\Delta(G, m) = 0$. The first step in our algorithm is to compute the *partial degrees* $\delta_U(v)$ for the vertices $v \in C$. The computation time to generate these degrees is $O(|V| + |E|)$. Let us assume that the clique vertices are ordered due to these degrees $\delta_U(v_1) \leq \ldots \leq \delta_U(v_{|C|})$. This order can be generated using bucket sort in $O(|V|)$ time.

Lemma 4.1 *Let G be a split graph with independent set U and clique C. Then, for each k with $1 \leq k \leq |C|$, $\Delta(G, |U| + k) = \delta_U(v_k) + k - 1$.*

Proof. Let us assume that we have a set $C' \cup U'$ with $C' \subset C$, $U' \subset U$ and $|C'| + |U'| = |U| + k$. Clearly, $|C'| \geq k$. The maximum degree $\Delta(G[C' \cup U'])$ in the graph induced by $C' \cup U'$ can be bounded as follows: $\Delta(G[C' \cup U']) \geq |C'| - 1 + \delta_{U'}(v)$ for $v \in C'$ where $\delta_{U'}(v)$ gives the number of neighbours of v in U'. Since $\delta_{U'}(v) + |U \setminus U'| \geq \delta_U(v)$ for $v \in C$, we get for each $v \in C'$: $\Delta(G[C' \cup U']) \geq |C'| - 1 + |U'| - |U| + \delta_U(v) = k - 1 + \delta_U(v)$. Since $|C'| \geq k$, we know that $\delta_U(v) \geq \delta_U(v_k)$ for at least one vertex $v \in C'$. It follows that $\Delta(G[C' \cup U']) \geq k - 1 + \delta_U(v_k)$.

On the other hand, the set $U \cup \{v_1, \ldots, v_k\}$ generates a subgraph with maximum degree $k - 1 + \delta_U(v_k)$. $\square$

The lemma above and the formula

$$\omega_k(G) = |U| + max\{h|1 \leq h \leq |C|, \Delta(G, |U| + h) \leq k\}$$

implies the following theorem.

Theorem 4.2 *A maximum k-dependent set for a split graph can be computed in linear time $O(|V| + |E|)$.*

Theorem 4.3 *The maximum f-dependent set problem for split graphs is NP-complete.*

The proof is omitted because of space considerations. It is by reduction from 3-SAT [5].

5 Graphs with bounded treewidth

In [5] we give a polynomial-time dynamic programming solution to the maximum f-dependent set problem restricted to graphs with constant treewidth. Our algorithm relies on the linear-time tree-decomposition algorithm for graphs with treewidth $\leq k$ from [1]. We also give a linear-time algorithm for the maximum f-dependent set problem restricted to trees. Because of space considerations we report only the main results here.

Theorem 5.1 *Given a graph with treewidth at most k, the maximum f-dependent set problem is solvable in $O(|V|^{2(k+1)})$ time.*

Corollary 5.2 *If all values $f(v) \leq k'$ with k' constant, a maximum f-dependent set in graph with constant treewidth can be generated in linear time.*

Theorem 5.3 *A maximum f-dependent set for a tree can be computed in linear time $O(|V|)$.*

6 Parallel algorithms

In this section, we outline processor efficient, poly-log versions of the sequential polynomial algorithms presented in the previous sections.

6.1 Cographs

Theorem 6.1 *Given a cograph G with n nodes and $k \in \mathbb{N}_0$, a corresponding cotree T, a maximum k-dependent set can be constructed in time $O(\log n \log k)$ on an EREW PRAM with $O(nk^2/\log n \log k)$ processors.*

Proof. The idea is to apply the optimal tree-contraction parallel algorithm from [13] and [12] to the cotree in order to recursively evaluate the values for each node of the tree as described in lemma 3.3. By the distributivity of min with respect to max and vice versa, and the distributivity of + with respect min, max, the vector of k maxmin+ expressions associated to the edges (corresponding to the arithmetic ones in [12]) can be kept within $O(k^2)$ total size. Therefore, each single SHUNT operation applied to a leaf could be implemented in time $O(\log k)$ using $O(k^2/\log k)$ EREW PRAM processors. The leaves, i.e. vertices of G to be included in the maximum k-dependent set can be found by recontraction. □

Note that if k is constantly bounded and the cotree is given then the above theorem yields an optimal logarithmic-time algorithm for maximum k-dependent graph in a cograph.

He has recently presented an $O(\log^2 n)$-time, $O(n+m)$ processor CRCW PRAM algorithm for recognizing cographs on n nodes and m edges and constructing their cotrees (non necessarily binary) [9]. As a non-binary cotree can be easily transformed into a binary one within the above time and processor bounds, Theorem 6.1 yields the following theorem.

Theorem 6.2 *Given a cograph G with n nodes and m edges, and $k \in \mathbb{N}_0$, a maximum k-dependent set can be constructed in time $O(\log^2 n)$ on an CRCW PRAM with $O(m + nk^2/\log n \log k)$ processors.*

We can generalize the above result to include maximum f-dependent set by parallelizing the sequential algorithm from the proof of Theorem 3.8. Again, we can use the tree-contraction technique. Now, with each edge a vector of polynomial number of disjunctions, each of polynomial length would be associated. We leave the details to the reader reporting only the following general result.

Theorem 6.3 *Given a cograph G where each node v has weight $f(v) \leq n$, a maximum f-dependent set can be constructed by an NC algorithm.*

6.2 Split graphs

Theorem 6.4 *A maximum k-dependent set for a split graph can be computed in time $O(\log n)$ on an EREW PRAM with $O((m + n)/\log n + I(n))$ processors, where $I(n)$ is the number of processors necessary for sorting n integers in $O(\log n)$ time on an EREW PRAM.*

Proof. Compute the degrees of the vertices and sort them in decreasing order with respect to degree; $\delta(v_1) \geq \ldots \geq \delta(v_{|V|})$. It is easy to see that, if we take a maximum clique C, the clique will contain exactly those nodes v_i where $\delta(v_i) \geq i - 1$. Find the largest j such that $\delta(v_j) \geq k + j$. Lemma 4.1 now tells us that the maximum k-dependent set consists of $C = \{v_{j+1}, \ldots, v_n\}$. To compute the degrees we need $O(\log n)$ time and $O((m + n)/\log n)$ processors using parallel list ranking in the adjacency lists. The sorting step can be performed in time $O(\log n)$. Also $O(\log n)$ time and $O(n/\log n)$ processors are sufficient for finding the required maximum. $\square$

6.3 Graphs with bounded treewidth

The general idea of parallelization of the polynomial-time algorithm for maximum f-dependent set in graphs with bounded treewidth is similar to that for cographs. First, we find a tree-decomposition of the input graph using the method of [2] or [14] and then apply the tree contraction method of [13]. In effect, we can report the following results.

Theorem 6.5 *Given a graph G with treewidth at most k where each node has weight $f(v) < n$, a maximum f-dependent set can be constructed by an NC algorithm.*

Theorem 6.6 *Given a tree G where each node v has weight $f(v) \leq n$, a maximum f-dependent set can be constructed in time $O(\log n)$ using an EREW PRAM with $O(n/\log n)$ processors.*

7 Overview of results

In the following table we give an overview about the complexity of the maximum k-dependent set and maximum f-dependent problem for several graph classes. The entry NPc means that the problem is NP-complete and the other entries correspond to our time upper bounds. The membership of the problem in the NC class is denoted by NC.

graph class	k-dependent	f-dependent
bipartite graphs	NPc	NPc
trees	$O(|V|), NC$	$O(|V|), NC$
planar graphs	NPc	NPc
split graphs	$O(|V| + |E|), NC$	NPc
cographs	$O(|V|^2), NC$	$O(|V|^3), NC$
$O(1)$ tree-width	$O(|V|^{2(K+1)}), NC$	$O(|V|^{2(K+1)}), NC$

References

[1] H.L. Bodlaender: A linear time algorithm for finding tree-decompositions of small treewidth, Proceedings of 25th STOC, 1993.

[2] H.L. Bodlaender: NC-algorithms for graphs with small treewidth, Proccedings of WG 88, Amsterdam, Netherlands (1988), Springer Verlag, LNCS 344, 1 – 8.

[3] S. Carlsson, Y. Igarashi, K. Kanai, A. Lingas, K. Miura, O. Petersson: Information disseminating schemes for fault tolerance in hypercubes, *IEICE Trans. Fundamentals* **75**, 1992, 255–260.

[4] D.G. Corneil, Y. Perl and L.K. Stewart: A linear recognition algorithm for cographs, *SIAM J. Comput.* **14**, 1985, 926–934.

[5] A. Dessmark, K. Jansen, A. Lingas: The complexity of maximum k-dependent and f-dependent set, Technical Report, LU-CS-TR:93-119, Dept. Computer Science, Lund University, Sweden, 1993.

[6] H. Djidjev, O. Garrido, C. Levcopoulos, A. Lingas: On the maximum k-dependent set problem, Technical Report, LU-CS-TR:92-91, Dept. Computer Science, Lund University, Sweden, 1992.

[7] K. Diks, O. Garrido, A. Lingas: Parallel algorithms for finding maximal k-dependent sets and maximal f-matchings, Proccedings of ISA 91, Taipei, Taiwan (1991), Springer Verlag, LNCS 557, 385-395.

[8] M.C. Golumbic: Algorithmic graph theory and perfect graphs, Academic Press, New York, 1980.

[9] X. He: Parallel algorithm for cograph recognition with applications, Proceedings of SWAT'92, Helsinki, Finland, Springer Verlag, LNCS 621, 94-105.

[10] J.E. Hopcroft and R.M. Karp: An $n^{2.5}$ algorithm for maximum matching in bipartite graphs, SIAM J. Comp. 2 (1973), 225-231.

[11] P.L. Hammer, B. Simeone: The splittance of a graph, *Combinatorica* **1** (3), 1981, 275–284.

[12] R.M. Karp and V. Ramachandran: Parallel algorithms for shared memory machines, J. van Leeuwen, ed., Handbook of Theoretical Computer Science, Vol. A (Elsevier Science Publishers, Amsterdam, 1990) 869-941.

[13] S.R. Kosaraju and A.L. Delcher: Optimal parallel evaluation of treestructured computation by ranking (extended abstract), Proc. AWOC 88, Springer-Verlag, N.Y., 1988, LNCS 319, 101-111.

[14] J. Lagergren: Efficient Parallel Algorithms for Tree-Decomposition and Related Problems. Proc. 31st FOCS, 1990.

[15] C.H. Papadimitriou and K. Steiglitz: Combinatorial Optimization, Algorithms and Complexity, Prentice-Hall, New Jersey, 1982.

Finding Shortest Non-Crossing
Rectilinear Paths in Plane Regions

Jun-ya Takahashi, Hitoshi Suzuki and Takao Nishizeki

Graduate School of Information Sciences
Tohoku University
Sendai 980, Japan

e-mail: junya@nishizeki.ecei.tohoku.ac.jp

Abstract. Let A be a plane region which is bounded by an outer rectangle and an inner one and has r rectangular obstacles inside the region. Let k terminal pairs lie on the outer and inner rectangular boundaries. This paper presents an efficient algorithm which finds k "non-crossing" rectilinear paths in the plane region A, each connecting a terminal pair without passing through any obstacles, whose total length is minimum. Non-crossing paths may share common points or line segments but do not cross each other in the plane. The algorithm runs in time $O(n \log n)$ where $n = r + k$.

1. Introduction

This paper deals with a rather unknown topic "non-crossing paths." Non-crossing paths should not cross each other in the plane but may share common points or line segments unlike "disjoint paths." The "non-crossing path problem" asks to find non-crossing paths of minimum total length, each connecting a specified terminal pair in two-dimensional plane regions or plane graphs. Our motivation comes from an application of the problem to VLSI layout design [1,2,3]. For the sake of simplicity, we assume that a VLSI routing region and internal blocks (obstacles) are all rectangular in shape. Usually such a routing region in a VLSI single-layer routing problem is represented by a two-dimensional plane region or a plane graph [3]. We wish to find "shortest" non-crossing paths (wires) in the region, each of which connects a terminal pair and does not pass through any blocks, and whose total length is minimum. Such shortest non-crossing paths give a routing which minimizes the area used by wires or the time-delay in VLSI circuits. (See Figure 1.)

Let k be the number of terminal pairs, let r be the number of rectangular obstacles, and let $n = k + r$. D. T. Lee *et al.* presented an algorithm which finds shortest k non-crossing rectilinear paths in a **plane region** in time $O((k^2!)n \log n)$ [2]. On the other hand, we gave an algorithm which finds shortest k non-crossing paths in a **plane graph** in time $O(n \log n)$ for the case when all $2k$ terminals lie on two specified face boundaries [1], where n is the number of vertices in the plane graph.

In this paper we consider a plane region A which is bounded by an outer rectangle and an inner one and has r rectangular obstacles inside the region, as depicted in Figure 1. We wish to find shortest non-crossing rectilinear paths passing through none of the obstacles in the plane region A. We give an algorithm

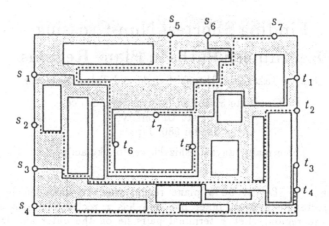

Fig. 1. Shortest non-crossing paths in plane region A.

which finds such shortest non-crossing paths in A in time $O(n \log n)$ for the case when all the terminals are located on the outer and inner boundaries, where $n = r + k$ and neither r nor k is assumed to be a constant.

Our idea is to reduce the problem on a plane region to a problem on a plane graph. However a straightforward reduction requires time $O(n^2 \log n)$ as follows. Construct from a given plane region A a plane graph G as depicted in Figure 2(a): add to G all corners of rectangles and all terminals as vertices, draw horizontal and vertical line segments from all these vertices until they hit a rectangle, and then add as vertices of G all the intersections of these line segments. One can observe that shortest non-crossing paths in the resulting **plane graph** G are shortest non-crossing **rectilinear** paths in the **plane region** A. Thus one can find shortest non-crossing rectilinear paths in A by applying the algorithm in [1] to G. Such a straightforward algorithm requires time $O(n^2 \log n)$ since G has $O(n^2)$ vertices.

Our algorithm in this paper improves the time complexity to $O(n \log n)$. The idea is to divide the plane region A into several subregions, in each of which shortest non-crossing paths may be assumed to run either horizontally or vertically, and is to construct a new plane graph G', as depicted in Figure 2(b), by drawing either horizontal or vertical line segments in each of the subregions. One can show that G' has $O(n)$ vertices and that shortest non-crossing paths in G' are shortest non-crossing rectilinear paths in A. Thus our algorithm runs in time $O(n \log n)$.

2. Preliminaries

In this section we give a formal description of the non-crossing path problem and define several terms. In this paper all rectangles are assumed *axis-parallel*, that is, all sides are parallel to x-axis or y-axis. We assume that there is an outmost rectangle R_O and that all rectangular obstacles are contained in R_O and do not overlap each other. Let A be a plane region which is inside R_O and outside the other rectangles, where A contains the boundaries of all rectangles. Let R_I be one of the rectangular obstacles in R_O. We call the boundary of R_O the *outer*

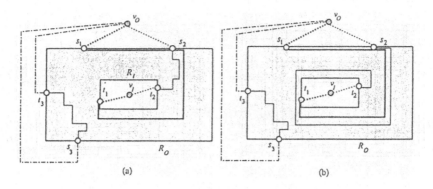

Fig. 2. Plane graphs G and G' constructed from plane region A.

boundary B_O of A, and call the boundary of R_I the *inner boundary* B_I of A. Let r be the number of rectangles except R_O and R_I. A pair of points s_i and t_i which we wish to connect by a path is called a *terminal pair* (s_i, t_i). Assume that all the terminals are located on the two boundaries B_O and B_I. Let k be the number of terminal pairs. In this paper neither r nor k is assumed to be a constant.

We define *non-crossing paths* in the plane region A as follows. Let $P_1, P_2, \cdots, P_k$ be paths in A connecting the k terminal pairs. Add a new point v_O outside R_O, and connect v_O to each terminal on B_O with a path. Similarly, add a new point v_I in R_I, and connect v_I to each terminal on B_I with a path. Let P'_i, $1 \leq i \leq k$, be a path (or a cycle) obtained from P_i by adding two new paths: one connecting s_i to v_O or v_I, and the other connecting t_i to v_O or v_I. We define paths $P_1, P_2, \cdots, P_k$ in region A to be *non-crossing* (for the boundaries B_O and B_I) if P'_i, $1 \leq i \leq k$, do not cross each other in the plane. This definition is appropriate for the VLSI single layer routing problem mentioned in Section 1. In Figure 3(a), paths P_1 and P_2 cross each other (for the boundaries B_O and B_I). On the other hand, the three paths P_1, P_2 and P_3 in Figure 3(b), do not cross each other.

In this paper we consider only rectilinear paths, that is, those consisting of axis-parallel line segments. The *length* of a rectilinear path is the sum of lengths of line segments in the path. Non-crossing paths $P_1, P_2, \cdots, P_k$ are *shortest* if the sum of lengths of $P_1, P_2, \cdots, P_k$ is minimum. Figure 1 depicts shortest non-crossing rectilinear paths in A. It is easy to check the existence of non-crossing paths in A. Therefore we assume that A necessarily contains non-crossing paths, and presents an efficient algorithm which finds the shortest non-crossing rectilinear paths in A.

We then define "monotone" paths. Let $x(p)$ be the x-coordinate of a point p, and let $y(p)$ be the y-coordinate of p. We recursively define an xy-path P starting at a point p, as follows [4]:

(a) P is the horizontal line $y = y(p)$ for $x \geq x(p)$ (not crossing any obstacle); or

(b) P follows $+x$ direction to the boundary of a rectangular obstacle, then follows $+y$ direction up to the upper left corner p' of the rectangular obstacle, and

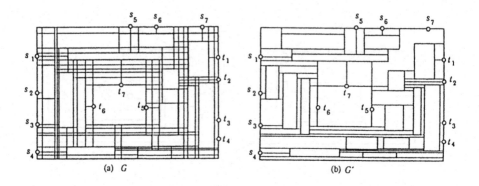

Fig. 3. Crossing and non-crossing paths.

then follows an xy-path starting at p'.

We similarly define $x(-y), -xy, -x(-y), yx, y(-x), -yx$ and $-y(-x)$-paths.

Define eight paths $\Pi_1, \Pi_2, \cdots, \Pi_8$ starting at the four corners of R_I in A as follows: the yx-path Π_1 and the xy-path Π_2 starting at the upper-right corner, the $x(-y)$-path Π_3 and the $-yx$-path Π_4 starting at the lower-right corner, the $-y(-x)$-path Π_5 and the $-x(-y)$-path Π_6 starting at the lower-left corner, and the $-xy$-path Π_7 and $y(-x)$-path Π_8 starting at the upper-left corner. The eight paths are drawn in thick lines in Figure 4(a). Let Π_i start at corner q_i and end at point p_i on B_O, then clearly each Π_i, $1 \leq i \leq 8$, is a shortest path in A connecting p_i and q_i.

We divide A into eight subregions $A_1, A_2, \cdots, A_8$ by the eight monotone paths Π_i, $1 \leq i \leq 8$. Let A_i be the subregion of A inside a cycle consisting of four paths: Π_i, the path going clockwise from p_i to p_{i+1} on B_O, Π_{i+1}, and the path going clockwise from q_i to q_{i+1} on B_I, where $i + 1 = 1$ if $i = 8$. Note that A_i includes the boundary of the cycle. The regions $A_1, A_2, \cdots, A_8$ are shaded in Figure 4(a).

3. Shortest non-crossing paths in the plane region

In this section we present an algorithm for finding shortest non-crossing rectilinear paths $P_1, P_2, \cdots, P_k$ in the plane region A, where each P_i, $1 \leq i \leq k$, represents a path connecting a terminal pair (s_i, t_i). We write $p \in B_O$ when point p lies on B_O. Similarly, we write $p \in B_I$ when p lies on B_I. One may assume without loss of generality that $s_i \in B_O$ and $t_i \in B_I$ if either s_i or t_i is on B_O and the other is on B_I. Then the set S of k terminal pairs can be partitioned into the following three subsets:

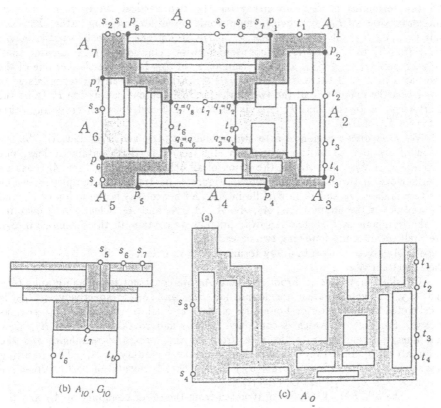

Fig. 4. Illustration of subregions and plane graphs for Case 1.

$$S_{IO} = \{(s_i, t_i) | s_i \in B_O \text{ and } t_i \in B_I\},$$
$$S_I = \{(s_i, t_i) | s_i, t_i \in B_I\}, \text{ and}$$
$$S_O = \{(s_i, t_i) | s_i, t_i \in B_O\}.$$

However one may assume, without loss of generality, $S_I = \phi$. Otherwise, one can decide all paths P_i for pairs $(s_i, t_i) \in S_I$ as follows. Let Q_i be a path which goes from s_i to t_i clockwise on B_I, and let Q_i' be a path which goes from s_i to t_i counterclockwise on B_I. Since we assume that there exist non-crossing paths in A, either Q_i or Q_i' passes through none of the terminals of S_{IO}. If $S_{IO} = \phi$, then let P_i be the shorter one of Q_i and Q_i'. If $S_{IO} \neq \phi$, let P_i be either Q_i or Q_i' that passes through none of the terminals in S_{IO}. All the paths P_i for S_I decided in this way do not cross each other and do not cross any paths for S_{IO} or S_O, since all these paths P_i traverse only the inner boundary B_I.

The set S_O is partitioned further into two subsets S_{Oa} and S_{Op}:

$$S_{Oa} = \{(s_i, t_i) \in S_O \mid s_i \text{ and } t_i \text{ are on the same side or on the two adjacent}$$
$$\text{sides of } R_O \},$$
$$S_{Op} = \{(s_i, t_i) \in S_O \mid s_i \text{ and } t_i \text{ are on the two parallel sides of } R_O \}.$$

All the terminals of S_{Op} lie either on the two vertical sides or on the two horizontal sides of R_O; otherwise, there would be no non-crossing paths. For each pair $(s_i, t_i) \in S_O$, all the terminals s_j of S_{IO} lie on one of the two paths in B_O going from s_i to t_i clockwise or counterclockwise. One may further assume that, for each pair $(s_i, t_i) \in S_{Oa}$, all the terminals s_j of S_{IO} lie on the shorter one of the two paths in B_O: if there is a pair $(s_i, t_i) \in S_{Oa}$ for which all the terminals s_j of S_{IO} lie on the longer one of the two paths in B_O, then one can decide P_i to be the shorter one in B_O because P_i is truly shortest in A and does not cross any other paths for S_O or S_{IO}.

We next divide region A into two subregions A_{IO} and A_O. Let A_{IO} be the union of B_I and all of the subregions $A_1, A_2, \cdots, A_8$ that contain at least one terminal s_j in S_{IO}. Let A_O be the union of all other subregions, then A_O consists of either one or two connected regions. Figure 4 illustrates an example, for which the terminals s_5, s_6 and s_7 in S_{IO} lie in A_8, and hence A_{IO} is the union of B_I and A_8, and A_O is the union of $A_1, A_2, A_3, A_4, A_5, A_6$ and A_7. Clearly A_{IO} contains all the terminals in S_{IO}, but A_O does not always contain all the terminals in S_O. We shall consider the following two cases.

Case 1: A_{IO} does not contain any terminal in S_O, and

Case 2: Otherwise.

Consider first Case 1. From A_{IO} and A_O we construct two plane graphs G_{IO} and G_O, in which paths are found for S_{IO} and S_O, respectively. G_{IO} is constructed from the outer boundary of A_{IO} and all the rectangles in A_{IO}, as follows. For each A_i which is contained in A_{IO} and contains a side of R_I, draw new lines, perpendicular to the side, from all the corners of rectangles and the terminals in A_i until they hit a rectangle or the boundary of A_i. The resulting plane graph is G_{IO}. (See Figure 4(b).) Since either horizontal or vertical lines are drawn in each A_i, G_{IO} has $O(n)$ vertices.

On the other hand, G_O is constructed from the outer boundary of A_O and the rectangles in A_O, as follows. Consider first the case in which $S_{Oa} = \phi$, that is, all the pairs of terminals in S_O lie on the two parallel sides of R_O. Then we draw new lines, perpendicular to the sides on which the terminals of S_O lie, from all terminals and all corners of rectangles in A_O until they hit a rectangle or the boundary of A_O. The resulting graph is G_O.

Consider next the case in which $S_{Oa} \neq \phi$. There is a pair in S_{Oa}, say (s_1, t_1), such that all the terminals of S_O lie on the longer one Q_1 of two paths on B_O going from s_1 to t_1 clockwise or counterclockwise, as shown in Figure 4(a). One may assume that Q_1 goes from s_1 to t_1 counterclockwise on B_O. Let S_O contain l pairs, then one may further assume that the $2l$ terminals $s_1, s_2, \cdots, s_{l-1}, s_l, t_l, t_{l-1}, \cdots, t_2, t_1$ in S_O appear on Q_1 in this order. Let $m = |S_{Oa}|$, then clearly

$$S_{Oa} = \{(s_i, t_i) | 1 \leq i \leq m\}, \text{ and}$$

$$S_{Op} = \{(s_i, t_i) | m + 1 \leq i \leq l\}.$$

(In the example depicted in Figure 4 $m = 2$.)

We then define three subregions A_O^1, A_O^2 and A_O^3 of A_O. Let Q_m be the path which goes from s_m to t_m counterclockwise. Let the end p_α of Π_α appear first on Q_m among the eight monotone paths. Let p_β appear last on Q_m. (In the example depicted in Figure 4, $\alpha = 7$ and $\beta = 3$.) Then $s_m, p_\alpha, p_\beta, t_m$ appear counterclockwise on Q_m in this order. Let A_O^1 be the subregion of A_O inside a cycle consisting of two paths: a shortest path in A_O which connects s_{m+1} and t_{m+1}; and the path on B_O which connects s_{m+1} and t_{m+1} counterclockwise. Let A_O^2 be the subregion of A_O inside a cycle consisting of two paths: a shortest path in A_O which connects s_m and q_α; and the path which goes from s_m to q_α clockwise on the outer boundary of A_O. Let A_O^3 be the subregion of A_O inside a cycle consisting of two paths: a shortest path in A_O which connects t_m and q_β; and the path which goes from t_m to q_β counterclockwise on the outer boundary of A_O. One may assume that A_O^1, A_O^2 and A_O^3 do not overlap each other. A_O^1, A_O^2 and A_O^3 are shaded in Figure 4(c).

We can derive the following two lemmas.

LEMMA 1. The subregion A_{IO} of A contains paths $P_{l+1}, P_{l+2}, \cdots, P_k$ which are included in a set of shortest non-crossing paths in A.

LEMMA 2. The subregion A_O of A contains paths $P_1, P_2, \cdots, P_l$ which are included in a set of shortest non-crossing paths in A.

By the two lemmas above, we can find shortest non-crossing paths in A as the union of two sets of non-crossing paths: one in A_O and the other in A_{IO}.

We are now ready to construct G_O. G_O is constructed from the outer boundaries of A_O, A_O^1, A_O^2 and A_O^3, all the corners of B_I, and all the rectangles in A_O as follows. For the subregion A_O^1, draw new lines, perpendicular to the side of R_O on which s_{m+1} lies, from all the corners of rectangles and the terminals in A_O^1 until they hit a rectangle or the boundary of A_O^1. For the subregion A_O^2, draw new lines, perpendicular to the side on which s_m lies, from all the corners of rectangles and the terminals in A_O^1 until they hit a rectangle or the boundary of A_O^2. For the subregion A_O^3, draw new lines, perpendicular to the side on which t_m lies, from all the corners of rectangles and the terminals in A_O^1 until they hit a rectangle or the boundary of A_O^3. The resulting plane graph is G_O. We can derive the following two lemmas.

LEMMA 3. The plane graph G_{IO} contains a set of shortest non-crossing paths for S_{IO} in A_{IO}.

LEMMA 4. The plane graph G_O contains a set of shortest non-crossing paths for S_O in A_O.

By Lemmas 3 and 4, we can observe that the union of two sets of shortest non-crossing paths, one in G_O and the other in G_{IO}, is a set of shortest non-crossing rectilinear paths in A.

Consider next Case 2. We assume, without loss of generality, that $S_O = \{(s_1, t_1), (s_2, t_2), \cdots, (s_l, t_l)\}$ and $S_{IO} = \{(s_{l+1}, t_{l+1}), (s_{l+2}, t_{l+2}), \cdots, (s_k, t_k)\}$. We define a partial ordering $\succ$ among all terminal pairs in S_O. Let $(s_i, t_i), (s_j, t_j) \in S_O$. There exist two paths Q_i and Q_i' on B_O connecting s_i and t_i, Q_i going from s_i to t_i clockwise and Q_i' counterclockwise. Either Q_i or Q_i' passes through both s_j and t_j. Let Q be such a path. Define $(s_i, t_i) \succ (s_j, t_j)$ if Q does not pass through any terminals in S_{IO}. Pair (s_i, t_i) is *maximal* when for each j, $1 \le j \le l$, either $(s_i, t_i) \succ (s_j, t_j)$ or there is no relation between (s_i, t_i) and (s_j, t_j). The assumption of location of the terminals in B_O implies that there exist one or two maximal terminal pairs if $S_O \ne \phi$. For each maximal terminal pair

(s_i, t_i), let P be a path connecting s_i and t_i on B_O and it contains none of the terminals in S_{IO}. Let p_α be the end of Π_α that appears first on P when traversing P from s_i to t_i. Let p_β be the end of Π_β that appears last on P when traversing P from s_i to t_i. We define Q_α and Q_β as follows. If s_i is contained in A_{IO}, let Q_α be a shortest path connecting s_i and q_α. Similarly, if t_i is contained in A_{IO}, let Q_β be a shortest path connecting s_i and q_β.

We partition the subregion A_{IO} into at most five subregions. When there is exactly one maximal terminal pair, we define A_{IO}^1 and A_{IO}^2 as follows. Let A_{IO}^1 be the subregion of A_{IO} inside a cycle consisting of three paths: Q_α, Π_α and the subpath of P which connects s_i and p_α. Let A_{IO}^2 be the subregion of A_{IO} inside a cycle consisting of three paths: Q_β, Π_β and the subpath of P which connects t_i and p_β. When there are exactly two maximal terminal pairs, we define A_{IO}^1 and A_{IO}^2 for one of the two pairs and define A_{IO}^3 and A_{IO}^4 for the other. Let A_{IO}^5 be the subregion of A_{IO} which is not contained in any A_{IO}^i ($1 \le i \le 4$). Let $A'_{IO} = A_{IO}^5$, and let A'_O be the union of A_O and A_{IO}^i ($1 \le i \le 4$). Figure 5 depicts an example. In this example there is exactly one maximal terminal pair (s_1, t_1). Two subregions A_{IO} and A_O are defined as in Case 1. (See Figures 5(b) and (c).) Since s_1 lies in A_{IO}, A_{IO} is partitioned into two subregions. A'_{IO} is the subregion of A_{IO} which does not contain s_1. All the terminals in S_{IO} lies in A'_{IO}. A_O and the subregion of A_{IO} which contains a terminal s_1 are merged into the subregion A'_O of A. (See Figures 5(d) and (e).)

We can derive the following two lemmas.

LEMMA 5. The subregion A'_{IO} of A contains paths $P_{l+1}, P_{l+2}, \cdots, P_k$ which are included in a set of shortest non-crossing paths in A.

LEMMA 6. The subregion A'_O of A contains paths $P_1, P_2, \cdots, P_l$ which are included in a set of shortest non-crossing paths in A.

From the two lemmas above, we can find shortest non-crossing paths in A as the union of two sets of non-crossing paths: one in A'_O and the other in A'_{IO}.

We next show how to construct two plane graphs G_{IO} and G_O which correspond to A'_{IO} and A'_O, respectively. (See Figures 5(d) and (e).) G_{IO} is constructed from A'_{IO} as in Case 1. G_O is constructed from the outer boundaries of A_O and A_{IO}^i ($1 \le i \le 4$), all the corners of B_I and all the rectangles in A'_O as follows. For the subregion A_{IO}^i ($1 \le i \le 4$), draw new lines, perpendicular to the side of R_O on which terminals in A_{IO}^i lie, from all the corners of rectangles and the terminals in A_{IO}^i until they hit a rectangle or the boundary of A_{IO}^i. For the subregion A_O, draw new lines as in Case 1. Let G_O be the resulting graph. Then we have the following two lemmas.

LEMMA 7. The plane graph G_{IO} contains a set of shortest non-crossing paths $P_{l+1}, P_{l+2}, \cdots, P_k$ for S_{IO} in A'_{IO}.

LEMMA 8. The plane graph G_O contains a set of shortest non-crossing paths for S_O in A'_O.

We can observe that the union of two sets of shortest non-crossing path, one in G_{IO} and the other in G_O, is a set of shortest non-crossing paths in A.

The discussion above leads to the following algorithm to find shortest non-crossing rectilinear paths in A.

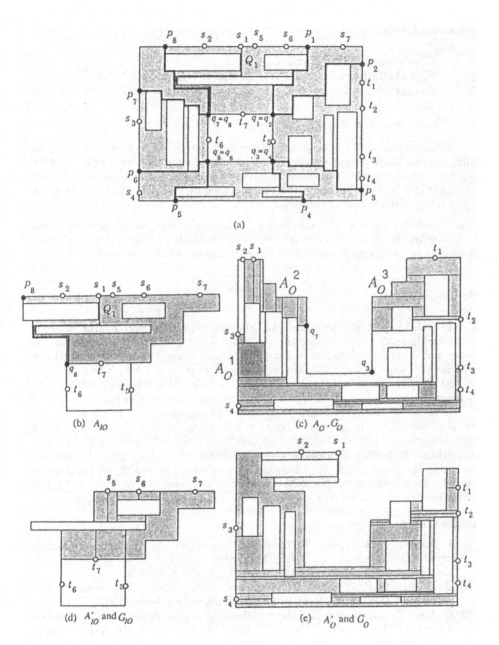

Fig. 5. Illustration of subregions and plane graphs for Case 2.

procedure PATH;
    begin
1.   Construct two plane graphs G_{IO} and G_O;
2.   Find shortest non-crossing paths for S_{IO} in G_{IO} and for S_O in G_O;
3.   Output the union of two sets of shortest non-crossing paths obtained above
    end.

One can construct two plane graphs G_{IO} and G_O in time $O(n \log n)$ by applying the plane sweep technique [4,5]. Since G_{IO} and G_O have $O(n)$ vertices, one can find the shortest non-crossing paths in G_{IO} and G_O in time $O(n \log n)$ by the algorithm in [1]. Therefore the running time of PATH is $O(n \log n)$ in total. Thus we have the following theorem.

THEOREM 1. Given a plane region A having r rectilinear obstacles inside and k terminal pairs on its outer and inner boundaries, one can find shortest non-crossing rectilinear paths in A in time $O(n \log n)$, where $n = r + k$.

4. Conclusion

In this paper we presented an efficient algorithm for finding shortest non-crossing rectilinear paths for the case when k terminal pairs are located on the outer and inner boundaries of a doughnut-shaped plane region with r obstacles. The running time of our algorithm is $O(n \log n)$ where $n = r + k$. The complexity is best possible since one needs $\Omega(n \log n)$ time to find a single shortest path between two points in a plane region with n rectangular obstacles [4].

It is rather straightforward to modify our sequential algorithm to an NC parallel algorithm which finds shortest non-crossing rectilinear paths in polylog time using a polynomial number of processors. Note that there are NC parallel algorithms for finding shortest non-crossing paths in a plane graph [1] and for executing the plane sweep [6]. There are some other future works, for example, to extend our algorithm to the case when terminals are located more than two boundaries or when the shape of obstacles are axis-parallel convex polygons.

References

[1] J. Takahashi, H. Suzuki, T. Nishizeki, "Algorithms for finding non-crossing paths with minimum total length in plane graphs," *Proc. of ISAAC '92*, Lect. Notes in Computer Science, vol. 650, pp. 400–409, 1992.
[2] D. T. Lee, C. F. Shen, C. D. Yang, and C. K. Wong, "Non-crossing path problems," *Manuscript*, Dept. of EECS, Northwestern Univ., 1991.
[3] W. Dai, T. Asano and E. S. Kuh, "Routing region definition and ordering scheme for building-block layout," *IEEE Trans. Computer-Aided Design*, vol. CAD-4, no. 3, pp. 189–197, July 1985.
[4] P. J. de Rezende, D. T. Lee, Y. F. Wu, "Rectilinear shortest paths in the presence of rectangular barriers," *Discrete & Comput. Geometry*, 4, pp. 41-53, 1989.
[5] F. P. Preparata and M. I. Shamos, *Computational Geometry*, Reading, M.A., Springer-Verlag, 1985.
[6] J. JáJá, *An Introduction to Parallel Algorithms*, Reading, M.A., Addison Wesley, 1992.

Treewidth of Circle Graphs *

T. Kloks**

Department of Mathematics and Computing Science
Eindhoven University of Technology
P.O.Box 513
5600 MB Eindhoven, The Netherlands

Abstract. In this paper we show that the treewidth of a circle graph can be computed in polynomial time. A circle graph is a graph that is isomorphic to the intersection graph of a finite collection of chords of a circle. The TREEWIDTH problem can be viewed upon as the problem of finding a chordal embedding of the graph that minimizes the clique number. Our algorithm to determine the treewidth of a circle graph can be implemented to run in $O(n^3)$ time, where n is the number of vertices of the graph.

1 Introduction

Consider a set of n chords of a circle. Associate with this set an undirected graph as follows. The vertex set is the set of chords and two vertices are adjacent if and only if the corresponding chords intersect. Such a graph is called a *circle graph* and we call a set of chords representing the graph a *circle model*. In this paper we do not distinguish between a circle graph and the circle model, i.e., we assume that we have a circle model of the graph. If the circle model is not given, it can be found in $O(ne)$ time [6, 7, 11, 22], where n is the number of vertices and e is the number of edges in the graph.

It is interesting to note that some important problems remain NP-complete when restricted to circle graphs. These problems include for example the CHROMATIC NUMBER problem [15], the COCHROMATIC NUMBER problem [12, 26] and the ACHROMATIC NUMBER problem [2]. There exists a heuristic for coloring circle graphs with performance guarantee of $O(\log n)$ [25]. On the other hand, some NP-complete problems are solvable in polynomial time, when restricted to circle graphs, for example the MAXIMUM INDEPENDENT SET, which can be solved in $O(n^2)$ time [24]. For more information on circle graphs and related classes of graphs we refer to [8, 13].

The treewidth of a graph is the minimum of the clique number minus one over all chordal embeddings of a graph. Since so many problems become solvable in polynomial time when restricted to the class of graphs with bounded treewidth, it is of importance to find a chordal embedding of a graph with a small clique number. Since the problem is NP-complete in general it is of interest to find fast algorithms for special graph classes. The TREEWIDTH can be solved in polynomial

* This research was partly supported by the foundation for Computer Science (S.I.O.N.) of the Netherlands Organization for Scientific Research (N.W.O.)

** Email: ton@win.tue.nl

time for example for cographs [5], permutation graphs [4, 16, 18], chordal bipartite graphs [16, 19, 18], circular arc graphs [23, 18], cotriangulated graph [16, 18] and for the class of graphs with bounded treewidth [3]. Recent results show that for graph classes with a polynomial number of minimal separators (including most of the above mentioned graph classes), general techniques can be applied to obtain a polynomial time algorithm for the treewidth problem (and the related minimum fill-in problem) [18, 20]. The treewidth problem remains NP-complete when restricted to bipartite and cobipartite graphs [1]. For more information on treewidth we refer to [16].

In this paper we give a simple and efficient algorithm to determine the treewidth of circle graphs. To illustrate the simplicity of the algorithm, and to whet the reader's appetite, we describe the algorithm here. Consider the circle model. Go around the circle in clockwise order and place a new vertex between every two consecutive end vertices of chords. Let Z be the set of these new vertices. Consider the polygon $\mathcal{P}$ with vertex set Z, and let T be a triangulation of this polygon. For a triangle in this triangulation define the weight as the number of chords in the circle model that cross the triangle. The weight of the triangulation T is the maximum weight of the triangles. The treewidth of the circle graph is the minimum weight minus one over all triangulations of the polygon $\mathcal{P}$. It is not hard to see that, using dynamic programming, the treewidth can be computed in $O(n^3)$ time.

Since the class of permutation graphs is properly contained in the class of circle graphs, our result generalizes some results of [4, 16] where an $O(nk)$ algorithm is given for the treewidth (and pathwidth) of permutation graphs, where k is the treewidth of the graph. (The PATHWIDTH problem was recently shown to be NP-complete for circle graphs [18, 21].)

2 Preliminaries

Definition 1. A *circle graph* is a graph for which one can associate a chord of a circle for each vertex such that two vertices are adjacent if and only if the corresponding chords have a nonempty intersection.

Without loss of generality we can assume that no two chords share an end vertex. A set of chords of a circle such that the graph is isomorphic with the intersection graph is called a *circle model* for the graph. Throughout this paper we identify a circle graph and a circle model of the graph, i.e., we assume that we have a circle model.

Definition 2. A graph is *chordal* if it has no induced cycle of length at least four.

Definition 3. A *triangulation* of a graph G is a graph H with the same vertex set as G such that G is a subgraph of H and such that H is chordal.

There are two problems concerned with triangulations of graphs that have drawn much attention because of the large number of applications. One is called the MINI-MUM FILL-IN problem. In this case one tries to find a triangulation with a minimum number of edges. The other problem is the TREEWIDTH problem. In this case the problem is to find a triangulation H such that $\omega(H)$ (i.e., the maximum number of vertices in a clique of H) is as small as possible. Both these problems are NP-complete [1, 27]. In this paper we concentrate on finding the treewidth of circle graphs.

Definition 4. The *treewidth* of a graph G is the minimum clique number $\omega(H)$ minus one over all triangulations H of G.

One of the main tools in this paper is a method to locate all minimal vertex separators in circle graphs quickly.

Definition 5. Let $G = (V, E)$ be a graph. A subset $S \subset V$ of vertices is an a, b-*separator* for non adjacent vertices a and b if the removal of S separates a and b in distinct connected components. If no proper subset of S is also an a, b-separator then S is called a *minimal* a, b-separator. A *minimal separator* S is a subset S of vertices for which there are non adjacent vertices a and b such that S is a minimal a, b-separator.

If $G = (V, E)$ is a graph and S is a subset of vertices then we write $G[S]$ for the subgraph of G induced by S. Minimal separators are easy to recognize using the following well-known result (see, e.g., [13]).

Lemma 6. *Let S be a separator of a graph $G=(V,E)$. Then S is a minimal separator if and only if there are two different connected components in $G[V - S]$ such that every vertex of S has a neighbor in both of these components.*

The following characterization of chordal graphs was found by Dirac [9].

Lemma 7. *A graph is chordal if and only if every minimal separator induces a complete subgraph.*

We use the following theorem which appeared in [4, 16].

Theorem 8. *Let $G = (V, E)$ be a graph with treewidth k. There exists a triangulation of G into a chordal graph H such that the following three statements hold:*

1. *$\omega(H) = k + 1$.*
2. *If a and b are non adjacent vertices in H then every minimal a, b-separator in H is also a minimal a, b-separator in G.*
3. *If S is a minimal a, b-separator in H and C is the vertex set of a connected component in $H[V - S]$ then C induces also a connected component in $G[V - S]$.*

Proof. Let H be a triangulation of G with $\omega(H) = k + 1$ and with a minimal number of edges. Let C be a minimal a, b-separator of H such that either C induces no minimal a, b-separator in G or the connected components of $H[V - C]$ are different from those of $G[V - C]$. Let $S \subseteq C$ be a minimal a, b-separator in G and let $C_1, \ldots, C_t$ be the connected components of $G[V - S]$. Define a chordal graph H' with vertex set V as follows. For each $1 \leq i \leq t$ take the chordal subgraph $H[C_i \cup S]$. There are no other edges in H'. Since S is a clique this defines a chordal graph H'. H' is a triangulation of G with treewidth k. The vertex sets of the connected components of $H'[V - S]$ are the same as in $G[V - S]$. We claim that the number of edges of H'

is smaller than the number of edges in H, which is a contradiction. Clearly H' is a subgraph of H.

First assume $S \neq C$, and let $x \in C \setminus S$. In H, x has a neighbor in the component containing a and a neighbor in the component containing b, since C is a minimal a, b-separator (Lemma 6). Not both these edges can be present in H'.

Now assume $S = C$. By assumption the vertex sets of the components of $H'[V-S]$ are different from those of $H[V-S]$. Then there must exist a connected component of $H[V-S]$ containing two connected components of $H'[V-S]$. This can only be the case if there is some edge between these components in $H[V-S]$ which is not there in $H'[V-S]$. This proves the claim. □

Definition 9. We call a triangulation of which the existence is guaranteed by Theorem 8 a *minimal triangulation*.

3   Scanlines

As mentioned before, we assume that no two chords of the circle model share an end vertex.

Definition 10. Place new points on the circle as follows. Go around the circle in clockwise order. Between every two consecutive end vertices of chords, place a new vertex. These new vertices are called *scanline vertices*.

If n is the number of vertices in the circle graph then there are $2n$ scanline vertices. We denote the set of scanline vertices by Z.

Definition 11. A *circle-scanline* is a chord of the circle of which the end vertices are scanline vertices.

Definition 12. Two circle-scanlines *cross* if they have a nonempty intersection but no end vertex in common.

Definition 13. Given two non crossing chords s_1 and s_2. A circle-scanline s is *between* s_1 and s_2 if every path from an end vertex of s_1 to an end vertex of s_2 along the circle passes through an end vertex of s.

If a and b are nonadjacent vertices of the circle graph then the corresponding chords in the circle model do not cross. Take a circle-scanline s which is between the chords of a and b. Clearly, the set of vertices, corresponding to chords that cross s, is an a, b-separator. The following lemma is a generalization of a result in [4].

Lemma 14. *Let $G = (V, E)$ be a circle graph and let a and b be non adjacent vertices. For every minimal a, b-separator S there exists a circle-scanline s between a and b such that the chords corresponding to vertices of S are exactly the chords crossing s.*

Proof. The proof is basically the same as in [4]. Let S be a minimal a, b-separator. Consider the connected components of $G[V - S]$. Let C_a and C_b be the components containing a and b respectively. These components must form 'connected' parts in the circle model. Notice that we can choose a circle-scanline crossing no chord of $G[V - S]$ and which is between the chords of a and b. All chords crossing this circle-scanline are elements of S. S is a minimal separator and hence all the chords corresponding with elements of S must cross the circle scanline, since such a chord of S intersects with a chord of C_a and a chord of C_b (Lemma 6). $\qquad\square$

Corollary 15. *There are at most $O(n^2)$ minimal vertex separators in a circle graph.*

4 Components and realizers

Let $G = (V, E)$ be a circle graph. Consider a circle model for G with the set Z of scanline vertices.

Definition 16. Let $Y \subseteq Z$ be a set of scanline vertices with at least three elements. Consider the convex polygon $\mathcal{P}(Y)$ with vertex set Y. The *component* $G(Y)$ is the subgraph of G induced by the set of vertices corresponding with chords in the circle model which have a non empty intersection with the interior region of $\mathcal{P}(Y)$.

Hence the edges of the polygon $\mathcal{P}(Y)$ are circle-scanlines. Notice that if $Y = Z$ then $G(Y)$ is simply the graph G.

Definition 17. Let Y be a set of at least three scanline vertices and consider the component $G(Y)$. For each circle-scanline that is an edge of the polygon $\mathcal{P}(Y)$, add edges between vertices of $G(Y)$ of which the corresponding chords cross that circle-scanline. In this way we obtain the *realizer* $R(Y)$ of the component $G(Y)$.

Hence each component is a subgraph of its realizer. We mention here that if Y is a set of three scanline vertices then

1. each chord corresponding with a vertex of $G(Y)$ intersects exactly two edges of $\mathcal{P}(Y)$ and hence
2. $R(Y)$ is a clique.

Lemma 18. *If $G(Y)$ is a component then the realizer $R(Y)$ is a circle graph.*

Proof. Let $s_1, s_2, \ldots, s_t$ be the circle-scanlines which are the edges of the polygon $\mathcal{P}(Y)$. Consider these circle-scanlines one by one.

Let $c_1, c_2, \ldots, c_\ell$ be the chords that cross some circle-scanline s_k. Call the end vertices of c_i, a_i and b_i. Choose a_i and b_i such that, when going along the chord from a_i to b_i first the circle-scanline s_k is crossed before the interior region of $\mathcal{P}(Y)$ is entered. Rearrange the order of the vertices $a_1, \ldots, a_\ell$ on the circle such that, afterwards, every pair of chords c_i and c_j cross. Notice that this only adds edges in the component between vertices with corresponding chords in $\{c_1, \ldots, c_\ell\}$.

In this way we obtain a circle model for the realizer. $\qquad\square$

We identify the realizer $R(Y)$ with a circle model for $R(Y)$ obtained as in the proof of Lemma 18.

Definition 19. Let $G(Y)$ be a component with realizer $R(Y)$. A circle-scanline s in the circle model for $R(Y)$ is *Y-nice* if the end vertices of s are elements of Y.

We now state one of our main results.

Lemma 20. *Let $R(Y)$ be a realizer of a component $G(Y)$. Let a and b be two non adjacent vertices in $R(Y)$ and let S be a minimal a, b-separator in $R(Y)$. Then there is a Y-nice circle-scanline s such that S consists of the vertices corresponding with the chords that cross s.*

Proof. Consider the circle model for $R(Y)$. Since a and b are not adjacent we know that there is a circle-scanline s (with end vertices in Z) between the chords of a and b such that the set of chords crossing s corresponds to S. Choose such a circle-scanline s with a minimum number of end vertices that are not in Y. If both end vertices of s are elements of Y then s is Y-nice. Assume this is not the case. Then s crosses with at least one circle-scanline s' which is an edge of the polygon $\mathcal{P}(Y)$. Let the end vertices of s be x and y chosen in such a way that, if we traverse s from x to y, then we first cross s' before entering the region of $\mathcal{P}(Y)$.

Let α and β be the end vertices of s' chosen such that α is on the same side of s as the chord a, and β is on the same side of s as the chord b.

Let s^* be the circle-scanline with end vertices y and α and let s^{**} be the circle-scanline with end vertices y and β.

Since a and b are non adjacent in $R(Y)$ the corresponding chords of a and b do not both cross the circle-scanline s'. Assume that the chord of a does *not* cross with s'. We now consider two cases.

b and s' do not cross. Then s^* and s^{**} are both circle-scanlines between a and b. Let S^* and S^{**} be the corresponding separators. We claim that either $S^* \subseteq S$ or $S^{**} \subseteq S$. This can be seen as follows. Assume there is a chord p in the realizer crossing with s^* but not with s and a chord q in the realizer crossing with s^{**} but not with s. Then p and q must both cross with s'. But this is a contradiction since p and q cannot cross without one of them also crossing s.

b and s' do cross. In this case s^* is a circle-scanline between a and b. We claim that $S^* \subseteq S$. Assume there is a chord p in $R(Y)$ that crosses with s^* but not with s. Then p and b cannot cross. But this is a contradiction, since both p and b cross with s'.

This shows that either s^* or s^{**} is a circle-scanline between a and b of which the corresponding separator is a subset of the separator of s. But both s^* and s^{**} both have one more end vertex in Y. This proves the theorem. $\square$

Definition 21. A component $G(Y)$ is k-feasible if the realizer $R(Y)$ has treewidth at most k.

In other words, the component $G(Y)$ is k-feasible if and only if there is a triangulation of the component such that each clique has at most $k + 1$ vertices and such that for every circle-scanline which is an edge of $\mathcal{P}(Y)$ the set of vertices that cross that circle-scanline induce a clique in the triangulation.

Lemma 22. *Let Y be a set of scanline vertices with at least three elements such that the component $G(Y)$ has at least $k + 2$ vertices. Then the following two statements hold.*

114

1. *If $G(Y)$ is k-feasible then there is a Y-nice circle-scanline, dividing the polygon $\mathcal{P}(Y)$ in two new polygons with vertex sets, say, Y_1 and Y_2, such that the components $G(Y_1)$ and $G(Y_2)$ both have less vertices than $G(Y)$ and such that both $G(Y_1)$ and $G(Y_2)$ are k-feasible, and*
2. *if there is a Y-nice circle-scanline, dividing the polygon $\mathcal{P}(Y)$ in two new smaller polygons with vertex sets Y_1 and Y_2, such that the components $G(Y_1)$ and $G(Y_2)$ are both k-feasible, then $G(Y)$ is also k-feasible.*

Proof. Assume that $G(Y)$ is k-feasible. Consider a minimal triangulation H of $R(Y)$ (Definition 9). Since the number of vertices is at least $k+2$ there must be a pair of non adjacent vertices a and b in H. Consider a minimal a,b-separator S in H. Since H is a minimal triangulation, S is also a minimal a,b-separator in G. Since H is chordal S induces a clique in H. By Lemma 20 there is a Y-nice circle-scanline s corresponding with S. s divides the polygon $\mathcal{P}(Y)$ into two new polygons. Let Y_1 and Y_2 be the vertex sets of these two new polygons.

We may assume that $G(Y_1)$ contains the chord corresponding with a and $G(Y_2)$ contains the chord corresponding with b. Let C_a and C_b be the vertex sets of $G(Y_1)$ and $G(Y_2)$ respectively. Then it follows that the induced subgraphs $H[C_a]$ and $H[C_b]$ are triangulations of $R(Y_1)$ and $R(Y_2)$, and hence $G(Y_1)$ and $G(Y_2)$ are k-feasible. Since a is not in $G(Y_2)$ and b not in $G(Y_1)$ it follows that both these components have less vertices than $G(Y)$.

Assume that there is a Y-nice circle-scanline s dividing the polygon $\mathcal{P}(Y)$ in two new smaller polygons with vertex sets Y_1 and Y_2, such that the components $G(Y_1)$ and $G(Y_2)$ are both k-feasible. Consider triangulations H_1 and H_2 of $R(Y_1)$ and $R(Y_2)$ respectively. Let S be the set of vertices corresponding with chords that cross s. Since S induces a clique in H_1 and in H_2, it follows that we can obtain a triangulation H of $R(Y)$ by identifying the vertices of S is H_1 and H_2. This shows that $G(Y)$ is k-feasible. □

5 Triangulating the polygon

Definition 23. Let $\mathcal{P}$ be a polygon with m vertices. A *triangulation* of $\mathcal{P}$ is a set of $m-3$ non crossing diagonals in $\mathcal{P}$ that divide the interior of $\mathcal{P}$ in $m-2$ triangles.

Definition 24. Let Y be a set of at least three scanline vertices. Consider a triangulation T of $\mathcal{P}(Y)$. The *weight* of a triangle is the number of chords in the circle model that have a non empty intersection with the triangle. The *weight of the triangulation*, $w(T)$, is the maximum weight of a triangle.

We can now state our main result.

Theorem 25. *Let Y be a set of at least three scanline vertices. A component $G(Y)$ is k-feasible if and only if there is a triangulation T with weight at most $k+1$.*

Proof. First assume that $G(Y)$ is k-feasible. If $G(Y)$ has at most $k+1$ vertices, then *any* triangulation of $\mathcal{P}(Y)$ has weight at most $k+1$. We proceed with induction on the number of vertices of $G(Y)$. Assume $G(Y)$ has more than $k+1$ vertices. By the first part of Lemma 22 there is a Y-nice circle-scanline which divides the polygon $\mathcal{P}(Y)$ in two new polygons $\mathcal{P}(Y_1)$ and $\mathcal{P}(Y_2)$ such that the components $G(Y_1)$ and $G(Y_2)$ both have less vertices than $G(Y)$ and such that both $G(Y_1)$ and $G(Y_2)$ are k-feasible. By induction there are triangulations T_1 of $\mathcal{P}(Y_1)$ and T_2 of $\mathcal{P}(Y_2)$ both with weight at most $k+1$. Then, clearly, $T = T_1 \cup T_2$ is a triangulation of $\mathcal{P}(Y)$.

Now assume T is a triangulation of $\mathcal{P}(Y)$ with weight at most $k+1$. If Y has only three vertices, $G(Y)$ has at most $k+1$ vertices, and hence $G(Y)$ is k-feasible. We now proceed with induction on the number of vertices of Y. Take any diagonal of T. This divides the polygon into two smaller polygons with vertex sets Y_1 and Y_2 say. Since there are triangulations of $\mathcal{P}(Y_1)$ and of $\mathcal{P}(Y_2)$ with weight at most $k+1$ we may conclude that both $G(Y_1)$ and $G(Y_2)$ are k-feasible. But then by the second part of lemma 22 also $G(Y)$ is k-feasible. □

6 Algorithm

The following theorem describes an algorithm to find the treewidth of a circle graph.

Theorem 26. *Given a circle graph G with n vertices. There exists an $O(n^3)$ algorithm to determine the treewidth of G.*

Proof. First compute a circle model for G. As mentioned earlier this step can be performed in $O(ne)$ time, where n is the number of vertices and e the number of edges of G. We may assume that we can decide whether two chords cross in $O(1)$ time.

Clearly we may assume that $n > 1$. Determine a set of scanline vertices Z. Since $n > 1$, Z has at least four vertices, and hence the polygon $\mathcal{P}(Z)$ is well defined. The algorithm we describe finds a triangulation with minimal weight for $\mathcal{P}(Z)$.

First, for each of the circle-scanlines compute the number of chords that cross the circle-scanline. Since there are $O(n^2)$ circle-scanlines, and the test if a circle-scanline and a chord cross can be performed in $O(1)$ time, this step costs $O(n^3)$ time.

Use dynamic programming to find an optimal triangulation for $\mathcal{P}(Z)$. Let the scanline vertices be $s_0, s_2, \ldots, s_{\ell-1}$ ordered clockwise. Let $P(i,t)$ be the polygon defined by $s_i, \ldots, s_{i+t-1}$, where indices are to be taken modulo ℓ. We define $w(i,t)$ as the minimum weight of a triangulation of the polygon $P(i,t)$. Let $c(i,j)$ be the number of chords crossing the circle-scanline with end vertices s_i and s_j. Then $w(i,t)$ can be determined in $O(n^3)$ time using the following. Set all $w(i,2)$ equal to 0. For $t = 3, \ldots, \ell$, compute for all i:

$$w(i,t) = \min_{2 \le j < t} \left(\max\left(w(i,j), w(i+j-1, t-j+1), F(i,j) \right) \right)$$

$$\text{where } F(i,j) = \frac{c(i, i+j-1) + c(i+j-1, i+t-1) + c(i, i+t-1)}{2}$$

Correctness follows from the fact that each chord crossing a triangle intersects exactly two sides of the triangle. The treewidth of G is $w(0, \ell) - 1$. □

In the rest of this section we show that it is also easy to find a triangulation of G with minimum clique number. Notice that the algorithm described above can

116

be easily adapted to return a triangulation T of the polygon $\mathcal{P}(Z)$ with minimum weight. Define the graph $H(T)$ with the same vertex set as G as follows. Two vertices are adjacent in $H(T)$ if there is a triangle such that the chords corresponding with the vertices intersect this triangle. Notice that G is a subgraph of $H(T)$. We show that $H(T)$ has a perfect elimination scheme. Consider a vertex z of $\mathcal{P}(Z)$ which is *not* incident with a diagonal of T. This vertex is incident with exactly one triangle Q. Consider a vertex x of which the corresponding chord intersects Q but no other triangle. Then the neighborhood of x, $N(x)$, is a clique, hence x is simplicial. Notice also that the number of vertices in $N(x)$ is the weight of Q minus one, showing that the number of vertices in the clique $\{x\} \cup N(x)$ is equal to the weight of Q. Remove x from the graph $H(T)$ and the corresponding chord from the circle model. If there is no chord left in the circle model which intersects Q but no other triangle, then remove z from Z. Repeating this process gives a perfect elimination scheme, showing that $H(T)$ is chordal. This also shows that the number of vertices in a maximum clique of $H(T)$ is equal to the weight of the triangulation.

7 Conclusions

In this paper we showed that the treewidth of circle graphs can be computed in polynomial time. A related problem is the PATHWIDTH problem. Recently it was shown that computing the pathwidth for circle graphs is NP-complete [18, 21]. In view of this it is interesting to notice that the pathwidth of permutation graphs can be computed in linear time [4]. In fact, for cocomparability graphs (a class of graphs properly containing the permutation graphs) it can be shown that the treewidth and pathwidth are equal [4]. (It was shown in [14] that this equality holds even for all AT-free graph.) Perhaps it is possible to find 'good' approximation algorithms for the pathwidth of a circle graph. Approximations (with approximation factor equal to the treewidth) for the treewidth and pathwidth of cocomparability graphs exist [17].

8 Acknowledgements

I like to thank M. de Berg, H. Bodlaender, A. Jacobs and H. Müller for valuable discussions.

References

1. Arnborg, S., D. G. Corneil and A. Proskurowski, Complexity of finding embeddings in a k-tree, *SIAM J. Alg. Disc. Meth.* 8, (1987), pp. 277–284.
2. Bodlaender, H. L., Achromatic number is NP-complete for cographs and interval graphs, *Information Processing Letter* 31, (1989), pp. 135–138.
3. Bodlaender, H., A linear time algorithm for finding tree-decompositions of small treewidth, Technical report RUU-CS-92-27, Department of Computer Science, Utrecht University, Utrecht, The Netherlands, (1992).
4. Bodlaender, H., T. Kloks and D. Kratsch, Treewidth and pathwidth of permutation graphs, Technical report RUU-CS-92-30, Department of Computer Science, Utrecht University, Utrecht, The Netherlands, (1992). To appear in *Proceedings of the 20th International colloquium on Automata, Languages and Programming.*

5. Bodlaender, H. and R. H. Möhring, The pathwidth and treewidth of cographs, *Proceedings 2nd Scandinavian Workshop on Algorithm Theory*, Springer Verlag, Lecture Notes in Computer Science 447, (1990), pp. 301–309.

6. Bouchet, A., A polynomial algorithm for recognizing circle graphs, *C. R. Acad. Sci. Paris, Sér. I Math.* 300, (1985), pp. 569–572.

7. Bouchet, A., Reducing prime graphs and recognizing circle graphs, *Combinatorica* 7, (1987), pp. 243–254.

8. Brandstädt, A., Special graph classes — a survey, Schriftenreihe des Fachbereichs Mathematik, SM-DU-199 (1991), Universität Duisburg Gesamthochschule.

9. Dirac, G. A., On rigid circuit graphs, *Abh. Math. Sem. Univ. Hamburg* 25, (1961), pp. 71–76.

10. Farber, M. and M. Keil, Domination in permutation graphs, *J. Algorithms* 6, (1985), pp. 309–321.

11. Gabor, C. P., W. L. Hsu and K. J. Supowit, Recognizing circle graphs in polynomial time, *26th Annual IEEE Symposium on Foundations of Computer Science*, (1985).

12. Gimbel, J., D. Kratsch and L. Stewart, On cocolourings and cochromatic numbers of graphs. To appear in *Disc. Appl. Math.*

13. Golumbic, M. C., *Algorithmic Graph Theory and Perfect Graphs*, Academic Press, New York, 1980.

14. Habib, M. and R. H. Möhring, Treewidth of cocomparability graphs and a new order theoretic parameter, Technical report 336/1992, Technische Universität Berlin, 1992.

15. Johnson, D. S., The NP-completeness column: An ongoing guide, *J. Algorithms* 6, (1985), pp. 434–451.

16. Kloks, T., *Treewidth*, PhD Thesis, Utrecht University, The Netherlands, (1993).

17. Kloks, T. and H. Bodlaender, Approximating treewidth and pathwidth of some classes of perfect graphs, *Proceedings Third International Symposium on Algorithms and Computation, ISAAC'92*, Springer Verlag, Lecture Notes in Computer Science, 650, (1992), pp. 116–125.

18. Kloks, T., H. Bodlaender, H. Müller and D. Kratsch, Computing treewidth and minimum fill-in: all you need are the minimal separators. To appear in *Proceedings of the First Annual European Symposium on Algorithms*, (1993).

19. Kloks, T. and D. Kratsch, Treewidth of chordal bipartite graphs, 10th *Annual Symposium on Theoretical Aspects of Computer Science*, Springer-Verlag, Lecture Notes in Computer Science 665, (1993), pp. 80–89.

20. Kloks, T. and D. Kratsch, Finding all minimal separators of a graph, Computing Science Note, 93/27, Eindhoven University of Technology, Eindhoven, The Netherlands, (1993).

21. Müller, H., Chordal graphs of domino type. Manuscript, (1993).

22. Naji, W., Reconnaissance des graphes de cordes, *Discrete Mathematics* 54, (1985), pp. 329–337.

23. Sundaram, R., K. Sher Singh and C. Pandu Rangan, Treewidth of circular arc graphs. To appear in *SIAM J. Disc. Math.*

24. Supowit, K, J., Finding a maximum planar subset of a set of nets in a channel, *IEEE Trans. Computer Aided Design* 6, (1987), pp. 93–94.

25. Supowit, K. J., Decomposing a set of points into chains, with applications to permutation and circle graphs, *Information Processing Letters* 21, (1985), pp. 249–252.

26. Wagner, K., Monotonic coverings of finite sets, *Journal of Information Processing and Cybernetics*, EIK, 20, (1984), pp. 633–639.

27. Yannakakis, M., Computing the minimum fill-in is NP-complete, *SIAM J. Alg. Disc. Meth.* 2, (1981), pp. 77–79.

A Framework for Constructing Heap-Like Structures In-Place

Jingsen Chen*

Division of Computer Science, Luleå University, S-971 87 Luleå, Sweden

Abstract. Priority queues and double-ended priority queues are fundamental data types in computer science. Several implicit data structures have been proposed for implementing the queues, such as heaps, min-max heaps, deaps, and twin-heaps. Over the years the problem of constructing these heap-like structures has received much attention in the literature, but different structures possess different construction algorithms. In this paper, we present a uniform approach for building the heap-like data structures in-place. The study is carried out by investigating hardest instances of the problem and developing an algorithmic paradigm for the construction. Our paradigm produces comparison- and space-efficient construction algorithms for the heap-like structures, which improve over those previously fast known algorithms.

1 Introduction

A *priority queue* is a set of elements on which two basic operations are defined: inserting a new element into the set; and retrieving and deleting the minimum element of the set. A *double-ended priority queue* (*priority deque*, for short) provides insert access and remove access to both the minimum and maximum elements efficiently at the same time. They have been useful in many applications [1,12]. Several implicit data structures have been developed for implementing priority queues and priority deques, namely the heap [17], the twin-heap ([12], p. 159), the min-max heap [2], and the deap [5]. The problem of constructing these heap-like structures has received much attention in the literature [2,5,6,8–10,12,15]. However, different structures possess different construction algorithms. In this paper, we present an algorithmic paradigm for the construction problems, which provides us with a simple and uniform way of designing comparison- and space-efficient construction algorithms. Our results also improve over those of previously fast known algorithms.

2 Data Structures

A k-ary (min-)heap [17] is a k-ary tree satisfying: (*i*) It has the *heap shape*: all levels are complete, except possibly the last level where all leaves occupy the leftmost positions; (*ii*) It is *min-ordered*: the key value associated with each node is not smaller than that of its parent. The minimum element is then at the root. A (max-)heap is defined similarly. Instead of 2-ary heap we say just heap.

---

* Part of the work was conducted while the author was with Lund University, Sweden.

A min-max heap [2] is a heap-shape binary tree with a *min-max ordering*: elements on even(resp. odd) levels are smaller(resp. greater) than or equal to their descendants, where the *level* of an element at location i is $\lceil \log i \rceil$. The minimum and maximum elements are at the root and one of its children, respectively. A max-min heap is similarly defined.

A deap [5] is a heap-shape binary tree with a hole at the position of the root. The left and the right subtree of the root is respectively a min-heap and a max-heap. Any node i in the min-heap is smaller than the corresponding element in the max-heap, $i+j$, if it exists; and $\lfloor \frac{i+j}{2} \rfloor$ otherwise, where j is $2^{\lfloor \log i \rfloor - 1}$. So the minimum element in a deap is stored in the position 2 and the maximum in position 3. Similar to the deap, the twin-heap [12] also keeps two heaps in a suitable way. The difference between a deap and a twin-heap is that the latter has two separate heaps of the same size. The upper bounds for the twin-heap construction can be derived from the corresponding results for the deap.

For the sake of simplicity, we shall assume that all the elements are distinct and use the term QUEUE to denote one of the following (binary or k-ary) structures: the heap or the min-max heap; and the term DEQUE for either the deap or the twin-heap. A k-ary QUEUE of size $\frac{k^{h+1}-1}{k-1}$ is called a *full k-ary* QUEUE. Similarly, a *full k-ary* DEQUE has size $\frac{2 \cdot k^h - 2}{k-1}$ while a *half-full k-ary* DEQUE is the one on $\frac{3 \cdot k^h - 2}{k-1}$ elements.

3   Hardest Instances

To seek efficient construction algorithms for heaps and the like, one is often concerned with finding an elegant method which has a good performance on the "hardest" instance of the construction problem. Examples include the construction algorithms for heaps [9] and for min-max heaps [6,15]. The "hardest" is in the sense that the construction complexity of the "hardest" instances gives an upper bound on the cost for building structures of arbitrary sizes. For the problem of constructing binary heaps or binary min-max heaps, it has been shown that the hardest instance for the problem occurs when the size of the input is $2^h - 1$ [9,15]. This can easily be generalized to the problem of constructing k-ary heaps, k-ary min-max heaps, k-ary deaps, or k-ary twin-heaps. In what follows, the problem of *merging* QUEUEs is to build a k-ary QUEUE starting from k disjoint k-ary QUEUEs plus one singleton element. For k-ary DEQUEs, the merge operation is to construct a k-ary DEQUE starting from k disjoint k-ary DEQUEs plus two singletons. When it is understood, we will not explicitly mention the singleton element(s).

Proposition 1. *If the (worst-case) cost to construct a k-ary full* QUEUE *of any size m is $f(m)$, then building an n-element k-ary* QUEUE *takes at most $f(n) + \mathcal{O}\left(k \cdot (\log_k n)^2\right)$ comparisons (in the worst case), where $f(n) \in \Theta(n)$.*

Proof. Consider first the merging operation on QUEUEs, which can also be stated as: Given a k-ary tree T of n nodes with root x and its subtrees $T_1, T_2, \cdots, T_k$ which are k-ary QUEUEs, we want to build a k-ary QUEUE on all the elements of T. The operation can be carried out in $\mathcal{O}(k \cdot \log_k n)$ comparisons by accessing the root x, which works as follows: If x were smaller than all its children, the tree would be a

k-ary QUEUE. If not, we simply swap the position of x with its smallest child, and repeat this step level by level downwards. Since each step needs k comparisons and the height of T is $\mathcal{O}(\log_k n)$, the merge operation uses $\mathcal{O}(k \cdot \log_k n)$ comparisons. We next present an algorithm for building any k-ary QUEUE of size n in time $f(n) + \mathcal{O}\left(k \cdot (\log_k n)^2\right)$. Notice that all subtrees hanging off the siblings of the elements lying on the path in a k-ary tree from the root to the last leaf are full k-ary trees, which can be converted into k-ary QUEUEs in $f(n) + o(f(n))$ comparisons. By performing merge operations in turn on these full k-ary subQUEUEs in a bottom-up fashion, which leads to an $\mathcal{O}\left(k \cdot (\log_k n)^2\right)$ additional term, we can successfully build a QUEUE on n elements in $f(n) + \mathcal{O}\left(k \cdot (\log_k n)^2\right)$ comparisons.

Observe that the upper bound on the construction complexity of full DEQUE can similarly be extended to that for building DEQUEs of arbitrary sizes without affecting the linear order term in the time complexity. More precisely, the merge operation is now referred to the construction of an n-element DEQUE, starting from a DEQUE of size n where the elements currently stored in the second and third positions of the DEQUE violate the DEQUE-ordering relation. Notice that here a DEQUE is implicitly viewed as an array of the elements. Analogous to Proposition 1, the following result can be derived immediately by checking the shape of subDEQUEs.

Proposition 2. *Let $f(m')$ and $g(m'')$ be the number of comparisons required to build a k-ary DEQUE of sizes $m' = \frac{2 \cdot k^{h'} - 2}{k-1}$ and $m'' = \frac{3 \cdot k^{h''} - 2}{k-1}$, respectively, where both $f(\cdot)$ and $g(\cdot)$ belong to $\Theta(n)$. Then the cost to construct a k-ary DEQUE of any size n is at most $\max\{f(n), g(n)\} + \mathcal{O}\left(k \cdot (\log_k n)^2\right)$. In particular, the cost to build a k-ary twin-heap of size n is at most $f(n) + \mathcal{O}\left(k \cdot (\log_k n)^2\right)$.*

4 Algorithmic Paradigm

From Propositions 1 and 2, the fullness of the structures enables us to use the divide and conquer techniques recursively in designing efficient construction algorithms that are simple and somewhat easier to be analyzed than those for non-full structures. We shall outline our paradigm only for constructing binary priority queue and priority deque structures. The algorithm is a variant of Floyd's method for heaps [8] in some sense, which proceeds by constructing small structures of the same sizes and consists of a sequence of merge operations on the sub-structures built precedingly.

Algorithm 3. *Suppose that n is the size of a full QUEUE or a full DEQUE (i.e., $n = 2^{h+1} - 1$ for QUEUEs and $n = 2^{h+1} - 2$ for DEQUEs).*

1. *Construct $2^{\lfloor \log n \rfloor - \lfloor \log n_0 \rfloor}$ full QUEUEs/DEQUEs on n_0 elements.*
2. *Insert the remaining elements into the QUEUEs/DEQUEs to create a QUEUE/DEQUE of size n by applying merge operations recursively on the smaller QUEUEs/DEQUEs built thus far.*

Let $\mathbb{F}(h)$ be the worst-case number of comparisons done by Algorithm 3 to build a full QUEUE/DEQUE of size n, where $h = \lfloor \log n \rfloor$ (called the *depth* of the QUEUE/DEQUE). Denote also the worst-case number of comparisons made by some algorithm to merge two full QUEUES/DEQUES of depth $h-1$ plus one/two singleton element(s) into one full QUEUE/DEQUE of depth h by $\mathbb{M}(h)$. Hence,

$$\mathbb{F}(i) \leq 2 \cdot \mathbb{F}(i-1) + \mathbb{M}(i) \quad \text{if } i > m; \qquad \text{and} \qquad \mathbb{F}(h_0) = f_0,$$

where $h_0 = \lfloor \log n_0 \rfloor$ is the depth of the smaller QUEUES/DEQUES which we intend to build with some particular fast methods. The recurrence above gives the cost of Algorithm 3 for constructing a QUEUE/DEQUE of depth h:

$$\mathbb{F}(h) \leq 2^{h-h_0} \cdot \mathbb{F}(h_0) + 2^h \cdot \sum_{i=h_0+1}^{h} \frac{\mathbb{M}(i)}{2^i} \tag{1}$$

Therefore, the worst-case complexity of Algorithm 3 depends on the efficiency of methods for building small structures and for merging the structures. The merge operation usually consists of two phases: *trickle-down* and *bubble-up*. In the stage of the trickle-down, a search path is found in a top-down fashion on which the singleton element(s) will be placed. For example, we start at the root of a min-heap find a path of minimum children using one comparison per node in the path. The bubble-up stage locates the correct position(s), for the element(s) to be inserted with respect to the required orderings of the QUEUE/DEQUE by binary or linear search, on the path staring from the bottom of the QUEUE/DEQUE. For instance, for a min-heap, the element is bubbled-up along with the path until it is larger than its parent. For different structures, different merge strategies will be employed.

5 Constructing Heap-Like Structures

5.1 Heaps

Much attention has been devoted to the problem of building heaps in the literature. The first algorithm was proposed by Williams [17], which requires $n \log n + \mathcal{O}(n)$ comparisons in the worst case to create an n-element heap. Floyd [8] reduced the cost to $2n + o(n)$ [12]. The fastest algorithm by Gonnet and Munro [9], called Algorithm GM, takes $1.625n + o(n)$ comparisons. The key step of Algorithm GM is to construct a full heap of size $2^{h+1} - 1$ from a binomial tree on 2^{h+1} elements recursively. To reduce the amount of space consumed, Carlsson [3] present a heap construction algorithm, referred to as Algorithm S, that makes $1.82n - \mathcal{O}(\log n)$ comparisons and does not use any extra space. In the following, we shall demonstrate that the algorithmic paradigm presented in the preceding section provides us with a simple way of reducing the space requirement of construction algorithms. First, we show how to apply the "mass production" technique [14] to heap construction.

Lemma 4. *Four heaps of size 7 can be constructed in at most 31 comparisons.*

Proof. After 14 comparisons (see Figure 1), we compare a with c. If $a \prec c$, then we compare c' with b; else a' is compared to d. With these two additional comparisons, we can create a heap of size 7 and three chains of size 3. Notice that a 7-element heap can be built by using 5 comparisons starting from a chain of size 3 plus 4 singleton elements. Hence, the total cost for constructing four heaps of size 7 is at most $14 + 2 + 3 \times 5 = 31$ comparisons. The lemma follows.

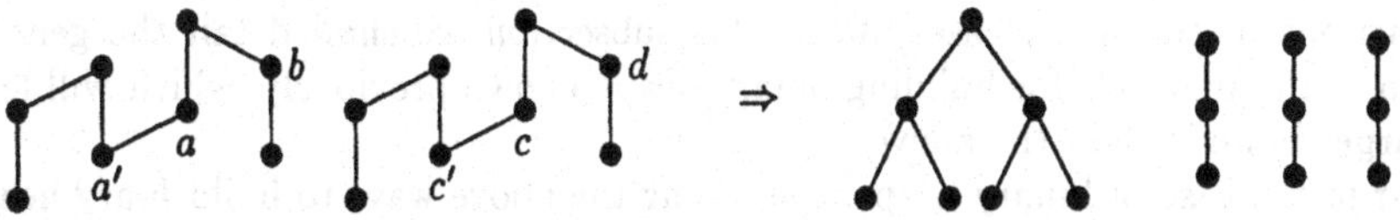

After 14 comparisons   After 16 comparisons

FIGURE 1: Build four heaps of size seven.

Since building a 7-element heap costs at least 8 comparisons, we have saved 1 comparison over four heaps of size 7. Therefore, from the algorithm in [3], we have:

Lemma 5. *An n-element heap can be built in at most* $1.78875n$ $(= 1.82n - \frac{n}{32})$ *comparisons in the worst case, using only 32 bits of extra space.*

The above upper bound can be further reduced by specifying two steps of Algorithm 3. In fact, combining Algorithm S [3] with Algorithm GM [9] yields

Theorem 6. *An n-element heap can be built in* $\left(1.945 - \frac{1}{2} \cdot \sum_{i=1}^{k} \frac{\lfloor \log i \rfloor}{2^i}\right) n + \mathcal{O}(\log^2 n)$ *comparisons in the worst case by using $m+1$ bits of extra space for any $m = 2^{k+1}-1$, where $k \geq 1$ and $m \leq n$.*

Proof. To achieve the desired time and space bounds, we shall employ Algorithm GM [9] to build smaller full heaps of size m in the first step of Algorithm 3. For the second step, Algorithm S [3] will be applyed to handle the remaining elements.
Consider the problem of building full heaps. Let $n = 2^{h+1} - 1$. Recall that Algorithm S [3] costs $\frac{3}{2} \cdot 2^{k+1} - (k+1) + 2^{k+1} \cdot \sum_{i=1}^{k} \frac{\lfloor \log i \rfloor}{2^{i+1}} - 2$ comparisons to build a heap of size $2^{k+1} - 1$, while Algorithm GM [9] uses only $\frac{13}{8} \cdot 2^{k+1} - (k+1) - 2$ comparisons. Hence, the worst-case number of comparisons done by Algorithm 3 is at most

$$1.82n - \mathcal{O}(\log n) - 2^{h-k} \cdot \left(2^{k+1} \cdot \sum_{i=1}^{k} \frac{\lfloor \log i \rfloor}{2^{i+1}} - \frac{1}{8} \cdot 2^{k+1}\right) \leq \left(1.945 - \frac{1}{2} \cdot \sum_{i=1}^{k} \frac{\lfloor \log i \rfloor}{2^i}\right) n$$

By Proposition 1, the theorem follows.

One specific case of the theorem occurs when $m = 1023$. That is,

Corollary 7. *A heap on n elements can be constructed in at most $1.6316n + \mathcal{O}(\log^2 n)$ comparisons in the worst case by using only a constant amount of extra space.*

5.2 k-Ary Heaps

Apart from the problem of constructing binary heaps, considerable attention has also been devoted to a more general problem of building k-ary heaps together with the sorting problem [7,11–13]. Without doubt analogous algorithms for building k-ary heaps can be obtained directly from those for binary heaps. For example, Floyd's algorithm [8] can be generalized to build k-ary heaps, which takes $\frac{k}{k-1} \cdot n + o(n)$

comparisons in the worst case [13]. In this subsection we shall detail the general version of the methods for building binary heaps known previously, which will lead to sharper bounds than the above.

Similar to the case of binary heaps, combining the above ways to build k-ary heaps and using the techniques in [4,9], we can reduce the number of comparisons required by k-ary heap construction. This can be done by first finding a path of minimum children from the root to a leaf position and then performing a binary search on this path to determine the final position of the element to be inserted.

Theorem 8. *A k-ary heap on n elements can be constructed in-place in at most $\left(\frac{k+1}{k} + \frac{k-1}{k} \cdot \sum_{i=1}^{h} \frac{\lfloor \log i \rfloor}{k^i}\right) \cdot n + o(n)$ comparisons in the worst case, where $h = \lfloor \log_k((k-1)n+1) - \log_k 2 \rfloor$ (the height of the heap).*

Proof. We only consider the case of full k-ary heaps, for otherwise the construction will need at most $\mathcal{O}(k \cdot (\log_k n)^2)$ additional comparisons from Proposition 1. Suppose now that $n = \frac{k^{h+1}-1}{k-1}$. The procedure of merging two full k-ary heaps of height i (whose complexity is denoted by $\mathbb{M}(i)$) is the same as that for the conventional heaps [3] except that $(k-1)i$ comparisons are needed to find the path of minimum children down to the leaf level. Since the number of elements on this path is i, we know that a binary search on the path will cost at most $\lfloor \log i \rfloor + 1$ comparisons in the worst case. Hence, $\mathbb{M}(i) \leq (k-1)i + \lfloor \log i \rfloor + 1$.

Notice that the number of sub-k-ary heaps of height i in a full k-ary heap of height h is $\frac{k^{h+1}}{k^{i+1}}$. Therefore, the total number of comparisons for building a k-ary heap of size $n = \frac{k^{h+1}-1}{k-1}$ is at most

$$k^h \cdot \sum_{i=1}^{h} \frac{(k-1)i + \lfloor \log i \rfloor + 1}{k^i} = \left(\frac{k+1}{k} + \frac{k-1}{k} \cdot \sum_{i=1}^{h} \frac{\lfloor \log i \rfloor}{k^i}\right) \cdot n + o(n).$$

Notice that $\sum_{i=1}^{h} \frac{\lfloor \log i \rfloor}{k^i} < \frac{1}{(k-1)^2}$ and thus $\left(\frac{k+1}{k} + \frac{k-1}{k} \cdot \sum_{i=1}^{h} \frac{\lfloor \log i \rfloor}{k^i}\right) < \frac{k}{k-1}$. Therefore, Theorem 8 makes an improvement in the upper bounds on the number of comparisons required for k-ary heap constructions.

5.3 Min-Max Heaps

It has been observed that algorithms for building min-max heaps often turn out to be analogous to those for traditional heaps, due to the similarity between these two data structures. For example, Floyd's heap construction algorithm [8], which uses smaller heaps as building blocks for larger ones, has been employed to the creation of min-max heaps. The complexity of this algorithm is roughly $2\frac{1}{3}n$ comparisons in the worst case [2]. As another example, the immediate utility of the idea behind the variants of Floyd's algorithm [3,9] results in a better min-max heap construction algorithm, which uses $2.15n$ comparisons (stated also in [2]). Moreover, the best known algorithm for building min-max heaps was developed by Strothotte et al. [15] with the help of binomial trees, which takes approximately $\frac{17}{9}n$ comparisons in the worst case. In the following, we shall show how to build an n-element min-max heap in at most $1.87153n$ comparisons in the worst case, which is slightly better than the known upper bound. Moreover, the amount of extra space required by our algorithm is only $\mathcal{O}(1)$. First, we need fast construction methods for special cases.

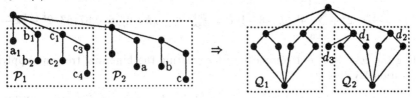

FIGURE 2: The construction of a min-max heap of size 15.

Lemma 9. *The worst-case number of comparisons for converting a binomial tree of size 16 into a min-max heap with 15 elements plus one extra element is at most 6.*

Proof. We shall actually show that 4 comparisons are enough for the transformation from $\mathcal{P}_1$ to $\mathcal{Q}_1$ while converting $\mathcal{P}_2$ into $\mathcal{Q}_2 + \{d_3\}$ costs 2 comparisons (see Figure 2). The conversion from $\mathcal{P}_2$ to $\mathcal{Q}_2 + \{d_3\}$ is simple: Comparing b with c and the larger one is then compared to a. The procedure of converting $\mathcal{P}_1$ into $\mathcal{Q}_1$ begins with a comparison between a_1 and c_2.

If a_1 is the larger, then we assume without loss of generality that a_1 is larger than c_4 (cost: one comparison) and compare again a_1 with b_2. Finally, b_1 is compared with c_1 if $a_1 \prec b_2$; else a comparison between b_1 and c_2 is made.

For the case when c_2 is the larger, we compare a_1 with b_1. If $a_1 \prec b_1$, then c_4 is compared with b_2 and the larger one is compared with c_2; else (for $a_1 \succ b_1$), a comparison between c_4 and c_2 is made and the loser is compared with b_2.

The correctness of the algorithm becomes obvious if we depict the sequence of comparisons graphically.

Lemma 9 improves the previously best known algorithm [15] for building a 15-element min-max heap, which takes 22 comparisons provided that an extra element is available. Notice that in the proof of Lemma 9, we actually construct a structure that is a min-max heap on 15 elements plus additional order relations $d_1 \prec d_2$ and $d_1 \prec d_3$ (see Figure 2). It can easily be checked that merging two such structures into a max-min heap of size 31 costs only 8 comparisons. Hence,

Lemma 10. *The number of comparisons sufficient to convert a (min-)binomial tree of size 64 into a 63-element min-max heap is at most 44 in the worst case.*

This can be done by first creating a binomial tree of size 64 and then converting it into a min-max heap of size 63. Now we are ready to describe our fast min-max heap construction algorithm. Let $n = 2^k$ (for even k). First, a (min-)binomial tree of size n is built. Then, the binomial tree is split recursively into four parts same as that done by the algorithm in [15]. Each part is converted into a min-max heap of size $2^{k-2} - 1$, recursively. The recursion ends up on binomial trees of size 64. Let $T(2^k)$ be the cost of our algorithm for converting a (min-)binomial tree of size 2^k into a min-max heap on $n = 2^k - 1$ elements plus one extra element. Then

$$
\begin{cases}
T(64) &= 44 \qquad \text{and} \\
T(2^k) &= 4 \cdot T(\tfrac{2^k}{4}) + k - 2 + 2(2(k-2) + 1) = 4 \cdot T(\tfrac{2^k}{4}) + 5k - 8,
\end{cases}
$$

where the $2(2(k-2)+1)$ term is the cost of merging two min-max heaps of size $2^{k-2}-1$ (for even k) [15]. The recurrence above exploits

$$T(2^k) = \tfrac{T(64)}{64}\cdot 2^k + \sum_{i=0}^{\frac{k}{2}-4} 4^i(5k-8-10i) = \tfrac{251}{288}\cdot 2^k - \tfrac{5}{3}k - \tfrac{16}{9} \doteq 0.87153\cdot 2^k - \tfrac{5}{3}k - \tfrac{16}{9}.$$

Adding the cost of 2^k-1 comparisons for creating the binomial tree and applying Proposition 1 yields:

Theorem 11. *A min-max heap on n elements can be constructed in at most $1.87153n + \mathcal{O}(\log^2 n)$ comparisons in the worst case.*

Moreover, we can apply Algorithm 3 to the construction method above as well.

Theorem 12. *A min-max heap on $2^{h+1}-1$ elements can be constructed in at most $2.149733n - \left(\tfrac{1}{2}\cdot\sum_{i=1}^{h_0}\tfrac{\lfloor\log i\rfloor}{2^i} + \tfrac{1}{2^{h_0+1}}\left(\tfrac{h_0}{6}-\tfrac{2}{9}\right) - \tfrac{11}{288}\right)\cdot n$ comparisons in the worst case, using $\mathcal{O}(2^{h_0})$ amount of extra space ($h_0 < h$ is even).*

Proof. If every min-max heap of size $2^{h_0+1}-1$ is created by the above construction method, then $\mathbb{F}(h_0) \le 1\tfrac{251}{288}\cdot 2^{h_0+1} - \tfrac{5}{3}(h_0+1) - \tfrac{16}{9}$. For the recursive merging procedure (i.e., the second step of Algorithm 3), we employ the algorithm in [2], which takes at most $2.14733n + \mathcal{O}(\log^2 n)$ comparisons in the worst case. Therefore, according to Equation (1), we know that the worst-case cost of the algorithm is

$$\mathbb{F}(h) \le 2^{h-h_0}\cdot\mathbb{F}(h_0) + 2^h\cdot\sum_{i=h_0+1}^{h}\frac{\mathbb{M}(i)}{2^i} = 2^h\cdot\sum_{i=1}^{h}\frac{\mathbb{M}(i)}{2^i} - 2^h\cdot\sum_{i=1}^{h_0}\frac{\mathbb{M}(i)}{2^i} + 2^h\cdot\frac{\mathbb{F}(h_0)}{2^{h_0}},$$

which is at most

$$2.149733n - 2^h\left(\tfrac{11}{3} + \tfrac{\frac{1}{3}}{4^{\lceil\frac{h_0}{2}\rceil}} - \tfrac{3h_0+8}{2^{h_0+1}} + \sum_{i=1}^{h_0}\frac{\lfloor\log i\rfloor}{2^i}\right) + \tfrac{2^h}{2^{h_0}}\left(1\tfrac{251}{288}\cdot 2^{h_0+1} - \tfrac{5}{3}(h_0+1) - \tfrac{16}{9}\right).$$

Evaluating Theorem 12 for $h_0 = 11$ and combining it with Proposition 1 establishes:

Corollary 13. *An n-element min-max heap ($n > 4096$) can be built in at most $1.87186n + \mathcal{O}(\log^2 n)$ comparisons using $\mathcal{O}(1)$ extra space in the worst case.*

5.4 Twin-Heaps and Full Deaps

The objective of this subsection is to exhibit sharper upper bound on the worst-case number of comparisons required for the twin-heap construction. As recognized in Proposition 2, only two special cases are needed to be investigated: full twin-heaps and half-full twin-heaps. Moreover, the relationship between the twin-heap and deap helps us to develop efficient algorithms to solve twin-heap construction problem. Carlsson [5] proposed a $2.5644n$-comparison algorithm that builds a deap by repeated pairwise merging smaller deaps. The merging stage uses the technique that starts by finding the path of minimum/maximum children when inserting new elements. This construction algorithm can be directly applied to twin-heaps by Proposition 2.

Lemma 14. *A twin-heap of size n can be built bottom-up by merging successive pairs of smaller twin-heaps into larger ones in at most $2.5664n + \mathcal{O}(\log^2 n)$ comparisons.*

We now giving an algorithm that has a worst-case complexity of $2.50n$ comparisons. The new twin-heap construction algorithm uses smaller twin-heaps as building blocks, which is motivated by the observation that two twin-heaps of size 6 can be built faster than constructing them separately. Notice that we can create a twin-heap of size 6 in 8 comparisons by using the algorithm in [5]. However, if we construct two twin-heaps of size 6 simultaneously by using information generated from comparisons more than once, we can save one comparison totally.

Lemma 15. *Two twin-heaps of size 6 can be constructed with at most 15 comparisons in the worst case.*

Proof. First, two binomial trees of size 4 are created with 6 comparisons. Next, by comparing elements x and y, we obtain the ordering relations shown in Figure 3.

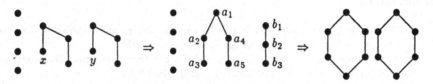

FIGURE 3: Build two 6-element twin-heaps.

One 6-element twin-heap will be built out of $\{a_1, \cdots, a_5\}$ plus one singleton element. We do this by first comparing a_3 with a_5 (assume without loss of generality that $a_3 \succ a_5$) and then inserting the singleton element into the chain $\{a_1 \prec a_2 \prec a_3\}$ with a binary search. The total cost of constructing the first twin-heap is 3 comparisons more. For building another twin-heap, we first compare two singletons, and then compare the smaller element with b_1. Similar to the case when constructing the first twin-heap, we can generate the second one with three additional comparisons. Hence, the overall cost is $6 + 1 + 3 + 2 + 3 = 15$ comparisons.

Theorem 16. *A twin-heap with n elements can be constructed in at most $2.5039n + \mathcal{O}(\log^2 n)$ comparisons in the worst case using $\mathcal{O}(1)$ extra space.*

Proof. Combining Algorithm 3 with Lemma 15 specifies:

1. Build $\frac{n+2}{8}$ twin-heaps with 6 elements using the method in Lemma 15.
2. The remaining elements are inserted into the twin-heap in the same way as the algorithm in [5].

We claim that the above algorithm takes $2.5039n$ comparisons. Notice first that the algorithm in [5] needs 16 comparisons to build two twin-heaps of size 6. However, we can saved one comparison for every two twin-heaps of size 6 from Lemma 15. By Lemma 14, we know that the number of comparisons needed by this new algorithm is $(2.5664 - \frac{1}{16})n$. From Proposition 2 the result follows.

6 Conclusions

This paper provides a systematic study of the problem of efficiently constructing various kinds of implicit priority queue and priority deque structures. In particular, the algorithmic paradigm presented enables us to develop new and fast construction methods, which improve upon the previously best known worst-case upper bounds for solving the construction problems and they are very close to being optimal. Remark that similar average-case results can be obtained by employing the above algorithmic paradigm as well. Our uniform approach for the construction problems also offers trade-off between the building time (evaluated by the number of comparisons used) and the space consumed. Interestingly, it is the technique, which builds many isomorphic copies of heap-like structures simultaneously, that leads to a better understanding of their construction complexities. This technique has previously been used for designing the fast known selection algorithm [14].

References

1. A. V. Aho, J. E. Hopocroft, and J. D. Ullman: *The Design and Analysis of Computer Algorithms*. Addison-Wesley, Reading, Massachusetts, 1974.
2. M. D. Atkinson, J.-R. Sack, N. Santoro, and Th. Strothotte: Min-max heaps and generalized priority queues. *CACM* **29** (10) (1986), 996-1000.
3. S. Carlsson: A variant of heapsort with almost optimal number of comparisons. *Information Processing Letters* **24** (4) (1987), 247-250.
4. S. Carlsson: Average-case results on heapsort. *BIT* **27** (1) (1987), 2-17.
5. S. Carlsson: The deap - A double-ended heap to implement double-ended priority queues. *Information Processing Letters* **26** (1) (1987), 33-36.
6. L. Draws, P. Eriksson, E. Forslund, L. Höglund, S. Vallner, and Th. Strothotte: Two new algorithms for constructing min-max heaps. In: *Proce. 1st SWAT* (1988), 43-50.
7. H. Erkiö: On heapsort and its dependence on input data. *Technical Report* A-1979-1. Department of Computer Science, University of Helsinki, Finland, 1979.
8. R. W. Floyd: Algorithm 245 - Treesort 3. *CACM* **7** (12) (1964), 701.
9. G. H. Gonnet and J. I. Munro: Heaps on heaps. *SIAM J. Comput.* **15** (1986), 964-971.
10. A. Hasham and J.-R. Sack: Bounds for min-max heaps. *BIT* **27** (3) (1987), 315-323.
11. D. B. Johnson: Efficient algorithms for shortest paths in sparse networks. *Journal of the ACM* **24** (1) (1979), 1-13.
12. D. E. Knuth: *The Art of Computer Programming. Vol. 3: Sorting and Searching.* Addison-Wesley, Reading, Massachusetts, 1973.
13. S. Okoma: Generalized heapsort. In: *Proce. 9th MFCS* (1980), 439-451.
14. A. Schönhage, M. Paterson, and N. Pippenger: Finding the median. *Journal of Computer and System Sciences* **13** (2) (1976), 184-199.
15. Th. Strothotte, P. Eriksson, and S. Vallner: A note on constructing min-max heaps. *BIT* **29** (2) (1989), 251-256.
16. J. Vuillemin: A data structure for manipulating priority queues. *CACM* **21** (1978), 309-314.
17. J. W. J. Williams: Algorithm 232: Heapsort. *CACM* **7** (6) (1964), 347-348.

Double-Ended Binomial Queues

C. M. Khoong[†] and H. W. Leong[‡]

Abstract. This paper introduces a highly efficient double-ended heap structure called *d-b-queues*, which are an extension of binomial queues. D-b-queues achieve the following amortized time bounds: $O(1)$ for *findmin, findmax, insert*, and *merge* operations, and $O(\log n)$ for *deletemin, deletemax, decreasekey, increasekey*, arbitrary *delete*, and *split* operations. An n-node d-b-queue can be constructed in $1.75n$ comparisons. Our results include a simple proof of $O(1)$ amortized time merging for ordinary binomial queues.

1 Introduction

We present an efficient pointer-based, double-ended heap data structure. The structure, which we call *d-b-queue*, is an extension of the well-known binomial queue [Vuil78]. The novel features used are systematic management of "matings" and "crisscrossings" across subheaps and the use of amortized analysis. D-b-queues achieve $O(1)$ time for *findmin, findmax, insert*, and *merge* operations, and $O(\log n)$ time for *deletemin, deletemax, decreasekey, increasekey*, arbitrary *delete*, and *split* operations. The time bounds are amortized. Furthermore, an n-node d-b-queue can be constructed in $1.75n$ comparisons. A d-b-queue requires the same amount of storage as an ordinary binomial queue of the same size.

We employ amortized time complexity analysis techniques similar to those we used in the design of efficient semi-implicit heap structures [KhLe93]. To our knowledge, d-b-queues are the first double-ended heap structure to achieve $O(1)$ time for merging and $O(\log n)$ time for splitting. Other known structures, such as [ASSS86, Carl87, OOW91], do not appear to support splitting, and require at least $O(\log n \log k)$ time for merging. Our results also include a proof of $O(1)$ amortized time merging for ordinary binomial queues. Previously, only an $O(\log n)$ amortized time bound was known.

2 Structure of D-b-queues

The d-b-queue data structure essentially consists of two binomial queues that are separate except for the fact that leaves of one binomial queue are linked to leaves in the other binomial queue. The following definition gives the basic structure of a d-b-queue.

Definition. A d-b-queue is a double-ended priority queue data structure that comprises two subheaps, namely a min-subheap and a max-subheap, such that: (i) The min-subheap (resp. max-subheap) is a binomial queue data structure with the *heap order* property: each node has a key that is no larger (resp. smaller) than that of

---

[†] *Information Technology Institute, National Computer Board, 71 Science Park Drive, Singapore 0511*

[‡] *Department of Information Systems and Computer Science, National University of Singapore, Lower Kent Ridge Road, Singapore 0511*

any of its children; (ii) Each leaf in the min-subheap (resp. max-subheap) is either *mated* or *free*. A leaf x is mated if there is another leaf y in the other subheap, such that $mate(x) = y$ and $mate(y) = x$. A leaf x is free if $mate(x) = $ nil. For any two nodes x and y where x is in the min-subheap and y is in the max-subheap, if $mate(x) = y$, then $key(x) \leq key(y)$.

Let the *rank* of a tree denote the number of children that its root has. The following structural invariants are maintained in d-b-queues in order to achieve efficiency in heap operations:

(i)   *Balance Invariant.* The sizes of the two subheaps differ by at most 1;

(ii) *Free Leaf Invariant.* A free leaf exists if and only if the larger subheap has a singleton tree, and there is at most one free leaf in the entire heap, which belongs to the singleton tree in the larger subheap. Furthermore, if the key of the free leaf is the smallest in the heap, it must be in the min-subheap; if its key is the largest in the heap, it must be in the max-subheap; otherwise its location does not matter.

(iii) *Covering Invariant.* We say that a tree T in one subheap *covers* another tree T' in the other subheap if the mates of leaves in T' are found entirely in T. Each tree in the larger subheap, except for the smallest-rank tree in that subheap, covers a tree in the smaller subheap of the same rank, and vice versa. Let the rank of the smallest-rank tree in the larger subheap be r. This tree covers all trees in the smaller subheap of ranks $0, 1, .., r-1$.

An example of a d-b-queue is shown in Figure 1a. A schematic of a d-b-queue illustrating the Covering Invariant is shown in Figure 1b. The following six structural scenarios for d-b-queues are distinguished. Case 1: Both subheaps have no singletons. Case 2: Both subheaps have singletons. Case 3: The larger subheap has a singleton, and the min-subheap is larger. Case 4: The larger subheap has a singleton, and the max-subheap is larger. Case 5: The smaller subheap has a singleton, and min-subheap is smaller. Case 6: The smaller subheap has a singleton, and the max-subheap is smaller. Each of these scenarios will be referred to by Case numbers in the sequel.

We mention some implementation details here. Each node has a key and 3 pointers for parent, first child, and sibling. The ranks (i.e. number of children) of root nodes in a subheap need not be stored, since the ranks are given implicitly by the binary expansion of the subheap's size. (For implementation convenience, it is assumed that the binary expansion is obtainable in $O(1)$ time. If this assumption is not reasonable, then $O(\log n)$ extra bits may be stored.) If the node is a leaf, its first child pointer is used as its mate pointer. If the node is a tree root, its sibling pointer is used to point to the tree root of the next higher rank; otherwise its sibling pointer points to a sibling of the next lower rank (if any).

The min-subheap (resp. max-subheap) has a minimum (resp. maximum) node pointer to identify the node with the smallest (resp. largest) key, and a minimum-rank root pointer to identify the root of the lowest-rank tree.

Due to the nature of some heap operations, matings in a heap may be crisscrossed. This means that the Hasse diagram of the heap may show crossed mating lines resulting from a *pairwise merge* with its parent. Crisscrossing complicates *split* operations, but fortunately *split* can still be done efficiently with the help of some structural information. Each node maintains a bit to indicate crisscrossing of mating lines, which we call the *c-bit*. The c-bit is basically used to disentangle crisscrossings when a tree needs to be split. Figure 2 illustrates the usage of c-bits. Apart from this extra bit, each node in a d-b-queue uses the same amount of storage as a node in an ordinary binomial queue.

3 Heap Algorithms

In d-b-queues, *findmin* and *findmax* are easy to do, since pointers to the minimum and maximum nodes are maintained. The algorithms for the other heap operations are given below. The following notation is used for convenience in discussion: $H(n)$ denotes a d-b-queue containing n nodes, $Hmin(n)$ denotes the min-subheap of $H(n)$, and $Hmax(n)$ denotes the max-subheap of $H(n)$. Define a *tree-pair* as a pair of equal-rank trees, one in each subheap, such that each tree covers the other. For instance, in Figure 1b, A and D forms one tree-pair.

3.1 Preliminaries

We first develop a set of primitives which are utilized in the algorithms for heap operations. These primitives are *shift up, pairwise merge, parallel pairwise merge, combine, parallel combine, recombine, long swap, migrate up*, and *migrate down*.

A *shift up* is used when a node violates heap order with respect to its ancestors. A *shift up* involves repeated comparisons and swaps along the path upwards to the tree root, until heap order is restored. Note that, when a leaf node is swapped with its parent $p(x)$, the former parent takes over the mate pointer of x, i.e. $mate(p(x)) := mate(x)$.

A *pairwise merge* links two trees of equal ranks into one, as shown for trees A and C in Figure 2. A *parallel pairwise merge* is the same as a *pairwise merge* except that two tree-pairs from different heaps are involved. This is illustrated for tree-pairs A-B and C-D in Figure 2.

Note that, in a *pairwise merge* or *parallel pairwise merge* involving singleton trees, the mate pointers need to be updated. Furthermore, heap order may now be violated between mates after the merge. This scenario is illustrated in Figure 3. Thus an extra comparison is needed to detect this situation. To restore heap order, a primitive called *long swap* is defined that swaps the affected mates, and then does a *shift up* in the relevant subheap. A *shift up* may be necessary since the mate of a singleton may belong to a tree of rank greater than zero in the other subheap.

A *combine* is a sequence of repeated *pairwise merges* in a subheap in increasing order of rank, until *pairwise merges* are not possible, after a new tree is introduced into the subheap. A *parallel combine* is a sequence of *parallel pairwise merges* of tree-pairs, after a tree-pair is introduced into the heap.

In *pairwise merge, parallel pairwise merge, combine*, and *parallel combine* primitives, one crucial additional detail is the maintenance of c-bits. This maintenance process is illustrated in Figure 2b for a *parallel pairwise merge*.

Other applicable primitives that will not be elaborated on here are *recombine* and *migrate up/down*. A *recombine* is used when a node violates heap order with respect to its descendents. A *recombine* breaks the node from its subtrees, treats its subtrees as a forest, and then *combines* the node with its subtrees. This effectively implements a *shift down* primitive (opposite of *shift up*), and restores heap order. A *migrate up/down* removes a node from one subheap and *combines* it into the other subheap. The node to be removed is the singleton (if the subheap has one) or the mate of the singleton from the other subheap.

A critical primitive is the disentanglement of crisscrossings. The need to disentangle crisscrossings arises when a covering tree is to be broken up into a forest of new trees. An example of a disentanglement operation is shown in Figure 4. The key problem in disentanglement is in finding the list L of nodes whose c-bits determine the history of pairwise merges that resulted in the current tree. The procedure $FindL$ below finds the list L for a tree rooted at x. The procedure $Disentangle(x, F)$ returns a forest F of disentangled forest of trees of ranks $0, ..., r-1$ plus an extra singleton, where r is the original rank of x. In the procedure $Disentangle(x, F)$, the parent of $L[r-1]$ is always the extra singleton in F because of the following reason. The rank of $L[r-1]$ is always 0, which implies that the cbit of $L[r-1]$ is always 0. So $L[r-1]$ is always assigned to be the root of the rank 0 tree in F at the last iteration of the **for**-loop, leaving its parent to be assigned as the extra singleton in F.

```
procedure FindL(x);                     procedure Disentangle(x, F);
begin                                   begin
  r := rank of x;                         for i := 0 to r-1 do
  y := x→firstSon;                          if L[i]→cbit = 0
  for i := 0 to r-1 do                      then
  begin                                         Let L[i] be the root of the rank
    L[i] := y;                                    (r-i-1) tree in F
    if y→cbit = 0                           else
      then y := y→sibling                     Let parent of L[i] be the root
      else begin                                of the rank (r-i-1) tree in F;
            x := y;                       Let parent of L[r-1] be the extra
            y := x→firstSon;                singleton in F;
          end;                          end;
end;
```

3.2 Makeheap

Heap creation is done as follows. Let the set of nodes be $x_1, x_2, .., x_n$. Break the set into pairs (x_{2i-1}, x_{2i}), where $1 \le i \le \lfloor n/2 \rfloor$. If n is odd, x_n is left unpaired. For each pair, the node with smaller key goes into $Hmin(n)$, and the other goes into $Hmax(n)$. Now if n is odd, x_n is placed into the correct subheap as follows. Pick any node x_j from $Hmin(n)$, and any node x_k from $Hmax(n)$. If $x_n < x_j$, x_n goes into $Hmin(n)$; if $x_n > x_k$, x_n goes into $Hmax(n)$; otherwise x_n may go into either subheap. The node pairs are treated as rank-0 tree pairs, and recursively *parallel pairwise merged* to form a d-b-queue. Note that the matings would be set up in the natural way, so that the Covering Invariant is maintained. The minimum and maximum nodes in $H(n)$ are identified via comparisons among tree roots.

3.3 Insert

The algorithm for *insert* is presented below. The subheap into which to insert x must be selected carefully to respect the structural invariants. If both subheaps are equal in size, x may become a free leaf. Also, the first *pairwise merge* in the *combine* may incur a *long swap*, for the following reason. Without loss of generality, assume that we are inserting x into $Hmin(n)$, and that $Hmin(n)$ has a singleton y. Thus the first *pairwise merge* in the *combine* involves x and y. If $key(x) > key(mate(y))$, then a *long swap* would be required.

Algorithm *insert(x, H(n))*:
begin
  if $H(n)$ is Case 1 or 2
  {** i.e. subheaps are equal in size **}
  then
    if $key(x) <$ (key of minimum node
      in $H(n)$)
    then
      Let $H^*(n)$ denote *Hmin(n)*

  else
    Let $H^*(n)$ denote *Hmax(n)*
  else
    Let $H^*(n)$ denote the smaller
    subheap in $H(n)$;
  combine(x, H*(n));
  Update the minimum or maximum
    node pointer in $H(n)$;
end;

3.4 Decreasekey, Increasekey, Deletemin, Deletemax, Delete

The algorithms for these operations are fairly easy generalizations of the ones on binomial queue structures; we summarize key ideas here. For *decreasekey* and *increasekey* , a *shift up* or *recombine* operation apply, depending on which subheap the item affected is in. For *deletemin* and *deletemax*, the main concerns are to restore the Balance and Free Leaf Invariants after the removal of the old minimum/maximum and when establishing the new minimum/maximum. The old minimum/maximum is basically replaced by the root of the smallest-rank tree in the same subheap. In so doing, should a tree be broken up into a forest, crisscrossings in mating lines must be disentangled to restore the Covering Invariant. An arbitrary *delete(x, H(n))* operation can be performed by doing a *decreasekey(x, ∞, H(n))* followed by a *deletemin(H(n))*.

3.5 Merge

A merge operation merges a heap $H'(k)$ into a heap $H(n)$, where $n \geq k$. To simplify the exposition, we first treat the case in which both n and k are even. In this case, the subheaps in each heap are structurally the same. It is thus possible to take a tree-pair at a time from $H'(k)$ and *parallel combine* the tree-pair into $H(n)$. Update the minimum and maximum node pointers in $H(n+k)$. We call this procedure *simpleMerge(H(n), H'(k))*.

To make *simpleMerge* efficient, the search for tree-pairs in $H(n)$ where each parallel combine begins has to be handled carefully. The idea is as follows. Perform the *parallel combines* in decreasing order of tree-pair size. For the first tree-pair from $H'(k)$, search along the $H(n)$ root list in ascending order of tree size to locate the start point of the *parallel combine*. The end point, say x, of this *parallel combine* is noted. For the next tree-pair from $H'(k)$, search along the $H(n)$ root list *backwards* starting at x to locate the start point of the *parallel combine*. The same logic applies for subsequent tree-pairs from $H'(k)$. As shown in §4, the cost of this search process can be amortized out.

If n and k are not both even, there are several ways to do the merge in the same time complexity. Here we provide transformations of these other cases to make use of *simpleMerge*, together with some extra insertions and deletions. In §4, it will be shown that the costs of the extra operations can be amortized out.

n even, k odd:
1. Delete an arbitrary node x_k from $H'(k)$.
2. *simpleMerge(H(n), H'(k–1))*.

n odd, k even:
1. Delete two arbitrary nodes x_{k1} and x_{k2} from $H'(k)$.
2. *insert(x_{k1}, H(n))*.

n and k both odd:
1. Delete an arbitrary node x_k from $H'(k)$.
2. *insert(x_k, H(n))*.
3. *simpleMerge(H(n+1),*

3. $insert(x_k, H(n+k-1))$. 3. $simpleMerge(H(n+1),$ $H'(k-1))$.
                                    $H'(k-2))$.
                               4. $insert(x_{k2}, H(n+k-1))$.

3.6 Split

A split operation removes k nodes from a heap $H(n)$ to form a new heap $H'(k)$, where $n \geq 2k$. As in *merge*, we first simplify the exposition by treating the case where n and k are both even. In this case, starting from the lowest rank, as many tree-pairs from $H(n)$ are removed as possible, totalling $k' \leq k$ nodes, and these tree-pairs are collectively called $H'(k')$. Then break the tree-pair of smallest rank remaining in $H(n-k')$ into forests. The breaking is done in such a way as to disentangle crisscrossings. It is easy to see that $k''=k-k'$ nodes may be obtained from a subset of the forest-pair without further subdivision. The forest-pair subset is collectively called $H''(k'')$. Finally, do a $merge(H'(k'), H''(k''))$. Also set the minimum and maximum node pointers in $H(n-k)$ and $H'(k)$. We call this splitting procedure $simpleSplit(H(n), k)$.

We provide here transformations of the cases when n and k are not both even to make use of *simpleSplit*, together with some extra insertions and deletions. In §4, it will be shown that the costs of the extra operations can be amortized out.

n even, k odd:
1. $simpleSplit(H(n), k-1)$.
2. Delete an arbitrary node x_n from $H(n-k+1)$.
3. $insert(x_n, H'(k-1))$.

n odd, k even:
1. Delete an arbitrary node x_n from $H(n)$.
2. $simpleSplit(H(n-1), k)$.
3. $insert(x_n, H(n-k-1))$.

n and k both odd:
1. Delete an arbitrary node x_n from $H(n)$.
2. $simpleSplit(H(n-1), k-1)$.
3. $insert(x_n, H'(k-1))$.

4 Analysis

The efficiency of the heap algorithms are analyzed from the amortized point of view, using a credit allocation approach. We distinguish two types of credits in the analysis of d-b-queues:

(i) *h-credits*, to pay for *combine* and *long swap* primitives. An h-credit covers the cost of one comparison-linking operation in a *pairwise merge* plus the cost of one comparison-swap step in a *long swap*.

(ii) *m-credits*, to pay for some of the work in *merge* operations. An m-credit pays for a constant amount of work in a root search process plus a constant amount of work done in a *delete* operation incurred during the *merge*.

The following analytic invariants are maintained in d-b-queues:

Invariant 1. Each tree in the heap has one h-credit available to it.

Invariant 2. The number of m-credits available to a heap $H(n)$ is at least $\lceil \log n \rceil$.

The efficiency of the heap operations is dependent on the primitives. It is easy to see that *shift up* and *long swap* both take (actual and amortized) time proportional to the height of the tree involved, which is at most $O(\log n)$. *Pairwise merge* and *parallel pairwise merge* both take $O(1)$ time (actual and amortized), plus the time required for a *long swap* if incurred. In the analysis of heap operations, *long swaps* would be accounted for separately.

To analyze *combine* and *parallel combine*, let us first ignore the possible occurence of the *long swap* in the first *pairwise merge*. Note that each tree has one h-credit available to it, including the tree just introduced into the heap. Thus each *pairwise merge* in the *combine* operation can be paid for by one h-credit, and after the last *pairwise merge* the resultant tree would still have one h-credit available to it. This implies that *combine* takes $O(1)$ amortized time. The same analysis for *parallel combine* implies that its amortized running time is also $O(1)$. Now add to these running times the time needed to perform the possible *long swap*.

4.1 Merging Binomial Queues in $O(1)$ Time

As part of our research on d-b-queues, a result that is important in itself is that ordinary binomial queues [Vuil78] can be merged in $O(1)$ time. This is a clear improvement over the previously known bound of $O(\log n)$. We state the proof here. The result generalizes to $O(1)$ merging for d-b-queues, since an binomial queue is conceptually one half (either the min- or max-subheap) of a d-b-queue, minus the sophistication of matings and c-bits.

Let Invariants 1 and 2 apply to the binomial queues to be merged: $H(n)$ and $H'(k)$. It is quite easy for the heap operations to maintain the invairants; we provide the details in §4.2 for d-b-queues. Consider the *merge* algorithm for d-b-queues (§3.6) applied to binomial queues: replace *parallel combine* of tree-pairs with *combine* of trees. We first analyze the complexity of *simpleMerge*, and then consider the costs of extra *inserts* and *deletes* in the algorithm. The m-credits are critical in the analysis: the $O(\log k)$ m-credits in $H'(k)$ (from Invariant 2) would pay for the entire root search process in *simpleMerge* and a constant number of extra *insert* and *delete* operations.

In *simpleMerge*, note that each *combine* for each tree from $H'(k)$ is paid for entirely by h-credits. The entire root search process for locating start points of *combines* require $O(\log k)$ time, since the first search involves stepping through at most $\lceil \log k \rceil$ roots, while the subsequent searches involve stepping through a total of at most $\lceil \log k \rceil$ roots in the reverse direction. Thus the search process is paid for by the m-credits in $H'(k)$. Updating of minimum and maximum node pointers can be done in $O(1)$ time. In summary, *simpleMerge* runs in $O(1)$ amortized time.

For general values of n and k, extra *deletes* from $H'(k)$ and *inserts* into $H(n)$ are incurred. The *inserts* cost $O(1)$ amortized time (since *combines* cost $O(1)$; see also §4.2) and the *deletes* $O(\log k)$ time. Again, this cost is covered by the m-credits available to $H'(k)$. To maintain Invariant 2, one new m-credit is allocated to $H(n+k)$. The total number of m-credits available to $H(n+k)$ is then at least $\lceil \log n \rceil + 1 \geq \lceil \log(n+k) \rceil$. The overall amortized time complexity of *merge* is thus $O(1)$.

4.2 Analysis of D-b-queues

We now turn to the analysis of the d-b-queue heap operations. First note that *findmin* and *findmax* run in $O(1)$ (actual and amortized) time, since pointers to the minimum and maximum nodes are maintained.

For *makeheap*, we note that $\lfloor n/2 \rfloor$ comparisons are used to divide the set of nodes into two subsets. If n is odd, 2 extra comparisons are incurred to place x_n into the right subset. At most $\lfloor n/2 \rfloor - 1$ parallel pairwise merges are done. If less than this number is done, the difference goes into comparisons to identify the minimum and maximum nodes. Each parallel pairwise merge incurs 2 comparisons, except the first iteration of $\lfloor n/4 \rfloor$ rank-0 merges, which cost 3 comparisons each. In total, at most $1.75n$ comparisons are done. To maintain Invariants 1 and 2, $\lceil \log n \rceil$ each of h-credits and m-credits are allocated to the heap.

For *insert*, one new h-credit and one new m-credit is allocated. The new h-credit, together with the h-credits of trees involved in the *combine* primitive, would pay for the *combine* primitive less the *long swap* primitive if it occurs. The m-credit is saved to help pay for work done in a future *merge* operation. We now turn to the payment of work done by the *long swap* in the *combine* primitive.

First note that the *long swap*, if it occurs, would be part of the first *pairwise merge* (of singleton trees) in the *combine* primitive. Let the new node be labeled c. We consider the various cases. If $H(n)$ has no singletons (Case 1), then the *long swap* would not occur. If $H(n)$ has a singleton only in the larger subheap (Cases 3 and 4) or if $H(n)$ has singletons in both subheaps (Case 2), then the *long swap* takes $O(1)$ actual time, since in the worst case c would need to be swapped with its singleton mate.

If $H(n)$ has a singleton only in the smaller subheap (Cases 5 and 6), then the *long swap* is similar to the scenario depicted in Figure 3d, in which a *shift up* for c may be required in the other subheap after swapping the mates. Let the tree in which the *shift up* occurs be labeled T_r, where r is its rank. From the Covering Invariant, it is known that T_r covers trees of ranks $0, 1, .., r-1$ in the other subheap; furthermore, note that the other subheap cannot have a tree of rank r. Thus the number of *pairwise merges* that would occur there due to the new node would be exactly r. The *shift up* in T_r would incur at most r comparison-swap steps. Thus the h-credits used to pay for the *pairwise merges* can also cover the cost of the *shift up*.

In summary, the overall amortized time complexity of *insert* is $O(1)$. Also note that Invariants 1 and 2 are maintained at the end of the operation.

The operations *decreasekey, increasekey, deletemin, deletemax,* and *delete* do not utilize any credits. The running time of each operation is bounded by the time for primitives such as *recombine, shift up., migrate up/down,* and/or disentangling of crisscrossings, all of which can be done in $O(\log n)$ time. Thus the operations *decreasekey, increasekey, deletemin, deletemax,* and *delete* run in $O(\log n)$ (actual and amortized) time.

The amortized time complexity of *merge* is clearly $O(1)$ from the analysis in §4.1.

For *split*, note that sizes of trees are given implicitly by the binary expansion of subheap sizes. Removing k' nodes requires $O(\log n)$ time (to scan the subheap root lists). The time to remove the remaining k'' nodes depends on the disentangling of crisscrossings at the tree-pair involved, which takes $O(\log n)$ time. The cost of the *merge* is amortized out using a sufficient pool of h-credits and m-credits. The extra *inserts* and *deletes* used to cater to general values of n and k contribute only $O(\log n)$ cost. Sufficient numbers ($O(\log n)$) of h-credits and m-credits are allocated to maintain Invariants 1 and 2. Thus the overall (actual and amortized) time complexity of *split* is $O(\log n)$.

We summarize our results in the following lemma:

Lemma 1. *D-b-queues achieve the following amortized times: $O(1)$ for findmin, findmax, insert, and merge operations, and $O(\log n)$ for deletemin, deletemax, delete, decreasekey, increasekey, and split operations. Furthermore, a d-b-queue can be constructed in $1.75n$ comparisons.*

5 Concluding Remarks

The heap algorithms for d-b-queues have been implemented in the C programming language. Simulation results [Khoo93] indicate that the constant factors in actual running times are small. An intriguing open problem would be to explore the possibility of removing the need to maintain some of the structural invariants or the c-bits.

References

[ASSS86] M. D. Atkinson, J.-R. Sack, N. Santoro, and T. Strothotte, Min-max heaps and generalized priority queues. *Comm. ACM* **29** (1986) 996–1000.

[Carl87] S. Carlsson, The deap – A double-ended heap to implement double-ended priority queues. *Inform. Proc. Lett.* **26** (1987) 33–36.

[GNT91] G. Gambosi, E. Nardelli, and M. Talamo, A pointer-free data structure for merging heaps and min-max heaps. *Theoret. Comput. Sci.* **84** (1991) 107–126.

[Khoo93] C. M. Khoong, The design and analysis of heap algorithms. M.Sc. Thesis, Department of Information Systems and Computer Science, National University of Singapore, Singapore, 1993.

[KhLe93] C. M. Khoong and H. W. Leong, Relaxed inorder heaps. Submitted for publication.

[OOW91] S. Olariu, C. M. Overstreet, and Z. Wen (1991), A mergeable double-ended priority queue. *Computer J.*, Vol. 34, pp. 423–427.

[Vuil78] J. Vuillemin, A data structure for manipulating priority queues. *Comm. ACM* **21** (1978) 309–315.

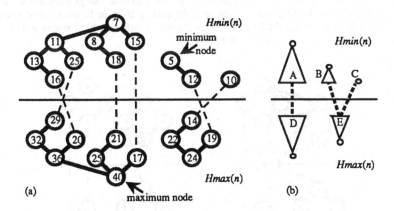

Figure 1. $Hmin(n)$ denotes the min-subheap and $Hmax(n)$ denotes the max-subheap. (a) Hasse diagram of a d-b-queue. Matings are indicated by the dashed lines. The occurrence of crisscrossings is indicated by intersecting dashed lines. (b) A schematic of the d-b-queue in (a) illustrating the Covering Invariant. Coverings are indicated by the dotted lines. For instance, A covers D and vice versa; E covers B and C.

137

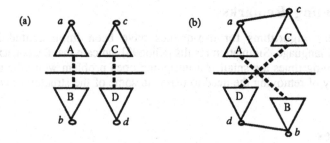

Figure 2. Illustration of crisscrossings. (a) Two tree-pairs A-B and C-D from different heaps, where A, B, C, and D are all of the same rank, are given. (b) The tree-pairs are *parallel pairwise merged*. It is assumed that $key(a) \geq key(c)$ and $key(d) \leq key(b)$. Now the c-bits of a and d are set to record a crisscrossing. The c-bits of c and b are irrelevant to the current crisscrossing.

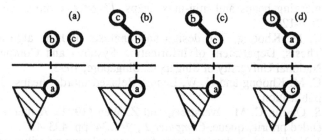

Figure 3. Intricacies in *pairwise merge* of singleton trees. (a) The singleton c is about to be *pairwise merged* with the singleton b. Node b is mated with node a, which would be a leaf in a tree of rank ≥ 0. (b) If $key(c) \leq key(b)$, then the *pairwise merge* does not affect the mating. (c) If $key(c) > key(b)$, then c takes over the mating from b; furthermore if $key(c) > key(a)$ then a *long swap* would be needed, as shown in (d), in which nodes a and c are swapped, and a *shift up* occurs starting at node c.

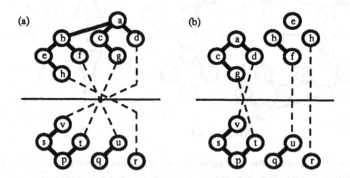

Figure 4. Disentanglement of crisscrossings. (a) A node is to be deleted from the covering tree of rank 3 in a tree-pair. Only the c-bits of nodes b, c, and e are set to 1. (b) The tree is broken up into a forest of ranks 2, 1, and 0 and an extra singleton (node c) in a way that disentangles crisscrossings, by checking the c-bits of descendents of the tree root. Each shaded set of nodes is a new tree in the disentangled forest.

A Simple Balanced Search Tree

with $O(1)$ Worst-Case Update Time [1]

by

Rudolf Fleischer [2]

ABSTRACT   In this paper we show how a slight modification of (a,b)-trees allows us to perform member and neighbor queries in $O(\log n)$ time and updates in $O(1)$ worst-case time (once the position of the inserted or deleted key is known). Our data structure is quite natural and much simpler than previous worst-case optimal solutions. It is based on two techniques : 1) *bucketing*, i.e. storing an ordered list of $2 \log n$ keys in each leaf of an (a,b) tree, and   2) *lazy splitting*, i.e. postponing necessary splits of big nodes until we have time to handle them. It can also be used as a finger tree with $O(\log^* n)$ worst-case update time.

1. Introduction

One of the most common (and most important) data structures used in efficient algorithms is the balanced search tree. Hence there exists a great variety of them in literature. Basically, they all store a set of n keys such that location, insertion and deletion of keys can be accomplished in $O(\log n)$ worst-case time.

In general, updates (insertions or deletions) are done in the following way : First, locate the place in the tree where the change has to be made; second, perform the actual update; and third, rebalance the tree to guarantee that future query times are still in $O(\log n)$. The second step usually takes only $O(1)$ time, whereas steps 1 and 3 both need $O(\log n)$ time. But there are applications which do not need the first step because it is already known where the key has to be inserted or deleted in the tree. In these cases we would like to have a data structure which can do the rebalancing step as fast as the actual update, i.e. in constant time. Good worst-case behaviour is especially important in real-time applications.

One such example are dynamic planar triangulations. In [M91] Mulmuley examined (among others) point location in dynamic planar Delauney triangulations. The graph of the triangulation is stored such that at each node of the planar graph the adjacent triangles are stored (sorted in clockwise radial order) in a balanced search tree. But now, whenever a point v is deleted or inserted, all points in the neighborhood of v can be affected by the retriangulation because their sequence of adjacent triangles might have changed. However, these changes are only local in the sense that one triangle (which is known at that time and has not to be searched in the radial tree) must be deleted or get some new neighbors (see [M91], 3.2 for details). To guarantee a worst-case update time for the triangulated point set which is proportional to the structural change (i.e. the number of deleted or newly created triangles) one needs search trees at the nodes which can handle updates in constant worst-case time.

It has been well known for a long time that some of the standard balanced search trees can achieve $O(1)$ amortized update time once the position of the key is known

---

[1] This work was supported by the ESPRIT II program of the EC under contract No. 3075 (project ALCOM)

[2] Max-Planck-Institut für Informatik, D66123 Saarbrücken, Germany, e-mail: rudolf@mpi-sb.mpg.de

([GMPR],[HM],[O82]). But for the worst-case update time the best known method had been an $O(\log^* n)$ algorithm by Harel ([[HL],[H80]). It has also been known that updates can be done with $O(1)$ structural changes (e.g. rotations) but the nodes to be changed have to be searched in $\Omega(\log n)$ time ([T83],[DSST]). Levcopoulos and Overmars ([LO]) have only recently come up with an algorithm achieving optimal $O(1)$ update time (similar results had been obtained by [DS] and [vE]). They use the *bucketing technique* of [O82] : Rather than storing single keys in the leaves of the search tree, each leaf (*bucket*) can store a list of several keys. Unfortunately, the buckets in [LO] have size $O(\log^2 n)$; so they need a 2-level hierarchy of lists to guarantee $O(\log n)$ query time within the buckets. They show that this bucket size is sufficient if after every $\log n$ insertions the biggest bucket is split into two halves and then the rebalancing of the search tree is distributed over the next $\log n$ insertions (for which no split occurs).

Our paper simplifies this approach considerably : We, too, distribute the rebalancing over the next $\log n$ insertions into the bucket which was split, but allow many buckets to be split at consecutive insertions (into different buckets). This seems fatal for internal nodes of the search tree : they may grow arbitrarily big because of postponed (but necessary) splits. But we show that internal nodes will never have more than twice the allowed number of children; hence queries can be done in $O(\log n)$ time. Furthermore, our buckets can grow only up to size $2\log n$, which means that we only need an ordered list to store the keys in a bucket. Also, the analysis of our algorithm seems simpler and more natural than in [LO].

Since the buckets are organized as a linear list, our data structure does not allow efficient finger searches, i.e. given a pointer to some known element and a key in distance d from this element we can not locate the key in time $O(d)$. However, iterating the construction, i.e. using our tree recursively in the buckets instead of linear lists, gives a data structure which allows efficient finger searches; but then the worst-case update time increases to $O(\log^* n)$. This matches the best previous bounds ([HL],[H80]).

We note that Dietz and Raman ([DR]) recently presented a variant of the Levcopolous/Overmars algorithm where the buckets are organized in a more complicated way such that efficient finger searches are possible with constant update time. However, their solution involves bitmanipulations and table lookup and therefore works only in the RAM model, whereas all other results mentioned in this paper, as well as our result, are achieved in the pointer machine model.

The paper is organized as follows. In Section 2 we define the data structure and give the algorithms for find and insert. In Section 3 we prove their efficiency. Then we conclude with some remarks in Section 4.

2 . The Data Structure

In this Section we will describe a simple data structure which maintains a set S of ordered keys and allows for operations *query*, *insert* and *delete*. Queries are the so-called *neighbor queries* : given a key K, if it is in the current set S report it, otherwise, report one of the two neighbors in S according to the given order. *Insert* and *delete* assume that we have previously located the key (to be deleted) or one of its neighbors (if we insert a new key) in the data structure. As was illustrated by the triangulation example in Section 1 where we have two nested data structures, this does not necessarily mean that we must perform a query in our data structure to locate this key. Our data structure is basically a balanced search tree, a variant of an (a, b)-tree ($4 \le 2a \le b$ and b even).

The main problem with update operations in a balanced search tree (and all other are still to be efficient. This means that we should rebalance the tree after each update. Unfortunately, this rebalancing can affect the whole path from the node to the root of the tree, which can be of length $\Omega(\log n)$. However, we will show in the next Section that in our search tree the rebalancing does not need to be performed immediately but can be distributed, step by step, over following updates which do not need a costly rebalancing. Thus we can guarantee constant worst-case update time.

In [OvL] and [O83], Overmars presented a very general method of handling deletions efficiently, the *global rebuilding technique*. We only give a short outline of this method here and refer to the original papers for details.

The idea is that a delete operation only deletes the node without doing any other operations, especially no rebalancing. This does not increase the query or insert time, but it does not decrease it either as it should. But since the optimal query time for $\frac{n}{c}$ keys is still $\log n - c$, we can afford being lazy for quite a long time before running into trouble (this is the reason why this method works only for deletions and not for insertions: $\frac{n}{c}$ insertions, all into the same position, can result in a path of length $\Omega(n)$ which would be disastrous for queries).

If there are too many deletions and the number of keys in the tree sinks below $\frac{n}{c}$ (here $c \geq 2$ is some constant), we start rebuilding the whole tree from scratch. Since this takes linear time, we distribute it over the next $\frac{n}{3c}$ operations, i.e. we still use and update our original (but meanwhile rather unbalanced) tree, but in parallel we also build a new tree, and we do this three times as fast. If the new tree is completed, we must still perform the updates which occured after we started the rebuilding. But again, we can do this in parallel during the next $\frac{n}{9c}$ operations. This continues until we finally reach a state where both trees store the same set of (at least $(\frac{1}{3} + \frac{1}{9} + \cdots)\frac{n}{c} = \frac{n}{2c}$) keys. Now we can dismiss the old tree and use the new tree instead. Since $c \geq 2$, we are always busy constructing at most one new tree. Hence the query and update time can only increase by a factor of 4.

This allows us to focus on a data structure which can only handle insertions; deletions can then be done using the global rebuilding technique. Now we give the details of our data structure. Assume that we initially have a set S_0 of n_0 keys (n_0 could be zero).

The Tree : Let $4 \leq 2a \leq b$ and b even. Then our tree T can be viewed as an $(a, 2b)$-tree, i.e. its internal nodes have between a and $2b$ children. However, each leaf does not store a single key but contains a doubly-linked ordered list of several keys; so we call the leaves *buckets*. Furthermore, each bucket B has a pointer r_B which points to some node within T (usually on the path from the root to B).

We remark that it is not really important that the keys in a bucket are stored in sorted order, but our algorithm automatically inserts new keys at the correct position of the list. For a node v of T let $size(v)$ denote the number of its children. We call v *small* if $size(v) \leq b$, otherwise *big*. We want to split big nodes into two small nodes whenever we encounter one. This makes our tree similar to an (a, b)-tree; the main difference is that we cannot afford to split big nodes immediately when they are created, but instead have to wait for a later insertion which can do the job.

Therefore we call this method *lazy splitting* : We only keep track of the fact that the tree may be unbalanced at a node (if seen as an (a, b)-tree) but we do not rebalance it until this part of the tree is used again later (otherwise rebalancing would be a waste of time). We note that in [T85] and [DSST] the different concept of *lazy recolouring* is used to guarantee a constant number of structural changes.

which could contain the key, and then search within the bucket. Insertions are done using Algorithm A.

Algorithm A : (A.1) Insert the new key into the bucket. Let this bucket be B with r_B pointing to node v.

(A.2) If v is big then split v into two small nodes.

(A.3) If v is the root of T then split B into two halves and let the r-pointer of both new buckets point to their common father. Otherwise, set r_B to $father(v)$, i.e. move it up one level in the tree.

Initially, we start with an (a,b)-tree T_0 (which is also an $(a,2b)$-tree) for the set S_0 such that each bucket contains exactly one key of S_0. Also, all pointers r_B point to their own bucket B. However, it is not clear at this point that Algorithm A really preserves the $(a,2b)$-property of T; one could think that some nodes could grow arbitrarily big because we could split the children of a node too often before testing and splitting the node itself in (A.2). But we will show in the next Section that this can not happen. Therefore it is reasonable to speak of splitting a big node into only two small nodes in (A.2). Nevertheless, this splitting can not be done arbitrarily but must follow some easy rules which will be given in the proof of Lemma 3.1 and in Lemma 3.3.

Furthermore, in (A.3), we want to split a bucket into two halves, and this should be done in constant time. Therefore we must design the buckets a little bit more complicated than just using a list. It is not really difficult, but it is technical : Each bucket has two headers (with five entries) to control the list (see Fig. 2.1). Initially, or after a split, *rightheader* controls an empty sublist. At each insertion, one key is added to the sublist of *rightheader* (either the inserted key or the rightmost key of the sublist of *leftheader*). Then we can easily split a bucket in (A.3) in constant time by changing only the header information and creating two new empty headers. This does not split the bucket exactly into two halves, but it is sufficient for our purpose as the next Lemma shows.

Lemma 2.1 Let n be the number of keys currently stored in T. Then $size(B) \leq 2\log n$ for all buckets B.

Proof : Easy by induction once we have proven that T is always an $(a,2b)$-tree of height at most $\log n$ (see Theorem 3.2). □

The r-pointers of the buckets always move up the tree from the leaves to the root, then starting again at the leaf-level. They usually follow the path from their own bucket to the root but if a node v is split (by an insertion into another bucket below v) then some r-pointers may point to the wrong sibling (if we create one new node as a sibling for v then all buckets in the subtree below the new node have their r-pointer still pointing to v which is not on their path to the root). But the analysis in Lemma 3.1 shows that this is not a problem. Hence it is not necessary to update the r-pointers pointing to v when v is split.

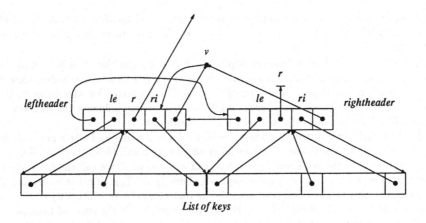

Fig. 2.1 A bucket

There is a second technical difficulty which we should be aware of. If we are given a new key k_1 to be inserted as the right neighbor of key k_2 which is the biggest key in bucket B_1 then we can not be sure that k_1 can be inserted into B_1. Let B_2 be the bucket to the right of B_1 with smallest key k_3, and let v be the lowest common ancestor of B_1 and B_2 in the tree. Then the search paths to B_1 and B_2 are split in v using a comparison with some key k_4 stored in v, $k_2 \leq k_4 < k_3$. Therefore k_1 belongs into bucket B_1 if and only if $k_1 \leq k_4$, otherwise into bucket B_2.

So we must maintain for each bucket B its left and right neighbor together with their respective lowest common ancestors. This can easily be done using a doubly linked list connecting alternately buckets and their lowest common ancestors (dotted line in Fig. 2.2), i.e. the usual preorder list. Splitting a node causes only local changes in this list which can be done in constant time.

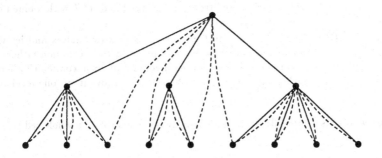

Fig. 2.2 The tree

3 . The Analysis

In this Section we prove that Algorithm A will never blow up an internal node of T, i.e. T actually always remains an $(a, 2b)$-tree. The idea is to show that the algorithm always splits a big node into only two small nodes, and that none of these two nodes can be split again before their common father has been tested and found to be small (or made small by a split). Since we start with an (a, b)-tree T_0, this proves that T is always an $(a, 2b)$-tree.

In order to prove this we extend Algorithm A to an Algorithm B which computes some additional attributes for the nodes and edges of the tree. These attributes allow us to conclude easily the desired properties of the tree T' constructed by Algorithm B. But since T and T' are identical (the additional attributes in no way influence the structure of the tree) the claim also holds for the tree T of Algorithm A.

First, we define the attributes which are just colours : Nodes and edges are coloured *blue* or *red*. A node is red if one of its incident edges is red, and blue otherwise. The red nodes together with the red edges form a forest of subtrees of T; we call the trees of this forest *red trees*. Leaves (buckets) are always red. Initially (in T_0), all other nodes and all edges are blue; this means that each bucket is the root of a red tree which only consists of the bucket itself. We call these red trees *trivial*. As we will see in the proof of Lemma 3.1, red edges can only be created by splitting a blue edge into two red edges; hence red edges always appear as pairs of two red edges. The red colour indicates that these edges can not be split (as long as they are red).

The edge connecting a node to its father is called *f-edge*, whereas the edges going to its children are called *c-edges*. We define the (non-existent) f-edge of $root(T)$ as always being blue. For a node v, let d_v (e_v) denote the number of its red (blue) c-edges. Obviously, $size(v) = d_v + e_v$. If $d_v + 2e_v \leq b$ then v is called *blocking*. In particular, blocking nodes are small and can not grow big (because red edges are never split and blue edges are always split into two red edges); hence they can not propagate a split to their father. For a leaf (bucket) B we define $d_B = e_B = 0$, i.e. leaves are blocking.

We now give the extended algorithm.

Algorithm B : (B.1) Insert the new key into the bucket. Let this bucket be B with r_B pointing to node v.

(B.2) If v is big then split v into small nodes and colour these nodes, their f-edges and their common father red (Lemma 3.1 shows that we always obtain only two small nodes). Otherwise, if v is the root of a nontrivial red tree (i.e. v is a red node of T with a blue f-edge) then *colour_blue(v)*.

(B.3) If v is the root of T then split B into two halves and let the r-pointer of both new buckets point to their common father; also, colour both buckets, their f-edges and their common father red. Otherwise, set r_B to $father(v)$, i.e. move it up one level in the tree.

colour_blue(v) is a subroutine which recursively recolours (blue) a whole red tree with root v.

colour_blue(v) : Colour v blue;
For all red c-edges $e = (v, w)$ colour e blue, and if w is not a leaf then *colour_blue(w)*;

Now we define invariant (I) which we will show to be true before and after every execution of Algorithm B.

(I) : (I.1) Blue nodes are small.

(I.2) Red edges always appear as pairs.

(I.3) T is an $(a, 2b)$-tree.

Let rT be any red subtree of T with root v.

(I.4) All nodes w of rT, $w \neq v$, are blocking.

(I.5) rT contains at least one bucket of T.

(I.6) If rT is a nontrivial red tree then, for all buckets B of rT, r_B points into rT.

(I.4) shows that no insert operation can make a node of a red tree big, i.e. red trees block the up-propagation of splits. (I.5) guarantees that all red trees are continuous, i.e. they always have at least one bucket as leaf. (I.6) shows that r-pointers pointing to a node v need not be updated to the correct sibling if v is split.

Lemma 3.1 Invariant (I) holds true after each step of Algorithm B. In particular, all nodes of T always have size at most $2b$, and we always split big nodes into only two small nodes.

Proof : By induction.

Initially, (I) is true because in T_0 all edges are blue and all nodes are small. So suppose that, after some time, we have a tree T and want to insert a new key into bucket B. Let r_B point to node v.

(B.1) This step does not affect (I). So (I) is true after (B.1) iff it was true before.

(B.2) We have to consider two cases.

<u>v is small</u> : If v is the root of a nontrivial red tree rT then rT is coloured blue in (B.2). This destroys one red tree and creates many trivial red trees (the buckets of rT). Therefore (I.3)-(I.6) hold after (B.2). (I.1) is true because we only change the colour of nodes of rT; but v was small by assumption and the other nodes were small by (I.4). And (I.2) is true because both edges of a pair of red edges must belong to the same red tree and hence are either both unaffected or both coloured blue.

And if v is not the root of a red tree then nothing happens in (B.2).

<u>v is big</u> : Then (B.2) does not affect (I.1). From (I.1) and (I.4) we conclude that v must be the root of a nontrivial red tree rT. v has at most $2b$ children by (I.3). Hence we can split v into only two small nodes; this, together with the fact that the f-edge of v must be blue, proves (I.2). However, we have to be careful about how to split v. Both new nodes must get at least a children to satisfy (I.3); on the other hand, both nodes must not get too many children because they must become blocking to satisfy (I.4). Lemma 3.3 below shows that it is always possible to split v such that both (I.3) and (I.4) are satisfied.

The red tree rT grows by the split; either $father(v)$ becomes its new root, or, if it was a red node of a red tree rT' before, rT becomes a subtree of the bigger tree rT'. In any case, (I.5) and (I.6) hold after the split.

(B.3) If v is the root of T before (B.3) then B is split. We now show that B was a trivial red tree before (B.3). Either B was a trivial red tree before (B.2) (and hence is also afterwards), or B was part of some red tree rT. Since v was the root of T, v was also the root of rT by (I.6). But v was not split in (B.2). Hence rT was coloured blue in (B.2) which left all its buckets as trivial red trees. Hence (I) holds after splitting B.

If v is not the root of T then r_B is moved up to $father(v)$. But this can only affect (I.6). With a similar argument as above we conclude that (I.6) is still true after (B.3) : either B was a trivial red tree before (B.3) (and therefore still is afterwards), or v was a node of the red tree rT containing B; if v was not coloured blue in (B.2) then $father(v)$ must also be in rT after (B.2). □

From this follows immediately

Theorem 3.2 *Algorithm A always maintains an $(a, 2b)$-tree T. It supports neighbor queries in time at most $(\lceil \log b \rceil + 2) \cdot \log n$ and insertions in time $O(1)$, once the position where the key is to be inserted in the tree is known. Also, deletions can be done in time $O(1)$ using the global rebuilding technique.*

Proof : We start with an (a, b)-tree T_0 and always maintain an $(a, 2b)$-tree T by Lemma 3.1. Hence the height of T is always bounded by $\log n$, and, doing a query, we can decide in time $\lceil \log b \rceil$ at which child of an internal node the search must be continued. And in the leaves, we can locate each key in time $2 \log n$ by Lemma 2.1. □

It remains to prove that we can always split a big node into two small nodes satisfying (I.3) and (I.4). This is an easy consequence from the following combinatorial Lemma (if blue edges are coded as $c_i = 1$ whereas pairs of red edges are coded as $c_i = 2$).

Lemma 3.3 Let $2a \leq b$, b even, $k \leq b$ and $c_1, \ldots, c_k \in \{1, 2\}$ with $b < \sum_{i=1}^{k} c_i$. Then an $j < k$ exists such that $a \leq \sum_{i=1}^{j} c_i$, $a \leq \sum_{i=j+1}^{k} c_i$, $2j \leq b$ and $2(k - j) \leq b$.

Proof : Let $j_1 := \min_j (\sum_{i=1}^{j} c_i \geq a)$ and $j_2 := \max_j (\sum_{i=j+1}^{k} c_i \geq a)$. From $\sum_{i=1}^{k} c_i \geq 2a+1$ and $c_i \in \{1, 2\}$ follows $j_1 \leq j_2$, $j_1 \leq \frac{b}{2}$ and $k - j_2 \leq \frac{b}{2}$. Let $j := \min(j_2, \frac{b}{2})$. Then $j_1 \leq j \leq j_2$, $2j \leq b$ and $2(k - j) \leq b$ (because $k \leq b$). □

4. Conclusions

We have seen how to implement a simple data structure which supports neighbor queries in time $O(\log n)$ and updates in time $O(1)$. However, as in [LO], our data structure can not be used as a finger tree. Hence, it remains an open problem to obtain a finger tree with only constant worst-case update time (which would have many more useful applications). However, using our tree recursively in the buckets (instead of the ordered lists), it is possible to obtain a data structure with $O(\log^* n)$ worst-case update time which also allows finger searches. This matches the best previous bounds ([HL],[H80]).

Another open problem is the question whether the query time can be reduced to $\log n$ (with factor 1); Andersson and Lai recently addressed this problem but could only find an optimal amortized solution ([AL]).

Our data structure seems to depend heavily on special properties of (a, b)-trees. It is not clear how to apply our techniques to other kinds of balanced search trees (e.g. BB[α], AVL,...). But we hope that it can be used in all kinds of efficient dynamic data structures (as in the triangulation example in Section 1).

Acknowledgements

We would like to thank Rajeev Raman for stimulating this research and Simon Kahan for reading a preliminary version of this paper.

References

[AL]   A. Andersson, T.W. Lai : "Comparison-efficient and write-optimal searching and sorting" *Proc. 2nd ISA* 1991, 273–282

[DR]   P.F. Dietz, R. Raman : "A constant update time finger search tree" *Advances in Computing and Information — ICCI '90, Lecture Notes in Computer Science*, Vol. 468, Springer 1990, 100–109 Also as Univ. of Rochester CS Dept. TR 321, December 1989

[DS]   P.F. Dietz, D.D. Sleator : "Two algorithms for maintaining order in a list" *Proc. 19th ACM STOC* 1987, 365–372

[DSST]   J.R. Driscoll, N. Sarnak, D.D. Sleator, R.E. Tarjan : "Making data structures persistent" *Journal of Computer and System Sciences* **38** *(1989)*, 86–124

[GMPR]   L. Guibas, E. McCreight, M. Plass, J. Roberts : "A new representation for linear lists" *Proc. 9th ACM STOC* 1977, 49–60

[H80]   D. Harel : "Fast updates with a guaranteed time bound per update" Technical Report 154, Dept. of ICS, University of California at Irvine, 1980

[HL]   D. Harel, G. Lueker : "A data structure with movable fingers and deletions" Technical Report 145, Dept. of ICS, University of Califonia at Irvine, 1979

[HM]   S. Huddleston, K. Mehlhorn : "A new data structure for representing sorted lists" *Acta Informatica* **17** *(1982)*, 157–184

[LO]   C. Levcopoulos, M.H. Overmars : "A balanced search tree with $O(1)$ worst-case update time" *Acta Informatica* **26** *(1988)*, 269–277

[M91]   K. Mulmuley : "Randomized multidimensional search trees : Dynamic sampling" *Proc. 7th Symposium on Computational Geometry* 1991, 121–131

[O82]   M.H. Overmars : "A $O(1)$ average time update scheme for balanced search trees" *Bull. EATCS* **18** *(1982)*, 27–29

[O83]   M.H. Overmars : "The design of dynamic data structures" *Lecture Notes in Computer Science*, Vol. 156, Springer 1983

[OvL]   M.H. Overmars, J. van Leeuwen : "Worst-case optimal insertion and deletion methods for decomposable searching methods" *Information Processing Letters* **12** *(1981)*, 168–173

[T83]   R.E. Tarjan : "Updating a balanced search tree in $O(1)$ rotations" *Information Processing Letters* **16** *(1983)*, 253–257

[T85]   A.K. Tsakalidis : "AVL-trees for localized search" *Information and Control* **67** *(1985)*, 173–194

[vE]   J. van der Erf : "Een datastructuur met zoektijd $O(\log n)$ en constante update-tijd (in Dutch)" Technical Report RUU-CS-87-19, Dept. of Computer Science, University of Utrecht 1987

MAPPING DYNAMIC DATA AND ALGORITHM STRUCTURES INTO PRODUCT NETWORKS

Sabine R. Öhring   and   Sajal K. Das *

Department of Computer Science   Department of Computer Science
University of Würzburg   University of North Texas
D 97074 Würzburg, Germany   Denton, TX 76203-3886, USA

Abstract. This paper presents optimal dynamic embeddings of dynamically growing or shrinking trees and three types of dynamically evolving grids into the de Bruijn graphs, and product networks such as (generalized) hypercube, hyper–de Bruijn, hyper Petersen, folded Petersen and product–shuffle networks.
Our results are important in mapping data and algorithm structures into multiprocessor interconnection networks. Tree embeddings can be used to maintain dynamic data structures such as quad–trees in image processing or data dictionaries, or to efficiently parallelize tree–based computations in divide–and–conquer or branch–and–bound algorithms. Dynamic embeddings of grids are used to parallelize solution methods for partial differential equations, for adaptive mesh refinement or hierarchical domain decomposition in approximation and interpolation of surfaces, image processing, or dynamic programming algorithms.

Keywords: de Bruijn, dynamic embedding, folded Petersen, grid, hypercube, hyper–de Bruijn, hyper Petersen, product–shuffle, tree.

1   Introduction

The performance of distributed–memory multiprocessors is profoundly affected by the underlying network topology. In this paper, we concentrate on a class of networks derived from the cartesian product operation, such as hypercubes, hyper–de Bruijn [1], product–shuffle [2], hyper Petersen [3], and folded Petersen network [4].
When the tasks of a parallel computation are mapped onto a multiprocessor system, the inter–task dependence structure is accomodated in the interconnection network. This mapping can be described graph–theoretically as embedding of a *guest* graph G, with the nodes representing the parallel processes or tasks and the edges representing the communication requirements among them, into a *host* graph H which represents the multiprocessor. An *embedding* of a graph $G = (V_G, E_G)$ into $H = (V_H, E_H)$ is formally defined by the tuple (f, g) with a node–mapping $f : V_G \rightarrow V_H$ and an edge–mapping $g : E_G \rightarrow \mathcal{P}_H$ (pathset of H), where $g(\{u, v\})$ connects $f(u)$ and $f(v)$ in V_H, for all $\{u, v\} \in E_G$. A *static* embedding maps the nodes of the task graph to the processors before the computation starts. Since task graphs often grow or shrink during computation, we present *dynamic* embeddings in this paper. In our context, *task allocation* means that all tasks remain on their host nodes during execution, contrary to *task migration* . While static embeddings of many data and algorithm structures in the de Bruijn and such product networks as hypercubes, hyper–de Bruijn, hyper Petersen and folded Petersen have been studied [4, 5, 6, 7, 8, 9, 10], dynamic embeddings of trees and dynamic grid

---

* This work is partially supported by Texas Advanced Research Program Grant under Award No. 003594003. The authors can be reached via E-mail at oehring@informatik.uni-wuerzburg.de and das@cs.unt.edu

structures in these networks (except dynamic tree embeddings in hypercubes [5]) have not gained attention, which motivates this paper. The *dilation* of the embedding (f, g) is the maximum distance in the host between the images of adjacent guest–nodes. The *expansion* is the ratio $\frac{|V_H|}{|V_G|}$, while the *load* is the maximum, over all host–nodes, of the number of guest–nodes mapped to a host node. The *edge–congestion* is the maximum number of edges of G that are routed by the edge–mapping g over a single edge of H.

2 Notations and Definitions

For $x = x_{n-1} \ldots x_0 \in \mathbb{Z}_2^n$, let $x(i) = x_{n-1} \ldots x_{i+1} \overline{x_i} x_{i-1} \ldots x_0$, $0 \le i \le n-1$, with $\overline{x_i}$ the complementary bit of x_i. $dist_G(x, y)$ is the distance of nodes x and y in graph G. The *binary hypercube* $Q_n = (V_n^Q, E_n^Q)$ of order n consists of the node–set $V_n^Q = \mathbb{Z}_2^n$ and there exists an edge $\{x, y\} \in E_n^Q$ between two nodes x and y iff their binary representations differ exactly in one bit. Fig. 1a) depicts Q_3. A *generalized hypercube* topology [11], of base b and dimension n is a graph $GQ_b^n = (V, E)$, where $V = \mathbb{Z}_b^n = \{x_{n-1} \ldots x_0 \mid x_i \in \mathbb{Z}_b, \ 0 \le i \le n-1\}$ and $E = \{(x_{n-1} \ldots x_0, y_{n-1} \ldots y_0) \mid \exists j, 1 \le j \le n, x_j \ne y_j \text{ and } x_i = y_i \text{ for } i \ne j\}$.

The undirected binary *de Bruijn* graph, $DG(2, n) = (V_n^D, E_n^D)$, of order n is defined by $V_n^D = \mathbb{Z}_2^n$ and $E_n^D = \{\{x_{n-1} \ldots x_0, x_{n-2} \ldots x_0 p\} \mid p, x_i \in \mathbb{Z}_2, 0 \le i \le n-1\}$. Fig. 1b) depicts $DG(2, 3)$. The *Shift–Left–neighbor* of $x = x_{n-1} \ldots x_0$ is defined as $SL(x) = x_{n-2} \ldots x_0 x_{n-1}$, while the *Shift–Exchange–Left–neighbor* as $SEL(x) = x_{n-2} \ldots x_0 \overline{x_{n-1}}$.

There are several graph–theoretic operations which enable us to aggregate large networks from smaller networks of known properties. One such operation, the so called *cartesian product* of two graphs $G_1 = (V_1, E_1)$ and $G_2 = (V_2, E_2)$, is defined as $G = G_1 \times G_2$ having the node–set $V = V_1 \times V_2$. There exists an edge $\{(u_1, u_2), (v_1, v_2)\}$ in the product network G iff either $u_1 = v_1$ and $(u_2, v_2) \in E_2$ or $u_2 = v_2$ and $(u_1, v_1) \in E_1$. The product of n graphs, for $n \ge 2$, can be defined analogously. By definition, the hypercubes $Q_n = Q_{n-1} \times K_2$ and $GQ_b^n = GQ_b^{n-1} \times K_b$, where K_b is a complete graph of order b, are product networks. The order–(m, n) *hyper–de Bruijn* graph [1] is the product graph $HD(m, n) = Q_m \times DG(2, n)$. For a node (x, y) of $HD(m, n)$, x denotes the hypercube–part–label and y the de Bruijn–part–label. The order–(m, n) *product–shuffle* graph [2] is defined as $PS(m, n) = DG(2, m) \times DG(2, n)$.

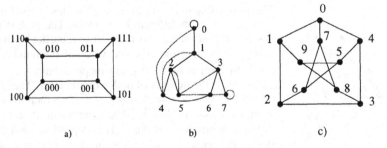

a)   b)   c)

Fig. 1. a) Q_3 b) $DG(2, 3)$ c) Petersen graph P

The *Petersen graph*, P, [12] with ten nodes has an outer and inner 5-cycle and five spokes joining them (see Fig. 1c)). The *n-folded Petersen graph*, $FP_n = P \times \ldots \times P = FP_{n-1} \times P$, introduced by us in [4] is the iterative cartesian product on the Petersen graph, while the *n-dimensional hyper Petersen network* is defined as $HP_n = Q_{n-3} \times P$ [3]. A generalization of both networks is the folded Petersen cube $FPQ_{n,k} = Q_n \times FP_k$, we proposed in [10].

3 Mapping Dynamic Trees

In many applications the tree structure is not complete, and may grow/shrink dynamically as the computation progresses in a way that is not known apriori. For example, a dictionary maintains a dynamic set of data elements, where the existing data can be deleted or new data can be inserted. Also, divide–and–conquer or branch–and–bound algorithms have dynamic tree structures, where a problem is recursively split into subproblems.

In this section, we describe dynamic task–allocation embeddings of dynamically evolving trees in various hosts under consideration. The embedding of dynamic trees for the hypercube and butterfly by a randomized flip–bit algorithm was first studied by Leighton et al. [5].
Following a similar technique, we show the results in Table 1. The phrase "Q is less than $\mathcal{O}(\beta)$ with high probability" expresses that for every c there exists a constant k independent of N such that the probability that Q exceeds $k\beta$ is less than N^{-c}.

Table 1: Dynamic Tree Embedding

Host	dil.	load (with high probability)
$DG(2,n)$	1	$\mathcal{O}(\frac{M}{2^n}+n)$
FP_n	1	$\mathcal{O}(\frac{M}{10^n}+n)$
$HD(m,n)$	2	$\mathcal{O}(\frac{M}{2^{m+n}}+m\cdot(m+n))$
HP_n	2	$\mathcal{O}(\frac{M}{2^n+2^{n-2}}+n^2)$
$PS(m,n)$	1	$\mathcal{O}(\frac{M}{2^{m+n}}+(m+n)^2)$
GQ_b^n	1	$\mathcal{O}(\frac{M}{b^n}+n)$

3.1 de Bruijn Host

Theorem 1. : *There exists a dynamic embedding of an arbitrary binary tree $T(M)$ with M nodes into an $N = 2^n$–node $DG(2,n)$ with dilation 1, and high probability load of at most $\mathcal{O}(\frac{M}{N}+\log^2 N)$ and edge–congestion of at most $\mathcal{O}(\frac{M-1}{2N}+\log^2 2N)$ [7].*

Proof : The node–mapping f starts with mapping the root of the dynamically evolving tree $T(M)$ to the processor $(0)^{n-1}1$ of $DG(2,n)$. In $T(M)$ of height h, for each node v a flip–bit, flip(v), is generated at random that has equally likely the value 0 or 1. If flip$(v) = 0$, then the left son(v) is mapped to $SL(f(v))$ and the right son(v) to $SEL(f(v))$. If flip$(v) = 1$, the children of node v are mapped in the reverse fashion. Thus, f has dilation 1. We divide the dynamic tree in subtrees T_j, $1 \leq j \leq t$, that are rooted at a node in level $i + k \cdot n$ and consist of all descendants between levels $k \cdot n + i + 1$ and $(k+1) \cdot n + i$ as sketched in Fig. 2.

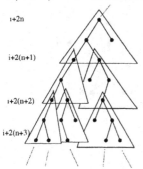

The embedding maps the at most 2^n nodes in level n of T_j, $0 \leq j \leq t$, on different host–nodes. The probability that an arbitrary node x is mapped to a particular host–node y is $\frac{1}{2^n}$, and the expected value of the number of nodes in level n of T_j that are mapped onto y is $\frac{m_j}{2^n}$, where m_j is the number of nodes in that level. Thus, the probability that a node in level n of T_j is mapped to a given processor, is also $\frac{m_j}{2^n}$. The assignment of nodes in different subtrees T_j is mutually independent (and thus Bernoulli events), because a host–node $f(x)$ is generated at random dependent only on the last n flip–bits, which are again picked independently in different T_j's.

Figure 2: Partition of $T(M)$

The proof is completed similar as in [5] with

Lemma Hoeffding : *For N independent Bernoulli trials with probabilities $p_1, \ldots, p_N$ such that $pN = \sum p_i$, the probability of at least m successes is at most $B(m,N,p) \leq (\frac{N \cdot p \cdot e}{m})^m$, where $m \geq pN + 1$ is an integer.*

3.2 Folded Petersen Host

Let f denote the required mapping of the nodes of the dynamically growing tree $T(M)$ in the folded Petersen network, FP_n. The root is mapped to the node $(0, \ldots, 0)$ of FP_n. Each tree–node x is assigned a track number, $t(x)$. If level$(x) = 0$, then $t(x) = 1$. And if $y = $ child(x), then $t(y) = t(x)$ if level$(y) = $ odd, otherwise $t(y) = (t(x) \bmod n) + 1$. Also x is assigned a random flip bit, $flip(x)$, that has equally likely the value 0 or 1. If x is mapped on $f(x)$ in FP_n, then its sons are embedded to neighbors of $f(x)$ as follows. For flip$(x) = 0$, $f($leftson$(x)) := (f(x)^{t(x)})'$ and $f($rightson$(x)) := (f(x)^{t(x)})''$. Otherwise, the assignments are interchanged. Thereby, $(f(x)^{t(x)})'$ and $(f(x)^{t(x)})''$, respectively, denote the first and second neighbors of $f(x)$ in the $t(x)$-th direction, where y_i is the element in direction i of $y = (y_1, \ldots, y_n)$. The first neighbor of $z = (z_1, \ldots, z_n)$ in the i–th direction is defined as $(z^i)' = (z_1, \ldots, z_{i-1}, z_i', z_{i+1}, \ldots, z_n)$. The second neighbor is defined similarly. In the Petersen graph P, the first and second neighbors (z_i' and z_i'')

of z_i, where $i = t(x)$, of a node $z = f(x)$, are chosen according to Table 2, iff $t(x) \neq t(y)$ and $y = $ father(x) in $T(M)$. Otherwise, we choose the two neighbors of z_i that are not equal to $(f(y))_i$, such that the first neighbor has the smaller label.

Table 2: First and second neighbors (v_1 and v_2) of nodes v in P

v	0	1	2	3	4	5	6	7	8	9
v_1	1	2	3	4	0	6	7	8	9	5
v_2	7	9	6	8	5	4	2	0	3	1

Theorem 2. : *Any dynamically evolving binary tree $T(M)$ can be dynamically embedded in FP_n with dilation 1 and load (with high probability) of at most $\mathcal{O}(\frac{M}{10^n} + n)$.*

Proof : We divide $T(M)$ in subtrees $T_1, \ldots, T_l$, such that the root of each T_j, $1 \le i \le l$, lies in a level $k \cdot n + i$, for k even and $0 \le k \le \frac{h-i}{n}$, where h denotes the height of $T(M)$. The subtrees consist of the root and all successors between levels $k \cdot n + i$ and $(k+2) \cdot n + i$. Each T_j has height $2n$ and m_j nodes. The flip–bit algorithm guarantees that the number of nodes in T_j placed on processor x is a random variable with value 0 or 1. The probability that any node from T_j is placed on x is $\frac{1}{10^n}$. Therefore, the probability that a node from T_j is mapped on x is $\frac{m_j}{10^n}$. Since the flip–bits in each subtree are generated independently, mapping nodes in different subtrees to processors in FP_n is mutually independent. The use of the Lemma of Hoeffding completes the proof. $\square$

3.3 Hyper–de Bruijn Host

Theorem 3. : *There exists a dynamic embedding of an arbitrarily evolving binary tree $T(M)$ in $HD(m, n)$ with dilation 2 and high probability load $\le \mathcal{O}(\frac{M}{2^{n+m}} + m \cdot (m+n))$.*

Proof : We first embed $T(M)$ dynamically in the intermediate host, $BF(m) \times DG(2, n)$, using the dynamic tree embedding in a butterfly $BF(m)$ by Leighton et al. [5], and in de Bruijn networks as in Section 3.1. Then $BF(m) \times DG(2, n)$ is embedded in $Q_m \times DG(2, n)$ with load m and dilation 1 by the mapping $\varphi((i, x, y)) = (x, y)$, where $0 \le i \le m-1$, $x \in \mathbb{Z}_2^m$ and $y \in \mathbb{Z}_2^n$. The node–mapping f of the dynamic tree $T(M)$ into the intermediate host starts with mapping the root into node $(0, (0)^m, (0)^{n-1}1)$. Again each tree–node x is assigned a random flip–bit, $flip(x) \in \{0, 1\}$. In each step of our algorithm, depending only on the level–number of x, its sons are mapped to either the de Bruijn– or butterfly–neighbors of $f(x) = (i, y, z)$, for $0 \le i \le m-1$, $y \in \mathbb{Z}_2^m$, $z \in \mathbb{Z}_2^n$, as given in Table 3. The following scheme describes for which level–numbers the sons of x are mapped to appropriate neighbors of $f(x)$. Assume that $n \le m$ and $s = \frac{m}{n} \ge 1$. (For $n > m$, exchange the role of de Bruijn and butterfly neighbors.)

Table 3. Mapping the sons of x

flip(x)	neighbor in ...	f(left son(x))	f(right son(x))
0	$BF(m)$	$(i+1 \bmod m, y, z)$	$(i+1 \bmod m, y(i), z)$
1	$BF(m)$	$(i+1 \bmod m, y(i), z)$	$(i+1 \bmod m, y, z)$
0	$DG(2, n)$	$(i, y, SL(z))$	$(i, y, SEL(z))$
1	$DG(2, n)$	$(i, y, SEL(z))$	$(i, y, SL(z))$

Algorithm : Selecting the type of neighbors
$i := 0$; $z_0 := 0$;
Repeat until all nodes of $T(M)$ are embedded
begin for $j := 1$ to n **do begin**
    Map the nodes y in level $i \cdot (m+n) + z_{j-1} + j$ to de Bruijn neighbors of $f(x)$,
    with $x = \text{father}(y)$ in $T(M)$;
    Map the nodes in levels $i \cdot (m+n) + z_{j-1} + j + 1$ through $i \cdot (m+n) + \lfloor j \cdot s \rfloor + j$
    to the butterfly neighbors; $z_j := \lfloor j \cdot s \rfloor$;
end;
$i := i + 1$
end.

Using an $(m+n)$–way level–balancing transformation on the dynamic tree [5], a tree $T'(M)$ is generated from $T(M)$ such that with high probability the total number W_i of vertices in levels $i+k \cdot (m+n)$, for $0 \le k \le \frac{h-i}{m+n}$, of $T'(M)$ is at most $\mathcal{O}(\frac{M}{m+n} + 2^{m+n})$. Next, we partition $T'(M)$ into subtrees $T_1, T_2, \ldots, T_l$, where each subtree is rooted at some vertex in level $k \cdot (m+n) + i$, for $0 \le i \le n+m-1$, and consists of all successors of that vertex between levels $k \cdot (m+n) + i + 1$ and $(k+1) \cdot (m+n) + i$ and complete the proof analogously to the embedding in de Bruijn graphs. $\square$
For the embedding in a Hyper Petersen, Product Shuffle or generalized hypercube GQ_d^n network we proceed along the line of dynamic tree embedding in Hyper de Bruijn or [5] to show the results in Table 1. For details we refer to [13].

4 Mapping Dynamic Grids

4.1 Type 1 Grids

This type of dynamic grids is a hierarchy of coarser grids and arises in multigrid algorithms for solving partial differential equations.

Definition 4. : A (two–dimensional) dynamic grid $dG_{m,n}^1$ of Type 1 having maximal side–lengths 2^m and 2^n in the x and y directions, respectively, is a hierarchy of grids starting with the grid $M^0(2^m, 2^n)$ of node–set $V = \{(x, y) \mid 0 \le x \le 2^m - 1, 0 \le y \le 2^n - 1\}$, and a sequence of successively coarser grids $M^{i+1}(2^{(m-i-1)+}, 2^{(n-i-1)+})$ that results from $M^i(2^{(m-i)+}, 2^{(n-i)+})$ by deleting every second point in the set of x and y coordinates, iff more than two points are left. ($k_+ = k$ for $k \ge 1$, otherwise $k_+ = 1$.)

Fig. 3 shows a dynamic grid $dG_{2,3}^1$ of Type 1. We have to consider that neighboring points in all M^j, $0 \le j \le \max\{n-1, m-1\}$ are mapped onto processors as close as possible. Chan and Saad [14] studied the mapping of such dynamic grids in hypercubes, showing that $dG_{m,n}^1$ can be embedded in the binary hypercube Q_{m+n} with load 1 and dilation 2. This type of grids can also be studied for maximal side lengths b^m and b^n, where only each bth node is taken for the next coarser grid. Using the generalized Gray codes due to [15]

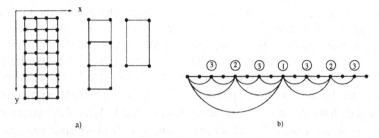

Fig. 3. a) Dynamic grid $dG_{2,3}^1$ of Type 1   b) Neighborhood–graph NG_x^{i3} for $dG_{4,4}^1$

defined for the generalized hypercubes GQ_b^k of base b, along with the fact that there exists a cyclic Gray code $G_b^{ik} = \{g_0, \ldots, g_{b^k-1}\}$ for even b satisfying $dist_{G_b^k}(g_i, g_{(i+b^j) \bmod b^k}) = 2$ for $0 < j < k$ and $dist_{G_b^k}(g_i, g_{(i+1) \bmod b^k}) = 1$, we can show

Theorem 5. : A k–dimensional dynamic grid of Type 1 with maximal side length b^{n_i}, $1 \leq i \leq k$, such that $\sum_{i=1}^{k} n_i = n$, b even, can be embedded dynamically in GQ_b^n with load 1 and dilation 2.

Theorem 6. : A k–dimensional dynamic grid $dG_{n_1,\ldots,n_k}^1$ can be embedded dynamically in HP_s with load 1 and dilation 2, where $\sum_{i=1}^{k} n_i = s$.

Proof : Let $k = 2$, $s = m + n$, and S_x^i be the set of x–coordinates in the i–th step of constructing $dG_{m,n}^1$, for $0 \leq i \leq m - 1$. It holds $S_x^i = \{0, 1, \ldots, 2^m - 1\} \cap \{k \cdot 2^i \mid 0 \leq k \leq \lfloor \frac{2^m-1}{2^i} \rfloor\}$, where $0 \leq i \leq m - 1$. Analogously, the set S_y^j of y–coordinates, for $0 \leq j \leq n - 1$, is defined. Similar to the approach in [8], the concept of *neighborhood–graphs* is used for mapping the grid–nodes onto processors with close proximity.

Definition 7. : The *neighborhood–graph* (for the x–coordinates) corresponding to a dynamic grid $(M^j, 0 \leq j \leq n)$ with the set S_x^{k} of x–coordinates in the k-th step, $0 \leq k \leq n$, is the graph $NG_x^{ik} = (V_{NG_x^k}, E_{NG_x^k})$. Here $V_{NG_x^k} = \bigcup_{k=0}^{n} S_x^k$ and $(x_{i_1}, x_{i_2}) \in E_{NG_x^k}$ iff x_{i_1} and x_{i_2} are distinct x–coordinates of grid–points (x_{i_1}, y_{i_1}) and (x_{i_2}, y_{i_2}) of M^k such that there exists an i, where $0 \leq i \leq k$, and $\{(x_{i_1}, y_{i_1}), (x_{i_2}, y_{i_2})\}$ is an edge in M^i. (Similary, the neighborhood graph NG_y^{ik} for the y–coordinates can be defined.)

Fig. 3 represents the graph NG_x^{i3} for $dG_{4,4}^1$. Our embedding uses the m–bit binary reflected Gray codes $\mathcal{G}_2^m = \{\mathcal{G}_2^m(0), \ldots, \mathcal{G}_2^m(2^m - 1)\} = \{g_0, \ldots, g_{2^m-1}\}$, recursively defined as $\mathcal{G}_2^1 = \{0, 1\}$, $\mathcal{G}_2^{m+1} = \{g_0 0, g_0 1, g_1 1, g_1 0, g_2 0, g_2 1, \ldots, g_{2^m-1} 1, g_{2^m-1} 0\}$. Since $HP_{m+n} = Q_m \times (Q_{n-3} \times P)$, we define the node–mapping $f((x, y)) = f_1(x) f_2(y)$ for a vertex (x, y) in $dG_{m,n}^1$. Let $f_1(x) = \mathcal{G}_2^m(x)$. Due to the properties of $\mathcal{G}_2^m$, the neighborhood graph NG_x^{m-1} is mapped into Q_m with load 1 and dilation 2. For f_2, we define the coding $C_{n-3} = \{c_{n-3}(0), \ldots, c_{n-3}(2^n - 1)\} = \{(g_0', 0), (g_0', 6), (g_0', 2), (g_0', 3), (g_0', 4), (g_0', 5), (g_0', 1), (g_0', 7), (g_1', 0), (g_1', 6), (g_1', 2), (g_1', 3), (g_1', 4), (g_1', 5), (g_1', 1), (g_1', 7), \ldots, (g_{2^{n-3}-1}', 0), \ldots, (g_{2^{n-3}-1}', 7)\}$, where $g_i' := \mathcal{G}_2^{n-3}(i)$ denotes the binary reflected Gray codes. One can easily prove that C_{n-3} maps the neighborhood–graph NG_y^{m-1} on $Q_{n-3} \times P$ with dilation 2 and load 1. Both dilation and load of 1 cannot be satisfied simultaneously, since HP_s does not contain a 3–cycle as subgraph.

Theorem 8. : A dynamic grid $dG_{m,n}^1$ can be dynamically embedded in the hyper–de Bruijn network, $HD(m, n)$, with load $\mathcal{O}(1)$ and dilation $\mathcal{O}(\log n)$, and in the product–shuffle network $PS(m, n)$, with load $\mathcal{O}(1)$ and dilation $\mathcal{O}(\max\{\log m, \log n\})$.

Proof : Here we provide the proof for $HD(m,n)$ as host. The neighborhood–graph NG_x^{m-1} is mapped in Q_m using $f_1(i) = \mathcal{G}_2^m(i)$ for $0 \le i \le 2^m - 1$. For all nodes v in NG_y^{n-1}, the degree $deg(v) \le 2n - 1$ since in each step i, that is in $M^i(2^{m-i}, 2^{n-i})$, for $0 \le i \le n-1$, an y–coordinate gets at most two new neighbored y–coordinates. The graph NG_y^{n-1} has an $(\frac{1}{2}, \frac{1}{2})$–node–separator (cf. [16]) of size 1. The deletion of the node numbered by 2^{n-1} from S_y^0 partitions NG_y^0 in two subgraphs with at most $\frac{1}{2}|S_y^0|$ nodes and each having an $(\frac{1}{2}, \frac{1}{2})$–node separator of size 1. Again these graphs can be partitioned by deleting the (2^{m-i})th smallest element in the corresponding sets of y–coordinates, for $2 \le i \le m-1$. In Fig. 3, a circled label on a node gives the time–step when it is used as a separator. With Theorem 9, it follows the existence of a mapping f_2 of NG_y^{n-1} in $DG(2,n)$ with load $\mathcal{O}(1)$ and dilation $\mathcal{O}(\log n)$. The desired embedding is $f((x,y)) = f_1(x)f_2(y)$. $\square$

Theorem 9. *[16]: An n–node graph with maxdegree–d and a node–separator of size $S(x)$ can be embedded in a de Bruijn graph with simultaneous load 1, dilation $\mathcal{O}(\log(d \cdot \sum_i S(\frac{n}{2^i})))$ and expansion $\mathcal{O}(1)$.*

4.2 Type 2 Grids

This type of dynamically evolving grids (*quasi–grids*) arises e.g. in *Adaptive Mesh Refinement* (AMR) algorithms which dynamically match the local resolution of a computational grid to the numerical solution being sought.

Definition 10. [17]: A dynamic grid of Type 2, denoted as dG^2, incrementally grows starting from the quasi-grid QG^0, bounded by the vertices $(0,0), (1,0), (1,1)$ and $(0,1)$, and by inserting division–points P_i (see Fig. 4). In the ith step, a division–point $P_i = (\frac{x_1+x_2}{2}, \frac{y_1+y_2}{2})$ is inserted in an *arbitrary* subsquare S_{i-1} of QG^{i-1}, bounded by the vertices $i_1 = (x_1, y_1)$, $i_2 = (x_1, y_2)$, $i_3 = (x_2, y_2)$ and $i_4 = (x_2, y_1)$. Thus, $j_1 = (x_1, \frac{y_1+y_2}{2}), j_2 = (\frac{x_1+x_2}{2}, y_2), j_3 = (x_2, \frac{y_1+y_2}{2})$ and $j_4 = (\frac{x_1+x_2}{2}, y_1)$ are inserted.

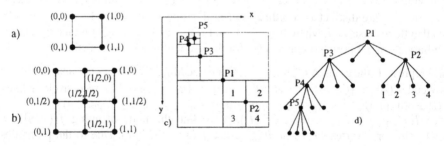

Fig. 4. a) QG^0　　b) QG^1　　　　c) Quasi–grid　　　　d) Quasi–tree for (c)

If a quasi–grid is considered as a region divided into subregions, the evaluation of the quasi–grid can be expressed by a quad–tree (cf. Fig. 4d)). QG^0 corresponds to the quad–tree consisting of a single node. If a division point P_i, $i \ge 1$, is inserted in a subsquare S_i of QG^{i-1}, then four sons are generated for the quadtree–node corresponding to S_i. The jth son, $1 \le j \le 4$, corresponds to subregion j as in Fig. 4 c) and d) for point P_2. With respect to the applications, the quad–tree is assumed to be height–balanced.

Definition 11. : A quad–tree $QT = (V_{QT}, E_{QT})$ is *height–balanced*, if $h(T_{v'}) = \Theta(\log |T_v|)$ for all $v \in V_{QT}$ and $\forall v' \in \mathcal{N}(v)$, the set of children of a node $v \in V_{QT}$. Here, $h(T)$ denotes the height of the tree and T_v the subtree of QT rooted at v.

Theorem 12. : *There exists a dynamic embedding f of a dynamically evolving quasi–grid QG with $|S_x|$ different x–coordinates and $|S_y|$ different y–coordinates (corresponding to a height–balanced quad–tree QT) in $HD(m,n)$ with load of at most $\mathcal{O}(\max\{1, 2^{\lceil \log|S_x|\rceil + \lceil \log|S_y|\rceil - m - n}\})$ and dilation of at most $\mathcal{O}(\log\log|S_y|)$. QG can be embedded dynamically in Q_n or HP_n with dilation of at most 2 and load of at most $\mathcal{O}(\max\{1, 2^{\lceil \log|S_x|\rceil + \lceil \log|S_y|\rceil - n}\})$.*

Proof : For $HD(m,n)$ as host, the x–coordinates of the grid points are embedded by the *binary reflected Gray codes* G_2^m, and the y–coordinates are mapped in $DG(2,n)$. Let S_y be the set of y–coordinates in the quasi–grid at any step, starting with $S_y^0 = \{y_1, y_2\}$, where $y_1 = 0$ and $y_2 = 1$. If the division–point $P_k = (\tilde{x}_k, \tilde{y}_k)$ is inserted in the k–th step, then $S_y^k = S_y^{k-1} \cup \{\tilde{y}_k\}$ for $\tilde{y}_k \notin S_y^{k-1}$, $S_y^k = S_y^{k-1}$ else. Fig. 5a) shows the neighborhood–graph NG_y^4 for the y–coordinates of the quasi–grid in Fig. 4c), where we use i instead of y_i. Let $h = \Theta(\log N)$ denote the height of the quad–tree, QT (having

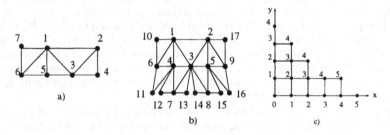

Fig. 5. a) Neighborhood–graph NG^4 b) NG_{max}^4 c) Dynamic grid of Type 3

$|V_{QT}| = N$ nodes), corresponding to the quasi–grid, QG, to be embedded. We construct the neighborhood–graph $NG_{max} = NG_{max}^h$ for the complete 4–ary, quad–tree of height h (cf. Fig. 5b)). Clearly, the neighborhood–graph NG of QT is isomorphic to a subgraph of NG_{max}. Also, each node $x \in NG_{max}$ has degree $deg(x) \leq 2h = \mathcal{O}(\log N)$ and NG_{max} has an $(\frac{1}{2}, \frac{1}{2})$–node separator of size $S(x) = 2$, if $x = |V_{NG_{max}}|$; otherwise $S(x) = 1$. Applying Theorem 9, NG_{max} and hence NG can be embedded in the de Bruijn host with load 1, dilation $\mathcal{O}(\log(h^2)) = \mathcal{O}(\log\log N)$, and expansion $\mathcal{O}(1)$. We can show that $|S_y| = cN^\alpha$ for $\alpha \leq 1$ and therefore the neighborhood–graph NG for the y–coordinates is embedded in the de Bruijn host $DG(2, \lceil \log|S_y|\rceil)$ with load $\mathcal{O}(1)$ and dilation $\mathcal{O}(\log\log|S_y|)$. Thus there exists an embedding f_2 of the neighborhood–graph NG with $|S_y|$ nodes into $DG(2,n)$ with simultaneous load $\mathcal{O}(\max\{1, 2^{\lceil \log|S_y|\rceil - n}\})$ and dilation $\mathcal{O}(\log\log|S_y|)$.

The mapping of the x–coordinates S_x starts with $f_1(0) = g_0, f_1(1) = g_0$, and $f_1(\frac{x_k + x_l}{2}) = g_{\lfloor \frac{i_k + i_l}{2}\rfloor}$, where $f_1(x_k) = g_{i_k}, f_1(x_l) = g_{i_l}$, and $G_2^m = \{g_0, \dots, g_{2^m - 1}\}$, the binary reflected Gray code. The embedding of the quasi grid in $HD(m,n)$ is then realized by $f(x,y) = f_1(x)f_2(y)$ together with a load balancing scheme. Details are given in [13].
□

4.3 Type 3 Grids

Definition 13. : The (2–dimensional) dynamic grid of Type 3, denoted as dG^3, starts with the node $\{(0,0)\}$ in dG_0^3. In each step $i \geq 1$, any of the nodes (x,y) with $x + y = i$ can be inserted simultaneously in the grid iff the nodes $(x-1, y), (x, y-1)$ and $(x-1, y-1)$ exist in dG_{i-1}^3. The resulting grid after step i is dG_i^3.

This type of grids arises in dynamic programming algorithms, which solve a problem by combining the solutions to subproblems, which in turn share the same subsubproblems. Every subproblem is solved just once and then its solution is stored in a table. Examples include optimization problems such as matrix–chain multiplication, polygon triangulations and longest common subsequence. Fig. 5b) shows a specific instance of dG^3. The labels on the nodes give the time steps of their generation.

Theorem 14. : A dynamic grid dG^3 can be embedded in Q_n, GQ_b^n, HP_n, FP_n and $HD(m,n)$ (for $m \leq n$) with dilation 1 and load of at most $\frac{10}{3}$ times the optimal load.

Proof : We give here only the proof for Q_n as host. Further results are shown similarly in [13]. The x and y coordinates of the grid are encoded by the Gray codes generated the by Gray–code transition sequences (GCTS) defined in Greenberg et al. [18]. Let $D = \{0, 1, \ldots, n-1\}$ be the set of bit–positions in the node labels of the hypercube Q_n, and $\mathcal{P} = \{p_0, p_1, \ldots, p_t\}$ be a set of pairwise disjoint bit–positions $0 \leq p_i \leq n-1$, $0 \leq i \leq t$. For any $r \leq t$, the rth GCTS specified by $\mathcal{P}$ and denoted by $GS[r; \mathcal{P}]$, is the sequence of integers of length $(2^r - 1)$ and defined inductively by : $GS[1; \mathcal{P}] = p_0$ and $GS[k+1; \mathcal{P}] = GS[k; \mathcal{P}], p_k, GS[k; \mathcal{P}]$. Let the ith element of $GS[r; \mathcal{P}]$ be denoted as $GS[r; \mathcal{P}]_i$, where the least value of i is 0. Starting with an arbitrary node y_0 in Q_n, we get a Gray code sequence $(y_0, \ldots, y_{2^r-1})$ of length 2^r when successively generating y_{i+1} by flipping bit–position $GS[r, D]_i$ of y_i for $0 \leq i \leq 2^r - 1$.

The node $(0, 0)$ of dG^3 is embedded by the node–mapping f to node $(0)^n$ of Q_n. Let D_i, $i \in \{0, 1\}$, denote the bit–positions, such that the Gray–code transition sequences $GCTS_i$ for encoding the nodes $(x, 0)$ and $(0, y)$, use only bits in D_0 and D_1, respectively. Algorithm (GCTS's for dG^3) computes the GCTS's. Let the nodes $(x-1, y)$, $(x, y-1)$ and $(x-1, y-1)$ be already embedded in Q_n by f, where $f((x-1, y-1))$ and $f((x, y-1))$ differ in bit–position i while $f((x-1, y-1))$ and $f((x-1, y))$ differ in bit–position j. $f((x, y))$ is then mapped to $f((x-1, y))(j)$, i.e. bit j is complemented.

```
Algorithm : GCTS's for dG³ ;
GCTS₀ = GCTS₁ = ∅;
D₀ = D₁ = {0, 1, . . . , n − 1}; k = 1; flag = 0;
While (nodes with new x or y coordinate are inserted) and
      (a new bit in GCTSᵢ, i ∈ {0, 1}, is needed) do
   begin   D̃ᵢ := Dᵢ; Dᵏᵢ := ∅;
      while D̃ᵢ ≠ ∅ do begin
            choose a new bit–position z ∈ D̃ᵢ;
            D̃ᵢ := D̃ᵢ − {z}; D̃₁₋ᵢ := D̃₁₋ᵢ − {z};
            Dᵏᵢ := Dᵏᵢ ∪ {z};
            GCTSᵢ = GCTSᵢ, z, GCTSᵢ;
      end;
      k = k + 1;
      if (flag = 0) then Dᵢ = Dᵢ¹ for i = 0, 1; flag = 1;
   end.
```

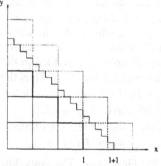

Figure 6: Rectangles with load 1

Each time $\tilde{D}_i$ gets emptyset for $i \in \{0, 1\}$, a new rectangle of grid nodes can be defined, on which the node–mapping f has load 1.

We assume that the dynamic grid that proceeds in diagonals is complete until step l, but step $l + 1$ is incomplete (cf. Fig. 6). Thus, for the number M of nodes in dG_3 holds : $(\sum_{i=1}^{l} i) \cdot N \leq M \leq (\sum_{i=1}^{l} i+1) \cdot N$. Let l be the largest number of the diagonal of rectangles completely filled with nodes of grid dG^3. The optimal load for embedding this grid is $L_{opt} \geq \sum_{i=1}^{l} i \geq 2l - 1$ whereas the real load is $L \leq L_{opt} + 2l + 3$. Therefore, we have : $\frac{L}{L_{opt}} \leq \frac{L_{opt}+2l+3}{L_{opt}} \leq 1 + \frac{2l+3}{2l-1} \leq \frac{10}{3}$, assuming $l \geq 2$. □

5 Conclusion

We presented optimal (within constant factors), dynamic task allocation embeddings of dynamically evolving binary trees and three types of grid structures into the de Bruijn graph and representative product networks – namely, (generalized) hypercubes, hyper–de Bruijn, hyper Petersen, folded Petersen and product–shuffle.
Dynamic trees arise in divide–and–conquer or branch–and–bound algorithms. The dynamic grid structures considered in this paper cover many applications, such as adaptive mesh refinement, hierarchical domain decomposition or dynamic programming methods. The mapping of dynamic quasi–grids for the case of an unbalanced quadtree or in the case that the growing pattern is known beforehand and the Type 3 grids that are arbitrary growing, will be analyzed in future research. A further direction is to search for other types of dynamic and irregular data–and–algorithm structures.

References

1. E. Ganesan and D.K. Pradhan. The hyper-de bruijn multiprocessor networks. In *Proc. 11th Conf. on Distributed Computing Systems*, pp. 492–499, Arlington, TX, May 1991.
2. A.L. Rosenberg. Product–shuffle networks : Towards Reconciling Shuffles and Butterflies. *Discrete Applied Mathematics* , vol. 37/38, July 1992.
3. S.K. Das and A.K. Banerjee. Hyper petersen network : Yet another hypercube–like topology. In *Proc. 4th Symp. on the Frontiers of Massively Parallel Computation (Frontiers' 92)*, pp. 270 – 277, McLean, Virginia, USA, Oct. 1992.
4. S. Öhring and S.K. Das. The Folded Petersen Network : A New Communication–Efficient Multiprocessor Topology. In *Proc. 1993 Int. Conf. on Parallel Processing., vol. I*, pp. 311 – 314, Aug. 1993.
5. F.T. Leighton, M.J. Newman, A.G. Ranade, and E.J. Schwabe. Dynamic Tree Embeddings in Butterflies and Hypercubes. *SIAM Journal on Computing*, 21(4):639 – 654, Aug. 1992.
6. S.K. Das and S. Öhring. *Embeddings of Tree–Related Topologies in Hyper Petersen Networks* . Technical Report CRPDC-92-16, Univ. North Texas, Univ. Wuerzburg, Sep. 1992.
7. S. Öhring. Dynamic Tree Embeddings into de Bruijn Graphs. In *Lecture Notes in Computer Science*, vol. 634, pp. 783–784. Springer, Sep. 1992.
8. S. Öhring and S.K. Das. Dynamic Embeddings of Trees and Quasi–Grids into Hyper–de Bruijn Networks,. In *Proc. 7th Int. Parallel Processing Symposium*, Newport Beach, CA, pp. 519–523, Apr. 1993.
9. S. Öhring and S.K. Das. The Folded Petersen Network : A New Versatile Multiprocessor Interconnection Topology. to appear in *Proc. 1993 Int. Workshop on Graph–Theoretic Concepts in Computer Science (WG'93)*, June 1993.
10. S. Öhring and S.K. Das. The Folded Petersen Cube Networks : New Competitors for the Hypercube. to appear in *Proc. Fifth IEEE Symp. on Par. and Dist. Computing*, Dec. 1993.
11. L. Bhuyan and D.P. Agrawal. Generalized Hypercubes and Hyperbus Structures for a Computer Network. *IEEE Transactions on Computers*, C-33:323–333, 1984.
12. G. Chartrand and R.J. Wilson. The Petersen Graph. *Graphs and Applications* (Eds. F. Harary and J.S. Maybee), pp. 69 –100, 1985.
13. S. Öhring and S.K. Das. Mapping Dynamic Data Structures on Product Networks. Technical Report CRPDC-93-5, Univ. Wuerzburg, Univ. North Texas, Mar. 1993.
14. T.F. Chan and Y. Saad. Multigrid Algorithms on the Hypercube Multiprocessor. In *IEEE Transactions on Computers*, vol. C–35, No. 11, pp. 969–977, Nov. 1986.
15. L. S. Barasch, S. Lakshmivarahan, and S.K. Dall. Generalized Gray Codes and Their Properties. In *Proc. 3rd Int. Conf. on Supercomputing*, vol. III, pp. 331–337, 1988.
16. S.N. Bhatt, F.R.K. Chung, J.-W. Hong, F.T. Leighton, B. Obrenic, A.L. Rosenberg, and E.J. Schwabe. Optimal Embeddings by Butterfly–Like Networks. COINS Technical Report 90-108, Computer and Information Science Department, University of Massachusetts, 1990.
17. A.L. Rosenberg. personal communication, May 1992.
18. D.S. Greenberg, L.S. Heath, and A.L. Rosenberg. Optimal Embeddings of Butterfly–Like Graphs in the Hypercube. *Math. Systems Theory*, 23:61–77, 1990.

Permutation Routing on Reconfigurable Meshes
(Extended Abstract)

J. C. Cogolludo and S Rajasekaran
Dept. of CIS, Univ. of Pennsylvania

Abstract. In this paper we present efficient algorithms for packet routing on the reconfigurable linear array and the reconfigurable two dimensional mesh. We introduce both worst case algorithms and algorithms that are better on average. The time bounds presented are better than those achievable on the conventional mesh and previously known algorithms.

We present two variants of the reconfigurable mesh. In the first model, M_r, the processors are attached to a reconfigurable bus, the individual edge connections being bidirectional. In the second model, M_{mr}, the processors are attached to two unidirectional reconfigurable buses. In this paper, we present lower bounds and nearly matching upper bounds for packet routing on these two models. As a consequence, we solve two of the open problems mentioned in [15].

I.   Introduction

Meshes with reconfigurable buses have become attractive models of computing in recent times as revealed by the vast number of papers that have been written on this model. There have been a number of interesting results obtained for the reconfigurable mesh (see e.g., [1-12,14,15,17]).

I.1   Packet   Routing

The problem of packet routing is this: Each node in a network (such as a linear array or a mesh) has one packet of information that it wishes to send to another node in the network. The task is to send all the packets to their correct destinations as quickly as possible such that at most one packet passes over any connection at any time. In particular, in the problem of partial permutation packet routing, each node is the origin of no more than one packet, and the destination of at most one packet as well. We deal with partial permutation routing problems in this paper.

I.2   Model   Definitions

A mesh is an n x n square grid with a processing element at each grid point. Meshes with reconfigurable buses were introduced by Miller et. al. in [11].

We identify two slightly different variations of the reconfigurable mesh model. The first model, which we will abbreviate as M_r, is the same as the one introduced by Rajasekaran and McKendall [15]. Here the processors are connected to a reconfigurable broadcast bus, that can be partitioned into subbuses at any time. One of the proessors connected to a bus can broadcast a message which is assumed to be readable by all the processors connected to the bus in the same time unit. The second model employed, denoted as M_{mr}, is perhaps more consistent with the MIMD model that has been widely employed for a conventional mesh ([13,16]). The only difference with respect to the M_r model is that in this case the processors are connected to two unidirectional reconfigurable buses instead of just one. These buses can also be partitioned in any way we desire, as long as they contain contiguous processors. This means that in one unit of time, we can send two packets along the same processors, one in each direction, whereas we could only send one with the M_r model.

I.3 Contributions of this Paper

Leung and Shende [9,10] have shown that routing on a linear array with a fixed bus requires at least (2n/3) steps in the worst case, and have presented an algorithm that runs in (2n/3) steps in the worst case.

In the case of the reconfigurable model Mr, Rajasekaran and McKendall [15] have presented an algorithm for routing a permutation in (3n/4) steps. The problem of optimal routing in this case still remains open. In this paper, we present an algorithm that runs in (2n/3) steps.

Rajasekaran and McKendall [15] have also presented a randomized routing algorithm for the n x n mesh that runs in time n + o(n) time steps. It was an open problem whether n was a lower bound on the routing time. In this paper we answer this question by presenting an algorithm that runs in time (17/18)n+o(n) with high probability.

On an n-node linear array M_{mr}, clearly, (n/2) is a lower bound on permutation routing time. We present a matching algorithm. The same lower bound applies for 2D mesh routing as well. We also present an algorithm for 2D mesh permutation routing that runs in time (2n/3) - (n/64) + o(n) with high probability (abbreviated as `w.h.p.' from hereon; the probability refered to here is $> 1-n^{-c}$, for any fixed c>0).

In this paper we also present algorithms that are better on average. The average behavior referred to here is as follows: there is a packet to start with at each node of the network; the destination of every packet is chosen randomly and uniformly. On a linear array M_r, our algorithm runs in time (7n/12) + o(n) w.h.p. For a 2D Mr, we could process random routing in time (2n/3) - (n/20)+ o(n) w.h.p. On the other hand, we show that for M_{mr}, random routing on an n-node linear array can be completed in time (n/3) - (n/64) + o(n) w.h.p. and on a 2D mesh the same task can be done in time (n/3) - (n/64) + o(n) w.h.p.

Problem	M_r (W.C.)	M_r (Avg)	M_{mr} (W.C.)	M_{mr} (Avg)
Normal Routing	$\frac{2n}{3}$+o(n)	$\frac{7n}{12}$+o(n)	$\frac{n}{2}$ √	$\frac{n}{3} - \frac{n}{64} + o(n)$
Partial Routing	$\frac{4n}{9}$+o(n)	$\frac{n}{3} - \frac{n}{40} + o(n)$	$\frac{n}{3}$ √	$\frac{n}{6} - \frac{n}{128} + o(n)$
2D Mesh Routing	$\frac{17n}{18} + o(n)$	$\frac{2n}{3} - \frac{n}{20} + o(n)$	$\frac{2n}{3} - \frac{n}{64} + o(n)$	$\frac{n}{3} - \frac{n}{64} + o(n)$

Table 1: Contributions of this paper. √ Indicates optimal algorithm

II. The Mr Model
II.1 Worst Case Linear Array Routing (Algorithm 1)

Rajasekaran and McKendall [15] have presented an algorithm for permutation routing on a linear array with a worst case run time of (3/4)n, whereas an obvious lower bound for this problem is only n/2. In this section we improve the result of [15] by describing a (2/3)n step algorithm.

Let A be the region of the first n/3 nodes of the linear array; B be the next n/6 nodes; C be the next n/6 nodes; D be the rest of the nodes.

Phase I Form $\frac{n}{2}$ groups of processors as follows: group 1 is composed of processor 1 only. Group i is composed of processors 2i - 2 & 2i - 1, for 1<i<(n/2+1). Processor n is not part of any group.

In parallel, for each group do:
• first, check if any of the two packets (or one packet in the case of group 1), has destination in region A and it currently lies to the right of its destination. If one packet satisfies this condition, attach to it the number of its group as its rank. If both packets satisfy the condition, then attach the rank to the packet with a destination further to the left.
• second, if neither of the packets satisfies the first condition, then check if either of them has destination in region B, and currently lies to the right of its destination. Again, attach the group number to this packet. If both packets satisfy this condition, then attach the number to the packet with destination further to the left.
• third, if both of the above conditions are violated, then do nothing.

When all the groups have finished performing these operations, then they are repeated, but this time the order of the groups is reversed, that is, processors 2 & 3 form group (n/2) and so on, up to processors (n/2 - 2) & (n/2 - 1), which form group 2 and processor n/2 which forms group 1. In this case, we repeat the steps listed above for each group, but now the roles of A and B will be played by C and D respectively. We also interchange "right" with "left" and vice-versa.

The ranks attached in the two cases are of different types.

Phase II Perform normal routing for n/2 steps. Note that after k steps (1 $\leq k \leq \frac{n}{2}$), processors 1 through k will not be sending any packets to the right. Similarly, processors n - k through n will not be sending any packets to the left. Thus, these processors will only be sending packets in one direction after the k steps. In particular, one can form a bus in these regions, and broadcast packets as they enter the bus, so that they reach their destination quickly. At step k, we form a bus from processors 1 to k and from n - k to n. At step k + 1, the left bus grows to processor k + 1 and the right bus grows to processor n - (k + 1). In addition, we are performing normal routing for those packets that are not part of any bus. (Here normal routing refers to routing using edge connections). On the left half, for instance, at any given step k, a maximum of two new packets can enter the growing bus: one because a packet is routed normally from node k + 1 to node k, and another because the bus has now grown to include node k + 1, and another packet may have also reached that node from node k + 2. This process continues for n/2 steps.

The packets that may enter the growing buses together correspond to groups formed in phase I. As we said, there may be two new packets after each step as the bus grows, and only one can be broadcast. At step i, the processor that contains the packet with rank i will be part of the corresponding bus, and the packet is broadcast at this step. If there is no packet with rank i, or if it has reached its destination, then no packet is broadcast at this step. This process continues for n/2 steps, after which all the packets have reached their correct half.

Phase III Let $|A_A|=$ number of packets now in region A with destination in A;

$|B_A|=$ number of packets now in region B with destination in region A;

$|B_B|$ = number of packets now in region B with destination in B;

There are no A_B packets.

Compute $|A_A|$, $|B_A|$, and $|B_B|$ using prefix computations in time $O(\log n)$.

Phase IV Form two buses, one with processors 1 through n/3 and the other with processors n/3 + 1 through n/2. At step i, each bus broadcasts packet with rank i to its correct destination. This phase goes on for $\max\{|A_A|, |B_B|\}$ steps. At the end of this phase, A_A and B_B packets have been properly routed.

Phase V Form one bus, with processors 1 through n/2. At step i, broadcast the packet with rank i to its correct destination. This goes on for $|B_A|$ steps, after which all the packets have been properly routed.

Time Analysis of the Algorithm

Phase I takes constant time; **Phase II** takes $\frac{n}{2}$ steps. **Phase III** performs several prefix computations and hence takes $O(\log n)$ time. **Phase IV** takes $\max\{|A_A|, |B_B|\}$ steps. **Phase V** takes $|B_A|$ steps. Thus, the total run time of the algorithm is $O(1) + n/2 + O(\log n) + \max\{|A_A|, |B_B|\} + |B_A|$. We show (see the following Lemma and [17]) that $\max\{|A_A|, |B_B|\} + |B_A| \leq \frac{n}{6}$.

Lemma II.1: $\mathrm{Max}\{|A_A|, |B_B|\} + |B_A| \leq \frac{n}{6}$.

Thus, the total time for the algorithm is $\leq$ $O(1) + O(\log n) + \frac{n}{2} + \frac{n}{6} = 2n/3 + o(n)$.

Theorem II.1: Permutation packet routing on the linear array can be performed in $\frac{2n}{3} + o(n)$ steps in the M_r model.

II.2 Worst Case Packet Routing with Packets on Alternate Nodes

We now consider a linear array of n processors, with a packet originating from every other node. No more than a packet is destined for any node. The goal is to route these packets. This problem is relevant to packet routing on a 2D mesh as will be shown later. In this section we present an algorithm that solves this problem of alternate node routing on a linear array in (4n/9) + o(n) steps.

Algorithm 2 - Packets on Alternate Nodes

There are three phases in the algorithm:

Phase I We will partition the array into two (maybe unequal) parts, say A and B. A 'good' packet is a packet whose origin and destination fall in the same part after we have partitioned the array. Obviously, any packet may or may not be 'good', depending exclusively on the partition we choose for the array.

We will partition the array in such a way that both parts will have the same number of 'good' packets. We show that such a partition point exists and it is unique, and furthermore, this point must fall somewhere between node n/3 and node 2n/3. (For a proof see [17].) In Phase I, we identify such a partition point. This is done by performing a binary search in the interval [n/3, 2n/3] in a total of $O(\log^2 n)$ time.

Phase II We now *compress* the packets originating from nodes closer to the two end points of the linear array. These packets will be compressed at some intermediate nodes near the partition point, one packet per node. We form buses to send these packets closer to the partition point, without interrupting the flow of packets coming from the other side through normal routing. Although the partition point is input dependent, there is no problem in transmitting the information necessary to create and modify the buses required by the algorithm.

When we are compressing the packets, we are sending them as close to the partition point as possible. Notice that the final destination of the packet might fall in the bus itself, in which case we would want to send the packet to the actual destination and not to its corresponding intermediate destination. But this should be no problem. In this case, there will be two processors claiming the packet. We can establish a priority scheme in which the final destination will always have priority.

After the smaller part (of A and B) has been totally compressed, packets coming from the other side will start using buses to reach their destination. At this point, the larger part will not have finished compressing (except if both parts are equal), and thus the packets coming from the smaller side cannot be sent to their destinations yet.

We will continue this procedure for n/3 steps, first compressing the two sides and performing normal routing for those packets that are not in any bus. After each side is compressed, the packets that are coming from the other side can be sent to their final destinations using buses. We claim that after n/3 steps, both streams have crossed each other completely, and we can form two independent parts separated by the node where the two streams finally separate. The proof of the following Lemma can be found in [17]:

Lemma II.2 In the above algorithm, the two streams of packets will be separated by step n/3.

Phase III After n/3 steps, we can subdivide the problem into two independent problems that can run in parallel. We also know when we can start forming buses to send packets to their destinations. In particular, the packets originating from each part will be sent to their destination through buses only after the other part has finished compressing its packets. We show that the buses will be busy from then on until step $\frac{n}{3}$ [17]. Thus, one packet can reach its destination at every step (on each side) since we start forming the buses to send packets to their final destination until step n/3.

Therefore, one group sends packets from step 2x/3 to step $\frac{n}{3}$. Thus, there are n/3 - 2x/3 packets sent. Therefore, there are (n/2 - x) - (n/3 - 2x/3) packets left = n/6 - x/3. (Remember here that n/6 ≤ x ≤ n/3). The other group sends packets from step 2/3(n/2 - x) to step n/3. Thus, it can send n/3 -2/3(n/2 - x) = 2x/3 packets. Thus, there are x - 2x/3 = x/3 packets left.

Therefore the total time bound is = n/3 + max{(n/6 - x/3), x/3}.

In both cases, the maximum time that phase III can take is n/9 (given that n/6 ≤ x ≤ n/3), thus making the worst case time bound for this algorithm = 4n/9 + o(n).

Theorem II.2: Worst case packet routing in the linear array with packets originating from alternate nodes can be performed in $\frac{4n}{9}$ +o(n) steps in the M_r model.

II.3 Worst Case Packet Routing on the 2D-Mesh

In this section we present a randomized algorithm for permutation routing on a 2D mesh whose run time is only $17n/18 + o(n)$. This run time reflects an improvement over that of [14]'s. Perhaps the importance of this algorithm lies in the fact that it rules out a lower bound of n for permutation routing on a 2D mesh.

Algorithm 3 - Packet Routing on the 2D Mesh

The algorithm we use is analogous to the one used in [14]. There are three phases in the algorithm. The mesh is partitioned into slices with $e\,n$ rows in each slice. The mesh is also partitioned into slices where there are $e\,n$ columns in each slice. Packets are colored red or blue at the beginning. The coloring is random and unbiased.

Red packets use the following algorithm while the blue packets execute an analogous algorithm but in orthogonal directions. Consider a red packet q whose origin is (i,j) and whose destination is (k,l). q chooses a random node i' in the column of its origin and the slice of its origin and traverses to (i',j) along column j. In phase II, it traverses along row i' up to the node (i',l). Finally in phase III, q travels along column l and reaches its actual destination. The blue packets will traverse along rows in the first phase, along columns in the second phase, and so on.

The expected number of red packets originating from any node is clearly 1/2. Thus the run time of phase I is $e\,n/2 + o(n)$ w.h.p. Run time of phase II is $4n/9 + o(n)$ w.h.p. (from an adoptation of Theorem II.2), and the run time of phase III is $n/2 + o(n)$ w.h.p. Put together, the total run time of the algorithm is $17n/18 + o(n)$ w.h.p.

Theorem II.3: Worst Case permutation packet routing on the 2D mesh can be performed in $17n/18 + o(n)$ steps w.h.p. in the M_r model.

II.4 Average Permutation Packet Routing on the Linear Array

We now show that on an n-node linear array if a packet originates from every node and the destination of each packet is random, then, we could perform routing in time less than $2n/3$ w.h.p.

Algorithm 4 - Permutation Packet Routing with Random Input

Partition the array into two equal halves, say A and B. The algorithm consists of log n phases. In the first phase we send all the packets that originate from A that are destined for B, followed by packets that originate from B which are destined for A. Each such packet is sent to its correct destination using a single broadcast configuring the whole array as a bus. Clearly this can be done in $n/2 + o(n)$ steps w.h.p. In phase 2, we partition A into two equal halves, say, A1 and A2, and process packets that originate from A1 that are destined for A2 and vice-versa. Similar processing is done independently in B as well. Phase 2 takes $n/8 + o(n)$ steps w.h.p.

In an analogous manner we execute the rest of the phases. The total run time of the algorithm will be $n/2 + n/8 + n/32 + \ldots + o(n) = 2n/3 + o(n)$, which is no better than the run time of the worst case permutation routing algorithm. However we can do better. The crucial fact used is that phase 1 of this algorithm can indeed be completed in $5n/12$ steps.

Consider phase 1, i.e., consider only packets that originate from A with a destination in B and vice-versa. The expected number of such packets originating from each node is 1/2. We can modify the algorithm using the idea of compressing the packets (similar to that of Algorithm 2), and can complete phase 1 in time $5n/12 + o(n)$ w.h.p. We only indicate the modifications necessary. We perform a prefix sums computation to identify the packets to be compressed. The rest of the algorithm can be retained as such. Thus random routing can be performed in time no more than 2n/3 - n/12 + o(n) = 7n/12 + o(n) w.h.p. Notice also that it is possible to show even better upper bounds on the run time, with a computation of similar gains in the run times of subsequent phases as well.

Theorem II.4: Average case packet routing on the linear array can be performed in $\frac{7n}{12}$ + o(n) steps w.h.p. in the M_r model.

II.5 Average Packet Routing with Packets on Alternate Nodes (Algorithm 5)

Consider an n-node linear array such that each node has an expected 1/2 packets. The destination of each packet is random. For this case, routing can be done in time n/3 - n/40 + o(n) w.h.p. The algorithm is very similar to the one presented in the previous section for the case with one packet per node. We can show the following

Theorem II.5: Average packet routing on the linear array with packets originating from alternate nodes can be done in time $\frac{n}{3} - \frac{n}{40}$ + o(n) w.h.p. on the M_r model.

II.6 Average Packet Routing on the 2D Mesh (Algorithm 6)

In this section, we consider the problem of random routing in the case of the 2D mesh M_r. We show that this problem can be done in time 2n/3 - n/20 + o(n) w.h.p. The algorithm to be used is quite simple and analogous to Algorithm 3, except that we perform only the last two phases.

In any phase, clearly, the expected number of blue (red) packets that originate from any node is 1/2. Since we also assume that each packet is destined for a random node, we could apply Theorem II.5 to analyze each phase. Thus, we can conclude that each phase of this algorithm runs in time n/3 - n/40 + o(n) w.h.p., resulting in the following theorem:

Theorem II.6: 2D routing can be done in time 2n/3 - n/20 + o(n) w.h.p. in the average case, using the M_r model.

III. The Mmr Model

In this model there are two unidirectional reconfigurable buses. For example, at a given point, we can have two buses formed, both for the whole linear array, but one that can send packets to the right only while the other can send packets to the left only. The model is very similar to the previous model and is perhaps closer to conventional MIMD mesh. In fact, we have a stronger model here, because Mr can be simulated by Mmr, while the converse is not necessarily true. We will now present two optimal algorithms for this model. These algorithms are quite simple to understand compared to the previous algorithms presented in the paper so far.

III.1 Worst Case Linear Array Packet Routing (Algorithm 7)

As in the previous model, the lower bound for worst case input in this model is n/2. We provide a matching algorithm here.

Phase I Divide the array into two equal parts. Note that if we have a packet on every node, then this division ensures that both parts have the same amount of 'good' packets, where 'good' packets were defined in section II.2.

In each half, form two buses, one to go in each direction. Send all the good packets in each half to their destination in parallel. If there are k 'good' packets in each half, then this phase can be completed in k steps.

Phase II

Now combine both parts into one, and form two buses (each composed of the whole array). Each half must now send $n/2 - k$ packets. All the packets that are sent now are going to the other half, so note that all the packets from the left half will utilize the same bus (namely the one going to the right), while all the packets from the right half will utilize the other bus only. Therefore, this phase can be completed in $n/2 - k$ steps.

Total Run Time $= k + (\frac{n}{2} - k) = \frac{n}{2} \ Steps$.

Theorem III.1: Worst case permutation packet routing on the linear array can be performed in $\frac{n}{2}$ steps in the M_{mr} model.

III.2 Worst Case Packet Routing with Packets on Alternate Nodes (Algorithm 8)

In this case, the lower bound is $\frac{n}{3}$. We also prove:

Theorem III.2: Worst case packet routing on the linear array with packets originating on alternate nodes can be performed in $\frac{n}{3}$ steps in the M_{mr} model.

III.3 Worst Case Packet Routing on the 2D Mesh (Algorithm 9)

In the model M_{mr} also, we could employ the 3 phase algorithm described for the model M_r. The first phase will run in e n steps; the second phase will take $n/3 + o(n)$ steps w.h.p. adopting Algorithm 8; Phase III will take $n/2 + o(n)$ steps w.h.p. Therefore, the algorithm will run in $5n/6 + e$ n $+ o(n)$ steps w.h.p. We could obtain a better run time by randomizing over the whole column (or row) in phase I. This modified algorithm takes $n/6 - n/128 + o(n)$ steps w.h.p. in phase I; the same amount of time for phase III; phase II can be completed in $n/3 + o(n)$ steps with high probability. Thus the overall run time will be $2n/3 - n/64 + o(n)$ w.h.p. Moreover, the queue length will be O(1) w.h.p.

Theorem III.3: Worst case permutation packet routing on the 2D mesh can be performed in $\frac{2n}{3} - \frac{n}{64} + o(n)$ steps in the M_{mr} model.

III.4 Average Permutation Packet Routing on the Linear Array

In a linear array of size n, if there is a packet to start with at each node and the destinations of the packets are random, then we could perform routing in time $n/3 - n/64 + o(n)$ w.h.p. We first describe an algorithm with a run time of $n/3 + o(n)$ and then indicate how to improve the run time further.

Algorithm 10 - Permutation Packet Routing on Array with Random Input

The algorithm is analogous to the one used in Section II.4. There are log n phases in the algorithm. In phase 1, we partition the array into two equal halves, say A and B and process packets that have to go from A to B and vice-versa. The number of packets that have to travel from A to B as well as the number of packets that have to traverse from B to A is easily seen (using Chernoff bounds) to be $n/4 + o(n)$ w.h.p. In the second model of our concern, realize that phase 1 of routing can be done in time $n/4 + o(n)$ w.h.p. Using an identical analysis, we see that phase 2 terminates in time $n/16 + o(n)$ w.h.p., and so on. Totally, the algorithm runs in time $n/3 + o(n)$.

The stated improvement in the run time is achieved as follows: Consider only phase 1. Let x be the node $3n/8$ and y be the node $5n/8$. Call the first $3n/8$ nodes of the array as zone C, the next $n/4$ nodes as zone D, the last $3n/6$ nodes as zone E. The number of packets that have to travel from zone C to zone B is clearly $3n/16$ and also the number of packets that have to travel from A to the right of node x to the right of node y is $3n/64$ w.h.p. This implies that after $3n/16 + 3n/64 = 15n/64$ steps there won't be any activity in regions C and E. There are $n/64$ more steps left in phase 1. In this time we could send some of the packets of phase 2 to their correct destinations. For instance, we could process some of the packets that originate from the first quarter of the array which are destined for the rest of the nodes in zone C and vice-versa. The number of such packets is $n/32 + o(n)$ w.h.p. and hence of these packets, we could send $n/64$ of them to their correct destinations. This means a reduction of $n/64$ steps in the run time of phase 2.

Similar reductions are achievable in the rest of the phases also. Even in phase 1, we could obtain a better gain by extending the above idea further.

Theorem III.4: Average case permutation packet routing on the linear array can be performed in $\frac{n}{3} - \frac{n}{64} + o(n)$ steps in the M_{mr} model.

III.5 Average Packet Routing with Packets on Alternate Nodes (Algorithms 11 and 12)

Now we consider the following problems: The expected number of packets originating from any node of an n-node linear array and on a 2D mesh is 1/2. The destination of each packet is random. Route these packets. We obtain the following:

Theorem III.5: Average case packet routing on the linear array with an expected 1/2 packet from each node can be performed in $\frac{n}{6} - \frac{n}{128} + o(n)$ steps in the M_{mr} model. Average case permutation packet routing on the 2D mesh can be performed in $\frac{n}{3} - \frac{n}{64} + o(n)$ steps.

References

[1] Y. Ben-Asher, D. Peleg, R. Ramaswami, and A. Schuster, The Power of Reconfiguration, Journal of Parallel and Distributed Computing, 1991, pp. 139-153.

[2] D.P. Doctor and D. Krizanc, Three Algorithms for Selection on the Reconfigurable Mesh, Manuscript, 1992.

[3] H. ElGindy and P. Wegrowicz, Selection on the reconfigurable Mesh, Proc. International Conference on Parallel Processing, 1991, pp. 26-33.

[4] E. Hao, P.D. McKenzie and Q.F. Stout, Selection on the Reconfigurable Mesh, Proc. Frontiers of Massively Parallel Computation, 1992, pp. 38-45.

[5] J. Jang, H. Park, and V.K. Prasanna, A fast Algorithm for Computing Histograms on a Reconfigurable Mesh, Proc. Frontiers of Massively Parallel Computation, 1992, pp. 244-251.

[6] J. Jang, H. Park, and V.K. Prasanna, An Optimal Sorting Algorithm on the Reconfigurable Mesh, Proc. International Parallel Processing Symposium, 1992, pp. 130-137.

[7]   J Jenq and S. Shani, Reconfigurable Mesh Algorithms for Image Shrinking, Expanding, Clustering, and Template Matching, Proc. International Parallel Processing Symposium, 1991, pp. 208-215.

[8]   J Jenq and S. Shani, Histogramming on a Reconfigurable Mesh Computer, Proc. International Parallel Processing Symposium, 1992, pp. 425-432.

[9]   J Y-T. Leung and S. M. Shende, Packet Routing on Square Meshes with Row and Column Buses, in Proc. IEEE Symposium on Parallel and Distributed Processing, Dallas, Texas, Dec. 1991, pp. 834-837.

[10]   J Y-T. Leung and S. M. Shende, On Multi-Dimensional Packet Routing for Meshes with Buses, to appear in Journal of Parallel and Distributed Computing, 1993.

[11] R. Miller, V.K. Prasanna-Kumar, D. Reisis, and Q.F. Stout, Meshes with Reconfigurable Buses, in Proc. 5th MIT Conference on Advanced Research in VSLI, 1988, pp. 163-178.

[12]   K. Nakano, D. Peleg, and A. Schuster, Constant Time Sorting on a Reconfigurable Mesh, Manuscript, 1992.

[13]   S. Rajasekaran, k-k Routing, k-k Sorting, and Cut Through Routing on the Mesh, Technical Report, Department of CIS, University of Pennsylvania, Philadelphia, PA 19104, October 1991.

[14]   S. Rajasekaran, Mesh Connected Computers with Fixed and Reconfigurable Buses: Packet Routing, Sorting, and Selection, to be presented in the First Annual European Symposium on Algorithms, October 1993.

[15]   S. Rajasekaran and T. McKendall, Permutation Routing and Sorting on the reconfigurable Mesh, Technical Report MS-CIS-92-36, Department of Computer and Information Science, University of Pennsylvania, May 1992.

[16]   S. Rajasekaran and Th. Tsantilas, Optimal Routing Algorithms for Mesh-Connected Processor Arrays, Algorithmica 8, 1992, pp. 21-38.

[17]   J.C. Cogolludo and S. Rajasekaran, Packet Routing Algorithms for Meshes with Reconfigurable Buses, Technical Report, Department of CIS, University of Pennsylvania, Philadelphia, PA 19104, June 1993.

Adaptive and Oblivious Algorithms for D-Cube Permutation Routing

Miltos D. Grammatikakis[1], D. Frank Hsu[2] * and Frank K. Hwang[3]

[1] Institute for Computer Science, F.O.R.T.H., Heraklion 711 10, Greece
[2] Computer and Information Science, Fordham University, Bronx, NY 10458
[3] AT&T Bell Laboratories, Murray Hill, NJ 07974

Abstract. Assuming d-port multiaccepting communication, we design an optimal locally adaptive permutation routing algorithm on the binary hypercube of dimension 7 (called 7-cube). We also prove an $\Omega(\sqrt{N}/\log N)$ lower bound for the class of deterministic restricted oblivious permutation routing algorithms. Finally, we design optimal deterministic oblivious permutation routing on the d-cube, $d \leq 6$.

1 INTRODUCTION

We consider the binary hypercube, with bidirectional edges between each pair of adjacent nodes. In the beginning, each node contains a message with a destination, and the $N = 2^d$, destinations are all distinct (permutation routing). The nodes are *synchronous* and the routing is divided into "moves". At each move a message either stays put, or moves to an adjacent node if the edge is free. In other words, at any move a message occupies either a node (if staying put), or an edge. As in the *multiple-accepting PE* scheme studied by Abraham and Padmanabhan [1] we allow multiple messages to occupy the same node. However, only one message can occupy an edge at any move. When several messages compete for the same edge (called a conflict), we assume that there is a register to decide which one occupies the edge. Finally, we assume that each node is capable of doing d *-port communication* , i.e., it can simultaneously receive and transmit up to d messages as long as no two are claiming the same edge.

Packet routing forms the kernel of many parallel algorithms, such as sorting, computing multidimensional FFTs, matrix transposition, and divide and conquer strategies. Online routing algorithms can be classified into either *oblivious* , or *adaptive* strategies. A routing algorithm is *deterministic oblivious* if the route of any packet depends only on its origin and destination, while it is *randomized oblivious* if the route of any packet is independently chosen according to a probability distribution which is a function of its origin and destination [2]. A deterministic oblivious algorithm belongs to the class of *deterministic restricted oblivious* routing algorithms if the route of a packet is uniquely determined from its current position and final destination. Lower bounds on the time delay for oblivious per-

---

* Research partially supported by an endowed Komatsu Chair Visiting Professorship at the Graduate School of Information Science, Japan Advanced Institute of Science and Technology

Table 1. Time delay for on-line permutation routing on the d-cube

Dimension (d)	Previous Best Algorithm [6]	Our algorithms (optimal)
4	5	4 *(both adaptive, and deterministic oblivious)*
5	7	5 *(both adaptive, and deterministic oblivious)*
6	9	6 *(both adaptive, and deterministic oblivious)*
7	11	7 *(adaptive)*

mutation routing have attracted much interest. Borodin and Hopcroft [3], considered the possibility that some permutations may cause conflicts, leading to large queuing delays. Their architectural model was a graph of N nodes and degree Δ. For the class of deterministic oblivious routing algorithms, they established better lower bounds of $\Omega(\sqrt{N}/\Delta^{3/2})$, More recently, Kaklamanis et al. [6] improved this lower bound to $\sqrt{N}/(\sqrt{2}\,\Delta)$, which specializes to an $\sqrt{N}/(\sqrt{2}\log N)$ lower bound for the binary hypercube. This lower bound is tight, as they also provided a deterministic oblivious permutation routing algorithm, based on a Hamiltonian cycle decomposition of the hypercube, which routed an arbitrary permutation in $2\sqrt{N}/(d-2\log d+2)+d/2 = O(\sqrt{N}/\log N)$ moves. They commented that their algorithm, was the most efficient one for all hypercubes with dimensions up to 14.

Other possible hypercube routing algorithms can be derived from Cypher and Plaxton's recent $O(\log N \log^2 \log N)$ hypercube adaptive sorting algorithm, called sharesort [4], or use Valiant and Brebner's $O(c \log N)$ probabilistic routing techniques [7]. However, since the constant involved in sorting is in the thousands, and also $c > 2$, both algorithms are not practical for small dimension hypercubes ($d \leq 14$).

First, in Section 2 of this paper, we provide an adaptive algorithm which completes arbitrary permutation routing on the binary 7 dimensional hypercube (7-cube) in 7 moves, the minimum number required for general permutations. Similar algorithms for routing on the d-cube, $1 \leq d \leq 6$ can be easily derived. The proposed routing is *local* since a message does not need to know the routes of other messages, only assuming that a message moving to a node can detect whether an adjacent edge is already occupied.

In Section 3, we prove a $\sqrt{N}/\log N$ lower bound for deterministic restricted oblivious permutation routing on the d-cube, under d-port communication. The multiplicative constant involved in this lower bound is improved to 1, from $1/\sqrt{2}$ in [6]. Based on the concept of a destination tree, we design deterministic oblivious routing algorithms which route an arbitrary permutation on the d-cube in d moves, $2 \leq d \leq 6$.

As shown in Table 1, our results improve on the previously best routing strategy on the d-cube, for any $4 \leq d \leq 7$. For example, while the previous best routing algorithm on the 7-cube requires 11 time-units, our algorithm routes all packets in the minimum possible time delay of only 7 units. It remains an interesting open problem whether a d-move adaptive permutation routing algorithm exists for arbitrary d.

2 ADAPTIVE ROUTING ALGORITHMS

In this Section we provide an optimal adaptive permutation routing algorithm for the 7-cube. The algorithm requires no preprocessing and offers distributed control of a packet route at each hypercube node. Optimal adaptive routing on the d-cube, for $d \leq 6$, are easily derived based on this algorithm [5]. We first state some useful lemmas and definitions.

Lemma 1. *A conflict cannot occur during the first or last move.*

Proof. At the first move no two messages can have the same starting node. At the last move no two messages can share the same destination. Hence no conflict can occur during these two moves. □

By *checking bit i* we mean a move in which a message stays put if bit i of its current node binary representation agrees with the corresponding destination bit, while it goes to the adjacent node with the correct bit i if not. A message is of *pattern B* , if B is its set of checked bits. A message of pattern B is called *maximal* if it disagrees with its final destination in every bit not in B .

We define *maximal checking* as follows. If the current message is maximal, we check a specified bit of the maximal message. If the message is not maximal, we check a bit which agrees with the destination of the message. Following our definitions, all non-maximal messages would stay put.

Lemma 2. *At each node there is at most one maximal message for each pattern.*

Proof. Two maximal messages at a node with the same pattern imply that they have the same destination, which is impossible. □

Theorem 3. *There exists a 7-move local permutation routing for the 7-cube.*

Proof. The 7-move routing algorithm proceeds as follows:

At the first move we check the first bit. No conflict occurs due to Lemma 1.

At the second move we check bit 2 if possible (which means that the dimension-2 edge has not been occupied by another message at move 2). If this is not possible, we check bit 3. Since each node has at most two messages at the start of move 2, either bit 2 or bit 3 is checked for every message and no conflict can occur.

After the second move, at any node there may be up to three messages of pattern $\{1, 2\}$, and only one message of pattern $\{1, 3\}$. We check bit 2 of the message with pattern $\{1, 3\}$. For the messages of pattern $\{1, 2\}$, at most two can have destinations whose third, fourth, fifth, and sixth bits are all 1. We check bit 3 of the first such message encountered, and check bit 4 of the second one if it appears. Any other message of pattern $\{1, 2\}$ stays put. Since we route at most three messages from a node, and all along different dimensions, no conflict can occur.

At the beginning of the fourth move, all messages have the patterns either $\{1, 2, 3\}$, or $\{1, 2, 4\}$. Among the messages of pattern $\{1, 2, 3\}$ ($\{1, 2, 4\}$), at most two can have destinations with 1 in bits 4, 5, and 6 (3, 5, and 6). We check bit 4 (3) for the first message and bit 5 (6) for the second. All other messages stay put. Since the checked bits are all distinct, no conflict can occur.

At the beginning of the fifth move, all messages are of patterns $\{1, 2, 3, 4\}$, $\{1, 2, 3, 5\}$, $\{1, 2, 4, 6\}$. Among the messages of pattern $\{1, 2, 3, 4\}$, use the maximal checking and check bit 7. Among the messages of pattern $\{1, 2, 4, 6\}$ ($\{1, 2, 3, 5\}$) at most two can have destinations with 1 in both bits 3 and 5 (4 and 6). We check bit 3 (4) for the first message and bit 5 (6) for the second. Since all checked bits are distinct, no conflict can occur.

At the sixth move we use the maximal checking; for pattern $\{1, 2, 3, 4, 5\}$ we check bit 6, for pattern $\{1, 2, 3, 4, 6\}$ we check bit 7, for pattern $\{1, 2, 3, 4, 7\}$ we check bit 5, for pattern $\{1, 2, 3, 5, 6\}$ we check bit 4, and for pattern $\{1, 2, 4, 5, 6\}$ we check bit 3.

Finally, no conflict can occur at move 7 by Lemma 1.   □

3   DETERMINISTIC OBLIVIOUS ROUTING ALGORITHMS

The following theorem proves a $\sqrt{N}/\log N$ lower bound for deterministic restricted oblivious routing on the binary d-cube of $N = 2^d$ nodes. For this special class of hypercube routing algorithm, Theorem 4 improves the lower bound of Kaklamanis et al. [6] by a constant multiplicative factor.

Theorem 4. *Any deterministic restricted oblivious algorithm for realizing permutations on the binary d-cube of $N = 2^d$ nodes takes at least a) $\Omega(\sqrt{N})$ time for 1-port communication, or b) $\Omega(\sqrt{N}/\log N)$ time for d-port communication.*

Proof. In accordance with [3], for every destination node (v_0) we define a *destination tree* consisting of all d-cube nodes, with edges defined by the following recursive process (the recursion stops when all nodes have been considered): a) the root node v_0 connects to its adjacent nodes, and b) each new node in the tree connects to any adjacent nodes which are not already in the tree. Figure 1 shows a destination tree for the 4-cube with root node $v_0 = 0000$. We define the *origins* of a node v, as all the descendants of v in the destination tree of v_0. If a packet directed to v_0 is placed at any of these origins, the packet will go through node v.

Take v to be a random node in the d-cube. Then, v will be in all the N destination trees. We are interested in counting the number of origins of v. We know that the number of nodes at distance i from a given node of a binary d-cube is $\binom{\log N}{i}$.

Therefore, in $DT = \begin{pmatrix} \log N \\ i \end{pmatrix}$ destination trees we have $OR = \sum_{l=i+1}^{\log N} \begin{pmatrix} \log N \\ l \end{pmatrix}$ origins of v.

Since DT becomes maximum at $i_{max} = \lceil \log N/2 \rceil$, we can say that (at $\tau = i_{max} - 1$) we have at least $\begin{pmatrix} \log N \\ \tau \end{pmatrix}$ origins of v in $\begin{pmatrix} \log N \\ \tau \end{pmatrix}$ destination trees.

Now, we can start a packet at each of these destination trees at a different node, such that the packets destined for distinct destinations will all go through v. Therefore, assuming 1-port communication, the delay must be at least:

$$\begin{pmatrix} \log N \\ \tau \end{pmatrix} \tag{3.1}$$

Under d-port communication, a maximum of $\log N$ packets might be diverted to different edges after reaching node v, thus causing a smaller overall queuing delay. Thus, assuming d-port communication, the delay must be at least:

$$\frac{\begin{pmatrix} \log N \\ \tau \end{pmatrix}}{\log N} \tag{3.2}$$

Suppose $n = \log N/2$. The quantity of interest (2) for d-port communication becomes:

$$\frac{\begin{pmatrix} 2n \\ n-1 \end{pmatrix}}{2\,n} \frac{2^n}{2\,n} = \frac{\sqrt{N}}{\log N}$$

Similarly, assuming $n = \log N/2$ the quantity of interest (1) for 1-port communication becomes:

$$\begin{pmatrix} 2n \\ n-1 \end{pmatrix} 2^n = \sqrt{N} \quad \square$$

The *greedy routing algorithm* sends each packet from its source to its destination by following a minimum distance path. Dimensions are examined in a left to right order, and packets which are not in their final destination move by correcting the next dimension for which the corresponding source and destination bits are different. Node conflicts (1-port communication), or edge conflicts (d-port communication) are commonly resolved with a farthest-first (or random) queuing discipline.

Suppose that a destination tree consists of d chains starting from the same root node v_0, with two nodes adjacent on a chain only if they are adjacent on the d-cube. A depth d destination tree, which contains all $N = 2^d$ nodes of the d-cube, is shown in Figures 2, 3 for $d = 2, 3$.

Remark. Destination trees with root node other than v_0 can be obtained by complementing appropriate dimensions of all nodes in the destination tree with root node v_0.

The following Lemmas 5-7 (8-10) deal with one to many routing (correspondingly many to one routing). By combining these lemmas, we provide in Theorem 11 a d-move deterministic oblivious permutation routing algorithm for the d-cube, $2 \leq d \leq 6$.

Lemma 5. *N packets initially located at distinct nodes on the d-cube ($d = \log N$, $d \geq 1$) can correct their first dimension without any edge conflicts.*

Proof. Trivial, since all N packets are initially located at distinct nodes. □

Lemma 6. *N packets initially located at distinct nodes on the d-cube ($d = \log N$, $d \geq 2$) can correct their first two dimensions without any edge conflicts.*

Proof. Packets move on a destination tree from a node with representation equivalent to its first two source bits, towards the root node v_0 equivalent to its first two destination bits, according to the following schedule. In the first step packets at distances 1 and 2 move, and in the second step packets at initial distance 2 move.

Consider two packets moving on the same destination tree. If the packets are initially located at different nodes, they can not conflict, since they move synchronously towards the root of the tree. If the packets are initially located at the same node (identical first two source bits), at least one of their last $d - 2$ source bits must be different, and therefore they move towards the root of the tree without conflict.

Now, consider two packets moving on different destination trees. Since packets are initially located at distinct nodes, and after correcting their first two dimensions they are located at different root nodes, no edge conflict can occur. □

Lemma 7. *N packets initially located at distinct nodes on the d-cube ($d = \log N$, $d \geq 3$) can correct their first three dimensions without any edge conflicts.*

Proof. Packets move synchronously on a destination tree from a node with representation equivalent to its first three source bits, towards the root node v_0 equivalent to its first three destination bits, as follows. In the first step packets at distances 1, 2, and 3 move, in the second step packets at initial distances 2, and 3 move, and in the third step packets at initial distance 3 move.

We omit the remaining part of proof, since it is similar to Lemmas 5, 6 above. The proof considers the possibility of a conflict between two packets which a) move on the same destination tree, and are either located at the same tree node, or at different tree nodes, b) move on different destination trees, and are initially at distance 3, distance 2, or distances 3 and 2 from the root node of the destination tree. □

Lemma 8. *N packets on the d-cube ($d = \log N$, $d \geq 1$) with distinct destinations, and their first $d - 1$ dimensions correct, can correct their last dimension without any edge conflicts.*

Proof. It is obvious, since all N packets have distinct destinations. □

Lemma 9. *N packets on the d-cube ($d = \log N$, $d \geq 2$) with distinct destinations, and their first $d - 2$ dimensions correct, can correct their last two dimensions without edge conflicts.*

Proof. Each packet moves on a destination tree from the root node with representation equivalent to its last two source bits, to a node equivalent to its last two destination bits, according to the following schedule. In the first step only packets at distance 2 move, and in the second step packets at initial distances 1 and 2 move.

Consider two packets moving on the same destination tree. If the packets are destined to different nodes, they can not conflict, since they move synchronously (one after the other) towards their destination node in the tree. If the packets are destined to the same node (identical last two destination bits), they must differ in at least one of their first $d - 2$ dimensions, and therefore they can progress towards their destination nodes without edge conflicts.

Now, consider two packets moving on different destination trees. Since they are initially located at different nodes, and also after correcting their last two dimensions they are still located at distinct nodes, no edge conflict can occur. □

Lemma 10. *N packets on the d-cube ($d = \log N$, $d \geq 3$) with distinct destinations, and their first $d - 3$ dimensions correct, can correct their last three dimensions without edge conflicts.*

Proof. Each packet moves on a destination tree from the root node with representation equivalent to its last three source bits, to a node equivalent to its last three destination bits, as follows. In the first step only packets at distance 3 move, in the second step packets at initial distances 2, or 3 move, and in the third step packets at initial distance 3 move.

Consider two packets moving on the same destination tree. If the packets are destined to different nodes, they can not conflict, since they move synchronously (one after the other) towards their destination node in the tree. If the packets are destined to the same node (identical last three destination bits), they must differ in at least one of their first $d - 3$ dimensions, and therefore they can progress towards their destination nodes without edge conflicts.

Next, consider two packets moving on different destination trees. Since packets are initially located at distinct root nodes, and after correcting their first three dimensions they are located at distinct destination nodes, no edge conflict can occur during the first and third step. An edge conflict may only arise during the second step. There are three cases to consider;

1) Both packets are initially at distance 3. Since both packets change their third dimension during the first step (Figure 3, Remark), they arrive at distinct nodes after the first step, and thus no edge conflict occurs.

2) Both packets are initially at distance 2. Since after the second step the packets arrive at distinct destinations, no edge conflict during the second step occurs.

3) One packet is initially at distance 3, and the other packet is initially at distance 2. The packet at distance 3 changes the third dimension during the first step, the second dimension during the second step, and the first dimension during the third step. For a conflict to occur the second packet must also change the second dimension during the second step. Let's suppose that this is true. Then, it follows from Figure 3, and our Remark that the second packet changes the third dimension during the first step. Therefore, both packets change the same dimension during the third step, just after the conflict. This means that both packets reach the same destination, a contradiction to our initial assumption.

□

By combining our routing techniques above we can easily prove,

Theorem 11. *There exists a d move deterministic oblivious routing algorithm for the d-cube, $2 \leq d \leq 6$.*

Proof. By applying Lemmas 5, and 8 with $d = 2$, Lemmas 5, and 9 with $d = 3$, Lemmas 6, and 9 with $d = 4$, Lemma 6, and 10 with $d = 5$, and Lemmas 7, and 10 with $d = 6$, we easily see that this theorem holds for $d = 2, 3, 4, 5, and 6$ respectively.

□

Remark. All permutation routing algorithms of Section 3 are deterministic oblivious, since the path of a packet does not depend on the presence (or absence) of other packets in its route.

Remark. All permutation routing algorithms presented are on-line. They can also be used to perform partial permutations, by treating nodes without packets in exactly the same way as nodes with packets already in their final destination.

4 SOME REMARKS

Although, a depth 4 destination tree, which contains all $N = 2^d$ nodes of the 4-cube can be constructed, we have not yet managed to route optimally on the 8-cube. By considering rearrangements of the nodes in the destination tree, and/or introduction of delays for certain packets in the schedule, efficient routing algorithms for higher dimensional hypercubes can be constructed. The routing method based on destination trees that we introduced can be extended to other graphs, such as the generalized hypercube and Cayley group graphs.

References

1. Abraham, S. and Padmanabhan, K.: Performance of the Direct Binary N-cube Network for Multiprocessors. IEEE Transactions on Computers, **C-38 (7)** 1989, 1000–1011.
2. W. A. Aiello, W. A., Leighton, F. T., Maggs, B. M., Newman, M.: Fast Algorithms for Bit Serial Routing on a Hypercube. Mathematical Systems Theory. **24, 1991** 253–271.

3. Borodin, A. and Hopcroft, J. E.: Routing, Merging, and Sorting on Parallel Models of Computation. Journal of Computer and System Science. **30**, **1985**, 130–145.
4. Cypher, R., Plaxton, C. G.: Deterministic Sorting in Nearly Logarithmic Time on the Hypercube and Related Computers. In Proc. 22nd ACM Symposium of Theory of Computing. **1990** 193–203.
5. Grammatikakis, M., Hsu, D. F., Hwang, F. K.: Universality of the D-cube, $D < 8$. In Proc. Parallel Computer Conference (PARCO 93), Advances in Parallel Computing, Elsevier Sience Publishers. **1993** to appear.
6. Kaklamanis, C., Krizanc, D., Tsantilas, A.: Tight Bounds for Oblivious Routing on the Hypercube. In Proc. 2nd Annual ACM Symposium of Parallel Algorithms and Architectures. **1990** 31–36.
7. Valiant, L. G., and Brebner, G. J.: Universal Schemes for parallel communication. In Proc. 13th ACM Symposium of Theory of Computing. **1981** 263–277.

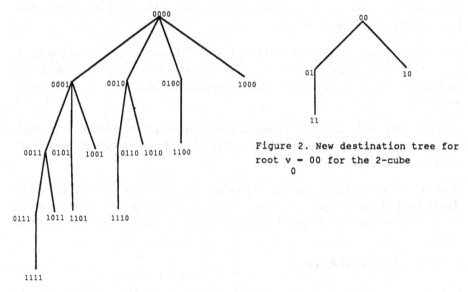

Figure 2. New destination tree for root $v_0 = 00$ for the 2-cube

Figure 1. Destination tree for root $v_0 = 0000$

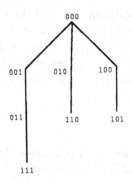

Figure 3. New destination tree for root $v_0 = 000$ for the 3-cube

On Quadratic Lattice Approximations

Anand Srivastav
Research Institute of Discrete Mathematics
University of Bonn

Peter Stangier
Institute for Computer Science
University of Cologne

Abstract

We consider the problem of approximating a system of linear and quadratic forms evaluated at a rational point by 0-1 vectors. When only linear forms are given this is the well known lattice approximation problem. We call the general version the quadratic lattice approximation problem. In this paper we construct via derandomization lattice points with small linear and quadratic discrepancies. Unfortunately the known derandomization methods do not apply to the quadratic variant of the lattice approximation problem. Therefore we develop a new derandomization technique, which captures non-linearity and dependencies among the random variables under consideration, extending the conditional probability/ pessimistic estimator method of Spencer and Raghavan. The essential new tool is an algorithmic version of Azuma's martingale inequality.

1 Introduction

We first discuss a new derandomization method, which we call the algorithmic Azuma inequality and then show how it can be applied to lattice approximations.

Algorithmic Azuma Inequality

The basic problem of the derandomization method of conditional probabilitites is the computation of such probabilities or of appropriate upper bounds, the so called pessimistic estimators in polynomial-time (Spencer [17], Ragahvan [15]). One key fact for the success of the conditional probability method is that pessimistic estimators can be constructed, whenever *linear* objective functions are involved, because then

Chernoff and Hoeffding type inequalities on the deviation of linear sums of independent random variables from their expectation are available. But the situation becomes quite complicated, if quadratic forms or other non-linear functions arise:

We are given the discrete n-dimensional cube $\Omega = \{0,1\}^n$, a rational vector $c \in [0,1]^n$ a rational vector and a rational, symmetric $n \times n$-matrix D. Troughout this paper let $f : \Omega \mapsto \mathbb{Q}$ denote a linear form $(f(x) = c^T x)$ or a quadratic form $(f(x) = x^T D x)$, $x \in [0,1]^n$ or a sum of both. Let $\mathbb{P}$ be a probability measure on Ω and for $\lambda \in \mathbb{R}$ let $F(\lambda)$ be the probability distribution

$$F(\lambda) = \mathbb{P}(|f - \mathbb{E}(f)| > \lambda).$$

If the x_i's are outcomes of completely independent Bernoulli trials, then $\mathbb{P}$ is the product measure induced by these Bernoulli trials and Hoeffding's inequality ([12], Corollary 5.2 (a)), which generalizes the Chernoff inequality in the Binomial case, shows for a linear form f

$$F(\lambda) \leq 2 \exp(-\frac{2\lambda^2}{n}).$$

In randomized rounding procedures inequalities of this type have been used to prove the correctness of the probabilistic method.

If all the c_i's are 0 or 1, Raghavan [15] first gave a derandomized counterpart of randomized rounding algorithms *on the RAM model* of computation. An examination of Raghavan's proof from the standpoint of probability theory shows that derandomization is the search for a vector $x \in \Omega$ which is concentrated around the expectation $\mathbb{E}(f)$, for example satisfies

$$|f(x) - \mathbb{E}(f)| \leq \lambda.$$

If f is no longer a linear form, for example $f(x) = x^T D x$, then we have a dependent sum of random variables and may loose the control on large deviations, i.e. do not have substitutes for the central limit theorem.

Fortunately, if f can be written as a martingale, Azuma [3] gave a generalized Chernoff type bound: Let $(\mathcal{F}_k)$ be an increasing sequence of sigma algebras,

$$\{\emptyset, \Omega\} = \mathcal{F}_o \subseteq \mathcal{F}_1 \subseteq \ldots \subseteq \mathcal{F}_n = \mathcal{P}(\Omega),$$

called finite filtration.

Denote by $f_k := \mathbb{E}(f|\ \mathcal{F}_k)$ the conditional expectation of f with respect to $\mathcal{F}_k$. The sequence $(f_k)_{k=0}^n$ is a Doob's martingale process with $f_0 = \mathbb{E}(f)$ and $f_n = f$. Denote by ϕ_k the martingale differences $\phi_k = f_k - f_{k-1}$. Let d_k be bounds on the martingale differences, i.e $\|f_k - f_{k-1}\|_\infty \leq d_k$ for all k. Then by Azuma's [3] inequality we have

$$F(\lambda) \leq 2\exp(-\frac{\lambda^2}{2\sum_{k=1}^n d_k^2}).$$

For certain values of λ the right hand side is less than 1 and the task of derandomization is to find a point $x \in \Omega$ with

$$|f(x) - \mathbb{E}(f)| \leq \lambda \qquad (1)$$

in deterministic polynomial time. We prove that this can be done. Let $0 < \delta < 1$ and set

$$\lambda(\delta) = \sqrt{2\sum_{k=1}^n d_k^2 \ln\frac{1}{\delta}}.$$

Then $F(\lambda) \leq \delta$ and the desired vector exists. Our first main result is the following algorithmic version of Azuma's inequality:

Theorem *A vector $x \in \Omega$ satisfying (1) can be constructed in $O(n^3 log\frac{1}{\delta})-$ time, if f is a quadratic form and in $O(n^2 log\frac{1}{\delta})-$ time, if f is a linear form.*

This result can be extended to $m \geq 2$ numbers of linear or quadratic form and events of the type

$$f_i(x) - \mathbb{E}(f_i) \leq \lambda_i \ (resp. \geq -\lambda_i)$$

for $i = 1, \ldots, m$. Then the running times are $O(mn^3 log\frac{m}{\delta})-$ resp. $O(mn^2 log\frac{m}{\delta})-$.

Quadratic Lattice Approximation

Let $c, p \in [0,1]^n$ be rational vectors, D a symmetric rational $n \times n$ matrix and C a rational $(r-1) \times n$ matrix with $0 \leq c_{ij} \leq 1$. By the *Quadratic Lattice Approximation* problem we address the problem of finding a lattice point $q \in \{0,1\}^n$ in polynomial-time such that the following conditions hold:

(a) $|c^T(p-q) + p^T Dp - q^T Dq|$ is small

(b) $\|C(p-q)\|_\infty$ is small.

Without the quadratic term (i.e. $D \equiv 0$) and taking the matrix $[C, c]$ instead of C this is exactly the (linear) lattice approximation problem.

Both the linear and quadratic approximation problems have interesting applications:

- When the matrix C represents the constraints and the objective of a $0-1$ integer linear program and p is a solution vector of the corresponding linear programming realaxtion, then lattice approximation is exactly the problem of rounding the vector p such that no constraint is violated too much and the objective function value of the rounded $0-1$ vector is close to the real objective function value.

- The same question arises in $0-1$ quadratic optimization, when Z is the objective function $Z(x) = c^T x + x^T Dx$.

- Linear lattice approximations are closely related to discrepancies of hypergraph colorings [17] and applications in computational geometry like the construction of small $\epsilon-$samples and nets [10].

Previous Work

Given $p \in [0,1]^n$, $q \in \{0,1\}^n$ and $x \in [0,1]^n$ let

$$\Delta_i = |\sum_{j=1}^n c_{ij}(p_j - q_j)|,$$

$$Z(x) = c^T x + x^T Dx$$

and

$$\Delta Z = |Z(p) - Z(q)|.$$

In the linear case Spencer [17] showed the existence of a lattice point with $\Delta_i \leq 6\sqrt{n}$ for all i, while Beck and Fiala [4] gave an algorithm constructing a lattice point with $\Delta_i \leq 2\sqrt{2n \ln 2r}$ for all $i = 1, \ldots, r$. Raghavan [15] gave a derandomized algorithm finding a vector $q \in \{0,1\}^n$ such that $\Delta_i \leq s_i D(s_i, \frac{1}{2r})$,

where
$s_i = \sum_{j=1}^n c_{ij} p_j$ and $D(s_i, \frac{1}{2r})$ is a function asymptotically better than the Beck-Fiala bound. Raghavan showed in the unweighted case a polynomial-time implementation of his algorithm in the RAM-model, whereas the same problem in the weighted case $0 \leq c_{ij} \leq 1$ remained open. As far as we know no positive results have been discovered for the more difficult quadratic version of the problem.

The Discrepancy Result

For a moment let us assume that $n = r$. This is not really a restriction and we will distinguish between r and n in our main statements. In the pure quadratic case ($C = 0, c = 0$) we can find by a greedy procedure a $x \in \{0,1\}^n$ with $\Delta Z \leq 2n$. This bound is sharp. The problem becomes interesting only if we wish to minimize the linear and quadratic discrepancies *simultaneously*. In the pure linear case the algorithmically reachable bound is $O(\sqrt{n \ln n})$ and from the above $O(n)$ is the greedy bound for the mixed problem. This gives us a feeling for good discrepancies. We wish to push ΔZ and Δ_i towards $O(\sqrt{n \ln n})$ as much as possible. Our main discrepancy result is:

Theorem *Let $d = 2\max_{1 \leq i \leq n} \sum_{j=1}^n |d_{ij}|$ and $\alpha \geq 0$ with $\text{trace}(D) \leq \alpha d \sqrt{n}$. Then a lattice point $x \in \{0,1\}^n$ can be constructed in $O(n^2 r \log r + n^3)$ time such that*

(a) $|c^T(p - q) + p^T D p - q^T D q| \leq 2\sqrt{n \ln 2r} + (3 + \alpha) d \sqrt{n}$

(b) $\|C(p - q)\|_\infty \leq 2\sqrt{n \ln 2r}$.

This result covers some interesting special cases:

- **Linear forms with rational c**
  If $D = 0$ and c rational, then we have an $O(n^2 r \log r)$-time algorithm.

- **Sparse Matrix Case**
  If $d = O(\log n)$ and $\text{trace}(D) = O(\log n \sqrt{n})$ we obtain $\Delta_i = O(\sqrt{n \log r})$ and $\Delta Z = O(\log n \sqrt{n})$ for all $i = 1, \ldots, r$.

- **Indefinite Matrices**
  If $\text{trace}(D) = 0$, D is indefinite and $\alpha = 0$. For small d, i.e. $d = O(n^{\frac{1}{2} - \epsilon})$ with $0 < \epsilon < \frac{1}{2}$ we get $\Delta Z = O(n^{1 - \epsilon})$ and $\Delta_i = O(\sqrt{n \log r})$.

At this moment we are not able to remove the assumption on the boundedness of the trace of D. It remains an interesting open problem whether or not better discrepancy bounds can be derived (see also the discussion at the end of this paper). The paper is organized as follows.

In the next two sections we construct pessimistic estimators for quadratic forms and derive an algorithmic version of Azuma's inequality. In the last section the application to lattice approximation is discussed.

The model of computation throughout this paper is the RAM-model (see [11]). By the size of an input we mean the number of datas in the description of the input. By the encoding length of the input we call the maximal binary encoding of numbers in the input. In the RAM-model an algorithm runs in polynomial-time (resp. strongly polynomial-time), if the number of elementary arithmetic operations (briefly called running time) is polynomially bounded in the size and the encoding length of the input (resp. *only* in the size of the input) *and* the maximal binary encoding length of a number appearing during the execution of the algorithm (briefly called space) is polynomially bounded in the size and encoding length of the input.

2 Pessimistic Estimators and Martingales

We start with a brief explanation of discrete martingales. Let $n \in \mathbb{N}$ and $\Omega = \{0,1\}^n$. Take the powerset $\mathcal{P}(\Omega)$ over Ω as the σ-algebra and let $\mathbb{P}$ be a probability measure defined by n independently Bernoulli trials where $\omega_i = 1$ with probability p_i and $\omega_i = 0$ with probability $1 - p_i$, $0 \leq p_i \leq 1$. Define a probability measure on Ω by

$$\mathbb{P}(\{\omega\}) = \prod_{i=1}^n p_i^{\omega_i}(1 - p_i)^{1 - \omega_i}, \quad (\omega \in \Omega).$$

Denote by $[n]$ the set $\{1, \ldots, n\}$. Let $E \subseteq \Omega$ be an event such that for a $0 < \delta < 1$, $\mathbb{P}(\bar{E}) \leq \delta$, where $\bar{E}$ is the complement of E. Our task is to find an $\omega \in E$. This can be done with Spencer's method of conditional probabilities described in the following procedure:

Definition 2.1 Algorithm WALK($\bar{E}$)

(a) For $i = 1, .., n$ do

    Compute ω_i with $\mathbb{P}(\bar{E}|\omega_1, \ldots, \omega_{i-1}, \omega_i)$

    $= \min\limits_{j=0,1} \mathbb{P}(\bar{E}|\omega_1, \ldots, \omega_{i-1}, j)$

(c) Output the vector $\omega = (\omega_1, .., \omega_n)$.

The striking observation is that $\omega \in E$ (see [17] or [15]).

The problem of efficient computation of the conditional probabilities can be removed in many examples using Raghavan's idea of "pessimistic estimators" [15]. The following definition formalizes this concept:

Let $i \in [n]$, $\omega_1, \ldots, \omega_{i-1} \in \{0, 1\}$,

$$C_{ij} = \{\omega' \in \Omega; \omega'_k = \omega_k \text{ for } k = 1, \ldots, i-1$$
$$\text{and} \quad \omega'_i = j\}$$

and $C_i = \{\omega' \in \Omega; \omega'_k = \omega_k \text{ for } k = 1, \ldots, i-1\}$.
Define $\mu_{ij}(\omega_1, \ldots, \omega_{i-1}) = \dfrac{\mathbb{P}(C_{ij})}{\mathbb{P}(C_i)}$.

Definition 2.2 *Let $U = \{U_{ij}(\omega_1, \ldots, \omega_{i-1}),$ $(i \in [n]$ and $j, \omega_1, \ldots, \omega_{i-1} \in \{0, 1\}\}$ be a family of real-valued functions with the convention that U_{1j} denote the j-th function on the first level. The family U is called a pessimistic estimator (resp. weak pessimistic estimator) for the event $\bar{E}$, if for each $i \in [n]$ and $\omega_1, \ldots, \omega_{i-1} \in \{0, 1\}$, the following conditions (i) - (iv) (resp. (i) - (iii)) are satisfied:*

(i) $\mathbb{P}(\bar{E}|\ \omega_1, \ldots, \omega_{i-1}, \omega_i = j)$

$$\leq U_{ij}(\omega_1, \ldots, \omega_{i-1})$$

(ii) U is $\mathbb{P}$-convex, that means

$$\sum_{j=0}^{1} \mu_{ij}(\omega_1, \ldots, \omega_{i-1}) U_{ij}(\omega_1, \ldots, \omega_{i-1})$$

$$\leq U_{i, \omega_{i-1}}(\omega_1, \ldots, \omega_{i-2}).$$

(iii) $\min\limits_{j=0,1} U_{1j} \leq \delta < 1$

(iv) U is computable in polynomial-time, that means each function $U_{ij}(\omega_1, \ldots, \omega_{i-1})$ can be computed in the RAM model of computation in time bounded by a polynomial in n and $\log \frac{1}{\delta}$.

Taking the pessimistic estimator instead of the conditional probabilities in the WALK algorithm we obtain indeed a polynomial-time algorithm.

Definition 2.3 *Let $E \subset \Omega$ be an event with $\mathbb{P}(\bar{E}) \leq \delta < 1$ and let U be a pessimistic estimator for $\bar{E}$. Then let $D - WALK(\bar{E})$ the algorithm defined as in Definition 2.1, but where the conditional probabilities have been replaced by the corresponding functions of the pessimistic estimator U.*

Combination of pessimistic estimators of different events are nothing else than sums of the corresponding families and $\mathbb{P}$-convexity implies

Proposition 2.4 *Let $E_i \subset \Omega$, $i = 1, \ldots, l$, be events with $\mathbb{P}(\bar{E}_i) \leq \delta_i < 1$ and $\delta_1 + \ldots + \delta_l < 1$. Let $U^{(i)}$ be a pessimistic estimators for $\bar{E}_i$ and let $U = U^{(1)} + \ldots + U^{(l)}$. Then U is a pessimistic estimator for the event $\bar{E}_1 \vee \ldots \vee \bar{E}_l$.*

Definition 2.5 *To shorten notation define for $\epsilon > 0$*

(i) the above event $E_a(\epsilon)$ as $f(\omega) \geq \mathbb{E}(f) + \epsilon$.

(ii) the below event $E_b(\epsilon)$ as $f(\omega) \leq \mathbb{E}(f) - \epsilon$.

(iii) the concentrated event $E_c(\epsilon)$ as
    $|f(\omega) - \mathbb{E}(f)| \leq \epsilon$.

Remark: it is very easy to find points *either* above or *below* the expectation, provided that the conditional expectations can be computed efficiently. But for combination of such events, for example concentrated points, the problem becomes difficult. Therefore we are urged to construct pessimistic estimator for the above and below type events separately, because then we will be able to analyse any combination of such events according to Proposition 2.4.

Before we can proceed, we must assure that the events under consideration hold with non-zero probability. This is done invoking Azumas martingale inequality. The martingale we are interested in is constructed as follows:

Say $\omega, \omega' \in \Omega$ are k-equivalent, i.e. $\omega \equiv_k \omega'$ if $\omega_j = \omega'_j$ for all $1 \leq j \leq k$. k-equivalency defines an equivalence relation on Ω and induces for each k a partition P_k of Ω where $P_n = \{\{\omega\}; \omega \in \Omega\}$ and $P_0 = \{\Omega\}$. Denote by $\mathcal{F}_k$ the σ-algebra generated by P_k. Then $\{\emptyset, \Omega\} = \mathcal{F}_o \subseteq \mathcal{F}_1 \subseteq \ldots \subseteq \mathcal{F}_n = \mathcal{P}(\Omega)$ and $(\mathcal{F}_k)_{k=0}^n$ is a finite filtering of σ-algebras.

Denote by $f_k = \mathbb{E}(f|\ \mathcal{F}_k)$ the conditional expectation of f with respect to $\mathcal{F}_k$. The sequence $(f_k)_{k=0}^n$ is a Doob's martingale process with $f_0 = \mathbb{E}(f)$ and $f_n = f$. Denote by ϕ_k the martingale differences $\phi_k = f_k - f_{k-1}$. Let $d_k := 2\sum_{j=1}^r |d_{kj}|$. Then

$$|f(x) - f(x')| \le d_k,$$

if $x_i = x_i'$ for all $i \ne k$ and as in the proof of Theorem 4.1 of [1] we have

Lemma 2.6 $\|\phi_k\|_\infty \le d_k$ for all k.

Let $\Delta := \sum_{k=1}^n d_k^2$, $0 < \delta < 1$ and $\epsilon_i := \sqrt{2\Delta \ln \frac{i}{\delta}}$ for $i = 1, 2$. Using Lemma 2.6 we infer from Azuma's inequality [3]

Proposition 2.7 .

(i) $\mathbb{P}(f - \mathbb{E}(f) \le -\epsilon_1(\delta)) \le \exp(-\frac{\epsilon_1^2}{2\Delta}) = \delta$

(ii) $\mathbb{P}(f - \mathbb{E}(f) \ge \epsilon_1(\delta)) \le \exp(-\frac{\epsilon_1^2}{2\Delta}) = \delta$

(iii) $\mathbb{P}(|f - \mathbb{E}(f)| \ge \epsilon_2(\delta)) \le 2\exp(-\frac{\epsilon_2^2}{2\Delta}) = \delta$

Proposition 2.7 shows the non emptiness of our events and we may proceed to the definition of the basic functions from which the pessimistic estimators are derived.

Definition 2.8 *For $i \in [n], j, \omega_1, \ldots, \omega_{i-1} \in \{0,1\}$ and parameters $\epsilon, t > 0$ define the families of functions $U^{(a)}, U^{(b)}$ and $U^{(c)}$ by*

(i) $U_{ij}^{(a)}(\omega_1, \ldots, \omega_{i-1}) :=$
$e^{-t(\epsilon+\mathbb{E}(f))} e^{\frac{1}{2}t^2(d_{i+1}^2+\ldots+d_n^2)} e^{t\mathbb{E}(f|\omega_1,\ldots,\omega_{i-1},j)}.$

(ii) $U_{ij}^{(b)}(\omega_1, \ldots, \omega_{i-1}) :=$
$e^{-t(\epsilon-\mathbb{E}(f))} e^{\frac{1}{2}t^2(d_{i+1}^2+\ldots+d_n^2)} e^{-t\mathbb{E}(f|\omega_1,\ldots,\omega_{i-1},j)}$

(iii) $U_{ij}^{(c)}(\omega_1, \ldots, \omega_{i-1}) :=$
$(U_{ij}^{(a)} + U_{ij}^{(b)})(\omega_1, \ldots, \omega_{i-1}).$

Unfortunately, the functions in Definition 2.8 are not efficiently computable. But the they have the nice property of a weak pessimistic estimator. We will use this fact later in the final step of the construction.

Theorem 2.9 *Let $0 < \delta < 1$, $\epsilon_i = \sqrt{2\Delta \ln \frac{i}{\delta}}$ and $t_i = \epsilon_i \Delta^{-1}$, $(i = 1, 2)$. The families $U^{(a)}$, $U^{(b)}$ and $U^{(c)}$ are weak pessimistic estimators for the events $\bar{E}_a(\epsilon_1), \bar{E}_b(\epsilon_1)$ and $\bar{E}_c(\epsilon_2)$.*

In the proof of Theorem 2.9 we need the following two lemmas.

Lemma 2.10 *For all $t > 0$ and $k \in \{1, \ldots, n\}$ we have*

$$\mathbb{E}(e^{t(f_k - f_{k-1})}| \mathcal{F}_{k-1}) \le \exp(\frac{td_k^2}{2}).$$

Lemma 2.11 *For $k \in \{1, \ldots, n\}$ let $C \in \mathcal{P}_k$ be a partition set. Then we have for all $t > 0$*
$\mathbb{E}(e^{tf_k}|C) = e^{t\mathbb{E}(f|C)}.$

Sketch of the Proof of Theorem 2.9:
We first consider the event $\bar{E}_a(\epsilon_1)$ which represents "$f - \mathbb{E}(f) > \epsilon_1$". The proofs for the other events are similar.
(i) Let $\epsilon > 0$ be arbitrary. For the upper bound condition we must show for all i and j :

$$\mathbb{P}(\bar{E}_a(\epsilon)|\omega_1, \ldots, \omega_{i-1}, j) \le U_{ij}^{(a)}(\omega_1, \ldots, \omega_{i-1}).$$

Let C_i and C_{ij} be as in Definition 2.2. Then using Lemma 2.10, induction on k and Lemma 2.11 we have for any $t > 0$

$$\begin{aligned} \mathbb{P}\ &(\bar{E}_a(\epsilon)|C_{ij}) = \mathbb{P}(f - \mathbb{E}(f) > \epsilon|C_{ij}) \\ \le\ & \exp(-t\epsilon) E\left(e^{t(f-\mathbb{E}(f))}|C_{ij}\right) \\ =\ & \exp(-t\epsilon) \\ \times\ & E\left[\mathbb{E}(1_{C_{ij}} e^{t(f-\mathbb{E}(f))}|\mathcal{F}_{n-1})\right] \cdot \mathbb{P}(C_{ij})^{-1} \\ \le\ & \exp(-t\epsilon)\exp(\frac{1}{2}t^2 \sum_{k=i+1}^n d_k^2) \\ \times\ & E\left(1_{C_{ij}} e^{t(f_i-\mathbb{E}(f))}\right) \mathbb{P}(C_{ij})^{-1} \\ =\ & U_{ij}^{(a)}(\omega_1, \ldots, \omega_{i-1}). \end{aligned}$$

(ii) $\mathbb{P}$-convexity of $U^{(a)}$ is proved as follows. Again the special choice $\epsilon = \epsilon_1$ is not needed, so let $\epsilon > 0$ be arbitrary and let $\mu_{ij} := \mu_{ij}(\omega_1, \ldots, \omega_{i-1})$ as in Definition 2.2 (ii). Then

$$\begin{aligned} & \sum_{j=0}^1 U_{ij}^{(a)}(\omega_1, \ldots, \omega_{i-1})\mu_{ij} \\ =\ & \exp(-t\epsilon)\exp(\frac{1}{2}t^2 \sum_{k=i+1}^n d_k^2) \\ \times\ & \sum_{j=0}^1 \mu_{ij} \exp(t[\mathbb{E}(f|C_{ij}) - \mathbb{E}(f)]) \\ =\ & (1). \end{aligned}$$

With Lemma 2.11 we get

$$\exp(t\mathbb{E}(f|C_{ij})) = E\left(e^{tf_i}|C_{ij}\right) .$$

With this and Lemma 2.10 we have the estimates

$$
\begin{aligned}
(1) \quad &= \quad \exp(-t\epsilon)\exp(\frac{1}{2}t^2\sum_{k=i+1}^{n}d_k^2) \\
&\quad \times \; E\left(e^{tf_i}|C_i\right)\cdot\exp(-t\mathbb{E}(f)) \\
&\leq \; U_{i-1,\omega_{i-1}}^{(a)}(\omega_1,\ldots,\omega_{i-2})
\end{aligned}
$$

(iii) Now we choose special values for ϵ and t. Let $\epsilon := \epsilon_1$ and $t := t_1 = \epsilon_1\Delta^{-1}$. We show the initial condition

$$
\min U_{1j}^{(a)} \leq \delta.
$$

Let $C_{1j} = \{\omega \in \Omega;\ \omega_1 = j\}$ and $p_{1j} = \mathbb{P}(C_{1j})$. Then

$$
\begin{aligned}
U_1^{(a)}(\omega_1) &= \min_{j=0,1}U_{1j}^{(a)} \\
&\leq \sum_{j=0}^{1}p_{1j}U_{1j}^{(a)} \\
&= \exp(-t\epsilon)\exp(-t\mathbb{E}(f))\exp(\frac{t^2}{2}\sum_{k=2}^{n}d_k^2) \\
&\quad \times \sum_{j=0}^{m-1}p_{1j}\exp(-t\mathbb{E}(f|C_{1j})) \\
&= \exp(-t\epsilon)\exp(\frac{t^2}{2}\sum_{k=2}^{n}d_k^2) \\
&\quad \times E\left(e^{t[\mathbb{E}(f|\mathcal{F}_1)-\mathbb{E}(f)]}|\mathcal{F}_0\right) \\
&\leq \exp(-t(\epsilon-\frac{t}{2}\sum_{k=1}^{n}d_k^2)) \quad \text{(Lemma 2.10)} \\
&= \delta < 1.
\end{aligned}
$$

$\square$

3 Taylor Polynomials and Approximate Pessimistic estimators

In this section we undertake the final step in the construction of the pessimistic estimators. The idea can be described as follows: Suppose a family $U^{(a)}$ is a weak pessimistic estimator. Suppose furthermore that the $U_{ij}^{(a)}$ can be approximated by polynomial-time computable functions up to a specified precision $\gamma \geq 0$. Let $q(n) \geq 1$ be a polynomial with $\delta + \frac{\delta}{q(n)} < 1$. Then with

$\gamma = \frac{\delta}{(4n-1)q(n)}$ we approximate $U_{ij}^{(a)}$ by functions $V_{ij}^{(a)}$ up to the absolute error γ and verify that a family of the form $(V^{(a)} + (4n-1)\gamma)$ is the desired pessimistic estimator in the RAM model.

We need the following central lemma, which assures that the exponential function can be approximated on the RAM model in an efficient way.

Lemma 3.1 [18] Let a, b, c, d, γ be rational numbers with encoding length at most L. Suppose that $b, d \geq 1$ and $0 < \gamma \leq 1$. Let T denote the Taylor polynomial of the exponential function. Then rational numbers x and y with

$$
|e^{a\ln b + \sqrt{c\ln d}} - T(x)T(y)| \leq \gamma
$$

can be computed in
$O(\max(n, |a| + |c|)\log(\frac{|b|+|d|}{\gamma}))$-time using at most $O(L[\max(n, |a|+|c|)\log(\frac{|b|+|d|}{\gamma})]^2)$ space.

The main theorem of this section is:

Theorem 3.2 Let f be a quadratic form or a linear form. Let $0 < \delta < 1$ and $\epsilon_i = \sqrt{2\sum_{k=1}^{n}d_k^2\ln\frac{2i}{\delta}}$, $(i = 1, 2)$. Then

(i) Pessimistic estimators for the events $\bar{E}_a(\epsilon_1)$, $\bar{E}_b(\epsilon_1)$, $\bar{E}_c(\epsilon_2)$ can be computed in $O(n^2\log\frac{1}{\delta})$-time.

(ii) The procedures $D-WALK(\bar{E}_a(\epsilon_1))$, resp. $D-WALK(\bar{E}_b(\epsilon_1))$, resp. $D-WALK(\bar{E}_c(\epsilon_2))$ find in $O(n^3\log\frac{1}{\delta})$ time a $x \in \Omega$ such that $f(x) \leq \mathbb{E}(f) + \epsilon_1$, resp. $f(x) \geq \mathbb{E}(f) - \epsilon_1$, resp. $|f(x) - \mathbb{E}(f)| \leq \epsilon_2$.

(iii) In the linear case ($D = 0$) the time for the computation of the pessimistic estimator is $O(n\log\frac{1}{\delta})$ and the running time for the D-WALK procedure is $O(n^2\log\frac{1}{\delta})$.

Remark As the running time in Theorem 3.2 is independent of the encoding length we have indeed a strongly polynomial algorithm.

Proof of Theorem 3.2: We analyse the running times and refer for space considerations to the full paper. Let f be a quadratic form. We consider the event $E_a(\epsilon_1)$. The argumentation for the other two events and for linear forms goes similar. By Definition 2.8

$$U_{ij}^{(a)}(\omega_1, ..., \omega_{i-1})$$

$$= \exp(-t_1[\epsilon_1 - \frac{t_1}{2}(d_{i+1}^2 + ... + d_n^2)$$

$$+ \mathbb{E}(f) - \mathbb{E}(f| \ \omega_1, ..., \omega_{i-1}, j)]).$$

With $\Delta := \sum_{k=1}^n d_k^2$, $a := \Delta^{-1} \sum_{k=i+1}^n d_k^2 - 2$, $b = \frac{2}{\delta}$ and

$$c := \frac{4}{\delta\Delta}(\mathbb{E}(f| \ \omega_1, ..., \omega_{i-1}, j) - \mathbb{E}(f))^2$$

the function $U_{ij}^{(a)}$ is rewritten in the simple form

$$U_{ij}^{(a)}(\omega_1, ..., \omega_{i-1}) = \exp(a \ln b + \sqrt{c \ln d}).$$

Invoking Lemma 3.1 we can compute a rational number y and a polynomial $S(y)$ with

$$|\exp(a \ln b + \sqrt{c \ln d}) - S(y)| \leq \frac{\gamma}{2}$$

in $O(n^2 \log \frac{1}{\delta})$ time. Now define the family $W^{(a)}$ by

$$W_{ij}^{(a)}(\omega_1, ..., \omega_{i-1}) = S(y) + \frac{(2n-i)\delta}{(4n-1)}.$$

Then by the weak pessimistic estimator property (Theorem 2.9) it is not difficult to show that the family $W^{(a)}$ is a pessimistic estimators for the event $\bar{E}_a(\epsilon_1)$.

We have to compute on each of the n levels the two functions functions $W_{i1}^{(a)}$ and $W_{i0}^{(a)}$, hence the running time of the D-WALK procedure is $O(n^3 \log \frac{1}{\delta})$ (resp. $O(n^2 \log \frac{1}{\delta})$ in the linear case) and the assertions (i) – (iii) of the theorem are proved. □

4 Lattice Approximation

An instance of the *quadratic lattice approximation* problem is a symmetric $n \times n$ matrix D, a $(r-1) \times n$ matrix C, rational vectors $c, p \in [0, 1]^n$ and an objective function $x \to c^T x + x^T Dx$ $(x \in [0, 1]^n)$. The problem is to find a lattice point $q \in \{0, 1\}^n$ in polynomial-time such that

(a) $|c^T(p - q) + p^T Dp - q^T Dq|$ is small

(b) $\|C(p - q)\|_\infty$ is small.

We assume that the entries of D, C are rational numbers and $0 \leq c_{ij} \leq 1$. Let L be the encoding length of (D, C, c, p). Let $d := 2 \max_{1 \leq i \leq r} \sum_{j=1}^n |d_{ij}|$. Derandomization gives the following result

Theorem 4.1 *Let* $\alpha \geq 0$ *and* $trace(D) \leq \alpha d\sqrt{n}$. *Then the procedure D-WALK finds in* $O(\ n^2 r \log r + n^3 \ )$*-time a vector* $q \in \{0, 1\}^n$ *such that*

(i) $|c^T(p - q) + p^T Dp - q^T Dq| \leq 2\sqrt{n \ln 2r} + (3 + \alpha)d\sqrt{n}$

(ii) $\|C(p - q)\|_\infty \leq 2\sqrt{n \ln 2r}.$

Proof. Define $\epsilon_1 = 2\sqrt{n \ln 2r}$ and $\epsilon_2 = 3d\sqrt{n}$. Let f be the function $f(x) = x^T Dx$, $x \in [0, 1]^n$. Denote by $\bar{E}_0$ the event

$$"|c^T(p - q)| > \epsilon_1",$$

by $\bar{E}_i$, $i = 1, ..., r - 1$ the events

$$"|\sum_{j=1}^n c_{ij}(p_j - q_j)| > \epsilon_1",$$

and by $\bar{E}_r$ the event

$$"|q^T Dq - \mathbb{E}(f)| > \epsilon_2".$$

By Theorem 3.2 (i) we can compute a pessimistic estimator $W^{(n)}$ for the event $\bar{E}_r$ in $O(n^2)$-time. By Theorem 3.2 (iii) pessimistic estimators $W^{(i)}$ for the events $\bar{E}_i$, $i = 0, ..., r - 1$, can be computed in $O(n \log r)$-time. Then by Proposition 2.4 $W := \sum_{i=0}^r W^{(i)}$ is a pessimistic estimator for the event $\bar{E}_0 \vee ... \vee \bar{E}_{r-1} \vee \bar{E}_r$ and can be computed in $O(rn \log r + n^2)$-time. Hence $D-WALK(\bar{E}_0 \vee ... \vee \bar{E}_{r-1} \vee \bar{E}_r)$ finds in $O(\ n^2 r \log r + n^3 \ )$-time a vector $q \in \{0, 1\}^n$ such that

(a) $|q^T Dq - \mathbb{E}(f)| \leq 3d\sqrt{n}$

(b) $\|[C, c](p - q)\|_\infty \leq 2\sqrt{n \ln 2r}.$

Let (ξ_i) be the Bernoulli trials under considerations defined trough $\mathbb{P}(\xi_i = 1) = p_i$ and $\mathbb{P}(\xi_i = 0) = 1 - p_i$. Then the expectation $\mathbb{E}(f)$ is:

$$\mathbb{E}(f) = \mathbb{E}(\sum_{i,j=1}^n d_{ij}\xi_i\xi_j) = \sum_{i \neq j} d_{ij}p_ip_j + \sum_{i=1}^n d_{ii}p_i.$$

183

But this together with

$$|p^T Dp - \mathbb{E}(f)| \leq \sum_{j=1}^{n} d_{jj}(p_j - p_j^2) \leq \alpha d\sqrt{n},$$

implies the theorem.

□

Remark Theorem 4.1 shows the similarity between the linear and quadratic discrepancy bounds ($n = r$). In the quadratic case we have an $O(d\sqrt{n})$ bound, while in the linear case the algorithmic reachable bound is $O(\sqrt{n \ln n})$ and the existence bound of Spencer is $O(\sqrt{n})$. It is known that Spencer's bound is sharp for Hadamard matrices. The interesting question arising here is whether the gap factor d reflects the quadratic behaviour and so is best possible or not. For small d, i.e.
$d = O(n^{\frac{1}{2}-\epsilon})$, $0 < \epsilon \leq \frac{1}{2}$, and if the trace of D is not too large trace our bound is good compared with the greedy bound $O(n)$, which is also the worst case discrepancy (attained for $D = (d_{ij}), d_{ij} = 1$ for all i, j and $p = (\frac{1}{2}, \ldots, \frac{1}{2})$). It would be interesting to exhibit more classes of matrices where lattice approximations beating the $O(n)$ greedy bound are possible.

In the weighted linear case Theorem 4.1 gives an $O(rn^2 \log r)$-time algorithm achieving discrepancies within $2\sqrt{n \ln 2r}$:

Corollary 4.2 *The procedure D-WALK finds in $O(rn^2 \log r)$-time a vector $q \in \{0,1\}^n$ such that*
$$\|[C, c](p - q)\|_\infty \leq 2\sqrt{n \ln 2r}.$$

□

Remark Raghavan improved the Beck-Fiala bound using Angluin-Valiant type inequalities. He showed a derandomized algorithm which achieves
$\|C(p - q)\|_\infty \leq \max_{1 \leq i \leq n} s_i D(s_i, \frac{1}{2r})$, where $s_i = \sum_{j=1}^{n} c_{ij} p_j$. Unfortunately in the weighted case the algorithm has no polynomial-time implementation, because the numbers $D(s_i, \frac{1}{2r})$ and $e^{c_{ij}}$ cannot be computed efficiently in the RAM model. But using the upper bounds of Raghavan on $D(s_i, \frac{1}{2r})$ (see [15], 1.13 and 1.14) and Taylor approximations the following result can be proved in a similar way as in Theorem 3.2:

Theorem 4.3 *Let $\Delta_i = ([C, c]p)_i$. Derandomization gives an $O(rn^2 \log r)$-time algorithm, which finds a vector $q \in \{0,1\}^n$ such that*

(i) $\Delta_i \leq 3\sqrt{s_i \ln 2r}$, if $s_i > \ln 4r$ for all i.

(ii) $\Delta_i \leq 6 \ln 2r$, if $s_i \leq \ln 4r$ for all i.

□

5 Concluding Remarks

(a) We have presented an algorithmic version of the classical martingale inequality of Azuma and showed its application to the quadratic lattice approximation problem. Since in the case of a 0-1 weighted linear form, m constraints and all deviation parameters being integers derandomization runs in $O(mn)$ time ([15], p.138), it is an interesting question, if the running time of $O(mn^2 \log \frac{m}{\delta})$ in the weighted case can be improved towards $O(mn)$.

(b) It would also be very interesting to exhibit other classes of matrices D of the quadratic lattice approximation problem with an $O(\sqrt{n \ln r} + d\sqrt{n})$ (or even better) discrepancy bound ?

(c) Our algorithms are sequential. The interesting question here is, whether one can parallelize them as Berger/Rompel [6] and Motwani/Naor and Naor [13] showed for some linear problems. This might be possible using estimates on the variation of functions with small martingale differences.

(e) With algorithmic counterparts of the inequalities of Chernoff, Hoeffding and Angluin/Valiant on large deviations of *rational weighted* sums of Bernoulli trials [12] developed in the companion paper [18] some new applications of randomized rounding and derandomization can be showed [19], [20]. It would be interesting to give more examples.

Acknowledgement We would like to thank Dr. Sachin Patkar for his help in preparing this paper.

184

References

[1] N. Alon, J. Spencer, P. Erdős; *The prababilistic method.* John Wiley & Sons, Inc. 1992.

[2] N. Alon; *A parallel algorithmic version of the Local Lemma.* Random Structures and Algorithms, Vol.2, No.4 (1991), 367-378.

[3] K. Azuma, *Weighted sums of certain dependent variables.* Tohoku Math. Journ. 3, (1967), 357-367.

[4] J. Beck, Y. Fiala; *Integer-making theorems.* Discrete Appl. Math. 3 (1991), 1-8.

[5] J. Beck, J. Spencer; *Balancing Matrices with line shifts.* Combinatorica 3, Vol. 3-4, (1983), 299-304.

[6] B. Berger, J. Rompel; *Simulating (log^{c}n)-wise Independence in NC.* Proceeding of FOCS 1989, IEEE Coputer Society Press, Los Alamitos, CA, 2-8.

[7] B. Bollobás; *The chromatic number of random graphs.* Combinatorica 8, (1988), 49-56.

[8] H. Černov; *A measure of asymptotic efficiency for test of a hypothesis based on the sum of observation.* Ann. Math. Stat. 23, (1952), 493-509.

[9] W. Hoeffding; *On the distribution of the number of success in independent trials.* Annals of Math. Stat. 27, (1956), 713-721.

[10] J. Matousek, E. Welzl, L. Wernisch; *Discrepancy and ϵ-approximations for bounded VC-dimension* Report B 91-06, Institute for Computer Science, FU Berlin, (1991). Sringer-Verlag (1984)

[11] K. Mehlhorn; *Data structures and algorithms 1: Sorting and Searching.* Sringer-Verlag (1984)

[12] C. McDiarmid; *On the Method of Bounded Differences.* Surveys in Combinatorics, 1989. J. Siemons, Ed.: London Math. Soc. Lectures Notes, Series 141, Cambridge University Press, Cambridge, England 1989.

[13] R. Motwani, J. Naor, M. Naor; *The probabilistic method yields deterministic parallel algorithms.* Proceedings 30the IEEE Conference on Foundation of Computer Science (FOCS'89), (1989), 8 -13.

[14] P. Raghavan, C. D. Thompson; *Randomized Rounding: A technique for provably good algorithms and algorithmic proofs.* Combinatorica 7 (4), (1987), 365-374.

[15] P. Raghavan; *Probabilistic construction of deterministic algorithms: Approximating packing integer programs.* Jour. of Computer and System Sciences 37, (1988), 130-143.

[16] E. Shamir, J. Spencer; *Sharp concentration of the chromatic number of random graphs $G_{n,p}$.* Combinatorica 7, (1987), 121-129.

[17] J. Spencer; *Ten lectures on the probabilistic method.* SIAM, Philadelphia (1987).

[18] A. Srivastav, P. Stangier; *Algorithmic Chernoff-Hoeffding inequalities in integer programming.* Bonn (1993), submitted to Random Structures & Algorithms.

[19] A. Srivastav, P. Stangier; *Integer multicommodity flows with reduced demands.* to appear in Springer Lecture Notes in Computer Science (1993) (Proceedings of the First Annual European Symposium on Algorithms (ESA'93), Bonn 1993).

[20] A. Srivastav, P. Stangier; *Weighted fractional and integral k-matching in hypergraphs.* to appear in Disc.Appl.Math. (1994).

A 2/3–Approximation of the Matroid Matching Problem

Toshihiro Fujito*

Department of Computer Science, The Pennsylvania State University
University Park, PA 16802, USA

Abstract. The computational complexity of the matroid matching problem, which generalizes the matching problem in general graphs and the matroid intersection problem, remains unresolved. Under the general assumption it can be shown to be exponential complexity. In this paper an approximation algorithm is given which achieves at least $\frac{2}{3}$ of the optima under the same assumption. The theorems behind analysis of the algorithm also shed some light on the structure of matroid matching.

1 Introduction

Suppose we are given a matroid $\mathcal{M}$ over a finite set S of elements and a family $\mathcal{F}$ of k–element subsets of S. We are interested in finding the maximum size subfamily $\mathcal{I}$ of $\mathcal{F}$ s.t. the union of sets in $\mathcal{I}$ forms an independent set in $\mathcal{M}$. While this problem is easily solvable by the greedy algorithm when $k = 1$, it becomes $\mathcal{NP}$–hard (actually MAX $\mathcal{SNP}$–hard) for any $k \geq 3$.

In this paper we study the case of $k = 2$. This version of the problem is called *matroid matching* (also known as *matroid parity*), and any subfamily the union of which is independent is called a *matching*. We conjecture that this problem shares the properties of the cases when $k \geq 3$ by not having polynomial–time exact solutions and the case of $k = 1$ by having a polynomial–time approximation schema.

Matroid matching is a generalization of the matching problem in general graphs, the matroid intersection problem, and many other problems. Consequently, it arises in many applications (see [Lov80a, Rec89]). Besides its practical value, matroid matching is theoretically interesting because it lies on the boundary of polynomial–time solvability and unsolvability. The aforementioned two problems, the graph matching and the matroid intersection, both are known to have good characterizations as well as poly–time algorithms for their optima. On the other hand, the computational complexity of the general matroid matching problem remains unresolved. In particular, it was shown, by Lovász [Lov81] and Jensen and Korte [JK82] independently, that exponential time is required if the matroid is given by the oracle computing its rank function. This shows that no polynomial–time algorithm can work for all underlying matroids

---

* This work was partially supported by NSF Grant CCR-9114545

(for the discussion on the oracle computation model, see [LP86]). Some special cases of this problem have been found to be polynomial–time solvable; most importantly, Lovász [Lov80b, Lov81] discovered a polynomial–time algorithm for the linear matroid matching problem, which was later improved and simplified [GS86, OVV90].

In this paper we consider the approximation of the general matroid matching problem under the oracle computation model. While it is easy to see that a greedy heuristic finds a matching of size $\frac{1}{2}$ of the optimal one, we show that a slightly more complicated heuristic achieves the performance ratio of $\frac{2}{3}$. At the end of this paper we discuss about a natural generalization of that algorithm and, as its consequence, the possibility of approximation schema for the general matroid matching.

A matroid $\mathcal{M}$ is represented throughout the paper as a pair of ground set S and rank function r. A *contraction* of $\mathcal{M} = (S, r)$ at $X \subseteq S$ is a matroid $\mathcal{M}_X = (S \setminus X, r_X)$, where $r_X(Y) = r(Y \cup X) - r(X)$ for all $Y \subseteq S \setminus X$. $sp(X)$ denotes the span of X for $X \subseteq S$ while sp_X denotes the span in $\mathcal{M}_X$. Without loss of generality we assume that a given set family $\mathcal{F}$ is a partition of matroid ground set S. Thus, each element $x \in S$ has a unique pair "mate", denoted $\hat{x}$, s.t. $\{x, \hat{x}\}$ is an element of $\mathcal{F}$. For $\mathcal{E} \subseteq \mathcal{F}$, $\overline{\mathcal{E}}$ denotes $\bigcup_{E \in \mathcal{E}} E$.

For a matching $\mathcal{I}$, a set $\mathcal{J} \subseteq \mathcal{F}$ such that $\mathcal{I} \oplus \mathcal{J}$ is a larger matching, is called an *improvement* of $\mathcal{I}$. If $\mathcal{J}$ has $k - 1$ elements in $\mathcal{I}$ and k elements in $\mathcal{F} \setminus \mathcal{I}$, we say that it is a *k–improvement*.

2 Main results and approximation

We now present an approximation algorithm for the matroid matching problem for $\mathcal{M} = (S, r)$ and $\mathcal{F}$, where $\mathcal{M}$ is given by the oracle computing r:

1. $\mathcal{I} \leftarrow \emptyset$
2. For every $\beta \in \mathcal{F}$
   If β is a 1–improvement of $\mathcal{I}$ then
   $\qquad \mathcal{I} \leftarrow \mathcal{I} + \beta$
3. $\Upsilon \leftarrow \{\{\alpha, \beta_1, \beta_2\} | \alpha \in \mathcal{I}, \beta_1, \beta_2 \in \mathcal{F} \setminus \mathcal{I}\}$
4. For every $\mathcal{J} \in \Upsilon$
   If $\mathcal{J}$ is a 2–improvement of $\mathcal{I}$ then
   $\qquad \mathcal{I} \leftarrow \mathcal{I} \oplus \mathcal{J}$

Theorem 1. *If a matching $\mathcal{I}$ does not have any 1– nor 2–improvement, and $\mathcal{N}$ is another matching, then $|\mathcal{I}| \geq \frac{2}{3}|\mathcal{N}|$.*

The proof of this theorem is given in the next section.

Theorem 2. *The algorithm given above yields a matching that does not have any 1– nor 2–improvement.*

From these two theorems it follows that the approximation ratio of the algorithm is $\frac{2}{3}$. Also note that the algorithm uses at most $|\mathcal{F}|^3$ oracle queries.

Proof of Theorem 2. Obviously the algorithm produces a matching. It is also clear that after step 2 $\mathcal{I}$ does not have any 1–improvement.

Claim 3. *Suppose $\mathcal{I}$ has no 1–improvement. If $\{\alpha, \beta_1, \beta_2\}$ is a 2–improvement for $\mathcal{I}$, $sp(\overline{\mathcal{I}}) \subseteq sp(\overline{\mathcal{I} \oplus \{\alpha, \beta_1, \beta_2\}})$.*

Proof. Assume otherwise. Then, $\alpha \not\subseteq sp(\overline{\mathcal{I} \oplus \{\alpha, \beta_1, \beta_2\}})$, which implies that $r(\overline{\mathcal{I} + \beta_1 + \beta_2}) > r(\overline{\mathcal{I} \oplus \{\alpha, \beta_1, \beta_2\}}) = r(\overline{\mathcal{I}}) + 2$, which in turn implies that either $\mathcal{I} + \beta_1$ or $\mathcal{I} + \beta_2$ is a matching. $\square$

This claim shows that an application of 2–improvement in step 4 does not introduce a new 1–improvement, and hence that the algorithm produces a matching which has no 1–improvement. Also implied is that once $\{\alpha, \beta_1, \beta_2\}$ is found not to be a 2–improvement in step 4, it remains not, throughout.

Claim 4. *Once $\{\alpha, \beta_1, \beta_2\}$ is applied as a 2–improvement in step 4, there is no 2–improvement involving $\alpha, \beta_1,$ or β_2 for $\mathcal{I}$ after step 4.*

Proof. Let $\mathcal{I}' = \mathcal{I} \oplus \{\alpha, \beta_1, \beta_2\}$ be a matching. Due to Claim 1, it suffices to show that $\mathcal{I}'$ does not have a 2–improvement involving $\alpha, \beta_1,$ or β_2. Suppose $\{\alpha', \beta_1', \beta_2'\}$ is a 2–improvement for $\mathcal{I}'$, and let $\mathcal{I}'' = \mathcal{I}' \oplus \{\alpha', \beta_1', \beta_2'\}$.
If $\alpha' = \beta_1$ then $\mathcal{I}'' = \mathcal{I} - \alpha \cup \{\beta_2, \beta_1', \beta_2'\}$.
If $\beta_1' = \alpha$ then $\mathcal{I}'' = \mathcal{I} - \alpha' \cup \{\beta_1, \beta_2, \beta_2'\}$.
    Either case implies the existence of 1–improvement followed by a 2–improvement on $\mathcal{I}$. Hence, it contradicts the assumption that $\mathcal{I}$ has no 1–improvement. $\square$

This claim concludes that there is no 2–improvement after step 4. $\square$

3 Proof of Theorem 1

The following base exchange theorem for matroids is useful in proving the disjoint attack lemma.

Base Exchange Theorem [Gre73, Woo74]. *If B_1, B_2 are bases of the matroid $\mathcal{M}$ and $X_1 \subseteq B_1$ then there exists $X_2 \subseteq B_2$ s.t. $(B_1 \setminus X_1) \cup X_2$ and $(B_2 \setminus X_2) \cup X_1$ are both bases of $\mathcal{M}$.*

In $\mathcal{M}$ over S, $\{\alpha, b_1, b_2\}$ is called an *attack* for a matching $\mathcal{I}$, where $\alpha \in \mathcal{I}$ and $b_1, b_2 \in S$, if $sp(\overline{\mathcal{I} \oplus (\alpha \cup \{b_1, b_2\})}) = sp(\overline{\mathcal{I}})$.

Lemma 5. *Given a matching $\mathcal{I}$ and an independent set I s.t. $I \subseteq sp(\overline{\mathcal{I}})$, there exist $|I| - |\mathcal{I}|$ disjoint attacks in $\mathcal{I} \cup I$.*

Proof by induction on $|\mathcal{I}|$. Take the matroids $\mathcal{M} = (\overline{\mathcal{I}} \cup I, r)$ and $\mathcal{M}_\alpha = (\overline{\mathcal{I} - \alpha} \cup I, r_\alpha)$ for any $\alpha \in \mathcal{I}$. Since $r_\alpha(\overline{\mathcal{I} - \alpha}) = r(\overline{\mathcal{I}}) - 2 = r(\overline{\mathcal{I} \cup I}) - 2 = r_\alpha(\overline{\mathcal{I} - \alpha} \cup I)$, $\overline{\mathcal{I} - \alpha}$ is a base of $\mathcal{M}_\alpha$. Also observe that an attack on $\overline{\mathcal{I} - \alpha}$ in $\mathcal{M}_\alpha$ is an attack on $\mathcal{I}$ in $\mathcal{M}$.

Case 1. $r_\alpha(I) = r(I)$. Then, I is still independent in $\mathcal{M}_\alpha$. Apply the induction hypothesis to $\mathcal{I} - \alpha$ and I in $\mathcal{M}_\alpha$

Case 2. $r_\alpha(I) < r(I)$. Extend I by $X \subseteq \overline{\mathcal{I}}$ so that $I \cup X$ is a base of $\mathcal{M}$. Apply the base exchange theorem to $\overline{\mathcal{I} - \alpha} \cup \alpha$ and $I \cup X$, and find $\beta \subseteq I \cup X$ s.t. both $\overline{\mathcal{I} - \alpha} \cup \beta$ and $(I \cup X) \setminus \beta \cup \alpha$ are bases.

Since $r(I \cup \alpha) \leq r(I) + 1$, $\alpha \not\subseteq X$ and hence, $\beta \not\subseteq X$.

Suppose $|\beta \cap X| = |\beta \cap I| = 1$, and let $b \in \beta \cap I$. Then, $I - b \cup \alpha \subseteq (I \cup X) \setminus \beta \cup \alpha$ is independent in $\mathcal{M}$, and so is $I - b$ in $\mathcal{M}_\alpha$. Thus, by the induction hypothesis there exist $|I - b| - |\mathcal{I} - \alpha| = |I| - |\mathcal{I}|$ many disjoint attacks for $\mathcal{I} - \alpha$ in $\mathcal{M}_\alpha \setminus b = (\overline{\mathcal{I} - \alpha} \cup (I - b), r_\alpha)$.

Now suppose $\beta = (b_1, b_2) \subseteq I$. Then, $\{\alpha, b_1, b_2\}$ is an attack for $\mathcal{I}$, and $I \setminus \beta$ is independent in $\mathcal{M}_\alpha$. By the induction hypothesis, there exist $|I \setminus \beta| - |\mathcal{I} - \alpha| = |I| - |\mathcal{I}| - 1$ many disjoint attacks for $\mathcal{I} - \alpha$ in $\mathcal{M}_\alpha \setminus \beta = (\overline{\mathcal{I} \setminus \alpha} \cup I \setminus \beta, r_\alpha)$. $\square$

Now we proceed to the proof of Theorem 1. Let $n = |\mathcal{N}|$ and $m = |\mathcal{I}|$. Assuming that a matching $\mathcal{I}$ has no 1-improvement, consider a matroid $\mathcal{M} = (S, r)$, where $S = \overline{\mathcal{I}} \cup \overline{\mathcal{N}}$. Also partition $\mathcal{N}$ into $\mathcal{N}_0 + \mathcal{N}_1$ s.t. $\mathcal{N}_0 = \{\beta | \beta \cap sp(\overline{\mathcal{I}}) = \emptyset\}$ and $\mathcal{N}_1 = \{\beta | \beta \cap sp(\overline{\mathcal{I}}) \neq \emptyset\}$

Take $X \subseteq \overline{\mathcal{N}_0}$ s.t. $\overline{\mathcal{I}} \cup X$ is a maximal independent subset of $sp(\overline{\mathcal{I}}) \cup \overline{\mathcal{N}_0}$. Notice that $\beta \not\subseteq X$ for every $\beta \in \mathcal{N}_0$.

Consider $\mathcal{M}_X = (S \setminus X, r_X)$ and observe now that $\overline{\mathcal{N}_0} \setminus X \subseteq sp_X(\overline{\mathcal{I}})$. From $\mathcal{N}_0$ extract a set $\mathcal{N}_0^1$ of singletons and a set $\mathcal{N}_0^2$ of pairs s.t. $\overline{\mathcal{N}_0} \setminus X = \mathcal{N}_0^1 \cup \overline{\mathcal{N}_0^2}$, $\mathcal{N}_0^1 = \{b | \hat{b} \in X\}$ and $\mathcal{N}_0^2 = \{\beta | \beta \cap X = \emptyset\}$.

To apply the disjoint attack lemma, take a "representative" element, from each pair of $\mathcal{N}_1 \cup \mathcal{N}_0^2$, which belongs to $sp_X(\overline{\mathcal{I}})$. Now inside $sp_X(\overline{\mathcal{I}})$ there are n elements of $\overline{\mathcal{N}}$, each from a distinct pair of $\mathcal{N}$.

Find $n - m$ disjoint attacks, $\{\alpha, b_1, b_2\}$, for $\mathcal{I}$ in $\mathcal{M}_X$.

Claim 6. *In each of the following cases, there exists a 2-improvement for $\mathcal{I}$ in $\mathcal{M}$.*

1. $b_1, b_2 \in N_0^1$
2. $b_1 \in N_0^1, b_2 \notin N_0^1$, and $r_X(\overline{\mathcal{I} - \alpha} \cup \{b_1, b_2, \hat{b}_2\}) = r_X(\overline{\mathcal{I}}) + 1$
3. $b_1 \notin N_0^1, b_2 \notin N_0^1$, and $r_X(\overline{\mathcal{I} - \alpha} \cup \{b_1, \hat{b}_1, b_2, \hat{b}_2\}) = r_X(\overline{\mathcal{I}}) + 2$

Proof. Each case claims that $\overline{\mathcal{I} - \alpha} \cup ((\beta_1 \cup \beta_2) \setminus X)$ is independent in $\mathcal{M}_X$, where $b_1(b_2)$ is a representative element of $\beta_1(\beta_2)$ from $\mathcal{N}$. Hence, $\overline{\mathcal{I} - \alpha} \cup (\beta_1 \cup \beta_2) \subseteq \overline{\mathcal{I} - \alpha} \cup ((\beta_1 \cup \beta_2) \setminus X) \cup X$ is independent in $\mathcal{M}$. $\square$

Thus, by the assumption that there is no 2-improvement for $\mathcal{I}$ from $\mathcal{N}$ in $\mathcal{M}$, each attack falls into one of the following two types:

1. $b_1 \in N_0^1, b_2 \notin N_0^1$, and $r_X(\overline{\mathcal{I} - \alpha} \cup \{b_1, b_2, \hat{b}_2\}) = r_X(\overline{\mathcal{I}})$
2. $b_1 \notin N_0^1, b_2 \notin N_0^1$, and $r_X(\overline{\mathcal{I} - \alpha} \cup \{b_1, \hat{b}_1, b_2, \hat{b}_2\}) \leq r_X(\overline{\mathcal{I}}) + 1$

Rearrange a set of $\mathcal{N}$-elements by taking $\mathcal{N}_2 = \mathcal{N}_1 \cup \mathcal{N}_0^2$ as a set of pairs, and N_0^1 as a set of singletons. Let $n_2 = |\mathcal{N}_2|$ and $n_1 = |N_0^1|$. We now consider how much the rank of $sp_X(\overline{\mathcal{I}})$ would be increased by adding elements of $\mathcal{N}_2$ and N_0^1 to it.

- no increase if it is from N_0^1.
- no increase if $\beta \in \mathcal{N}_2$ is involved in an attack of type 1.
- if $\beta_1, \beta_2 \in \mathcal{N}_2$ are involved in an attack of type 2, they together can increase it by at most 1.
- increase of at most 1 if $\beta \in \mathcal{N}_2$ is not involved in any attack.

Thus, the total increase is at most $n_2 - (n - m)$, and hence, $r_X(\overline{\mathcal{I}} \cup \overline{\mathcal{N}_2} \cup N_0^1) \leq r_X(\overline{\mathcal{I}}) + n_2 - (n - m)$. Meanwhile, $r_X(\overline{\mathcal{N}_2} \cup N_0^1) = 2n_2 + n_1$. Therefore, $2n_2 + n_1 \leq 2m + n_2 - (n - m)$, and it follows that $2n \leq 3m$. This completes the proof of Theorem 1.

4 Discussion

One natural and perhaps most important question arising from this paper's results is whether the performance ratio of $\frac{k}{k+1}$ (or even just the one better than $\frac{2}{3}$) can be achieved by sequential application of i-improvements for i running from 1 upto k. Even if this approach fails to generalize Theorem 2, a matching without any i-improvement for $1 \leq i \leq k$ can be obtained in polynomial-time since there are only polynomial many such improvements. Therefore, what is actually needed to have an approximation scheme for the general matroid matching problem is a generalization of Theorem 1:

Conjecture 7. *If a matching $\mathcal{I}$ does not have any i-improvement for $1 \leq i \leq k$ and $\mathcal{N}$ is another matching, then $|\mathcal{I}| \geq \frac{k}{k+1}|\mathcal{N}|$.*

Moreover, this conjecture will be proven true if the following conjecture is true (which can be couched, quite properly, as a base exchange theorem for 2-polymatroids):

Conjecture 8. *For a matroid $\mathcal{M} = (S, r)$ let $\mathcal{B}_1$ be a set of pairs only and $\mathcal{B}_2$ be a set of pairs and singletons over S s.t. $\overline{\mathcal{B}_1}$ and $\overline{\mathcal{B}_2}$ are both bases of $\mathcal{M}$. Then there exist $\mathcal{X}_1 \subseteq \mathcal{B}_1$ and $\mathcal{X}_2 \subseteq \mathcal{B}_2$ s.t. $\overline{(\mathcal{B}_1 \setminus \mathcal{X}_1) \cup \mathcal{X}_2}$ and $\overline{(\mathcal{B}_2 \setminus \mathcal{X}_2) \cup \mathcal{X}_1}$ are both bases of $\mathcal{M}$ and $\mathcal{X}_2$ contains at most two singletons.*

Another important open problem is whether the approximation algorithm given in this paper generalizes for the *weighted* matroid matching problem. The poly-time solvability of this problem, even for the case of linear matroids, is completely unsettled.

5 Acknowledgments

We wish to thank P. Berman and V. Ramaiyer for helpful suggestions and discussions.

References

[Gre73] C. Greene. A multiple exchange property for bases. In *Proc. Amer. Math. Soc. 39*, pages 45–50, 1973.

[GS86] H.N. Gabow and M. Stallman. An augmenting path algorithm for linear matroid parity. *Combinatorica*, 6:123–150, 1986.

[JK82] P. Jensen and B. Korte. Complexity of matroid property algorithms. *SIAM Journal on Computing*, 11:184–190, 1982.

[Lov80a] L. Lovász. Matroid matching and some applications. *J. of Combinatorial Theory (B)*, 28:208–236, 1980.

[Lov80b] L. Lovász. Selecting independent lines from a family of lines in a space. *Acta Scientiarum Mathematicarum*, 42:121–131, 1980.

[Lov81] L. Lovász. The matroid matching problem. In *Algebraic Methods in Graph Theory, Vol. 2*, pages 495–518. North–Holland, 1981.

[LP86] L. Lovász and M.D. Plummer. *Matching Theory*, chapter 11, pages 413–415. North–Holland, 1986.

[OVV90] J.B. Orlin and J.H. Vande Vate. Solving the linear matroid parity problem as a sequence of matroid intersection problems. *Mathematical Programming*, 47:81–106, 1990.

[Rec89] A. Recski. *Matroid Theory and its Applications*, volume 6 of *Algorithms and Combinatorics*. Springer-Verlag, 1989.

[Woo74] D.R. Woodall. An exchange theorem for bases of matroids. *J. Combinatorial Theory (B)*, 16:227–229, 1974.

Extended Abstract

Using Fractal Geometry for Solving Divide-and-Conquer Recurrences

Simant Dube

Department of Mathematics, Statistics and Computing Science

University of New England

Armidale NSW 2351

Australia

Email: simant@germain.une.edu.au

Abstract

A relationship between the fractal geometry and the analysis of recursive (divide-and-conquer) algorithms is investigated. It is shown that the dynamic structure of a recursive algorithm which might call other algorithms in a mutually recursive fashion can be geometrically captured as a fractal (self-similar) image. This fractal image is defined as the attractor of a mutually recursive function system. It then turns out that the Hausdorff-Besikovitch dimension D of such an image is precisely the exponent in the time complexity of the algorithm being modeled. That is, if Hausdorff D-dimensional measure of the image is finite then it serves as the constant of proportionality and the time complexity is of the form $\Theta(n^D)$, else it implies that the time complexity is of the form $\Theta(n^D \log^p n)$ where p is an easily determined constant.

Keywords: Fractals, fractal dimension, Hausdorff-Besikovitch dimension, iterated function systems, divide-and-conquer recurrences.

1   Introduction

The analysis of the time complexity of algorithms is of fundamental importance to computer scientists. A great number of useful algorithms use Divide-and-Conquer approach, in which the original problem is reduced to a number of smaller problems [1, 6]. In this paper, we consider a new fractal geometry based approach to analyze such algorithms, in which the size of a smaller problem is related to that of the original problem by a multiplicative factor.

The problem of analysis of such recursive algorithms reduces to solving divide-and-conquer recurrence relations. A number of methods have been developed for solving such recurrence relations, and also for general recurrence relations [11, 12, 15].

In [6] the Master method to solve divide-and-conquer recurrences is discussed. The Master Method is based on the Master Theorem, which is adapted from [5]. In past literature, mutual recurrence relations of more general nature have been considered.

```
Algorithm A (B[1...n])
Array B;
if n=1 then ..., print("hello");
   else call A(B[1...n/2]),
        call A(B[n/4...3n/4]),
        call A(B[n/2...n]);
end A;
```

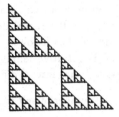

Figure 1: Dynamic Structure of a Recursive Algorithm Captured by Sierpinski Triangle

In [15], such recurrence relations are called multi-dimensional linear first order recurrences. Divide-and-conquer recurrences are called extended first order recurrences. A divide-and-conquer recurrence can be reduced to a (single-dimensional) linear first order recurrence and a secondary recurrence [15].

In this paper, we present an altogether new approach to solve mutual divide-and-conquer recurrences, which gives more general results in a far simpler manner.

Surprisingly, we make use of the recent developments in fractal geometry, which has gained a remarkable popularity among scientists and mathematicians since it was shown by B. Mandelbrot in [13] that many natural objects possess fractal (self-similar) geometries. If one magnifies one of the parts of a self-similar object then it resembles the whole. Clouds, mountains, trees, human circulatory system are examples of fractal objects.

An important step in the development of "computational fractal geometry" is taken by M. Barnsley in [2]. He has introduced Iterated Function Systems (IFS). For image generation and compression purposes, IFS are generalized to Mutually Recursive Function Systems (MRFS) in [7] and are also studied in [9]. MRFS are related to Recurrent IFS introduced in [3]. An interesting special case of MRFS is studied in [8].

An MRFS consists of n variables defined in a mutually recursive fashion as unions of each others under affine transformations. IFS is a special case of MRFS when $n = 1$.

Results on Hausdorff-Besikovitch dimension of objects defined by MRFS are shown in [14] which generalize those in [3]. In this paper, we will be using the results from [14] to build a relationship between fractal geometry and analysis of recursive algorithms.

At a conceptual level, the notion of self-similarity is not limited solely to images but can be used to describe many natural phenomena like distribution of noise on a channel, Brownian motion of particles in air [4]. In this paper, we show that a divide-and-conquer algorithm is also "self-similar" as it is made of its smaller "copies." Here self-similarity is temporal while in case of a natural object it is spatial.

For example, consider a recursive algorithm A as shown in Fig. 1. The algorithm A calls itself 3 times and at each recursive call the input size is halved. The system of recurrence relations is:

$$T(n) = 3T(n/2)$$
$$T(1) = O(1).$$

Note that we assume that all computation is done at the "trivial-case" $n = 1$. Therefore, $T(n) = \Theta(n^{\log_2 3})$.

Now A can be modeled by an IFS which generates the well-known Sierpinski Triangle, see Fig. 1. Its fractal dimension is $\log_2 3$. This is no coincidence as can be intuitively deduced as follows:

Consider the intuitive definition of fractal dimension of a self-similar image $\mathcal{O}$ which implies that if $\mathcal{O}$ has fractal dimension D then

$$\text{(number of self-similar copies)} \approx C(\text{magnification factor})^D \qquad (1)$$

where C is some positive constant. Now consider the recursive algorithm A such that at each of its recursive calls the size of the input is reduced by a multiplicative factor. Each such calls creates a "copy" of A on smaller input. Since we assume that the only computation is done at the "trivial-case" when the size of the input is 1, the total time taken by A on an input of size n is the total number of recursive calls made with input size equal to 1. How many such trivial-case recursive calls are made? For this, we rewrite (1) as

$$\text{(number of recursive calls)} \approx C(\text{magnification factor})^D \qquad (2)$$

In our case

$$\text{magnification factor} = \frac{\text{size of the original input}}{\text{size of the trivial-case input}} = \frac{n}{1} = n.$$

Therefore, from (2) the time complexity of the algorithm A is

$$T(n) \approx Cn^D.$$

Now this interrelationship between divide-and-conquer recurrences and fractals can be generalized: to a system of recurrences and in which the multiplicative factor is any real number. Also one can easily handle the case in which computation is done at other recursion levels besides the trivial-case. This generalization is the aim of this paper.

The main result in this paper is the theorem stated in Section 4.

Consider a group $\mathcal{A}$ of n mutually recursive algorithms and let one algorithm be distinguished as the main algorithm (main "routine" in the terminology of programming languages) which is called first. An algorithm may call itself or any other algorithm. In such a recursive call the size of input is reduced by a multiplicative factor.

Now $\mathcal{A}$ can be modeled as an MRFS $\mathcal{M}$ defined on n variables, such that the execution of $\mathcal{A}$ corresponds "graphically" with the sequence of images generated while executing the Deterministic Algorithm on $\mathcal{M}$. Let the attractor of $\mathcal{M}$ (the fractal image defined by $\mathcal{M}$) be $\mathcal{O}$. Then the theorem states the mathematical relationship between the Hausdorff-Besikovitch dimension D and the Hausdorff D-dimensional measure of $\mathcal{O}$ and the time complexity of $\mathcal{A}$. This theorem is interesting because:

1. It provides a mathematically rigorous method to analyze mutually recursive algorithms.

2. It generalizes the known methods to solve recurrence relations.

3. It provides another link between discrete mathematics and continuous mathematics.

2 Preliminaries

For the definitions of concepts used in this paper see [2, 10]. Let A be an "image" in $X = R^n$ i.e. A is a topologically closed and bounded subset of X. The set of all images in X is denoted by $\mathcal{H}(X)$. Let $\epsilon > 0$. Let $B(x, \epsilon)$ denote the closed ball of radius ϵ and center at a point $x \in X$. That is,

$$B(x, \epsilon) = \{y \in X | d(x, y) \le \epsilon\}.$$

Let $\mathcal{N}(A, \epsilon)$ be the least number of closed balls of radius ϵ needed to cover A. That is,

$$\mathcal{N}(A, \epsilon) = \text{ smallest integer } M \text{ such that } A \subseteq \bigcup_{n=1}^{M} B(x_n, \epsilon),$$

for some set of distinct points $\{x_n | n = 1, 2, \ldots, M\} \subseteq X$.

Let $f(\epsilon)$ and $g(\epsilon)$ be real valued functions of the positive real variable ϵ. Then $f(\epsilon) \approx g(\epsilon)$ means that

$$\lim_{\epsilon \to 0} \frac{\ln(f(\epsilon))}{\ln(g(\epsilon))} = 1.$$

The intuitive idea behind the definition of fractal dimension is that a set A has fractal dimension D if

$$\mathcal{N}(A, \epsilon) \approx C \epsilon^{-D}$$

for some positive constant C. Mathematically we define it as the limit

$$D = \lim_{\epsilon \to 0} \frac{\ln \mathcal{N}(A, \epsilon)}{\ln(1/\epsilon)}$$

if it exists. The notation $D = D(A)$ is used to say that D is the *fractal dimension* of A. Another important property of a fractal image is its *Hausdorff-Besikovich dimension* and *Hausdorff p-dimensional measure*. For the definition of these concepts, refer to [2]. In this paper, for all the fractal images, the fractal dimension and Hausdorff-Besikovich dimension are same and we would use them interchangeably.

3 Recursive Algorithms as Fractals

3.1 Mutually Recursive Algorithms

Consider a group $\mathcal{A} = \{A_1, A_2, \ldots, A_n\}$ of N algorithms. One algorithm is called first and therefore is distinguished as the *main algorithm*. The time complexity of $\mathcal{A}$ is therefore the time complexity of the main algorithm. We assume that there are no unreachable algorithms in $\mathcal{A}$ i.e. it should be possible to call every algorithm during the execution of $\mathcal{A}$ on some input.

The structure of each $A \in \mathcal{A}$ is as follows.

1. The algorithm A can call any algorithm $B \in \mathcal{A}$ such that the size of the input is changed by a multiplicative factor. If the size of original input is n and the size of the input in a recursive call is changed to n/b where $1/b$ is the multiplicative factor, then we interpret n/b as either $\lceil n/b \rceil$ or as $\lfloor n/b \rfloor$. This is because the size of the input must be an integer.

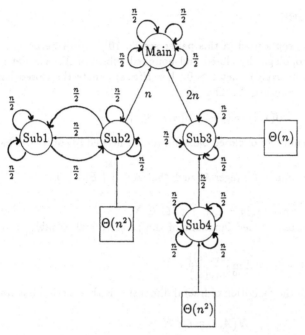

Figure 2: A Group of Mutually Recursive Algorithms

2. The algorithm A can do some additional computation taking $\Theta(n^D)$ steps, where D is a nonnegative real number. If $D = 0$ then this computation can be ignored.

3. The algorithm A performs a constant amount of computation at the trivial case, when the input size is equal to 1.

If there is a chain of recursive calls $B_1, B_2, \ldots, B_m$ where $B_i \in \mathcal{A}, i = 1, 2, \ldots, n$, B_i calls B_{i+1} and $B_1 = B_m$ then the size of the input in call to B_m should be strictly less than the size of the input in call to B_1. In other words, over every possible loop, the size of the input is "contracted" by a multiplicative factor.

Example 1 Consider Fig. 2 which shows a group of mutually recursive algorithms. One algorithm is distinguished as the main routine. The main routine and other "subroutines" call each other recursively. In total, there are four other subroutines. If a routine calls other on input size n/b then there is an appropriately labeled arc from the called routine to the calling routine. The "additional" computation performed by the routines is indicated by nodes drawn as squares. Note that each of the subroutines 2 and 4 perform $\Theta(n^2)$ additional computation and the subroutine 3 performs $\Theta(n)$ additional computation. □

3.2 Mutually Recursive Function Systems (MRFS)

Let $X = R^k$, the k-dimensional Euclidean space, be the underlying metric space.

A *variable* V is a symbol whose "value" at any time is an image. Recall that the set of all images is denoted by $\mathcal{H}(X)$.

Let C be any given image and suppose the fractal dimension of C is known. Then we call C to be a *condensation image*. Define a mapping $W : \mathcal{H}(X) \to \mathcal{H}(X)$ such that for all $A \in \mathcal{H}(X)$, $W(A) = C$. Then W is called the *condensation mapping* defining C.

Now we generalize Mutually Recursive Function Systems (MRFS) as studied in [7, 9], to *condensation MRFS* just to facilitate our discussion.

Informally, in a condensation MRFS a variable is defined as the union of the values of some other variables mapped under affine transformations and some condensation images.

A *condensation MRFS* $\mathcal{M}$ is a triple $(\mathcal{V}, \mathcal{C}, G)$ where:

$\mathcal{V} = \{V_1, V_2, \ldots, V_n\}, n \geq 1$ is a nonempty finite set of variables,

$\mathcal{C} = \{C_1, C_2, \ldots, C_m\}, m \geq 0$ is a finite set of condensation images,

G is the *underlying labeled digraph*.

The n variables in $\mathcal{V}$ are defined mutually recursively as

$$V_i = (w_{i_1}(V_{j_1}) \cup w_{i_2}(V_{j_2}) \cup \ldots \cup w_{i_r}(V_{j_r})) \cup (C_{k_1} \cup C_{k_2} \cup \ldots \cup C_{k_t}), \; i = 1, 2, \ldots, n$$

for some variables $V_{j_1}, V_{j_2}, \ldots, V_{j_r}$ (which need not be distinct) in $\mathcal{V}$, and for some affine transformations $w_{i_1}, w_{i_2}, \ldots, w_{i_r}$, and for some condensation images $C_{k_1}, C_{k_2}, \ldots, C_{k_t}$ in $\mathcal{C}$, and where t and r depend on i.

The labeled digraph G has $n+m$ nodes representing the n variables and m condensation images and labeled arcs of G represent the interrelationships among variables and between variables and condensation images, see [10].

For many interesting fractal images generated by MRFS see [7, 9]. One can apply either the Deterministic Algorithm or the Chaos Game Algorithm to generate the images [2, 7].

3.3 Modeling Algorithms by MRFS

A group $\mathcal{A} = \{A_1, A_2, \ldots, A_n\}$ of mutually recursive algorithms can be *modeled* by a condensation MRFS $\mathcal{M} = (\mathcal{V}, \mathcal{C}, G)$ where

$$\mathcal{V} = \{V_1, V_2, \ldots, V_n\}$$

and the variable V_i represents the algorithm $A_i, i = 1, 2, \ldots, n$, and

$$\mathcal{C} = \{C_1, C_2, \ldots, C_n\}$$

where the set C_i represents the additional computation done by the algorithm A_i. C_i is a condensation image with Hausdorff-Besikovitch dimension equal to D_i if the algorithm A_i does additional computation of $\Theta(n^{D_i})$ steps.

The underlying labeled digraph G of $\mathcal{M}$ represents the interrelationships between algorithms and the additional computation performed by them.

Consider each algorithm $A_i \in \mathcal{A}$. Suppose the algorithm A_i calls recursively the algorithms $A_{j_1}, A_{j_2}, \ldots, A_{j_r}$ and at these recursive calls the input size is contracted by a factor $s_1, s_2, \ldots, s_r$, respectively. Then, the variable V_i representing the algorithm A_i is mutually recursively defined as

$$V_i = w_{i_1}(V_{j_1}) \cup w_{i_2}(V_{j_2}) \cup \ldots \cup w_{i_r}(V_{j_r}) \cup C_i$$

where w_{i_k} is a similitude with contractivity factor equal to s_k, $k = 1, 2, \ldots, r$.

If one needs to generate the actual fractal which geometrically captures the working of the algorithm, then one needs to choose the affine transformations and the condensation images. These have to be chosen so that the condensation MRFS $\mathcal{M}$ is nonoverlapping [10]. For definition of nonoverlapping MRFS see [10, 14]. However if one needs to only determine the time complexity of $\mathcal{A}$ then one needs only the contractivities of the affine transformations and the Hausdorff-Besikovitch dimensions of the condensation images.

Example 2 Refer to Example 1 and Fig. 2. This group of recursive algorithms can be modeled by an MRFS $\mathcal{M}$ with 5 variables M, S1, S2, S3 and S4 representing the 5 routines and 3 condensation sets C_1, C_2 and C_3 representing the additional computation performed by subroutines Sub2, Sub4 and Sub3, respectively. The MRFS is specified by:

$$
\begin{aligned}
M &= w_1(M) \cup w_2(M) \cup w_3(M) \cup w_4(M) \cup u(S2) \cup v(S3) \\
S1 &= w_5(S1) \cup w_6(S1) \cup w_7(S2) \cup w_8(S2) \\
S2 &= w_9(S1) \cup w_{10}(S2) \cup w_{11}(S2) \cup C_1 \\
S3 &= w_{12}(S3) \cup w_{13}(S3) \cup w_{14}(S3) \cup w_{15}(S4) \cup C_3 \\
S4 &= w_{16}(S4) \cup w_{17}(S4) \cup w_{18}(S4) \cup w_{19}(S4) \cup C_2
\end{aligned}
$$

The transformations $w_1, w_2, \ldots, w_{19}$ have contractivity factors equal to $1/2$, and u and v have equal to 1 and 2 respectively. The fractal dimensions of the condensation sets C_1, C_2 and C_3 are 2, 2 and 1 respectively. The transformations and these sets can be appropriately chosen in 4-dimensional Euclidean space so that the MRFS defines a particular fractal [10]. □

4 The Main Result

Given a digraph G let $\mathrm{SC}(G)$ denote the set of strongly connected components of G. Also for a strongly connected component $H \in \mathrm{SC}(G)$ let α_H denote the Hausdorff-Besikovitch dimension of the attractor of the MRFS represented by H. In the underlying digraph of a condensation MRFS, *each condensation image with its single self-loop is a strongly connected component*. Therefore, if C is a condensation image with Hausdorff-Besikovitch dimension equal to D, then we have a strongly connected component H containing C such that $\alpha_H = D$.

Let $\mathcal{M} = (\mathcal{V}, \mathcal{C}, G)$ be a condensation MRFS. From the Mauldin-Williams Theorem [14], the Hausdorff-Besikovitch dimension of the attractor of $\mathcal{M}$ is

$$\alpha = \max\{\alpha_H | H \in \mathrm{SC}(G)\}.$$

Consider all those strongly connected components of G with Hausdorff-Besikovitch dimension equal to α,

$$\text{SCMAX}(G) = \{H \in \text{SC}(G) | \alpha_H = \alpha\}.$$

Now from G we construct a reduced digraph G' by collapsing elements of $\text{SCMAX}(G)$ into nodes. For each H in $\text{SCMAX}(G)$ we create a node v_H in G'. If there is path P in G from H to K where $H, K \in \text{SCMAX}(G)$ and H and K are distinct, such that the path P does not pass through any other component in $\text{SCMAX}(G)$ then we place an arc from v_H to v_K.

Then G' so obtained is a directed acyclic graph (DAG) and is called the *order structure DAG* of $\mathcal{M}$.

Theorem 1 *Let $\mathcal{A}$ be a group of mutually recursive algorithms. Let $\mathcal{M}$ be a condensation MRFS modeling $\mathcal{A}$ and having $\mathcal{O}$ as its attractor. Let the Hausdorff-Besikovitch dimension of $\mathcal{O}$ be D and its Hausdorff D-dimensional measure be k. Let $T(n)$ be the time complexity of $\mathcal{A}$ on input of size n. Then,*

$$T(n) = \Theta(n^D \log^p n)$$

where p is the length of the longest path in the order structure DAG of $\mathcal{M}$. Furthermore, if $p = 0$ (equivalently, $k < \infty$) then k is the constant of proportionality by which two algorithms with same value of D can be compared.

The proof follows from the way the MRFS $\mathcal{M}$ is constructed to model $\mathcal{A}$, definition of Hausdorff-Besikovitch dimension and the Mauldin-Williams Theorem as stated in [14]. For details of the proof see [10].

Remark 1: In the above theorem we have assumed that an algorithm can perform an additional computation (which for example may involve the linear time taken to read the input) taking $\Theta(n^D)$ steps where D is nonnegative real number. In [10] the above theorem is extended to the case when this additional computation has time complexity of the form $\Theta(n^D f(n))$ where $f(n)$ belongs to a special class of functions called as *logarithmically slow functions* which includes powers of the logarithmic function i.e. $\log^k n$.

Remark 2: A special case of the above theorem is the Master Theorem as stated in [6]. Here MRFS is a condensation IFS i.e. it has one variable and one condensation set. Master Theorem is an easy corollary of Theorem 1. Compare this with its long proof in [6].

Example 3 Refer to Examples 1 and 2. How can we determine the time complexity of the algorithms? We need to solve the following set of divide-and-conquer recurrence relations:

$$
\begin{aligned}
T_M(n) &= 4T_M(n/2) + T_{S2}(n) + T_{S3}(2n) \\
T_{S1}(n) &= 2T_{S1}(n/2) + 2T_{S2}(n/2) \\
T_{S2}(n) &= T_{S1}(n/2) + 2T_{S2}(n/2) + \Theta(n^2) \\
T_{S3}(n) &= 3T_{S3}(n/2) + T_{S4}(n/2) + \Theta(n) \\
T_{S4}(n) &= 4T_{S4}(n/2) + \Theta(n^2)
\end{aligned}
$$

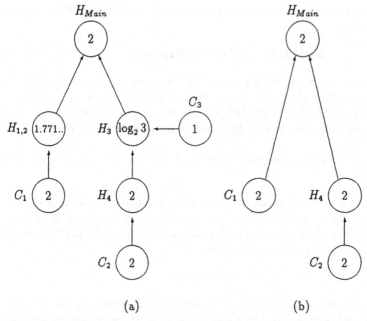

Figure 3: (a) Fractal dimension of strongly connected components of MRFS modeling algorithms in Fig. 2 (b) Order structure DAG with length of the longest path equal to 2

Conventional approaches, such as the Master Theorem, seem inadequate [6, 15]. However, using Theorem 1 one can easily solve the recurrence relations. In Fig. 3(a), we show the strongly connected components of the MRFS $\mathcal{M}$ in Example 2. In total there are 7 such components. Each of these components is an MRFS by itself and defines an image. Using the results in [14, 16] one quickly computes the fractal dimension of these images. For example, the fractal dimension of the image defined by the component $H_{1,2}$ is $\log_2(2 + \sqrt{2})$ where $2 + \sqrt{2}$ is obtained as the largest eigen-value of the matrix:

$$C = \begin{bmatrix} 2 & 1 \\ 2 & 2 \end{bmatrix}$$

The number inside a circle in Fig. 3(a) is the fractal dimension of the image defined by the corresponding component. Since the maximum of the fractal dimensions is 2, we construct the order structure DAG as shown in Fig. 3(b) by keeping only the components with fractal dimension equal to 2. The length of the longest path in the DAG is 2. Therefore, substituting $D = 2$ and $p = 2$ in Theorem 1 we obtain the solution

$$T_M(n) = \Theta(n^2 \log^2 n).$$

$\square$

As a final remark, we mention that the fractal image associated with recursive algorithms with time complexity $\Theta(n)$ would be a line segement, with $\Theta(n \log n)$ would be infintely many such line segments $L_0, L_1, L_2, \ldots$ which become closer and closer, e.g.

L_i is a line segment between 2-D points $(0, (2^i - 1)/2^i)$ and $(1, (2^i - 1)/2^i)$. Similarly, a quadratic algorithm is encoded by a filled square, and an algorithm with time complexity $T(n^2 \log n)$ by a 3-D image consisting of infinitely many such squares which become closer and closer [10].

References

[1] A. V. Aho, J. E. Hopcroft and J. D. Ullman, *The Design and Analysis of Computer Algorithms*, Addison-Wesley, 1974.

[2] M. F. Barnsley, *Fractals Everywhere*, Academic Press, 1988.

[3] M. F. Barnsley, J. H. Elton and D. P. Hardin, "Recurrent Iterated Function Systems," *Constructive Approximation*, 5, 3-31 (1989).

[4] M. F. Barnsley, R. L. Devaney, B. B. Mandelbrot, H-O. Peitgen, De Saupe, and R. F. Voss, *Science of Fractal Images*, Springer-Verlag, 1988.

[5] J. L. Bentley, D. Haken and J. B. Saxe, "A General Method for Solving Divide-and-Conquer Recurrences," *SIGACT News*, 12, 36-44 (1980).

[6] T. H. Cormen, C. E. Leiserson and R. L. Rivest, *Introduction to Algorithms*, MIT Press, 1990.

[7] K. Culik II and S. Dube, "Affine Automata and Related Techniques for Generation of Complex Images," *Theoretical Computer Science* 116, 373-398 (1993).

[8] K. Culik II and S. Dube, "Rational and Affine Expressions for Image Synthesis." *Discrete Applied Mathematics* 41, 85-120 (1993).

[9] K. Culik II and S. Dube, "Balancing Order and Chaos in Image Generation," Preliminary version in Proc. of *ICALP'91*. Lecture notes in Computer Science 510, Springer-Verlag, pp. 600-614. *Computer and Graphics*, to appear.

[10] S. Dube, "Using Fractal Geometry for Solving Divide-and-Conquer Recurrences," Technical Report 93-70, Dept of Math, Stat and Comp Sci, University of New England at Armidale, Australia.

[11] R. L. Graham, D. E. Knuth and O. Patashnik, *Concrete Mathematics*, Addison-Wesley, 1989.

[12] D. H. Greene and D. E. Knuth, *Mathematics for the Analysis of Algorithms*, Birkhäuser Boston, 1982.

[13] B. Mandelbrot, *The Fractal Geometry of Nature*, W. H. Freeman and Co., San Francisco, 1982.

[14] R. D. Mauldin and S. C. Williams, "Hausdorff Dimension in Graph Directed Constructions," *Transactions of American Mathematical Society*, 309, 811-829 (1988).

[15] P. W. Purdom, Jr. and C. A. Brown, *The Analysis of Algorithms*, Holt, Rinehart and Winston, 1985.

[16] L. Staiger, "Quadtrees and the Hausdorff Dimension of Pictures," Workshop on Geometrical Problems of Image Processing (GEOBILD'89), *Math. Research No. 51*, Akademie-Verlag, Berlin, pp. 173-178 (1989).

Simple Combinatorial Gray Codes Constructed by Reversing Sublists

Frank Ruskey[1]

Department of Computer Science, University of Victoria,
Victoria, B.C., V8W 3P6, Canada, e-mail:fruskey@csr.uvic.ca

Abstract. We present three related results about simple combinatorial Gray codes constructed recursively by reversing certain sublists. First, we show a bijection between the list of compositions of Knuth and the list of combinations of Eades and McKay. Secondly, we provide a short description of a list of combinations satisfying a more restrictive closeness criteria of Chase. Finally, we develop a new, simply described, Gray code list of the partitions of a set into a fixed number of blocks, as represented by restricted growth sequences. In each case the recursive definition of the list is easily translatable into an algorithm for generating the list in time proportional to the number of elements in the list; i.e., each object is produced in $O(1)$ amortized time by the algorithm.

1 Introduction

Frank Gray patented the Binary Reflected Gray Code (BRGC) in 1953 for use in "pulse code communication", but the underlying construction of the code existed for centuries as the solution of a puzzle that is known in the West as the "Chinese Rings". The BRGC has since found use in the minimization of logic circuits, in hypercube architectures, in the construction of multikey hashing functions, and has even been proposed for use in the organization of books on library shelves! For the purposes of this paper, we regard the BRGC as a way of listing all 2^n bitstrings of length n so that successive strings differ by a single bit or, equivalently, as a way of listing all subsets of an n-set so that successive subsets differ by a single element.

It is natural to take a more abstract view and look for listings of other combinatorial objects so that successive objects differ by a very small amount; such lists have come to be known as *combinatorial Gray codes*. In this paper, we shed new light on some previously known lists of combinations and compositions, and develop new lists for combinations and set partitions. In each case the list has a very simple recursive description. The simplicity of the recursion is mirrored in simple proofs and simple efficient algorithms for generating the lists. These algorithms generate the objects so that only a constant amount of computation is done between successive objects, in an amortized sense.

* Research supported in part by the Natural Sciences and Engineering Research Council of Canada under Grant A3379.

There is a well-known simple recursive construction for the BRGC. This construction involves the concatenation of two recursively defined smaller sublists, one of which is reversed, as shown below.

$$\mathbf{B}(n) = \begin{cases} \emptyset & \text{if } n = 0 \\ \mathbf{B}(n-1) \cdot 0 \circ \overline{\mathbf{B}(n-1)} \cdot 1 & \text{if } n > 0 \end{cases}$$

This simple recursive construction is easily proven to be correct and can be implemented in an efficient manner to actually generate the list. In the construction of other combinatorial Gray codes, it is natural to try to invent similar recursive definitions. In general, the recursion defining the combinatorial Gray code should mirror some simple recursion for counting the combinatorial objects, in this case $a_n = 2a_{n-1}$ with $a_0 = 1$.

Very simple constructions have been obtained in the cases of combinations (e.g., Reingold, Nievergelt, and Deo [ReNiDe] or Eades and McKay [EaMc]), compositions (e.g., Wilf [Wi]), and well-formed parentheses (e.g., Ruskey and Proskurowski [RuPr]). Less simple recursive constructions, that are nevertheless based on reversing sublists, have been obtained for numerical partitions (Savage [Sa]), and set partitions (Fill and Reingold [FiRe], Ehrlich [Eh]).

We regard the objects being listed as strings over some alphabet. If L is a list of strings and x is a symbol, then $L \cdot x$ denotes the list of strings obtained by appending an x to each string of L. For example if $L = 01, 10$ the $L \cdot 1 = 011, 101$. If L and L' are lists then $L \circ L'$ denotes the concatenation of the two lists. For example, if $L = 01, 10$ and $L' = 11, 00$, then $L \circ L' = 01, 10, 11, 00$.

For a list L, let $first(L)$ denote the first element on the list and let $last(L)$ denote the last element on the list L. If L is a list $l_1, l_2, \ldots, l_n$, then $\overline{L}$ denotes the list obtained by listing the elements of L in reverse order; i.e., $\overline{L} = L^{-1} = l_n, \ldots, l_2, l_1$. Note the obvious equations $first(\overline{L}) = last(L)$ and $last(\overline{L}) = first(L)$.

By a k-combination of n we mean a bitstring of length n with exactly k ones. The set of all k-combinations of n is denoted $\mathbf{B}(n, k)$; i.e.,

$$\mathbf{B}(n, k) = \{b_1 b_2 \cdots b_n \mid b_i \in \{0, 1\}, \Sigma b_i = k\}$$

By a k-composition of n we mean a solution $(x_1, x_2, \ldots, x_k)$, in non-negative integers, to the equation $x_1 + x_2 + \cdots + x_k = n$.

A partition of $[n] = \{1, 2, \ldots, n\}$ into k blocks is a collection of k non-empty disjoint sets $B_0, B_1, \ldots, B_{k-1}$ whose union is $[n]$. Assuming that the B_i are ordered by their smallest elements, a convenient representation of the partition is as a sequence $x_1 x_2 \cdots x_n$ where $x_i = j$ if $x_i \in B_j$. For example, the sequence corresponding to the partition $\{1, 3, 4\}, \{2, 6\}, \{5, 7\}$ is 0100212. Such sequences are known as restricted growth sequences (or RG sequences) in the combinatorial literature (e.g., Milne [Mi]) and are characterized by the following property, which explains the name. If $i = 1$ then $x_i = 0$, and if $i > 1$, then

$$0 \le x_i \le 1 + \max\{x_1, x_2, \ldots, x_{i-1}, k - 2\}.$$

In this note we will omit all proofs. In each case they are straightforward inductive arguments. The real difficulties lie in coming up with the recursive definitions and the statements of the lemmata.

2 The equivalence of two well-known lists

The following recursively defined list of k-combinations of n is due to Eades and McKay [EaMc].

$$
\mathbf{C}(n,k) = \begin{cases}
0^n & \text{if } k = 0 \\
10^{n-1} \circ 010^{n-2} \circ \cdots 0^{n-1}1 & \text{if } k = 1 \\
\mathbf{C}(n-1,k) \cdot 0 & \\
\quad \circ \overline{\mathbf{C}(n-2,k-1)} \cdot 01 & \\
\quad \circ \overline{\mathbf{C}(n-2,k-2)} \cdot 11 & \text{if } 1 < k < n \\
1^n & \text{if } k = n
\end{cases}
\tag{1}
$$

This list has a number of interesting properties which are embodied in the next lemma.

Lemma 1. *The list* $\mathbf{C}(n,k)$ *satisfies the following properties.*

1. *Successive combinations differ by a transposition of two bits and all bits between those that are transposed are zeroes.*
2. $first(\mathbf{C}(n,k)) = 1^k 0^{n-k}$.
3. $last(\mathbf{C}(n,k)) = 0^{n-k} 1^k$.

Another list, this one of k-compositions of n, is attributed in Wilf [Wi] to Donald Knuth. Consider the list, $\mathbf{Comp}(n,k)$, of all k-compositions of n, defined as follows. If $k = 1$, then $\mathbf{Comp}(n,1) = n$ and if $k > 1$, then $\mathbf{Comp}(n,k)$ is given by list (a) below.

$\mathbf{Comp}(n, k-1) \cdot 0\circ$
$\mathbf{Comp}(n-1, k-1) \cdot 1\circ$
$\mathbf{Comp}(n-2, k-1) \cdot 2\circ$
$\mathbf{Comp}(n-3, k-1) \cdot 3\circ$

$$\vdots$$

$\mathbf{Comp}(0, k-1) \cdot n$

(a)

$\mathbf{Comb}(n, k-1) \cdot 01^0\circ$
$\mathbf{Comb}(n-1, k-1) \cdot 01^1\circ$
$\mathbf{Comb}(n-2, k-1) \cdot 01^2\circ$
$\mathbf{Comb}(n-3, k-1) \cdot 01^3\circ$

$$\vdots$$

$\mathbf{Comb}(0, k-1) \cdot 01^n$

(b)

The basic properties of this list are embodied in the next lemma.

Lemma 2. *The list* $\mathbf{Comp}(n,k)$ *satisfies the following properties.*

1. *Successive compositions differ by one in exactly two positions.*
2. $first(\mathbf{Comp}(n,k)) = n0^{k-1}$.
3. $last(\mathbf{Comp}(n,k)) = 0^{k-1}n$.

Every k-composition of n corresponds to a element of $\mathbf{B}(n + k - 1, n)$. This correspondence is obtained by turning each x_i into a run of x_i 1's and separating the k runs of 1's by $k - 1$ 0's. For example $300123 \in \mathbf{Comp}(9, 6)$ corresponds to $11100010110111 \in B(14, 9)$. Applying this correspondence to the list definition (a) yields the list $\mathbf{Comb}(n, k)$ of elements of $\mathbf{B}(n + k - 1, n)$ as shown in (b). The following lemma states that the Eades-McKay and Knuth lists are essentially identical.

Lemma 3. $\mathbf{Comb}(n, k) = \mathbf{C}(n + k - 1, n)$.

3 A more restrictive closeness condition

In the Eades-McKay list the two bits that are transposed could be quite far apart. The elements of $\mathbf{B}(n, k)$ cannot be listed so that successive strings differ by the transposition of two *adjacent* bits, unless n is even and k is odd, or $k \in \{0, 1, n - 1, n\}$ (e.g., Buck and Wiedemann [BuWi], Eades, Hickey and Read [EaHiRe]). Let us say that two distinct bitstrings are *two-close* if they differ by a transposition of two bits that are either adjacent or have a single 0 between them. Chase [Ch] showed that the elements of $\mathbf{B}(n, k)$ could always be listed so that successive strings are two-close. He also presented a loopfree algorithm for producing such a listing. However, his construction takes far more explanation than the one we will now describe.

The two-close condition can be handled by the same list-reversing technique that we used previously, but we add an additional boolean parameter. Let $\mathbf{R}(n, k, 1)$ denote a listing of the bitstrings in $\mathbf{B}(n, k)$ that starts at $0^{n-k-1}1^k0$ and ends at $0^{n-k}1^k$ and satisfying the two-close condition. Let $\mathbf{R}(n, k, 0)$ be a similar list, except starting at $0^{n-k-2}1^k00$. The list $\mathbf{R}(n, k, 1)$ exists only if $0 \leq k < n$ and the list $\mathbf{R}(n, k, 0)$ exists only if $0 \leq k < n - 1$. These lists may be defined recursively as shown below.

$$
\mathbf{R}(n, k, p) = \begin{cases}
0^n & \text{if } k = 0 \\
1^{n-1}0 \circ \mathbf{R}(n - 1, k - 1, 1) \cdot 1 & \text{if } p = 1 \text{ and } k = n - 1 \\
\begin{aligned} &\overline{\mathbf{R}(n - 1, k, 1)} \cdot 0 \\ &\circ \mathbf{R}(n - 1, k - 1, 0) \cdot 1 \end{aligned} & \text{if } p = 1 \text{ and } 0 < k < n - 1 \\
\begin{aligned} &\mathbf{R}(n - 1, k, 1) \cdot 0 \\ &\circ \mathbf{R}(n - 1, k - 1, 1) \cdot 1 \end{aligned} & \text{if } p = 0 \text{ and } 0 < k < n - 1
\end{cases}
$$

These lists have a number of interesting properties which are embodied in the next lemma.

Lemma 4. *The lists* $\mathbf{R}(n, k, 0)$ *and* $\mathbf{R}(n, k, 1)$ *satisfy the following properties.*

1. *Successive bitstrings are two-close.*
2. $first(\mathbf{R}(n, k, 0)) = 0^{n-k-2}1^k00$
3. $first(\mathbf{R}(n, k, 1)) = 0^{n-k-1}1^k0$
4. $last(\mathbf{R}(n, k, 0)) = last(\mathbf{R}(n, k, 1)) = 0^{n-k}1^k$

4 Set partitions

Previous works have defined two set partitions to be *close* if they differ by taking an element from one block and moving it into another block. Gray codes under this criteria have been developed by Knuth (see Wilf [Wi]) for the set of all partitions, and by Ehrlich [Eh] and Hansche (see Fill and Reingold [FiRe]) for partitions into a fixed number of blocks. It is easy to see that two RG sequences that differ in one position have corresponding partitions that are close, but not vice-versa. We will show how to list RG sequences representing partitions into a fixed number of blocks so that successive sequences differ in exactly one position. Previous lists for partitions into a fixed number of blocks did not have this property, nor were they as simple as what we present here.

We define two lists of RG sequences representing partitions of $[n]$ into k blocks, denoted $\mathbf{S}(n, k, 0)$ and $\mathbf{S}(n, k, 1)$. These definitions have similarities to the $\mathbf{R}(n, k, \cdot)$ lists defined earlier. Successive RG sequences on these lists have the property that they differ in exactly one position.

Recall the basic recurrence relation for the Stirling numbers of the second kind $\left\{ {n \atop k} \right\}$, which count partitions of $[n]$ into exactly k blocks.

$$\left\{ {n \atop k} \right\} = \left\{ {n-1 \atop k-1} \right\} + k \left\{ {n-1 \atop k} \right\}$$

This recurrence relation may be proven in terms of RG sequences $x_1 x_2 \cdots x_n$ by classifying them according to the value of x_n. If $x_n = k - 1$ then $x_1 x_2 \cdots x_{n-1}$ has largest element equal to $k-2$ or $k-1$. If $0 \leq x_n < k-1$, then $x_1 x_2 \cdots x_{n-1}$ must have largest value equal to $k - 1$.

The construction of the lists depends upon the parity of k as shown in the tables below, first for k even and then for k odd. If $k = 1$ then $\mathbf{S}(n, 1, 0) = \mathbf{S}(n, 1, 1) = 0^n$ and otherwise use the tables below. Here's the lists for k even.

$\mathbf{S}(n, k, 0)$ even k	$\mathbf{S}(n, k, 1)$ even k
$\mathbf{S}(n-1, k-1, 0) \cdot (k-1)\circ$	$\mathbf{S}(n-1, k-1, 1) \cdot (k-1)\circ$
$\mathbf{S}(n-1, k, 1) \cdot (k-1)\circ$	$\mathbf{S}(n-1, k, 1) \cdot (k-1)\circ$
$\mathbf{S}(n-1, k, 1) \cdot (k-2)\circ$	$\mathbf{S}(n-1, k, 1) \cdot (k-2)\circ$
$\vdots$	$\vdots$
$\mathbf{S}(n-1, k, 1) \cdot 1\circ$	$\mathbf{S}(n-1, k, 1) \cdot 1\circ$
$\mathbf{S}(n-1, k, 1) \cdot 0$	$\mathbf{S}(n-1, k, 1) \cdot 0$

Here's the lists for k odd.

$S(n,k,0)$ odd k	$S(n,k,1)$ odd k
$S(n-1,k-1,1)\cdot(k-1)\circ$	$S(n-1,k-1,0)\cdot(k-1)\circ$
$S(n-1,k,1)\cdot(k-1)\circ$	$S(n-1,k,1)\cdot(k-1)\circ$
$S(n-1,k,1)\cdot(k-2)\circ$	$S(n-1,k,1)\cdot(k-2)\circ$
$\vdots$	$\vdots$
$S(n-1,k,1)\cdot1\circ$	$S(n-1,k,1)\cdot1\circ$
$S(n-1,k,1)\cdot0$	$S(n-1,k,1)\cdot0$

These lists have a number of interesting properties which are embodied in the following lemma.

Lemma 5. *The lists* $S(n,k,0)$ *and* $S(n,k,1)$ *satisfy the following properties.*

1. *Successive RG sequences differ in exactly one position.*
2. $first(S(n,k,0)) = first(S(n,k,1)) = 0^{n-k}012\cdots(k-1)$
3. $last(S(n,k,0)) = 0^{n-k}12\cdots(k-1)0$
4. $last(S(n,k,1)) = 012\cdots(k-1)0^{n-k}$

5 Algorithmic Considerations

We will now illustrate how the recursive definitions given earlier can be translated into efficient procedures for generating the objects. The proof of Lemma 4 leads to the consideration of the following table.

$p=1$	$p=0$	
$0^{n-k-2}01^{k-1}1\ 0$	$0^{n-k-2}1^{k-1}10\ 0$	
$\vdots$	$\vdots$	$\binom{n-1}{k}$
$0^{n-k-2}1^{k-1}10\ \underline{0}$	$0^{n-k-2}01^{k-1}\underline{1}\ 0$	
$0^{n-k-2}1^{k-1}00\ \underline{1}$	$0^{n-k-2}01^{k-1}\underline{0}\ 1$	
$\vdots$	$\vdots$	$\binom{n-1}{k-1}$
$0^{n-k-2}001^{k-1}1$	$0^{n-k-2}001^{k-1}1$	

This table reveals that the transposition between the two sublists occurs in positions $n-2$ and n if $p=1$ and is in positions $n-1$ and n if $p=0$. This revelation in turn leads directly to the Pascal procedure **gen(n,k,p)** of Algorithm 1 which generates the list $C(n,k,p)$. There is also a symmetric procedure **neg(n,k,p)** which generates the reversed list $\overline{C(n,k,p)}$. The procedure **swap(i,j)** swaps the bits in positions i and j (and does whatever processing of the bitstring is required).

What is the running time of **gen(n,k,p)**? We need only observe the following.

```
procedure gen ( n,k,p : integer );
begin
  if p = 1 then begin
    if k = n-1 then begin
      for i := n-1 downto 1 do swap( i,i+1 );
    end else
    if (0 < k) and (k < n-1) then begin
      neg( n-1, k, 1 );  swap( n-2, n );  gen( n-1, k-1, 0 );
    end
  end else {p = 0} begin
    if (0 < k) and (k < n-1) then begin
      gen( n-1, k, 1 );  swap( n-1, n );  gen( n-1, k-1, 1 );
    end
  end;
end {of gen};
```

Fig. 1. Two-close generation of Combinations.

- Each recursive call to **gen** or **neg** can be charged to a call to **swap** so that any call to **swap** is charged at most twice.
- The total number of calls to **swap** is one less than $\binom{n}{k}$.
- The for loop makes no recursive calls, and each iteration of the for loop produces one new combination.

Thus total amount of computation is proportional to the number of recursive calls, which is $O(\binom{n}{k})$.

Analogous algorithmic development may be carried out for the other lists of this paper. In each case the proof reveals the change that occurs between successive sublists, and the costs of the recursive calls can be charged to the strings so that no string gets charged more than a constant amount.

References

[BuWi]   M. Buck and D. Wiedemann, *Gray Codes with Restricted Density*, Discrete Mathematics, 48 (1984) 163-171.

[Ch]   P.J. Chase, *Combination Generation and Graylex Ordering*, Congressus Numerantium, 69 (1989) 215-242.

[EaHiRe] P. Eades, M. Hickey and R. Read, *Some Hamilton Paths and a Minimal Change Algorithm*, Journal of the ACM, 31 (1984) 19-29.

[EaMc]   P. Eades and B. McKay, *An Algorithm for Generating Subsets of Fixed Size with a Strong Minimal Change Property*, Information Processing Letters, 19 (1984) 131-133.

[Eh]   G. Ehrlich, *Loopless Algorithms for Generating Permutations, Combinations and Other Combinatorial Configurations*, Journal of the ACM, 20 (1973) 500-513.

[FiRe]   J.A. Fill and E.M. Reingold, *Solutions Manual for Combinatorial Algorithms: Theory and Practice*, Prentice-Hall, 1977.

[Mi]   S.C. Milne, *Restricted Growth Functions, Rank Row Matchings of Partition Lattices, and q-Stirling Numbers*, Advances in Mathematics, 43 (1982) 173-196.

[ReNiDe] E.M. Reingold, J. Nievergelt and N. Deo, *Combinatorial Algorithms*, Prentice-Hall, Inc., Englewood Cliffs, New Jersey, 1977.

[RuPr]   F. Ruskey and A. Proskurowski, *Generating Binary Trees by Transpositions*, J. Algorithms, 11 (1990) 68-84.

[Sa]   C.S. Savage, *Gray Code Sequences of Partitions*, J. Algorithms, 10 (1989) 577-595.

[Wi]   Herbert S. Wilf, *Combinatorial Algorithms: An Update*, SIAM CBMS-55, 1989.

Time Space Tradeoffs
(Getting Closer to the Barrier?)

Allan Borodin

Department of Computer Science, University of Toronto
Toronto, Ontario, Canada M5S 1A4

(presently on sabbatical leave at Laboratory for Computer Science,
Massachusetts Institute of Technology, Cambridge, MA 02139, U.S.A.)

Abstract. We survey the current status of the time-space complexity of
various basic computational problems. For time-space (or space alone)
lower bounds, Boolean branching programs are the "ultimate model".
We consider restricted and unrestricted branching program models as
well as certain "structured models" which are appropriate to the particular problems being studied. Recent results on graph connectivity are
especially noteworthy.

1 Introduction

Time-space tradeoffs have been a topic of interest in Computer Science since
the beginning of the discipline. Early compiler writers understood that trying
to economize on the use of registers often seemed to require recomputation of
partial results. We will see that this topic has been prominent in complexity
theory.

Cobham [29] pioneered the theoretical study of time T-space S tradeoffs.
He showed that on any computational model with restricted input access (i.e.,
sequential access as in a single head Turing machine input tape) certain Boolean
problems required $TS = \Omega(n^2)$ where n is the length of the input and that this
was essentially optimal for all S in the range $\log n \leq S \leq n$. One could say that
the next thirty years have been spent trying to prove similar results for random
access input models.

In spite of substantial progress, there is still no lower bound for a decision
problem on an unrestricted Boolean branching program. Indeed, establishing
such a result is what we would call the "smallest fundamental complexity barrier"; namely, find an explicit (i.e., in NP) Boolean decision problem for which
$TS = O(n \log n)$ is not possible.

In the next sections, we will provide a brief history and reflect on whether or
not we are approaching the smallest complexity barrier. We mention a number
of open problems.

2 A Brief Review of Tradeoffs on the Straight Line Program (= Circuit Pebbling) Model

One natural model for which interesting tradeoffs have been derived is the Straight Line Program Model which is equivalent to a pebbling game played on a circuit which realizes the function being considered. One can naturally consider either Boolean problems on Boolean circuits or arithmetic problems on arithmetic circuits. In either case, one considers any circuit for realizing the desired function. Viewing the circuit as an unlabelled DAG, a pebbling of the DAG attempts to pebble each output node by the following rules:

R1: a pebble can always be removed from a node

R2: if the predecessors (at most two for Boolean and arithmetic circuits) of a node v are both pebbled, then v can be pebbled by a new pebble or by using the pebble from one of its predecesors; as a special case, a pebble can always be placed on an input node

The pebbling literature and its application to time-space tradeoffs is extensive and thorough discussions can be found in the comprehensive surveys by Pippenger [50] and Savage [55]. Hence we will only mention relatively few results in this section. In this model, each circuit or DAG admits a number of different pebblings. For a particular pebbling, space is measured by the maximum number of pebbles used at any point of time (corresponding to register use) and time by the number of pebble placements according to rule R2 (corresponding to operations executed). The first result in this context concerned the best pebbling for certain DAGs (not necessarily related to any particular circuit). These results can be thought of as time-space tradeoffs as would be encountered by optimizing compilers which only work with the DAG of straight line program segments. Obviously, any n node DAG can be pebbled in time n using space n . What is surprising is that Hopcroft,Paul and Valiant [36] show that every n node DAG can be pebbled in $O(n/\log n)$ pebbles but using $2^{\Omega(n/\log n)}$ time. Paul,Tarjan and Celoni [49] show that $\Omega(n/\log n)$ pebbles are needed in general. It then remained to prove how much time was needed when the space (ie. the number of pebbles) is less than linear. Paul and Tarjan [48] show that there are n node graphs which can be pebbled in space $c_2\sqrt{n}$ and time n but for which any pebbling using only space $c_1\sqrt{n}$ requires time $2^{c_3\sqrt{n}}$ for appropriate constants $c_1 < c_2$ and c_3. For larger space bounds, perhaps the ultimate results in this regard are by Lengauer and Tarjan [42], following results by Pippenger [51], and Reischuk [54]. Lengauer and Tarjan show that there are DAGs for which any $O(n/\log n)$ space pebbling requires exponential time. Furthermore they show that $\Theta(n/\log\log n)$ is a breakpoint for the ability to pebble in polynomial time. That is, for appropriate constants $c_1 < c_2$ with space exceeding $c_2 n/\log\log n$, any n node graph can be pebbled in polynomial time but this is no longer true for space less than $c_1 n/\log\log n$.

For more natural DAGs (i.e., DAGs which correspond to known algorithms) the tradeoffs are not as dramatic but perhaps more important. Furthermore, for a given function, we must look at the best time-space performance over all circuits realizing the function. Here follow some illustrative results. Savage and Swamy [56] considered the FFT (Fast Fourier Transform) algorithms for the n point DFT (Discrete Fourier Transform) and showed a $TS = \Omega(n^2)$ tradeoff for this widely used algorithm. Grigoriev [35] and independently Tompa [60] then showed that a number of common algebraic problems (eg. polynomial and matrix multiplication, convolution, the DFT) have inherent time space tradeoffs in this model. For example, Tompa shows that any algorithm for the DFT has the same $TS = \Omega(n^2)$ lower bound as holds for the FFT. Tompa [61] also considered a class of algorithms for directed s-t connectivity, namely algorithms based on "repeated squaring" as in the Savitch method. Tompa showed that any such algorithm has the property that sublinear space forces non polynomial time.

A couple of important results for the circuit pebbling model have been derived via the branching program model and hence their precise statements might best be delayed until later. Briefly stated, the following results have been established. In either the Boolean or arithmetic setting, any polynomial size formula can be realized by a circuit pebbling which uses only a constant number of pebbles and polynomial time. This very surprising result was first established in the Boolean case by Barrington [14] and later extended to the arithmetic case by Ben Or and Cleve [18]. Finally, the previous lower bounds did not establish an inherent lower bound for a decision problem. Alon and Maass [5] do establish a tradeoff bound for a Boolean decision problem via results for oblivious branching programs.

3 Some Turing Machine Results

As mentioned in the introduction, Cobham [29] began the formal study of time-space tradeoffs. Cobham considered an abstraction of bounded space Turing machines; namely, while retaining the single head-read only sequential input tape, he abstracted the finite control and the storage structures to be any set Q of states and measured space by $\log_2 \#Q$, which he called the capacity $C(n)$; i.e., if $Q(n)$ is the set of all distinct configurations obtained on inputs of length n, then $C(n)$ provides a lower bound on space. Using a crossing sequence argument, Cobham established $T \cdot S = \Omega(n^2)$ bounds for a number of problems, including recognition of the set $\{w\#w \mid w \in \{0,1\}^*\}$.

Clearly the above set is recognizeable in "constant space" and linear time by a Turing machine model which allows two heads for the input. Duris and Galil [31] consider a k head nondeterministic Turing machine model and show that an explicit Boolean function has a lower bound of $T^2 \cdot S = \Omega(n^3)$ on this model. Space for a k input head machine is defined as $k \log n + C(n)$ where $C(n)$ is again the capacity. Karchmer [39] uses the same techniques to show the identical bound for the Boolean version of the element distinctness problem. These proofs, like Cobham's, only depend on the size of the storage structure and not how it is organized.

A general construction of relevance to this topic is Hopcroft, Paul and Valiant's [36] result that any function computable by a time T Turing machine can be computed by a space $T/\log T$ Turing machine. This construction relies substantially on their pebbling result mentioned in the previous section. The pebbling lower bounds give evidence that no general simulation can reduce space without dramatically increasing time but do not provide a proof that a tradeoff must hold for any given function.

Finally, although we are mainly concerned in this survey with "easy problems" (i.e., all the explicit problems mentioned here are easily computable within polynomial time), we should note the paper by Bruss and Meyer [26] where they consider the time-space complexity for the provably hard (i.e., exponential time) decision problem for the theory of real addition. Here they show that any nondeterministic decision procedure either uses $2^{\Omega(n)}$ space or time $2^{\Omega(n^2)}$. Without any space restriction, the best known time lower bound is $2^{\Omega(n)}$ but as Bruss and Meyer point out, this merely leaves open the possibility of a time space tradeoff.

4 Branching Programs and Progress Arguments

One can use the techniques employed for circuit pebbling or for sequential access Turing machines to prove lower bounds on a problem such as sorting on these models (Tompa [60]). However, these models are not sufficiently general to distinguish between merging (which is computable in $O(\log n)$ space, linear time) and sorting. To this end, Borodin,Fischer,Kirkpatrick,Lynch and Tompa [22] considered sorting on a comparison branching program. Comparison branching programs are similar to comparison trees except that "equivalent nodes" are coalesced to form a DAG rather than a tree. Thus we think of a node of the branching program as a state or configuration of the computation and we label each node by the comparison that takes place in this state. For our purposes, we can assume without loss of generality that branching programs are levelled so that level i corresponds to the possible configurations at time i. It is perhaps suprising that comparison branching programs do not appear to have been considered earlier since the Boolean branching program model has a long and diverse history both as a basic computation model (see Razborov [53] for a comprehensive survey of branching program size result including the early Russian literature) and as a representation for Boolean functions (see Lee [41] for an early advocate of such use). Following Cobham again, we can define the space of a branching program (hereafter called a BP) to be log(size of the BP). Borodin et al. [22] show that in the comparison BP model, sorting requires $TS = \Omega(n^2)$. The proof technique is to look at a sufficiently small "stage" of the computation and argue that during this stage most inputs cannot make too much "progress" (e.g., we can't correctly determine too many ranks).

Borodin and Cook [19] introduce a useful extension of Boolean branching programs whereby inputs are over some finite set R rather than just $R = \{0,1\}$. A BP node is now labelled by a particular input and there is a branch for each of the $\#R$ possible values. Borodin and Cook show that sorting (more precisely, the somewhat easier problem of "ranking") elements in the range $[1, n^2]$ requires $TS = \Omega(n^2/\log n)$. Beame [15] greatly simplies this argument and shows that sorting integers in the range $[1, n]$, or outputting the unique elements in a list of such integers, requires $TS = \Omega(n^2)$ which is optimal.

The progress based BP argument has been exploited for a number of multi output functions where progress can be naturally associated with the number of outputs produced thus far. Of particular note, is that on the R-way BP model, Abrahamson [1] establishes strong lower bounds for the expected space $\overline{S}$ and time $\overline{T}$ used for a number of algebraic problems, in many cases matching (and sometimes improving) on the bounds proved on the straight line or circuit pebbling model. Here, the inputs are assumed to come from a field and elements are chosen randomly from an appropriate finite subset D of the field. For example, he shows $\overline{TS} = \Omega(n^2 \log \delta)$ for n element convolution and $\overline{T}^2\overline{S} = \Omega(n^6 \log \delta)$ for $n \times n$ matrix multiplication where $\delta = \#D$. The latter result improves the $TS = \Omega(n^3)$ result of Yesha [63] which held for sufficiently large fields. Another result by Abrahamson [2] shows that the tradeoff for $n \times n$ Boolean matrix multiplication experiences a sharp break at $\overline{T} = \Theta(n^{2.5})$; namely, for appropriate constants $c_1 < c_2$ if $\overline{T} \leq c_1 n^{2.5}$ then $\overline{TS} = \Omega(n^{3.5})$ and if $\overline{T} > c_2 n^{2.5}$ then $\overline{TS} = \Omega(n^3)$, both bounds optimal to within a $\log n$ factor. Mansour, Nisan and Tiwari [43] prove that any implementation of a universal hashing function from n bit strings to m bit strings requires $TS = \Omega(nm)$; for example, the function $x + y \cdot z$ requires $TS = \Omega(n^2)$ where x, y, z are in a field F and $n = \log(\#F)$.

When considering decision problems, it is no longer as clear what the notion of progress should be. By recalling that the element distinctness problem (hereafter called ED) is equivalent to sorting in the comparison tree model, Borodin, Fich, Meyer auf der Heide, Upfal and Wigderson [21] associate progress with the number of adjacent (in the sorted order) elements that have been compared. For the comparison BP model, they are able to show that for ED, $T^2S = \Omega(n^3)$. By using a more involved analysis on the amount of progress that can be made in a stage when the space is constricted throughout the stage (unlike the analysis in the previous arguments which only restrict space at the beginning of a stage), Yao [62] shows that ED requires $TS = \Omega(n^{2-\epsilon(n)})$ where $\epsilon(n) = O(\log n)^{-1/2}$. A number of further such TS tradeoffs for set operations are shown in Patt and Pelag [47]. For decision problems, no analogous results are known for the Boolean BP model.

5   Restricted Branching Programs

While the situation for unrestricted Boolean BPs remains wide open (at least for decision problems), there are a number of well-motivated restricted BP models relative to which many interesting bounds have been derived.

The most severe space restriction that one might consider is that of a constant width BP. It is easy to see that every Boolean function can be computed in width 2 but possibly using exponential time. In fact, Yao [62] (improving on a result of Borodin, Dolev, Fich and Paul [20]) shows that the majority function requires non polynomial time relative to width 2 BP. Shearer [58] shows that any width 2 BP computing the "zero mod 3" function must have exponential length. The first results giving time (= length) lower bounds for arbitrary constant width were due to Chandra,Furst and Lipton [27]. At present, the best time lower bound for computing majority by a constant width BP is $\Omega(n \log n)$ and this was proven independently by Babai, Pudlak, Rodl and Szemerdi [8] and Alon and Maass [5]. This proof also applies to the "exactly $n/2$ out of n function" and more precisley the bound is $\Omega(n \log n / \log w)$ which for width $w \geq \log n$ can be seen to be optimal by computing the function modulo primes of size around w. It is reasonable to conjecture that for majority an exponential time lower bound should hold for any constant width BP. In a very surprising result, Barrington [14] shows that every balanced formula of size t can be realized by a width 5 BP of length $O(t^2)$. (Note that Kosaraju [40] shows that any length t formula can be converted to a balanced formula of size t^2) . Since majority can be realized by polynomial size formulas it follows that majority can be computed by a polynomial length width 5 BP. Ben Or and Cleve [18] generalize this result to arithmetic computation over any ring. Furthermore Cleve [28] shows how width (register space) can be traded for length (time) in realizing any formula of length t; namely he shows that any balanced formula of size t can be realized by a width w BP of length $t^{1+O(1/w)}$.

Another useful restriction is that of oblivious BP. A BP is oblivious if at each level, every node queries the same input. Oblivious computation is well motivated in complexity theory and oblivious BP lower bounds can be applied to the circuit pebbling model. Also it is clear that any constant width BP can be made oblivious with the same width and increasing the length by a constant factor. Alon and Maass [5] consider the Boolean set equality problem and the sequence equality problem over 0,1,2. For example they show that any oblivious 3-way BP computing sequence equality having width $2^{n/2^{h(n)}}$ requires time $\Omega(n h(n))$ for any $h(n) \leq \log n/4$; ie. space $o(n)$ implies non linear time. The strongest lower bound concerning oblivious BP computation is due to Babai, Nisan and Szegedy [7]. They exhibit an explicit Boolean function for which $o(n \log^2 n)$ time implies space $(= \log_2 $ BP size$)$ $\Omega(n)$. This result then also gives a lower bound for constant width but not the best that is known since the technique of Neciporuk can be used to derive BP size lower bounds for explicitly defined functions (see,for example,the discussion in Razborov [53]). Indeed, Beame and Cook [17] show that the Neciporuk method can be used to derive a $\Omega(n^2/\log^2 n)$ lower bound on the BP size for a Boolean version of the element distinctnes problem. The Alon and Maass, and Babai et al. results rely on two-party (respectively, multi-party as introduced in Chandra,et al [27]) communication complexity results.

One final restricted model is that of read-once and, more generally, read-k times BP. Here we can think of the read-once (read-k) restriction as a restriction on time (say time $\leq kn$). There is a long history of results concerning read-once BP starting with the work of Masek [44] and Zak [64]. For example, Babai, Hajnal, Szemeredi and Turan [6] show that a read-once program for counting mod 2 the triangles of a m node graph presented as a $\binom{m}{2}$ bit adjacency matrix, requires space $\Omega(m)$. Borodin, Razborov and Smolensky [23] and independently, Okolnishnikova [46] proved exponential size lower bound (i.e., n^ϵ space bounds) for syntactic read k BP. In particular, Borodin et al. [23] show that for some explicit Boolean function, a syntactic read-k nondeterministic BP requires space $\Omega(n/4^k k^3)$.

6 Graph Connectivity Problems

Graph connectivity, in particular undirected st connectivity USTCON and directed st connectivity STCON, is one area in which substantial progress has been made for both upper and lower bounds. The Savitch [57] algorithm which still remains the best deterministic or randomized space bound for STCON uses $n^{O(\log n)}$ time. For over a decade it remained open whether or not there exists a deterministic or randomized polynomial time algorithm for USTCON or STCON which uses less than linear (in n, the number of nodes) space.

Aleliunas, Karp, Lipton, Lovasz and Rackoff [4] showed that by using random walks on undirected graphs, USTCON was solvable in a randomized (1-sided error) log space, and time $O(mn)$. For regular graphs, Kahn, Linial, Nisan and Saks [38] show that $O(n^2)$ time is sufficient. This immediately yields upper bounds for non uniform log space; in particular, we can state such bounds in terms of universal traversal sequences (i.e., a single string of edge directions which traverses every n node connected graph insuring that every node will be visited) or the JAG model of Cook and Rackoff [30]. A JAG is a structured model for graph traversal problems in which pebbles are used to remember nodes and such pebbles can be moved along an edge or jumped to another pebble. This model is sufficiently powerful so as to simulate the standard breadth or depth first searches, as well as Savitch's algorithm and random walks or universal traversal sequences. In the JAG model, space is counted as $p \log n + \log \#Q$ where p is the number of pebbles and Q is the set of states of the JAG (ie. $\log \#Q$ is again Cobham's Capacity measure). Universal traversal sequences (UTSs) can be viewed as one pebble JAGs (when the pebble is placed on the source and the sink is a distinguished node recognizable by the JAG). Aleliunas et al. show the existence of a length $O(m^2 n \log n)$ UTS (or by Kahn et al. [38], length $O(mn^2 \log n)$ for regular graphs).

The generalization of the 1 pebble walk to randomized p pebble traversal of graphs was first considered in Broder, Karlin, Raghavan and Upfal [25] and further improved in Barnes and Feige [12] and Feige [34]. These algorithms all place pebbles randomly on the graph (but in different ways) and then build connected components by observing when the pebbles collide. Using the "soft oh" $\tilde{O}$ notation to hide $(\log n)^{O(1)}$ factors, Broder et al. show $T \cdot S = \tilde{O}(m^2)$, which is improved to $\tilde{O}(m^{3/2} n^{1/2})$ in Barnes and Feige and to $\tilde{O}(mR)$ by Feige where $R = \sum_u 1/\deg(u)$ is the virtual resistance. For regular graphs, the Feige bound becomes $TS = \tilde{O}(n^2)$ which (up to the $\log n$ factors) seems optimal. For deterministic algorithms, Barnes and Ruzzo [13] show that a $S = O(n^{1/c})$, $T = n^{O(c)}$ tradeoff can be realized for USTCON. Nisan [45] then shows how to use pseudo random generators to simulate any random log space computation (in particular, a random walk for USTCON) in $O(\log^2 n)$ space and polynomial time. For STCON, the situation is not nearly so positive, the only progress being the result of Barnes, Buss, Ruzzo and Schieber [10] that sublinear space (specifically space $O(n/2^{\sqrt{\log n}})$ and polynomial time is possible. All the previously mentioned algorithms can be realized on randomized JAGs.

Turning to lower bounds, one would expect that it should be easier to derive bounds for STCON. Cook and Rackoff [30] introduced the JAG model and showed that in this model, STCON requires $\Omega(\log^2 n/\log\log n)$ space independent of time. Time space lower bounds for the JAG begin with lower bounds on the length of universal traversal sequences (UTS). For UTS results we usually assume the graphs are regular. The first published UTS lower bound was obtained by Bar-Noy, Borodin, Karchmer, Linial and Werman [9]. In particular, they showed that degree 2 connected graphs (= cycles) required $\Omega(n \log n)$ length sequences and degree $d \leq n/2$ graphs required $d(n-d)$. Borodin, Ruzzo and Tompa [24] showed that degree d graphs $(3 \leq d \leq \frac{n}{3}-2)$ required $\max(\Omega(m^2), \Omega(dn^2 \log n))$. Their linear algebra based proof is applied in Beame, Borodin, Raghavan, Ruzzo and Tompa [16] to show that for p pebble JAGs where $p-1$ pebbles cannot be moved after initial placement, the time is $\Omega(n^2)$ independent of the number of states. This result is substantially improved in Edmonds [33] where he uses the linear algebra proof as the base case for a recursive construction of graphs for which any $p = O(\log n/\log\log n)$ pebble probabilistic JAG (with no restrictions—i.e., all pebbles can move and jump) must use time $n \times 2^{\Omega(\log n/\log\log n)}$ as long as the number of states is bounded by $2^{(\log n)^{O(1)}}$. That is, in the probabilistic JAG model any $O(\log^2 n/\log\log n)$ space machine must use time exceeding $n(\log n)^{O(1)}$.

For STCON, Edmonds [32] introduces an elegent class of "comb graphs" with outdegree d for any $d \geq 2$ and shows that independent of the number of states that $Tp^{1/2} = \Omega(mn^{1/2})$. Edmonds also considers the extended NNJAG model of Poon [52] where nodes are given labels which the pebbles can identify. (Poon [52] gives evidence that the model may be more powerful by showing how the Immerman [37]-Szelepcsényi [59] construction can be simulated on nondeterministic NNJAG's.) For probabilistic 1-sided error, Edmonds (see also Barnes and Edmonds [11]) obtains the bound $TS^{1/3} = \Omega(m^{2/3}n^{2/3})$ for the complement of STCON (i.e., st non connectivity). This obviously yields the same bound for deterministic NNJAG's accepting STCON. Barnes and Edmonds [11] improve the bound to $TS = \Omega(n^2/\log n)$ for the JAG model(using a variant of the comb graphs). It is interesting to note that the NNJAG result is obtained by viewing the JAG computation as a branching program computation with an assumption inherited from the JAG computation as to how the program must make progress. Then the progress based arguments mentioned in section 4 can be applied.

7   Conclusions and Open Problems

It is, of course, never a priori clear how close "partial results" bring us to breaking a long range goal such as the complexity barrier stated in the introduction. In fact, to paint the bleakest situation, it appears that the following may be open:

find an explicit Boolean decision problem for which it can be proved that $O(\log n)$ space requires time $n + 1$ on an unrestricted Boolean BP. (If this turns out to be easy to show then consider a slightly bigger time bound such as $n + \log n$ or $2n$). In spite of this bleak portrayal, there are too many important and substantial results to not consider time-space tradeoffs as one of the more successful aspects of complexity theory (e.g., compare with circuit size bounds or even depth size tradeoffs for circuits).

Many "partial results" as well as "major results" remain open. We list a few partial results as suggestive topics.

A) Concerning the circuit pebble model.
1. Can the algorithmic assumptions in Tompa's [61] result be weakened by using ideas from (say) Barnes and Edmonds [11]?
2. Is there a function which exhibits a tradeoff breakpoint for the model as in Abrahamson's [2] result for branching programs?

B) Concerning restricted BPs.
1. Can the syntactic read-k results be extended to semantic read-k branching programs? (A BP is semantic read k if only consistent computation paths are subject to the read k restriction.)
2. Can better lower bounds be obtained for branching programs which are both oblivious and read-k?

C) Concerning graph connectivity questions.
1. What is the smallest space bound for which STCON can be solved in polynomial time? That is, improve on the Barnes et al. [10] result or give a non polynomial lower bound for a space restricted JAG
2. Can the USTCON lower bound for JAGs be improved?
3. Extend the USTCON lower bounds to the NNJAG models.
4. Can it be shown that a non-deterministic log space NNJAG requires more than linear time for accepting the complement of STCON?

D) Find other natural classes of problems and appropriate structured models in which to study time-space tradeoffs.

Acknowledgements: I thank Paul Beame, C.K. Poon and Martin Tompa for many valuable comments on the first draft of this paper.

References

1. K. Abrahamson: Time-space tradeoffs for algebraic problems on general sequential machines. JCSS 43, 269–289 (1991)
2. K. Abrahamson: A time-space tradeoff for Boolean Matrix multiplication, In Proc. 31st FOCS, 412-419 (1990)
3. M. Ajtai, L. Babai, P. Hajnal, J. Komlos, P. Pudlak, V. Rodl, E. Szemeredi, Gy. Turan: Two lower bounds for branching programs, In Proc. 18th ACM STOC, 30-38, (1986)

4. R. Aleliunas, R.M. Karp, R.J. Lipton, L. Lovasz, C. Rackoff: Random walks, universal traversal sequences, and the complexity of maze problems. In Proc. 20th Annual Symposium on Foundations of Computer Science, 1979, 218-223
5. N. Alon, W. Maass: Meanders and their applications in lower bounds arguments. JCSS 37, 118-129 (1988)
6. L. Babai, P. Hajnal, E. Szemeredi, G. Turan: A lower bound for read-once branching program. JCSS 35, 153-162 (1987)
7. L. Babai, N. Nisan, M. Szegedy: Multiparty protocols, pseudorandom generators for logspace and time-space trade-offs. JCSS 45, 204-232 (1992)
8. L. Babai, P. Pudlák, V. Rödl, E. Szemerédi, E. Lower bounds to the complexity of symmetric Boolean functions Theoretical Computer Science, vol 74, 313-324 (1990)
9. A. Bar-Noy, A. Borodin, M. Karchmer, N. Linial, M. Werman: Bounds on universal sequences. SIAM Journal on Computing 18, 268-277 (1989)
10. G. Barnes, J.F. Buss, W.L. Ruzzo, B. Schieber: A sublinear space, polynomial time algorithm for directed s-t connectivity. In Proc. 7th Annual Conference on Structure in Complexity Theory, 1992, 27-33
11. G. Barnes, J. Edmonds: Time-space trade-off lower bounds for directed ST-connectivity on JAGs and stronger models. To appear in Proc. 34th Annual Symposium on Foundations of Computer science, 1993
12. G. Barnes, U. Feige: Short random walks, In ACM STOC93, 728-737 (1993)
13. G. Barnes, W.L. Ruzzo: Deterministic algorithms for undirected s-t connectivity using polynomial time and sublinear space. In Proc. 23rd Annual ACM Symposium on the Theory of Computing, 1991, 43-45
14. D.A. Barrington: Bounded-width polynomial-size branching programs recognize exactly those languages in NC^1, In Proc. 18th ACM STOC, 1986, 1-5
15. P. Beame: A general time-space tradeoff for finding unique elments. SIAM Journal on Computing 20, 270-277 (1991)
16. P. Beame, A. Borodin, P. Raghavan, W.L. Ruzzo, M. Tompa: Time-space tradeoffs for undirected graph connectivity. In Proc. 31st Annual Symposium on Foundations of Computer Science, 1990, 429-438
17. P. Beame, S. Cook: unpublished manusrcipt
18. M. Ben-Or.R. Cleve: Computing Algebraic Formulas Using a Constant Number of Registers, In SIAM Journal on Computing 21, 54-58, (1992)
19. A. Borodin, S. Cook: A time-space tradeoff for sorting on a general sequential model of computation. SIAM Journal on Computing 11, 287-297 (1982)
20. A. Borodin, D. Dolev, F. Fich, W. Paul: Bounds for width two branching programs. SIAM Journal on Computing 15, 549-560 (1986)
21. A. Borodin, F. Fich, F. Meyer auf der Heide, E. Upfal, A. Wigderson: A time-space tradeoff for element distinctness. SIAM Journal on Computing 16, 97-99 (1987)
22. A. Borodin, M.J. Fischer, D.G. Kirkpatrick, N.A. Lynch, M. Tompa: A time-space tradeoff for sorting on non-oblivious machines. JCSS 22, 351-364 (1979)
23. A. Borodin, A. Razborov, R. Smolensky: On lower bounds for read-k times branching programs. Computational Complexity 3, 1-18 (1993)
24. A. Borodin, W.L. Ruzzo, M. Tompa: Lower bounds on the length of universal traversal sequences. JCSS 45, 180-203 (1992)
25. A.Z. Broder, A.R. Karlin, P. Raghavan, E. Upfal: Trading space for time in undirected s-t connectivity. In Proc. 21st Annual ACM Symposium on the Theory of Computing, 1989, 543-549
26. A.R. Bruss, A.R. Meyer: On time-space classes and their relation to the theory of real addition. Theoretical Computer Science 11, 59-69 (1980)

27. A. Chandra, M. Furst, R. Lipton: Multiparty protocols, In Proc. 15th ACM STOC, 94-99 (1983)

28. R. Cleve: Towards Optimal Simulations of Formulas by Bounded-Width Programs, Computational Complexity 1, 91-105, (1991)

29. A. Cobham: The Recognition problem for the set of perfect squares. IBM Watson Research Center, Research paper RC-1704 (1966)

30. S.A. Cook, C.W. Rackoff: Space lower bounds for maze threadability on restricted machines. SIAM Journal on Computing 9, 636-652 (1980)

31. P. Duris, Z. Galil: A time-space tradeoff for language recognition. In Proc. 22nd Annual Symposium on Foundations of Computer Science, 1981, 53-57

32. J. Edmonds: Time-space lower bounds for undirected and directed ST-connectivity on JAG models. Ph.D. thesis, University of Toronto, 1993

33. J. Edmonds: Time-space trade-offs for undirected st-connectivity on a JAG. In Proc. 25th Annual ACM Symposium on the Theory of Computing, 1993, 718–727

34. U. Feige: A randomized time-space tradoff of $\hat{0}(m\mathring{R})$ for USTCON: To appear in FOCS 93

35. D. Grigoriev: An Application of separability and independence notions for proving lower bounds of circuit complexity. Notes of Scientific Seminars, 60, Leningrad Department, Steklov Mathematical Institute, 1976, 38–48 (in Russian)

36. J.E. Hopcroft, W.J. Paul, L.G. Valiant: On time versus space and related problems, JACM vol 24,number 2,332-337 (1977)

37. N. Immerman: Nonterministic space is closed under complementation. SIAM Journal on Computing 17, 935–938 (1988)

38. J.D. Kahn, N. Linial, N. Nisan, M.E. Saks: On the cover time of random walks on graphs. Journal of Theoretical Probability 2, 121–128 (1989)

39. M. Karchmer: Two time-space tradeoffs for element distinctness. Theoretical Computer Science 47, 237–246 (1986)

40. S.R. Kosaraju: Parallel Evaluation of Division-Free Arithmetic Expressions, Proc. of 18th Annual ACM Symp. on Theory of Computing, 1986, pp. 231-239.

41. C.Y. Lee: Representation of switching circuits by binary-decision programs. Bell System Technical Journal 38, 985–999 (1959)

42. T. Lengauer, R.E. Tarjan: Asymptotically tight bounds on time-space trade-offs in a pebble game. JACM 29, 1087–1130 (1982)

43. Y. Mansour, N. Nisan, P. Tiwari: The computational complexity of universal hashing. In Proc. 22nd Annual ACM Symposium on the Theory of Computing, 1990, 235–243

44. W. Masek: A fast algorithm for the string editing problem and decision graph complexity. M.Sc. thesis, Massachusetts Institute of Technology, Cambridge, 1976

45. N. Nisan: $RL \subseteq SC$. In Proc. 24th Annual ACM Symposium on the Theory of Computing, 1992, 619–623

46. E.A. Okolnishnikova: Lower bounds for branching programs computing characteristic functions of binary codes. Metody Discretnogo Analiza 51, 61–83 (1991) (in Russian)

47. B. Patt-Shamir, D. Pelag: Time-space tradeoffs for Set Operations, Theoretical Computer Science 110, 99-129 (1993)

48. W.J. Paul, R.E. Tarjan: Time-space trade-offs in a pebble game. Acta Informatica 10, 111–115 (1978)

49. W.J. Paul,R.E. Tarjan,J.R. Celoni Space Bounds for a Game on Graphs, Mathematical Systems Theory, vol 10, number 3, 239-251 (1977) (See also Correction in vol 11,number 1,page 85,1977)

50. N. Pippenger: Pebbling. In Proc. 5th IBM Symposium on Mathematical Foundations of Computer Science, 1980
51. N. Pippenger: A time-space trade-off, J. ACM 25, 509-515 (1978)
52. C.K. Poon: Space bounds for graph connectivity problems on node-named JAGs and node-ordered JAGs. To appear in Proc. 34th Annual Symposium on Foundations of Computer science, 1993
53. A. Razborov: Lower bounds for deterministic and nondeterministic branching programs. In Proc. 8th FCT, Lecture Notes in Computer Scienc, 529, New York/Berlin, 1991, Springer-Verlag, 47-60
54. R. Reischuk: Improved bounds on the problem of time-space trade-off in the pebble game, J. ACM 27, 839-850 (1980)
55. J.E. Savage: Space-time tradeoffs - A survey. In Proc. 3rd Hungarian Computer Science Conference, 1981
56. J.E. Savage, S. Swamy: Space-time tradeoffs on the FFT algorithm. IEEE Trans. Inform. Theory 24, 563-568 (1978)
57. W.J. Savitch: Relationships between nondeterministic and deterministic tape complexities. JCSS 4, 177-192 (1970)
58. J. Shearer: unpublished manuscript
59. R. Szelepcsényi: The method of forcing for nondeterministic automata. BullaEuropean Assoc. Theoret. Comput. Sci. 33, 96-100 (1987)
60. M. Tompa: Time-space tradeoffs for computing functions, using connectivity properties of their circuits. JCSS 20, 118-132 (1980)
61. M. Tompa: Two Familiar Transitive Closure Algorithms which Admit no Polynomial Time, Sublinear Space Implementations, Siam Journal on Computing vol 11,no. 1, 130-137 (1982)
62. A.C. Yao: Near-optimal time-space tradeoff for element distinctness. In Proc. 29th Annual Symposium on Foundations of Computer Science, 1988, 91-97. Accepted for publication in Siam Journal on Computing.
63. Y. Yesha: Time-space tradeoffs for matrix multiplication and the discrete Fourier transform on any general sequential random-access computer. SIAM Journal on Computing 29, 183-197 (1984)
64. S. Zak: An exponential lower bound for one-time-only branching programs. In Proc. 11th MFCT, Lecture Notes in Computer Science, 176, New York/Berlin, 1984, Springer-Verlag, 562-566

Separating Exponentially Ambiguous NFA from Polynomially Ambiguous NFA *

Hing Leung

Department of Computer Science, New Mexico State University,
Las Cruces, NM 88003, U.S.A.
hleung@nmsu.edu

Abstract. We resolve an open problem raised by Ravikumar and Ibarra on the succinctness of representations relating to the types of ambiguity of finite automata. We show that there exists a family of nondeterministic finite automata $\{A_n\}$ over a two-letters alphabet such that, for any positive integer n, A_n is exponentially ambiguous and has n states, whereas the smallest equivalent deterministic finite automaton has 2^n states and any smallest equivalent polynomially ambiguous finite automaton has $2^n - 1$ states.

1 Introduction

In [RI89], questions relating the type of ambiguity of finite automata to the succinctness in their number of states are studied. The following five classes of finite automata have been considered: DFA (deterministic finite automata), NFA (nondeterministic finite automata), UFA (unambiguous NFA), FNA (finitely ambiguous NFA) and PNA (polynomially ambiguous NFA).

Let C_1 and C_2 be any two of the above five classes of finite automata. We say that C_1 can be polynomially converted to C_2 (written $C_1 \leq_P C_2$) if there exists a polynomial p such that for any finite automaton in C_1 with n states, we can find an equivalent finite automaton in C_2 with the number of states being at most $p(n)$. C_1 is said to be polynomially related to C_2 (written $C_1 =_P C_2$) if $C_1 \leq_P C_2$ and $C_2 \leq_P C_1$. C_1 is said to be separated from C_2 if $C_1 \neq_P C_2$. Furthermore, we write $C_1 <_P C_2$ if $C_1 \leq_P C_2$ and $C_1 \neq_P C_2$.

It is immediate that DFA $\leq_P$ UFA, UFA $\leq_P$ FNA, FNA $\leq_P$ PNA and PNA $\leq_P$ NFA. The following results have been obtained: DFA $<_P$ NFA ([MF71], [Mo71]), DFA $<_P$ UFA ([Sc78], [SH85], [RI89]), UFA $<_P$ FNA ([Sc78], [RI89]), and UFA $<_P$ NFA ([SH85]). It is unknown whether FNA $<_P$ NFA. It is conjectured by [RI89] that FNA $<_P$ PNA and PNA $<_P$ NFA.

In this paper, we prove that PNA $<_P$ NFA which immediately implies that FNA $<_P$ NFA. The other conjecture that FNA $<_P$ PNA still remains open.

Specifically, we show that there exists a family of NFAs $\{A_n \mid n \geq 1\}$ over a two-letters alphabet such that, for any positive integer n, A_n is exponentially ambiguous and has n states, whereas the smallest equivalent DFA has 2^n states and any smallest equivalent PNA has $2^n - 1$ states.

---

* This research is supported by an Alexander von Humboldt research fellowship. It was done while the author was visiting the Univerisity of Frankfurt, Germany.

Our results show that any PNA equivalent to A_n cannot do better in the number of states than the smallest equivalent DFA obtained by the subset construction except for the saving of the dead state.

Another way to interpret our results is as follows: let us first define that an NFA is strongly ambiguous if there is a useful state q (that is, q can be reached from some starting state and can reach some final state) and there is a string w such that M can process w starting from state q and ending also with state q in more than one ways. Then by a characterization in [IR86], our results show that A_n is strongly ambiguous with n states, whereas the smallest equivalent DFA has 2^n states and any smallest equivalent NFA that is not strongly ambiguous has $2^n - 1$ states.

Section 2 presents the definitions and some basic results. Section 3 presents the family of NFAs $\{A_n\}$ and proves the main result of this paper by a series of lemmas.

2 Preliminaries

We assume that the reader is familiar with the basic definitions, results and notations in finite automata theory [HU79] and graph theory [Ha69].

Let w be a string. Then w^R denotes the reverse of w. For a language L, L^R is the set of strings w^R where $w \in L$.

Throughout this paper, we assume a model of NFA that is slightly more general than the one defined in [HU79] in that we allow a set of starting states instead of only one starting state. Thus, an NFA M is a 5-tuple $(Q, \Sigma, \delta, Q_I, Q_F)$ where Q is the set of states, Σ is the alphabet set, $\delta : Q \times \Sigma \longrightarrow 2^Q$ is the transition function, Q_I is the set of starting states and Q_F is the set of final states.

Given an NFA M, we define the ambiguity of a string w to be the number of different accepting paths for w in M. Note that a string w is in the language of M if and only if the ambiguity of w is not zero. The ambiguity function $amb_M : \mathbb{N}_0 \longrightarrow \mathbb{N}_0$ is defined such that $amb_M(n)$ is the maximum of the ambiguities of strings that are of length n or less. Remark: amb_M is nondecreasing.

M is called unambiguous if the ambiguity of any string is either zero or one. M is called finitely (respectively: polynomially, exponentially) ambiguous if amb_M can be bounded by a constant (respectively: polynomial, exponential) function f; that is for all $n \in \mathbb{N}_0$, $amb_M(n) \leq f(n)$.

It is easy to see that $amb_M(n) \leq s_1 s^n$ where s_1 is the cardinality of Q_I and s is the cardinality of Q. Thus, every NFA must be exponentially ambiguous.

M is called strictly exponentially ambiguous ([IR86]) if M is exponentially ambiguous but not polynomially ambiguous. It is known ([IR86]) that M is strictly exponentially ambiguous if and only if there is a useful state q and there is a string w such that M can process w starting from state q and ending also with state q in more than one ways.

3 Main result

For any positive integer n, we define an NFA $A_n = (Q, \Sigma, \delta, \{q_1\}, \{q_1\})$ where $Q = \{q_1, q_2, \ldots, q_n\}$, q_1 is the only starting state and the only final state, $\Sigma = \{0, 1\}$ and δ (see Figure 1) is defined as follows: $\delta(q_1, 0) = \{q_1, q_2\}$, $\delta(q_i, 0) = \{q_{i+1}\}$ for $2 \le i \le n - 1$, $\delta(q_n, 0) = \{q_1\}$, $\delta(q_1, 1) = \emptyset$ and $\delta(q_i, 1) = \{q_i\}$ for $2 \le i \le n$.

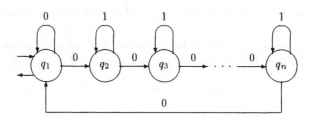

Fig. 1. Transition diagram of A_n.

We denote the language of A_n by L_n, which is $(0 + (01^*)^{n-1}0)^*$ in regular expression. It is easy to see that $L_n = L_n^R = L_n^*$.

First, we present some definitions. Given a language L and a string x,

$$prefix(L) \overset{\text{def}}{=} \{w \mid \exists w', \ ww' \in L\}$$

and

$$x^{-1}(L) \overset{\text{def}}{=} \{w \mid xw \in L\}.$$

The two operations *prefix* and x^{-1} commute since both $x^{-1}(prefix(L))$ and $prefix(x^{-1}(L))$ equal to $\{w \mid \exists w', \ xww' \in L\}$.

Let $kill\_non\_q1$ denotes $0(10)^{n-1}$. Then for any $P \subseteq Q$, $\delta(P, kill\_non\_q1) = P - \{q_2, \ldots, q_n\}$.

Let *accept* denotes 0^{n-1} and *reset* denotes *accept kill_non_q1*. For any nonempty subset $P \subseteq Q$, observe that $q_1 \in \delta(P, accept)$ and $\delta(P, reset) = \{q_1\}$. Equivalently, for any $x \in prefix(L_n)$, $x \ accept \in L_n$ and $(x \ reset)^{-1}(L_n) = L_n$.

For any $P \subseteq Q$, let $w_P \in \Sigma^*$ be $w_1 0 w_n 0 w_{n-1} 0 \ldots 0 w_1$ where $w_i = \epsilon$ if $q_i \in P$, and $w_i = 1$ otherwise; and let $u_P \in \Sigma^*$ be $0^{n-1} w_P$.

Lemma 1. *For any $P \subseteq Q$, we have*
(1) $\delta(P, w_P) = P$
(2) for any $q \in P$, $\delta(q, w_P) \supseteq \{q\}$
(3) $\delta(Q - P, w_P) = \emptyset$

Proof. Given a state $q_i \in Q$, we define $shift(q_i)$ to be q_{i+1} if $1 \le i \le n - 1$, and q_1 if $i = n$. Given $P \subseteq Q$, we extend the definition of *shift* such that $shift(P) = \{shift(q) \mid q \in P\}$. Moreover, for any $q \in Q$ and $P \subseteq Q$, we define $shift^0(q) = q$ and $shift^0(P) = P$. Note that $shift^n(q) = q$ and $shift^n(P) = P$.

(1) Observe that $\delta(P, w_1) = P$. This is because if $q_1 \in P$, then $w_1 = \epsilon$ and $\delta(P, w_1) = \delta(P, \epsilon) = P$. Otherwise if $q_1 \notin P$, then $w_1 = 1$ and $\delta(P, w_1) = \delta(P, 1) = P$ since $q_1 \notin P$.

Therefore, $\delta(P, w_P) = \delta(P, 0w_n0w_{n-1} \ldots 0w_1)$.

We want to show by induction that $\delta(P, 0w_n0w_{n-1} \ldots 0w_{n-i+1}) = shift^i(P)$ for $0 \leq i \leq n$. Then we are done since $\delta(P, 0w_n0w_{n-1} \ldots 0w_1) = shift^n(P) = P$.

Base. $i = 0$. $\delta(P, \epsilon) = P = shift^0(P)$.

Induction hypothesis: Assume that the statement is true for $0 \leq k \leq n - 1$.

Induction step. By induction, we have $\delta(P, 0w_n0w_{n-1} \ldots 0w_{n-k+1}0w_{n-k})$ $= \delta(\delta(P, 0w_n0w_{n-1} \ldots 0w_{n-k+1}), 0w_{n-k}) = \delta(shift^k(P), 0w_{n-k})$. The induction proof is completed if we can verify that $\delta(shift^k(P), 0w_{n-k}) = shift^{k+1}(P)$.

Case 1. $(q_{n-k} \in P)$

Then $w_{n-k} = \epsilon$ and $q_n \in shift^k(P)$. Thus, $\delta(shift^k(P), 0w_{n-k}) = \delta(shift^k(P), 0)$ $= shift^{k+1}(P)$ since $q_n \in shift^k(P)$.

Case 2. $(q_{n-k} \notin P)$

Then $w_{n-k} = 1$ and $q_n \notin shift^k(P)$. Thus, $\delta(shift^k(P), 0w_{n-k}) = \delta(shift^k(P), 01) = shift^{k+1}(P)$ since $q_n \notin shift^k(P)$.

(2) Observe that $\delta(q, w_1) = \{q\}$. This is because if $q_1 \in P$, then $w_1 = \epsilon$ and $\delta(q, w_1) = \delta(q, \epsilon) = \{q\}$. Otherwise if $q_1 \notin P$, then $w_1 = 1$ and $\delta(q, w_1) = \delta(q, 1) = \{q\}$ since $q \neq q_1$ by the facts that $q_1 \notin P$ and $q \in P$.

Therefore, $\delta(q, w_P) = \delta(q, 0w_n0w_{n-1} \ldots 0w_1)$.

We want to show by induction that $\delta(q, 0w_n0w_{n-1} \ldots 0w_{n-i+1}) \supseteq shift^i(\{q\})$ for $0 \leq i \leq n$. Then we are done since $\delta(q, 0w_n0w_{n-1} \ldots 0w_1) \supseteq shift^n(\{q\})$ $= \{q\}$.

Base. $i = 0$. $\delta(q, \epsilon) = \{q\} = shift^0(\{q\})$. Thus, $\delta(q, \epsilon) \supseteq shift^0(\{q\})$.

Induction hypothesis: Assume that the statement is true for $0 \leq k \leq n - 1$.

Induction step. By induction, we have $\delta(q, 0w_n0w_{n-1} \ldots 0w_{n-k+1}0w_{n-k}) = \delta(\delta(q, 0w_n0w_{n-1} \ldots 0w_{n-k+1}), 0w_{n-k}) \supseteq \delta(shift^k(\{q\}), 0w_{n-k})$. The induction proof is completed if we can verify that $\delta(shift^k(\{q\}), 0w_{n-k}) \supseteq shift^{k+1}(\{q\})$.

Case 1. $(q_n = shift^k(q))$

Then $w_{n-k} = \epsilon$ since $q_{n-k} = q \in P$. Thus, $\delta(shift^k(\{q\}), 0w_{n-k}) = \delta(shift^k(\{q\}), 0) = \delta(q_n, 0) = \{q_1\} = shift(\{q_n\}) = shift(shift^k(\{q\})) = shift^{k+1}(\{q\})$. Hence, $\delta(shift^k(\{q\}), 0w_{n-k}) \supseteq shift^{k+1}(\{q\})$.

Case 2. $(q_n \neq shift^k(q))$

Then no matter whether $w_{n-k} = \epsilon$ or $w_{n-k} = 1$, we always have $\delta(shift^k(\{q\}), 0w_{n-k}) \supseteq shift^{k+1}(\{q\})$.

(3) We want to show that $\delta(Q - P, w_10w_n0w_{n-1}0 \ldots 0w_2) = \emptyset$. Hence, $\delta(Q - P, w_P) = \delta(Q - P, w_10w_n0w_{n-1}0 \ldots 0w_20w_1) = \delta(\delta(Q-P, w_10w_n0w_{n-1}0 \ldots 0w_2), 0w_1) = \delta(\emptyset, 0w_1) = \emptyset$.

First observe that $\delta(Q - P, w_1) = Q - P - \{q_1\}$. This is because if $q_1 \in P$ then $w_1 = \epsilon$ and $q_1 \notin Q - P$. Thus $\delta(Q - P, w_1) = \delta(Q - P, \epsilon) = Q - P = Q - P - \{q_1\}$ since $q_1 \notin Q - P$. Otherwise if $q_1 \notin P$ then $w_1 = 1$. Thus $\delta(Q - P, w_1) = \delta(Q - P, 1) = Q - P - \{q_1\}$.

Next we want to show by induction that for $1 \leq i \leq n$, $\delta(Q - P - \{q_1\}, 0w_n0w_{n-1} \ldots 0w_{n-i+2}) = shift^{i-1}(Q - P) - \{q_1, q_2, \ldots, q_i\}$. Then we are done since by taking $i = n$, we have $\delta(Q-P-\{q_1\}, 0w_n0w_{n-1}\ldots 0w_2) = shift^{n-1}(Q-P) - \{q_1, q_2, \ldots, q_n\} = \emptyset$.

Base. $i = 1$. $\delta(Q - P - \{q_1\}, \epsilon) = Q - P - \{q_1\} = shift^0(Q-P) - \{q_1\}$.

Induction hypothesis: Assume that the statement is true for $1 \leq k \leq n-1$.

Induction step. By induction, we have $\delta(Q - P - \{q_1\}, 0w_n0w_{n-1}\ldots$ $0w_{n-k+2}0w_{n-k+1}) = \delta(shift^{k-1}(Q-P) - \{q_1, q_2, \ldots, q_k\}, 0w_{n-k+1})$. The induction proof is completed if we can verify that $\delta(shift^{k-1}(Q-P) - \{q_1, q_2, \ldots, q_k\}, 0w_{n-k+1}) = shift^k(Q-P) - \{q_1, q_2, \ldots, q_k, q_{k+1}\}$.

Case 1. ($q_{n-k+1} \in P$)

Then $w_{n-k+1} = \epsilon$ and $q_n \notin shift^{k-1}(Q-P)$. Thus, $shift^{k-1}(Q-P) = shift^{k-1}(Q-P) - \{q_n\}$. Hence, $\delta(shift^{k-1}(Q-P) - \{q_1, q_2, \ldots, q_k\}, 0w_{n-k+1}) = \delta(shift^{k-1}(Q-P) - \{q_n, q_1, q_2, \ldots, q_k\}, 0) = shift^k(Q-P) - \{q_1, q_2, \ldots, q_k, q_{k+1}\}$.

Case 2. ($q_{n-k+1} \notin P$)

Then $w_{n-k+1} = 1$. Thus, $\delta(shift^{k-1}(Q-P) - \{q_1, q_2, \ldots, q_k\}, 0w_{n-k+1}) = \delta(shift^{k-1}(Q-P) - \{q_1, q_2, \ldots, q_k\}, 01) = shift^k(Q-P) - \{q_1, q_2, \ldots, q_k, q_{k+1}\}$. □

Corollary 2. *For any $P \subseteq Q$, $\delta(q_1, u_P) = P$.*

Proof. $\delta(q_1, u_P) = \delta(q_1, 0^{n-1} w_P) = \delta(Q, w_P) = \delta(P, w_P) \bigcup \delta(Q - P, w_P) = \delta(P, w_P) = P$ by parts (1) and (3) of lemma 1. □

Corollary 3. *For any $P, P' \subseteq Q$, $u_P w_{P'} \in prefix(L_n)$ if and only if $P \bigcap P' \neq \emptyset$.*

Proof. Suppose $P \bigcap P' \neq \emptyset$. Let $q \in P \bigcap P'$. Then $\delta(q_1, u_P w_{P'}) = \delta(P, w_{P'}) \supseteq \delta(q, w_{P'}) \supseteq \{q\} \neq \emptyset$ by corollary 2 and part (2) of lemma 1. Since all states in A_n are useful, $u_P w_{P'} \in prefix(L_n)$.

Suppose $P \bigcap P' = \emptyset$. Then $P \subseteq Q - P'$ and $\delta(q_1, u_P w_{P'}) = \delta(P, w_{P'}) \subseteq \delta(Q - P', w_{P'}) = \emptyset$ by corollary 2 and part (3) of lemma 1. Thus, $u_P w_{P'} \notin prefix(L_n)$. □

Lemma 4. *The smallest DFA recognizing L_n has 2^n states.*

Proof. By corollary 2, all subsets of states can be realized in the subset construction. Next, we want to show that any two different subsets of states are not equivalent, then we are done by the Myhill-Nerode theorem. Let $P \neq P' \subseteq Q$. Let q_i be the state with the largest subscript such that q_i belongs only to exactly one of P and P'. That is, for any q_j where $i + 1 \leq j \leq n$, either q_j belongs to both P and P' or q_j does not belong to any one of P and P'. If $i = 1$, then P and P' can be distinguished by the empty string. Assume that $1 < i \leq n$. Then P and P' can be distinguished by $(10)^{n+1-i}$. □

Let M_n be a $2^n - 1$ by $2^n - 1$ matrix over the field of characteristic 2 with rows and columns indexed by the nonempty subsets of Q such that $M_n(P, P') = 1$ if $u_P w_{P'}$ accept $\in L_n$, and $M_n(P, P') = 0$ otherwise. By corollary 3 and the property of accept, $M_n(P, P') = 1$ if $P \bigcap P' \neq \emptyset$ and $M_n(P, P') = 0$ otherwise.

Lemma 5. *The rank of M_n is $2^n - 1$.*

Proof. The proof will be given in the full version of the paper. $\square$

Lemma 6. *A smallest UFA recognizing L_n has $2^n - 1$ states.*

Proof. First, by removing the dead state from the DFA obtained by the subset construction, we have a UFA with $2^n - 1$ states. Next we are going to use a technique introduced in [Sc78] to show that any UFA would require at least $2^n - 1$ states.

Let U be a UFA recognizing L_n with the finite set of states denoted by K. Let R be a matrix over the field of characteristic 2 with rows indexed by K and columns indexed by the nonempty subsets of Q such that $R(k, P) = 1$ if U can reach a final state starting from state $k \in K$ on consuming w_P *accept*, and $R(k, P) = 0$ otherwise.

We claim that any row in M_n is a linear combination of the rows in R. Given a nonempty subset P of Q, let $K' \subseteq K$ be the set of states reached by U from the set of starting states on consuming u_P. For any $k_1 \neq k_2 \in K'$ and for any nonempty subset P' of Q, $R(k_1, P')$ and $R(k_2, P')$ cannot both have value one; that is, at most one of $R(k, P')$, for $k \in K'$, is 1. Otherwise, there are two different accepting paths for $u_P w_{P'}$ *accept* in U, which contradicts to the assumption that U is unambiguous. Thus, the row indexed by P in M_n is the sum of the rows indexed by K' in R.

Therefore, the rank of M_n is less than or equal to the rank of R. Hence, K must have at least $2^n - 1$ states so that the rank of R is at least $2^n - 1$. $\square$

Lemma 7. *Any UFA recognizing L such that $prefix(L) = prefix(L_n)$ requires at least $2^n - 1$ states.*

Proof. Since L is regular, there exists a finite set of strings $\{\gamma_1, \ldots, \gamma_h\} \subseteq \Sigma^*$ such that for any $z \in prefix(L)$, exactly one of $z\gamma_i$, for $1 \leq i \leq h$, is in L.

Let X be a $2^n - 1$ by $h(2^n - 1)$ matrix over the field of characteristic 2 with rows indexed by the nonempty subsets of Q and columns indexed by $\{(P, i) \mid \emptyset \neq P \subseteq Q, 1 \leq i \leq h\}$ such that $X(P, (P', i)) = 1$ if $u_P w_{P'} \gamma_i \in L$, and $X(P, (P', i)) = 0$ otherwise.

We claim that the rank of X is $2^n - 1$. We define X' to be a $2^n - 1$ by $2^n - 1$ matrix over the field of characteristic 2 with rows and columns indexed by the nonempty subsets of Q such that the column indexed by P is the sum of the h columns in X indexed by $\{(P, i) \mid 1 \leq i \leq h\}$.

We want to show that X' is the same matrix as M_n. That is, we want to show that $X'(P, P') = 1$ if $P \cap P' \neq \emptyset$ and $X'(P, P') = 0$ otherwise.

Suppose $P \cap P' \neq \emptyset$. By corollary 3, $u_P w_{P'} \in prefix(L_n) = prefix(L)$. By the definition of γ_i's, exactly one of $u_P w_{P'} \gamma_i$, for $1 \leq i \leq h$, is in L. That is, exactly one of $X(P, (P', i))$, for $1 \leq i \leq h$, is 1. Hence, $X'(P, P') = 1$.

Suppose $P \cap P' = \emptyset$. By corollary 3, $u_P w_{P'} \notin prefix(L_n) = prefix(L)$. Therefore, $u_P w_{P'} \gamma_i \notin L$ for $1 \leq i \leq h$. Hence, $X(P, (P', i)) = 0$ for $1 \leq i \leq h$ and $X'(P, P') = 0$.

By lemma 5, the rank of X' is $2^n - 1$. Since each column of X' is obtained by a linear combination of the columns of X, the rank of X must be bigger than or equal to the rank of X'. Thus, the rank of X is at least $2^n - 1$. Moreover since the number of rows in X is $2^n - 1$, the rank of X is at most $2^n - 1$. Therefore, the rank of X is $2^n - 1$.

Finally, since the rank of X is $2^n - 1$ and by using the same technique as in the proof of lemma 6, a smallest UFA for L must have at least $2^n - 1$ states. $\square$

The following lemma is a generalization of lemma 7. This is because we can obtain lemma 7 by letting x to be the empty string in the following lemma.

Lemma 8. *Let U be a UFA with the number of states less than $2^n - 1$ which accepts L such that $prefix(L) \subseteq prefix(L_n)$. Then for all $x \in prefix(L_n)$,*

$$x^{-1}(prefix(L)) \subset x^{-1}(prefix(L_n)) .$$

Proof. From $prefix(L) \subseteq prefix(L_n)$, it is immediate that $x^{-1}(prefix(L)) \subseteq x^{-1}(prefix(L_n))$ for any arbitrary x. We want to show that $x^{-1}(prefix(L)) \neq x^{-1}(prefix(L_n))$ for any $x \in prefix(L_n)$.

Suppose the contrary that $x^{-1}(prefix(L)) = x^{-1}(prefix(L_n))$ for some $x \in prefix(L_n)$. Thus, $z^{-1}(prefix(L)) = z^{-1}(prefix(L_n))$ where $z = x$ reset. Since z^{-1} and $prefix$ commute, $prefix(z^{-1}(L)) = prefix(z^{-1}(L_n))$. Let L' be $z^{-1}(L)$. Then we obtain $prefix(L') = prefix(L_n)$ since $z^{-1}(L_n) = L_n$ by the property of $reset$.

We are going to construct a UFA U' with less than $2^n - 1$ states to recognize L'; hence by lemma 7, a contradiction.

The transition diagram for U' is the same as that of U. The set of starting states for U' is defined to be the set of states reached by U on consuming z from the set of starting states of U. The set of final states for U' is again the same as that of U. It is clear that the language accepted by U' is $L' = z^{-1}(L)$. Also, U' cannot be ambiguous otherwise U is also ambiguous. Moreover, the number of states in U' is the same as the number of states in U; therefore, it is less than $2^n - 1$. $\square$

Theorem 9. *A smallest PNA recognizing L_n has $2^n - 1$ states.*

Proof. By removing the dead state from the DFA obtained by the subset construction, we have a UFA, which is polynomially ambiguous, with $2^n - 1$ states.

Let M be a PNA for L_n with the smallest number of states. Then every state in M must be useful. Consider the transition diagram of M. Let us look at one strongly connected component, denoted T, that cannot be reached from other strongly connected components.

We claim that T must have at least $2^n - 1$ states. Suppose the contrary that it has less than $2^n - 1$ states.

T must have some starting states in it. Otherwise it is not useful which contradicts to the definition of M. Let the set of starting states of M that appear in T be $\{p_1, \ldots, p_k\}$.

Let $1 \le i \le k$. We define an NFA T_{p_i} such that p_i is now the only starting and final state, and the transition diagram for T_{p_i} is T. We want to check that lemma 8 can be applied to the language of T_{p_i}. First, T_{p_i} has less than $2^n - 1$ states. Next, T_{p_i} is a UFA; otherwise by the characterization given in section 2, M is strictly exponentially ambiguous, a contradiction. Let $w \in prefix(L(T_{p_i}))$. Then p_i must reach a nonempty subset of states in T on consuming w. Since all states in M are useful, w is therefore in $prefix(L_n)$. Hence, $prefix(L(T_{p_i})) \subseteq prefix(L_n)$.

Consider the set of UFAs $\{T_{p_i} \mid 1 \leq i \leq k\}$. Let $x_0 = \epsilon \in prefix(L_n)$. For $1 \leq i \leq k$, we define x_i to be a string chosen arbitrarily from

$$(x_0 x_1 \ldots x_{i-1})^{-1}(prefix(L_n)) - (x_0 x_1 \ldots x_{i-1})^{-1}(prefix(L(T_{p_i}))) .$$

Thus, $x_0 x_1 \ldots x_i \in prefix(L_n)$ for $0 \leq i \leq k$. The existence of x_i, for $1 \leq i \leq k$, is then guaranted by lemma 8. Let x be $x_1 \ldots x_k$. By the way x_i's are defined, $x \notin prefix(L(T_{p_i}))$ for $1 \leq i \leq k$. Thus, each UFA T_{p_i}, $1 \leq i \leq k$, reaches the empty set from the starting state p_i on consuming x since all states in T_{p_i} are useful.

Let $z = x$ reset. Since $x \in prefix(L_n)$, then $z^{-1}(L_n) = L_n$ by the property of reset.

Let us consider M again. From the set of starting states, M reaches a subset of states, denoted P, on consuming z. Since $z \in prefix(L_n)$ and $L(M) = L_n$, P is not empty. Moreover, by the previous discussions, P does not include any state in T.

We define another NFA M' by removing the set of states in T from the state set of M and let P be the new set of starting states, whereas the set of final states is the set of final states of M minus the set of states in T. By the facts that M accepts L_n and $z^{-1}(L_n) = L_n$, M' must also accept L_n. But this is a contradiction since M' is now a PNA accepting L_n with a smaller number of states than M.

Therefore, T cannot have less than $2^n - 1$ states. Hence, M has at least $2^n - 1$ states. $\square$

Acknowledgements. I gratefully thank Andreas Weber, Jonathan Goldstine and Detlef Wotschke for their valuable discussions.

References

[Ha69] F. Harary, *Graph theory*, Addison-Wesley, Reading, MA, 1969.

[HU79] J. Hopcroft and J. Ullman, *Introduction to automata theory, languages and computation*, Addison-Wesley, Reading, MA, 1979.

[IR86] O. Ibarra and B. Ravikumar, *On sparseness, ambiguity and other decision problems for acceptors and transducers*, Proc. 3rd Annual Symposium on Theoretical Aspects of Computer Science, 1986, pp. 171–179.

[MF71] A. Meyer and M. Fischer, *Economy of description by automata, grammars, and formal systems*, Proc. 12th Symposium on Switching and Automata Theory, 1971, pp. 188–191.

[Mo71] F. Moore, *On the bounds for state-set size in the proofs of equivalence between deterministic, nondeterministic, and two-way finite automata*, IEEE Trans. Comput., 20 (1971), pp. 1211–1214.

[Re77] C. Reutenauer, *Propriétés arithmétiques et topologiques de séries rationnelles en variables non commutatives*, Thèse troisième cycle, Université Paris VI, 1977.

[RI89] B. Ravikumar and O. Ibarra, *Relating the type of ambiguity of finite automata to the succinctness of their representation*, SIAM J. Comput., 18 (1989), pp. 1263–1282.

[Sc78] E. Schmidt, *Succinctness of descriptions of context-free, regular, and finite languages*, Ph.D. Thesis, Cornell University, 1978.

[SH85] R. Stearns and H. Hunt, *On the equivalence and containment problems for unambiguous regular expressions, regular grammars and finite automata*, SIAM J. Comput., 14 (1985), pp. 598–611.

[WS91] A. Weber and H. Seidl, *On the degree of ambiguity of finite automata*, Theoret. Comput. Sci., 88 (1991), pp. 325–349.

Threshold Computation and Cryptographic Security[*]

Yenjo Han[1], Lane A. Hemaspaandra[1], Thomas Thierauf[2][**]

[1] Dept. of Computer Science, University of Rochester, Rochester, NY 14627 USA
[2] Abteilung für Theoretische Informatik, Universität Ulm, 89069 Ulm, Germany

Abstract. Threshold machines [21] are Turing machines whose acceptance is determined by what portion of the machine's computation paths are accepting paths. Probabilistic machines [11] are Turing machines whose acceptance is determined by the probability weight of the machine's accepting computation paths. Simon [21] proved that for unbounded-error polynomial-time machines these two notions yield the same class, PP. Perhaps because Simon's result seemed to collapse the threshold and probabilistic modes of computation, the relationship between threshold and probabilistic computing for the case of bounded error has remained unexplored.

In this paper, we compare the bounded-error probabilistic class BPP with the analogous threshold class, BPP_{path}, and, more generally, we study the structural properties of BPP_{path}. We prove that BPP_{path} contains both NP^{BPP} and $\text{P}^{\text{NP}[\log]}$, and that BPP_{path} is contained in $\text{P}^{\Sigma_2^p[\log]}$, BPP^{NP}, and PP. We conclude that, unless the polynomial hierarchy collapses, bounded-error threshold computation is strictly more powerful than bounded-error probabilistic computation.

We also consider the natural notion of secure access to a database: an adversary who watches the queries should gain no information about the input other than perhaps its length [5]. We show, for both BPP and BPP_{path}, that if there is *any* database for which this formalization of security differs from the security given by oblivious [9] database access, then $\text{BPP} \neq \text{PP}$. It follows that if any set lacking small circuits can be securely accepted, then $\text{BPP} \neq \text{PP}$.

1 Introduction

In 1975, Simon [21] defined threshold machines. A threshold machine is a non-deterministic Turing machine that accepts a given input if more than half of all computation paths on that input are accepting paths. Gill [11] defined the class PP as the class of sets for which there exists a probabilistic polynomial-time Turing machine that accepts exactly the members of the set with probability greater than 1/2. Simon [21] showed that the class of sets accepted by polynomial-time

---

[*] Research supported in part by grants NSF-CCR-8957604, NSF-CCR-9057486, and NSF-INT-9116781/JSPS-ENG-207, and by DFG Postdoctoral Stipend Th 472/1-1.
[**] Work done while visiting the University of Rochester and Princeton University.

threshold machines characterizes the unbounded-error probabilistic class PP. This equality notwithstanding, in some sense the threshold paradigm is more natural for programming than the probabilistic model. In the threshold model, one does not have to be concerned with depth and probability weight normalization, and thus "ragged" computation trees present no problem.

In this paper, we extend the notion of threshold computation to bounded-error probabilistic classes, and we study the degree to which threshold and probabilistic database ("oracle") computations hide information from observers.

In particular, we introduce BPP_{path} and R_{path} as the threshold analogs of BPP and R [11]. We give evidence that, unlike the case for PP, these threshold classes are different from their probabilistic counterparts. Section 3 studies the properties of the class BPP_{path} and its relationship to other complexity classes. For example, we show—in contrast to the BPP case—that BPP_{path} is self-low (i.e., $BPP_{path}{}^{BPP_{path}} = BPP_{path}$) only if the polynomial hierarchy collapses. We also show that BPP is low for BPP_{path}, that there is a relativized world in which BPP_{path} does not contain the smallest reasonable counting class, and that BPP_{path} has many closure properties. The diagram on the right gives an overview of the inclusion relations we establish between BPP_{path} and other complexity classes; in particular, note that, though contained in the polynomial hierarchy, BPP_{path} contains NP and coNP.

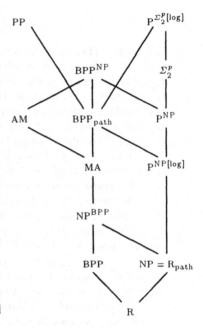

Section 4 studies, for threshold and probabilistic computations that have Turing (that is, adaptive) access to a database, the degree to which the input can be hidden from an observer. In particular, we consider the least restrictive possible notion ensuring that a powerful observer should gain no information about the input other than its length [5]. For the cases of unbounded-error probabilistic and threshold computation, we note that this optimal degree of security can be achieved in all cases. For the cases of bounded-error probabilistic and threshold computations, we prove the following result: If there exists any database D to which secure access yields more power than oblivious access (a notion in which the querying machine—until finished querying—is wholly denied access to the input other than the length of the input [9]), then $BPP \neq PP$.

2 Definitions and Discussion

Throughout this paper, we use the alphabet $\Sigma = \{0, 1\}$. For a string $x \in \Sigma^*$, $|x|$ denotes the length of x. For a set $A \subseteq \Sigma^*$, $A(x)$ denotes the characteristic function of A, $A^{=n}$ denotes $\{ y \mid y \in A$ and $|y| = n \}$, $A^{\leq n}$ denotes $\{ y \mid y \in A$ and $|y| \leq n \}$, and $\|A\|$ denotes the cardinality of A.

Let $(\cdot, \cdot)_b : \Sigma^* \times \Sigma^* \to \Sigma^*$ be a polynomial-time computable, polynomial-time invertible, one-to-one, onto function. For any string z, let $z + 1$ denote the string that lexicographically follows z, and for any string $z \neq \epsilon$, let $z - 1$ denote the string that lexicographically precedes z. Let k_s be the lexicographically kth string in Σ^*. We define our (multi-arity, onto) pairing function by $(x_1, x_2, \cdots, x_k)$ equals (a) $(\epsilon, \epsilon)_b$ when $k = 0$, (b) $(\epsilon, x_1 + 1)_b$ when $k = 1$, and (c) $(k_s, (x_1, (x_2, (\cdots (x_{k-1}, x_k)_b \cdots)_b)_b)_b)_b$ when $k \geq 2$.

We use P and FP, respectively, to denote the class of polynomial-time computable sets and polynomial-time computable total functions. NP is the class of sets that are computable by nondeterministic polynomial-time Turing machines. $P^{NP[\log]}$ denotes the class of languages accepted by some deterministic polynomial-time Turing machine that makes $\mathcal{O}(\log n)$ queries to its NP oracle. P/poly [14] denotes the class of sets having small circuits. We use the standard notions of truth-table reductions [17], and the standard notion (that is, giving all machines access to the oracle) of relativization [3].

For a nondeterministic polynomial-time Turing machine M, let $acc_M(x)$ $(rej_M(x))$ denote the number of accepting (rejecting) paths of M on input x and let $total_M(x)$ denote the total number of paths of M on input x.

We use the standard model of nondeterministic Turing machines, in which the nondeterministic branching degree is at most two. M is polynomial-normalized (henceforward denoted *normalized*) if there is a polynomial p such that on every input x the machine M makes exactly $p(|x|)$ nondeterministic moves on each computation path.

2.1 Probabilistic and Threshold Computation

A *probabilistic polynomial-time Turing machine* [11] is a nondeterministic polynomial-time Turing machine M such that M chooses with equal probability each of the nondeterministic choices at each choice point. $\Pr[M(x) = 1]$ denotes the probability weight of those paths on which M accepts x and $\Pr[M(x) = 0]$ denotes the probability weight of those paths on which M rejects x.

We now define some complexity classes in terms of probabilistic polynomial-time Turing machines.

Definition 1. [Probabilistic Classes]

1. PP [11] is the class of all sets L such that there exists a probabilistic polynomial-time Turing machine M such that for all $x \in \Sigma^*$ it holds that $\Pr[M(x) = L(x)] > \frac{1}{2}$.

2. BPP [11] is the class of all sets L such that there exist a probabilistic polynomial-time Turing machine M and an $\epsilon > 0$ such that for all $x \in \Sigma^*$ it holds that $\Pr[M(x) = L(x)] > \frac{1}{2} + \epsilon$.

3. R [11] is the class of all sets L such that there exists a probabilistic polynomial-time Turing machine M such that for all $x \in \Sigma^*$ it holds that $x \in L \Longrightarrow \Pr[M(x) = 1] > \frac{1}{2}$ and $x \notin L \Longrightarrow \Pr[M(x) = 0] = 1$.

It is well known that $R \subseteq BPP \subseteq PP$ [11].

By looking at the portion of accepting paths rather than the probability weight of the accepting paths, we now introduce the threshold analogs of the above probabilistic classes. Let $\#[M(x) = 1]$ denote $acc_M(x)$ and let $\#[M(x) = 0]$ denote $rej_M(x)$.

Definition 2. [Threshold Classes]

1. PP_{path} [21] is the class of all sets L such that there exists a nondeterministic polynomial-time Turing machine M such that for all $x \in \Sigma^*$ it holds that $\#[M(x) = L(x)] > \frac{1}{2} total_M(x)$.

2. BPP_{path} is the class of all sets L such that there exist a nondeterministic polynomial-time Turing machine M and an $\epsilon > 0$ such that for all $x \in \Sigma^*$ it holds that $\#[M(x) = L(x)] > (\frac{1}{2} + \epsilon) total_M(x)$.

3. R_{path} is the class of all sets L such that there exists a nondeterministic polynomial-time Turing machine M such that for all $x \in \Sigma^*$ it holds that $x \in L \implies acc_M(x) > \frac{1}{2} total_M(x)$ and $x \notin L \implies rej_M(x) = total_M(x)$.

It is easy to see that $R_{path} \subseteq BPP_{path} \subseteq PP_{path}$. For all threshold classes in this paper, as a notational convenience we will place oracles above the word "path" (e.g., BPP_{path}^{BPP} denotes $(BPP_{path})^{BPP}$).

It is known that R, BPP, and PP sets can be accepted via normalized probabilistic polynomial-time Turing machines: just extend each computation path of a given machine up to a fixed polynomial length and, on each new path, accept if the path that was extended accepted, and reject otherwise. The modified machine has the same acceptance probability as the original one. Observe that for normalized machines, the probabilistic interpretation of the machine accepts the same set as the threshold interpretation of the machine. Thus, each of the probabilistic classes is contained in the corresponding threshold class, i.e., $PP \subseteq PP_{path}$, $BPP \subseteq BPP_{path}$ and $R \subseteq R_{path}$.

In fact, Simon [21] has already shown that PP_{path} is not bigger than PP.

Theorem 3. [21] $PP_{path} = PP$.

Interestingly, this equivalence between probabilistic and threshold classes cannot hold for R and BPP unless the polynomial hierarchy [19, 23] collapses to its second level. This follows from the fact that NP is contained in R_{path} and thus is also contained in BPP_{path}.

Proposition 4. $R_{path} = NP$.

Corollary 5. $NP \subseteq BPP_{path}$.

It follows that if BPP_{path} has small circuits or BPP_{path} has normalized machines, then, by the result of Karp, Lipton, and Sipser (see [14]), the polynomial hierarchy collapses. We will strengthen Corollary 5 in Section 3.

2.2 Secure Computation

In this subsection and in Section 4, we study notions of secure adaptive access to databases in the presence of a powerful spying observer. We give below what we feel are the most natural definitions. In these definitions, we obtain security by requiring that an observer (seeing a path drawn uniformly from all the machine's paths) should learn nothing about the input string other than perhaps its length. For threshold computation, this notion is new. For probabilistic computation, this definition is equivalent (For a proof, see the full version [12] of this paper.) to the notion of "one-oracle instance-hiding schemes that leak at most the length of their inputs" [5].

Definition 6. [Secure Threshold Computation] For any set D, a set A is said to be in secureBPP$_{\text{path}}^D$ (that is, is said to be "securely accepted by a bounded-error threshold polynomial-time machine via access to database D") if there is a nondeterministic polynomial-time Turing machine N such that:

1. [$A \in$ BPP$_{\text{path}}^D$ via machine N] There exists an $\epsilon > 0$ such that for all $x \in \Sigma^*$ it holds that $\#[N^D(x) = A(x)] > (\frac{1}{2} + \epsilon) \, total_{N^D}(x)$ (see Part 2 of Definition 2).
2. [The queries of N^D reveal no information to an observer other than perhaps the length of the input] For every $k \in \{0, 1, 2, \cdots\}$, and every vector $\widetilde{v} = (v_1, v_2, \cdots, v_k)$, $v_1, v_2, \cdots, v_k \in \Sigma^*$, and every pair of strings $x \in \Sigma^*$ and $y \in \Sigma^*$ such that $|x| = |y|$, it holds that

$$\frac{path\text{-}occurances_{N^D(x)}(\widetilde{v})}{total_{N^D}(x)} = \frac{path\text{-}occurances_{N^D(y)}(\widetilde{v})}{total_{N^D}(y)},$$

where $path\text{-}occurances_{N^D(z)}(\widetilde{v}) = ||\{p \mid p$ is a path of $N^D(z)$ on which $\widetilde{v}$ is the sequence of queries asked to the oracle (in the order asked, possibly with duplications if the same query is asked more than once)[3]$\}||$.

Definition 7. [Secure Probabilistic Computation] For any set D, a set A is said to be in secureBPP^D (that is, is said to be "securely accepted by a bounded-error probabilistic polynomial-time machine via access to database D") if there is a probabilistic polynomial-time Turing machine N such that:

1. [$A \in$ BPP^D via machine N] There exists an $\epsilon > 0$ such that for all $x \in \Sigma^*$ it holds that $\Pr[N^D(x) = A(x)] > \frac{1}{2} + \epsilon$ (see Part 2 of Definition 1).
2. [The queries of N^D reveal no information to an observer other than perhaps the length of the input] For every $k \in \{0, 1, 2, \cdots\}$, and every vector $\widetilde{v} = (v_1, v_2, \cdots, v_k)$, $v_1, v_2, \cdots, v_k \in \Sigma^*$, and every pair of strings $x \in \Sigma^*$ and $y \in \Sigma^*$ such that $|x| = |y|$, it holds that $\Pr[$the query vector of $N^D(x)$ is $\widetilde{v}] = \Pr[$the query vector of $N^D(y)$ is $\widetilde{v}]$.

Oblivious self-reducibility was discussed in [9], and we now define complexity classes capturing the notion of oblivious access.

Definition 8. [Oblivious Probabilistic and Threshold Classes] For any set D, a set A is said to be in obliviousBPP$_{\text{path}}^D$ (respectively, obliviousBPP^D) if

---

[3] Henceforward, we'll refer to this as a *query vector*.

there is a nondeterministic (respectively, probabilistic) polynomial-time Turing machine N such that:

1. $[A \in \text{BPP}^D_{\text{path}}$ (respectively, $A \in \text{BPP}^D$) via machine $N]$ There exists an $\epsilon > 0$ such that for all $x \in \Sigma^*$ it holds that $\#[N^D(x) = A(x)] > (\frac{1}{2} + \epsilon)\, total_{N^D}(x)$ (respectively, $\Pr[N^D(x) = A(x)] > \frac{1}{2} + \epsilon$).

2. N is an oblivious machine in the sense that on an input z it initially is given access to a "pre-input" tape on which $0^{|z|}$ is written. N then performs its adaptive queries to D. Then, after making all queries to D, machine N is given access to z.

It is clear that, for every D, $\text{BPP}^D_{\text{path}} \supseteq \text{secureBPP}^D_{\text{path}} \supseteq \text{obliviousBPP}^D_{\text{path}}$, and $\text{BPP}^D \supseteq \text{secureBPP}^D \supseteq \text{obliviousBPP}^D$.

Are these inclusions proper? In other words, does using security against observers as the *definition* of secure computation ($\text{secureBPP}^D_{\text{path}}$, secureBPP^D) yield a more flexible notion of security than does blinding the machine to its input ($\text{obliviousBPP}^D_{\text{path}}$, obliviousBPP^D)? Formally, is $\text{obliviousBPP}^D_{\text{path}} \neq \text{secureBPP}^D_{\text{path}}$ or $\text{obliviousBPP}^D \neq \text{secureBPP}^D$? Our intuition says that both inequalities hold. However, Section 4 shows that establishing that "yes" is the answer implies that $\text{P} \neq \text{PSPACE}$ (and even implies the stronger result that $\text{BPP} \neq \text{PP}$). Since it is commonly believed that $\text{P} \neq \text{PSPACE}$, this does not provide evidence that equality holds; rather, it merely suggests that witnessing a separation will be hard with current techniques. We note that results (such as Theorem 23 and Corollary 24) that connect the existence of an oracle separation to the existence of a real-world separation (see, e.g., the survey [7]) usually occur in cases in which the oracle is tremendously restricted (e.g., to the class of tally sets or the class of sparse sets [4, 18]); in contrast, Section 4 provides such a relativization result that applies without restriction of the database D.

There is no point in defining security classes for unbounded-error computation, as it is easy to see that, for every D, $\text{PP}^D = \text{securePP}^D = \text{obliviousPP}^D = \text{PP}^D_{\text{path}} = \text{securePP}^D_{\text{path}} = \text{obliviousPP}^D_{\text{path}}$.

Finally, we note that all sets that are accepted by an oblivious machine relative to some database D have small circuits. Let obliviousBPP^* denote $\bigcup_{D \in 2^{\Sigma^*}} \text{obliviousBPP}^D$. We have the following result.

Proposition 9. $\text{obliviousBPP}^* = \text{P/poly}$.

Corollary 10. $(\exists L)\,[L \notin \text{obliviousBPP}^L]$.

Though for most common classes $\mathcal{C}$ it holds that $(\forall L)\,[L \in \mathcal{C}^L]$, Corollary 10 should not be surprising; it is natural that weak machines, when accepting a hard set via a hard database, may leak some information to an observer. Interestingly, there is a similar result regarding the existence of NP-hard sets that are securely accepted by a bounded-error probabilistic polynomial-time machine: that is, for any set D, no NP-hard set is in secureBPP^D unless the polynomial hierarchy collapses [1].

3 BPP_path

In this section, we study BPP_{path} in more detail. Although the previous section implies that BPP and BPP_{path} differ unless the polynomial hierarchy collapses, these classes nonetheless share certain properties. For example, as is also the case for BPP [27], BPP_{path} has a strong amplification property.

Theorem 11. Let L be in BPP_{path}. For each polynomial q, there is a nondeterministic polynomial-time Turing machine M such that for all $x \in \Sigma^*$ it holds that $\#[M(x) = L(x)] > \left(1 - 2^{-q(|x|)}\right) \, total_M(x)$.

The proof is analogous to the corresponding proof for BPP.

BPP is closed under Turing reductions [15, 25]. For BPP_{path}, we can prove closure under truth-table reductions.

Theorem 12. BPP_{path} is closed under polynomial-time truth-table reductions.

Corollary 13. BPP_{path} is closed under complementation, intersection, and union.

Corollary 14. $P^{NP[\log]} \subseteq BPP_{path}$.

It is natural to hope that Theorem 12 can be improved to the case of Turing reductions. However, no relativizable proof can establish the closure of BPP_{path} under Turing reductions, as Beigel [6] constructed an oracle set A such that $P^{NP^A} \not\subseteq PP^A$ and thus, relative to this oracle, BPP_{path} is not closed under Turing reductions, i.e., $BPP_{path}^A \neq P^{BPP_{path}^A}$.

It is known that BPP is low for PP [16] and for itself [15, 25], i.e., $PP^{BPP} = PP$ and $BPP^{BPP} = BPP$. We show in the next theorem that BPP is also low for BPP_{path}. Observe that relative to the above oracle A, BPP_{path} is not low for PP, i.e., $PP^{BPP_{path}^A} \neq PP^A$. As a consequence of Corollary 5 and Theorem 18 below, BPP_{path} is not low for BPP_{path} unless the polynomial hierarchy collapses.

Theorem 15. $BPP_{path}^{BPP} = BPP_{path}$.

Corollary 16. $NP^{BPP} \subseteq BPP_{path}$.

Babai [2] introduced the Arthur-Merlin classes MA and AM. It is known that $NP^{BPP} \subseteq MA \subseteq AM \subseteq BPP^{NP}$ [2, 26].[4] Below, we strengthen Corollary 16 to show that even MA is contained in BPP_{path}. This improves the recent result of Vereshchagin [24] that $MA \subseteq PP$.

Theorem 17. $MA \subseteq BPP_{path}$.

---

[4] It is not known whether $NP^{BPP} = MA$. Since BPP is closed under truth-table reductions, it is immediate that $NP^{BPP} = \exists \cdot BPP$. So the interesting question is whether $\exists \cdot BPP = MA$. The *machines* for both classes can be thought of as consisting of two stages: nondeterministic guess followed by probabilistic computation. $\exists \cdot BPP$ requires that the probability of the latter stage must be bounded away from half for *all* (whether *correct* or *wrong*) nondeterministic guesses of the earlier stage; in contrast, MA requires that the promise of bounded probability be kept only for correct guesses. It is an open question whether the weakened requirement lets MA accept more languages than $\exists \cdot BPP$.

It is an open question whether AM is contained in BPP_{path}. Vereshchagin [24] recently constructed a set A such that relative to A the class AM is not a subset of PP, i.e., $\text{AM}^A \not\subseteq \text{PP}^A$. Thus, AM is not a subset of BPP_{path} relative to that set A, i.e., $\text{AM}^A \not\subseteq \text{BPP}^A_{\text{path}}$. On the other hand, BPP_{path} is not a subset of AM unless the polynomial hierarchy collapses. This follows from the result of Boppana, Håstad, and Zachos [8] that if $\text{coNP} \subseteq \text{AM}$ then the polynomial hierarchy collapses to its second level. Since $\text{coNP} \subseteq \text{BPP}_{\text{path}}$, we get the same consequence from the assumption that BPP_{path} is contained in AM.

Sipser [22] located BPP within the second level of the polynomial hierarchy by showing that $\text{BPP} \subseteq \text{R}^{\text{NP}}$. We show that BPP_{path} is included in the third level of the polynomial hierarchy. The proof applies Sipser's Coding Lemma for universal hashing [22].

Theorem 18. $\text{BPP}_{\text{path}} \subseteq \text{P}^{\Sigma_2^p[\log]} \cap \text{BPP}^{\text{NP}}$

Corollary 19. $\text{BPP}^{\text{BPP}_{\text{path}}}_{\text{path}} = \text{BPP}_{\text{path}} \implies \text{PH} = \text{P}^{\Sigma_2^p[\log]}$.

Zachos [26] has shown that $\text{NP} \subseteq \text{BPP}$ implies $\text{PH} = \text{BPP}$. Since this result relativizes (i.e., for all A, $\text{NP}^A \subseteq \text{BPP}^A$ implies $\text{PH}^A = \text{BPP}^A$), we obtain the following corollary from Theorem 18.

Corollary 20. $\Sigma_2^p \subseteq \text{BPP}_{\text{path}} \implies \text{PH} = \text{BPP}^{\text{NP}}$.

Ogiwara and Hemachandra [20] and Fenner, Fortnow, and Kurtz [10] independently defined the counting class SPP as follows.

Definition 21. [20, 10] SPP is the class of all sets L such that there exist a nondeterministic polynomial-time Turing machine M and an FP function f such that for all $x \in \Sigma^*$ it holds that $x \in L \implies acc_M(x) = f(x) + 1$ and $x \notin L \implies acc_M(x) = f(x)$.

Fenner, Fortnow, and Kurtz [10] argue that SPP is, in some sense, the smallest class that is definable in terms of the number of accepting and rejecting computations. In particular, SPP is low for $\text{PP}, \text{C}_=\text{P}$, and $\oplus\text{P}$ [10]. Though it is an open question whether SPP is contained in BPP_{path}, there is an oracle relative to which this is not the case.

Theorem 22. There is an oracle A such that $\text{SPP}^A \not\subseteq \text{BPP}^A_{\text{path}}$.

4   If Secure and Oblivious Computation Differ, then P $\neq$ PSPACE

Section 2, and in particular Section 2.2, provide the class definitions and motivation for this section. We show, for both threshold and probabilistic computation, that secure computation is more powerful than oblivious computation only if $\text{BPP} \neq \text{PP}$ (which would resolve in the affirmative the important question of whether polynomial time differs from polynomial space).

Very informally summarized, the proof decomposes a secure computation (query vector by query vector), uses the power of counting ($\#\text{P}$) to allow an oblivious machine to mimic a secure machine, and via the BPP $\neq$ PP clause and the fact that BPP computations are easy for BPP_{path} machines to incorporate (essentially as proved in Theorem 15) puts #P-like computational power into the hands of our oblivious BPP_{path} machine.

Theorem 23. If there is a database D such that $secureBPP^D_{path} \neq$ $obliviousBPP^D_{path}$, then $BPP \neq PP$.

Corollary (to the Proof) 24. If there is a database D such that $secureBPP^D \neq obliviousBPP^D$, then $BPP \neq PP$.

Recall that P/poly is the class of languages that have "small circuits."

Corollary 25. If there is a database D such that $secureBPP^D \not\subseteq P/poly$, then $BPP \neq PP$.

5 Open Problems

There are several open problems regarding BPP_{path}. Is BPP_{path} contained in Σ^p_2 or even in R^{NP}? It seems that the proof technique of Theorem 18 doesn't suffice to establish either of these relationships. Does BPP_{path} have complete sets? There is a relativized world in which BPP lacks complete sets [13]; we conjecture that the same holds for BPP_{path}.

Regarding secure computation, does there exist a structural condition that completely characterizes the conditions under which $(\forall D)\,[secureBPP^D = obliviousBPP^D]$ or that completely characterizes the conditions under which $(\forall D)\,[secureBPP^D_{path} = obliviousBPP^D_{path}]$?

Acknowledgments

For helpful discussions, we are grateful to F. Ablayev, G. Brassard, J. Cai, F. Green, J. Seiferas, and S. Toda. We thank S. Homer for informing us of the close relationship between our notion of secure probabilistic access and the notion of 1-oracle instance hiding schemes.

References

1. M. Abadi, J. Feigenbaum, and J. Kilian. On hiding information from an oracle. *Journal of Computer and System Sciences*, 39:21–50, 1989.
2. L. Babai. Trading group theory for randomness. In *Proceedings of the 17th ACM Symposium on Theory of Computing*, pages 421–429, April 1985.
3. T. Baker, J. Gill, and R. Solovay. Relativizations of the P=?NP question. *SIAM Journal on Computing*, 4(4):431–442, 1975.
4. J. Balcázar, R. Book, and U. Schöning. The polynomial-time hierarchy and sparse oracles. *Journal of the ACM*, 33(3):603–617, 1986.
5. D. Beaver and J. Feigenbaum. Hiding instances in multioracle queries. In *Proceedings of the 7th Annual Symposium on Theoretical Aspects of Computer Science*, pages 37–48. Springer-Verlag *Lecture Notes in Computer Science #415*, 1990.
6. R. Beigel. Perceptrons, PP, and the polynomial hierarchy. In *Proceedings of the 7th Structure in Complexity Theory Conference*, pages 14–19. IEEE Computer Society Press, June 1992.
7. R. Book. Restricted relativizations of complexity classes. In J. Hartmanis, editor, *Computational Complexity Theory*, Proceedings of Symposia in Applied Mathematics #38, pages 47–74. American Mathematical Society, 1989.

8. R. Boppana, J. Håstad, and S. Zachos. Does co-NP have short interactive proofs? *Information Processing Letters*, 25:127–132, 1987.

9. J. Feigenbaum, L. Fortnow, C. Lund, and D. Spielman. The power of adaptiveness and additional queries in random-self-reductions. In *Proceedings of the 7th Structure in Complexity Theory Conference*, pages 338–346. IEEE Computer Society Press, June 1992.

10. S. Fenner, L. Fortnow, and S. Kurtz. Gap-definable counting classes. In *Proceedings of the 6th Structure in Complexity Theory Conference*, pages 30–42. IEEE Computer Society Press, June/July 1991.

11. J. Gill. Computational complexity of probabilistic Turing machines. *SIAM Journal on Computing*, 6(4):675–695, 1977.

12. Y. Han, L. Hemachandra, and T. Thierauf. Threshold computation and cryptographic security. Technical Report TR-461, University of Rochester, Department of Computer Science, Rochester, NY, 1993.

13. J. Hartmanis and L. Hemachandra. Complexity classes without machines: On complete languages for UP. *Theoretical Computer Science*, 58:129–142, 1988.

14. R. Karp and R. Lipton. Some connections between nonuniform and uniform complexity classes. In *Proceedings of the 12th ACM Symposium on Theory of Computing*, pages 302–309. April 1980.

15. K. Ko. Some observations on the probabilistic algorithms and NP-hard problems. *Information Processing Letters*, 14(1):39–43, 1982.

16. J. Köbler, U. Schöning, S. Toda, and J. Torán. Turing machines with few accepting computations and low sets for PP. *Journal of Computer and System Sciences*, 44(2):272–286, 1992.

17. R. Ladner, N. Lynch, and A. Selman. A comparison of polynomial time reducibilities. *Theoretical Computer Science*, 1(2):103–124, 1975.

18. T. Long and A. Selman. Relativizing complexity classes with sparse oracles. *Journal of the ACM*, 33(3):618–627, 1986.

19. A. Meyer and L. Stockmeyer. The equivalence problem for regular expressions with squaring requires exponential space. In *Proceedings of the 13th IEEE Symposium on Switching and Automata Theory*, pages 125–129, 1972.

20. M. Ogiwara and L. Hemachandra. A complexity theory for feasible closure properties. In *Proceedings of the 6th Structure in Complexity Theory Conference*, pages 16–29. IEEE Computer Society Press, June/July 1991.

21. J. Simon. *On Some Central Problems in Computational Complexity*. PhD thesis, Cornell Univeristy, Ithaca, N.Y., January 1975. Available as Cornell Department of Computer Science Technical Report TR75-224.

22. M. Sipser. A complexity theoretic approach to randomness. In *Proceedings of the 15th ACM Symposium on Theory of Computing*, pages 330–335, 1983.

23. L. Stockmeyer. The polynomial-time hierarchy. *Theoretical Computer Science*, 3:1–22, 1977.

24. N. Vereshchagin. On the power of PP. In *Proceedings of the 7th Structure in Complexity Theory Conference*, pages 138–143. IEEE Computer Society Press, June 1992.

25. S. Zachos. Robustness of probabilistic complexity classes under definitional perturbations. *Information and Control*, 54:143–154, 1982.

26. S. Zachos. Probabilistic quantifiers and games. *Journal of Computer and System Sciences*, 36:433–451, 1988.

27. S. Zachos and H. Heller. A decisive characterization of BPP. *Information and Control*, 69:125–135, 1986.

On the Power of Reading and Writing Simultaneously in Parallel Computations*

Rolf Niedermeier and Peter Rossmanith**

Fakultät für Informatik, Technische Universität München
Arcisstr. 21, 80290 München, Fed. Rep. of Germany

Abstract. In the standard model of Cook, Dwork, and Reischuk [1] for the derivation of lower bounds, one computation step of a CREW-PRAM consists of a read phase, internal computation, and a write phase. We investigate the case when one step comprises only internal computation and *one* communication phase instead of two, but we have *either* reading *or* writing and internal computation, and thus allow the mixed occurrence of reading and writing (of different memory cells) at the same time. We show that simultaneous reading and writing saves communication phases. In detail, we show that the computation of the OR-function requires exactly $\log n$ "single communication" steps instead of $0.72 \log n$ "double communication" steps on the standard model. We provide a general lower bound of $\log(\deg(f))$ in terms of the degree of a function f. We obtain a lower bound of $0.76 \log n$ for the computation of critical functions and present a critical function that can be computed in $0.90 \log n$ steps. Finally we demonstrate a tight correspondence between CROW-PRAM's of the modified form and the decision tree complexity.

1 Introduction

The derivation of upper bounds has seen immense progress. The matter with lower bounds, however, seems to be more intricate. In general, only little is known about the time it takes at minimum to solve a given problem on a specific computational model. In this paper we continue and complement work concerning upper and lower bounds for concurrent read, exclusive write parallel random access machines (CREW-PRAM's), which Cook, Dwork, and Reischuk [1] initiated in their seminal paper. The most expensive aspect in parallel computation is communication. Often the number of necessary communcations completely dominates the time required by a parallel algorithm, while the costs of the internal computations play only a minor rôle.

In Cook, Dwork, and Reischuk's setting [1], a computation step of a CREW-PRAM consists of, first, a pure read instruction of all processors, then some internal computation of each processor, and finally a pure write instruction of all processors. A CREW-PRAM is made up of an unlimited number of processors and shared memory cells. A computation consists of T synchronous steps that are performed by all processors parallel in lock-step. Kutyłowski [8] showed that it takes exactly $\phi(n)$

* Research supported by DFG-SFB 0342 TP A4 "KLARA."
** Email: niedermr/rossmani@informatik.tu-muenchen.de

steps on a CREW-PRAM to compute the OR of n variables. The function $\phi(n)$ is defined as $\phi(n) = \max\{ t \mid F_{2t+1} \leq n \} \approx 0.72 \log n$***, where F_k is the kth Fibonacci number. Since each step comprises a pure read *and* a pure write phase, this means that actually $1.44 \log n$ communication phases are necessary. As Kutyłowski showed, this cannot be improved. One crucial property of a CREW-PRAM is that all processors read and write in alternating manner. Here the question arises what happens if a computation step consists of only one communication phase that, however, may be *either* one write *or* one read instruction. Subsequently we will refer to this model as a PRAM *with simultaneous reading and writing,* because a write may take place exactly at the same time (i.e., simultaneously) as a read. Clearly, counting the number of computation steps as defined above, our model is at least half as fast as a traditional CREW-PRAM and it never is faster. If we count the two communication phases of a computation step in the standard model as two time units, because in our model a computation step only needs one time unit, our model appears to be faster.

In Section 3 we show that our variant of the CREW-PRAM can compute the OR of n variables much faster than in $1.44 \log n$ steps. Actually, if we denote by $\overline{\text{CREW}}(f)$ the number of steps needed to compute a Boolean function f by a CREW-PRAM with simultaneous reading and writing, $\overline{\text{CREW}}(\text{OR}) = \lceil \log n \rceil + 1$ holds for $n > 1$. This means that a CREW-PRAM can take advantage of the ability to read and write at the same time. Recently, Dietzfelbinger, Kutyłowski, and Reischuk [2] came up with a more general method (compared to Kutyłowski [8]) that delivers lower bounds for Boolean functions in terms of their *degree* [2, 12]. Their main result is $\text{CREW}(f) \geq \phi(\deg f)$. Since almost all functions have degree n and every Boolean function can be computed in $\phi(n) + 1$ steps [2, 11], nearly all functions have time complexity $\phi(n)$ plus an *additive* constant of at most one. We show the corresponding result $\overline{\text{CREW}}(f) \geq \lceil \log(\deg f) \rceil + 1$ by a similar proof technique.

Critical functions [1] (that are those functions for which there exists an input such that flipping any single input bit changes the output bit) represent a class of Boolean functions for which up to now in general no matching lower and upper bounds are known. Cook, Dwork, and Reischuk [1] gave a lower bound of $0.44 \log n$ and presented a critical function that can be computed in $0.64 \log n$ steps. Parberry and Yan [10] improved these results. Their lower bound is $0.5 \log n$ and they gave a critical function that can be computed in $0.57 \log n$ steps. In Section 4 a careful adaption of Parberry and Yan's methods yields a lower bound of $0.76 \log n$ for simultaneous reading and writing and an upper bound of $0.90 \log n$. Again the upper bound refers to some particular function. We summarize lower and upper bounds up to constant, additive terms in Table 1.

Whereas it is open whether there exists a general relationship between the computational power of CREW-PRAM's with and without simultaneous reading and writing, such a relation exists for CROW-PRAM's [3]. Fich and Wigderson [5] noted that $\text{CROW}(f) = \log D(f)$, where D denotes the decision tree complexity of f. In Section 5 we show that $\overline{\text{CROW}}(f) \approx 1.44 \log D(f)$, yielding the general relationship $\overline{\text{CROW}}(f) \approx 1.44 \, \text{CROW}(f)$.

We define the speedup $S_{EW}(f) := 2\,\text{CREW}(f)/\overline{\text{CREW}}(f)$ (resp. $S_{OW}(f) :=$

*** All logarithms in this paper are taken to base 2 if not stated otherwise.

Type of functions	Lower bounds		Upper bounds	
	CREW	$\overline{\text{CREW}}$	CREW	$\overline{\text{CREW}}$
OR	$0.72 \log n$ [8]	$\log n$	$0.72 \log n$ [1]	$\log n$
critical (easiest)	$0.5 \log n$ [10]	$0.76 \log n$	$0.57 \log n$ [10]	$0.90 \log n$
Boolean (general)	$0.72 \log(\deg(f))$ [2]	$\log(\deg(f))$	$0.72 \log n$ [1]	$\log n$

Table 1. Lower and upper bounds for CREW-PRAM's

$2 \text{CROW}(f)/\overline{\text{CROW}}(f))$ as the factor how much simultaneous read and write access speeds up the computation of f in terms of communication phases. We have $S_{EW}(\text{OR}) \approx 1.44$ and our results also show that the speedup $S_{EW}(f)$ for critical functions f is at most 1.89. We even can derive from our results that for almost all Boolean functions we have $S_{EW} \approx 1.44$. Eventually, our results suggest the conjecture $1.39 \, \text{CREW}(f) = \overline{\text{CREW}}(f)$ for all functions f, which will probably be very hard to prove. If, however, the conjecture holds, there will be some consequences: By exploiting some specialties of simultaneous read and write access we can go beyond the possibilities of Parberry and Yan's lower bound proof and obtain $\overline{\text{CREW}}(f) \geq 0.76 \log n$ for each critical function f. A general relationship $1.39 \, \text{CREW}(f) = \overline{\text{CREW}}(f)$ would imply $\text{CREW}(f) \geq 0.55 \log n$ for all critical functions f and this tentative lower bound nearly coincides with the upper bound $0.57 \log n$ of Parberry and Yan [10]. For CROW-PRAM's life is easier: $S_{OW}(f) \approx 1.39$ for all functions f.

Due to the lack of space several proofs had to be omitted. They appear in the full paper.

2 Preliminaries

In this section we outline some of the necessary background. For a more general introduction to PRAM's (and also lower bounds for them) we refer to JáJá [6] and Karp and Ramachandran [7]. Let $\mathbb{B}_n$ denote the set of all Boolean functions mapping $\{0,1\}^n$ to $\{0,1\}$ and let $f \in \mathbb{B}_n$. We abbreviate the minimal number of steps necessary to compute f on a CREW-PRAM as $\text{CREW}(f)$. Here one step consists of reading *and* writing. If one step consists *either* of reading *or* writing, we call the minimal number of steps $\overline{\text{CREW}}(f)$. Any number of processors can read from the same memory cell simultaneously, but at most one processor must write into a memory cell in each step. Furthermore, if one processor writes into a memory cell, then no processor is allowed to read from it at the same time.

There is an arbitrary number of different states for each processor. In the beginning of a computation, each processor is in its initial state. The memory cells can hold arbitrary values. Initially, the first n cells contain the n input bits. In the standard CREW-PRAM model [1] each step consists of three parts. First, a processor reads from a memory cell that is determined by the current state of the processor. Then the processor changes its state according to the value read and its old state. Finally, the processor writes into a memory cell according to its new state. See the work of Cook, Dwork, and Reischuk [1] for more details. We study a model whose computation steps only consist of two parts: *either* read *or* write and a state transition.

There are lower bounds in terms of the *degree* of a Boolean function [2], a generalization of Kutyłowski's [8] tight bounds for the computation of the OR-function. Smolensky [12] expressed any Boolean function $f(\vec{x})$ as a polynomial of $\vec{x} = (x_1,\ldots,x_n)$ over the reals as follows: Let $S \subseteq \{1,\ldots,n\}$ and $m_S := \prod_{i \in S} x_i$. Every function f is a unique linear combination of monomials:

$$f(\vec{x}) = \sum_{S \subseteq \{1,\ldots,n\}} \alpha_S(f)\, m_S(\vec{x}),$$

where $\alpha_S(f)$ is a positive integer. We define the degree of f as

$$\deg(f) := \max\{\, |S| \mid \alpha_S(f) \neq 0 \,\}.$$

Dietzfelbinger, Kutyłowski, and Reischuk [2] proved $\mathrm{CREW}(f) \geq \phi(\deg(f))$.

3 Lower bounds in terms of the degree and the complexity of the OR-function

A naïve algorithm for the OR-function needs $\log n$ steps. Cook, Dwork, and Reischuk [1] found a quicker solution. In this section we show $\overline{\mathrm{CREW}}(\mathrm{OR}) = \lceil \log n \rceil + 1$. This may be a hint for the naturalness of simultaneous reading and writing. We start with a lower bound in terms of the degree of a function similar to Dietzfelbinger, Kutyłowski, and Reischuk [2]. We also represent Boolean functions as polynomials, but instead of using partitions of the input space and the "full information" assumption (i.e., each processor remembers everything it has read so far and when it writes, the symbol written includes all information the processor has accumulated so far as well as the processor identification number and the current time step), we use sets that contain the characteristic functions of memory cells or processors at a given point of time.

Theorem 1. $\overline{\mathrm{CREW}}(f) \geq \lceil \log \deg(f) \rceil + 1$ *for* $f \in \mathbb{B}_n$ *and* $\deg(f) > 1$.

Proof. For $t \geq 0$ let S_t and M_t be sets of Boolean functions defined via the system of recurrences

$$S_0 = \{0,1\},$$
$$M_0 = \{0,1,x_1,\ldots,x_n,1-x_1,\ldots,1-x_n\},$$
$$S_{t+1} = \sum M_t S_t \cap \mathbb{B}_n,$$
$$M_{t+1} = S_t + M_t(1-S_t) \cap \mathbb{B}_n.$$

In the preceding equations the intersections with $\mathbb{B}_n$ guarantee that we do not leave the set of Boolean functions. By $M_t S_t$ we understand the set $\{\, g \cdot h \mid g \in M_t,\ h \in S_t \,\}$ and by $\sum M_t S_t$ the set of all sums consisting of elements of $M_t S_t$. By induction we get $\deg(M_t), \deg(S_t) \leq 2^{t-1}$ for $t > 0$. So any function $f \in M_t$ fulfills $\lceil \log \deg(f) \rceil + 1 \leq t$. In the following we show that any Boolean function that can be computed in t steps is contained in M_t. From this the result follows.

We use the following notation. The Boolean function $s_{p,t,q}(\vec{x})$ has value 1 iff processor p is in state q after step t. Similarly, $m_{k,t,a}(\vec{x})$ has value 1 iff global memory cell k contains value a after step t. Thus $s_{p,t,q}$ and $m_{k,t,a}$ denote the characteristic functions of processor p and memory cell k on input $\vec{x}$. We prove $s_{p,t,q} \in S_t$ for all p and q and $m_{k,t,a} \in M_t$ for all k and a. The new state of a processor depends on its old state and the value read during the step. So $s_{p,t,q}(\vec{x}) = 1$ if there exist q', k, and a such that $s_{p,t-1,q'}(\vec{x}) = 1$ and $m_{k,t-1,a}(\vec{x}) = 1$, and processor p reads from memory cell k if in state q' and changes to state q if the value read is a. So we can express $s_{p,t,q}$ as

$$s_{p,t,q} = \sum_{(q',k,a)} s_{p,t-1,q'} \cdot m_{k,t-1,a},$$

where the sum is taken over appropriate (q', k, a) as described above. Clearly, $s_{p,t,q}$ is in $\mathbb{B}_n$ and $s_{p,t-1,q'} \in S_{t-1}$ and $m_{k,t-1,a} \in M_{t-1}$ follow from the induction hypothesis. By definition, the left part of the equation is in S_t.

There are two reasons why memory cell k contains a after step t: (i) Some processor writes the value in step t. This happens if there exists a processor p that is in state q and q is a state that makes p write into k. A Boolean function $f_1 = \sum_{(p,q)} s_{p,t-1,q}$, where we sum over appropriate (p, q), computes whether this is the case. The set S_{t-1} contains f_1, because S_{t-1} is closed under summation of functions with disjoint support sets and each $s_{p,t-1,q} \in S_{t-1}$. (ii) No processor writes into cell k and k already contained a after step $t-1$. The function $f_2 = 1 - \sum_{(q',p)} s_{p,t-1,q'}$ computes the first condition; here q' is a state causing some processor p to write into k. Function $f_3 = m_{k,t-1,a}$ computes the second condition. Again we know $1 - f_2 \in S_{t-1}$ and $f_3 \in M_{t-1}$. Combining (i) and (ii) we get $m_{k,t,a} = f_1 + f_2 f_3$, which is contained in M_t. We can add since the cases (i) and (ii) are disjoint. □

Corollary 2. $\overline{\mathrm{CREW}}(\mathrm{OR}) \geq \lceil \log n \rceil + 1$.

Proof. This is a direct consequence of $\deg(\mathrm{OR}) = n$. □

We now give an algorithm for the OR-function to provide a matching upper bound.

Lemma 3. $\overline{\mathrm{CREW}}(\mathrm{OR}) \leq \lceil \log n \rceil + 1$.

Proof. To compute the OR of n values, we use two procedures to compute the OR of some parts of the input. Procedure $M[i,j]$ computes $x_i \vee \cdots \vee x_j$ in the ith memory cell. Procedure $P[i,j]$ computes $x_i \vee \cdots \vee x_j$ such that the ith processor knows the result. For $M[i,i]$ there is nothing to do because the ith memory cell already contains the result x_i. The computation of $P[i,i]$ needs one step. Processor i simply reads x_i from the ith memory cell. Now we state how to perform $M[i,j]$ and $P[i,j]$ in general. Let $k = \lfloor (i+j)/2 \rfloor$. To do $M[i,j]$, first do in parallel $M[i,k]$ and $P[k+1,j]$. Now the ith memory cell contains $x_i \vee \cdots \vee x_k$ and the $(k+1)$st processor knows $y := x_{k+1} \vee \cdots \vee x_j$. In the next step processor $k+1$ writes 1 into the ith memory cell if $y = 1$, and does nothing otherwise. To compute $P[i,j]$, first perform $P[i,k]$ and $M[k+1,j]$ in parallel. Then processor i reads from memory cell $k+1$ and thus knows $x_i \vee \cdots \vee x_j$. A short analysis shows that $M[1,n]$ takes exactly $\lceil \log n \rceil + 1$ steps. $\qquad\qquad\qquad\qquad\qquad\qquad\qquad\qquad\qquad\qquad\qquad\qquad\qquad\qquad\qquad$ □

Since the upper bound of Lemma 3 exactly matches the lower bound from Corollary 2, we have determined the time complexity of the OR-function precisely.

Corollary 4. $\overline{\mathrm{CREW}}(\mathrm{OR}) = \lceil \log n \rceil + 1$ *for* $n > 1$.

In the same way we can compute the AND-function. Using the conjunctive normal form for the representation of Boolean functions, we can compute *any* Boolean function with just two more steps than the AND-function itself.

Theorem 5. $\overline{\mathrm{CREW}}(f) \leq \lceil \log n \rceil + 3$ *for all* $f \in \mathbb{B}_n$.

Proof. (Sketch) For every possible input $\vec{w}$, there will be one processor that knows after $\lceil \log n \rceil + 2$ steps whether $\vec{x} = \vec{w}$. This processor writes the result in step $\lceil \log n \rceil + 3$ into the first memory cell. $\qquad\qquad\qquad\qquad\qquad\qquad\qquad$ □

Almost all Boolean functions have degree n [2, 11]. We conclude that $\overline{\mathrm{CREW}}(f) = \log n + c$ for some $c \in \{1, 2, 3\}$ and for almost all Boolean functions f.

4 Lower and upper bounds for critical functions

A *critical* function is a Boolean function for which an input I exists such that flipping any single bit of I changes the output. The OR-function is a well-known critical function whose *critical input* is the all-zero word. For traditional CREW-PRAM's Parberry and Yan [10] proved a lower bound of $0.5 \log n$ for all critical functions. They also presented a particular critical function and showed how to compute it in $0.57 \log n$ steps. In the following we give a corresponding lower bound of $0.76 \log n$ for simultaneous reading and writing. For this model, we also exhibit a particular critical function and demonstrate how to compute it in $0.90 \log n$ steps.

Definition 6. [1]

1. Input index i *affects* processor p *at time* t *with* I if the state of p at time t with input I differs from the state of p at t with input $I(i)$. Input index i *affects* memory cell k *at time* t *with* I if the contents of k at time t with input I differs from the contents of k at t with input $I(i)$.
2. The set of indices that affect processor p at time t on input I is denoted by $P(p,t,I)$. The set of indices that affect memory cell k at time t on input I is denoted by $M(p,t,I)$.
3. $P_t := \max_{p,I} |P(p,t,I)|, \quad M_t := \max_{c,I} |M(c,t,I)|$.

We will subsequently figure out the values of P_t and M_t. Note that we can't compute a function with a critical input of length n in time T unless $M_T \geq n$ [1, 10].

As Cook, Dwork, and Reischuk [1] did, we start with an investigation of *oblivious* PRAM's. A PRAM is oblivious if the points of time and addresses of all writes depend only on the input length, but not on the input itself.

Theorem 7. *If an oblivious CREW-PRAM with simultaneous reading and writing computes a critical function in T steps, then $n \leq F_T$.*

To prove Theorem 7, we derive the recurrence $P_0 = 0$, $M_0 = 1$, $P_1 = 1$, $M_1 = 1$, $P_{t+1} = P_t + M_t$, $M_{t+1} = P_t$, whose solution is $M_t = F_t$. In contrast to oblivious PRAM's, for *semi-oblivious* PRAM's the decision whether or not it is written into a cell may depend on the input. Again Cook, Dwork, and Reisschuk's [1] techniques yield the following theorem.

Theorem 8. *A semi-oblivious CREW-PRAM with simultaneous reading and writing needs at least $\lceil \log n \rceil + 1$ steps to compute a critical function. This lower bound cannot be improved.*

The semi-oblivious algorithm of Lemma 3 computes the OR-function in $\lceil \log n \rceil + 1$ steps. Since the OR-function is critical, the above lower bound $\lceil \log n \rceil + 1$ is optimal. Finally, we come to critical functions on general CREW-PRAM's with simultaneous reads and writes. We have to rework the intricate proof of Parberry and Yan [10] to fit our model.

Theorem 9. *Every CREW-PRAM with simultaneous reading and writing requires at least $0.76 \log n$ steps to compute a critical function.*

Proof. (Sketch) A careful analysis of Parberry and Yan's proof [10, Theorem 4.7] yields the following recurrence: $P_0 = 0$, $P_1 = 1$, $P_2 = 2$, $M_0 = 1$, $M_1 = 1$, $M_2 = 2$, $P_{t+1} = P_t + M_t$, and $M_{t+1} = 2P_t + P_{t-2} + M_t$. The lower bound resulting from this recurrence is $\log_\alpha n$, where $\alpha \approx 2.47$. $\square$

Theorem 10. *There is a critical function $f \in \mathbb{B}_n$ that can be computed by a CREW-PRAM with simultaneous reading and writing in less than $0.90 \log n$ steps.*

Proof. Let us call a pair consisting of a memory cell and a processor simply a *pair*. We say that a pair computes a Boolean function $f(\vec{x})$ in step t, if after step t the processor is in a distinguished state iff $f(\vec{x}) = 1$ and the memory cell contains $f(\vec{x})$. A pair may compute the OR of two bits in two steps. Starting from this point, we compute bigger and bigger critical functions by pairs and by plain memory cells. If M_t is the size of a critical function computed by a memory cell in t steps and P_t is the corresponding size for a pair, then we already know $M_2 = P_2 = 2$. Now let us assume we have three pairs (p_1, m_1), (p_2, m_2), and (p_3, m_3), each of them computing functions of size P_t after step t. Step $t+1$ consists of the following actions: Processor p_1 reads from m_2, p_2 reads from m_3, and p_3 reads from m_1. In step $t+2$ each of these three processors writes 1 into a memory cell m_4, if the processor computed itself 1, but has read 0. Cell m_4 computes a function of size M_{t+1} after step $t+1$, so m_4 computes a function of size $M_{t+1} + 3P_t$ after step $t+2$. This gives us the first recurrence we need: $M_{t+2} = M_{t+1} + 3P_t$, for $t > 1$.

The second recurrence reads $P_{t+1} = P_t + M_t$ and is established in the following way: Assume that processors p and p' compute the same function. Processor p reads from a copy of m' and simultaneously p' writes 1 into m', if p' has computed 1. The original pair was, say, (p, m), but now we consider the pair (p, m'). It computes a critical function of size $P_t + M_t$, if we assume that (p, m) computed a function of size P_t and m' computed a function of size M_t. The solution of the recurrence $M_2 = P_2 = 2$, $M_{t+2} = M_{t+1} + 3P_t$, $P_{t+1} = P_t + M_t$ is

$$M_t = \alpha^t + \beta^t \cos(at + b) \quad \text{for } t \geq 2,$$
$$\text{where } \alpha = \frac{2}{3} + \frac{\sqrt[3]{2}}{3\sqrt[3]{79 + 9\sqrt{77}}} + \frac{\sqrt[3]{79 + 9\sqrt{77}}}{3\sqrt[3]{2}} \text{ and } \beta < \alpha.$$

This means that our algorithm takes $\log_\alpha(n) + \Theta(1) \approx 0.89 \log n$ steps to compute the described critical function of n variables. We do not examine the function any further. It suffices that it has the critical input $\vec{0}$ and that there are no write conflicts. $\square$

5 CROW-PRAM's

For CREW-PRAM's we conjectured a speedup of 1.44 for all Boolean functions. We couldn't, however, prove it and the reasons that support the conjecture are rather weak. For CROW-PRAM's we can prove that all Boolean functions have the same speedup. Dymond and Ruzzo [3] introduced CROW-PRAM's (concurrent read, owner write) as a frequently occurring subclass of CREW-PRAM's. Here each global memory cell has a unique "write-owner" that is the only processor allowed to write into it. Nisan [9] showed that CREW- and CROW-PRAM's need the same amount of time up to a constant factor to compute a function on a full domain. Therefore, $\text{CREW}(f) = \Theta(\text{CROW}(f))$. A *decision tree* computes a Boolean function by repeatedly reading input bits until the function can be determined from the bits accessed. The decision of which bit to read at any time may depend on the previous bits read. The complexity measure $D(f)$ is the minimal height of a decision tree that computes f [9, 13]. There is a tight connection between CROW-PRAM's and decicion trees: $\text{CROW}(f) = \log(D(f)) + \Theta(1)$ [4, 5]. In this section we prove a corresponding relation for CROW-PRAM's with simultaneous reading and writing.

Theorem 11. $\overline{\text{CROW}}(f) = \lceil \log_\alpha D(f) \rceil - O(1) \approx 1.44 \log D(f)$, *where* $\alpha = \frac{1}{2}(1 + \sqrt{5})$.

Proof. "$\geq$": Here we assume that the state of a processor reflects its whole history including all values read and written. The contents of a cell encodes the value and the time of the latest write and the state of the processor that has written. Instead of accepting with a 1 in the first cell of global memory and rejecting with a 0, we agree upon accepting on a value in the first memory cell that is a member of an arbitrary "set of accepting values." An ordinary CROW-PRAM can simulate such a modified accepting mechanism with at most two additional steps. We err in the right direction, hence we can assume this full information assumption without loss of generality. We introduce two types of Boolean functions. First, we have $m_{k,t,i}(\vec{x})$ that evaluates to 1 if on input $\vec{x}$ memory cell k contains value i at time t. Second, we have $s_{p,t,q}(\vec{x})$ which evaluates to 1 if on input $\vec{x}$ processor p is in state q at time t. Furthermore, let M_t denote the maximum height necessary for decision trees to compute all functions of type $m_{k,t,i}(\vec{x})$ and let P_t be the corresponding maximum height for decision trees to compute all functions $s_{p,t,q}(\vec{x})$. By induction we prove the recurrence $P_0 = 0$, $M_0 = 1$, $M_1 = 1$, $M_t = P_{t-1}$, and $P_t = P_{t-1} + M_{t-1}$. The solution is $M_t = F_{t-1}$, from which $\overline{\text{CROW}}(f) \leq \log_\alpha(D(f))$ follows. The basis is immediate. We start directly with the induction step.

First we show $M_t = P_{t-1}$ by proving $m_{k,t,i}(\vec{x}) \in P_{t-1}$. Due to the full information assumption, we can uniquely determine the point of time $t' \leq t - 1$ when the last write took place and the state q of the processor p which has written into k. Thus we know that cell k at time t has contents i if and only if processor p at time t' is in state q. This implies $m_{k,t,i}(\vec{x}) = s_{p,t',q}(\vec{x}) \in P_{t-1}$. Next, we show $P_t = M_{t-1} + P_{t-1}$. We have to show the existence of a decision tree that computes whether a processor p is in state q at time t. The height of this decision tree must be at most $M_{t-1} + P_{t-1}$. From q we determine that processor p read last time the value a from memory cell k at time t'. We also conclude from q that processor p was in state r at time $t' - 1$. The reverse is also true: $s_{p,t,q}(\vec{x}) = s_{p,t'-1,r}(\vec{x}) \wedge m_{k,t',a}(\vec{x})$. By induction hypothesis there are decision trees that compute $s_{p,t'-1,r}(\vec{x})$ and $m_{k,t',a}(\vec{x})$ since $t' \leq t - 1$. Placing one of these decision trees on top of the other yields a decision tree that computes $s_{p,t,q}(\vec{x})$.

"$\leq$": Now we come to the reverse direction of the proof, where a CROW-PRAM simulates a decision tree. The basic idea is "pointer doubling" on decision trees [9]. Without modificaton this approach leads to an algorithm with $2 \log(D(f))$ steps. First, for each node of the decision tree a processor sets up a pointer to the successor determined by the variable's value. Then we have to show how to perform pointer doubling on a list of length n in $1.44 \log n$ steps. The standard algorithm for pointer doubling, where the processor reads the successor's sucessor and writes it in its own memory cell, requires $2 \log n$ steps. We improve this to $1.44 \log n$. We avoid alternating statements of the form read, write, read, write, and so on. In lieu, a processor reads repeatedly followed by a final write. This works as follows. Instead of having only one tree where the computation takes place, we initially make $T \approx 1.44 \log n$ copies of that tree. We associate one processor with each element in each tree. We will use the first tree only in the first step, the second tree only in first two steps, and so on. In the tth step a processor determines its successor's successor by reading from the approporiate place *in the t-th copy*. Unless the processor belongs to the $(t + 1)$st copy it does not write the value back, but proceeds reading. The owner write property of the above algorithm is obvious.

We prove the running time of approximately $1.44 \log D(f)$ as follows. Let P_t denote the distance a pointer, known to a processor, spans after step t. Similarily, M_t denotes the distance a pointer, contained in a memory cell, spans after step t. All writes are oblivious, so $M_t = P_{t-1}$. The knowledge of a processor that always reads is composed of the knowledge of the processor at time $t - 1$ and the value read at time $t - 1$. A processor computes its new pointer by combining its old pointer with a pointer read from a memory cell, so $P_t = P_{t-1} + M_{t-1}$. $\square$

6 Conclusion

CREW-PRAM's and CROW-PRAM's with simultaneous reading and writing are a reasonable alternative to the standard model of Cook, Dwork, and Reischuk. In general, simultaneous reading and writing saves time. The speedup lies, of course, always between one and two. For CROW-PRAM's we showed that *all* functions have the same speedup 1.39. Whether CREW-PRAM's have also a fixed speedup for all functions, remains an open question. We conjecture a fixed speedup of 1.44. Almost all functions indeed have a speedup of 1.44, but there still could exist functions with speedups of one or two. A fixed, general speedup for CREW-PRAM's would be nice, because results for simultaneous reading and writing would translate to the traditional model. A proof of our conjecture would immediately improve Parberry and Yan's results for critical functions. We used proof techniques already known for the standard model, but also introduced new techniques.

Acknowledgement We thank Klaus-Jörn Lange for many helpful discussions and comments.

References

1. S. A. Cook, C. Dwork, and R. Reischuk. Upper and lower time bounds for parallel random access machines without simultaneous writes. *SIAM Journal on Computing*, 15(1):87–97, 1986.
2. M. Dietzfelbinger, M. Kutyłowski, and R. Reischuk. Exact time bounds for computing boolean functions on PRAMs without simultaneous writes. In *Proc. of 2d SPAA*, pages 125–135, 1990.
3. P. Dymond and W. L. Ruzzo. Parallel RAMs with owned global memory and deterministic language recognition. In *Proc. of 13th ICALP*, number 226 in LNCS, pages 95–104. Springer-Verlag, 1986.
4. F. E. Fich. The complexity of computation on the parallel random access machine. In J. H. Reif, editor, *Synthesis of Parallel Algorithms*, chapter 20, pages 843–900. Morgan Kaufmann Publishers, 1993.
5. F. E. Fich and A. Wigderson. Toward understanding exclusive read. *SIAM Journal on Computing*, 19(4):718–727, 1990.
6. J. JáJá. *An Introduction to Parallel Algorithms*. Addison-Wesley, 1992.
7. R. M. Karp and V. Ramachandran. A survey of parallel algorithms for shared-memory machines. In J. van Leeuwen, editor, *Algorithms and Complexity*, volume A of *Handbook of Theoretical Computer Science*, chapter 17, pages 869–932. Elsevier, 1990.
8. M. Kutyłowski. Time complexity of boolean functions on CREW PRAMs. *SIAM Journal on Computing*, 20(5):824–833, 1991.
9. N. Nisan. CREW PRAMs and decision trees. *SIAM Journal on Computing*, 20(6):999–1007, 1991.
10. I. Parberry and P. Yuan Yan. Improved upper and lower time bounds for parallel random access machines without simultaneous writes. *SIAM Journal on Computing*, 20(1):88–99, 1991.
11. R. L. Rivest and J. Vuillemin. On recognizing graph properties from adjacency matrices. *TCS*, 3:371–384, 1976.
12. R. Smolensky. Algebraic methods in the theory of lower bounds for boolean circuit complexity. In *Proc. of 19th STOC*, pages 77–82, 1987.
13. I. Wegener. *The Complexity of Boolean Functions*. Wiley-Teubner, 1987.

Relativizing Complexity Classes
With Random Oracles*

Ronald V. Book

Department of Mathematics, University of California, Santa Barbara CA 93106 USA
(e-mail: book@math.ucsb.edu)

Abstract. It is known that for almost every oracle set A, $P(A) \neq NP(A)$
and $PH(A) \neq PSPACE(A)$; but there are no known results that relate these
facts to the P =?NP or PH =?PSPACE problems. Here the following result
is shown (Theorem 6):

> If for almost every oracle set A, the polynomial-time hierarchy relative to
> A collapses (to some finite level), then the unrelativized polynomial-time
> hierarchy collapses.

Also, a new characterization (Theorem 2) of classes of the form ALMOST-$\mathcal{R} =$
$\{A \mid$ for almost every B, $A \in \mathcal{R}(B)\}$, where $\mathcal{R}$ is any appropriate reducibility, is established. This new characterization, *The Random Oracle Charac-
terization*, is based on algorithmically random oracle sets, and shows that for
any appropriate reducibility $\mathcal{R}$, ALMOST-$\mathcal{R}$ is precisely the recursive part
of $\mathcal{R}(B)$ for every algorithmically random language B. The paper itself is
part of a development of the properties of classes of the form ALMOST-$\mathcal{R}$,
extending the results of Book, Lutz, and Wagner [4].

There has been a great deal of interest in the (potential) differences between
deterministic and probabilistic computation. In structural complexity theory one
approach to this problem is the study of the question of whether the class BPP is the
same as the class P; if one compares nondeterministic and probabilistic computation,
then the corresponding problem is that of whether the class AM is the same as the
class NP.

In this paper these (and other) problems are studied by considering classes of
the form ALMOST-$\mathcal{R} = \{A \mid$ for almost every B, $A \in \mathcal{R}(B)\}$, where $\mathcal{R}$ is any
"appropriate" reducibility. (All of the reducibilities commonly used in complexity
theory are appropriate, e.g., $\leq_m^P$, $\leq_{btt}^P$, $\leq_T^P$, $\leq_T^{NP}$, $\leq_T^{PH}$.) Well-known examples of
ALMOST-$\mathcal{R}$ include ALMOST- $\leq_{btt}^P=$ P, ALMOST- $\leq_T^P=$ BPP, ALMOST- $\leq_T^{NP}=$
AM, ALMOST- $\leq_T^{PH}=$ PH, and ALMOST- $\leq^{PQH}=$ PSPACE, where $A \leq^{PQH} B$

---
* This work was supported in part by the National Science Foundation under Grants
CCR-8913584 and CCR-9302057.

if and only if $A \leq_T^{\text{PH}} QBF \oplus B$ (see [1, 11]). The starting point of the present work is the class of "algorithmically random languages," denoted by RAND; this is the class of languages whose characteristic sequences are algorithmically random in the sense of Martin-Löf [9]. The class RAND has been characterized in terms of Kolmogorov complexity (a characterization far different from the original definition): a language A is in RAND if and only if there is a constant $c > 1$ such that for all but finitely many n, the Kolmogorov complexity of the initial prefix of length n of the characteristic sequence of A is greater than $n - c$. It follows that almost every language is in RAND but no recursively enumerable language (hence, no recursive language) is in RAND.

An important result relating classes of the form ALMOST-$\mathcal{R}$ and RAND is the following:

Proposition 1. [4] *For every appropriate reducibility* $\mathcal{R}$, ALMOST-$\mathcal{R} = \mathcal{R}(\text{RAND}) \cap$ REC, *where* REC *denotes the class of recursive languages.*

Thus, for any appropriate reducibility $\mathcal{R}$, a recursive language A is in ALMOST-$\mathcal{R}$ if and only if there exists a language B in RAND such that $A \in \mathcal{R}(B)$. The examples stated above yield P $=$ P$_{\text{btt}}$(RAND) $\cap$ REC, BPP $=$ P$_T$(RAND) $\cap$ REC, AM $=$ NP$_T$(RAND) $\cap$ REC, PH $=$ PH$_T$(RAND) $\cap$ REC, and PSPACE $=$ PQH(RAND) $\cap$ REC. The characterization of ALMOST-$\mathcal{R}$ as $\mathcal{R}$(RAND) $\cap$ REC shows that ALMOST-$\mathcal{R}$ is precisely the *recursive part of* $\mathcal{R}$(RAND). In the study of computational complexity, one goal is to develop a theory that will allow for the classification of the intrinsic complexity of functions and predicates that are recursive; thus, showing that each class of the form ALMOST-$\mathcal{R}$ is the recursive part of $\mathcal{R}$(RAND) should be considered an important step in classifying the intrinsic computational complexity of such classes.

Proposition 1 shows that for every algorithmically random language B, $\mathcal{R}(B) \cap$ REC $\subseteq$ ALMOST-$\mathcal{R}$. This raises the question of whether for every B in RAND, this inclusion is proper. Using a result similar to a theorem proved by Kautz in his (as yet) unpublished doctoral dissertation [5], it can be shown that in fact this inclusion is *not* proper for any B in RAND.

Theorem 2. (The Random Oracle Characterization) *For every appropriate reducibility* $\mathcal{R}$, ALMOST-$\mathcal{R}$ = $\mathcal{R}(B) \cap$ REC *for every language B in* RAND.

This result implies that AM = NP(B) $\cap$ REC for every language B in RAND, PH = PH(B)$\cap$REC for every language B in RAND, and that PSPACE = PQH(B)$\cap$ REC for every language B in RAND. There are other useful corollaries.

Corollary 3. *For every choice of B and C in* RAND, $\mathcal{R}(B) \cap$REC $= \mathcal{R}(C) \cap$REC.

The reader should not be misled: it is not hard to see that for almost every pair B and C in RAND, $\mathcal{R}(B) \neq \mathcal{R}(C)$; Theorem 2 shows only that the recursive part of $\mathcal{R}(B)$ is equal to the recursive part of $\mathcal{R}(C)$.

(To the best of the author's knowledge, there is only one previously known example of Theorem 2 that is nontrivial; this is the characterization of BPP as $P_T(B) \cap$ REC for every B in RAND. This characterization is due to Lutz [7, 8]; the choice of the name "The Random Oracle Characterization" for Theorem 2 follows Lutz's usage for his characterization of BPP.)

Theorem 2 leads to a number of other results concerning well-studied open problems about the relationships between certain complexity classes.

Theorem 4. *Let* $\mathcal{R}$ *be an appropriate reducibility. For any class* **D** *of recursive languages, the following are equivalent:*

(a) **D** $\subseteq$ ALMOST-$\mathcal{R}$;
(b) there exists $B \in$ RAND *such that* **D** $\subseteq \mathcal{R}(B)$;
(c) for every $B \in$ RAND, **D** $\subseteq \mathcal{R}(B)$;

From Theorem 2 and results in [4], it follows that **D** $\subseteq$ ALMOST-$\mathcal{R}$ if and only if for almost every language A, **D** $\subseteq \mathcal{R}(A)$.

Theorem 4 has the following corollaries.

Corollary 5. *(a) Let* **C** *be any class in* {BPP, NP, AM, PP, PH, P(PP), PSPACE, **C**$_=$P,...}. *Then* **C** = P *if and only if there exists a language B in* RAND *such that* **C** $\subseteq$ P$_{btt}$(B) *if and only if for every language* $B \in$ RAND, **C** $\subseteq$ P$_{btt}$(B) *if and only if for almost every language A,* **C** $\subseteq$ P$_{btt}$(A).

(b) Let **C** be either P(PP) or PSPACE. Then **C** = PH if and only if there exists a language $B \in$ RAND such that **C** $\subseteq$ PH(B) if and only if for every language $B \in$ RAND, **C** $\subseteq$ PH(B) if and only if for almost every language A, **C** $\subseteq$ PH(A).

(c) Let **C** be any class in {AM, PP, PH, P(PP), PSPACE}. Then **C** = BPP if and only if there exists a language $B \in$ RAND such that **C** $\subseteq$ P$_T$(B) if and only if for every language $B \in$ RAND, **C** $\subseteq$ P$_T$(B) if and only if for almost every language A, **C** $\subseteq$ P$_T$(A).

(d) Let **C** be any class in {PP, PH, P(PP), PSPACE}. Then **C** = AM if and only if there exists a language $B \in$ RAND such that **C** $\subseteq$ NP$_T$(B) if and only if for every language $B \in$ RAND, **C** $\subseteq$ NP$_T$(B) if and only if for almost every language A, **C** $\subseteq$ NP$_T$(A).

(e) PSPACE = DTIME(2^{poly}) if and only if there exists a language $B \in$ RAND such that DTIME(2^{poly}) $\subseteq$ PQH(B) if and only if for every language $B \in$ RAND, DTIME(2^{poly}) $\subseteq$ PQH(B) if and only if for almost every language A, DTIME(2^{poly}) $\subseteq$ PQH(A).

(f) P(PP) = PH if and only if PP $\subseteq$ PH if and only if there exists a language $B \in$ RAND such that PP $\subseteq$ PH(B) if and only if for every language $B \in$ RAND, PP $\subseteq$ PH(B) if and only if for almost every language A, PP $\subseteq$ PH(A).

Recall that it is not known whether the polynomial-time hierarchy has infinitely many levels or collapses to some finite level. In addition, it is not known whether the collapse of the polynomial-time hierarchy implies the existence of a random language relative to which the polynomial-time hierarchy collapses. Here we show that the converse is true, that is, if there exists a random language relative to which the polynomial-time hierarchy collapses, then the polynomial-time hierarchy collapses.

Theorem 6. If there exists a language $B \in$ RAND such that the polynomial-time hierarchy relative to B collapses, then the (unrelativized) polynomial-time hierarchy collapses. Thus, if for almost every language A, the polynomial-time hierarchy relative to A collapses, then the (unrelativized) polynomial-time hierarchy collapses.

From results in [4] and Theorem 2, it follows that exactly one of two things is true:

I. For almost every language A, the polynomial-time hierarchy relative to A collapses. This is equivalent to the existence of a language $A \in \text{RAND}$ such that the polynomial-time hierarchy relative to A collapses ([4]). It follows from Theorem 2 that this is equivalent to showing that the polynomial-time hierarchy relative to A collapses for every language $A \in \text{RAND}$.

II. For almost every language A, the polynomial-time hierarchy relative to A does not collapse. By results in [4], this is equivalent to the existence of a language $A \in \text{RAND}$ such that the polynomial-time hierarchy relative to A does not collapse. This, in turn, is equivalent to showing that the polynomial-time hierarchy relative to A does not collapse for every language $A \in \text{RAND}$.

If I. happens, then it follows from Theorem 6 that the (unrelativized) polynomial-time hierarchy collapses.

It is not known whether the converse of Theorem 6 is true. It does not appear likely that the techniques used in the present paper would allow one to settle this question. This conclusion rests on the observation that these techniques appear to yield information only about classes of recursive languages.

Theorem 6 relates the relativized situation to the unrelativized situation for the question "does the polynomial-time hierarchy collapse?" For problems such as $P = ?NP$ and $PH = ?PSPACE$, there are known separation results that hold for almost every oracle; but in these cases no result relating the relativized situation to the unrelativized situation is known, that is, for those problems no result similar to Theorem 6 is known.

While the Random Oracle Characterization is the principal result of this paper, Theorem 6 is probably the interesting result for most researchers since almost everyone working in theoretical computer science understands that the question of whether the polynomial-time hierarchy collapses is an important open question.

The reader will have noticed that many of the individual results in Corollary 5 are very similar in form to known results about the comparison of complexity classes via sparse oracles; this is particularly true in the case of questions such as $P = ?NP$, $P = ?PSPACE$, $PH = ?PSPACE$, and "does the polynomial-time hierarchy collapse?". Thus, it is appropriate to compare the power of sparse oracles with that of random oracles in the context of comparing complexity classes. Here one such comparison is developed. While there have been several studies of the power of sparse languages used as oracles, this appears to be the first time that the power of sparse oracles and that of algorithmically random oracles are compared.

Let SPARSE denote the class of all sparse languages.

Consider Corollary 5(a) when the reducibility $\mathcal{R}$ is chosen to be either $\leq_m^P$ or $\leq_{btt}^P$. It is known (from the literature) that with this choice of $\mathcal{R}$, if the class RAND is replaced by the class SPARSE throughout, then with the exception of $C \in \{BPP, AM\}$, the result remains the same. The same thing is true for the remaining parts of Corollary 5.

It is known (from the literature) that the polynomial-time hierarchy collapses if and only if there is a language S in SPARSE such that the polynomial-time hierarchy relative to S collapses if and only if for every language S in SPARSE, the polynomial-time hierarchy relative to S collapses. The proof of this fact can be easily modified to show that it holds for the recursive part of the polynomial-time hierarchy relative to a sparse language. This yields the following fact.

Theorem 7. *The following are equivalent:*

(a) *the polynomial-time hierarchy collapses;*

(b) *there exists $S \in$ SPARSE such that the recursive part of the polynomial-time hierarchy relative to S collapses;*

(c) *for every $S \in$ SPARSE, the recursive part of the polynomial-time hierarchy relative to S collapses.*

Examination of the known proofs about the comparison of complexity classes by using sparse oracles with the proofs in Corollary 5 will show that they are quite different in essentially all respects. The fact that they give similar results for comparing these nonrelativized classes does not imply that the class RAND is related to the class SPARSE (in fact, RAND $\cap$ SPARSE $= \emptyset$) or that for any appropriate reducibility $\mathcal{R}$, the class ALMOST-$\mathcal{R}$ is related to the class $\mathcal{R}$(SPARSE) $\cap$ REC. In fact, these results themselves are very different as can be seen by the following.

Theorem 8. *If $\mathcal{R}$ is an appropriate reducibility, then $\mathcal{R}(\mathrm{SPARSE}) \cap \mathrm{REC}$ is not included in ALMOST-$\mathcal{R}$.*

Theorem 8 and the remarks preceding it suggest that the languages in RAND and also the languages in SPARSE do not have great expressive power when used as oracles. There are two very different ways of making this intuition formal.

One way to compare the expressive power of the classes RAND and SPARSE is based on Kolmogorov complexity. For every language A and integer $n > 0$, define the Kolmogorov complexity $K(A_{\leq n})$ of the finite language $A_{\leq n}$ to be the size of the shortest program Π such that on input (Π, n), a (previously fixed) optimal universal machine U outputs the $2^{n+1} - 1$ bit characteristic sequence of $A_{\leq n}$ and halts.

This notion of the Kolmogorov complexity of a finite language is considered to be a measure of the "information content" of that language. It is well known (for example, see [6]) that there exist a constant c (depending only on U) such that for every A and n, $K(A_{\leq n})$ is no greater than $2^{n+1} + c$.

Martin-Löf [10] showed that every language in RAND has very high Kolmogorov complexity, that is, if B is in RAND, then for all but finitely many n, $K(B_{\leq n}) > 2^{n+1} - 2n$. This justifies the view that RAND is the class of those languages having very high information content. (Lutz [7] has shown that there exist languages D that are not in RAND but have very high Kolmogorov complexity, that is $K(D_{\leq n}) > 2^{n+1} - 2n$.)

In contrast, every language in SPARSE is considered to have low information content since its census function is bounded above by a polynomial. In terms of Kolmogorov complexity, this implies that if S is sparse, then for all but finitely many n, $K(S_{\leq n}) < \mathrm{poly}(\log n)$. Thus, when Kolmogorov complexity is used to describe the information content of a language, then the class of algorithmically random languages and the class of sparse languages are very different, even though the remarks above and Theorems 6 and 7 show that these classes have similar power when used as oracle sets.

Another way to compare the expressive power of the classes RAND and SPARSE is to formulate a method for comparing the "useful information content" of such languages when used as oracles.

It seems reasonable to take the view that for information content to be useful, some reducibility or algorithm must be able to exploit it in order to perform a useful task. For simplicity, we restrict attention to tasks that are decision problems, that is, problems of membership in languages. Retaining "exploit" and "useful" as variable parameters to be specified by a reducibility $\mathcal{R}$ of the type normally considered in structural complexity theory and a class $\mathbf{C}$ of languages, say that *language A contains at least as much $\mathcal{R}$-useful information about C as language B if $\mathcal{R}(B) \cap C \subseteq \mathcal{R}(A) \cap C$.* Notice that this is a criterion for comparing the useful information content of A with that of B; it is *not* a measure of useful information.

Since sparse languages have very low information content (as measured by their generalized Kolmogorov complexity), it is reasonable to say that language A contains *negligible $\mathcal{R}$-useful information about* $\mathbf{C}$ if $\mathcal{R}(A) \cap \mathbf{C} \subseteq \mathcal{R}(\text{SPARSE}) \cap \mathbf{C}$, or equivalently, if $\mathcal{R}(A) \cap \mathbf{C} \subseteq \mathcal{R}(\text{SPARSE})$.

If $\mathbf{C}$ is taken to be the class REC of all recursive languages, then the situation changes since, as shown in Theorem 8, $\mathcal{R}(\text{SPARSE}) \cap \text{REC}$ is not included in ALMOST-$\mathcal{R}$.

More specifically, there is the following result.

Corollary 9. *Every algorithmically random language contains negligible P_m-useful, negligible P_{btt}-useful, negligible P_{tt}-useful, neglibible P_T-useful, and negligible PH-useful information about the class REC of all recursive languages.*

All of the results in this paper support the thesis that very little of the information encoded in algorithmically random languages (or in sparse languages) is computationally useful when the reducibilities described here are used to retrieve that information.

References

1. K. Ambos-Spies. Randomness, relativizations, and polynomial reducibilities, *Proc. 1st Conf. Structure in Complexity Theory*, Lecture Notes in Computer Sci. 223, Springer-Verlag, 1986, 23–34.
2. J. Balcázar, J. Díaz, and J. Gabarró. *Structural Complexity II*, Springer-Verlag, 1990.
3. C. Bennett and J. Gill. Relative to a random oracle, $P^A \neq NP^A \neq co\text{-}NP^A$ with probability 1, *SIAM J. Computing* 10 (1981), 96–113.

4. R. Book, J. Lutz, and K. Wagner. An observation on probability versus randomness with applications to complexity classes, *Math. Systems Theory* 26 (1994), to appear. See also *Proc. STACS 92, Lecture Notes in Computer Sci.* 577, Springer-Verlag, 319–328.

5. S. Kautz. Degrees of Random Sets, Ph.D. Dissertation, Cornell University, 1991.

6. M. Li and P. Vitanyi. Kolmogorov complexity and its applications, in J. van Leeuwen (ed.), *Handbook of Theoretical Computer Science*, vol. A, Elsevier Sci. Publishers (1990), 187–254.

7. J. Lutz. Almost everywhere high nonuniform complexity, *J. Comput. System Sci.* 44 (1992), 220–258.

8. J. Lutz, A pseudorandom oracle characterization of BPP, *SIAM J. Computing*, 24 (1993), to appear. See also *Proc. Sixth IEEE Conf. on Structure in Complexity Theory*, 1991, 190–195.

9. P. Martin-Löf. On the definition of random sequences, *Info. and Control* 9 (1966), 602–619.

10. P. Martin-Löf. Complexity oscillations in infinite binary sequences, *Zeitschrift für Wahrscheinlichkeitstheory und Verwandte Gebiete* 19 (1971), 225–230.

11. N. Nisan and A. Wigderson. Hardness vs. randomness, *Proc. 29th IEEE Symp. Found. of Comput. Sci.* (1988), 2–11.

An Introduction to Perpetual Gossiping

Arthur L. Liestman[1] and Dana Richards[2]

[1] School of Computing Science, Simon Fraser University,
Burnaby, B.C. V5A 1S6, Canada
[2] National Science Foundation, 1800 G St NW,
Washington, D. C. 20550, U.S.A

Abstract. In this paper, we introduce a new information dissemination problem in which gossiping is to occur continuously but with restricted use of the network. In this problem, information continues to be generated by each member of the network and, thus, the gossip process must be ongoing. However, in order to allow the network to be used for other purposes, the communications used by the gossip process are limited to k calls per time unit. We present some preliminary results on this new problem.

1 Introduction

Gossiping refers to the information dissemination problem that exists when each member of a network knows a unique piece of information and must transmit it to every other member. Gossiping is accomplished by a sequence of phone calls made between pairs of individuals, such that during each call the two people involved exchange all of the information they know at that time; and such that at the end of the sequence of calls, everybody knows everything. Numerous papers have considered various aspects of gossiping including the minimum number of calls required to gossip and the minimum time required to gossip.

Broadcasting is a related problem in which one member of the network has a message which is to be transmitted to all of the other members by a series of calls over the network. Each of these calls requires one unit of time; any member of the network can participate in at most one call per time unit; and any member can only call an adjacent member. Gossiping can be viewed as n simultaneous broadcasts. In fact, the term *all-to-all broadcasting* is sometimes used in place of the term *gossiping*. For a survey of work on gossiping, broadcasting, and related problems, see [2].

In previous work in this area, gossiping was assumed to be a one-shot problem, that is, a unique piece of information was known to each member and the goal was to produce a calling scheme to make all of the information known to all of the members. Perhaps a more realistic setting is to assume that new information arises at each member from time to time which must then be communicated to all of the other members. One way to accomplish this is to have the members gossip at fixed intervals. However, gossip schemes may require a large amount of information flow and this periodic interruption may significantly effect the performance of the overall network at these times. If the gossip scheme is spread out over time and then repeated, the problem of degrading the performance of the network may be alleviated but the result may be that the network members are not informed of new information in a timely fashion. We suggest an alternate approach, *perpetual gossiping*, where gossiping is considered to be an ongoing process which uses a limited amount of the network in each time unit but which ensures that information is disseminated in a timely fashion.

In Section 2, we define this problem more precisely. We discuss some preliminary results on paths, cycles, and hypercubes in Sections 3, 4, and 5, respectively.

2 Definitions

A graph $G = (V, E)$ represents a communications network in which the n vertices in V correspond to the members of the network and the edges of E correspond to communication links connecting pairs of members. The network is synchronous with each unit of time divided into two sections. A unique new piece of information may *arise* at any vertex in section 1 of any time unit; in the same time unit many vertices can simultaneously have their own individual piece of information arise. Communications (calls) are allowed during section 2 of any time unit. (These two sections are considered only to clarify that a piece of information which arises at vertex u at or before time t may be sent to a neighbor of u in time unit t.) We assume that time begins at time unit 1.

A *perpetual gossip scheme* is an infinite schedule of calls which ensures that any information arising at any vertex of G is eventually communicated to every other vertex of G. A *k-call perpetual gossip scheme* is a perpetual gossip scheme in which at most k calls are used in any given time unit.

Let P be a k-call perpetual gossip scheme for G. If w is the smallest integer such that for all times $i + 1$, all of the information which has arisen in G by section 1 of time $i + 1$ is known to all members of G by time $i + w$, then we say that P has a *gossip window* of size w. We are interested in determining the minimum gossip window size which is possible for a given graph G and a given call constraint k and in finding a perpetual gossip scheme that achieves this window size. Among all k-call perpetual gossip schemes for G, a scheme with the smallest gossip window size is called an *optimal k-call perpetual gossip scheme*. We use $W_k(G)$ to denote the minimum gossip window size for any k-call perpetual gossip scheme for G.

In the k-call perpetual gossip schemes described below, calls are made periodically according to a predetermined schedule. These schemes are represented by labeling the edges of G with integers $1, ..., m$ so that no more than k edges receive any particular label. The edges labeled i are used at all times t where $i \equiv t \pmod m$. We note that non-periodic schemes exist and may, in fact, be optimal.

We use the term *bide time* to denote the number of consecutive time units between the time that a message arises and the time that it is sent. For example, if the message arises in section 1 and is sent out in section 2 of the same time unit then the bide time is 0. We use the term *edge idle time* to denote the number of consecutive time units in which an edge is unused.

3 Paths

Let P_n denote the path of n vertices and $n - 1$ edges. We will use A and B to denote the left and right end vertices, respectively, of P_n. The edges are consecutively numbered $1, 2, ..., (n-1)$ with 1 incident on A and $(n-1)$ incident on B.

Theorem 1. $W_1(P_n) = 3n - 6$ *for* $n \geq 3$.

Proof. To show that $W_1(P_n) \geq 3n - 6$, assume that a scheme exists with a gossip window of size $x < 3n - 6$. Without loss of generality, there must be a time $i + 1$ such that a message α arises at A and arrives at B at time $i + x$. To reach B, α must be transmitted in at least $n - 1$ calls. A message β arising at time i at B must also be transmitted by at least $n - 1$ calls to reach A at a time no later than $i + x - 1$. The sequence of calls used to transmit β from B to A can share at most one call with the sequence of calls used to transmit α from A to B. Thus, $n - 2$ of the calls transmitting β are distinct from those transmitting α. A message γ arising at A at time i (1 unit before α) must reach B no later than time $i + x - 1$. The message γ can share at most one call with β but it can share no calls with α, otherwise either γ would arrive too late at B or α would continue to follow the calls used by γ and arrive prior to $i + x$, contradicting the assumption. So, in the $x + 1$ time steps from time i to time $i + x$, there are at least $(n - 1) + (n - 2) + (n - 2)$ distinct calls and $x \geq 3n - 6$.

To show that $W_1(P_n) \leq 3n - 6$, consider the following labeling of the edges of P_n. Beginning at the left end, label the first edge with "1", the ith edge with "$i, 2n - 2 - i$" for $i = 2, 3, ...n - 2$, and the last edge with "$n - 1$". It is simple to verify that, using this scheme, any message arising at time i at any vertex will be transmitted to all other vertices of P_n by time $i + 3n - 6$. □

Before we consider multiple call perpetual gossip schemes for paths, let us consider the scheme proposed above for $k = 1$. In this scheme, consecutive edges of the path are used to "push" all information from one end of the path to the other and then, reversing the order, to "push" the information in the other direction. We will use the idea of an information *pusher* to describe the schemes for larger k. These schemes can be thought of as having k pushers acting simultaneously on different parts of the path. A complication can arise in that these pushers might, at some time i, attempt to send messages on adjacent edges. This situation, which we call a *collision*, is not allowed since the vertex incident on both edges cannot participate in two calls at the same time. However, two pushers may *share* an edge.

Theorem 2. $\lceil \frac{n-1}{k} \rceil + n - 2 \leq W_k(P_n) \leq 2\lceil \frac{n-2}{k} \rceil + n - 2$ *for* $n \geq 3$ *and* $2 \leq k \leq \lceil \frac{n-1}{2} \rceil$.

Proof. Let $w = (n-1) + b$ be the gossip window size for some k-call perpetual gossip scheme. We claim that the maximum edge idle time for any edge is b.

To prove this claim, let edge j have idle time $b + x$ beginning at time i. Let α arise at A at time $i - j + 1$. Then, α cannot cross edge j until time $i + b + x$ and will not arrive at B until time $i + b + x + n - 1 - j$, for a total of $(n-1) + b + x$ time steps. This proves the claim.

Since the maximum edge idle time is b, each edge is used at least $\frac{1}{b+1}$ of the time. Hence $\frac{n-1}{b+1} \le k$ and $b \ge \lceil \frac{n-1}{k} \rceil - 1$, which establishes the lower bound.

We construct a k-call perpetual gossip scheme for P_n as follows: Schedule one pusher to begin at A at time 1. This pusher returns to A (and begins its next cycle) at time $(n-1) + (n-2) = 2n - 3$. We can add other pushers so that they cause no collisions provided that these other pushers also begin at A at odd time units between 1 and $2n - 3$. There are $n - 2$ such odd start times $(3, 5, ..., 2(n-2) + 1)$. To space the pushers as evenly as possible (thus keeping vertex bide time small), we add $k - 1$ other pushers as follows: Partition $n - 2$ into k parts, i of size $\lceil \frac{n-2}{k} \rceil$ and $k - i$ of size $\lfloor \frac{n-2}{k} \rfloor$ where i is the largest value such that $n - 2 = i\lceil \frac{n-2}{k} \rceil + (k - i)\lfloor \frac{n-2}{k} \rfloor$. Unless $\lceil \frac{n-2}{k} \rceil = \lfloor \frac{n-2}{k} \rfloor$, the next i pushers begin at times $1 + 2\lceil \frac{n-2}{k} \rceil$, $1 + 4\lceil \frac{n-2}{k} \rceil$, ... $1 + 2i\lceil \frac{n-2}{k} \rceil$. The remaining $k - i - 1$ pushers begin at times $1 + 2i\lceil \frac{n-2}{k} \rceil + 2\lfloor \frac{n-2}{k} \rfloor$, $1 + 2i\lceil \frac{n-2}{k} \rceil + 4\lfloor \frac{n-2}{k} \rfloor$, ... $1 + 2i\lceil \frac{n-2}{k} \rceil + 2(k - i - 1)\lfloor \frac{n-2}{k} \rfloor$. If $\lceil \frac{n-2}{k} \rceil = \lfloor \frac{n-2}{k} \rfloor$, then $k = i$ and the other $k - 1$ pushers begin at times $1 + 2\lceil \frac{n-2}{k} \rceil$, $1 + 4\lceil \frac{n-2}{k} \rceil$, ... $1 + (2k - 2)\lceil \frac{n-2}{k} \rceil$. (As described, this scheme is not in full swing until the k pushers are all active; effectively the scheme is operational starting at time $2n - 3$.)

The maximum idle time for any vertex is $b = 2\lceil \frac{n-2}{k} \rceil - 1$ (for example, for a message arising at A at time 2). Some of the interior vertices of the path do have smaller maximum idle times. At each path vertex a pusher will pass going left and a pusher will pass going right (possibly simultaneously if an edge is shared, such as at A and B). Note that after waiting at most b time units at least one pusher will pass going in each direction. Let x be the maximum number of time steps after a message has arisen before it has encountered a pusher in each direction. Clearly $x \le b$ for any message α and α will be broadcast in at most $x + (n-1)$ time units. □

4 Cycles

Let C_n denote the path of n vertices and n edges. Label the vertices $1, 2, ...n$ clockwise and let edge i connect vertex i with vertex $i + 1$. Note that operations on vertex numbers are performed modulo n, adopting the convention that n is used in place of 0.

Theorem 3. $W_1(C_n) = 2n - 3$ *for* $n \ge 3$.

Proof. For $n = 3$ or $n \ge 5$, $W_1(C_n) \ge 2n - 3$. This follows from a result of Bumby [1] which shows that any graph which does not contains a cycle of length 4 requires at least $2n - 3$ calls to complete gossip. If information arises at every vertex in C_n at time i, this information cannot be received by all of the other vertices before time $i + 2n - 3$ if at most one call is allowed per time unit.

For $n = 4$, observe that in any 1-call perpetual gossip scheme, some vertex B must be involved in two consecutive calls at times $i + 1$ and $i + 2$ for some i. (Otherwise, calls simply alternate between two opposite edges and gossip can never be completed.) Thus, vertex A (opposite B) is idle during times $i + 1$ and $i + 2$. Since each of the other vertices must receive A's message in separate calls, A's information cannot be known to all of the vertices before time $i + 5$ and $W_1(C_4) \geq 5$.

For the upper bounds, label the edges of the cycle consecutively with $1, 2, ..., n$. With this labeling, the edge idle time is $n - 1$ and each vertex bide time is $n - 2$. Any message arising at a vertex j at time $i + 1$ is sent to vertex $j - 1$ no later than at time $i + n - 2$. The message is sent to j at time no later than $i + n - 1$ and arrives at $j - 2$ no later than time $i + 2n - 3$. □

We note that the above scheme may be viewed as a single pusher moving clockwise around the cycle. In k-call perpetual gossip schemes for $k > 1$, we will use multiple pushers moving clockwise.

Theorem 4. $\frac{n-1}{\sqrt{k}} - 1 \leq W_k(C_n) \leq n - k + \lceil \frac{n}{k} \rceil - 2$ for $n \geq 3$ and $2 \leq k \leq \lfloor \frac{n}{2} \rfloor$.

Proof. To establish a lower bound on the minimum window size, consider the progress of each piece of information as calls are made in the cycle. A message arising at a particular vertex may spread both clockwise and counterclockwise through the cycle but can reach at most one new vertex with any given call. We refer to the furthest vertices (in each direction) which know the message as the *perimeter* of the vertex where the message arose. Thus, $n - 1$ calls are required to move that particular message. However, any call may move several messages.

Let us assume that a new piece of information arises at every vertex at time 1. At time i (for small i), any call can advance the perimeters of several vertices but it can only advance the perimeters of vertices at distance at most $i - 1$ from the call. Thus, a call made at time i (for small i) can only add to the perimeters of at most $2i$ vertices. Since at most k calls can be made in any time unit, in l time units, at most $\sum_{i=1}^{l} k \cdot 2i$ vertices can be added to various perimeters. To complete gossip, each of n messages must reach $n - 1$ new vertices. That is, the total number of advancements of the n perimeters must be $n(n - 1)$. So, if gossip is completed in l time units, then $\sum_{i=1}^{l} k \cdot 2i \geq n(n - 1)$. Simple algebra gives the lower bound; the computed bound given can be tightened slightly but we conjecture the true lower bound will be much larger.

We construct a k-call perpetual gossip scheme for C_n as follows: Partition n into k parts, i of size $\lceil \frac{n}{k} \rceil$ and $k - i$ of size $\lfloor \frac{n-2}{k} \rfloor$ where i is the largest value such that $n = i \lceil \frac{n-2}{k} \rceil + (k - i) \lfloor \frac{n-2}{k} \rfloor$.

We schedule k pushers to begin at vertex 1 at various times t. Each pusher uses a sequence of adjacent calls proceeding clockwise around the cycle. A pusher calls from vertex 1 to vertex 2 at time t, from vertex 2 to vertex 3 at time $t + 1$, ..., and finally calls from vertex n to vertex 1 at time $t + n - 1$. This pusher repeats this process beginning at time $t + n$.

Unless $\lceil \frac{n}{k} \rceil = \lfloor \frac{n}{k} \rfloor$, schedule i pushers to begin at vertex 1 at times $1, 1+\lceil \frac{n}{k} \rceil$, ..., $1+(i-1)\lceil \frac{n}{k} \rceil$ and the remaining $k-i$ pushers to begin at vertex 1 at times $1+i\lceil \frac{n}{k} \rceil, 1+i\lceil \frac{n}{k} \rceil + \lfloor \frac{n}{k} \rfloor, \ldots 1+i\lceil \frac{n}{k} \rceil + (k-i-1)\lfloor \frac{n}{k} \rfloor$. If $\lceil \frac{n}{k} \rceil = \lfloor \frac{n}{k} \rfloor$, then $k = i$ and the k pushers begin at vertex 1 at times $1, 1+\lceil \frac{n}{k} \rceil, 1+2\lceil \frac{n}{k} \rceil, \ldots, 1+(k-1)\lceil \frac{n}{k} \rceil$. (As described, this scheme is not in full swing until the k pushers are all active; effectively the scheme is operational starting at time $n+1$.)

From the point of view of any vertex on the cycle, a pusher arrives at some time i from the counterclockwise neighbor and leaves at time $i+1$ to the clockwise neighbor. The next pusher arrives either at time $i+\lceil \frac{n}{k} \rceil$ or at time $i+\lfloor \frac{n}{k} \rfloor$. Thus, the maximum vertex bide time is $\lceil \frac{n}{k} \rceil - 2$.

Information flows around the cycle in two directions, i.e., the perimeter advances in both directions. The first pusher to encounter a specific message is the *primary* pusher for that message, and it advances the perimeter clockwise at a rate of 1 edge per time unit. Further, as each new pusher arrives at the counterclockwise perimeter, the perimeter is advanced 1 edge counterclockwise. We refer to the collective contribution of the counterclockwise perimeter as the *backwash*.

Assume that a message has arisen at, say, vertex 1 and after biding for 0 or more time units encounters a pusher connecting it to vertex n, its counterclockwise neighbor. This will be the primary pusher for this message, but this initial call is considered to be the first call of the backwash for that message. At some point this message is brought to a vertex i by the primary pusher such that the same message has arrived at vertex $i+1$ through the backwash. At this point every pusher must have contributed to the backwash exactly once so $i = n - k - 1$ and, including the initial call of the primary pusher, the broadcast of that message takes at most $n - k$ time units.

The remaining case is when a message, after biding for 0 time units, first encounters a call connecting it to its clockwise neighbor. However its broadcast time is bounded by the broadcast time of a hypothetical message arising at the other end of that call at the same time. By the above analysis that hypothetical message is broadcast in $n - k$ time.

Summing this time with the maximum bide time gives the upper bound. $\square$

5 Hypercubes

Let Q_d denote the d-dimensional hypercube of 2^d vertices and $d \cdot 2^{d-1}$ edges.

Theorem 5. $W_2(Q_3) = 7$.

Proof. Consider a piece of information α that is first sent during time unit i. In each subsequent time unit at most two new vertices can learn α. Thus, α cannot be known to all 8 vertices prior to time $i + 3$. In fact, in order for all 8 vertices to learn α by time $i + 3$, α must be sent to two new vertices at time $i + 1$. This means that, in this case, the two calls at time $i+1$ must involve the two vertices involved in the time i call, that is, the three calls form a path with the time i

call in the middle. Further, at time $i + 1$ both calls must have an endpoint on this path and an endpoint off of it.

To show that $W_2(Q_3) \geq 7$, we need only show that some vertex X must be idle for at least 3 consecutive time units in any 2-call perpetual gossiping scheme for Q_3. If so, then a piece of information arising at X at time $j + 1$, the first of the three idle time units, cannot be sent from X until time $j + 4$ and cannot be known to all of the vertices until time $(j + 4) + 3 = j + 7$, giving the lower bound.

It is easy to show that some vertex must be idle for at least 2 consecutive time units in such a scheme. Otherwise, the 4 vertices which are not involved in calls at time i must be involved in calls at time $i + 1$. Similarly, the 4 vertices which are involved in calls at time i must again be involved in calls at time $i + 2$. Thus, for all vertices to avoid being idle for 2 consecutive time units, the calls do not create a spanning subgraph of Q_3 and cannot complete gossip.

Let us assume that $W_2(Q_3) \leq 6$ and consider a 2-call perpetual gossiping scheme for Q_3 with a window size of no more than 6. Note that such a scheme can have no vertices which are idle for 3 consecutive time units.

Choose a vertex A which is idle at times $i+1$ and $i+2$. A piece of information α arising at A at time $i + 1$ must be sent to a second vertex at time $i + 3$ and to two additional vertices at time $i + 4$ in order to be known to all vertices by time $i + 6$. Two of the four vertices which have learned α by time $i + 4$ must call uninformed vertices, Y and Z, at time $i + 5$.

There are two vertices, U and V, which were not involved in the path through A created at times $i + 3$ and $i + 4$, nor involved at time $i + 5$. To avoid a bide time of 3, U and V must have been connected by a call at time $i + 3$. Since Y and Z are idle for two consecutive time units, at times $i + 3$ and $i + 4$, their messages (arising at time $i + 3$) must both be sent to two new vertices at time $i+6$. This can only happen if the time $i+5$ and $i+6$ calls form a cycle of length 4. Thus, U and V are both idle at times $i + 4$, $i + 5$, and $i + 6$, a contradiction.

Thus, there must be a vertex which is idle for 3 consecutive time units and the lower bound follows.

$W_2(Q_3) \leq 7$ follows from the scheme of Figure 1. $\qquad \square$

Theorem 6. $\lceil \frac{2^{d-1}}{k} \rceil + \lceil \log_2 k \rceil + \lceil \frac{2^d - 2^{\lceil \log_2 k \rceil}}{k} \rceil \leq W_k(Q_d) \leq (d+1) \cdot \lceil \frac{2^{d-1}}{k} \rceil - 1$ for $d \geq 2$ and $1 \leq k \leq d$.

Proof. Given that there are 2^d vertices in Q_d and at most $2k$ of them can be involved in calls during any given time unit, there must be some vertex A which is not involved in calls for $\lceil \frac{2^{d-1}}{k} \rceil$ consecutive time units. If A's information (arising at the beginning of this bide time) is first sent to another vertex at time $i + 1$, it can be known to k vertices no sooner than at time $i + \lceil \log_2 k \rceil$ since the number of informed vertices can at most double in each time unit. Since at least $2^d - 2^{\lceil \log_2 k \rceil}$ vertices remain to be informed after time $i + \lceil \log_2 k \rceil$, and no more than k can be informed in any time unit, another $\lceil \frac{2^d - 2^{\lceil \log_2 k \rceil}}{k} \rceil$ time units are required.

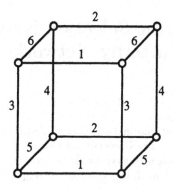

Fig. 1. an optimal 2-call perpetual gossip scheme for the 3-cube

Consider the following scheme: Phase i corresponds to dimension i, where $i \in 1, ..., d$, and consists of $\lceil \frac{2^{d-1}}{k} \rceil$ time units. In phase i, place calls on the 2^{d-1} edges of that dimension in some order, calling on k edges in each time unit except the last time unit of the phase (in which $2^{d-1} - k \cdot (\lceil \frac{2^{d-1}}{k} \rceil - 1)$ calls are made). The scheme is periodic and each call is repeated every $d \cdot \lceil \frac{2^{d-1}}{k} \rceil$ time units.

The ordering of the calls in this scheme guarantees that gossip is completed by any d consecutive phases. Thus, any message that arises during one phase is guaranteed to be known to all vertices after the next d complete phases. A piece of information that arises in the first time unit of a phase will be known to all vertices by the end of the d phases beginning with that time unit. If a piece of information arises later in the phase, it may not be involved in a call until the following phase. Thus, the gossip window for this scheme is no more than $d \cdot \lceil \frac{2^{d-1}}{k} \rceil + \lceil \frac{2^{d-1}}{k} \rceil - 1 = (d + 1) \cdot \lceil \frac{2^{d-1}}{k} \rceil - 1$ time units.

$\square$

6 Conclusions

We studied perpetual gossip in paths, cycles, and hypercubes. We were able to find optimal 1-call perpetual gossip schemes for paths of length n and for cycles of n vertices. We gave nontrivial lower and upper bounds on the size of the minimum gossip window for k-call perpetual gossip in paths and cycles for larger k. We were able to find an optimal 2-call perpetual gossip scheme for the 3-cube and gave non-trivial upper and lower bounds on the size of the minimum gossip window for k-call perpetual gossip in d-cubes for larger k.

References

1. Bumby, R.: A Problem with Telephones. *SIAM J. Alg. Discr. Meth.* **2** (1981) 13-18.
2. S. T. Hedetniemi, S. M. Hedetniemi, and A. L. Liestman: A Survey of Broadcasting and Gossiping in Communication Networks. *Networks* **18** (1988) 319-349.

A Probabilistic Selection Network with Butterfly Networks

Takahiro Ikeda

Department of Information Science, Faculty of Science,
The University of Tokyo
7-3-1 Hongo, Bunkyo-ku, Tokyo-to 113, Japan
e-mail: ike@is.s.u-tokyo.ac.jp

Abstract. This paper constructs a probabilistic selection network with size $0.5n \log n + O(n^{0.822} \log n)$ and depth $5.62 \log n + O(\log \log n)$. This paper further researches the practical aspects of a butterfly network and shows some useful properties of a butterfly network, which improves in the size and the depth of the selection network.

1 Introduction

A selection network is the comparator network classifying values into two classes in such a way that each value in one class is at least as large as all of those in the other. Especially the network classifying n values into classes with k values and $n - k$ values respectively is called an (n, k)-selector. A sorting network is the comparator network which sorts a set of values. Especially the network sorting n values is called an n-sorter. Since a sorting network is a selection network as it is, the results for sorting networks are applicable to selection networks as well.

In 1983, Ajtai, Komlós, and Szemerédi showed the existence of n-sorters with size $O(n \log n)$ and depth $O(\log n)$ based on utilization of expanders [1]. However, the constant factor in the depth bound is enormous and even the improvement by Paterson have not brought it below 1,000 [6].

Pippenger has reduced the constant factor in the size bound to two for $(n, n/2)$-selectors [7]. Moreover Jimbo and Maruoka have improved this construction and have achieved smaller size of $1.89 \log n + O(n)$ [3]. However, there still remains a distance to the lower bound $\frac{1}{2} n \log n$ given by Alekseev [2], and an enormous constant is hidden in the term $O(n)$.

On the other hand, Leighton and Plaxton have constructed much simpler sorting network using other approach. They have constructed a probabilistic n-sorter with depth $7.44 \log n + O(\log \log n)$ [5]. In addition to its small depth, their result has significance from the point of view that they have been concerned in the implementation of the network by using practical butterfly networks, at the expense of allowing the network to fail for a small number of input sets.

This paper applies their approach to construction of a selection network and builds a probabilistic (n, k)-selector with size $0.5n \log n + O(n^{0.822} \log n)$ and depth $5.62 \log n + O(\log \log n)$. Using only butterfly networks except for the last stage, this selection network achieves smaller size and depth than those of previously known selection networks. Especially its size is remarkable for its constant factor of the leading term which equals to that of the lower bound given by Alekseev. This paper further researches the practical aspects of a butterfly network and shows that it can approximately sort all input elements with extremely high probability and can sort almost sorted elements with less tolerance. From these properties, above probabilistic (n, k)-selector is improved in its size and depth.

This selection network is a probabilistic one which allows failure for few cases. For such a probabilistic algorithm, the estimation in the failure probability is also significant from its basic concept. With regard to the network provided in this paper, the probability to fail in selection is bounded by $O(2^{-2^{c\sqrt{\log n}}})$ where c is some small constant. This bound is sufficiently small for large n.

In concluding this section, some preliminaries are described. The *rank* of an element x in an ordered set X is defined as the number of elements in X smaller than x. On the probability that a probabilistic algorithm gives a correct result, the phrase *with high probability*, *with very high probability*, and *with extremely high probability* are used if the algorithm fails with probability $O(n^{-c})$, $O(2^{-2^{c\sqrt{\log n}}})$, and $O(2^{-n^c})$ respectively, where c is some positive constant and n is the input size for the algorithm. Throughout this paper, the log function refers to the base two logarithm.

2 A Strong Ranking Property of a Butterfly Network

An n-input butterfly network, defined if n is exponent of two, is a comparator network constructed as follows. In the first level, n input lines are connected to $n/2$ comparators. In the next level, the upper outputs and the lower outputs of the first level comparators are collected each other, and each set of them is connected to an $n/2$-input butterfly network recursively. This recursion is ended with 2-input butterfly networks, which are merely comparators. The depth of an n-input butterfly network is $\log n$ and the size is $\frac{1}{2}n \log n$. In this paper, assume that each comparator in a butterfly network is oriented to put out the larger value into its upper output and the smaller value into its lower output, and give an $\log n$-bits binary number to each output from $0\cdots0$ to $1\cdots1$ in order from the bottom to the top respectively (see Figure 1).

Leighton and Plaxton have proved that the approximate rank of the element coming to each output of a butterfly network is usually decidable. The following lemma, describing this property, is obtained directly from their result in [5].

Lemma 1. *Let $n = 2^m$ where m is some non-negative integer, and $X = \{0,1\}^m$ be the output set of an n-input butterfly network. Then there exists a fixed bijection $\pi : X \to \{0,\ldots,n-1\}$, a fixed subset Y of X such that $|Y| = n^\gamma$ and the following statement is satisfied with extremely high probability: the rank of the element i is in the range $[\pi(i) - n^\gamma, \pi(i) + n^\gamma]$ for all i in $X - Y$ where γ is a positive constant strictly less than unity.*

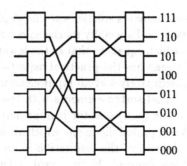

Figure 1. An 8-input butterfly network and binary numbers corresponding to its outputs, where each box denotes a comparator

Leighton and Plaxton have also provided an algorithm to compute π and Y described in above lemma and have shown that the value of γ is less than $0.822 + c$ where c is the constant appeared in the definition of the probability.

This lemma indicates that a butterfly network can give approximate ranks to most elements except for an extremely small fraction, namely $O(2^{-n^c})$, of $n!$ possible input assignments. By applying a random permutation to the input set, a butterfly network satisfies the useful probabilistic property of Lemma 1.

3 A Probabilistic Selection Network

In this section, a probabilistic (n, k)-selector will be constructed with butterfly networks. For simplicity, n is confined to an exponent of two. Then the main result of this section is stated as follows.

Theorem 1. *There exists a probabilistic (n, k)-selector with very high probability such that the size is $0.5n \log n + 2.47n^{0.822} \log n + 294n^{0.822} + O(1)$ and the depth is $5.62 \log n + 51.4 \log \log n + O(1)$.*

In order to prove this theorem, first a strong selection property of a butterfly network will be established in the following lemma.

Lemma 2. *Let $X = \{0,1\}^{\log n}$ be the output set of an n-input butterfly network, and k be an arbitrary integer satisfying $0 \leq k \leq n - 1$. Then there exists fixed subsets L and H of X such that each element in L has a rank below k in X with extremely high probability and each element in H has a rank above or equal to k in X with extremely high probability, and $|X - L - H| \leq 3n^\gamma$ where γ is a positive constant strictly less than unity.*

Proof. Define L and H with Y, π, and γ in Lemma 1 as follows: $L = \{x \in X - Y \mid \pi(x) < k - n^\gamma\}$, $H = \{x \in X - Y \mid \pi(x) \geq k + n^\gamma\}$. These L and H satisfy the conditions on ranks mentioned in this lemma from Lemma 1. Furthermore, the facts that inequations $k - n^\gamma \leq \pi(x) < k + n^\gamma$ are satisfied for arbitrary $x \in X - Y - L - H$ and that $\pi(x)$ is a distinct integer for every x imply that $|X - Y - L - H| \leq 2n^\gamma$. This concludes that $|X - L - H| \leq 3n^\gamma$. $\quad\square$

This lemma shows that an n-input butterfly network is a probabilistic partial (n, k)-selector in the sense that it classifies the input set except some fixed elements into two classes with extremely high probability. This probabilistic property requires a random permutation for the input sets similar to Lemma 1.

Then, consider the following recursive procedure to construct a very high probability (n, k)-selector. Let U and V be the empty sets. They will finally be two classes as the outputs of a very high probability (n, k)-selector.

1. Let X be the input set for this procedure. If $|X| > 2^{\epsilon\sqrt{\log n}}$ where ϵ is an arbitrary constant such that $\epsilon\sqrt{\log n} \geq \frac{1}{1-\gamma}\log 6 + 1$, then go to step 3.
2. Apply Batcher's odd-even sorting network to X. Add the elements which have ranks below $k - |U|$ in X to U and the others to V. Then halt.
3. Apply a random permutation and a butterfly network to X. From Lemma 2, X is separated into three disjoint subsets L, H, and M in such a way that each element in L has a rank below $k - |U|$ in X with extremely high probability and each element in H has a rank above or equal to $k - |U|$ in X with extremely high probability and M consists of the rest elements in X. Move arbitrary elements in H to M until $|M|$ becomes an exponent of two. Add the elements in L to U and the elements in H to V.
4. Call this procedure recursively with M being a new input set.

In the rest of this section, it will be proven that the network constructed by this recursive procedure is the probabilistic (n, k)-selector stated in Theorem 1. Let X_i be the input set for the i-th recursion of this procedure and n_i be the size of X_i. Then, X_0 denotes the original input set and n_0 equals to n.

Claim 1. $n_i \leq 6^{(1-\gamma^i)/(1-\gamma)}n^{\gamma^i}$ for $i \geq 0$.

Proof. This is clear for $i = 1$. Consider the j-th recursion. After applied a butterfly network, $|M|$ equals $2^{\gamma\log n_j + \log 3}$ and increases up to $2^{\lceil \gamma\log n_j + \log 3\rceil} < 2^{\gamma\log n_j + \log 3 + 1} = 6n_j^\gamma$ in step 3. Then, if the claim follows for $i = j$, n_{j+1} is less than $6n_j^\gamma = 6^{(1-\gamma^{j+1})/(1-\gamma)}n^{\gamma^{j+1}}$. This shows that the claim is satisfied for $i = j + 1$. Hence, the claim follows for all $i \geq 0$ by induction for i. $\quad\square$

Claim 2. Let $d = \log(\log n - \frac{1}{1-\gamma}\log 6)/\log\gamma^{-1}$. If $\log n > \frac{1}{1-\gamma}\log 6$, then the construction ends after at most d times recursion.

Proof. From Claim 1, n_d is less than $6^{(1-\gamma^d)/(1-\gamma)}n^{\gamma^d} = 2^{\frac{1}{1-\gamma}\log 6 + 1}$. Since n_d does not satisfy the condition in step 1, the claim follows. $\quad\square$

Claim 3. *Suppose that each butterfly network can classify the elements with probability 1 in step 3. Then the proposition that each element in U has a rank below k in X_0 and each element in V has a rank above or equal to k in X_0 is satisfied at any point in the recursion.*

Proof. This proposition is obviously satisfied at the beginning. Assume that it is satisfied at the entrance of the i-th recursion. Then exact $k - |U|$ elements have ranks below k in X_0. The elements added to U in step 3 have ranks below $k - |U|$ in X_i, and have ranks below k in X_0. Similarly, the elements added to V in step 3 have ranks above or equal to k in X_0. The same property is regarded for the operation in step 2. Then the claim follows by induction for i. $\square$

This claim shows that U and V finally become the outputs of (n, k)-selector, ignoring the success probability.

Claim 4. *The network constructed by the procedure fails in selection with probability $O(2^{-2^{\epsilon c}\sqrt{\log n}})$ if the butterfly network in each recursion fails in partial selection with probability $O(2^{-n^c})$.*

Proof. The failure probability for a butterfly network is at most $O(2^{-2^{\epsilon c}\sqrt{\log n}})$ because the input size is at least $2^{\epsilon}\sqrt{\log n}$. Since the failure probabilities for other butterfly networks are absorbed by this failure probability, the probability that the network fails in selection is bounded by $O(2^{-2^{\epsilon c}\sqrt{\log n}})$. $\square$

Finally the size and the depth of the network will be calculated.

Claim 5. *Let $S(n)$ be the size of the network constructed by the procedure. Then the upper bound for $S(n)$ is $0.5n \log n + 2.47n^{0.822} \log n + 294n^{0.822} + O(1)$.*

Proof. Consider the network due to step 3 in the procedure, and let $S_0(n)$ be its size. Since the size of an n_i-input butterfly network is $\frac{1}{2}n_i \log n_i$,

$$S_0(n) \le \frac{1}{2}n_0 \log n_0 + \frac{1}{2}n_1 \log n_1 + \frac{1}{2}\sum_{i=2}^{d-1} n_i \log n_i$$

$$\le \frac{1}{2}n \log n + 3n^\gamma (\gamma \log n + \log 6)$$

$$+ \frac{1}{2}6^{\frac{1-\gamma'}{1-\gamma}}\left(\log n - \frac{1-\gamma^{1-d}}{1-\gamma}\log 6\right)\sum_{i=2}^{d-1}\gamma^i n^{\gamma'} .$$

Then the following inequation is obtained by estimating the summation with the corresponding integral, where e denotes the base of the natural logarithm:

$$S_0(n) \le \frac{1}{2}n \log n + 3\gamma n^\gamma \log n + 3\left\{\log 6 + \frac{(\log e)^2}{\log \gamma^{-1}}\left(1 + \frac{\gamma}{1-\gamma}\log 6\right)\right\}n^\gamma$$

$$- \frac{1}{2}\frac{(\log e)^2}{\log \gamma^{-1}}\left(1 + \frac{\gamma}{1-\gamma}\log 6\right)2^{1/\gamma} \cdot 6^{\frac{1}{1-\gamma}} .$$

Finally the following inequation is established with $\gamma = 0.822$:

$$S_0(n) < 0.5n \log n + 2.47n^{0.822} \log n + 294n^{0.822} - O(1) .$$

The size of the final odd-even sorting network is at most $2^{\epsilon\sqrt{\log n}-2}(\epsilon^2 \log n - \epsilon\sqrt{\log n} + 4) - 1$. Then the additional size is asymptotic to $O(1)$ by bringing ϵ asymptotic to $(\frac{1}{1-\gamma}\log 6 + 1)/\sqrt{\log n}$. Hence, the upper bound for $S(n)$ is $0.5n \log n + 2.47n^{0.822} \log n + 294n^{0.822} + O(1)$. ☐

Claim 6. *Let $D(n)$ be the depth of the network constructed by the procedure. Then the upper bound for $D(n)$ is $5.62 \log n + 51.4 \log\log n + O(1)$.*

Proof. Consider the network due to step 3 in the procedure, and let $D_0(n)$ be its depth. Since the depth of an n_i-input butterfly network is $\log n_i$,

$$D_0(n) \le \sum_{i=0}^{d-1} \log n_i \le \sum_{i=0}^{d-1} \log\left(6^{\frac{1-\gamma^i}{1-\gamma}} n^{\gamma^i}\right)$$

$$\le \frac{1}{1-\gamma}\log n + \frac{1}{1-\gamma}\frac{\log 6}{\log \gamma^{-1}}\log\log n - \frac{1}{1-\gamma}\left(1 + \frac{1}{1-\gamma}\log 6\right).$$

Finally the following inequation is obtained with $\gamma = 0.822$:

$$D_0(n) < 5.62 \log n + 51.4 \log\log n + O(1).$$

The depth of the final odd-even sorting network is at most $\frac{1}{2}\epsilon^2 \log n + \frac{1}{2}\epsilon\sqrt{\log n}$. Then the additional depth is asymptotic to $O(1)$ by bringing ϵ asymptotic to $(\frac{1}{1-\gamma}\log 6 + 1)/\sqrt{\log n}$. Hence, the upper bound for $D(n)$ is $5.62 \log n + 51.4 \log\log n + O(1)$. ☐

Thus, a probabilistic selection network with size $0.5n \log n + 2.47n^{0.822} \log n + 294n^{0.822} + O(1)$ and depth $5.62 \log n + 51.4 \log\log n + O(1)$ has been obtained and Theorem 1 follows. This network only consists of butterfly networks and an odd-even sorting network, and is simple. Especially its size is remarkable for its constant factor of the leading term which equals to that of the lower bound.

However, there remains a problem how to perform random permutations efficiently. If inefficient random permutations are executed, the complexity of random permutations may bound that of the whole selection network. This problem will be reconsidered in Section 5.

4 A Better Strong Ranking Property

This section pays an attention to the set Y in Lemma 1, which refers to the subset of X such that the error in ranking for each element exceeds n^γ. Leighton and Plaxton have shown in [5] that $\pi(i)$ is the rank of $f_i(1/2)$ in the set $\{f_x(1/2) \mid x \in X\}$ and Y is the set $\{x \in X \mid \Delta(f_x(2^{-n^c}), f_x(1 - 2^{-n^c}) > 2n^{\gamma-1}\}$, where $\Delta(x,y) = \log\frac{y(1-x)}{(1-y)x}$ and f is defined recursively as follows:

$$f_\emptyset(z) = z \ , \qquad \begin{cases} f_{0x}(z) = 1 - \sqrt{1 - f_x(z)} \ , \\ f_{1x}(z) = \sqrt{f_x(z)} \ . \end{cases}$$

Table 1. The relation between $d_0(x)$ and $d_1(x)$ for all x in $\{0,1\}^4$

x	$d_0(x) \times 10^4$	$d_1(x) \times 10^3$	x	$d_0(x) \times 10^4$	$d_1(x) \times 10^3$
0000	2.30	8.18	1000	4.98	5.88
0001	7.65	5.38	1001	8.30	4.79
0010	6.53	5.41	1010	8.16	4.80
0011	8.02	4.97	1011	5.63	5.58
0100	5.63	5.58	1100	8.02	4.97
0101	8.16	4.80	1101	6.53	5.41
0110	8.30	4.79	1110	7.64	5.38
0111	4.98	5.88	1111	2.30	8.18

In fact, they have proved that the rank of $x \in X$ is in the range $[f_x(2^{-n^c})n, f_x(1-2^{-n^c})n]$ with extremely high probability and that at most n^γ elements in X satisfy the inequality $\Delta(f_x(2^{-n^c}), f_x(1-2^{-n^c})) > 2n^{\gamma-1}$, where γ is a positive constant less than $0.822 + c$. This concludes that $|Y| \le n^\gamma$ since the inequality $y - x \le \Delta(x, y)$ is satisfied for any $x < y$. Let $d_0(x)$ denote $f_x(1-2^{-n^c}) - f_x(2^{-n^c})$ and $d_1(x)$ denote $\Delta(f_x(2^{-n^c}), f_x(1-2^{-n^c}))$ for convenience. Then, $d_1(x)$ does not seem to be appropriate to bound $d_0(x)$ because the former is not small whenever the latter is sufficiently small. Consider the case that the input size $n = 16$ for example. Table 1 shows the values of $d_0(x)$ and $d_1(x)$ for each $x \in \{0,1\}^4$ where $c = 0.001$. This table indicates the tendency that $d_1(x)$ is smaller when $d_0(x)$ is larger and $d_1(x)$ is larger when $d_0(x)$ is smaller. The similar result is obtained for other input size, 32, 64, 128, etc., from actual calculation.

If this result is satisfied for any input size, it is strongly expected that $d_0(x)$ is sufficiently small for such x that $d_1(x) > 2n^\gamma$ and the error in ranking by π is at most n^γ for any output. Then the following conjecture can be stated.

Conjecture 1. *Let $n = 2^m$ where m is some non-negative integer, and $X = \{0,1\}^m$ be the output set of an n-input butterfly network. Then there exists a fixed bijection $\pi : X \to \{0, \ldots, n-1\}$ such that the rank of the element i is in the range $[\pi(i) - n^\gamma, \pi(i) + n^\gamma]$ with extremely high probability for all i in X where γ is a positive constant strictly less than unity.*

This conjecture represents Y is the empty set in Lemma 1. If this conjecture is satisfied, the size and the depth of the (n, k)-selector provided in Section 3 are improved to $0.5n \log n + 1.65n^{0.822} \log n + 155n^{0.822} + O(1)$ and $5.62 \log n + 39.8 \log \log n + O(1)$ respectively.

Since this conjecture has not been proved analytically, in order to verify this conjecture the computational experiment has been performed. The probability that the error in ranking with π exceeds $n^{0.822+c}$ is calculated by actual application of a butterfly network for random sequence of input assignments.

Table 2 shows the result of the experiment in the case that the length of the sequence is 100,000 and $c = 0.001$. F denotes the number of the elements for whom the error in ranking exceeds $n^{0.822+c}$ in $100,000n$ elements and $P =$

Table 2. The probability that errors in ranking exceed $n^{0.823}$

n	32	64	128	256	512	1024	2048	4096
2^{-n^c}	0.4988	0.4986	0.4983	0.4981	0.4978	0.4976	0.4974	0.4971
F	2	9	63	134	270	529	1145	2265
$P \times 10^6$	0.63	1.41	4.92	5.23	5.27	5.17	5.60	5.53

$F/(100,000n)$ corresponds to the required probability. From this result, F seems to be proportional to n for larger n. This implies P is asymptotic to some small constant. Hence, the probability that the error in ranking exceeds $n^{0.822+c}$ is less than the failure probability bound 2^{-n^c} until this failure probability bound decreases to the small constant. This constant is sufficient small for practical size n, and then Conjecture 1 follows for practical size n.

5 Ranking for an Almost Sorted Input Set

The (n, k)-selector in Section 3 includes some waste operations which discard given information on ranks. In other words, in spite that an input set on the way of recursion have been approximately sorted by a butterfly network, a random permutation in step 3 deserts its sortedness. If there exists an appropriate permutation to let a butterfly network utilize the sortedness of an input set, it will be a solution for this problem. In this case, it is further expected that a butterfly network has a better ranking property than described in Lemma 1.

In a butterfly network, a comparator in earlier level brings its two inputs to farther outputs each other and a comparator in later level brings them to nearer outputs each other. This suggests in substance that earlier comparators decide rough ranks of elements and later comparators decide strict ranks of elements. Then it is reasonable for each input line of a butterfly network to be given an element in such a way that the elements which have farther ranks are compared with each other in earlier level and the elements which have nearer ranks are compared with each other in later level. This idea is based on the concept analogous to seeding for a tournament, which is a principle preventing superior players to meet in the first battle in a tournament.

In order to represent this idea, give a $\log n$-bit number to each input line of an n-input butterfly network from $0 \cdots 0$ to $1 \cdots 1$ in order from the bottom to the top respectively similar to the numbering to the outputs. Let $X = \{x_0, \ldots, x_{n-1}\}$ be the input set and the rank of x_i be in the range $[i-n/2, i+n/2]$ with extremely high probability. This assumption corresponds to the property satisfied by the input set $X - L - H$, which is the set of the elements whose errors in ranking exceed $n^{0.822} + c$, if Conjecture 1 follows. Consider the bijection $\sigma : \{0, \ldots, n-1\} \to \{0, 1\}^{\log n}$ defined as follows:

$$\sigma(0) = 0 \ , \quad \sigma(i) = \begin{cases} 0\sigma(\lceil i/2 \rceil) & \text{if } i \text{ is even,} \\ 1\sigma(\lceil i/2 \rceil) & \text{if } i \text{ is odd.} \end{cases}$$

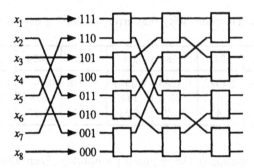

Figure 2. The assignment with the permutation by σ for an 8-input butterfly network represented as arrows

If x_i is assigned to the input $\sigma(i)$ of a butterfly network, then this assignment realizes the stated idea. Consider the case that $n = 8$ for example (see Figure 2). Then, x_0 is compared with x_4 in the first level, and is compared with either x_2 or x_6 in the secondary level, and is finally compared with an element among x_1, x_5, x_3, and x_7 in the last level. The similar observation shows that this assignment with σ satisfies the stated property. Thus, the following conjecture is stated.

Conjecture 2. *Let $X = \{x_0, \ldots, x_{n-1}\}$ be the input set for an n-input butterfly network and the rank of x_i be in the range $[i - n/2, i + n/2]$ with extremely high probability. Then a butterfly network can give an approximate rank to each output with less tolerance when elements in X are assigned to input lines of the butterfly network with the permutation by σ than with a random permutation.*

Since this conjecture has not been proved analytically, in order to verify this conjecture, the computational experiment has been performed. The average errors in ranking for the case that random input assignments are used and for the case that the input assignment with σ is used are calculated by actual application of a butterfly network for the input set $X' = \{x'_i\}$ such that the rank of x'_i is random value following the normal distribution $N(i, (n/5)^2)$. This X' approximately corresponds the input set X in above conjecture because x'_i distributes in the range $[i - n/2, i + n/2]$ with high probability 0.9876 from the property of the normal distribution. In this experiment, although the function π is used for random assignments to give an approximate rank to each output, the function obtained statistically from the result is used for the assignment with σ, since the appropriate function has not been found for this case.

Table 3 shows the result of the experiment for 100,000 cases. c also equals 0.001 in this experiment. E_r and E_σ denote the required average errors for random assignments and for the assignment with σ respectively. This result indicates that E_σ is about 75 % of E_r. This suggests that Conjecture 2 follows if an appropriate ranking function for an approximately sorted input set is given.

Table 3. The average errors in approximate ranks

n	32	64	128	256	512	1024	2048	4096
E_r	2.0	3.7	6.3	10.8	18.3	30.7	51.4	85.6
E_σ	1.5	2.7	4.8	8.1	13.8	23.1	38.7	64.4

If Conjectures 1 and 2 are satisfied, inner random permutations in the selection network in Section 3 can be removed and improvement in tolerance of approximate ranks enables probabilistic selection with less size and less depth.

6 Conclusion

This paper has built a very high probability (n, k)-selector with size $0.5n \log n + O(n^{0.822} \log n)$ and depth $5.62 \log n + O(\log \log n)$. Using only butterfly networks except for the last stage, it achieves smaller size and depth than that of previously known selection networks. Especially its size is remarkable for its constant factor of the leading term which equals to that of the lower bound. However, there remains a problem that this selection network is useful only for large input size. The construction of the simpler selection network surpassing the sorting network with $O(n(\log n)^2)$ size and $O((\log n)^2)$ depth such as an odd-even sorting network is an important subject.

This paper has further researched the practical aspects of a butterfly network and has shown some useful properties of a butterfly network. These properties imply that above probabilistic (n, k)-selector can be improved in its size and depth. Although analytical proofs of these properties have been remained, it will be accomplished with more minute research on a butterfly network. This research is expected to solve the problem on small input size and to be the basis to construct more practical selection network.

References

1. Ajtai, M., Komlós, J., and Szemerédi, E., "Sorting in $c \log n$ Parallel Steps," *Combinatorica, 3,* 1983, pp.1–19.
2. Alekseev, V. E., "Sorting Algorithms with Minimum Memory," *Kibernetica, 3,* 1969, pp.99–103.
3. Jimbo, M., and Maruoka, A., "Selection Networks with $8n \log_2 n$ Size and $O(\log n)$ Depth," *ISAAC '92, Lecture Notes in Computer Science 650,* 1992, pp.165–174.
4. Knuth, D. E., *The Art of Computer Programming, 3* Addison-Wesley, 1973.
5. Leighton, T., and Plaxton, C. G., "A (fairly) Simple Circuit that (usually) Sorts," *Proc. 31st Annual Symp. on Foundations of Computer Science, 1,* 1990, pp.264–274.
6. Paterson, M. S., "Improved Sorting Networks with $O(\log n)$ Depth," *Algorithmica, 5,* 1990, pp.75–92.
7. Pippenger, N., "Selection Networks," *SIAM Journal on Computing, 20,* 1991, pp.878–887.

Optimal Group Gossiping in Hypercubes Under Wormhole Routing Model

Satoshi Fujita, Masafumi Yamashita and Tadashi Ae

Department of Electrical Engineering
Faculty of Engineering, Hiroshima University
Kagamiyama 1-4-1, Higashi-Hiroshima, 724 Japan

Abstract. This paper introduces a new gossiping problem of exchanging tokens among specified nodes. We call this problem the *group gossiping problem*. The group gossiping problem is a generalization of the usual gossiping problem, which has been investigated extensively during the last decade. In this paper, we consider the gossiping problem in n-cubes under the *wormhole routing model*, which is a model of the communication by the wormhole switching. We propose an asymptotically optimal group gossiping algorithm for n-cubes under the model.

Keywords: parallel algorithm, gossiping, wormhole routing, optimal time bound, n-cubes

1 Introduction

This paper introduces a new gossiping problem of exchanging tokens among specified nodes. We call this problem the *group gossiping problem*. The group gossiping problem is a generalization of the usual gossiping problem, which has been investigated extensively during the last decade [1, 3, 6, 7]. Let $G = (V, E)$ be an undirected graph representing a processor network and U ($\subseteq V$) be the specified subset of nodes among which tokens are requested to exchange. The gossiping problem assumes $U = V$. Since $U \subseteq V$, any gossiping algorithm can correctly solve the group gossiping problem, as well.

Suppose that in each step, a node u can communicate with another node v only if $\{u, v\} \in E$. We call it the *link communication assumption*. Under the link communication assumption, in the worst case, group gossiping requires time greater than or equal to the diameter of the network. On the other hand, for each of many communication topologies, there is a gossiping algorithm which achieves a gossiping time close to the diameter of the network, even if each link in the network can carry tokens only in one direction in each step (i.e., half–duplex) [8][1]. Hence, under the link communication assumption, the group gossiping problem can be solved efficiently by gossiping algorithms in the sense of the worst case analysis. The link communication assumption models the communication by packet switching using the store–and–forward technique.

---

[1] If we further assume that each node can participate at most one communication in each step, in many cases, we can find lower bounds strictly greater than the diameter of the network[7, 8].

Recently, an alternative technique called wormhole switching was proposed for efficient internode communication, and has been realized in commercial parallel processors such as iPSC/2 or NCUBE-2 [10]. The wormhole switching is a hybrid of the packet switching and the circuit switching. It routes a message as follows:

- When a message is sent from a node to another node, which may not be adjacent with, the initial part of the message containing the communication route between the nodes is sent in the packet switching mode, and the route for the communication is reserved.
- The rest of the message flows through the reserved route without the need for store–and–forward delay.

An important property of the wormhole switching is that the message latency is considered to be a constant regardless of the location of the communicating nodes. As for iPSC/2, it is reported that the message latency is improved by 3–10 times over the iPSC/1, which adopts the store–and–forward technique [10].

This paper assumes that the communication time is a constant, and that each message can carry any number of tokens as in standard literatures (see [6]). Although each communication is made in a constant time regardless of the locations of the communicating nodes by the wormhole switching, it is required to reserve a route connecting them for the communication. When k pairs (u_i, v_i) $(1 \leq i \leq k)$ of nodes are communicating with each other simultaneously, k edge-disjoint routes connecting those pairs are reserved. A group gossiping algorithm requests many nodes to send tokens in parallel to achieve a good gossiping time, but such parallelism is obviously bounded by, e.g., the edge–connectivity of the network G, and moreover, the given subset U may be distributed inadequately for parallel communication. The main theme of this paper is how to extract parallelism under those conditions. In this paper, we consider the group gossiping under the wormhole routing model.

As the underlying communication topology G, we will adopt n-cubes, since there have been built several n-cube systems adopting the wormhole switching technique (e.g., iPSC/2 and NCUBE-2). We will propose two group gossiping algorithms for n-cubes. Let $|U| = 2^{\epsilon n}$ for $\epsilon \leq 1$. The first algorithm completes the gossiping provided that $\epsilon < 1$, and asymptotically achieves the lower bound when $1/(1-\epsilon) = o(n)$. The second algorithm, on the other hand, asymptotically achieves the lower bound when $1/(1 - \epsilon) = \Omega(n)$. By combining these two algorithms, we obtain an asymptotically optimal group gossiping algorithm for any U $(\subseteq V)$.

Under the wormhole routing model, several routing problems have been investigated [2, 4, 5, 9, 11]. To the authors' knowledge, however, no investigation has been carried out for the gossiping problem.

This paper is organized as follows. In Section 2, we introduce some notations, and define the wormhole routing model and the group gossiping problem formally. Sections 3, 4 and 5 are concerned with the first gossiping algorithm for $1/(1 - \epsilon) = o(n)$. In Section 3, we show an outline of the algorithm. The algorithm consists of three phases. Sections 4 and 5 describe efficient implementations of these phases. Section 6 proposes the second gossiping algorithm for $1/(1-\epsilon) = \Omega(n)$. The algorithm also uses the procedure proposed in Section 5.

2 Preliminaries

2.1 Notations

Let $\mathcal{H} = (V, E)$ be an (undirected) n-cube, where $V = \{0,1\}^n$, and for any $u, v \in V$, $\{u, v\} \in E$ iff u and v differ in exactly one bit. An element of V is called a *node*, and an element of E is called an *edge*. If u and v differ in the ith bit and $\{u, v\} \in E$, we denote as $u = \oplus_i v$. Note that if $u = \oplus_i v$ then $v = \oplus_i u$. An edge $\{u, v\}$ is called the ith edge of u if $u = \oplus_i v$. Let $E_i \subseteq E$ be the set of the ith edges of $\mathcal{H}$. An ordered pair (u, v) of nodes is called a link. If $\{u, v\} \in E_i$, then (u, v) is called the ith link of u. Let $L_i = \{(u, v), (v, u) : \{u, v\} \in E_i\}$ be the set of ith links. Finally, let $L = \bigcup_{1 \le i \le n} L_i$ be the set of all links.

Definition 1. Let m be an integer in $\{0, 1, \ldots, n\}$. For each $x \in \{0, 1\}^m$, let $\mathcal{H}_x$ be the subcube of $\mathcal{H}$ induced by the set of nodes $\{yx : y \in \{0, 1\}^{n-m}\}$. □

Definition 2. Let v be a node in V and m be an integer in $\{0, 1, \ldots, n\}$. For each $x \in \{0, 1\}^m$, $v(x)$ is defined as $v(x) = yx$, where y is the prefix of v with length $n - m$. For given $U \subset V$, $U(x)$ is defined as $U(x) = \{v(x) : v \in U\}$. □

Example 1. Consider the case of $n = 4$. $\mathcal{H}_{01}$ is the 2-cube induced by set $\{0001, 0101, 1001, 1101\}$. If $v = 1111$ and $x = 01$, then $v(x) = 1101$, since $v(x)$ has the same prefix 11 with v. □

Definition 3. Given $G = (V, E)$, a path in G connecting nodes u and v ($\in V$) is a sequence of links, $P = \ell_1, \ell_2, \ldots, \ell_m$ such that $\ell_i = (w_{i-1}, w_i) \in L$ for $1 \le i \le m$, $w_0 = u$, and $w_m = v$. If w_i's are distinct, path P is said to be *simple*. Since we only consider simple paths in this paper, a path means a simple path. If a path P contains a link (u, v) and another path Q contains a link (u, v) or (v, u), they are said to *share* edge $\{u, v\}$. A set of paths are said to be (mutually) *edge-disjoint* if any two paths in the set share no edges.

Let U and W be two disjoint subsets of V. We say that U and W are *connected by edge-disjoint paths*, if there is a set $\mathcal{P}$ of edge-disjoint paths, each connecting a node in U and a node in W, such that each node in U (resp. in W) is connected with a node in W (resp. in U) by some path in the set. If $\mathcal{P}$ can be taken so that it contains a path connecting u and v for any $u \in U$ and $v \in W$, the U and W are said to be *fully connected by edge-disjoint paths*. When $\{u\}$ and U are connected by edge-disjoint paths, we say that u and U are connected by edge-disjoint paths. □

2.2 Wormhole Routing Model

Each node of $\mathcal{H}$ communicates with another node of $\mathcal{H}$ by sending a message along with a path connecting them in $\mathcal{H}$. $\mathcal{H}$ has a global clock, and all nodes synchronously execute their operations according to the global clock.

In this paper, we make the following assumptions on the communication.

- Each message can carry any number of tokens in a step.
- When u wishes to send a message to v, u invokes communication by specifying a path connecting u and v. Let $\mathcal{P}$ be the set of paths selected by the nodes who wish to send messages in a step. If a path in $\mathcal{P}$ is edge–disjoint with the other paths in $\mathcal{P}$, then the message is sent through the path in the step. If more than one paths share an edge, one of them is selected arbitrarily, and the communication using the selected path occurs in the step.
- A node u can send out messages to *all* edges incident on u simultaneously. Moreover, those messages can be distinct. However, u can't receive any message from an edge to which it sends out a message, since the paths must be edge–disjoint. The message sent is received only by the destination node at the other end of the path. Other intermediate nodes in the path can't receive the message; namely, they merely relay the message.

We call this model the *wormhole routing model*. Note that under this model, each edge is *half-duplex*.

2.3 Group Gossiping Problem

Let $G = (V, E)$ be a graph. Each node $u \in V$ has a token $t(u)$. For a given subset $U \subseteq V$, consider the following operation: broadcast $\{t(u) : u \in U\}$ to all nodes in U. We call this problem the *group gossiping problem* for U. When $|U| = 2$, it is the point–to–point routing problem, and when $U = V$, it is the usual gossiping problem [6]. In what follows, let $|U| = 2^{\epsilon n}$ for $0 \le \epsilon \le 1$.

A lower bound on the group gossiping time under the wormhole routing model is derived as follows. Under the model, a node $v \in V$ can send any number of tokens to at most n nodes in a step since the degree of v is n. (Some of the n receivers may not be adjacent with v.) Hence, it requires at least $\lceil \log_2(|U| - 1)/\log_2 n \rceil$ steps to broadcast token $t(v)$ to all nodes in $U \setminus \{v\}$. Then, we have the following theorem.

Theorem 4. *The group gossiping for U requires at least $\lceil \log_2(|U| - 1)/\log_2 n \rceil$ steps.* □

3 Outline of The First Algorithm *GROUP_GOSSIP1*

This section describes an outline of the first group gossiping algorithm *GROUP_GOSSIP1* which asymptotically achieves the lower bound in Theorem 4 provided $1/(1-\epsilon) = o(n)$. In the following three sections, we assume that $1/(1-\epsilon) = o(n)$.

The following lemma provides the basis for distributing (and collecting) tokens to (and from) h other nodes in an h-cube in one step.

Lemma 5. *Let h be any natural number and let $\mathcal{H} = (X, A)$ be an h-cube. Then any set $Y \subseteq X$ of size h and any node v ($\in X \backslash Y$) are connected by edge–disjoint paths.* □

Algorithm *GROUP_GOSSIP1* consists of three phases. Recall that each node u in U initially holds a distinct token $t(u)$. Phase 1 moves those $|U|$ tokens to a set W of $2^d (\leq |U|)$ nodes, called *intermediate nodes*, in one step. Here d is an integer determined from $|U|$ and n. This move uses the edge-disjoint paths guaranteed in Lemma 5. At the end of Phase 1, each intermediate node ($\in W$) keeps at least $\lfloor |U|/2^d \rfloor$ tokens. In Phase 2, the nodes in W exchange the tokens they collected in Phase 1 by repeatedly applying a communication scheme proposed in Section 5. As we will see later, it takes $\log_2 |U|/\log_2 n + o(\log_2 |U|/\log_2 n)$ steps, and when Phase 2 finishes, every node in W holds the set of all tokens $T = \{t(u) : u \in U\}$. Finally in Phase 3, the nodes in W broadcast T to U in one step by using the paths used in Phase 1, in the reverse direction.

Algorithm *GROUP_GOSSIP1* is described as follows.

Algorithm *GROUP_GOSSIP1* (For $U = 2^{\epsilon n}$ satisfying $1/(1 - \epsilon) = o(n)$.)

Phase 1: Each node $u \in U$ sends $t(u)$ to a node in W in one step.

Phase 2: All nodes in W exchange their tokens in $\log_2 |U|/\log_2 n + o(\log_2 |U|/ \log_2 n)$ steps.

Phase 3: Each node in U receives the set of all tokens from a node in W in one step. □

The following two sections show concrete implementations of these three phases.

4 Communication with Intermediate Nodes

This section firstly determines the set of intermediate nodes, W, then constructs edge–disjoint paths connecting U and W. These edge–disjoint paths are used as the communication routes both in Phases 1 and 3: in Phase 1, tokens flow from U to W, and in Phase 3, from W to U, through the routes.

4.1 Intermediate Nodes W

Given U ($\subset V$), let d be the smallest non-negative integer which satisfies $\lceil |U|/2^d \rceil \leq n - d$. We partition U into 2^d subsets as even as possible, as follows. Note that since $1/(1 - \epsilon) < n$, d ($< n$) is well defined for any $n \geq 1$. Consider a partition $\mathcal{U}$ of U generated by the following procedure.

procedure *PARTITION*(m, U_x):

Step 1: If $m = n - d$, then output U_x and stop.

Step 2: Partition U_z into the following three subsets:

$$X_0 = \{w0y \in U_z : w1z \in U_z, |w| = m - 1 \text{ and } y, z \in \{0,1\}^{n-m}\},$$
$$X_1 = \{w1y \in U_z : w0z \in U_z, |w| = m - 1 \text{ and } y, z \in \{0,1\}^{n-m}\}, \text{ and}$$
$$Y = U_z \setminus (X_0 \cup X_1).$$

Note that for any element of X_0, there is an element having the same prefix with length $m - 1$ in X_1, and vice versa.

Step 3: Let Y_0 be an arbitrary subset of Y of size $\lceil |Y|/2 \rceil$. Let $Y_1 = Y \setminus Y_0$.

Step 4: $U_{0z} := X_0 \cup Y_0$ and $U_{1z} := X_1 \cup Y_1$.

Step 5: Call $PARTITION(m - 1, U_{0z})$ and $PARTITION(m - 1, U_{1z})$. □

Let $\mathcal{U}$ be the set of subsets generated by $PARTITION(n, U)$. The following lemma shows that $PARTITION$ partitions a given set U into subsets with almost equal sizes.

Lemma 6. *For any two subsets $U_x, U_y \in \mathcal{U}$, $|U_x|$ and $|U_y|$ differ at most 1.* □

By Lemma 6, we immediately obtain the following corollary, since the maximum level of recursion is d and hence $|\mathcal{U}| = 2^d$.

Corollary 7. *The size of each subset obtained by $PARTITION$ is at most $\lceil |U|/2^d \rceil$.* □

Using the partition $\mathcal{U}$, we define W as follows: $W = \{0^n(x) : U_x \in \mathcal{U}\}$. Recall the definition of notation $v(x)$ given in Definition 2.

4.2 Edge–Disjoint Paths Connecting U and W

This subsection describes how to connect U and W by edge–disjoint paths.

Let v be an arbitrary node in U. Suppose that v is in $U_x \in \mathcal{U}$. Let $u = 0^n(x)$ be a node in W. We will connect v and u by a path $P_v = Q_v R_v$ which is determined as follows:

- Q_v is the shortest path connecting v and $v(x)$ in such a way that if in Q_v, a link in L_x occurs before a link in L_y then $x > y$. In short, Q_v uses links in $L_n, L_{n-1}, \ldots, L_{n-|x|+1}$ in this order.
- R_v connects $v(x)$ and u. Notice that both $v(x)$ and $u (= 0^n(x))$ are in $\mathcal{H}_x$. Since $|\{v(x) : v \in U_x\}| \leq \lceil |U|/2^d \rceil \leq n - d$ by Corollary 7, there are edge-disjoint paths connecting u and each of $v(x)$'s by Lemma 5. We take the path connecting $v(x)$ and u as R_v.

Let $\Pi_1 = \{Q_v : v \in U\}$, $\Pi_2 = \{R_v : v \in U\}$, and $\Pi = \{Q_v R_v : v \in U\}$.

Theorem 8. *Any two paths $P_u, P_v \in \Pi$ are edge-disjoint.* □

Theorem 8 is immediate by the following lemmas.

Lemma 9. *Any two paths $Q_u \in \Pi_1$ and $R_v \in \Pi_2$ are edge-disjoint.* □

Lemma 10. *Any two paths $R_u, R_v \in \Pi_2$ are edge-disjoint.* $\square$

Lemma 11. *Any two paths $Q_u, Q_v \in \Pi_1$ are edge-disjoint.* $\square$

By using the edge-disjoint paths Π connecting U and W, Phases 1 and 3 are executed as follows. Let $\mathcal{U}$ be the partition of U obtained by $PARTITION(n, U)$.

Phase 1: Each $u \in U_x \in \mathcal{U}$ sends $t(u)$ to $w = 0^n(x) \in W$ through path P_u in Π connecting u and w. $\square$

Phase 3: Each $u \in U_x \in \mathcal{U}$, receives $T = \{t(v) : v \in U\}$ from $w = 0^n(x) \in W$ through path P_u in the reverse direction. $\square$

Since Π is a set of edge-disjoint paths, we have the following theorem.

Theorem 12. *Each of Phases 1 and 3 completes in one step.* $\square$

5 Exchange Tokens among Intermediate Nodes

Recall that set W of intermediate nodes satisfies that $|W| = 2^d$ and $d < n$. At the beginning of Phase 2, for each $\mathcal{H}_x$, the tokens of the nodes in U_x are held by node $0^{n-d}x$ in $\mathcal{H}_x$. Let $r = n - d - 1$, and $s = \lfloor \log_2 r \rfloor$ $(= \lfloor \log_2(n - d - 1) \rfloor)$. In this section, we assume that $d \leq n - 3$ without loss of generality. If $d \geq n - 2$, let $Y = \{0^3 x : x \in \{0,1\}^{n-3}\}$, which is a subset of W. Then we regard Y as W at the expense of at most 6 steps, since all nodes $yx \in W$ can send tokens to $0^3 x \in Y$ in 3 steps (for each x in parallel), and $0^3 x$ can broadcast T to all nodes yx in 3 steps.

The basic idea for quick execution of Phase 2 is to exchange tokens among 2^s nodes in one step by using a communication scheme similar to complete bipartite graphs. For each scheme, edge-disjoint paths are selected in such a way that they pass through disjoint subcubes.

5.1 Basic Communication Pattern

Fix any $w \in \{0,1\}^{d-s}$, and consider the following two subsets: $S_0 = \{00^r xw : x \in \{0,1\}^s\}$, and $S_1 = \{10^r xw : x \in \{0,1\}^s\}$. For each pair (u, v) in $S_0 \times S_1$, we give a path P_{uv} connecting u and v in $\mathcal{H}_w$, and show that the set of paths $\Gamma = \{P_{uv} : u \in S_0 \text{ and } v \in S_1\}$ is edge-disjoint. For each $x \in \{0,1\}^*$, $\bar{x}$ denotes the integer whose binary representation is x. By $0^{(r;i_1,i_2,\ldots,i_j)}$, we denote the bit sequence $b_1 b_2 \ldots b_r \in \{0,1\}^r$ such that $b_k = 0$ iff $k \neq i_1, i_2, \ldots, i_j$.

Let $u = 00^r yw \in S_0$ and $v = 10^r zw \in S_1$. If $s = 1$, there apparently exists a set Γ of edge-disjoint paths fully connecting S_0 and S_1. In the following, we consider the case of $s \geq 2$. When $s \geq 2$, the path P connecting u and v is determined as follows:

1. If $\bar{z} = \bar{y}$, we take $P_{uv} = (u, v)$.

2. If $\bar{y} > \bar{z}$, the path P_{uv} consists of four subpaths P_1, P_2, P_3, and P_4. The first part P_1 of P_{uv} is given as follows: $P_1 = (00^r yw, 00^{(r;\bar{z}+1)} yw), (00^{(r;\bar{z}+1)} yw,$ $00^{(r;\bar{z}+1,\bar{y}+1)} yw)$. Note that $\bar{z} + 1 < \bar{y} + 1 \leq r$, since $s = \lfloor \log_2 r \rfloor$. The second part P_2 of P_{uv} consists of a link in L_1: $P_2 = (00^{(r;\bar{z}+1,\bar{y}+1)} yw, 10^{(r;\bar{z}+1,\bar{y}+1)} yw)$. The third part P_3 of P_{uv} connects nodes $10^{(r;\bar{z}+1,\bar{y}+1)} yw$ and $10^{(r;\bar{z}+1,\bar{y}+1)} zw$ by the shortest path which uses links in $L_{r+2}, L_{r+3}, \ldots, L_{r+s+1}$ in this order. The last part P_4 of P_{uv} is given as follows: $P_4 = (10^{(r;\bar{z}+1,\bar{y}+1)} zw, 10^{(r;\bar{y}+1)} zw),$ $(10^{(r;\bar{y}+1)} zw, 10^r zw)$.

3. If $\bar{y} < \bar{z}$, the first, second and fourth parts P_1, P_2, P_4 of P_{uv} are given in the way same as the case of $\bar{y} > \bar{z}$. The third part P_3 of P_{uv} connects $u_1 = 10^{(r;\bar{z}+1,\bar{y}+1)} yw$ and $u_2 = 10^{(r;\bar{z}+1,\bar{y}+1)} zw$ as follows.

   (a) If for each $1 \leq i \leq s$, the ith bits of y and z differ, i.e., the Hamming distance between u_1 and u_2 is s $(= |y|)$, then P_3 is the shortest path connecting u_1 and u_2 using links in $L_{r+2}, L_{r+3}, \ldots, L_{r+s+1}$ in this order, i.e., P_3 is the same as in the case of $\bar{y} > \bar{z}$.

   (b) If not, let k be an integer in $[r+2, r+s+1]$ such that the kth bits of u_1 and u_2 take the same value. Then, P_3 is given as $P_3 = (u_1, \oplus_k u_1), P_3', (\oplus_k u_2, u_2)$ where P_3' is a shortest path connecting $\oplus_k u_1$ and $\oplus_k u_2$. □

Let $\Gamma = \{P_{uv} : u \in S_0 \text{ and } v \in S_1\}$. It is obvious that each path P_{uv} in Γ correctly connects $u \in S_0$ and $v \in S_1$ using edges in $\mathcal{H}_w$. The following three lemmas guarantee that Γ is a set of edge–disjoint paths.

Lemma 13. *No edge in E_1 is shared by plural paths in Γ.* □

Let $\mathcal{E}_1 = \bigcup_{2 \leq i \leq r+1} E_i$ and $\mathcal{E}_2 = \bigcup_{r+2 \leq i \leq r+s+1} E_i$.

Lemma 14. *No edge in $\mathcal{E}_1$ is shared by plural paths in Γ.* □

Lemma 15. *No edge in $\mathcal{E}_2$ is shared by plural paths in Γ.* □

By Lemmas 13, 14, and 15, we have the next theorem.

Theorem 16. *Sets S_0 and S_1 are fully connected by set Γ of edge–disjoint paths in $\mathcal{H}_w$.* □

5.2 Exchange Tokens among All Nodes in W

Let σ be an integer in $\{0, 1, \ldots, d-1\}$. Fix any $w_1 \in \{0,1\}^\sigma$ and $w_2 \in \{0,1\}^{d-(\sigma+s)}$, and consider the following two subsets $W_{w_1,w_2} = \{00^r w_1 x w_2 : x \in \{0,1\}^s\}$ and $\overline{W}_{w_1,w_2} = \{10^r w_1 x w_2 : x \in \{0,1\}^s\}$. Then, by Theorem 16, W_{w_1,w_2} and $\overline{W}_{w_1,w_2}$ are fully connected by edge–disjoint paths in the subcube induced by $\{yw_1 x w_2 : y \in \{0,1\}^{r+1} \text{ and } x \in \{0,1\}^s\}$ because of the symmetricity with respect to each dimension. Let $\Gamma_{w_1,w_2} = \{Q_{uv} : u \in W_{w_1,w_2} \text{ and } v \in \overline{W}_{w_1,w_2}\}$ be a set of edge–disjoint paths. By using Γ_{w_1,w_2}'s, Phase 2 is executed as follows in $\lceil d/s \rceil$ steps.

procedure *TOKEN_EXCHANGE*

Step 1: Let $\sigma = 0$ and $b = 0$.

Step 2: Repeat Steps from 3 to 5, $\lceil d/s \rceil$ times.

Step 3: For each $w_1 \in \{0,1\}^\sigma$ and $w_2 \in \{0,1\}^{d-(\sigma+s)}$, let $W_{w_1,w_2} = \{00^r w_1 x w_2 : x \in \{0,1\}^s\}$. Then, $\mathcal{W} = \{W_{w_1,w_2} : w_1 \in \{0,1\}^\sigma \text{ and } w_2 \in \{0,1\}^{d-(\sigma+s)}\}$ forms a partition of W and $|\mathcal{W}| = 2^{d-s}$. Let $\overline{W}_{w_1,w_2} = \{1x \in V : 0x \in W_{w_1,w_2}\}$. Note that W_{w_1,w_2} and $\overline{W}_{w_1,w_2}$ are fully connected by a set of edge–disjoint paths Γ_{w_1,w_2}.

Step 4: If $b = 0$ (resp. $b = 1$), for all $w_1 \in \{0,1\}^\sigma$ and $w_2 \in \{0,1\}^{d-(\sigma+s)}$, each u in W_{w_1,w_2} (resp. $\overline{W}_{w_1,w_2}$) sends all tokens it holds to every node in $\overline{W}_{w_1,w_2}$ (resp. W_{w_1,w_2}) through paths in Γ_{w_1,w_2}.

Step 5: $\sigma := \sigma + s$ and $b := b \oplus 1$.

Step 6: After the repetition, if $\lceil d/s \rceil$ is even, every node in W holds the set of all tokens $T = \{t(v) : v \in U\}$, and if $\lceil d/s \rceil$ is odd, every node in $\overline{W}$ holds T. Hence, if $\lceil d/s \rceil$ is odd, each node u in $\overline{W}$ sends T to node $\oplus_1 u$ in W by link $(u, \oplus_1 u) \in L_1$. $\qquad \square$

Theorem 17. *TOKEN_EXCHANGE correctly completes the gossiping among all nodes in W.* $\qquad \square$

Since any two paths $P \in \Gamma_{w_1,w_2}$ and $Q \in \Gamma_{w_1',w_2'}$ are edge–disjoint, each round of *TOKEN_EXCHANGE* completes in one step. By definition, $s = \lfloor \log_2(n - d - 1) \rfloor$. Hence $s = \lfloor \log_2(n - d - 1) \rfloor \geq \log_2(n - d) - 2 \geq \log_2 n + \log_2(1 - \epsilon) - 2$. Since $\lceil |U|/2^d \rceil \leq n - d$ holds for $d = \epsilon n$ and $\epsilon n - 1$ (because $1/(1 - \epsilon) = o(n)$), $d \leq \epsilon n = \log_2 |U|$. Hence we have

$$\lceil d/s \rceil \leq \left\lceil \frac{\log_2 |U|}{\log_2 n + \log_2(1 - \epsilon) - 2} \right\rceil \leq \frac{\log_2 |U|}{\log_2 n + \log_2(1 - \epsilon) - 2} + 1.$$

Now since $1/(1 - \epsilon) = o(n)$, $\log_2 n + \log_2(1 - \epsilon) = \log_2 n - o(\log_2 n)$. That is, $\lceil d/s \rceil \leq \log_2 |U|/\log_2 n + o\,(\log_2 |U|/\log_2 n)$.

Each of Phases 1 and 3 takes one step by Theorem 12. When $d \geq n - 2$, we need at most 6 more steps. Consequently, we have the following theorem.

Theorem 18. *Let $|U| \leq 2^{\epsilon n}$ for $\epsilon < 1$. If $1/(1-\epsilon) = o(n)$, then GROUP_GOSSIP1 solves the group gossiping problem for U in $\log_2 |U|/\log_2 n + o\,(\log_2 |U|/\log_2 n)$ steps, which is asymptotically optimal.* $\qquad \square$

6 The Second Algorithm *GROUP_GOSSIP2*

This section proposes an algorithm which achieves the lower bound in Theorem 4, provided $1/(1 - \epsilon) = \Omega(n)$. If $1/(1 - \epsilon) = \Omega(n)$, $\epsilon = 1 - O(1/n)$, i.e., $|U| = 2^{\epsilon n} = 2^{n-O(1)}$. Let δ be an integer in $\{0, 1, \ldots, n - 3\}$. Algorithm *GROUP_GOSSIP2* is described as follows.

Algorithm *GROUP_GOSSIP2*

Phase 1: Let $W = \{0^{n-\delta}x : x \in \{0,1\}^{\delta}\}$. For each $x \in \{0,1\}^{\delta}$, consider the spanning tree, in which every node in $\mathcal{H}_x$ is connected with $0^n(x)$ by a shortest path. (The shortest paths are not necessarily edge–disjoint.) For each $x \in \{0,1\}^{\delta}$ in parallel, collect all tokens of nodes in $\mathcal{H}_x$ to $0^n(x)$ in δ steps through the path in the spanning tree.

Phase 2: Apply *TOKEN_EXCHANGE* to W to exchange tokens among all nodes in W. Since $\delta \leq n-3$, it correctly completes the operation in $\log_2 |U| / \log_2(\delta - 1)$ steps.

Phase 3: Each node $u = 0^n(x)$ in W, in parallel, sends the set of all tokens to all nodes of $\mathcal{H}_x$ in δ steps through the paths in the spanning tree. $\square$

The correctness of the algorithm is clear. The running time of the algorithm is $2\delta + \log_2 |U| / \log_2(\delta - 1)$. Since $|U| = \Theta(|V|)$, by selecting δ to satisfy $\delta = o(n/\log_2 n)$ and $\delta = \omega(n^{\epsilon})$ (e.g., $\delta = n/(\log_2 n)^2$ satisfies this condition), *GROUP_GOSSIP2* asymptotically achieves the lower bound in Theorem 4.

Theorem 19. *Let $|U| = 2^{\epsilon n}$ for $\epsilon \leq 1$. If $1/(1 - \epsilon) = \Omega(n)$, GROUP_GOSSIP2 correctly solves the group gossiping problem for U in $\log_2 |U| / \log_2 n + o(\log_2 |U| / \log_2 n)$ steps, which is asymptotically optimal.* $\square$

References

1. A. Bagchi, S. L. Hakimi, J. Mitchem, and E. Schmeichel. Parallel algorithms for gossiping by mail. *Inform. Process. Lett.*, 34:197–202, April 1990.
2. W. J. Dally and C. L. Seitz. Deadlock–free message routing in multiprocessor interconnection network. *IEEE Trans. Comput.*, 36(5):547–553, May 1987.
3. R. C. Entringer and P. J. Slater. Gossips and telegraphs. *J. Franklin Inst.*, 307:353–360, 1979.
4. A. M. Farley. Minimum-time line broadcast networks. *Networks*, 10:59–70, 1980.
5. C. J. Glass and L. M. Ni. Adaptive routing in mesh–connected networks. In *Proc. ICDC '92*, pages 12–19. IEEE, 1992.
6. S. M. Hedetniemi, S. T. Hedetniemi, and A. L. Liestman. A survey of gossiping and broadcasting in communication networks. *Networks*, 18:319–349, 1988.
7. D. W. Krumme. Fast gossiping for the hypercube. *SIAM J. Comput.*, 21(2):365–380, April 1992.
8. D. W. Krumme, G. Cybenko, and K. N. Venkataraman. Gossiping in minimal time. *SIAM J. Comput.*, 21(1):111–139, February 1992.
9. A. Sengupta and S. Bandyopadhayay. Deadlock-free routing in k-ary hypercube network in presence of processor failures. *Inform. Process. Lett.*, 34:323–328, May 1990.
10. A. Trew and G. Wilson, editors. *Past, Present, Parallel: A Survey of Available Parallel Computer Systems*, chapter 4, pages 125–147. Springer-Verlag, 1991.
11. H. Xu, P. K. McKinley, and L. M. Ni. Efficient implementation of barrier sychronization in wormhole–routed hypercube multicomputers. In *Proc. ICDC '92*, pages 118–125. IEEE, 1992.

Optimal linear broadcast routing with capacity limitations

(Extended Abstract)

S. Bitan [1] and *S. Zaks* [2]

Department of Computer Science
Technion, Haifa, Israel

Abstract

We study the problem of broadcast routing in high-speed networks with special switching hardware. In such networks replication of a packet in each internal vertex (as done in traditional broadcast algorithms) is inefficient. The method that is used in such networks is *linear broadcast routing*, in which the packets travel along linear routes, determined in the header of the packet. In this routing several packets travel simultaneously through one communication line and one switching subsystem. Each edge has a communication cost, and the total cost of the routing is determined by either the number of the packets or by the total cost of edges traversed.

The problem that arises in realistic networks is that of linear broadcast routing with limited bandwidth, that corresponds to edge capacity in the underlying graph. We show that the problem of determining the linear broadcast routing, with either smallest number of packets or with the smallest cost of edges utilization (for a given number of packets), for either bounded or unbounded headers, and with capacity limitations, is NP-hard for general graphs. We present polynomial-time algorithms for tree networks, for all of the above cases, in which the header is unbounded.

---

[1] Email address: SARAB@CS.TECHNION.AC.IL.
[2] Email address: ZAKS@CS.TECHNION.AC.IL .

1 Introduction

An *asynchronous distributed network* consists of a set of processors connected by communication lines. *Broadcasting* in such a network is the action of propagating information from a *source* processor r to all others. Traditional broadcast algorithms in packet-switching networks usually propagate information using a spanning tree [Seg83].

In high-speed networks with switching hardware such as PARIS [CG88], the processing time in the node is not negligible, so the replication of the packet in the intermediate node is inefficient. In such networks each node contains a general purpose processor, and a switching hardware. In this networks another broadcast technique called *linear broadcast routing* (see [CG89]) is used. This routing technique creates only linear routes, i.e. the packet is replicated only by the broadcasting node, and all the other processors propagate the packets to their destination. Thus the only work done by all the processors except for r is switching the packet from an incoming edge to an outgoing one. Since the communication in such networks is much faster, the switching functions are implemented automatically in hardware. It is customary thus to assume that a packet can go through a route between two processors in the network in one time unit (see [ICK88]), thus linear broadcasting has a cost of one time unit.

In linear broadcast routing several copies of the same packet go through one communication line simultaneously. In realistic networks the bandwidth of the communication lines is bounded, and we model it by edge capacities. Each of the packets contain a header that includes the switching information for its route. The header length thus corresponds to the number of vertices through which the packet travels. In addition, there might be distinct costs associated with each communication line.

We study the linear broadcast routing problem in general networks, and specifically in the commonly-used tree networks, with bandwidth and header length limitations. The problem of linear broadcast routing was suggested in [CG89]. They studied the problem in general graphs and in trees, with bounded and unbounded header. They however ignored the bandwidth limitations. Further study of the problem was done in [BZ93], where algorithms to solve the problem of linear broadcast routing in trees with a bounded number of packets were presented.

We discuss the problem of minimal cost linear broadcast routing with bandwidth limitations. We study the cost functions of number of packets and total communication. and show that linear broadcast routing with unbounded header is NP-Complete for general graphs, and with bounded header is NP-Complete even for trees. We present algorithm to solve it on trees with bandwidth limitations and a given number of packets.

In Section 2 we present definitions and notations. In Section 3 we summarize our results. One of the NP-completeness reductions is sketched in Section 4, and one of the algorithms is sketched in Section 5. Most proofs are omitted in this extended abstract.

2 Preliminaries

A communication network is viewed as an undirected graph $G = (V, E)$, where the vertices in V represent the processors in the network, and the edges in E represent the communication lines between the processors. A *capacity function* $c : E \rightarrow N = \{1, 2, 3, ...\}$ indicates for every edge $e \in E$ the number of packets that can go simultaneously through e. The capacity function corresponds to the bandwidth limitations of the communications lines. A *weight function* $w : E \rightarrow N$ indicates for every edge $e \in E$ the communication cost.

In order to solve the linear routing problem, we want to cover the graph by a set of paths (each corresponding to one packet), all starting at a given vertex r. Such a set will be termed *broadcast set*. It is emphasized that a path is not necessarily *simple*; *e.g.*, it can go through a given edge more than once (see [Eve79]).

A *linear broadcast problem with capacity limitations* $GLBC = (G, c, r)$ is to determine a broadcast set with the smallest number of paths, such that, for every edge e, at most $c(e) \geq 2$ paths go through e.

A *weighted linear broadcast problem with capacity limitations* is to determine a broadcast set of a total minimal cost, where the number of packets is either unbounded - $WGLBC = (G, c, w, r)$, or bounded by k - $WGLBCS = (G, c, w, r, k)$; in both cases, we require that for every edge e, at most $c(e) \geq 2$ paths go through e.

Example: Consider the graph $G_1(V, E)$ with weights as shown in Figure 1.

The path $p_0 = < e_6, e_5 >$ is a simple path connecting f and d. The path $p_1 = < e_{11}, e_8 >$ also connects d and f. $w(p_0) = 3$, while $w(p_1) = 11$. The path that consists of a concatenation of p_0, p_1, e_{12} and e_{12} is an example of a path that starts and ends in f, and goes through the vertices c, d, e, f and h, with a total weight of 18. Such a path is denoted by $p_0 \cdot p_1 \cdot e_{12} \cdot e_{12}$.

Consider the capacity function c_0 that assigns capacity 2 to all the edges in G_1. The broadcast set $\psi_0 = \{e_1 \cdot e_2 \cdot e_4 \cdot e_5 \cdot e_9 \cdot e_{10} \cdot e_{11} \cdot e_{12} \cdot e_{13}\}$ is a broadcast set for (G_1, r, c_0, w), with a weight of 31. The broadcast set $\psi_1 = \{e_3 \cdot e_2 \cdot e_4 \cdot e_5 \cdot e_8 \cdot e_{10} \cdot e_8 \cdot e_5 \cdot e_6 \cdot e_{12}\}$ is an optimal weight broadcast set containing one trail for (G_1, rc_0, w), with a weight of 15.

Consider the same graph and weight function, this time with capacity function c_1 that assigns capacity 2 to all the edges except for e_2, e_5 and e_{10}, whose capacity is 1. The broadcast set ψ_0 is still a feasible broadcast set for (G_1, r, c_1, w), but ψ_1 is no longer feasible. The optimal weight broadcast set this time is $\psi_2 = \{e_1, e_3 \cdot e_4 \cdot e_5 \cdot e_8 \cdot e_{10} \cdot e_9 \cdot e_7 \cdot e_6 \cdot e_{12}\}$, with a weight of 29.

A *tree linear broadcast problem with capacity limitations* $TLBC = (T, c, r)$, a *weighted tree linear broadcast problem with capacity limitations* $WTLBC = (T, c, w, r)$ and a *weighted tree linear broadcast problem with simultaneous send limitations and capacity limitations*, $WTLBCS = (T, c, w, r, k)$ are specified in ways similar to that of the corresponding $GLBC$, $WGLBC$ and $WGLBCS$

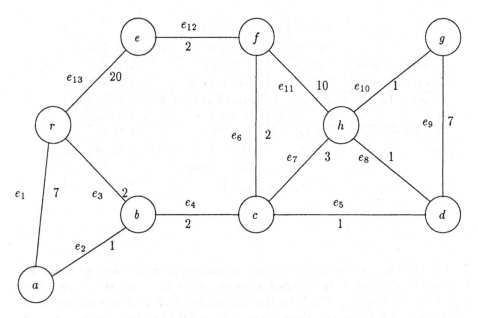

Figure 1:   An example

problems, where a tree $T(V, E)$ replaces the general graph $G(V, E)$.

3 Summary of results

We study the above four problems (GLBC, WGLBC, TLBC, WTLBC) with the additional three parameters :

1. The maximal path length - bounded or unbounded. This corresponds to the amount of space needed for the packet header.
2. The capacity function - either for each $e \in E, c(e) = 1$, or for each $e \in E, c(e) \geq 2$.
3. The bound on the number of paths in the broadcast set - bounded or unbounded.

Notation: WGLB will be the problem like WGLBC but without capacity limitations, and Bounded WGLB will be like WGLB but with a bound on the header length (and similarly for the other three problems).

Following is a brief summary of the corresponding four problems without the capacity limitations (i.e $GLB = (G, r)$ instead of GLBC, and similarly the problems $WGLB$, TLB and $WTLB$).

Linear Broadcast Routing			
	Unweighted	Weighted	Weighted, k trails
Unbounded trail, General Graph	*GLB* DFS	*WGLB* NPC[CG89]	*WGLBS* NPC[CG89]
Bounded Trail, Tree	*Bounded TLB* NPC[CG89]	*Bounded WTLB* NPC[CG89]	*Bounded WTLBS* NPC[CG89]
Unbounded trail, Tree	*TLB* DFS	*WTLB* Algorithm Tree Routing [CG89]	*WTLBS* Algorithms A1 A2 [BZ93]

Table 1

The GLB problem without capacity limitations is trivial, because when the maximal trail length is unbounded a broadcast set that consists of one *DFS* path is clearly optimal. The WGLB and Bounded TLB problems are shown in [CG89] to be NP-Complete. Since the problem is NP-Complete even for a tree, it is obvious that the corresponding bounded GLB and bounded WGLB problems are also NP-Complete.

The only problems in this table that can be polynomially solved are WTLB [CG89] and WTLBS [BZ93].

We now summarize - in Tables 2 and 3 - the results of this paper, that deal with linear broadcast with capacity limitations.

Notation: WTLBC1 will be the problem like WTLBC but with capacity $c(e) = 1$ for all $e \in E$ (and similarly for the other problems).

Table 2 contains four versions of the GLBC1 problem (The TLBC1 problem can clearly be solved only for trees with at most one vertex of degree > 2). In the next section we present a sketch of proof for one of the NP-Complete results.

Linear Broadcast Routing with Capacity Function $C : E \to N$			
General Linear Broadcast Routing $c(e) = 1, GLBC1$			
	Unweighted	Weighted	Weighted, k trails
Unbounded trail, General Graph	*GLBC 1* NPC	*WGLBC 1* NPC-[CG89]	*WGLBCS1* NPC[CG89]
Bounded trail, General Graph	*Bounded GLBC 1* NPC - Theorem 1	*Bounded WGLBC 1* NPC	*Bounded WGLBCS1* NPC

Table 2

Table 3 summarizes our results for the TLBC problems. As for the GLBC problem, the fact that capacity limitations are added does not affect the results, so the situation remains the same as in Table 1.

Tree Linear Broadcast Routing with Capacity Function $C: E \to N$			
Tree Linear Broadcast Routing $c(e) \geq 2$, *TLBC*			
	Unweighted	Weighted	Weighted, k trails
Bounded trail, Tree	*Bounded TLBC* NPC[CG89]	*Bounded WTLBC* NPC[CG89]	*Bounded WTLBCS* NPC[CG89]
Unbounded Trail, Tree	*TLBC* DFS	*WTLBC* Algorithm B	*WTLBC* Algorithm C

Table 3

As can be seen from the table, the situation in Bounded TLBC and Bounded WTLBC remains the same with and without capacity limitations. The solution for TLBC is trivial; namely the *DFS* set, with the capacity of all the edges is 2. In Section 5 we give a brief description of algorithm C.

4 NP-completeness results

We prove NP-completeness of the GLBC1, Bounded GLBC1 and Bounded WGLBC1 problems. The GLBC1 problem is shown to be NP-Complete by a reduction to the Hamiltonian circuit problem [GJT76], and the Bounded GLBC1 and Bounded WGLBC1 problems are shown to be NP-Complete in the strong sense by a reduction to the 3-Partition problem [GJ79]. Here we present the construction for the Bounded WGLBC1 problem.

Bounded GLBC1

Instance : An instance (G, c, r) of GLBC, with $c(e) = 1$ for all $e \in E$, bounds $M, L \in N$.

Question : Is there a broadcast set ψ for (G, c, r) under the above capacity function c, such that for all $t \in \psi, l(t) \leq L$ and $n(\psi) \leq M$?

Theorem 1. *The Bounded GLBC1 problem is NP-Complete in the strong sense.*

Proof: It can be easily seen that Bounded GLBC1 $\in$ NP. To show completeness we show a reduction to the following NP-Complete problem:

3-Partition :

Instance : Set A of $3m$ elements, a bound $B \in N$, and a size $s(a) \in N$ for each $a \in A$ such that $B/4 < s(a) < B/2$ and such that $\sum_{a \in A} s(a) = mB$.

Question : Can A be partitioned into m disjoint set $A_1, A_2, ..., A_m$ such that, for $1 \leq i \leq m, \sum_{a \in A_i} s(a) = B$?

Construction: Given an instance of 3-Partition we construct the following instance (G, c, r) of Bounded GLBC1:

$G = (V, E)$ where $V = \{r\} \cup V_m \cup U \cup W$ and $V_m = \{v_i | 1 \leq i \leq m\}$,

$U = \{u_{i,j} | 1 \leq i \leq 3m, 1 \leq j \leq a_j\}$

$W = \{w_1\} \cup \{w_{i,j} | 1 \leq i \leq m, 1 \leq j \leq B\}$

and $E = E_r \cup E_{V_m} \cup E_{U_1} \cup E_{U_2} \cup E_{U_3} \cup E_{w_1} \cup E_{w_2}$ where

$E_r = \{(r, v_i) | 1 \leq i \leq m\}$

$E_V = \{(v_i, u_{j,1}) | 1 \leq i \leq m, 1 \leq j \leq 3m\}$

$E_{U_1} = \{(u_{i,1}, u_{j,1}) | 1 \leq i, j \leq 3m, i \neq j\}$, i.e., the vertices $\{u_{i,1}\}$ form a clique.

$E_{U_2} = \{(u_{i,j}, u_{i,j+1}) | 1 \leq i \leq 3m, 1 \leq j < j < a_i\}, 3m$ chains, each contains a_i vertices.

$E_{U_3} = \{(u_{i,a_i}, u_{j,a_j}) | 1 \leq i, j \leq 3m, i \neq j\}$, namely the last vertices in the chains form a clique.

$E_{w_1} = \{(u_{i,a_i}, w_1 | 1 \leq i \leq 3m\} \cup \{(w_1, w_{i,1}) | 1 \leq i \leq m\}$

$E_{w_2} = \{(w_{i,j}, w_{i,j+1}) | 1 \leq i \leq m, 1 \leq j \leq B\}$.

In addition we set the bounds $M = m$ and $L = 2B + 2$, i.e. we ask if a broadcast set containing m trails whose maximal length is $2B + 2$ can be constructed under the above conditions. In Figure 2 the construction for $A = \{6, 5, 4, 4, 3, 2\}$, $m = 2$ and $B = 12$ is illustrated.

We note that the same reduction can be used to proof NP-completeness of the Bounded GLBC and the Bounded WGLBC problems, since we don't use anywhere the fact that we can go through each edge at most once, and the structure of the graph forces us to do so. But the NP-Completeness of the GLBC and WGLBC follows directly from the NP-completeness of Bounded TLBC and Bounded WTLBC.

5 Algorithm C for WTLBC

In this section we briefly sketch Algorithm C for the WTLBC problem, with time complexity $O(N^2)$. Its correctness proof and analysis, and the discussion of Algorithm B, are omittd here.

Lemma 2. *[CG89] : For every optimal broadcast set for a broadcast problem (T, w, r), and for any edge e in T, either there exists one path going up e and one going down e, or no path is going up e, and at least one is going down e.*

Although the lemma refers to the WTLB problem, it is very easy to see that it holds for the WTLBC problem. According to Lemma 1, the edges in E can be partitioned into two disjoint subsets with respect to an optimal broadcast set ψ. We say that an edge e or a vertex v is *red* w.r.t. to ψ if one path is going up through it, and *blue* w.r.t. to ψ otherwise.

With each broadcast set ψ we associate the set $blue_l(\psi)$ of its blue leaves. We say that two broadcast sets for the same tree are *equivalent* if their blue leaves sets are equal; namely, $\psi \sim \psi'$ if $blue_l(\psi) = blue_l(\psi')$. Note that two broadcast sets are equivalent if and only if their sets of blue and red edges are equal, so the above relation is indeed an equivalence. We use the following:

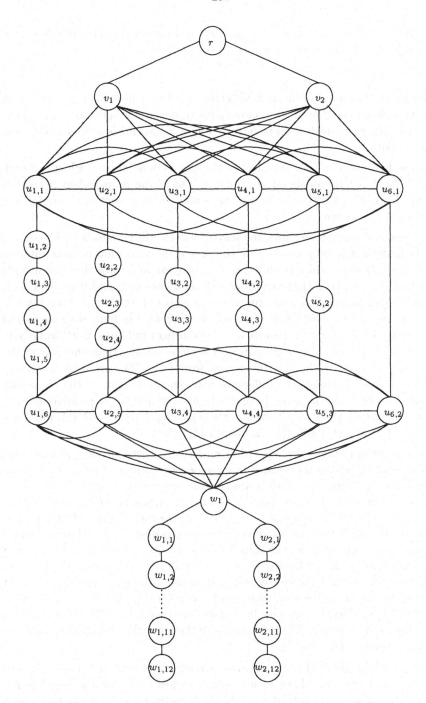

Figure 2: The construction for the NP-completeness proof of GLBC1

Lemma 3. *[BZ93] Let ψ and ψ' be two equivalent broadcast sets for a broadcast problem (G, w, r), then $w(\psi) = w(\psi')$.*

Algorithm C gets as an input a weighted tree linear broadcast problem with capacity limitations (T, c, w, r), and outputs an optimal weight broadcast set ψ for it. The algorithm is divided into three stages. Following is a brief description of the algorithm.

In the first stage a *DFS* traversal of the tree is done, during which for each node v, *depth* (v) is calculated, and also the *id* of the deepest leaf in *subtree* (v) (*max-leaf* (v)), and the *id* of v's son whose subtree contains *max-leaf* (v) (*max-son* (v)) is found.

The second stage determines the leaves that should be added to *blue-leaves* (ψ). To understand the process, note that when the algorithm is in an internal node $b \in p(r, l)$ whose color is blue there are two optional trails that can include l, the first one is a trail that arrives at b, from b goes to l, and then back to b, and the second is a trail that arrives at b, and ends in l. In the first trail l is red, and in the second l is blue. Note that in order to color l in blue, we have to add a new blue path from r to b, and then to l. Coloring l in blue is worthwhile only if the cost of the new blue section from r to b, is smaller than the cost of the section we that turned blue, i.e. the section from l to b. The profit from coloring l in blue, when b is blue is the difference between the weights of the above two sections, and the coloring is profitable only if the difference is positive; namely $w(p(r, b)) < w(p(l, b))$ or *depth* (b) - *depth* (r) < *depth* (l) - *depth* (b), but since *depth* $(r) = 0$, we get depth $(l) - 2 \cdot$ depth $(b) > 0$.

In this second stage, the algorithm recursively visits all the optional blue vertices, and calculates the profits from coloring leaves in blue. It starts from the broadcasting vertex r, which is always blue, goes to the maximal leaf in each of r's son subtrees, which of course gives the maximal profit in each subtree. Going down to a leaf adds a new blue path from the root to the leaf. After a leaf is reached, the algorithm climbs to its father (who is blue now), and proceed with the same procedure with all the sons, besides the one he is coming from. After the profits of coloring in blue the leaves in *subtree*(v) is calculated, the capacity limitations are checked. Since we said that in an optimal weight broadcast set all the trails end in a blue leaf, there can be at most c *(father* $(v), v)$ blue leaves in *subtree* (v) so that the capacity limitations are maintained. The best c *(father* $(v), v)$ leaves are selected. The second pass of the algorithm terminates when the algorithm traversed all r's sons.

In the third stage of the algorithm we actually constructs the trails. For each blue node v, it first goes down to *max-son* (v) when it reaches the blue leaf $l = $ *max-leaf* (v), it adds a new trail $p(r, l)$ to the broadcast set. From the leaf it goes up, to *father* (v). When it returns to a node v, if v has additional blue sons, that were not searched yet, it goes down to them in a similar way, otherwise, it picks a trail from the trails that go through v, and inserts a *DFS* traversal of all the remaining red subtrees to the trail.

References

[BZ93]   S. Bitan and S. Zaks.   Optimal linear broadcast.   *Journal of Algorithms*, 14:288–315, March 1993.

[CG88]   I. Cidon and I. S. Gopal. Paris: An approach to private integrated networks. *Journal of Analog and Digital Cabled Systems*, June 1988.

[CG89]   C. T. Chou and I. S. Gopal. Linear broadcast routing. *Journal of Algorithms*, 10(4):490–517, 1989.

[Eve79]   S. Even. *Graph Algorithms*. Computer Science Press, 1979.

[GJ79]   M. R. Garey and D. S. Johnson. *Computers and Intractability*. Freeman, 1979.

[GJT76]   M. R. Garey, D. S. Johnson, and R. E. Tarjan. The planar hamiltonian circuit problem is np-complete. *SIAM Journal of Computing*,

[ICK88]   I. S. Gopal I. Cidon and S. Kutten. New models and algorithms for future networks. In *Proceedings of the 7'th Annual ACM Symposium on Principles of Distributed Computing*, pages 75–89, Toronto, CANADA, August 1988.

[Seg83]   A. Segall. Distributed networks protocols. *IEEE Trans. on Information Theory*, IT-29(1):23–35, January 1983.

Multicommodity Flows: A Survey of Recent Research * (Extended Abstract)

Baruch Awerbuch[2] and Tom Leighton[1,2]

[1] Department of Mathematics and
[2] Laboratory for Computer Science
Massachusetts Institute of Technology
Cambridge, MA 02139

1 Introduction

The *multicommodity flow problem* consists of shipping several different commodities from their respective sources to their sinks through a common network so that the total flow going through each edge does not exceed its capacity. Associated with each commodity is a *demand*, which is the amount of that commodity that we wish to ship through the network. Given a multicommodity flow problem, we would like to know if there is a *feasible flow*, i.e. a way of shipping the commodities that satisfies the demands as well as the capacity constraints. More generally, we might also like to know the maximum value z such that at least z percent of each demand can be shipped without exceeding the capacity constraints. The latter problem is known as the *concurrent flow problem*, and is equivalent to the problem of determining the minimum ratio by which the capacities must be increased in order to satisfy 100% of all of the demands.

Multicommodity flow problems arise in a wide variety of contexts and have been extensively studied during the past several decades. For example, many product distribution, traffic planning, and scheduling problems can be expressed and solved as a multicommodity flow problem. In addition, it has recently been discovered [9, 7] that a wide variety of NP-hard problems (such as graph partitioning, minimum feedback arc set, minimum cut linear arrangement, minimum 2D area layout, via minimization, and optimal matrix arrangement for nested disection) can be approximately solved using multicommodity flow algorithms. Many packet routing and communication problems can also be expressed as multicommodity flow problems but since local-control, on-line algorithms for multicommodity flow were only recently discovered [1], flow techniques have not been commonly used to solve such problems.

Not surprisingly, the literature on flow problems is extensive. Much of the work centers on the much simpler problem of 1-commodity flow (also known as the *max-flow* problem). A survey of the many 1-commodity algorithms can

* Portions of this paper were taken from [1]. This work was supported in part by Air Force Contract AFOSR F49620-92-J-0125 and DARPA Contracts N00014-91-J-1698 and N00014-92-J-1799.

be found in [5]. Most of these algorithms rely on finding augmenting paths to increase the flow from source to sink. An exception is the recent algorithm of Goldberg and Tarjan [5] (which is based on an algorithm of Karzanov [6]). The latter algorithm maintains a preflow on the network and pushes local flow excess toward the sink along what is estimated to be a shortest path. The best of these algorithms run in $\tilde{O}(NM)$ steps, where N is the number of nodes in the network and M is the number of edges in the network.

There has been much less progress on the multicommodity flow problem, perhaps because handling K commodities seems to be much more difficult than handling one commodity. All exact algorithms for multicommodity flow are based on linear programming, all have horrendous running times (even though polynomial), and none are used for large networks in practice.

The situation is somewhat better for approximation algorithms, however. In particular, Vaidya [14] developed a $(1 + \varepsilon)$-approximation algorithm for the multicommodity flow problem based on linear programming that uses (roughly) $O(K^{\frac{7}{2}} N M^{\frac{5}{2}} \log(DU \varepsilon^{-1}))$ steps where K is the number of commodities, N and M are as before, D is the largest demand, and U is the largest edge capacity. More recently, Leighton et al [10] discovered a purely combinatorial $(1 + \varepsilon)$-approximation algorithm based on 1-commodity min-cost flows that runs in $O(K^2 N M \varepsilon^{-2} \log K \log^3 N)$ steps.

By using randomization, the running time of the latter algorithm can be improved by a factor of K. The resulting time bound $\tilde{O}(KNM)$ is very nice because it means that approximately solving a K-commodity flow problem is not much harder than solving K disjoint 1-commodity flow problems. In fact, the Leighton et. al. algorithm works by routing (and, later, rerouting) one commodity at a time. Each commodity is routed in a min-cost fashion where the cost of an edge is an exponential in the ratio of the flow currently passing through the edge to the capacity of the edge. In other words, the more congested an edge is, the more expensive it is to use. After rerouting $\tilde{O}(K)$ single commodity flows, it can be shown that [10] no edge capacity is exceeded by more than a $(1 + \varepsilon)$-factor.

2 A Simpler Algorithm

The preceding multicommodity flow approximation algorithms are fairly complicated to describe (and to analyze) in detail and none is well-suited for applications in contexts requiring local control and/or local decision-making. For example, none of the algorithms are amenable to implementation on a fault-prone distributed network. In fact, of the many 1-commodity flow algorithms known, only the Goldberg-Tarjan algorithm is implementable in a distributed network, and even there, the algorithm needs to maintain shortest path information and it does not tolerate faults or dynamic changes in the network structure.

Recently, Awerbuch and Leighton [1] have discovered a very simple approximation algorithm for the multicommodity flow problem. The algorithm is based on a simple "edge balancing" technique. In particular, the algorithm attempts to send a commodity across an edge $e = (u, v)$ if there is more of the commodity queued at u than there is queued at v. Contention for capacity is resolved by shipping the commodity which has the largest disparity in queue size across e. No attempt is made to find augmenting paths, shortest paths, min-cost paths, or even any path from a node to a sink. Commodities are simply entered at their respective sources, according to their demands, emptied from their sinks when present, and otherwise locally balanced across each edge.

To simplify presentation, the algorithm can be described in terms of "parallel rounds". At the start of each round, d_i units of commodity i are added to the source for commodity i for $1 \leq i \leq K$. Then flow is pushed across each edge in an attempt to balance each commodity across each edge (up to the limits imposed by capacity). Then any commodity present at the appropriate sink is removed from the network.

The key to the success of the algorithm is that the amount of flow contained in all the queues stays bounded over time (provided that there exists a feasible flow with demands of $(1 + \varepsilon)d_i$ for each commodity i). Thus, when the algorithm is run for a large number of rounds, the flow that remains in the system will become very small compared to the flow that was pumped into the system and we can compute an approximate solution to the original flow problem by seeing where the flow went during the course of the algorithm. In distributed network applications, we never need to compute the approximate solution explicitly – the commodities just flow through the network to their destinations on their own (without having any idea where the destination is in the network) and we are guaranteed to get near-optimal throughput.

The algorithm is similar in spirit to a physical network containing fluids. For each fluid, there is a source and a drain. As fluid enters at the source, pressure builds and the fluid spreads through the network. Eventually, each fluid reaches its sink and a fluid path is set up between source and sink. The key to the analysis is to show that no fluid can be blocked by the others for any serious period of time, and to show that the cumulative heights of all the fluids stays bounded over time. It is intuitively clear that these properties hold for one-commodity flow problems, but much less clear that they should hold for multicommodity flow problems. (In fact, there is a simple 4-commodity problem for which the maximum concurrent flow is 25%, but for which no flow ever reaches the correct sink. Hence, the threat of deadlock or blocking must be taken seriously.)

The algorithm is also similar in spirit to an algorithm proposed by Dennis [2] in 1964. In particular, Dennis showed how to model a multicommodity flow problem as a collection of tightly coupled electrical networks (one for each commodity). He then argued that the optimal flow could be found by measuring the current flows in the various networks. An important part of this work is the claim that the electrical system converges to a steady state (although no time bounds for convergence are established). Although the construction in Dennis' paper results in a highly non-linear system of equations, Dennis also observed that the linear approximation to this system of equations suggests an algorithm for multicommodity flow in which some sort of "edge-balancing" technique is used to find the optimal flow. The running time of this algorithm was not made clear, however.

Overall, the Awerbuch-Leighton algorithm can be shown to run in $\tilde{O}(LM)$ rounds on a distributed network or in $\tilde{O}(KLM^2)$ steps on a sequential machine, and it finds a feasible flow (provided that there exists a feasible flow when the demands are increased by a factor of $1 + \varepsilon$), where L is the length of the longest flow path.[3] The algorithm can also be used to find a $(1 - \varepsilon)$-optimal solution to the concurrent flow problem by using binary search.

It is certainly worth noting that the upper bound on running time just stated is asymptotically *inferior* to the bound for the multicommodity flow approximation algorithm of [10]. Nevertheless, the algorithm is still of interest since:

1. We suspect that the true running time of the algorithm (or a close variation) is much better—we just haven't proved it yet. In fact, we believe that the true running time may eventually be shown to be competitive with the algorithms developed in [10], at least for certain classes of flow problems.

2. The algorithm is very simple and intuitive. It is even simpler than the known algorithms for 1-commodity flow. Understanding this algorithm would seem to be an important step towards a better understanding of flow problems in general.

3. The algorithm appears to work well empirically. In particular, Mark Tsimelzon [13] has run simulation tests for DIMACS test problems and has found that several variations of the algorithm are comparable or superior in running time to the best known algorithms. Hence, the new algorithm may eventually prove to be of practical value.

4. Unlike all the other algorithms for multicommodity flow, the Awerbuch-Leighton algorithm can be used for routing in environments where global control is not possible and where routing decisions need to be made locally. Hence, multicommodity flow techniques can now be brought to bear on problems like message routing in networks.

5. The algorithm works in dynamically changing environments (i.e. in environments where demands and edge capacities can vary from step to step. As long as there exists a feasible flow over time, the algorithm is guaranteed to find it.)

3 Max-flow Min-cut Theorems

It has long been known that the maximum flow through a network is limited by the capacity of the minimum cut separating the sources from the sinks. In the case of 1-commodity flow problems, the classic max-flow min-cut theorem [3] states that the maximum flow equals the capacity of the minimum cut separating the source from the sink. It has also long been known that no such result holds for multicommodity flow problems with more than one commodity.

---

[3] The time bound proved in [1] is somewhat larger. The improved bound stated here can be obtained using an exponential potential function instead of the quadratic function used in [1].

In recent years, there has been a substantial amount of progress in the development of *approximate* max-flow min-cut theorems for multicommodity flow problems. For example, Leighton and Rao [9] proved that for uniform multicommodity flow problems, the ratio of the min-cut to the max-flow is at most $O(\log N)$. (In a *uniform* flow problem, the demand between every pair of nodes is the same. Uniform flow problems arise in a wide variety of applications [9].) In other words, there is always a concurrent flow of size $\Theta(1/\log N)$ of the limit imposed by the worst cut. Leighton and Rao also proved that the $\Theta(\log N)$ bound is tight since there are simple examples for which the min-cut max-flow ratio is $\Theta(\log N)$.

Klein, et. al., [7] generalized the previous result by showing that the min-cut max-flow ratio for any undirected flow problem is at most $O(\log C \log D)$ where C is the sum of the capacities and D is the sum of the demands. This bound was later improved to $O(\log N \log D)$ by Tragoudas [12], then to $O(\log K \log D)$ by Garg, Vazirani, and Yannakakis [4], and then to $O(\log^2 K)$ by Plotkin and Tardos [11]. Whether or not the true min-cost max-flow ratio is $\Theta(\log K)$, $\Theta(\log^2 K)$, or somwhere in between remains a very interesting and challenging open question. As a special case, Klein, Plotkin, and Rao [8] have shown that the ratio (for both uniform and undirected multicommodity flow problems) can be improved by a $\log K$ factor for graphs with a forbidden minor of constant size (e.g., for planar graphs).

References

1. Baruch Awerbuch and Tom Leighton. A Simple local-control approximation algorithm for multicommodity flow. In *Proc. 34th IEEE Symp. on Foundations of Computer Science*, November, 1993. To appear.
2. Jack B. Dennis. Distributed solution of network programming problems. In *IEEE Transactions of the Professional Technical Group on Communications Systems*, volume CS-12, number 2, pages 176–184, June 1964.
3. L.R. Ford and D.R. Fulkerson. Flows in Networks. Princeton University Press, Princeton, NJ, 1962.
4. N. Garg, V. Vazirani, and M. Yannakakis. Approximate max-flow min-(multi)cut theorems and their applications. In *Proc. 25th ACM Symp. on Theory of Computing*, pages 698–707, May 1993.
5. A.V. Goldberg and R. E. Tarjan. Solving minimum-cost flow problems by successive approximation. *Mathematics of Operations Research*, 15(3):430–466, 1990.
6. A. V. Karzanov. Determining the maximal flow in a network by the method of preflows. *Soviet Math. Dokl.*, 15:434–437, 1974.
7. P. Klein, A. Agrawal, R. Ravi, and S. Rao. Approximation through multicommodity flow. In *Proc. 31st IEEE Symp. on Foundations of Computer Science*, pages 726–727, 1990.
8. P. Klein, S. Plotkin, and S. Rao. Excluded minors, network decomposition, and multicommodity flow. In *Proc. 25th ACM Symp. on Theory of Computing*, pages 682–690, May 1993.
9. F. T. Leighton and Satish Rao. An approximate max-flow min-cut theorem for uniform multicommodity flow problems with applications to approximation algo-

rithms. In *29th Annual Symposium on Foundations of Computer Science, IEEE*, pages 422–431, 1988.

10. T. Leighton, F. Makedon, S. Plotkin, C. Stein, E. Tardos, and S. Tragoudas. Fast approximation algorithms for multicommodity flow problem. In *Proc. 23rd ACM Symp. on Theory of Computing*, pages 101–111, May 1991.

11. S. Plotkin and E. Tardos Improved bounds on the max-flow min-cut ratio for multicommodity flow. In *Proc. 25th ACM Symp. on Theory of Computing*, pages 691–697, May 1993.

12. S. Tragoudas. VLSI partitioning approximation algorithms based on multicommodity flows and other techniques. PhD thesis, University of Texas at Dallas, 1991.

13. Mark Tsimelzon. Bachelor thesis, MIT Lab. for Computer Science, 1993.

14. P.M. Vaidya. Speeding up linear programming using fast matrix multiplication. In *Proc. 30th IEEE Symp. on Foundations of Computer Science*, pages 332–337, 1989.

Parallel Construction of Canonical Ordering and Convex Drawing of Triconnected Planar Graphs

Xin He[1] Ming-Yang Kao[2]

1 Introduction

The problem of "nicely" drawing planar graphs on the plane has received increasing attention due to theoretical interests and practical applications (See [3] for a survey). A new concept, *canonical ordering* for planar triangular graphs, was recently developed for solving this class of problems [4]. By using canonical ordering, it was shown that every planar graph has a straight-line embedding on a $(2n - 4) \times (n - 2)$ grid which can be found in $O(n \log n)$ time [4]. The algorithm was improved to run in linear time [2]. This concept is generalized by Kant to general triconnected planar graphs [8]. By using this tool, Kant developed linear time algorithms for solving several interesting problems. In particular, Kant showed that every triconnected planar graph can be drawn with straight lines on a $(2n-4) \times (n-2)$ grid such that all faces are convex. Although this problem has been extensively studied in the past [11, 15], Kant's result is the first one showing such an embedding can be drawn on a small grid.

The canonical ordering is defined by induction and seems hard to construct in parallel. The *realizer* of planar triangular graphs is another newly developed concept [12]. By using it, Schnyder developed a linear time algorithm for embedding planar graphs on an $(n - 2) \times (n - 2)$ grid with straight lines [13]. A parallel algorithm for constructing realizer and such an embedding were given in [5], which takes $O(\log n \log \log n)$ time with $O(n/ \log n \log \log n)$ processors.

In this paper, we show that the canonical ordering and the realizer are closely related. This connection makes parallel construction of canonical ordering possible. In order to obtain such an algorithm, we define a *bidirectional realizer* of an extended graph G of G, and show a canonical ordering of G can be obtained from such a realizer. To find a bidirectional realizer of G, we show there exists an edge set E' that can be *contracted* from G while maintaining the triconnectivity. The algorithm is recursively applied to the resulting smaller graph. Then a bidirectional realizer of G is obtained by *expanding E'*. We also present a parallel convex drawing algorithm. Our main results are:

Theorem 1.1 *A canonical ordering of a triconnected planar graph G can be constructed in $O(\log^4 n)$ time with $O(n^2)$ processors on a CREW PRAM.*

Theorem 1.2 *Given a canonical ordering of a triconnected planar graph G, a convex drawing of G on a $(2n - 4) \times (n - 2)$ grid can be found in $O(\log n)$ time with $O(n)$ processors on a CREW PRAM.*

The present paper is organized as follows. In §2 we introduce the definitions of canonical ordering and realizer and discuss their connections. In §3 and §4, we present the parallel canonical ordering algorithm. §5 presents our convex drawing algorithm. Due to space limitation, most proofs are omitted.

[1]Department of Computer Science, State University of New York at Buffalo, Buffalo, NY 14260. E-mail xinhe@cs.buffalo.edu. Research supported in part by NSF grant CCR-9205982.
[2]Department of Computer Science, Duke University, Durham, NC 27706. Research supported in part by NSF Grant CCR-9101385.

2 Canonical Ordering and Realizer

Let $G = (V, E)$ be a planar graph with n vertices and m edges. For each $v \in V$, $N(v)$ denotes the set of v's neighbors. We assume G is equipped with a fixed embedding which can be constructed in $O(\log n)$ time with $O(n)$ processors [10]. If all faces of G are triangles, it is a *planar triangular graph*. For an interior vertex v of G, $N(v) = \{u_1, \ldots, u_k\}$ form a cycle denoted by $cycle(v)$. Define $star(v) = \{(v, u_i) \mid 1 \leq i \leq k\}$. The following concept is defined in [12, 13].

Definition 2.1 *Let G be a planar triangular graph with three exterior vertices v_1, v_2, v_n in counterclockwise order. A realizer of G is a partition of the interior edges of G into three sets T_1, T_2, T_n together with an orientation of the interior edges of G such that the following hold:*

1. For $i \in \{1, 2, n\}$, all interior edges incident to v_i are in T_i and directed to v_i.

2. The edges incident to each interior vertex u appear counterclockwise as:
- *one edge in T_1 leaves u; a set (maybe empty) of edges in T_n enters u;*
- *one edge in T_2 leaves u; a set (maybe empty) of edges in T_1 enters u;*
- *one edge in T_n leaves u; a set (maybe empty) of edges in T_2 enters u.*

For $i \in \{1, 2, n\}$, T_i is a tree rooted at v_i containing v_i and all interior vertices of G [12, 13]. For each interior vertex v, $p_i(v)$ denotes the parent of v in T_i.

Definition 2.2 *[4] Let $G = (V, E)$ be a planar triangular graph with three exterior vertices u, v, w. A canonical ordering of G is a numbering of V by $v_1 = u, v_2 = v,, v_3, \ldots, v_n = w$ such that the following hold for every $3 < k \leq n$:*

1. The subgraph G_{k-1} of G induced by $v_1, v_2, \ldots, v_{k-1}$ is biconnected and the boundary of its exterior face is a cycle C_{k-1} containing the edge (v_1, v_2).

2. v_k is in the exterior face of G_{k-1} and its neighbors in G_{k-1} form an (at least 2-elements) interval of the path $C_{k-1} - \{(u, v)\}$. If $k < n$, v_k has at least one neighbor in $G - G_k$.

Every planar triangular graph G has a realizer [12] and a canonical ordering [4]. Fig 1 (1) shows a realizer and a canonical ordering of G. The following lemmas establish the connection between these two concepts.

Lemma 2.1 *Let $\{v_1, \ldots, v_n\}$ be a canonical ordering of a planar triangular graph G. Assign direction to interior edges of G and partition them into three sets $\{T_1, T_2, T_n\}$ as follows. For $3 \leq k < n$, let $c_l, c_{l+1}, \ldots, c_r$ be the neighbors of v_k on the exterior face C_{k-1} of G_{k-1} in counterclockwise order. Put $v_k \to c_l$ in T_1; $v_k \to c_r$ in T_2; and $c_i \to v_k$ in T_n for each $l < i < r$. All interior edges incident to v_n are in T_n and directed to v_n. Then $\{T_1, T_2, T_n\}$ is a realizer of G.*

Lemma 2.2 *Let $\{T_1, T_2, T_n\}$ be a realizer of a planar triangular graph G. Direct the edges of G as follows. For each interior edge (u, v), direct u to v if and only if either $u \to v \in T_n$, or $u - v \in T_1$, or $u \leftarrow v \in T_2$. Direct the three exterior edges as: $v_1 \to v_2$, $v_2 \to v_n$, $v_1 \to v_n$. Then: (1) The resulting graph $\bar{G}$ is an acyclic graph. (2) Any topological ordering R of $\bar{G}$ is a canonical ordering of G.*

[8] extends the canonical ordering to triconnected planar graphs as follows. Without loss of generality, we assume the exterior face of G is a triangle.

Definition 2.3 *Let $G = (V, E)$ be a triconnected planar graph with three exterior vertices u, v, w. A canonical ordering of G is an ordering O of V by $\{v_1 = u, v_2 = v, v_3 \ldots, v_n = w\}$ such that the following hold for for every $k \geq 3$:*

1. Either v_k is on the exterior face of G_k and has at least two neighbors on the exterior face C_{k-1} of G_{k-1}. G_k is biconnected. If $k < n$, then v_k has at least one neighbor in $G - G_k$.

2 Or there exists an $l \geq 1$ such that $v_k, \ldots, v_{k+l}$ is a chain on the exterior face of G_{k+l} and has exactly two neighbors in G_{k-1}, which are on the exterior face C_{k-1} of G_{k-1}. G_{k+l} is biconnected. Every vertex $v_k, \ldots, v_{k+l}$ has at least one neighbor in $G - G_{k+l}$.

Every triconnected planar graph has a canonical ordering [8]. Fig 1 (2) shows an example. A canonical ordering O can be interpreted as adding vertices of G in stages. In stage k, we add either a vertex v_k, or a face F implied by the added vertices $v_k, \ldots, v_{k+l}$. For each vertex x added at stage k, we write $O(x) = k$ and the O-value of x is k. In applications, it is not necessary to use integer stage numbers. In the following, we also use fractional stage numbers.

An *independent set* of a graph $G = (V, E)$ is a subset $I \subset V$ such that no two vertices in I are adjacent. I is *maximal* if I is not contained in larger independent sets of G. A *k-coloring* of G is a partition of V into k independent sets.

Lemma 2.3 *[6] Given a planar graph G, we can find in $O(\log^2 n)$ time and $O(n)$ processors: (a) An independent set I of G with $|I| \geq n/5$; (b) A 5-coloring of G.*

3 Bidirectional realizer and canonical ordering

Let $\{T_1, T_2, T_n\}$ be a realizer of a planar triangular graph G and v an interior vertex of G with $u_1 = p_1(v)$, $u_2 = p_2(v)$, and $u_n = p_n(v)$. Let $cycle_{[1,2]}(v)$ denote the path on $cycle(v)$ between u_1 and u_2 in counterclockwise order. $Vcycle_{[1,2]}(v)$ denotes the vertices of $cycle_{[1,2]}(v)$ including u_1 and u_2. Similarly $Vcycle_{(1,2)}(v) = Vcycle_{[1,2]}(v) - \{u_2\}$. The notations $cycle_{[2,n]}(v)$, $Vcycle_{[2,n]}(v)$ $cycle_{[n,1]}(v)$, $Vcycle_{[n,1]}(v)$ are analogous. From the definition of realizer, it is easy to prove:

Lemma 3.1 *Let G be a planar triangular graph with a realizer $\{T_1, T_2, T_n\}$. Let v be an interior vertex of G. For any interior edge $e \in cycle(v)$: (1) If $e \in cycle_{[1,2]}(v)$, then either $e \in T_2$ directed counterclockwise; or $e \in T_1$ directed clockwise. (2) If $e \in cycle_{[2,n]}(v)$, then either $e \in T_n$ directed counterclockwise; or $e \in T_2$ directed clockwise. (3) If $e \in cycle_{[n,1]}(v)$, then either $e \in T_1$ directed counterclockwise; or $e \in T_n$ directed clockwise.*

Definition 3.1 *The extended graph $\hat{G}$ of a triconnected planar graph G is obtained as follows: For each interior face F of G, add a face vertex v_F in F and connect v_F to all vertices of G on the boundary of F. The added edges are called dummy edges. The edges of G are called real edges.*

We next define *bidirectional realizer* and *special realizer* of $\hat{G}$ and discuss how to use them to obtain a canonical ordering of G.

Definition 3.2 *A realizer $\hat{G}$ is called a bidirectional realizer if, for each face vertex v_F of $\hat{G}$ with $u_1 = p_1(v_F)$, $u_2 = p_2(v_F)$, $u_n = p_n(v_F)$, the following hold:*

1. $cycle_{[1,2]}(v_F)$ is partitioned into two parts: a path in T_2 starting at u_1 directed counterclockwise; and a path in T_1 starting at u_2 directed clockwise.

2. $cycle_{[2,n]}(v_F)$ is partitioned into two parts: a path in T_n starting at u_2 directed counterclockwise; and a path in T_2 starting at u_n directed clockwise.

3. $cycle_{[n,1]}(v_F)$ is partitioned into two parts: a path in T_1 starting at u_n directed counterclockwise; and a path in T_n starting at u_1 directed clockwise.

Any path in above definition maybe empty. Fig 2 shows an example.

Definition 3.3 *A bidirectional realizer $\{T_1, T_2, T_n\}$ of $\hat{G}$ is a special realizer if for each face vertex v_F the following hold: 1. Either $cycle_{[2,n]}(v_F) \subset T_2$; or $cycle_{[2,n]}(v_F)$ consists of exactly one edge which is in T_n. 2. Either $cycle_{[n,1]}(v_F) \subset T_1$; or $cycle_{[n,1]}(v_F)$ consists of exactly one edge which is in T_n.*

Given a bidirectional realizer $\{T_1, T_2, T_n\}$ of $\hat{G}$, if a face vertex v_F does not satisfy Definition 3.3, we say v_F is *bad*. We can convert $\{T_1, T_2, T_n\}$ to a special realizer of $\hat{G}$ by redirecting the edges in $cycle(v_F) \cup star(v_F)$ for each bad v_F.

Lemma 3.2 *A bidirectional realizer of $\hat{G}$ can be converted to a special realizer of $\hat{G}$ in $(\log^2 n)$ time with $O(n)$ processors on a PRAM.*

Proof: Let G^* be the dual graph of G. Let $\{I_1^*, \ldots, I_5^*\}$ be a 5-coloring of G^*. Since the face vertices in I_j^* do not share edges in G, we can perform the redirection for bad face vertices in I_j^* in parallel. All bad face vertices are removed after 5 stages. G^* and its 5-coloring can be constructed in $O(\log^2 n)$ time with $O(n)$ processors by Lemma 2.3 (b). The redirection of the edges around v_F is done locally. It takes $O(1)$ time with $O(n)$ processors. $\square$

Let $\hat{O}$ be the canonical ordering of $\hat{G}$ induced by a special realizer $\{T_1, T_2, T_n\}$ of $\hat{G}$. We can obtain a canonical ordering O of G from $\hat{O}$ as follows. For each face F of G, the face vertex v_F of $\hat{G}$ is responsible for assigning O-value to the vertices in $Vcycle_{[n,1)}(v_F) \cup Vcycle_{(2,n]}(v_F)$.

Case 1: All edges in $cycle_{[n,1]}(v_F)$ are in T_1 and all edges in $cycle_{[2,n]}(v_F)$ are in T_2. For all $x \in P = Vcycle_{[n,1)}(v_F) \cup Vcycle_{(2,n]}(v_F)$, set $O(x) = \hat{O}(v_F)$.

Case 2: Either $cycle_{[n,1]}(v_F)$ or $cycle_{[2,n]}(v_F)$ consists of exactly one edge which is in T_n. Let $x = p_n(v_F)$. Let $F_1, \ldots, F_t$ be all face vertices of $\hat{G}$ with $x = p_n(v_{F_i})$ in counterclockwise order. Then $cycle_{[n,1]}(v_{F_i}) \subset T_1$ and $cycle_{[2,n]}(v_{F_i}) \subset T_2$. (Fig 3). Let $P_1 = Vcycle_{(n,1)}(v_{F_i})$ and $P_2 = Vcycle_{(2,n)}(v_{F_i})$. Let $k = \min\{\hat{O}(v_{F_i}) \mid 1 \le i \le t\}$. Set $O(x) = k$.
Case 2a: $t > 2$. Set $O(y) = k + 1/3$ for $y \in P_1$ and $O(z) = k + 2/3$ for $z \in P_2$.
Case 2b: $k = 2$. Set $O(y) = k$ for $y \in P_1$ and $O(z) = k + 1/3$ for $z \in P_2$.

Theorem 3.3 *The ordering O defined above is a canonical ordering of G.*

Theorem 3.4 *Given a bidirectional realizer of the extended graph $\hat{G}$ of a triconnected planar graph G, a canonical ordering of G can be found in $O(\log^2 n)$ time with $O(n)$ processors on a PRAM.*

4   Constructing a bidirectional realizer

4.1   Edge contraction

We present an algorithm for finding a bidirectional realizer of the extended graph $\hat{G}$ of G. The basic technique is *edge contraction* and *edge expansion*. Let $e = (u, v)$ be an interior edge of G. Let F_1 and F_2 be the two faces of G with e on their boundary. The operation of *contracting* e merges u and v into a new *contracted vertex* o_e. If F_1 (or F_2) is a triangle, the two edges of F_1 (or F_2) are merged into one edge in G'. G/e denotes the resulting *contracted graph*. If G/e is triconnected, we say e is *contractible* and v is a *contractible neighbor* of u.

We discuss the effects of contracting e on $\hat{G}$. Let $e_1 = (v_{F_1}, u)$, $e_1' = (v_{F_1}, v)$, $e_2 = (v_{F_2}, u)$, $e_2' = (v_{F_2}, v)$, where the face vertices v_{F_1} and v_{F_2} correspond to F_1 and F_2. After contracting e, the edges e_1 and e_1' are replaced by a new edge (v_{F_1}, o_e). e_2 and e_2' are replaced by a new edge (v_{F_2}, o_e) (Fig 4). The edges (v_{F_1}, o_e), (v_{F_2}, o_e) are the *residue edges* of e, denoted by $Res(o_e)$. The four edges in $\hat{G}$ corresponding to the two residue edges are the *surrounding edges* of e.

Lemma 4.1 *Every vertex u has at least two contractible neighbors.*

4.2 Edge expansion

Let G be a triconnected planar graph and $e = (u, v)$ be a contractible edge of G. Let $G' = G/e$. Let G and G' be the extended graph of G and G', respectively. Let $\{T_1', T_2', T_n'\}$ be a bidirectional realizer of $\hat{G}'$. We can *expand* e to obtain a bidirectional realizer $\{T_1, T_2, T_n\}$ of $\hat{G}$ as follows.

Let $Res(o_e) = \{(y, o_e), (z, o_e)\}$ be the two residue edges of e in $\hat{G}'$. Let $e_1 = (y, u), e_1' = (y, v), e_2 = (z, u), e_2' = (z, v)$ be the four surrounding edges of e in $\hat{G}$. For any edge e' of G such that $e' \notin \{e, e_1, e_1', e_2, e_2'\}$, the label of e' with respect to $\{T_1, T_2, T_n\}$ is the same as its label with respect to $\{T_1', T_2', T_n'\}$. We need to specify labels of e, e_1, e_1', e_2, e_2' with respect to $\{T_1, T_2, T_n\}$. The four edges in $star(o_e)$ in G' that are adjacent to (o_e, y) (o_e, z) are called *essential edges* of o_e. They are denoted by $Ess(o_e) = \{g_1, g_2, g_3, g_4\}$.

Fig 5 shows several cases. If a residue edge is a real edge, a triangle face F of G is created by the expansion. We add a new face vertex v_F and three dummy edges and assign labels to these dummy edges (not shown in Fig 5). Note that we get a bidirectional realizer G after the expansion in all cases.

4.3 Parallel construction of bidirectional realizers

The realizer of G is constructed as follows: Find a large set E' of contractible edges of G and construct the contracted graph $G' = G/E'$; recursively find a bidirectional realizer of the expanded graph $\hat{G}'$ of G'; then expand E' to get a bidirectional realizer of G. In order to do so, we require (i) G/E' is triconnected; and (ii) all edges of E' can be expanded simultaneously.

Definition 4.1 *A set E' of contractible edges of G is compatible if:*
1. G/E' is triconnected.
2. For any two $e_1, e_2 \in E'$, $Res(o_{e_1}) \cap (Res(o_{e_2}) \cup Ess(o_{e_2})) = \emptyset$ in $\hat{G}/E'$.

We use an independent set I of G, as in Lemma 2.3 (a), to guide the construction of E'. By Lemma 4.1, each $v \in I$ has at least two contractible neighbors in G. Define: $E(I) = \{(v, u) | v \in I$ and u is a contractible neighbor of $v\}$.

Pick a contactable neighbor u_1 of the first vertex $v_1 \in I$ and contract $e_1 = (v_1, u_1) \in E(I)$ from G. We still use u_1 to denote the contracted vertex o_{e_1}. Since $G_1 = G/e_1$ is triconnected, the second vertex $v_2 \in I$ has at least two contractible neighbors in G_1. Since the edge contraction cannot create new contractible neighbors, any contractible neighbor of v_2 in G_1 is also a contractible neighbor of v_2 in G. (However, the contraction of e_1 may destroy the contractability of some neighbors of v_2). Pick a contractible neighbor u_2 of v_2 and contract $e_2 = (v_2, u_2) \in E(I)$ from G_1. Continuing this process, we can select a contractible neighbor u_i for each $v_i \in I$. Let E_1 be the set of the selected edges. Then $E_1 \subset E(I)$ and $G' = G/E_1$ is triconnected. We need to parallelize this process.

Definition 4.2 *Define the conflict graph $CG = (V_{CG}, E_{CG})$ of G as follows: $V_{CG} = E(I)$. Two nodes (v_1, u_1) and (v_2, u_2) of CG are adjacent in CG if and only if either (i) $v_1 = v_2$; or (ii) the graph $G/\{e_1, e_2\}$ is not triconnected.*

Lemma 4.2 *Let I_{CG} be a maximal independent set of CG. Then (a) $|I_{CG}| = |I|$; (b) the graph G/I_{CG}, (I_{CG} viewed as an edge set of G), is triconnected.*

The set I_{CG} satisfies the first but not the second condition of Definition 4.1. By removing some edges that violate the second condition, we can proof:

Lemma 4.3 *There exists a compatible set* $E' \subset I_{CG}$ *of contractible edges such that* $|E'| \geq n/25$.

We are now ready to describe:
Algorithm 1: Bidirectional Realizer
0. If G has no interior vertices, return the trivial bidirectional realizer of $\hat{G}$.
1. Find an independent set I of G as in Lemma 2.3 (a).
2. Construct the conflict graph CG.
3. Find a maximal independent set I_{CG} of CG.
4. Construct the set E' of contractible edges of G as in Lemma 4.3.
5. Construct the contracted graph $\hat{G}' = \hat{G}/E'$.
6. Recurvely find a bidirectional realizer $\{T_1', T_2', T_n'\}$ of $\hat{G}'$.
7. Expand E' to get a bidirectional realizer of $\hat{G}$.

The complexity of Algorithm 1 is dominated by step 3. Since CG contains $O(n)$ nodes and $O(n^2)$ edges, I_{CG} can be found in $O(\log^3 n)$ time with $O(n^2)$ processors [7]. By Lemma 4.3, the recursion depth is $O(\log n)$. This gives the following theorem which, combined with Theorem 3.4, proves Theorem 1.1.

Theorem 4.4 *A bidirectional realizer of the extended graph of a triconnected planar graph can be found in* $O(\log^4 n)$ *time with* $O(n^2)$ *processors on a PRAM.*

5 Convex drawing algorithm

We first present a sequential convex drawing algorithm. Let G be a triconnected planar graph with a canonical ordering $O = \{v_1, \ldots, v_n\}$. Let G_k denote the subgraph induced by $v_1, \ldots, v_k$ and C_k the exterior face of G_k. The x- and y-coordinates of a point p are denoted by $x(p)$ and $y(p)$. Given two points p_1 and p_2, let $\mu(p_1, p_2)$ denote the crossing point of the line of slope $+1$ from p_1 and the line of slope -1 from p_2. The algorithm proceeds in K stages, where K is the maximum O-value of the vertices. At stage k, the vertices with O-value k are added. The drawing of G_k satisfies the following:
 (a) v_1 is drawn at $(0, 0)$, v_2 is drawn at $(2k - 4, 0)$.
 (b) If $C_k = \{v_1 = c_1, c_2, \ldots, c_q = v_2\}$ (clockwise), then $x(c_1) < \ldots < x(c_q)$.
 (c) For $1 \leq i < q$, the edge (c_i, c_{i+1}) has slope either $+1$, or -1, or 0.
Case 1: One vertex v_k is added at stage k. Let $c_{i_1}, \ldots, c_{i_s}$ be the vertices on C_{k-1} adjacent to v_k. Let F_l $(1 \leq l < s)$ be the face formed by the edges (v_k, c_{i_l}), $(v_k, c_{i_{l+1}})$ and the path B_l of C_{k-1} between c_{i_l} and $c_{i_{l+1}}$. It is shown in [8] that each B_l has the following pattern:
 • From c_{i_l} to some vertex c_{α_l}, a sequence D_l, $|D_l| \geq 1$, of vertices with decreasing y-coordinate.
 • Two vertices $c_{\alpha_l}, c_{\beta_l}$ with same y-coordinate (possibly $c_{\alpha_l} = c_{\beta_l}$).
 • From c_{β_l} to $c_{i_{l+1}}$, a sequence U_l of vertices with increasing y-coordinate.
We shift c_j $(\alpha_1 \leq j \leq \beta_{s-1})$ to right by 1 and c_j $(\beta_{s-1} + 1 \leq j \leq q)$ to right by 2. (When c_j is shifted, some other vertices "below" c_j are also shifted by the same amount). v_k is drawn at the point $\mu(c_{i_1}, c_{i_s})$. It is shown in [8] that all faces F_l $(1 \leq l < s)$ are convex. When adding a vertex at a later stage, v_k (hence c_{i_s}) maybe shifted to right while c_{i_1} is not. In this case, we shift c_j $(\alpha_1 \leq j < i_s)$ to right by the same amount. Then the convexity of the faces F_l $(1 \leq l < s)$ will be maintained. Similarly, at a later stage, c_{i_s} maybe shifted to right while v_k is not. To maintain the convexity of F_{s-1}, we shift c_j $(\beta_{s-1} + 1 \leq j < i_s)$ to right by the same amount.

Case 2: A face F_1, with vertices $v_k^1, v_k^2, \ldots, v_k^t$, is added at stage k. Let c_{i_1} and c_{i_2} be the two vertices of C_{k-1} adjacent to v_k^1 and v_k^t. Let B_1 be the path of C_{k-1} between c_{i_1} and c_{i_2}. B_1 also has the above pattern. We shift c_{α_1} and c_{β_1} to right by 1 and c_j ($\beta_1 + 1 \leq j \leq q$) to right by $2t$. v_k^1 is drawn on the line of slope $+1$ from c_{i_1} with y-coordinate $y(c_{i_2}) + \delta y$ (δy to be determined). $c_k^1, c_k^2, \ldots, c_k^t$ are drawn on the same horizontal line with distance 2 apart. The edge (v_k^t, c_{i_2}) has slope -1. When adding vertices at later stages, the convexity of F_1 can be maintained as in Case 1.

[8] showed that the convexity of F_l ($1 \leq l \leq s - 1$, $s = 2$ in Case 2) can be maintained by adding dummy edges into G as follows: In F_1, add dummy edges (c_{i_1}, c_j) for $i_1 + 1 < j \leq \alpha_1$. In F_{s-1}, add dummy edges (c_{i_s}, c_j) for $\beta_{s-1} \leq j < i_s - 1$. For simplicity, the resulting *modified graph* is still denoted by G. The algorithm is performed on this modified graph. After the drawing is done, the dummy edges are removed. The faces of the original graph remain convex.

In order to make parallel implementation possible, we use a new $(\bar{x}, y)$ coordinate system. For a point $p = (x(p), y(p))$, define $\bar{x}(p) = x(p) - y(p)$. If two points p_1 and p_2 are on the same line of slope $+1$, then $\bar{x}(p_1) = \bar{x}(p_2)$. Given two points $p_1 = (\bar{x}_1, y_1)$ and $p_2 = (\bar{x}_2, y_2)$, the $\bar{x}$- and y-coordinate of the crossing point $q = \mu(p_1, p_2)$ is given by: $(\bar{x}(q), y(q)) = (\bar{x}_1, y_2 + \frac{1}{2}[\bar{x}_2 - \bar{x}_1])$.

At stage k, the vertices $v_k^1, \ldots, v_k^t$ are added. (Case 1 corresponds to $t = 1$. Case 2 corresponds to $t > 1$). Let $c_{i_1}, \ldots, c_{i_s}$ be the vertices of C_{k-1} adjacent to v_k^1 and v_k^t. The vertices c_j ($i_1 < j < i_s$) are called *internal vertices*, c_{i_1} and c_{i_s} are the *leftvertex* and the *rightvertex* of stage k, resp. If $i_1 + 1 < i_s$, c_{i_1+1} is the *leader* of stage k. Observe that when we draw v_k^j, it is not necessary to know the exact positions of c_{i_1} and c_{i_s}. Consider Case 1 where one vertex v_k^1 is added. If we know the $\bar{x}$-offset of c_{i_s} relative to c_{i_1}, namely $\delta \bar{x} = \bar{x}(c_{i_s}) - \bar{x}(c_{i_1})$, then the position of v_k^1 relative to c_{i_1} and c_{i_s} is given by:

$$(\bar{x}(v_k^1), y(v_k^1)) = \mu((0, y_1), (\delta \bar{x}, y_2)) = (0, y_2 + \frac{\delta \bar{x}}{2})$$

In Case 2, the $\bar{x}$- and y-offsets of v_k^j ($1 \leq j \leq t$) can also be easily determined.

A tree T rooted at v_1 is constructed during the algorithm. At any stage k, the path from v_1 to v_2 of C_k is the rightmost branch of T. For each v let:

- $Left(v) =$ the left T-son of v; $Right(v) =$ the right T-son of v;
- $\delta \bar{x}(v) =$ the $\bar{x}$-offset of v relative to its parent in current T;
- $\delta y(v) =$ the y-offset of v relative to the rightvertex c_{i_s} of stage $k = O(v)$.

The following is our sequential algorithm. It has four phases.

Phase I: Construct T, compute $\delta \bar{x}$- and δy-values for the vertices of G.

At the beginning, T has two vertices v_1, v_2 with root v_1. Initialize the variables: $Right(v_1) = v_2$, $Left(v_1) = Right(v_2) = Left(v_2) = nil$, $\delta \bar{x}(v_2) = 0$.

At stage k ($3 \leq k \leq K$), the vertices $v_k^1, \ldots, v_k^t$ are added. Let $c_{i_1}, \ldots, c_{i_s}$ be the vertices of C_{k-1} adjacent to v_k^1 and v_k^t. The following actions are taken:

(1) Adjusting T: The path $c_{i_1} \to v_k^1 \ldots \to v_k^t \to c_{i_s}$ becomes part of the rightmost branch of T; the links $c_{i_1} \to c_{i_1+1}$ and $c_{i_s-1} \to c_{i_s}$ are broken; the leader c_{i_1+1} becomes the left son of v_k^1.

(2) Update $\delta \bar{x}$- and δy-values to reflect the changes of T.

- $\delta \bar{x}(v_k^1) = 0$; $\delta \bar{x}(v_k^j) = 2$ for $2 \leq j \leq t$. (Since v_k^1 and c_{i_1} have the same $\bar{x}$-coordinate and c_{i_1} is the parent of v_k^1, $\delta \bar{x}(v_k^1) = 0$. Since v_k^{j-1} is the parent of v_k^j and they are on the same horizontal line with distance 2 apart, $\delta \bar{x}(v_k^j) = 2$).

- $\delta\bar{x}(c_{i_s}) = 2 + \sum_{j=i_1+1}^{i_s} \delta\bar{x}(c_j)$. (The $\bar{x}$-offset of v_{i_s} relative to c_{i_1} is $a = 2t + \sum_{j=i_1+1}^{i_s} \delta\bar{x}(c_j)$. The $\bar{x}$-offset of v_k^t relative to c_{i_1} is $b = 2(t-1)$. Since v_k^t is the parent of c_{i_s}, $\delta\bar{x}(c_{i_s}) = a - b$).
- If $i_1 + 1 < i_s$, $\delta\bar{x}(c_{i_1+1}) = 1 + \delta\bar{x}(c_{i_1+1})$. (Since c_{i_1+1} is shifted to right by 1 and its old parent c_{i_1} and the new parent v_k^1 has the same $\bar{x}$-coordinate, $\delta\bar{x}(c_{i_1+1})$ is increased by 1).
- $\delta y(v_k^1) = \ldots = \delta y(v_k^t) = \frac{1}{2}\delta\bar{x}(c_{i_s})$. (The y-offsets of v_k^j ($1 \le j \le t$) relative to the rightvertex c_{i_s} is one-half of the $\bar{x}$-offset of c_{i_s}).

Phase II: Compute $\bar{x}$-coordinate. For each vertex v of G, $\bar{x}(v)$ is the sum of the $\delta\bar{x}$-values of the vertices on the path of T from the root v_1 to v.

Phase III: Compute y-coordinate. Define a tree DY as follows. The node set is $\{a_2, a_3, \ldots, a_K\}$. For each $3 \le k \le K$, if the right vertex c_{i_s} of stage k has O-value k', add an edge from a_k to $a_{k'}$ in DY. DY is a tree rooted at a_2. Assign a_k a weight $\delta y(v_k^t)$. Let $y(a_k)$ be the total weight of the nodes on the path from a_k to a_2 in DY. Then $y(a_k)$ is the y-coordinate of the vertices $v_k^1, \ldots, v_k^t$.

Phase IV: Compute x-coordinate, which is simply $x(v) = \bar{x}(v) + y(v)$.

Fig 6 (1) shows a graph G with dummy edges $(6, 3_2), (7, 3_1), (8, 1)$. The final tree T is shown in (2). The integer near a node v is the value $\delta\bar{x}(v)$. The tree DY is shown in (3). The integer near a node k is the weight of a_k. The $\delta\bar{x}$-values of the vertices at stage k are shown in Table (4). The column labeled *stage $k = 10$* shows the final $\delta\bar{x}$-values. The vertices with a $*$ in this column are the leaders of various stages. The δy-, $\bar{x}$-, y- and x-values are also shown in the table. (5) shows the drawing of G produced by the algorithm.

Due to space limitation, we omit the parallel implementation. We only note that all phases of the algorithm are based on computation on certain trees. These trees can be constructed in parallel, and the computation can be done by using the pre-sum and the tree contraction [1, 9] within the bounds of Theorem 1.2.

References

[1] K. Abrahamson, N. Dadoun, D. G. Kirkpatrick, and T. Przytycka, A simple tree contraction algorithm, *Journal of Algorithms* 10 (1989), pp. 287–302.

[2] M. Chrobak and T. Payne, A linear time algorithm for drawing planar graphs on a grid, UCR-CS-90-2, Dept. of Math. and CS, UC at Riverside, 1990.

[3] P. Eades and R. Tamassia, Algorithms for Automatic Graph Drawing: An Annotated Bibliography, Dept. of CS, Brown Univ., TR CS-89-09, 1989.

[4] H. de Fraysseix, J. Pach and R. Pollack, How to draw a planar graph on a grid, *Combinatorica* 10 (1990), pp. 41–51.

[5] M. Fürer, Xin He, Ming-Yang Kao and B. Raghavachari, $O(n \log \log n)$-work parallel algorithm for straight-line grid embedding of planar graphs, in *Proc. 4th ACM SPAA*, San Diego, June 1992, pp. 100-109.

[6] A. Goldberg, S. Plotkin, and G. Shannon, Parallel symmetry-breaking in sparse graphs, *SIAM J. on Discrete Mathematics* 1 (1988), pp. 434-446.

[7] M. Goldberg and T. Spencer, Constructing Maximal Independent Set in Parallel, *SIAM J. Disc. Math.* Vol 2(3), 1989, pp. 322-328.

[8] G. Kant, Drawing planar graphs using the *lmc*-ordering, in *Proc. 33th IEEE FOCS*, Pittsburgh, 1992, pp. 101-110.

[9] G. Miller and J. Reif, Parallel tree contractions and its applications, in *Proc. 26th IEEE FOCS*, 1985, pp. 478–489.

[10] V. Ramachandran and J. H. Reif, An optimal parallel algorithm for graph planarity, in *Proc. 30th IEEE FOCS*, 1989, pp. 282–287.

[11] R. Read, A new method for drawing a planar graph given the cyclic order of the edges at each vertex, *Congressus Numerantium* 56 (1987), pp. 31–44.

[12] W. Schnyder, Planar graphs and poset dimension, *Orders* 5 (1989), pp. 323–343.

[13] ——, Embedding planar graphs on the grid, in *Proceedings of the 1st Annual ACM-SIAM Symposium on Discrete Algorithms*, 1990, pp. 138–148.

[14] R. Tamassia and J. S. Vitter, Parallel transitive closure and point location in planar structures, *SIAM J. Comput.* 20 (2), 1991, pp. 708-725.

[15] W. Tutte, How to draw a graph, *London Math. Soc.* 13, 1963, pp. 743–768.

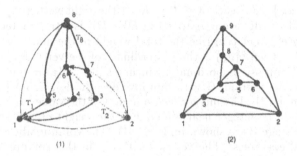

Figure 1: Realizer and canonical ordering of planar graphs

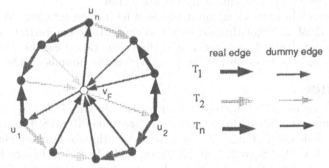

Figure 2: Edges around a face vertex v_F in a bidirectional realizer of $\hat{G}$

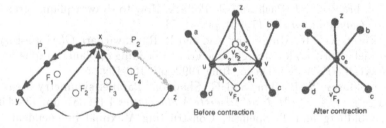

Figure 3: Assigning O values   Figure 4. Edge contraction

312

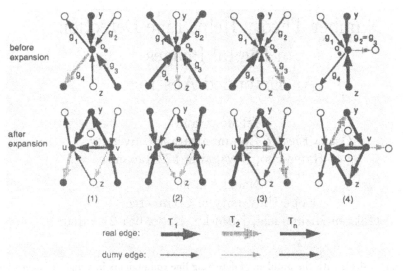

before
expansion

after
expansion

(1)   (2)   (3)   (4)

real edge: T₁   T₂   Tₙ

dumy edge:

Figure 5: Edge expansion

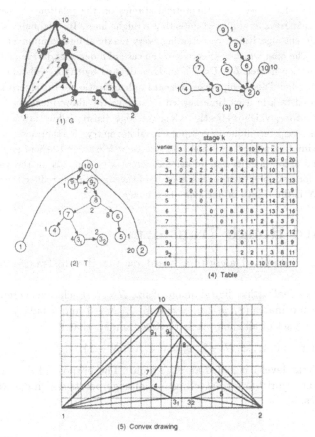

Figure 6: An example of the convex drawing algorithm

Number Theory Helps Line Detection in Digital Images
An Extended Abstract

Tetsuo Asano

Osaka Electro-Communication University,
Hatsu-cho, Neyagawa, 572 Japan

Naoki Katoh

Kobe University of Commerce,
Gakuen-Nishimachi, Nishi-ku, Kobe, 651-21 Japan

This paper deals with the problem of detecting line components in a digital image. For this purpose, the Hough Transform, which is based on voting in the dual plane, is widely used. However, there have been few theoretical studies on the relationship between its computational complexity and ability of detecting straight lines. In this paper we present two completely different algorithms for detecting every maximal line component contained in a digital image. The one, which is effective in the case of a dense digital image, is based on a new transformation named $L_1-Dual\ Transform$ defined by the L_1-distance between points and lines. Number Theory supports efficient implementation of the algorithm. It can complete the required task in least time needed to achieve the above-mentioned ability of line detection. The other, which is effective when the edge density is low, attains efficiency by using the plane sweep technique in computational geometry. Furthermore, we present an efficient approximation algorithm which can detect at least $\alpha \times 100\%$ of any maximal line component and show that its computational complexity depends on the value of α. Choosing $\alpha = 0.5$, for example, the time complexity of the algorithm is reduced from $O(N^4)$ to $O(N^3)$, where N is the length of one side of an image.

1 Standard Hough Transform

We first describe a standard algorithm based on the Hough Transform ([H62], [DH72]).

In this paper we deal with a digital image of size $N \times N$. Each pixel is represented by its (integral) coordinates (x_i, y_i), $x_i, y_i = 0, 1, \ldots, N - 1$ and a binary gray level (0 or 1). Let G be a set of all such pixels (lattice points), that is,

$$G = \{(x_i, y_i)|\ x_i = 0, 1, \ldots, N - 1,\ y_i = 0, 1, \ldots, N - 1\}.$$

A pixel with gray level 1 is called an edge point. The purpose of the paper is to develop an efficient algorithm for detecting sets of edge points which correspond to some straight lines.

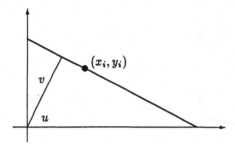

Figure 1: $u - v$ transform.

Let v be the distance from the origin to a line of an angle $u + \pi/2$ which passes through an edge point (x_i, y_i) (see Figure 1). Then, v is represented by

$$v = x_i \cos u + y_i \sin u. \tag{1}$$

This can be regarded as a transformation from a point (x_i, y_i) in the $x - y$ plane to a curve $v = x_i \cos u + y_i \sin u$ in the $u - v$ plane. Then, when the curve $v = x_i \cos u + y_i \sin u$ for an edge point (x_i, y_i) intersects the curve $v = x_j \cos u + y_j \sin u$ for (x_j, y_j) at (u', v'), the line passing through these two points is expressed by the equation:

$$x \cos u' + y \sin u' = v'. \tag{2}$$

In general, if m such curves intersect at a point (u', v'), its corresponding m points lie on the line (2). To find such intersections in the $u - v$ plane, the $u - v$ plane is partitioned into small equal regions (called buckets) in standard algorithms based on the Hough Transform. Then, following a curve $v = x_i \cos u + y_i \sin u$, we put a vote in each bucket passed by the curve. Finally, we enumerate all buckets having more votes than a predefined threshold and report the lines corresponding to these buckets.

The algorithm described above will be referred to as the algorithm based on the Hough Transform, or simply as Algorithm 1. To implement the algorithm we must determine in advance (1) a sequence $(u_1, u_2, \ldots, u_M)$ which specifies the angles of lines to be detected, (2) a function $q(u, v)$ to quantize v values, and (3) a threshold t determining the minimum cardinality of a set of edge points to give a digital line. The sequence of angles is related to both of the time complexity and the ability of line detection. The quantization function $q(u, v)$ is also important since it determines the size of the array for buckets. However, there have been very few theoretical studies on optimality of these parameters.

2   Evaluation of Line Detection Ability

Given a point p and a line l in the plane, the *vertical and horizontal distances* between them are the lengths of the vertical and horizontal segments, respectively, from p to l. Then, the $L_1 - distance$ between them is the minimum of the two distances.

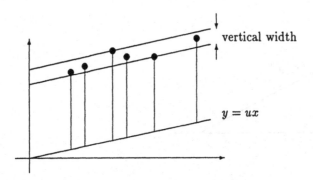

Figure 2: The vertical width of a point set at a slope u.

For a set P of edge points and a slope u, the *vertical and horizontal widths of P at slope u* are defined to be the maximum differences of vertical and horizontal distances, respectively, from points of P to the line $y = ux$ (see Figure 2). The $L_1-width$ of P for a slope u is defined to be the smaller one among the vertical and horizontal widths. Here note that which is larger the vertical or horizontal distance from a point to a line depends on the slope of the line. That is, the vertical distance is shorter if the slope is between -1 and 1 and the horizontal distance is shorter otherwise. Throughout this paper we consider only those lines whose slopes are between 0 and 1, and hence the L_1-distance is determined by the vertical distance. It might be easy to see that other cases can be dealt with in similar ways, for example, by exchanging $x-$ and $y-$coordinates.

Formally, we define a (signed) L_1-distance $d_1(x, y, a, b)$ between a point (x, y) and a line $y = ax + b$ by

$$d_1(x, y, a, b) = \begin{cases} y - ax - b & \text{if } |y - ax - b| \leq |x - (y - b)/a| \\ x - (y - b)/a & \text{otherwise.} \end{cases}$$

With these definitions, given a line $y = ux + v$ its corresponding set $G(u, v)$ of lattice points is defined to be all lattice points whose L_1-distances to the line are between -0.5 and 0.5. More precisely, we define the set by

$$G(u, v) = \{(x_i, y_i) \in G| \ -0.5 \leq d_1(x_i, y_i, u, v) \leq 0.5\}.$$

A set P of edge points corresponding to some line is called a *digital line* or *line component*. When there is no line component that properly includes P, the set P is called a *maximal line component*. In this paper we do not find a set of edge points for a given line but find a set of edge points which corresponds to some line and then calculate an equation of the line that fits the set best.

With the definitions above we have the following lemmas.

[Lemma 1] *A set P of edge points is a line component if and only if there is some slope u at which the $L_1-width$ of P is at most 1.*

Based on the preparation above, we shall describe the criterion on the line detection ability of algorithms.

[Definition 1] *An algorithm for detecting lines in a digital image is said to be perfect if it can detect every maximal line component of size not less than t for any digital image and a threshold t.*

[Definition 2] *An algorithm is said to be α−perfect if for each maximal line component in a given digital image it can detect α × 100% of its constituent edge points.*

Defining the line detection ability of an algorithm, the following lemma might be obvious.

[Lemma 2] *The algorithm based on the Hough Transform is not perfect.*

In the next section we shall present a perfect algorithm that can detect every maximal line component in a given digital image. The algorithm is based on a new transformation based on the L_1−distance which is in principle the same as the duality transform.

3 L_1−Dual Transform

The Duality Transform which is commonly used in Computational Geometry maps a point (a, b) into a line $y = ax + b$ and a line $y = cx + d$ into a point $(-c, d)$. It is well known that the transformation preserves the vertical relation between points and lines. The intersection point of the line $y = a_1x + b_1$ for a point (a_1, b_1) and the line $y = a_2x + b_2$ for a point (a_2, b_2) corresponds to the line passing through the two points. Therefore, if we transform edge points into lines and enumerate all the intersections at which t or more lines intersect, we have a set of line components which correspond to those intersections. However, this approach has a serious disadvantage that we cannot restrict the range in which intersections exist, in other words, intersections may be scattered in a very wide range. This means that the "voting" method which was used in the algorithm based on the Hough Transform is not valid. This is the reason why the original idea by Hough based on the Duality Transform has been considered to be "impractical" [DH72]. Notice that the Hough Transform based on the equation (1) ($v = x_i \cos u + y_i \sin u$) allows us to restrict the search space to a small rectangle defined by the inequalities $0 \leq u \leq \pi$ and $|v| \leq 2N$ (more precisely $|v| \leq \sqrt{2}N$ for a square image). This compactness of the search space is considered to be the key to the practical success of the Hough Transform. However, if we are careful in the line detection ability, the algorithm based on the Hough Transform is not the best. In this section we design a perfect algorithm along the definitions in the previous section.

The algorithm to be presented in this section is based on a transformation which is very similar to the duality transform. In the Hough Transform we computed the (perpendicular) distance v from the origin to a line passing through an edge point (x_i, y_i) whose angle is $\pi/2 - u$. In our transform we compute the L_1−distance. Therefore, we shall call it L_1−*Dual Transform*. For simplicity, throughout this paper we concentrate ourselves to detection of line components whose corresponding lines are of slopes between 0 and 1. So, the L_1−distance v from an edge point (x_i, y_i) to a line $y = ux$ is given by

$$v = y_i - u \cdot x_i. \tag{3}$$

Then, we sort edge points by their v values. If two points (x_i, y_i) and (x_j, y_j) have the same v value, that is, if we have

$$y_i - u \cdot x_i = y_j - u \cdot x_j = v,$$

then, we can see that the line $y = ux + v$ passes through these two points. This leads to the following observation: Given a set of edge points, compute their corresponding v values by Equation (3) and then sort them to find maximal subsets of edge points whose v values differ by at most one. Then, those subsets are the digital lines required. Thus, we have the following algorithm.

[Algorithm 2: Algorithm Based on L_1−Dual Transform]

Array: (1) bit $g[N][N]$; /* digital image */

 (2) struct point{ int x, y;} p[n]; /* array to store coordinates of edge points */

 (3) double v[n]; /* an array to store v values */

 (4) int $\sigma[n]$; /* an array to store labels of edge points */

 /* n is the total number of edge points */

Predetermined Parameters:

 (1) $(u_1, u_2, \ldots, u_M)$: /* a sequence of slopes of lines */

 (2) a threshold t determining the minimum cardinality of a set of edge points to give a digital line.

procedure:

 Store coordinates of all the edge points in the structured array p[.].

 for $i = 1$ to n $\sigma[i] = i$; /* Initialization of the array $\sigma[.]$ */

 for $i = 1$ to M /* for each slope */

  for $j = 1$ to n /* for each edge point */

   compute $v[j] = p[j].y - u_i \times p[j].x$;

  Sort v values so that we have $v[\sigma[1]] \le v[\sigma[2]] \le \cdots \le v[\sigma[n]]$;

  Based on the sorted sequence, find maximal line components;

A procedure for detecting a maximal line component:

 For the input of the sorted list of v values we want to find all maximal subsets such that their v values are different by at most one and they consist of t or more points. Since we have the sorted list, linear scan over the list suffices to find all those subsets.

This algorithm is different from the algorithm based on the Hough Transform in the following respects.

(1) The equation for $v[j]$ is different.

(2) Each u_i is not an angle but a slope of a line.

(3) The v values are not quantized but sorted to find maximal subsets having width at most one.

(4) The threshold specifying the allowance on v values is fixed to be 1.

[Lemma 3] *The algorithm based on the L_1−Dual Transform is perfect only if the length M of the slope sequence is $\Omega(N^2)$.*

Proof: Suppose that a digital image consists only of the four lattice points $A(0, 0)$, $B(0, 1)$, $C(a, b)$ and $D(a, b + 1)$. Then, the set of those lattice points is obviously a maximal line component corresponding to the line $y = (b/a)x + 0.5$. Therefore, the fraction b/a must be included in the sequence of slopes to detect this maximal line component when the threshold t is four. Since there are $\Omega(N^2)$ fractions[K81] of the form $b/a, b = 0, 1, \ldots, N - 1$, $a = 1, 2, \ldots, N - 1$, $b \leq a$, we have the lemma. $\square$

Let $(u_0 = 0, u_1, u_2, \ldots, u_M)$ be an increasing sequence of irreducible fractions of the form y/x, $0 \leq y < N$, $1 \leq x < N$, $y < x$. This sequence is known as a *Farey series*[GKP89]. The Farey series of order k, denoted by F_k, is the increasing sequence of all reduced fractions between 0 and 1 whose denominators are less than or equal to k.

[Lemma 4] *Let $(u_0, u_1, \ldots, u_M)$ be the Farey series of order $N - 1$. Given an arbitrary digital image, each of its maximal line components can be detected by the algorithm based on the L_1-Dual Transform.*
Proof: Let P be an arbitrary maximal line component. Without loss of generality we assume that it corresponds to a line of slope between 0 and 1. The vertical width of P is defined as before. Then, it is minimized when u is some slope of an edge of the convex hull of P. Since each such edge connects two edge points in P, the slope of the edge must be included in the Farey series. $\square$

The above discussions lead to the following theorem.

[Theorem 1] *The algorithm based on the L_1-Dual Transform is perfect with the sequence of slopes which is equal to the Farey series of length $\Theta(N^2)$.*

4 Efficient Approximation Algorithm

We have presented a perfect algorithm which can detect every maximal line component in a digital image using a slope sequence of length $\Theta(N^2)$. In this section we present an approximation algorithm which is more efficient at the expense of the line detection ability. The algorithm itself is just the same as the algorithm based on the L_1-Dual Transform (referred to simply as Algorithm 2, hereafter) except that it uses the slope sequence of length $O(N)$ rather than that of $O(N^2)$.

We shall show that careful design of a slope sequence of length $\Theta(N)$ makes Algorithm 2 1/2−perfect. We omit the proof of the validity of this approximation algorithm due to space limit. Such a slope sequence is defined as follows.

$$u_i' = \frac{i}{N-1}, \quad i = 0, 1, \ldots, N - 1. \tag{4}$$

[Theorem 2] *Algorithm 2 is 1/2−perfect with the above slope sequence of length $\Theta(N)$.*

The above method may yield a better accuracy than 1/2− perfectness. But, it is not the case in general. The reason is as follows: We can easily construct an example such that $|S_A| \approx |S|/2$ and all points of S are very close to L_A', while $|S_B| \approx |S|/2$ and all points of S are very close to L_B', and there does not exist a line with slope $k/(N - 1)$ for which the vertical distances from both a point of S_A and a point of S_B to that line are simultaneously less than or equal to one.

In the above method with 1/2−perfectness, we use a slope sequence such that the difference between each consecutive two slopes is equal to $1/(N - 1)$. It is reasonable to imagine that, if we use a finer sequence such that the length of the interval between each consecutive slopes is less than $1/(N - 1)$, we can obtain an

α—perfect algorithm with $\alpha > 1/2$. However, this is not the case in general because we can easily obtain an example in the manner similar to the above such that at most half of points in S have the vertical distance less than or equal to one to a line with any slope among those chosen.

In spite of this fact, if we are concerned with only dense digital line components, we can devise an α—perfect algorithm for any $\alpha > 1/2$.

A slope sequence for this purpose is defined as follows: Let and $u_k = b_k/a_k, k = 1, 2, \ldots, M$, where b_k/a_k is an irreducible fraction. For i, j with $0 \le i \le j \le M$, let

$$s_{ij} = \sum_{k=i}^{j} \frac{N-1}{a_k}.$$

Roughly speaking, s_{ij} approximates the number of lattice points in the triangle formed by three straight lines, $y = u_i x, y = u_j x$, and $x = N - 1$.

We then compute a subsequence $(u_0'(= 0), u_1', \ldots, u_{M'}'(= 1))$ of $(u_0(= 0), u_1, \ldots, u_M)$ in this order by the following formula.

$$u_{i+1}' := \min\{u_j \mid u_j > u_i' \text{ and } s_{i'+1,j} \ge (N-1)(1-\alpha)/3\} \tag{5}$$

Here i' satisfies $u_{i'} = u_i'$.

[Theorem 3] *When the slope sequence defined by (5) is used and the threshold t is changed to αt, Algorithm 2 is α—perfect.*

[Theorem 4] *The length of the slope sequence defined by (5) is $\Theta(N/(1-\alpha))$.*

From the above theorem, if we typically choose $\alpha = 1 - 1/\log N$, the length of the slope sequence becomes $\Theta(N \log N)$.

5   Optimality of the Algorithm

The algorithm based on the L_1—Dual Transform can report every maximal line component in $O(nN^2)$ time. We shall show that it is optimal within a constant factor when the density of edge points is high, more exactly when the number of edge points is of the order of the image size $O(N^2)$. If $n = O(N^2)$ then the time complexity is $O(N^4)$. Thus, it suffices to prove that there are $\Omega(N^4)$ maximal line components.

[Theorem 5] *An $N \times N$ digital image contains $\Theta(N^4)$ distinct maximal line components.*

Proof We are interested in the total number of maximal line components in an $N \times N$ digital image when every lattice point is an edge point. Let P be an arbitrary maximal line component. The vertical width of the convex hull of the maximal line component is exactly one. It is also easy to see that there at least two x—coordinates at which the vertical distance is one. Let x_1 and x_2 be the smallest and second smallest such x—coordinates. P has two lattice points at each of them. Let those points be$(x_1, y_1), (x_1, y_1 + 1), (x_2, y_2), (x_2, y_2 + 1)$. Then, the set of lattice points lying in the region bounded by the line passing through the two points (x_1, y_1) and (x_2, y_2) and that through $(x_1, y_1 + 1)$ and $(x_2, y_2 + 1)$ coincides with the set P. This implies that P can be completely specified only by the two points (x_1, y_1) and (x_2, y_2). So, these two points are referred to as the characteristic points of the maximal line component.

Now, how many ways to choose such two points? The point (x_1, y_1) is the leftmost point that gives the vertical distance one and (x_2, y_2) is the second leftmost such point. Here, we let $a = x_2 - x_1$ and $b = y_2 - y_1$. When $y_1 < b$, (x_1, y_1) can not be the leftmost characteristic point unless $x_1 \leq N - a$. Therefore, the number of possible (x_1, y_1) is at least

$b \times (N - a) + a \times (N - 2b) = (a + b)N - 3ab$

and at most

$b \times (N - a) + a \times (N - b) = (a + b)N - 2ab$.

The slope of the line corresponding to P is given by b/a, that is, $(y_2 - y_1)/(x_2 - x_1)$. This fraction is irreducible due to the definition of x_1, x_2, y_1 and y_2.

The discussion above means that we can evaluate the number of maximal line components by counting the number of ways to choose two characteristic points as above. Thus, it is given by computing the summation of $(a + b)N - 2ab$ and $(a + b)N - 3ab$ over all fractions b/a out of the Farey series of order $N - 1$. It is not so easy to have the exact summation. Considering the fact that two integers less than N are relatively prime at probability $6/\pi^2$, and more precisely the length of the Farey series of order $N - 1$ is $3/\pi^2 N^2 + O(N \log N)$, the exact summation can be bounded by the summation taken for every pair of integers less than N by some constant factor. The theorem follows from

$$\sum_{a=1}^{N-1} \sum_{b=1}^{N-1} ((a + b)N - 3ab) = \Theta(N^4).$$ □

6 Issues on Computational Complexities

In this section we evaluate the computational complexities of the algorithms presented so far. First of all, we redefine the parameters to specify the problem.
(1)N : image size (the number of lattice points is N^2).
(2)n : the number of edge points.
(3)M : the length of a slope sequence.

[Lemma 5] *The time complexity of the algorithm based on the standard Hough Transform is $O(N^2 + NM + nM)$ and it requires the space $O(NM)$ in addition to $O(N^2)$ space for a digital image.*

The algorithm based on L_1–Dual Transform (Algorithm 2) requires $O(nM \log n)$ time if $O(n \log n)$ time is needed to sort v values. Do we always need $O(n \log n)$ time for sorting? The answer is "No". In fact, we can show that the sorting can be done in $O(n)$ time for each slope by utilizing the integrality of coordinates. In addition, the standard Hough Transform requires complex computation using sin and cos while Algorithm 2 needs no complex calculations. This is considered to be some improvement from a practical point of view.

Now, we shall show how to sort v values in $O(n)$ time for each slope u_i. Based on this fact we present an efficient algorithm which detect every maximal line component from a set P of n edge points in an $N \times N$ digital image in $O(nM)$ time.

Let S be a list of increasing distinct slopes of lines passing through two lattice points in the $N \times N$ lattice plane (here only those slopes between 0 and 1 are considered). Let $M = |S|$ and u_i be the i–th element of S ($u_0 = 0$).

In the algorithm we maintain a one-dimensional array $\sigma[n]$ of length n in which edge points are sorted in the non-increasing order of vertical distances to the line of slope u_i passing through the origin. Suppose that $\sigma[j]$ stores a number of edge points whose vertical distance is the j-th smallest. The basic idea of the algorithm is as follows.

(1) The vertical distance from each edge point to the line $y = u_0 x (= 0)$ is determined by its y-coordinate. If there is any tie, we arrange them in the decreasing order of their x-coordinates.

(2) Given the content of the array $\sigma[n]$ for the slope u_i, that for the next slope u_{i+1} is obtained by modifying it as follows.

Here, we assume that the current array $\sigma[n]$ stores edge point numbers in the non-decreasing order of the vertical distances to the line of slope u_i passing through the origin. In case of ties the ones with larger x-coordinates are preceded. Let the vertical distance from a point $p_{\sigma[j]}$ to the line $y = u_{i+1} x$ be $d_{\sigma[j]}$.

Case 1: Points $p_{\sigma[j]}$ and $p_{\sigma[j+1]}$ have the same vertical distance to the line $y = u_i x$:

In this case, since the x-coordinate of $p_{\sigma[j]}$ is greater than that of $p_{\sigma[j+1]}$ by the definition, we have $d_{\sigma[j]} < d_{\sigma[j+1]}$.

Case 2: The vertical distance from point $p_{\sigma[j]}$ to $y = u_i x$ is smaller than that from $p_{\sigma[j+1]}$ to $y = u_i x$:

Since there is no slope of line that passes through two lattice points between u_i and u_{i+1}, we have $d_{\sigma[j+1]} \geq d_{\sigma[j]}$. If $d_{\sigma[j+1]} > d_{\sigma[j]}$ holds, we do nothing. On the other hand, in the case of $d_{\sigma[j+1]} = d_{\sigma[j]}$ we find every lattice point whose vertical distance is equal to it. This is done by extracting points from $\sigma[j+2]$ in order while checking the vertical distance to $y = u_{i+1} x$. Suppose that as the result those points from $\sigma[j]$ to $\sigma[j']$ have the same vertical distance to $y = u_{i+1} x$. Then, the x-coordinates of these points increase in the order of $\sigma[j] \ldots \sigma[j']$. This is because the vertical distances to $y = u_i x$ monotonically increase in that order. As for the vertical distances to $y = u_{i+2} x$, they monotonically increase in the reverse order. Thus, we should reverse the order from $\sigma[j]$ to $\sigma[j']$ to prepare for the next iteration.

It should be noted that we do not report each line component as a set of edge points but an equation of the line that fits the component best. It is not so hard to modify Algorithm 2 so as to report such equations.

[Lemma 6] *The algorithm based on the L_1-Dual Transform runs in $O(nM)$ time and $O(n + N)$ space.*

7 Line Detection in a Sparse Digital Image

For practical applications a digital image usually contains as many edge points as 10% or 15% of the whole pixels. Thus, in such cases the number n of edge points is proportional to the number of pixels or lattice points. In other words, $n = O(N^2)$ holds in many cases.

In this case the algorithms described thus far run in $O(nM) = O(N^4)$ time. On the other hand, for sparse digital images such that the density of edge points is very low, say $n = O(N)$, it looks very expensive to examine $O(N^2)$ different slopes to make an algorithm perfect. In this section we present an algorithm for detecting line components based on combinatorial observations rather than integral properties of coordinates of edge points.

The basic idea is a plane sweep which is often used in computational geometry. For an edge point $p = (x_p, y_p)$ let p^T and p^L be the lattice points just above and just left of p, respectively. More exactly, $p^T = (x_p, y_p + 1)$, $p^L = (x_p - 1, y_p)$. Then, while changing the slope of the parallel lines passing through the points p and p^T from -1 to $+1$ in a counterclockwise way, we maintain a set of edge points contained in the region bounded by the two lines. After that, while changing the slope of the parallel lines passing through the points p and p^L from $+1$ to ∞ and from ∞ to -1 also in a counterclockwise way, we maintain a set of edge points contained in the region bounded by the two lines.

If $O(n^2)$ space is allowed we can construct an arrangement of lines which are dual of edge points in $O(n^2)$ time. Using the arrangement of lines the angular sorting around each edge point can be done in $O(n)$ time and each insertion and deletion of an edge point can be done in $O(1)$ time. Therefore, we can reduce the total time complexity to $O(n^2)$.

8 Conclusions

In this paper we have considered the problem of detecting line components in a digital image and presented an efficient algorithm which can detect every maximal line component without missing any one. Although we did not achieve considerable improvement over the existing line detection algorithms, one remarkable advantage of the proposed algorithm is in the guarantee that it can detect every maximal line component. Although the considerations in this paper have been restricted to line detection, detection of circles and ellipses are also important. The problem of detecting such complicated curves is left for a future subject.

References

[H62] P. V. C. Hough: "Method and Means for Recognizing Complex Patterns", U.S. Patent 3069654, December 18, 1962.

[DH72] R. O. Duda and P.E. Hart: "Use of the Hough Transformation to Detect Lines and Curves in Pictures", Comm. of the ACM, 15, January 1972, pp.11-15.

[MK89] T. Matsuyama and H. Koshimizu: "Hough Transform and Pattern Matching in Computer Vision", Joho-Shori, 20, 9, 1989, pp.1035-1046.

[FWM91] Fujii, Wada and Matsuyama: "How to Construct a Parameter Space without any Distortion in Hough Transform", Proc. 1991 Domestic Conference of IEICE of Japan, D-562 (1991).

[K81] D. E. Knuth: "The Art of Computer Programming", Vol. II (Seminumerical Algorithms), p.324, Addison-Wesley, 1981.

[GKP89] R. L. Graham, D. E. Knuth, and O. Patashnik: "Concrete Mathematics", Addison-Wesley, 1989.

Optimally Computing the Shortest Weakly Visible Subedge of a Simple Polygon*
(Preliminary Version)

Danny Z. Chen[†]

University of Notre Dame

Abstract

Given an n-vertex simple polygon P, the problem of computing the shortest weakly visible subedge of P is that of finding a shortest line segment s on the boundary of P such that P is weakly visible from s (if s exists). In this paper, we present new geometric observations that are useful for solving this problem. Based on these geometric observations, we obtain optimal sequential and parallel algorithms for solving this problem. Our sequential algorithm runs in $O(n)$ time, and our parallel algorithm runs in $O(\log n)$ time using $O(n/\log n)$ processors in the CREW PRAM computational model. Using the previously best known sequential algorithms to solve this problem would take $O(n^2)$ time. We also give geometric observations that lead to extremely simple and optimal algorithms for solving, both sequentially and in parallel, the case of this problem where the polygons are rectilinear.

1 Introduction

In this paper, we consider a *weak visibility* problem. Weak visibility deals with visibility problems in which the "observers" are of the shape of line segments. An important class of weak visibility problems studies the case where the "opaque" objects are the boundaries of simple polygons. For a point p in a polygon P and a line segment s, p is said to be *weakly visible* from s iff p is visible from some point on s (depending on p). Polygon P is said to be *weakly visible* from a line segment s iff every point $p \in P$ is weakly visible from s. Many sequential algorithms [1, 2, 3, 4, 7, 8, 9, 10, 12, 13, 15, 18, 19, 20, 21, 22, 23] and parallel algorithms [5, 6, 11, 14] for solving various weak visibility problems on simple polygons have been discovered.

We consider the problem of computing the shortest weakly visible subedge of a simple polygon (called it the SWVS problem). That is, given an n-vertex simple polygon P, we would like to find a line segment s on the boundary of P such that (i) P is weakly visible from s (if s exists), and (ii) the length of s is the shortest among all such line segments on the boundary of P (it is possible that s is a single point on the boundary of P). Intuitively, if P represents a house whose interior is that of a simple polygon, then s is the shortest portion of any wall of P by which a guard has to patrol back and forth in order to keep the inside of P completely under surveillance.

There is related work on the SWVS problem. Avis and Toussaint [1] considered the problem of detecting the weak visibility of a simple polygon (that is, deciding whether a polygon P is weakly visible from an edge e of P, and reporting all such edges e for P); they presented a sequential linear time algorithm for the case of checking whether P is weakly visible from a *specified* edge e of P. Another sequential linear time algorithm for this case was given in [10]. Sack and Suri [20] and Shin [21] independently gave optimal linear time algorithms for solving the problem of detecting the weak visibility of a simple polygon. Chen [5] came up with

---

*This research was partially done when the author was with the Department of Computer Sciences, Purdue University, West Lafayette, Indiana, and was supported in part by the Office of Naval Research under Grants N00014-84-K-0502 and N00014-86-K-0689, the National Science Foundation under Grant DCR-8451393, and the National Library of Medicine under Grant R01-LM05118.

[†]Department of Computer Science and Engineering, University of Notre Dame, Notre Dame, IN 46556, U.S.A. E-mail: chen@cse.nd.edu.

an optimal parallel algorithm for this problem; Chen's algorithm runs in $O(\log n)$ time using $O(n/\log n)$ CREW PRAM processors.

Several problems on computing weakly visible line segments with respect to a simple polygon have been studied. Ke [15] and Doh and Chwa [8] gave $O(n \log n)$ time algorithms for computing a line segment in a polygon from which the polygon is weakly visible (such a segment can be in the interior of the polygon); in particular, Ke's algorithm finds such a line segment of shortest length. Lee and Chwa [19] designed a linear time algorithm for computing all the maximal convex chains or all the maximal reflex chains on the boundary of a polygon from which the polygon is weakly visible. Bhattacharya et al. [3] presented a linear time algorithm for computing a shortest line segment (not in the interior of a polygon) from which the boundary of the polygon is weakly visible (or externally visible). Ching et al. [7] showed that, if a polygon is weakly visible from a specified edge e, then the shortest weakly visible subedge on e can be computed in linear time by using the algorithm in [1]. The problem of computing in parallel the shortest weakly visible subedge on a specified polygon edge was solved optimally by Chen [6], in $O(\log n)$ time using $O(n/\log n)$ CREW PRAM processors.

The SWVS problem, obviously, is a natural generalization of the weak visibility problem first studied by Avis and Toussaint [1] and then by Sack and Suri [20] and Shin [21]. A straightforward sequential solution to the SWVS problem based on these known algorithms consists of the following steps: (1) Compute all the edges of P from each of which P is weakly visible, by using [20, 21]. (2) For every edge so obtained, compute the shortest weakly visible subedge on that edge, by using [1, 7]. (3) Among all the weakly visible subedges computed in step (2), find the one with the shortest length. Such an algorithm certainly solves the SWVS problem. However, because a simple polygon can have $O(n)$ edges from each of which the polygon is weakly visible, and because computing the shortest weakly visible subedge on a specified edge in general requires $O(n)$ time [1, 7], the above algorithm takes $O(n^2)$ time.

In this paper, we present new geometric observations that are useful for solving the SWVS problem. Based on these geometric observations, we obtain efficient sequential and parallel algorithms for solving the SWVS problem. Our sequential algorithm runs in $O(n)$ time, and our parallel algorithm runs in $O(\log n)$ time using $O(n/\log n)$ CREW PRAM processors. These algorithms are obviously optimal. We also give geometric observations that lead to extremely simple and optimal algorithms for solving, both sequentially and in parallel, the case of the SWVS problem where the polygons are rectilinear (i.e., the edges of the polygons are either vertical or horizontal).

2 Preliminaries

The input consists of an n-vertex simple polygon P, and the output is s, the shortest weakly visible subedge of P (if s exists). Polygon P is specified by a sequence $(v_1, v_2, \ldots, v_n)$ of its vertices, in the order in which they are visited by a counterclockwise walk along the boundary of P, starting from vertex v_1. Without loss of generality (WLOG), we assume that no three vertices of P are collinear.

The edge of P joining v_i and v_{i+1} is denoted by $e_i = \overline{v_i v_{i+1}}$ $(= \overline{v_{i+1} v_i})$, with the convention that $v_{n+1} = v_1$. The boundary of P is denoted by $bd(P)$, and the polygonal chain from v_i counterclockwise to v_j along $bd(P)$ is denoted by $C(i,j)$. The size of a chain C is the number of line segments on C, denoted by $|C|$.

An edge e of P from which P is weakly visible is called a weakly visible edge of P. We denote the set of all the weakly visible edges of P by WVE. Note that, for an arbitrary simple polygon of n vertices, the set of its weakly visible edges can be computed optimally, in $O(n)$ time sequentially [20, 21], and in $O(\log n)$ time using $O(n/\log n)$ CREW PRAM processors in parallel [5]. WLOG, we assume that $WVE \neq \emptyset$ (because if $WVE = \emptyset$, then P is not weakly visible from any of its edges and hence the shortest weakly visible subedge s on $bd(P)$ does not exist). For each edge $e \in WVE$, we denote the shortest weakly visible subedge of P on e by $s(e)$. Let $WVE = \{we_1, we_2, \ldots, we_m\}$, where $m = |WVE|$. Note that m can be $O(n)$. WLOG, we assume that $m > c$ for some constant integer $c \geq 1$ (c will be decided in Section 3). This is because if $m \leq c$, then s is one of the $m = O(1)$ $s(e)$'s, where $e \in WVE$. The $O(1)$ $s(e)$'s can be computed optimally, both sequentially and in parallel, by respectively applying the algorithms in [1, 6] to every edge $e \in WVE$.

We label the we_i's of WVE in such a way that $we_1 = e_1$ and that, when walking along $bd(P)$ counterclockwise by starting at v_1, we visit the we_i's in increasing order of their indices. In the rest of this paper, we use the following convention for the indices of the we_i's: For every integer $i = 1, 2, \ldots, m$, $we_{i+m} = we_i$, and for every integer $j = 0, 1, \ldots, m-1$, $we_{-j} = we_{m-j}$.

For an edge $we_i = e_j \in WVE$, we call v_j (resp., v_{j+1}) the *first vertex* (resp., *last vertex*) of we_i, and denote it by $fv(we_i)$ (resp., $lv(we_i)$). For two consecutive edges we_i and we_{i+1} of WVE, where $we_i = e_j$ and $we_{i+1} = e_k$, we denote by C_i the chain on $bd(P)$ from $lv(we_i)$ counterclockwise to $fv(we_{i+1})$ *excluding* $lv(we_i)$ and $fv(we_{i+1})$. Note that $C_i = (e_{j+1}, e_{j+2}, \ldots, e_{k-1}) - \{v_{j+1}, v_k\}$, and that C_i contains no edge in WVE. C_i can be $\emptyset$ for some i (when $lv(we_i) = fv(we_{i+1})$). Obviously, the we_i's and C_i's together form a partition of $bd(P)$.

A point p in the plane is represented by its x-coordinate and y-coordinate, denoted by $x(p)$ and $y(p)$, respectively. A vertex v_i is *convex* if the interior angle of P at v_i is $< \pi$. An edge e_i is *convex* if both v_i and v_{i+1} are convex. For any edge $we_i \in WVE$, if we_i is convex, then for any subchain $C(j,k)$ of $C(lv(we_i), fv(we_i))$, the (directed) shortest path from v_j to v_k inside P goes through only the vertices on $C(j,k)$, and the shortest path makes only right turns (this fact is shown in [1, 10]). Hence, we call such a shortest path the *internal convex path* between v_j and v_k along $C(j,k)$.

3   Useful Geometric Observations

The idea of our algorithms is to compute the shortest weakly visible subedge $s(we_i)$ on every edge $we_i \in WVE$. Because $|WVE|$ can be $O(n)$ and because computing each $s(we_i)$ in general requires $O(n)$ operations, the algorithms based on this idea appear to take $O(n^2)$ operations. The following lemmas are crucial to the optimality of our algorithms.

Lemma 1 *Suppose that* $|WVE| \geq 7$. *Then for every edge* $we_i \in WVE$, *the following are true:*

(1) *The vertex* $fv(we_i)$ *is visible from every point on the chain along* $bd(P)$ *from vertex* u' *clockwise to vertex* v', *where* $u' = fv(we_{i-2})$ *if* $C_{i-2} \neq \emptyset$ *and* $u' = fv(we_{i-3})$ *otherwise, and* $v' = lv(we_{i+1})$ *if* $C_i \neq \emptyset$ *and* $v' = lv(we_{i+2})$ *otherwise.*

(2) *The vertex* $lv(we_i)$ *is visible from every point on the chain along* $bd(P)$ *from vertex* u'' *counterclockwise to vertex* v'', *where* $u'' = lv(we_{i+2})$ *if* $C_{i+1} \neq \emptyset$ *and* $u'' = lv(we_{i+3})$ *otherwise, and* $v'' = fv(we_{i-1})$ *if* $C_{i-1} \neq \emptyset$ *and* $v'' = fv(we_{i-2})$ *otherwise.*

Proof. Note that, because $|WVE| \geq 7$, the chains defined in (1) and (2) both do not contain we_i. We only prove (1) (the proof for (2) is symmetric).

We first prove the case where C_{i-2} and C_i are both nonempty. Let p be an arbitrary point on the chain along $bd(P)$ from $fv(we_{i-2})$ clockwise to $lv(we_{i+1})$. To prove that p is visible from $fv(we_i)$, we need to show that the following are true: (i) The chain along $bd(P)$ from $fv(we_i)$ *clockwise* to p does not block the view between $fv(we_i)$ and p, and (ii) the chain along $bd(P)$ from $fv(we_i)$ *counterclockwise* to p does not block the view between $fv(we_i)$ and p.

Case (i) Let q be a point on C_{i-2}. If the view between $fv(we_i)$ and p were blocked by the chain along $bd(P)$ from q counterclockwise to $fv(we_i)$, then $fv(we_i)$ would have not been weakly visible from we_{i-2} (see Figure 1 (a)), a contradiction. If the view between $fv(we_i)$ and p were blocked by the chain along $bd(P)$ from p counterclockwise to q, then p would have not been weakly visible from we_{i-1} (see Figure 1 (b)), again a contradiction.

Case (ii) Let q be a point on C_i. If the view between $fv(we_i)$ and p were blocked by the chain along $bd(P)$ from $fv(we_i)$ counterclockwise to q, then $fv(we_i)$ would have not been weakly visible from we_{i+1} (see Figure 2 (a)), a contradiction. If the view between $fv(we_i)$ and p were blocked by the chain along $bd(P)$ from q counterclockwise to p, then p would have not been weakly visible from we_i (see Figure 2 (b)), again a contradiction.

Suppose that $C_i = \emptyset$. We need to show that the chain along $bd(P)$ from $fv(we_i)$ counterclockwise to $lv(we_{i+2})$ does not block the view between $fv(we_i)$ and $lv(we_{i+2})$. If the view were blocked by the chain along $bd(P)$ from $fv(we_i)$ counterclockwise to $fv(we_{i+2})$ excluding $fv(we_{i+2})$, then $fv(we_i)$ would have not been weakly visible from we_{i+2} (see Figure 3 (a)), a contradiction. If the view were blocked by we_{i+2} itself, then $lv(we_{i+2})$ would have not been weakly visible from we_i (see Figure 3 (b)), again a contradiction. The proof for other points on the chain along $bd(P)$ from $fv(we_{i-2})$ clockwise to $lv(we_{i+2})$ is similar to the proof of **Cases** **(i)** and **(ii)** above (with edge we_{i+1} playing the role of C_i).

The case where $C_{i-2} = \emptyset$ is proved similarly to **Cases (i)** and **(ii)**, because the chain along $bd(P)$ from $lv(we_{i-3})$ counterclockwise to $fv(we_{i-1})$ is nonempty, and hence it can play the role of C_{i-2} in the above proof. For an example of $fv(we_{i-2})$ not visible from $fv(we_i)$ when $C_{i-2} = \emptyset$, see Figure 4. □

Lemma 2 *Suppose that* $|WVE| \geq 7$. *Then for each edge* $we_i \in WVE$, we_i *is completely visible from every point on the chain along* $bd(P)$ *from vertex* u *clockwise to vertex* v, *where* $u = fv(we_{i-2})$ *if* $C_{i-2} \neq \emptyset$ *and* $u = fv(we_{i-3})$ *otherwise, and* $v = lv(we_{i+2})$ *if* $C_{i+1} \neq \emptyset$ *and* $v = lv(we_{i+3})$ *otherwise.*

Proof. Let C_{vu}^i be the chain along $bd(P)$ from u clockwise to v. Because $|WVE| \geq 7$, C_{vu}^i does not contain we_i. By Lemma 1, every point p on C_{vu}^i is visible from both endpoints $fv(we_i)$ and $lv(we_i)$ of we_i. Hence it is easy to see that p is visible from every point on we_i (see Figure 5). □

For every $we_i \in WVE$, let C_{vu}^i denote the chain along $bd(P)$ from vertex u clockwise to vertex v as defined in Lemma 2. The computational consequence of Lemma 2 is that, when computing $s(we_i)$ on every edge $we_i \in WVE$, we can simply ignore the effect of all the points on C_{vu}^i. This is because, by Lemma 2, edge we_i is completely visible from every point on C_{vu}^i. The points in P that we need to consider when computing $s(we_i)$, therefore, are all on the following two disjoint subchains of $bd(P)$:

(a) The chain from u counterclockwise to $fv(we_i)$, denoted by LC_i, and

(b) the chain from v clockwise to $lv(we_i)$, denoted by RC_i.

In summary, for every $we_i \in WVE$, the computation of $s(we_i)$ is based only on chains LC_i and RC_i.

Note that chain LC_i contains at most two nonempty chains C_j, where $j \in \{i-1, i-2, i-3\}$, and that RC_i contains at most two nonempty chains C_k, where $k \in \{i, i+1, i+2\}$. We only discuss the computation of $s(we_i)$ with respect to the points on RC_i (the computation of $s(we_i)$ with respect to LC_i is similar).

WLOG, we assume for the rest of this section that $|WVE| \geq 7$. Note that, based on the lemmas in this section, the integer parameter c of our algorithms (c was introduced in Section 2) is chosen to be 7.

The next lemma greatly reduces our effort in computing $s(we_i)$ with respect to the points on chain RC_i: It enables us to "localize" the computation to RC_i.

Lemma 3 *For a point* p *on* RC_i *and every point* q *on* we_i, *the chain along* $bd(P)$ *from* p *counterclockwise to* q *does not block the view between* p *and* q.

Proof. Suppose that the chain C_{pq} along $bd(P)$ from p counterclockwise to q did block the view between p and q. Let $ICP(C_{pq})$ be the internal convex path between p and q that passes only the vertices of C_{pq}, and let $\overline{pq'}$ be the line segment on $ICP(C_{pq})$ that is adjacent to p (see Figure 6). Since $|WVE| \geq 7$, there must be at least one edge $we_j \in WVE$ such that (1) we_j is not adjacent to q', and (2) we_j is either on the subchain of C_{pq} from p counterclockwise to q' or on the subchain of C_{pq} from q' counterclockwise to q. If we_j is on the subchain of C_{pq} from p counterclockwise to q', then q would have not been weakly visible from we_j, a contradiction. If we_j is on the subchain of C_{pq} from q' counterclockwise to q, then p would have not been weakly visible from we_j, again a contradiction. □

By Lemma 3, for every point p on RC_i and every point q on we_i, the view between p and q can be blocked only by the chain along $bd(P)$ from q counterclockwise to p.

We now consider the computation of $s(we_i)$ with respect to the points on RC_i. We further partition RC_i into two subchains: (a) The chain from the endpoint v of RC_i (as defined in Lemma 2) *clockwise* to $lv(we_{i+1})$ excluding $lv(we_{i+1})$, denoted by RC_i^r, and (b) the chain from $lv(we_{i+1})$ *clockwise* to $lv(we_i)$, denoted by RC_i^l. The following lemmas are useful in computing $s(we_i)$.

Lemma 4 *For a point p on RC_i, if $fv(we_i)$ is not visible from p, then $lv(we_i)$ is visible from* p.

Proof. By Lemma 3, the view between p and $fv(we_i)$ cannot be blocked by the chain along $bd(P)$ from p counterclockwise to $fv(we_i)$. So if $fv(we_i)$ is not visible from p, the view must be blocked by the chain C' along $bd(P)$ from p *clockwise* to $fv(we_i)$. But if the view between p and $lv(we_i)$ were also blocked by C', then p would have not been weakly visible from we_i, a contradiction. □

Corollary 1 *For a point p on RC_i, if $lv(we_i)$ is not visible from p, then $fv(we_i)$ is visible from* p.

Proof. An immediate consequence of Lemma 4. □

Corollary 2 *Let p be a point on RC_i^l. If p is not visible from $lv(we_i)$, then the subchain of RC_i^l from p clockwise to $lv(we_i)$ defines a point p' on we_i such that the segment $\overline{lv(we_i)p'}$ is the maximal segment on we_i that is not visible from p.*

Proof. An immediate consequence of Lemma 3 and Corollary 1. □

Lemma 5 *Let p be a point on RC_i^r, and let we_j be the edge of WVE such that we_j does not contain p and that we_j is the first edge encountered among the edges of WVE when walking along RC_i^r from p clockwise to $lv(we_{i+1})$. Then if p is not visible from $lv(we_i)$, then a vertex of we_j must define a point p' on we_i such that the segment $\overline{lv(we_i)p'}$ is the maximal segment on we_i that is not visible from p.*

Proof. Let p' be the point on we_i such that segment $\overline{lv(we_i)p'}$ is the maximal segment on we_i that is not visible from p. Note that p' can be $lv(we_i)$ because $lv(we_i)$ is not visible from p. By Lemma 3, the view between p and every point q on $\overline{lv(we_i)p'}$ can be blocked only by the chain along RC_i from p clockwise to q. If p' were defined by a point on the chain along RC_i from p clockwise to $lv(we_j)$ excluding $lv(we_j)$, then p would have not been weakly visible from we_j, a contradiction. If p' were defined by a point on the chain along RC_i from $fv(we_j)$ clockwise to p' excluding $fv(we_j)$, then $lv(we_i)$ would have not been weakly visible from we_j, again a contradiction. Hence only the vertices of we_j can define p' on we_i for p. □

Note that in Corollary 2 and Lemma 5, point $p \in RC_i$ is visible from every point on the segment $\overline{fv(we_i)p'} \subset we_i$ (this follows from Corollary 1). For every point p on RC_i^r, by Lemma 5, point p' on we_i can be easily computed. For every point p on RC_i^l, by Corollary 2, point p' on we_i can be found out if the line segment that is on the internal convex path from $lv(we_i)$ to p along RC_i^l and that is adjacent to p is known.

We define a total order on the points of we_i, as follows: For every two points q' and q'' on we_i, $q' \leq q''$ iff segment $\overline{fv(we_i)q'}$ is contained by segment $\overline{fv(we_i)q''}$. Let edge we_i correspond to the interval $[fv(we_i), lv(we_i)]$. For every vertex v_k of RC_i, let $[lp_k, rp_k]$ be the interval on we_i such that segment $\overline{lp_k rp_k}$ is the maximal segment on we_i that is visible from v_k. We denote $[lp_k, rp_k]$ by I_k. Note that it is possible that $lp_k = rp_k$. For example, if the only point on we_i from which v_k is visible is $lv(we_i)$, then $I_k = [lv(we_i), lv(we_i)]$. The intervals I_k have the following property:

Lemma 6 *For every pair of consecutive vertices v_k and v_{k+1} of RC_i, $I_k \cap I_{k+1} \neq \emptyset$.*

Proof. There are three cases to consider. If I_{k+1} is equal to $[lv(we_i), lv(we_i)]$, then I_k is also equal to $[lv(we_i), lv(we_i)]$, by Lemma 3 (otherwise, v_k would have not been weakly visible from we_i). If I_k is equal to $[lv(we_i), lv(we_i)]$ but I_{k+1} is not, then I_{k+1} must be equal to $[fv(we_i), lv(we_i)]$ (this also follows from Lemma 3). If both I_k and I_{k+1} are not equal to $[lv(we_i), lv(we_i)]$, then they must both contain $fv(we_i)$. □

From the intervals I_k of the vertices v_k on RC_i, we define a set of intervals on we_i, called the *characteristic intervals*, as follows: For every edge e_j on RC_i, let

$$CI_j = I_j \cap I_{j+1},$$

and call CI_j the *characteristic interval* of e_j. The next lemma illustrates the relation between $s(we_i)$ and the characteristic intervals for the edges of RC_i.

Lemma 7 *The shortest weakly visible subedge $s(we_i)$ on we_i must contain at least one point on interval CI_j, for every edge e_j on RC_i.*

Proof. This follows from the fact that edge e_j is completely visible from every point on interval CI_j (see Figure 7). □

4 Algorithms for the SWVS Problem

We need some simple notation for presenting the algorithms (we only give that with respect to the chain RC_i). If vertex $lv(we_i)$ is nonconvex, then let r_i be the ray originating from $fv(we_i)$ and passing $lv(we_i)$, and let r_i first hit $bd(P) - we_i$ at point h_i. Denote the chain along $bd(P)$ from $lv(we_i)$ counterclockwise to h_i by RP_i (called the *right pocket* of we_i). The following properties of RP_i are easily seen to be true:

- The chain $RP_i - \{lv(we_i), h_i\}$ can intersect at most two edges of WVE (i.e., we_{i+1} and we_{i+2}), and if this is the case, then $C_i = \emptyset$.

- Point h_i is contained in chain RC_i and is the first point on RP_i that intersects ray r_i, where r_i is viewed as a half-line (otherwise, some point on RP_i would have not been weakly visible from we_i).

- The only point on we_i from which every point on $RP_i - \{lv(we_i), h_i\}$ is visible is $lv(we_i)$.

For every vertex v_k of $RC_i - lv(we_i)$, let ICP_i^k be the internal convex path connecting $lv(we_i)$ and v_k, and let w_k be the line segment on ICP_i^k that is adjacent to v_k. Let r_i^k be the ray originating from v_k and containing w_k. Note that r_i^k must intersect we_i. Let the intersection point of r_i^k and we_i be ip_i^k.

The general procedure for solving the SWVS problem is as follows.

Algorithm SWVS.

Input. A simple polygon P of n vertices.

Output. The shortest weakly visible subedge s of P.

(1) Compute WVE for P. If $|WVE| < 7$, then compute $s(we_i)$ on every edge $we_i \in WVE$, find s from these $s(we_i)$'s, and stop.

(2) For every $we_i \in WVE$, perform the following computation on chain RC_i:

   (2.1) Vertex $lv(we_i)$ is convex. For every vertex v_k on RC_i^l, compute the segment w_k on ICP_i^k (by Corollary 2) and the intersection point ip_i^k between r_i^k and we_i; let interval I_k be $[fv(we_i), ip_i^k]$. For every vertex v_k on RC_i^r, compute ip_i^k by Lemma 5, and let I_k be $[fv(we_i), ip_i^k]$.

   (2.2) Vertex $lv(we_i)$ is nonconvex. For every vertex v_k on $RP_i - lv(we_i)$, let interval I_k be $[lv(we_i), lv(we_i)]$. For every vertex v_k on $RC_i^l - RP_i$ (resp., $RC_i^r - RP_i$), compute I_k as in Step (2.1), by using Corollary 2 (resp., Lemma 5).

   (2.3) Compute the characteristic interval CI_j for every edge e_j on RC_i. Let the set of characteristic intervals so obtained be I_i^R.

(3) For every $we_i \in WVE$, perform computation similar to Step (2) on chain LC_i. Let the set of characteristic intervals so obtained be I_i^L.

(4) For every $we_i \in WVE$, compute $s(we_i)$ as follows: Let

$$\alpha_i = \max\{lp_k \mid [lp_k, rp_k] \in I_i^R \cup I_i^L\},$$

and

$$\beta_i = \min\{rp_k \mid [lp_k, rp_k] \in I_i^R \cup I_i^L\}.$$

If $\alpha_i \leq \beta_i$, then let $s(we_i)$ be *any* point on interval $[\alpha_i, \beta_i]$; otherwise, let $s(we_i) = [\beta_i, \alpha_i]$.

(5) Let $s = s(we_j)$, where

$$|s(we_j)| = \min\{|s(we_i)| \mid we_i \in WVE\}.$$

Lemma 8 Algorithm SWVS *can be implemented sequentially in $O(n)$ time, and in parallel in $O(\log n)$ time using $O(n/\log n)$ CREW PRAM processors.*

Proof. The sequential implementation of **Algorithm SWVS** is done by using [1, 7, 20, 21]. The parallel implementation is done by using [5, 6]. The details are given in the full paper. □

5 The Rectilinear Case

In this section, we study the special case of the SWVS problem where the polygons are rectilinear (i.e., each edge of the polygons is either vertical or horizontal). We present very simple and optimal algorithms to solve this case, both sequentially and in parallel. As for the general SWVS problem, we also give interesting geometric observations for the rectilinear case. These geometric observations enable us to design extremely simple solutions to this case. The parallel computational model we use in this section is the EREW PRAM. Due to the space limitation, the proofs in this section are omitted and are left to the full paper.

In the rest of this section, we let P be a rectilinear simple polygon of n vertices. For a subchain $C(i, i+3) = (e_i, e_{i+1}, e_{i+2})$ of $bd(P)$, we call $C(i, i+3)$ a *concave chain* of P and call edge e_{i+1} the *center edge* of $C(i, i+3)$ iff e_{i+1} is nonconvex (i.e., the interior angles of P at v_{i+1} and v_{i+2} are both greater than π). Let the line containing an edge e_j be denoted by $l(e_j)$. We say that a concave chain $C(i, i+3)$ is *upward* (resp., *downward, leftward, rightward*) if e_{i+1} is horizontal (resp., horizontal, vertical, vertical) and if no point on $C(i, i+3)$ is strictly above (resp., below, to the left of, to the right of) line $l(e_{i+1})$.

For every vertex v_i of P, if v_{i+1} (resp., v_{i-1}) is nonconvex, then let r_i^+ (resp., r_i^-) denote the ray starting at v_i and containing e_i (resp., e_{i-1}). If ray r_i^+ (resp., r_i^-) is associated with v_i, then let $h(r_i^+)$ (resp., $h(r_i^-)$) denote the point on $bd(P) - e_i$ (resp., $bd(P) - e_{i-1}$) that is first hit by r_i^+ (resp., (r_i^-)).

A subchain C of $bd(P)$ is said to be *x-monotone* (resp., *y-monotone*) iff the intersection between C and every vertical (resp., horizontal) line is a single connected component. A subchain C' of $bd(P)$ is said to be a *staircase* iff C' is both x-monotone and y-monotone. Polygon P is said to be *x-monotone* (resp., *y-monotone*) iff $bd(P)$ can be partitioned into two x-monotone (resp., y-monotone) chains.

Let $C(i, i+3)$ be an upward concave chain (the other cases are similar). Then the following properties of $C(i, i+3)$ can be easily seen to hold (see Figure 8).

- The only possible weakly visible edge of P on $C(i, i+3)$ is the center edge e_{i+1} of $C(i, i+3)$.

- If $e_{i+1} \in WVE$, then the following are true:

  1. The subchain of $bd(P)$ from $h(r_{i+3}^-)$ counterclockwise to $h(r_i^+)$ is x-monotone.
  2. Both vertices v_i and v_{i+3} are convex.
  3. The subchain of $bd(P)$ from v_i (resp., v_{i+3}) clockwise (resp., counterclockwise) to $h(r_{i+2}^-)$ (resp., $h(r_{i+1}^+)$) is a staircase, and the subchain of $bd(P)$ from $h(r_{i+2}^-)$ (resp., $h(r_{i+1}^+)$) clockwise (resp., counterclockwise) to $h(r_i^+)$ (resp., $h(r_{i+3}^-)$) is a staircase, as shown in Figure 8.
  4. $s(e_{i+1}) = e_{i+1}$.

The following lemmas are useful for our algorithms.

Lemma 9 *If polygon P has two concave chains $C(i, i+3)$ and $C(j, j+3)$, where $C(i, i+3)$ is either upward or downward and $C(j, j+3)$ is either leftward or rightward, then P is not weakly visible from any of its edges.*

Corollary 3 *If polygon P is neither x-monotone nor y-monotone, then P is not weakly visible from any of its edges.*

Lemma 10 *Suppose that polygon P has only upward and downward (resp., leftward and rightward) concave chains. Let $C(i, i+3)$ be such a concave chain. Then WVE consists of at most two edges of P: (1) the center edge e_{i+1} of $C(i, i+3)$, and (2) the edge e_j such that e_j contains both the points $h(r_i^+)$ and $h(r_{i+3}^-)$ (if such an edge e_j exists).*

Lemma 11 *Suppose that polygon P is x-monotone and has two distinct upward (resp., downward) concave chains $C(i, i+3)$ and $C(j, j+3)$. Then P can possibly be weakly visible from at most one edge e_k such that e_k contains all the points $h(r_i^+)$, $h(r_{i+3}^-)$, $h(r_j^+)$, and $h(r_{j+3}^-)$ (if such an edge e_k exists).*

Lemma 12 *Let C be a staircase chain on $bd(P)$. Then if C has more than four edges, then no edge on C can be in WVE.*

Let e_j be an edge in WVE such that e_j is on a staircase of $bd(P)$ and that e_j is not the center edge of any concave chain of P. Then there are two possible cases for e_j: Either both the vertices of e_j are convex or exactly one vertex of e_j is convex. We consider first the case where exactly one vertex of e_j is convex. WLOG, let v_j be convex and v_{j+1} be nonconvex (the case where v_j is nonconvex and v_{j+1} is convex is symmetric). It is easy to see that the following properties with respect to e_j hold:

- Vertices v_{j-1} and v_{j+2} are both convex. Therefore, e_j must be adjacent to an ending edge of a maximal staircase of $bd(P)$.

- Suppose that line $l(e_j)$ is horizontal (the other case is similar). Then the subchain of $bd(P)$ from $h(r_{j+2}^-)$ counterclockwise to v_{j-1} is x-monotone.

- The subchain of $bd(P)$ from v_{j+2} counterclockwise to $h(r_j^+)$ is a staircase, and the subchain of $bd(P)$ from $h(r_j^+)$ counterclockwise to $h(r_{j+2}^-)$ is a staircase.

- Let $H_j = \{h(r_k^-) \mid v_k$ is on the subchain of $bd(P)$ from $h(r_{j+2}^-)$ counterclockwise to v_{j-1} and v_{k-1} is nonconvex$\}$. If $H_j = \emptyset$, then $s(e_j) = v_{j+1}$. Otherwise, let α_j be the point in H_j that is closest to v_j among all the points in H_j; then $s(e_j) = \overline{\alpha_j v_{j+1}}$.

In the case where both the vertices of $e_j \in WVE$ are convex, the following properties with respect to e_j hold:

- Vertices v_{j-1} and v_{j+2} are both convex. Hence e_j is an ending edge of a maximal staircase of $bd(P)$.

- The subchain of $bd(P)$ from v_{j+2} counterclockwise to v_{j-1} (i.e., $C(j+2, j-1)$) is monotone with respect to line $l(e_j)$.

- Let $RH_j = \{h(r_k^-) \mid v_k$ is on $C(j+2, j-1)$ and v_{k-1} is nonconvex$\}$, and let $LH_j = \{h(r_k^+) \mid v_k$ is on $C(j+2, j-1)$ and v_{k+1} is nonconvex$\}$. Let α_j (resp., β_j) be the point in RH_j (resp., LH_j) that is closest to v_j (resp., v_{j+1}) among all the points in RH_j (resp., LH_j). If both α_j and β_j do not exist, then $s(e_j)$ can be any point on e_j. If exactly β_j (resp., α_j) does not exist, then $s(e_j)$ can be any point on the segment $\overline{\alpha_j v_j}$ (resp., $\overline{\beta_j v_{j+1}}$). If both α_j and β_j exist, then $s(e_j) = \overline{\alpha_j \beta_j}$.

Lemma 13 *For polygon P, if $WVE \neq \emptyset$, then $|WVE| = O(1)$.*

Our results on solving the rectilinear case of the SWVS problem are summarized in the following lemma.

Lemma 14 *Given a rectilinear polygon P, there are extremely simple and optimal algorithms for computing, both sequentially and in parallel, the following information: (i) WVE, and (ii) the shortest weakly visible subedge s of P. The sequential algorithm runs in $O(n)$ time, and the parallel algorithm runs in $O(\log n)$ time using $O(n/\log n)$ EREW PRAM processors.*

331

References

[1] D. Avis and G. T. Toussaint. "An optimal algorithm for determining the visibility polygon from an edge," *IEEE Trans. Comput.*, C-30 (12) (1981), pp. 910–914.

[2] B. K. Bhattacharya, D. G. Kirkpatrick, and G. T. Toussaint. "Determining sector visibility of a polygon," *Proc. 5-th Annual ACM Symp. Computational Geometry*, 1989, pp. 247–254.

[3] B. K. Bhattacharya, A. Mukhopadhyay, and G. T. Toussaint. "A linear time algorithm for computing the shortest line segment from which a polygon is weakly externally visible," *Proc. Workshop on Algorithms and Data Structures (WADS'91)*, 1991, Ottawa, Canada, pp. 412–424.

[4] B. Chazelle and L. J. Guibas. "Visibility and intersection problems in plane geometry," *Discrete and Computational Geometry*, 4 (1989), pp. 551–581.

[5] D. Z. Chen. "An optimal parallel algorithm for detecting weak visibility of a simple polygon," *Proc. of the Eighth Annual ACM Symp. on Computational Geometry*, 1992, pp. 63–72.

[6] D. Z. Chen. "Parallel techniques for paths, visibility, and related problems," Ph.D. thesis, Technical Report No. 92-051, Dept. of Computer Sciences, Purdue University, July 1992.

[7] Y. T. Ching, M. T. Ko, and H. Y. Tu. "On the cruising guard problems," Technical Report, 1989, Institute of Information Science, Academia Sinica, Taipei, Taiwan.

[8] J. I. Doh and K. Y. Chwa. "An algorithm for determining visibility of a simple polygon from an internal line segment," *Journal of Algorithms*, 14 (1993), pp. 139–168.

[9] H. ElGindy. "Hierarchical decomposition of polygon with applications," Ph.D. thesis, McGill University, 1985.

[10] S. K. Ghosh, A. Maheshwari, S. P. Pal, S. Saluja, and C. E. V. Madhavan. "Characterizing weak visibility polygons and related problems," Technical Report No. IISc-CSA-90-1, 1990, Dept. Computer Science and Automation, Indian Institute of Science.

[11] M. T. Goodrich, S. B. Shauck, and S. Guha. "Parallel methods for visibility and shortest path problems in simple polygons (Preliminary version)," *Proc. 6-th Annual ACM Symp. Computational Geometry*, 1990, pp. 73–82.

[12] L. J. Guibas, J. Hershberger, D. Leven, M. Sharir, and R. E. Tarjan. "Linear time algorithms for visibility and shortest paths problems inside triangulated simple polygons," *Algorithmica*, 2 (1987), pp. 209–233.

[13] P. J. Heffernan and J. S. B. Mitchell. "Structured visibility profiles with applications to problems in simple polygons," *Proc. 6-th Annual ACM Symp. Computational Geometry*, 1990, pp. 53–62.

[14] J. Hershberger. "Optimal parallel algorithms for triangulated simple polygons," *Proc. 8-th Annual ACM Symp. Computational Geometry*, 1992, pp. 33–42.

[15] Y. Ke. "Detecting the weak visibility of a simple polygon and related problems," manuscript, Dept. of Computer Science, The Johns Hopkins University, 1988.

[16] C. P. Kruskal, L. Rudolph, and M. Snir. "The power of parallel prefix," *IEEE Trans. Comput.*, C-34 (1985), pp. 965–968.

[17] R. E. Ladner and M. J. Fischer. "Parallel prefix computation," *Journal of the ACM*, 27 (4) (1980), pp. 831–838.

[18] D. T. Lee and A. K. Lin. "Computing the visibility polygon from an edge," *Computer Vision, Graphics, and Image Processing*, 34 (1986), pp. 1–19.

[19] S. H. Lee and K. Y. Chwa. "Some chain visibility problems in a simple polygon," *Algorithmica*, 5 (1990), pp. 485–507.

[20] J.-R. Sack and S. Suri. "An optimal algorithm for detecting weak visibility of a polygon," *IEEE Trans. Comput.*, C-39 (10) (1990), pp. 1213–1219.

[21] S. Y. Shin. "Visibility in the plane and its related problems," Ph.D. thesis, University of Michigan, 1986.

[22] G. T. Toussaint. "A linear-time algorithm for solving the strong hidden-line problem in a simple polygon," *Pattern Recognition letters*, 4 (1986), pp. 449–451.

[23] G. T. Toussaint and D. Avis. "On a convex hull algorithm for polygons and its applications to triangulation problems," *Pattern Recognition*, 15 (1) (1982), pp. 23–29.

332

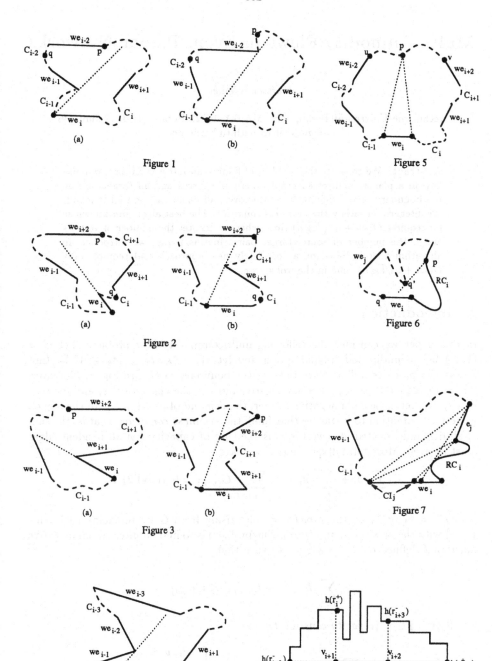

Figure 1

Figure 5

Figure 2

Figure 6

Figure 3

Figure 7

Figure 4

Figure 8

Multicommodity Flows in Even, Planar Networks

Karsten Weihe

Technische Universität Berlin, Sekr. MA 6-1, 10623 Berlin, Str.d. 17.Juni 136
e-mail: weihe@math.tu-berlin.de

Abstract. We consider the problem of finding an integral multicommodity flow in a planar, undirected graph where all sources and all targets are on the boundary of the infinite face. Moreover, all capacities and all demands are integer and satisfy the *evenness condition*. The best algorithm known so far requires $\mathcal{O}(kn + n^2\sqrt{\log n})$ time, where n denotes the number of vertices and k the number of source-target pairs. In this paper, we introduce an algorithm that is based on a completely new approach and requires only $\mathcal{O}(kn + n\sqrt{\log n})$ time in the worst case.

1 Introduction

In this paper we consider the following multicommodity flow problem. Let $G = (V, E)$ be an undirected, planar graph and let $s_1, \ldots, s_k, t_1, \ldots, t_k \in V$ be (not necessarily pairwise different) vertices on the boundary of G. See Fig. 1. Moreover, for each edge $e \in E$ let C_e be a nonnegative integer, the *capacity* of e, and for each pair $\{s_i, t_i\}$ let d_i be a nonnegative integer, the *demand* of pair $\{s_i, t_i\}$. In the sequel, we restrict attention to instances that fulfill the *evenness condition*, that is, the sum of all demands located at a vertex *plus* the sum of capacities of all incident edges must be even. More formally, for $v \in V$ we have

$$\sum_{i:s_i=v} d_i + \sum_{i:t_i=v} d_i + \sum_{w:\{v,w\}\in E} C_{\{v,w\}} = 0 \pmod 2 \ .$$

Let $G^{\rightarrow} = (V, E^{\rightarrow})$ be the *directed* graph arising from G by replacing each edge $\{v, w\}$ with the arcs (v, w) and (w, v). The problem is to find an integral, nonnegative function f defined on $E^{\rightarrow} \times \{1, \ldots, k\}$ such that

$$\sum_{i=1}^{k} (f_{(v,w),i} + f_{(w,v),i}) \leq C_{\{v,w\}}$$

for all $\{v, w\} \in E$ (*capacity constraints*) and

$$\sum_{w:\{v,w\}\in E} (f_{(v,w),i} - f_{(w,v),i}) = \begin{cases} d_i, & v = s_i, \\ -d_i, & v = t_i, \\ 0, & \text{else}, \end{cases}$$

for all $i = 1, \ldots, k$ and $v \in V$ (*flow conservation conditions*).

The problem can be solved in strongly polynomial time by means of linear programming [10], since the polytope of all real-valued solutions is integral. On the other hand, Okamura and Seymour proved a *purely combinatorial duality theorem*,

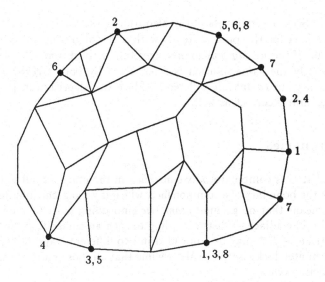

Fig. 1. An instance with eight pairs of terminals.

which states that an instance is solvable if and only if no cut is oversaturated [9]. That is, for no cut does the sum of all demands of source-target pairs separated by this cut exceed the capacity of this cut. All purely combinatorial algorithm known so far rely on this theorem. Hassin was the first to give a purely combinatorial algorithm, which requires $\mathcal{O}(n^4)$ time [5]. The fastest algorithm in the literature is due to Matsumoto, Nishizeki and Saito [8]. This algorithm uses decomposition techniques of Frederickson [1] and requires $\mathcal{O}(kn + n^2\sqrt{\log n})$ time.

A completely new approach to such problems, which does not make use of the Theorem of Okamura and Seymour (nor of any other strong duality type theorem) has been introduced in [11]. There only the special case is considered where $C_{\{v,w\}} = 1$ for each edge $\{v, w\} \in E$. In other words, the problem is to find edge-disjoint paths from s_i to t_i, respectively, with the additional restriction that the augmented graph $(V, E + \{s_1, t_1\} + \cdots + \{s_k, t_k\})$ be Eulerian. This special case is there solved in linear time. Clearly, that result immediately yields a weakly polynomial algorithm for the general capacitated case, simply by replacing each edge $\{v, w\} \in E$ with $C_{\{v,w\}}$ many copies.

In this paper now, we will show that a *strongly* polynomial algorithm for the general case can be derived from that approach as well. This algorithm requires $\mathcal{O}(kn + n\sqrt{\log n})$ time and $\mathcal{O}(kn)$ space. Note that the algorithm is asymptotically optimal for sufficiently large k, that is $k \geq \text{const} \cdot \sqrt{\log n}$, since the size of the output is obviously $\Theta(kn)$ in the worst case. Further note that, in particular, the relaxation of this problem to rational input and output data can be solved within the same time and space.

The paper is organized as follows. In Sect. 2 we will give some useful terminology; in Sect. 3 we will introduce the new algorithm and prove the claimed worst case bound; and in Sect. 4 correctness of the algorithm is sketched.

The main idea for the proof is to show that the algorithm introduced in this paper only fails if the *weakly* polynomial algorithm derived from [11] fails, and that otherwise exactly the same solution is constructed by both algorithms, although these algorithms run *completely* differently. Since the algorithm in [11] is correct, the new algorithm is correct as well.

2 Preliminaries

We assume that G is combinatorially embedded in the plane s.t. all sources and all targets lie on the boundary of a common face, which is w.l.o.g. the infinite face. *Combinatorially* means that no genuine *geometric* embedding must be known, only the sorting of all cyclic adjacency lists according to such an embedding. Note that each pair (v, w), $(w, v) \in E^\rightarrow$ may be embedded in two different ways: either as a clockwise or as a counterclockwise cycle. We assume that all those pairs are embedded as counterclockwise cycles.

We now give a "standard form" for input instances, which will simplify notation in what follows. Let s_{i_0} be an arbitrary s-terminal, which is fixed in the sequel. If, for $i \neq i_0$, the counterclockwise ordering of s_{i_0}, s_i, and t_i around G is given by $s_{i_0} \preceq t_i \prec s_i$, then we exchange s_i with t_i. Moreover, if the counterclockwise ordering of s_{i_0}, t_i, and t_j around G is given by $s_{i_0} \preceq t_i \prec t_j$ for $i > j$, we exchange the i-th pair with the j-th pair. Therefore, when we "walk" around the boundary of G counterclockwise, starting with s_{i_0}, then we do not pass t_i before s_i for $i = 1, \ldots, k$, and we do not pass t_{i+1} before t_i for $i < k$.

W.l.o.g. we assume that all s-terminals are disjoint with all t-terminals, and that all terminals have degree 1. This can be ensured as follows. For each vertex $v \in V$ where s-terminals (resp., t-terminals) are located, add a new vertex w to V and an edge $\{v, w\}$ to E, embedded in the infinite face, and shift all those terminals from v to w. The capacity of the new edge $\{v, w\}$ is the sum of all demands of terminals at w. Clearly, that does not change the asymptotic input size. Moreover, we assume that $G \setminus \{s_1, \ldots, s_k, t_1, \ldots, t_k\}$ is biconnected, because otherwise we could alternatively solve one auxiliary subproblem for each biconnected component. See Fig. 2.

Let p and q be two different *directed* paths in G or $G^\rightarrow$. Assume that p and q do not cross themselves nor each other and have the same startvertices and endvertices, respectively. Moreover, assume that both the startvertex and the endvertex lie on the boundary of the infinite face. Then p and q divide the rest of the graph into two *sides*, the *left side* and the *right side*, w.r.t. their orientations. In this case, p is said to be *more right than* q, if the right side of p is a proper subset of the right side of q.

Remark. It is very important to realize that a path which does not cross itself need not be simple. But the vertex where a cycle of such a path is closed is **no** crossing of this path. Moreover, in $G^\rightarrow$ such a path may even contain cycles of length 2, that is, pairs (v, w), $(w, v) \in E^\rightarrow$. Whether or not such a pair may be inserted in a non-crossing path without enforcing a crossing, highly depends on the common embedding of the pair. Remember that all such pairs in $E^\rightarrow$ are assumed to be commonly embedded such that they form counterclockwise cycles. □

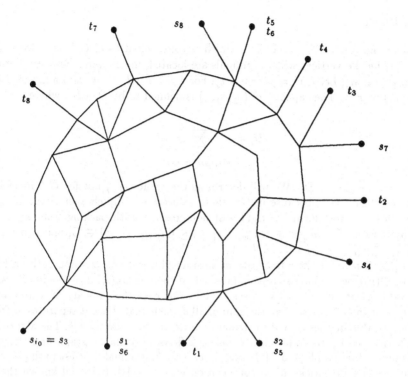

Fig. 2. An instance that is completely equivalent to the instance of Fig. 1, but satisfies all additional restrictions imposed in Sect. 2.

A *non-crossing path packing* in $G^\rightarrow$ from s_i to t_j, say, is an (s_i, t_j)-flow that can be decomposed into (s_i, t_j)-paths such that no path crosses itself and no two paths cross each other. The *rightmost non-crossing path packing from s_i to t_j with flow value d* is the unique feasible flow f constructed as follows (if possible): Let f initially be the zero flow. While the total value of f is smaller than d, increase f by one unit along the rightmost non-crossing, augmenting (s_i, t_j)-path. Obviously, the resulting flow is actually a non-crossing path packing.

3 The Algorithm

Basicly, our algorithm consists of two phases. In the first phase, we determine a new capacity $c_{(v,w)}$ for each arc $(v, w) \in E^\rightarrow$. These values will fulfill $c_{(v,w)} + c_{(w,v)} \le C_{\{v,w\}}$. In the second phase, we solve the *directed* multicommodity flow problem in $G^\rightarrow$ with the original sources s_i, targets t_i, and demands d_i, and with the new capacities $c_{(\cdot,\cdot)}$. The second phase produces a correct solution, unless the input instance itself is unsolvable. Clearly, otherwise a solution to this directed problem immediately yields a solution to the original problem.

First Phase

Consider the graph $G^{\Rightarrow} = (V, E^{\Rightarrow})$ which is constructed from $G^{\rightarrow}$ as follows. Let $v_1, \ldots, v_l$ be the vertices where terminals are located, in counterclockwise ordering, starting with s_{i_0}. Let $n(v) = j$ for $v = v_j$. For $j = 1, \ldots, l-1$, we add a directed arc (v_j, v_{j+1}) to $E^{\rightarrow}$. The capacity of (v_j, v_{j+1}) is defined to be $D - D_j$, where

$$D_j = \sum_{\substack{i=1 \\ n(s_i) \leq j, \, n(t_i) > j}}^{k} d_i$$

and $D = \max_{j=1,\ldots,k} D_j$. We will determine the values $c_{(\cdot,\cdot)}$ not for $G^{\rightarrow}$ itself but for its extension $G^{\Rightarrow}$, because in $G^{\Rightarrow}$ these values are easily characterized: In $G^{\Rightarrow}$, the values $c_{(\cdot,\cdot)}$ just form the rightmost non-crossing path packing from s_{i_0} to t_k with flow value D, subject to $c_{(v,w)} + c_{(w,v)} \leq C_{\{v,w\}}$, $\{v,w\} \in E$, and subject to $c_{(v_j,v_{j+1})} \leq C_{(v_j,v_{j+1})}$, $j = 1, \ldots, l-1$.

In [12], it is shown how to construct such a rightmost non-crossing path packing in time $\mathcal{O}(n\sqrt{\log n})$. This algorithm fails only if there is no (s_{i_0}, t_k)-flow in $G^{\Rightarrow}$ with flow value D at all. It is easy to see that in this case the capacity of a minimum (s_{i_0}, t_k)-cut in $G^{\rightarrow}$ is less than the sum of all d_i such that this cut separates s_i from t_i. Clearly, this implies that the original input instance is unsolvable. The algorithm in [12] is based on the interrelation between flows in a planar graph and shortest paths in the dual graph [4, 6, 7]. In principle, there first a dual shortest path problem is solved using Frederickson's decomposition technique [1]. It is well known that a maximum flow can be easily constructed in linear time from the resulting labeling of all faces. However, in general this maximum flow is not the rightmost non-crossing path packing. In order to construct this particular flow, first the dual labelings are modified slightly, before the translation into a primal flow may take place. We refer to [12] for further details.

Second Phase

In the second phase, we now consider the *directed* multicommodity flow problem formed by the directed graph $G^{\rightarrow}$, the terminals $s_1, \ldots, s_k, t_1, \ldots, t_k$, the demands $d_1, \ldots, d_k$, and the new capacities $(c_e)_{e \in E^{\rightarrow}}$. More precisely, the second phase works on $G_c = (V, E_c^{\rightarrow})$, where $E_c^{\rightarrow} = \{e \in E^{\rightarrow} : c_e > 0\}$. The second phase consists in a certain, very simple, variant on the preflow-push algorithm of Goldberg and Tarjan [3], which is tailored to the problem considered here and does not need any kind of distance labeling. All commodities are pushed in parallel, which enables us to touch each arc in $E_c^{\rightarrow}$ at most once. A formal description of the algorithm is given in Table 1. In the sequel, we now explain this algorithm informally.

For simplicity, we assume that the output of the algorithm consists of an integer array $f[\cdot, \cdot]$ with range $E_c^{\rightarrow} \times \{1, \ldots, k\}$, where $f[e, i]$ means the amount of the *i-th* commodity pushed through arc e. In principle, these arrays could be replaced by lists in order to reduce the space required. However, this complicates the description of the algorithm significantly, but would not reduce the *asymptotic* space requirement of $\Theta(kn)$, which is obviously optimal in the worst case.

Table 1. Formal description of the second phase

PROCEDURE *initialize* ();

    FOR $v \in V$ DO

        divide all arcs entering v into maximal subintervals of the adjacency list of v;

        initialize the union-find structure of v with these subintervals;

    FOR each subinterval X AND

        $i = 1, \ldots, k$ DO $excess_X[i] := 0$;

    FOR $i = 1, \ldots, k$ AND $e \in E_c^{\rightarrow}$

        DO $f[e, i] := 0$;

    make list L empty;

    FOR $i = 1, \ldots, k$ DO

        $f[(s_i, w_i), i] := d_i$;

        $X := find((s_i, w_i))$;

        $excess_X[i] := d_i$;

        IF $all\_touched(X)$

            THEN $L.insert(X)$;

PROCEDURE push_flow

    ($e \in E_c^{\rightarrow}$; VAR exc : excess array);

    $v := tail(e)$;

    $w := head(e)$;

    $X := find(e)$;

    FOR $i := 1$ TO k DO

        $y := \min\{c_e, exc[i]\}$;

        $f[e, i] := f[e, i] + y$;

        $exc[i] := exc[i] - y$;

        $excess_X[i] := excess_X[i] + y$;

        $c_e := c_e - y$;

    IF w is no t-terminal AND

    $all\_touched(X)$

    THEN $L.insert(X)$;

PROCEDURE handle

    (X : arc set entering $v \in V$);

    $e :=$ the counterclockwise next arc after X in the adjacency list of v;

    REPEAT

        $push\_flow(a, excess_X)$;

        $e :=$ the counterclockwise next arc after e in the adjacency list of v;

    UNTIL e enters v;

    IF $X \neq next\_in(X)$

        THEN $unite(X, next\_in(X))$;

(* Main Routine *)

    initialize ();

    REPEAT

        $X := L.remove()$;

        $handle(X)$;

    UNTIL $L.empty()$;

    IF the values $f[\cdot, \cdot]$ define a correct solution

    THEN return $f[\cdot, \cdot]$

    ELSE return "unsolvable";

For each $v \in V$, we define a *union-find structure* on all arcs in $E_c^{\rightarrow}$ entering v [2]. In the sequel, the sets in the union-find structure of v are called the *arc sets entering* v or simply *arcs sets*, if this does not lead to ambiguities. Note that the set of all arcs entering v can be decomposed into maximal subintervals such that there is no arc in $E_c^{\rightarrow}$ leaving v within such an interval. This is the initial collection of sets in the union-find structure of v. For an arc set X entering v, let $next\_in(X)$ denote the counterclockwise next arc set entering v in the adjacency list of v (Fig. 3). Each union operation unites an arc set X with $next\_in(X)$, and the name of the union is the name of $next\_in(X)$, which means that X is dropped. Of course, X and $next\_in(X)$ are united only if $X \neq next\_in(X)$, that is, more than one arc set entering v has "survived" the union operations at vertex v up to now.

Rather than excesses of vertices as in the usual preflow-push algorithm, we manage the excess of each arc set, that is, the amount of flow already pushed through this arc set and not pushed further yet. For this aim, we manage an array $excess_X$ with range $1, \ldots, k$ for each arc set X. The i-th component, $excess_X[i]$, stores the amount of the i-th commodity already pushed through the arcs of X. Hence, whenever we unite an arc set X with $next\_in(X)$, the excess array of the union is the component-wise sum of $excess_X$ and $excess_{next\_in(X)}$.

Throughout the second phase of the algorithm, we manage a list L of arc sets. An arc set X is inserted in L, once all arcs in X are touched (indicated by $all\_touched(X)$ = TRUE, for short). The core of the algorithm is a loop where in each iteration we remove an arc set X, say, from L and perform Procedure $handle(X)$, which we describe next.

Let X be an arc set entering $v \in V$ and let $next\_out(X)$ denote the set of all arcs in the adjacency list of v that appear counterclockwise after X and before $next\_in(X)$. Then all arcs in $next\_out(X)$ leave v. See Fig. 3. Procedure $handle(X)$ pushes as much of the excess of X as possible through the arcs of $next\_out(X)$. More precisely, $handle(X)$ first pushes as much as possible of the excess of the first commodity, then as much as possible of the excess of the second commodity, and so on. The arcs of $next\_out(X)$ are considered counterclockwise, after one another.

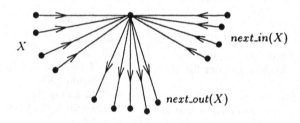

Fig. 3. An initial arc set X entering v and the related sets $next\_in(X)$ and $next\_out(X)$.

After that, Procedure $handle(X)$ unites X with $next\_in(X)$. This maintains as an invariant that, at each stage of the second phase, the current arc sets entering v just form the "coarsest" partition such that all untouched arcs leaving v are located between two such arc sets. In particular, all arcs leaving v that are enclosed in an arc set entering v are already touched.

In [11] it was shown how to apply the union-find structure of Gabow and Tarjan [2] to a similar situation. This data structure would work properly only if the adjacency lists were acyclic. (In general, it can be applied to tree structured union-find sequences, not only to linear ones.) Therefore, in the beginning we break the adjacency list of v between two arbitrary but subsequent arc sets entering v. If these two arc sets are later to be united, we do not really perform a union operation, but keep from then on "in mind" that, virtually, these two arc sets are merely two halves of one single arc set.

For $i = 1, \ldots, k$, let w_i denote the unique vertex incident to s_i. Initially, for all $i = 1, \ldots, k$, the arc (s_i, w_i) has flow $f[(s_i, v_i), i] = d_i$. For all other arcs $e \in E_c^{\rightarrow}$ we set $f[e, i] = 0$. For consistency, we set $excess_X[i] = d_i$ for the arc set X that contains (s_i, w_i). Moreover, $X = find((s_i, w_i))$ is initially inserted in L, if $all\_touched(X) =$ TRUE already, that is $X = \{(s_i, w_i)\}$.

See Table 1 for further details. It is not hard to see that using the technique of Gabow and Tarjan, this algorithm can be implemented so as to run in linear time.

4 Correctness (Sketched)

Remember that in [11] a linear time algorithm is introduced for the special case where $C_{\{v,w\}} = 1$ for each edge $\{v, w\} \in E$. Moreover, it was noted in the Introduction that this algorithm can be used for the general capacitated problem. Clearly, this algorithm is only weakly polynomial and, hence, not our method of choice. But we can use it for proving correctness of the algorithm introduced in this paper, namely by showing that our algorithm provides a solution, whenever the weakly polynomial algorithm does (exactly the same in fact). We use correctness of that algorithm because in our opinion a correctness proof "from scratch" would be much harder and would, in particular, have to repeat all *combinatorial* and *topological* arguments for the weakly polynomial algorithm in a *much more difficult situation*. The algorithm in [11] consists of two phases, too. The result of the first phase is a set of directed paths in the underlying graph, which are called *auxiliary paths*. The following Lemma 1 is an immediate consequence of [11].

Lemma 1. *The auxiliary paths constructed in the first phase of the algorithm in [11] form the rightmost non-crossing path packing from the s-terminals to the t-terminals with respect to unit edge capacities.*

When we replace each edge $\{v, w\} \in E$ by $C_{\{v,w\}}$ many copies and apply the algorithm in [11] to the resulting problem, exactly this is the result of the first phase. Let $e_1, \ldots, e_l$ be the copies of $\{v, w\} \in E$, embedded in *this* order, such that e_1 is the leftmost edge when we regard them as being oriented from v to w. Because of the "rightmost property" stated in Lemma 1, there are $i \in \{0, \ldots, l\}$ and $j \in \{i + 1, \ldots, l + 1\}$ such that $e_1, \ldots, e_i$ and $e_j, \ldots, e_l$ belong to auxiliary paths and $e_{i+1}, \ldots, e_{j-1}$ do not. Moreover, $e_1, \ldots, e_i$ are oriented by their auxiliary paths from w to v, and $e_j, \ldots, e_l$, from v to w. Therefore, when we merge $e_1, \ldots, e_i$ into a single arc (w, v) and $e_j, \ldots, e_l$ into a single arc (v, w), then (v, w) and (w, v) form a counterclockwise cycle. Hence, the result of the first phase translates into the rightmost non-crossing path packing from the s-terminals to the t-terminals in $G^{\rightarrow}$. Moreover, this flow extends to the rightmost non-crossing path packing from s_{i_0} to t_k in $G^{\Rightarrow}$. This already proves the following Lemma 2.

Lemma 2. *If the input instance is solvable, the first phase of the algorithm yields the same result as the first phase of the weakly polynomial algorithm resulting from [11].*

Now we turn our attention to the second phase of the algorithm. We have to show that, if the input instance is solvable at all, the result of the second phase equals the result of the second phase of the weakly polynomial algorithm resulting from [11].

The second phase of that weakly polynomial algorithm consists in a loop where in each iteration exactly one *solution path* is drawn. More precisely, first all d_1 many paths from s_1 to t_1 are drawn, then all d_2 many paths from s_2 to t_2, and so on. (From Sect. 2 recall the renumbering of all pairs (s_i, t_i) with respect to the ordering of the t-terminals.)

The solution paths provided by the weakly polynomial algorithm are uniquely characterized as follows. Let p be a solution path from s_i to t_i, say, and let $\mathcal{P}$ be the set of all solution paths drawn before p. Then p is the rightmost non-crossing path from s_i to t_i such that, for $(v, w) \in E_c^{\rightarrow}$, the number of paths in $\mathcal{P} \cup \{p\}$ containing (v, w) does not exceed $c_{(v,w)}$. Thus, we obtain the following characterization of the "merged" solution of the weakly dual algorithm, when applied to a general, capacitated input instance.

Lemma 3. *For each $i = 1 \ldots, k$, the flow f_i from s_i to t_i constructed by the weakly polynomial algorithm is the rightmost non-crossing path packing from s_i to t_i w.r.t. the residual capacities $c - (f_1 + \cdots + f_{i-1})$.*

We have to show that the second phase always sets *exactly these* flow values and, while there is still excess left at some non-terminal vertex, does not get stuck because list $L = \emptyset$. The proof of this is sketched only briefly (and rather metaphorically than formally).

For the correctness of all particular flow settings, it is first important to note that the greedy-like rule for "spreading" $excess_X [1], \ldots, excess_X [k]$ over $next\_out (X)$ completely conforms to Lemma 3, provided that actually all flow that has to be pushed through $next\_out (X)$ stems from $excess_X$. Therefore, we only have to show that the latter is true. To see this, we formulate an invariant, which is not hard to prove and might be insightful for its own.

Lemma 4. *At any stage of the second phase, there is a continuous, non-crossing, directed path $\mathcal{R}$ in the plane, from one infinite point to another one, such that all arcs not yet touched completely belong to the left side of $\mathcal{R}$, all touched arcs contributing to the excesses of their heads are crossed by $\mathcal{R}$ from right to left, and all other arcs belong to the right side of $\mathcal{R}$.*

Now suppose some flow has to be pushed through $next\_out (X)$ from anywhere else but X. At the current stage, all this flow ends with arcs crossed by the current path $\mathcal{R}$, because of Lemma 4. Focus on a particular unit of this flow and let a be the arc where this flow currently ends. Moreover, let p be the way of this unit from a to $next\_out (X)$ in the correct solution we want to obtain. This unit corresponds to a path constructed by the weakly polynomial algorithm in the second phase. Hence, it follows from Lemma 3 that this unit never crosses its own trace on the way from its s-terminal to its t-terminal.

Now consider the cycle $\mathcal{C}$ formed by p and a subpath (or reverse subpath) of $\mathcal{R}$. If $next\_out (X)$ belongs to the interior of $\mathcal{C}$, our unit of flow must later on leave $\mathcal{C}$ in order to reach its t-terminal finally. Since it does not cross its own trace, it cannot leave $\mathcal{C}$ by crossing p. On the other hand, it cannot leave $\mathcal{C}$ by crossing $\mathcal{R}$ either, since by induction the flow through all arcs on the right side of C is already correct. Hence, we have arrived at a contradiction. However, if $next\_out$ (X) belongs to the *exterior* of C, one can prove a contradiction with a certain property of $G_c^{\rightarrow}$, which results from [11], too: There is no clockwise cycle in G_c. Because of lack of space, the proof of the latter case is left out completely.

It remains to show that list L never becomes empty until all flow has been pushed to the t-terminals. For this aim, we characterize the members of L. We define a binary relation on all currently existing arc sets in all union-find structures: $X \prec Y$ if and only if there is an auxiliary path constructed in the first phase of the weakly polynomial algorithm which contains arcs of both sets, but the arc in X precedes the arc in Y on this path. One can show that this relation is a strict partial ordering. Now consider the relative minima in the set of all X with $all\_touched\,(X) \neq TRUE$ yet. In particular, let X be a such a relative minimum. Then one can prove that at least one immediate predecessor of X belongs to L (and conversely, each member of L is an immediate predecessor of such an X). Again, all further details are left out.

References

1. G.N. Frederickson (1987): *Fast algorithms for shortest paths in planar graphs with applications.* SIAM J. Comput. **16**, 1004-1022.
2. H.N. Gabow and R.E. Tarjan (1985): *A linear-time algorithm for a special case of disjoint set union.* J. Comp. Syst. Sci. **30**, 209-221.
3. A.V. Goldberg and R.E. Tarjan (1988): *A new approach to the maximum flow problem.* J. ACM **35**, 921-940.
4. R. Hassin (1981): *Maximum flow in (s, t) planar networks.* Inf. Proc. Letters **113**, p. 107.
5. R. Hassin (1984): *On multicommodity flows in planar graphs.* Networks **14**, 225-235.
6. T.C. Hu (1969): *Integer programming and network flows.* Addison-Wesley.
7. A. Itai and Y. Shiloach (1979): *Maximum flow in planar networks.* SIAM J. Comput. **8**, 135-150.
8. K. Matsumoto, T. Nishizeki and N. Saito (1985): *An efficient algorithm for finding multicommodity flows in planar networks.* SIAM J. Comp. **14**, 289-302.
9. H. Okamura and P.D. Seymour (1981): *Multicommodity flows in planar graphs.* Proc. London Math. Soc. **42**, 178-192.
10. É. Tardos (1985): *A strongly polynomial algorithm to solve combinatorial linear programs.* Operations Research **34**, 250-256.
11. D. Wagner and K. Weihe (1992): *A linear-time algorithm for edge-disjoint paths in planar graphs.* Proc. 1st Europ. Symp. Algorithms (ESA '93), Springer Lect. Notes Comp. Sci.
12. K. Weihe (1993): *Non-crossing path packings in planar graphs with applications.* Preprint no. 358/1993, Fachbereich Mathematik, Technische Universität Berlin.

Linear Time Algorithms for Disjoint Two-Face Paths Problems in Planar Graphs *

Heike Ripphausen-Lipa, Dorothea Wagner, Karsten Weihe

Technische Universität Berlin, MA 6–1, Straße des 17. Juni 136, 10623 Berlin, Germany
e-mail: ripphaus@, wagner@, and weihe@math.tu-berlin.de

Abstract. In this paper we present a linear time algorithm for the vertex-disjoint Two-Face Paths Problem in planar graphs, i.e., the problem of finding k vertex-disjoint paths between pairs of terminals which lie on two face boundaries. The algorithm is based on the idea of finding *rightmost* paths with a certain property in planar graphs. Using this method, a linear time algorithm for finding vertex-disjoint paths of a certain homotopy is derived. Moreover, the algorithm can be modified to solve the more general linkage problem in linear time as well.

1 Introduction

The problem of finding disjoint Steiner trees or paths in planar graphs is a core problem in VLSI design. In this paper we consider a special version of the k-disjoint paths problem, i.e., the problem to determine k vertex-disjoint paths between k pairs of vertices. This problem is $\mathcal{NP}$-hard, even for planar graphs [1]. If k is assumed to be fixed, then the problem can be solved in linear time [2]. But in this case the order constant grows tremendously in k. The Two-Face Paths Problem considered in this paper is defined as follows:

Problem 1. Let $G = (V, E)$ be a planar graph and let F^o, F^i be two different faces. $t_1^i, \ldots, t_k^i \in V$ are pairwise different vertices incident with F^i, and $t_1^o, \ldots, t_k^o \in V$ pairwise different vertices incident with F^o. The *Two-Face Paths Problem* (TFPP) is to connect the corresponding vertices t_j^i and t_j^o by pairwise vertex-disjoint paths.

Throughout the paper, we consider a fixed embedding of G in the plane where F^o is the outer face, and we call F^i the inner face. The adjacency lists of vertices are ordered counter-clockwise with respect to the fixed embedding. The vertices $t_1^i, \ldots, t_k^i$ are called *inner terminals*, the vertices $t_1^o, \ldots, t_k^o$ *outer terminals*. We assume that $t_1^i, \ldots, t_k^i$ resp. $t_1^o, \ldots, t_k^o$ appear in this order clockwise around the boundary of F^i and F^o, respectively. Upper indices i resp. o always indicate that the corresponding object is incident to the *inner* resp. *outer* face.

The TFPP is a special case of the vertex-disjoint Two-Face Steiner Trees Problem. This problem consists in finding vertex-disjoint Steiner trees in planar graphs. Again, the position of the terminals is restricted to two designated face boundaries.

---

* This research has been supported by the *Technische Universität Berlin* under grant FIP 3/1.

Suzuki, Akama, and Nishizeki have developed an algorithm for solving this problem in $O(\min\{n \log n, kn\})$ running time [5]. All parts of their algorithm have linear running time except a procedure for the TFPP. However, they claim that they can solve the TFPP in linear time if the *Menger Problem* can be solved in linear time.

Problem 2. Let $G = (V, E)$ be a planar graph, and $s, t \in V$, $s \neq t$. The *Menger Problem* is to find as many undirected pairwise vertex-disjoint (s, t)-paths in G as possible.

In [3] a linear time algorithm is presented for the vertex-disjoint Menger Problem in planar graphs. Using this algorithm, we now develop a new algorithm for solving the TFPP in linear time. This algorithm is much easier than the algorithm of Suzuki et al., and is therefore not difficult to implement. Furthermore, it can be used to solve the p–Homotopic Two-Face Paths Problem which is defined as follows:

Problem 3. Let p be an arbitrary path in G between t_1^o and t_1^i. The *p-Homotopic Two-Face Paths Problem* (p–HTFPP) is the Two-Face Paths Problem with the additional condition that the path which connects t_1^o and t_1^i is homotopic to p.

Moreover, the algorithm can easily be extended for solving a problem similar to the p–HTFPP, the *Linkage Problem*. For the Linkage Problem the pairs of terminals to be connected are not fixed.

The algorithms for solving the TFPP, the p–HTFPP, and the Linkage Problem use two sets of paths S^o and S^i which can be determined by the Menger Algorithm of [3]. The path set S^o consists of pairwise vertex-disjoint paths starting with outer terminals and ending with arbitrary vertices on the inner face boundary. Analogously the path set S^i starts with inner terminals and ends with arbitrary vertices on the outer face boundary. Paths starting with outer resp. inner terminals are denoted by *outer* resp. *inner paths*. The paths determined by the Menger Algorithm are in some sense *rightmost*. Therefore corresponding inner and outer paths have to intersect at least once if there exists a solution to the TFPP. The main idea is to concatenate segments of an outer path with segments of the corresponding inner path in order to connect an outer terminal with the corresponding inner terminal.

In Sect. 2 we introduce the *band model*, which turns out to be very useful for formulating some properties of paths and for the proofs. A binary relation *more right* on paths and path sets is defined in Sect. 3 in order to characterize *rightmost* paths (as those determined by the Menger Algorithm). In Sect. 4, a sufficient condition for the existence of a solution to the TFPP is given. Algorithms for determining solutions to the TFPP and the p–HTFPP are developed in Sect. 5.

2 The Band Model

Proofs become much easier if the graph and the paths are translated into the so-called *band model*. Informally, the band model is constructed as follows: Embed a graph with two designated faces on a cylinder such that the designated face boundaries lie on the boundaries of the cylinder. The band model is built by rolling off the cylinder infinitely often in the plane in both directions.

More formally, choose an arbitrary simple path $p^* = v_1, \ldots, v_s$ starting with a vertex v_1 incident to F^i and ending with a vertex v_s incident to F^o. In the sequel, p^* is called the *cutting path*. Then a graph G_{p^*} is constructed by replacing every vertex v_i $(1 \leq i \leq s)$ on p^* by two vertices v_i^l and v_i^r. In G_{p^*} edges to the right side of p in G are incident to v_i^l and edges to the left side of p are incident to v_i^r, if p is traversed from the inner to the outer boundary. Edges between v_i and v_{i+1} $(1 \leq i < s)$ are incident to v_i^l and v_{i+1}^l. In this construction the cyclic order of the remaining edges incident to v_i^l resp. v_i^r is maintained. Now we take an infinite number of copies of G_{p^*}. The j-th copy of G_{p^*} is denoted by $C_j(G_{p^*})$ (j integer). These copies are glued together by identifying vertices v_i^r in $C_j(G_{p^*})$ with v_i^l in $C_{j+1}(G_{p^*})$ $(1 \leq i \leq s)$. The resulting infinite graph is denoted by G^*. This construction is illustrated in Fig. 1.

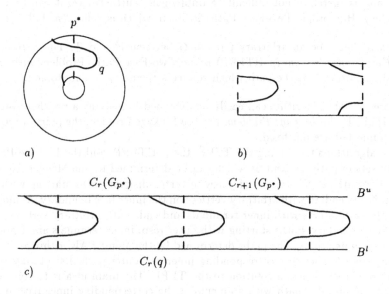

Fig. 1. a) The graph G, b) a copy of G_{p^*}, c) the band graph G^* corresponding to G

Notice, G^* does not depend on the choice of the cutting path p^*. We call G^* the *band graph w.r.t.* G. The boundary in G^* corresponding to the outer resp. inner face boundary in G is denoted by *upper boundary* B^u resp. *lower boundary* B^l. A relation *more right* can be defined in an obvious way on the vertices of the upper resp. lower boundary. We will even say that a vertex is *more right* than another vertex if they are identical. This relation is denoted by "$\succeq$".

Now we consider the translation of paths resp. sets of paths of G into G^* with cutting path p^*. The copy of vertex v in $C_r(G_{p^*})$ is denoted by $C_r(v)$. Let p be a path in G. We assume that p starts with vertex t incident to the inner face, and ends with a vertex incident to the outer face. In every copy $C_r(G_{p^*})$, there is a copy $C_r(p)$ of p which starts with $C_r(t)$.

Assume that in G the path p crosses the cutting path in vertex v from the left to the right side. Then the path $C_r(p)$ switches over from the current copy of G_{p^*} to the neighbouring copy of G_{p^*} to the right (cf. Fig. 1). Crossings from the right to the left side of the cutting path are handled analogously. Notice that the path $C_r(p)$ can end in a copy different from the copy it has started.

Consider intersections between two paths p and q in G. In G^* these intersections are represented by intersections of copies of p and q. In G^* every intersection between p and q is represented by an infinite number of intersections between pairs of copies of p and q. In order to find a representative pair of paths for every intersection of a path p with other paths, consider a fixed copy of p, say $C_r(p)$. Every intersection between p and q is represented by an intersection between $C_r(p)$ and a copy of q. See Fig. 2: the first and fourth intersection of p with q are represented by the pair $(C_r(p), C_s(q))$ $(s = r + 1)$. The second and third intersection is represented by the pair $(C_r(p), C_{s+1}(q))$. Every pair of copies $(C_r(p), C_s(q))$ can represent several intersections.

The same intersections are also represented by pairs $(C_{r+l}(p), C_{s+l}(q))$, (l integer). In the sequel we will only say that a pair $(C_r(p), C_s(q))$ *represents an intersection set*, if we can conclude that the corresponding paths have to intersect because of the constellation of their endvertices (independent of their detailed course). That is, startvertex of $C_r(p)$ is *more right* than startvertex of $C_s(q)$, but endvertex of $C_s(q)$ is *more right* than endvertex of $C_r(p)$.

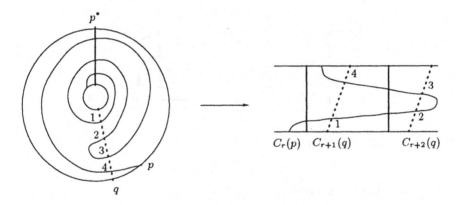

Fig. 2. Intersections represented by different pairs of copies of paths

3 Rightmost and Leftmost Paths

First we will define and prove some properties of paths resp. path sets in G^* (the band graph), before we will turn our attention to paths resp. path sets in G.

Definition 4. Let p and p' be two paths in G^* with endvertices s resp. s' on the lower boundary and t resp. t' on the upper boundary. Path p' is *more right* than p $(p' \succeq p)$ iff $s' \succeq s$ and $t' \succeq t$.

Let $P = \{p_1, \ldots, p_k\}$ and $P' = \{p'_1, \ldots, p'_k\}$ be two vertex-disjoint path sets in G^* where each path has one endvertex on the lower, and one on the upper boundary. P' is *more right* than the path set P $(P' \succeq P)$ iff $p'_j \succeq p_j$ for all j with $1 \leq j \leq k$.

Notice, if p and p' start and end with the same vertices, then $p \succeq p'$ and $p' \succeq p$. In this case p and p' are homotopic, but not necessarily equal. That is, the relation "$\succeq$" on paths is a partial order on homotopy classes. For concatenating segments of pairs of paths which are incomparable we need the following definition:

Definition 5. Let p and p' be two simple paths in G^* between the lower and the upper boundary. The *right [left] hull* p'' of p and p' is a concatenation of segments of p and p', such that p'' builds a simple path between the lower and the upper boundary. Furthermore all segments of p and p' lie to the left [right] side of p'' or on p''. (Figure 3 illustrates this notion.)

We can prove that such right resp. left hulls always exist. The algorithm for determining the right [left] hull of two paths p and p' uses a right-first search [left-first search] strategy: This is a depth-first search where in each step the edges which leave the current vertex are searched "from the right [left] to the left [right]". Let $\{v, w\}$ be the last edge added to a path and w the last vertex of this path. Then the next edge in right-first search order [left-first search order] is the next [previous] edge after [before] $\{v, w\}$ in the adjacency list of w.

$$p \quad p' \qquad\qquad \text{——— } p'' \text{ (right hull)}$$

Fig. 3. Right hull of two paths

Right [Left] Hull Algorithm

{Let p, p' be two paths between B^l and B^u which start with s, s' on B^l}
{w.l.o.g. $s \succeq s'$ $[s' \succeq s]$}
$s'' := s$; $p'' := s''$;
WHILE B^u is not reached by p''
  DO { let v be the last vertex of p'' }
    $p'' := p'' \cup \{v, w\}$;
    { where $\{v, w\}$ is the next edge in right[left]-first order of p or p' to the last edge of p'' in the adjacency list of v }

Note that only if v is a common vertex of p and p', there is more than one possibility for choosing the next edge for p''. If the algorithm should start on the upper boundary (this case occurs in Sect. 5), then a left[right]-first order must be used instead of a right[left]-first order. It can easily be proved:

Lemma 6. *(Proof omitted.) The Right [Left] Hull Algorithm determines the right [left] hull of two simple paths p and p'.*

In the sequel we denote a segment of a path p between vertices v and w by $p(v, w)$.

Lemma 7. *(Proof omitted.) Let $P = \{p_1, \ldots, p_k\}$ and $P' = \{p'_1, \ldots, p'_k\}$ be two sets of vertex-disjoint simple paths in G^* connecting vertices $t_1, \ldots, t_k$ resp. $t'_1, \ldots, t'_k$ incident with one boundary of G^* with some vertices incident with second boundary. Moreover, let $t_j \succeq t_i$ resp. $t'_j \succeq t'_i$ for $j > i$.*
  Let P'' be the set of right [left] hulls of P and P', that is, $P'' = \{p''_1, \ldots, p''_k\}$ where p''_j is the right [left] hull of p_j and p'_j. Then $P'' \succeq P$ and $P'' \succeq P'$ $[P \succeq P''$ and $P' \succeq P'']$.

Notice, that all definitions and statements about finite path sets in G^* can be generalized to infinite path sets. But since G is finite, we can find a finite part of G^* corresponding to G. Now we are able to define relations on sets of paths in G, and some important properties of these path sets.

Definition 8. Let $P = \{p_1, \ldots, p_k\}$ and $P' = \{p'_1, \ldots, p'_k\}$ be two sets of vertex-disjoint paths in G, both starting with vertices $t_1, \ldots, t_k$ incident with F^d, $d \in \{i, o\}$. Let p^* be an arbitrary cutting path, and $C_r(P)$ and $C_r(P')$ be copies of P and P' which start in the same copy $C_r(G_{p^*})$ of G^*. Then P' is *more right* than P iff $C_r(P')$ is *more right* than $C_r(P)$.

Notice, that the relation *more right* on sets of paths in G is independent of the cutting path.

Theorem 9. *(Proof omitted.) There exists (at least) one path set P_r in G consisting of vertex-disjoint paths starting with $t_1, \ldots, t_k$ on F^d, $d \in \{i, o\}$, which is more right than all other vertex-disjoint path sets P in G starting with the same vertices.*

The proof is based on a translation of paths from G to G^* and an application of Lemma 7.
  In the sequel, S^i and S^o are the sets of paths determined by the Menger Algorithm starting with $t^i_1, \ldots, t^i_k$ and $t^o_1, \ldots, t^o_k$, respectively. Now the useful properties of the path sets S^o and S^i can be formulated more precisely:

Lemma 10. *(Proved in [3].) The path set S^i is more right than all other path sets in G starting with $t^i_1, \ldots, t^i_k$, and the path set S^o is more right than all other path sets in G starting with $t^o_1, \ldots, t^o_k$.*

4 Solvability of the Two-Face Paths Problem

In this section we characterize solvable instances of the TFPP. Consider two path sets $P^o = \{p^o_1, \ldots, p^o_k\}$ starting with $t^o_1, \ldots, t^o_k$ and $P^i = \{p^i_1, \ldots, p^i_k\}$ starting with $t^i_1, \ldots, t^i_k$. Let p be a simple path between t^o_1 and t^i_1. The path sets P^o and P^i fulfill the *rightness property w.r.t. p*, if the following holds: Let G^* be the band graph w.r.t. G with cutting path p. Then for every pair $(C_r(p_j^o), C_r(p_j^i))$ in G^*, the path $C_r(p_j^o)$ ends to the right of (or on) $C_r(t_j^i)$, and $C_r(p_j^i)$ ends to the right of (or on) $C_r(t_j^o)$.

Theorem 11. *The TFPP is solvable iff there exist two sets of vertex-disjoint paths $P^o = \{p_1^o, \ldots, p_k^o\}$ starting with $t_1^o, \ldots, t_k^o$, $P^i = \{p_1^i, \ldots, p_k^i\}$ starting with $t_1^i, \ldots, t_k^i$, and a simple path p between t_1^i and t_1^o such that P^o and P^i fulfill the rightness property w.r.t. p.*

Proof. First, assume that a solution $S = \{p_1, \ldots, p_k\}$ to the TFPP exists. Obviously, $P^o := S$ and $P^i := S$ fulfill the rightness property w.r.t. p.

For proving the converse direction, consider path sets P^o and P^i which fulfill the rightness property w.r.t. some path p. Because of the Jordan Curve Theorem every path p_j^i intersects the corresponding path p_j^o at least once. We concatenate segments of p_j^o with segments of p_j^i, in order to connect the corresponding terminals. That is, a sequence of intersection vertices must be found to concatenate alternating segments of p_j^o and p_j^i.

In order to find a suitable sequence of intersection vertices, we translate the path sets P^o and P^i into paths in G^* (w.r.t. cutting path p). The pairs $(C_r(p_j^o), C_r(p_j^i))$ $(1 \leq j \leq k$, r integer) fulfill the assumption of Lemma 7. Because of the rightness property of the path sets P^o and P^i, the pairs $(C_r(p_j^o), C_r(p_j^i))$ $(1 \leq j \leq k)$ do not only represent sets of intersections, but also the left hulls of these pairs connect $C_r(t_j^o)$ with $C_r(t_j^i)$. This means that there exist vertex-disjoint paths between $C_r(t_j^o)$ and $C_r(t_j^i)$ in G^*, for all integer r. Hence, vertex-disjoint paths between t_j^o and t_j^i $(1 \leq j \leq k)$ exist in G which consist of segments of P^o and P^i. $\square$

If there are two path sets which fulfill the rightness property w.r.t. p, the path sets S^o and S^i fulfill the rightness property as well, since they are *rightmost*. This means, these sets of paths are our candidates for solving the TFPP. Notice, that Theorem 11 indeed gives a sufficient condition for the existence of a solution to the TFPP. However, in general we do not know a suitable cutting path p in advance.

5 The Algorithms

First we restrict our attention to the problem of finding paths of a certain homotopy, that is, the p–HTFPP (Problem 3). Let $I_1, \ldots, I_p$ be some forbidden faces in a planar graph, and p_1, p_2 two paths with the same endvertices. p_1 is *homotopic* to p_2, if p_1 can be moved continuously to p_2 without moving the endvertices and without crossing the forbidden faces.

In our problem the outer face F^o and the inner face F^i are the forbidden faces. Paths in G can be interpreted as curves in the plane. Obviously, for this special case the homotopy of one path in the solution determines the homotopy of all other paths. A characterization for the existence of a solution to the p–HTFPP can easily be given:

Lemma 12. *The p–HTFPP is solvable iff there exist two sets of vertex-disjoint paths $P^o = \{p_1^o, \ldots, p_k^o\}$ and $P^i = \{p_1^i, \ldots, p_k^i\}$ starting with $t_1^o, \ldots, t_k^o$ resp. $t_1^i, \ldots, t_k^i$ which fulfill the rightness property w.r.t. p*

Proof. The necessity of the rightness property w.r.t. p is trivial.

Assume that path sets P^o and P^i exist which fulfill the rightness property w.r.t. a path p. Theorem 11 implies that a solution to the TFPP exists. For the p–HTFPP just consider paths in G^*, where p is used as cutting path. Then the left hulls p_j of the pairs $(C_r(p_j^o), C_r(p_j^i))$ connect t_j^o and t_j^i. Obviously, p_1 is homotopic to p. ☐

If the path sets S^i and S^o determined by the Menger Algorithm do not fulfill the rightness property w.r.t. p, then there exists no solution to the p–HTFPP.

Lemma 13. *A solvable p–HTFPP can be solved in linear time.*

Proof. Assume that P^o and P^i fulfill the rightness property w.r.t. p. Such path sets can be determined in linear time by the Menger Algorithm. The proof of Lemma 12 shows that suitable intersection vertices for concatenating paths are represented by pairs $(C_r(p_j^o), C_r(p_j^i))$ in G^* (with cutting path p). In G, these intersection vertices correspond to intersection vertices v with the following property: For these vertices v, the number of times $p_j^o(t_j^o, v)$ and $p_j^i(t_j^i, v)$ surround the inner face in relation to p is equal. These vertices can easily be recognized and marked in linear time. Then the left hulls corresponding to the pairs $(C_r(p_j^o), C_r(p_j^i))$ are determined. The correct intersection vertices for concatenating (cf. the Right/Left Hull Algorithm) are just the marked intersection vertices. For this purpose every path of P^o and P^i is traversed at most once. That is, the algorithm has linear running time. ☐

Notice, a p–HTFPP is unsolvable if one of the left hulls does not end with the correct terminal.

Now, we consider the general TFPP. Lemma 10 shows that there are *rightmost* and *leftmost* path sets among all path sets starting with fixed vertices on one face boundary, and ending with arbitrary vertices on the second face boundary. Analogously, it can be proved that there exist *rightmost* and *leftmost* solutions to the TFPP.

Lemma 14. *A rightmost [leftmost] solution to a solvable TFPP can be determined in linear time.*

Proof. We assume that paths are directed from the inner to the outer terminals. First, *rightmost* solutions are constructed. Again, we will concatenate segments of paths from $S^o = \{p_1^o, \ldots, p_k^o\}$ and $S^i = \{p_1^i, \ldots, p_k^i\}$, the path sets determined by the Menger Algorithm.

Let us first consider these paths in the band graph G^* w.r.t. G. We will determine pairs $(C_r(p_j^o), C_s(p_j^i))$ in G^* such that the left hulls of these pairs induce vertex-disjoint paths between corresponding terminals in G. W.l.o.g. we can assume that the cutting path of G^* is the path p_1 of a solution to the TFPP.

The path p_j^i is obviously *more right* than the path p_j of a *rightmost* solution to the TFPP. Consider a path $C_s(p_j^i)$ in an arbitrary copy of G. Take a pair $(C_r(p_j^o), C_s(p_j^i))$ which represents a set of intersections between p_j^o and p_j^i such that r is as large as possible. Then the left hull corresponding to the pair $(C_r(p_j^o), C_s(p_j^i))$ is taken for connecting t_j^i with t_j^o. Obviously, a path in a *rightmost* solution to the TFPP cannot be more right than this left hull.

Such a pair is easier to determine if the paths are traversed in the opposite direction. Consider a path $C_r(p_j^o)$. Take a pair $(C_r(p_j^o), C_s(p_j^i))$ which represents a set of intersections between p_j^o and p_j^i such that s is as small as possible. That is, $C_s(p_j^i)$ is the copy of p_j^i which ends next to the right of $C_r(t_j^o)$. However, if we apply this procedure to all pairs of outer and inner paths, the resulting left hulls are not necessarily vertex-disjoint.

The endvertex of $C_r(p_j^i)$ on the upper boundary is denoted by $C_r(e_j^o)$. Notice, here we deviate from our general notation, since we do not know in which copy the path $C_r(p_j^i)$ ends.

Claim 1: (Proof omitted.) Let $(C_r(p_j^o), C_s(p_j^i))$ and $(C_{r'}(p_l^o), C_{s'}(p_l^i))$ represent intersection sets for connecting the corresponding inner and outer terminals such that $C_s(e_j^o) \succeq C_r(t_j^o)$ and $C_{s'}(e_l^o) \succeq C_{r'}(t_l^o)$. W.l.o.g. assume that $C_{r'}(t_l^o) \succeq C_r(t_j^o)$. The left hulls of these pairs are vertex-disjoint paths in G^* iff $C_{s'}(e_l^o) \succeq C_s(e_j^o)$.

That is, if $C_s(e_j^o) \succeq C_{s'}(e_l^o) \succeq C_{r'}(t_l^o) \succeq C_r(t_j^o)$ then the left hulls of the two pairs are not vertex-disjoint. We call this constellation of terminals and endvertices *forbidden constellation*. Obviously, there cannot exist a third pair of copies of paths $(C_{r''}(p_m^o), C_{s''}(p_m^o))$ with $C_{r''}(t_m^o)$ and $C_{s''}(e_m^o)$ lie in this order between $C_{r'}(t_l^o)$ and $C_{s'}(e_l^o)$. This would lead to a contradiction to the vertex-disjointness of the path sets.

Claim 2: (Proof omitted.) Let $(C_r(p_j^o), C_s(p_j^i))$ and $(C_{r'}(p_l^o), C_{s'}(p_l^i))$ be two pairs of paths such that $C_s(p_j^i)$ resp. $C_{s'}(p_l^i)$ are copies of p_j^i resp. p_l^i which end immediately to the right of $C_r(t_j^o)$ resp. $C_{r'}(t_l^o)$. Assume that $C_s(e_j^o) \succeq C_{s'}(e_l^o) \succeq C_{r'}(t_l^o) \succeq C_r(t_j^o)$. Then the pair $(C_{r'}(p_l^o), C_{s'+1}(p_l^i))$ represents a set of intersection.

This means, in the case of forbidden constellations there exist suitable pairs of paths representing intersection sets such that the left hulls of these pairs of paths are vertex-disjoint. Consider the copies of all outer paths which start in the same copy, say $C_r(G_{p_1})$. W.l.o.g. assume that there exists a forbidden constellation, as described above with $j = 1$.

Procedure Intersection Sets;
$C_{s_1}(p_1^i) := C_s(p_1^i);$
FOR $m := 2$ TO k DO
      $C_{s_m}(p_m^i) :=$ the copy of p_m^i which lies immediately to the right of $C_s(p_1^i);$
$\{ (C_r(p_j^o), C_{s_j}(p_j^i)) \text{ represent sets of intersections } \}$

That is, the copy $C_s(p_1^i)$ and the next $k-1$ copies of paths immediately to the right of $C_s(p_1^i)$ and all copies of outer paths in $C_r(G_{p_1})$ are taken.

Claim 3: (Proof omitted.) The pairs $(C_r(p_j^o), C_{s_j}(p_j^i))$ $(1 \leq j \leq k)$ determined by Procedure Intersection Sets represent sets of intersections such that the corresponding left hulls are pairwise vertex-disjoint.

We now complete the proof of Lemma 14. Because of the construction, the set of left hulls of these pairs is a *rightmost solution* to the TFPP.

It remains to determine these suitable sets of intersection vertices represented by $(C_r(p_j^o), C_{s_j}(p_j^i))$ in G, instead of G^*. The forbidden constellation can be recognized during a clockwise traversal of the outer boundary: t_i^o and e_i^o lie in this order between t_j^o and e_j^o. In a second scan of the outer boundary the set of intersections which are suitable for concatenating paths for all pairs of paths p_m^o and p_m^i can be determined. This is either the first, or the second possible intersection set represented by pairs of copies of paths, when traversing the outer paths from the outer to the inner boundary. After the correct sets of intersection vertices has been determined, we proceed just as in the algorithm for solving the p-HTFPP. However now the determination of the left hulls starts on the outer boundary.

In order to obtain a *leftmost* solution, the first or second possible intersection set for determining the left hull is taken, when inner paths are traversed from the inner to the outer boundary. This problem can be handled symmetric to the problem of finding a rightmost solution.   □

Again, there is no solution, when at least one left hull does not end with the correct terminal.

We call a *leftmost* Two-Face Paths Solution a solution of *homotopy class 0*. A solution which surrounds the inner face boundary s-times more than the leftmost solution, is called solution of *homotopy class s*. It can easily be shown:

Corollary 15. *All homotopy classes between the* rightmost *and the* leftmost *solution to the TFPP exist.*

Corollary 16. *The number of different homotopy classes can be determined in linear time.*

Corollary 15 has been shown in "Graph minors, VI" by Robertson and Seymour [4] as well. Precisely, they have considered a more general problem: the *Linkage Problem*. A small modification of the algorithm for the p-HTFPP suffices for solving the Linkage Problem in linear time resp. for finding the extreme linkages.

References

1. J.F. Lynch. *The equivalence of theorem proving and interconnection problem.* ACM SIGDA Newsletter 5, 31 – 65, 1975.
2. B. Reed, N. Robertson, A. Schrijver, P.D. Seymour. *Finding disjoint trees in graphs on surfaces.* To appear.
3. H. Ripphausen-Lipa, D. Wagner, K. Weihe. *The vertex-disjoint Menger problem in planar graphs.* Preprint no. 324, Technische Universität Berlin, 1992. (Extended abstract in: Proceedings of the fourth annual ACM-SIAM Symposium on Discrete Algorithms, 112 – 119, 1993.)
4. N. Robertson and P.D. Seymour. *Graph minors. VI. Disjoint paths across a disc.* Journal of Combinatorial Theory, B, Vol. 41, 115 – 138, 1986.
5. H. Suzuki, T. Akama, T. Nishizeki. *Finding Steiner forests in planar graphs.* Proceedings of the first annual ACM-SIAM Symposium on Discrete Algorithms, 444 – 453, 1993.

Robot Mapping: Foot–Prints vs Tokens

(Extended Abstract)

Xiaotie Deng * and Andy Mirzaian *

Abstract: We are interested in the problem of robot exploration from the approach of competitive analysis, where the cost of an online–strategy is compared with the minimum cost carrying out the same task with perfect information.

Our world model is restricted to graph maps. Within this model, there are several differences dealing with different robot sensors. Most often robots are assumed to have perfect vision. In this paper, however, we consider two different sensors: tokens and foot-prints. For the former, robots cannot recognize nodes or edges of the unknown graph under exploration but can drop some tokens which can be recognized if it returns to nodes where tokens are dropped. In the latter case, the robot has the power of knowing whether a node or an edge has been visited before, though it may not remember exactly when and where it was visited (similar to a traveler lost in the desert who recognizes its foot–print, or a robot smells its own trace.)

With competitive analysis, we want to minimize the ratio of the total number of edges traversed for mapping the graph divided by the optimum number of edge traversals for verifying the map. In particular, we call a strategy competitive if this ratio is constant. As a first step, we have developed a competitive strategy to map an unknown embedded planar graph with pure foot-prints. Then we apply this technique to obtain an algorithm using n identical tokens to competitively map unknown planar embedded graphs. We also give a lower bound of competitive ratio $\Omega(n)$ for mapping general embedded graphs with a single token, when robot strategies are slightly restricted. This is tight since there is an algorithm of competitive ratio $O(n)$ [DJMW].

1 Introduction

We are interested in designing algorithms for a robot to build a map of the environment, based on its own observations in exploring the world. Graph maps are used here to represent the world of robots. Thus, the robot can traverse edges to move from nodes to nodes in the graph. Exploration strategies are evaluated by the ratio of the cost of building the map, where we know nothing

*Department of Computer Science, York University, North York, Canada, M3J 1P3. {deng, andy}@cs.yorku.ca. Authors' research was partly supported by NSERC grants.

about the world initially, and the cost of verifying the map, where we have a map of the world and the initial position–orientation of the robot in the map, but still want to verify the correctness of the given information. This approach, called *competitive analysis* [ST], is used as a measure of information–efficiency of algorithms, for robot navigations and explorations [BCR, BBFY, BRS, DKP, DP, FFKRRV, Kl, KP, KRR, PY]).

Different built–in sensors result in different perceptions of the world by the robot. Usually, it is assumed that nodes or edges traversed before can be completely recognized. In contrast, it is assumed in [RS] that nodes are divided into a small number of classes, e.g., white and black colors, and can only be recognized as such. They take the approach of Valiant learning model to design algorithms which correctly infer the environment with high probability. An even more extreme assumption is made in [DJMW] that nodes cannot be recognized at all, as a result of errors in robot's estimation of its location, in sensor reading, in estimation of distance moved between measurements. (See also [KB, LL] for some other qualitative formulations of exploration problems.) To recognize the environment, tokens are used in [DJMW] for the robot to drop at nodes or pick up from nodes while exploring. Thus, by dropping one token at a node, we can recognize the node when we return to find the token there. We may also consider the mapping problem from a desert traveler's viewpoint: We may not recognize a node (or an edge) when we return but we know whether it was previously visited by recognizing our own foot–print.

We want to design an efficient strategy for a robot to obtain the map of an unknown embedded graph $G = (V, E)$. The embedding establishes a (circular) order of the edges incident to a node. Using the entrance edge, this circular order can be broken into a linear one. Each edge may lead to a new node not yet visited. Nodes can be given names the first time they are reached. However, when reaching a node that is already visited, we have to find out its name which was given the first time the node was reached. We also need to identify which node each edge incident to a given node leads to. We call this the *mapping* problem. Since the graph is unknown, the cost of mapping is dependent on both the strategy and the unknown graph. The competitive ratio of a strategy is defined as the maximum ratio of the number of traversed edges to establish the map divided by the minimum number of edges traversed for verifying the map. Obviously, we want to design a strategy with the minimum competitive ratio. Notice that, for robots with perfect sensors, mapping is the same as the graph traversal problem. However, when this ultimate power is deprived of, mapping a graph becomes much more different from the traversal problem of the graph. We may not know the graph even after all the edges are traversed!

There are some connections between these different models. For example, for a graph of n nodes and m edges, if we have $n + m$ distinct tokens, we can put one token at one node or edge. This would reduce the token model to the perfect sensor model. Also we can use $\binom{n}{2} + \binom{m}{2}$ identical tokens to simulate distinct tokens by using i identical tokens in place of a distinct token with label i. It is well–known that, with perfect sensors, we can explore a general graph without traversing an edge more than twice by depth–first search on the edges. Therefore, $\binom{n}{2} + \binom{m}{2}$ identical tokens would be enough for us to explore a general graph of n nodes and m edges competitively. Also, it is not difficult to see that the foot-print model is reducible to the identical token model with $n + m$ tokens, one for each node or edge. In this paper, we are able to develop a competitive algorithm for exploring embedded planar graphs using foot–prints. Furthermore, with some nontrivial modifications, we are able to apply this method to obtain a competitive algorithm for exploring embedded planar graphs of n nodes using n identical tokens.

Besides embedded planar graphs, we also consider graphs embedded in general surfaces. This can be viewed as an abstraction of bridges, tunnels, etc. Thus, at each node of the graph, there is a local surface homeomorphic to a disk centered at the node, on which edges are embedded. We call it a general embedded graph.

To illustrate tasks involved in the above model, we first introduce an algorithm with $O(mn)$ traversals for mapping general embedded graphs in the pure foot–print model, i.e., no tokens are used. Initially the robot is placed at an arbitrary node u in the graph. It can see all the edges incident to u. The spatial orientation of edges gives their circular order around u. When the robot first reaches a new node v from node u, it can remember the edge $e = (u, v)$ from which it arrived at v. Thus, edges incident to v can be (clockwisely) ordered with respect to the edge (u, v). If the robot moves back along (u, v), it knows it comes back to node u and its map shows the local orientation of edges it has made for node u before it went to node v. However, if the robot comes back to u from an unknown edge, it cannot distinguish this node from all other nodes with the same local adjacency configuration. Thus, the local maps of a node are different when the robot reaches the node from two different edges and the robot may not knows how to match the two linear orders of incident edges from these two maps!

To deal with these difficulties, at each stage the robot has explored a part of the graph, it draws a *partial graph* $P = (V', s, E', E'')$, where $V' \subset V$, $E' \subset E \cap V'^2$, E'' are edges with just one endpoint belonging to V'. Edges in E' are all the edges the robot has traversed and mapped up to this point. The robot is currently at node s, and know how to match edges incident to s with edges in the partial map it has at hand. Thus, when it moves out of node s along an edge in E', it knows which node it will arrive by consulting the map. The ability to remember this edge, combined with the map, enables the robot to obtain the matching between the edges it sees at that node with the edges in the map. It is easy to conclude that the robot can go to any edge in E' and any node in V' correctly by comparing with the map. However, the robot does not know where the edges in E'' will lead to. When E'' is empty, we know the whole graph. We call edges in E' the old edges and others new edges. Also call nodes in V' the old nodes and others new nodes.

Suppose now E'' is not empty. The robot will take a closest edge $e = (a, x) \in E''$, with $a \in V'$, and move to a and then move to the other endpoint along e, and continue until an old node t is reached (by recognizing a foot print). Let's denote this chain of new edges by S. Denote by (s, t) the last edge along S. We want to find out which node of V' is t. The robot can come back along S to node a. Since a and t are the only two nodes in P incident to an increased number of foot-printed edges, the robot can go through all the nodes in V' along edges in E' to find out which node in V' is t. Then the partial graph can be updated by $V' \leftarrow V' \cup \{V(S)\}$, $E' \leftarrow E' \cup \{E(S)\}$.

We notice that this algorithm may, in the worst case, exhaust all the nodes for validating a single edge. Thus the total number of traversals is $O(mn)$. Similar results using the same approach are obtained for the pure token model in [DJMW], with a single token. Both algorithms give a competitive ratio $O(n)$. In Section 2, we show this result is tight for the pure single token model, for a restricted class of strategies. We notice that, in both the token model and the foot-print model, it is a nontrivial task to design a general optimal algorithm for verifying a map. Fortunately, for the particular lower bound example, we are able to evaluate the number of traversals needed for verifying a map.

In Section 3, we develop a competitive strategy for the mapping problem on embedded planar graphs for the pure foot-print model. traversal of edges. Moreover, this DFS tree is obtained by ordering edges incident to a node counterclockwisely. Thus, we call it the *leftmost DFS tree*. In Section 4, we show that this strategy can be applied to the mapping problem in the token model with n tokens, by carefully studying the structure of the leftmost DFS tree. Notice that planar graph traversals have also been important in algorithms testing planarity of graphs, in which case a planar embedding is usually constructed for the graphs, (see, for example, [BL, CNA, HT].) Our problem is quite different, an unknown embedding of a planar graph is already given. Our task is to construct the map by collecting information while traversing the embedding. Section 5 concludes the paper with discussions on several related problems.

2 Mapping General Graphs with One Token

In this section, we show that there is no constant competitive ratio for the mapping problem with a single token. Moreover, we prove that the upper bound of competitive ratio $O(n)$ for the pure token model is tight under a restricted class of strategies. We want to build a graph of $O(n)$ edges, for which mapping takes $\Omega(n^2)$ traversals, and verification takes only $O(n)$ traversals. The restriction is to allow our strategy to put the token at the first unknown node reached, and try to find out if it is among the old nodes. We call such a strategy *depth-one search*.

We construct a star-shape graph of $2n+1$ nodes, one node of degree $2n$, and $2n$ nodes of degree two. There is a perfect matching between these $2n$ nodes of degree two, to be determined by the adversary, adapt to our strategy. We call the degree $2n$ node *the origin*, denoted by $< 0 >$. Then, we name all other nodes according to the circular order of edges incident to the origin $< 0 >$, and denote them by $< 0, i >$, $i = 1, 2, \cdots, 2n$. Each node $< 0, i >$, is matched to another node of degree two, which is called the matched node of $< 0, i >$ and denoted by $M(i)$. When we move to $< 0, i >$ from the origin, we denote the other adjacent node of $< 0, i >$ by $< 0, i, M >$. We can move from $< 0, i, M >$ back to $< 0, i >$, or move to the origin via another edge incident to $< 0, i, M >$. In the latter case, we don't know which node $< 0, j >$ is the same as $< 0, i, M >$. Thus, the task of mapping is to find out the matching $M(i) = j$ for all $i = 1, 2, \cdots, 2n$. Also, *depth–one search* algorithms use tests of type $M(i) = j$. First consider the following algorithm. For each $i = 1, 2, \cdots, 2n$, we find $M(i)$ by moving from $< 0 >$ to $< 0, i >$, then to $< 0, i, M >$. Then, put the token at $< 0, i, M >$ and move back to $< 0 >$ via $< 0, i >$ (in the reversal order). Then probe each of $< 0, j >$, $j = 1, 2, \cdots, 2n$ until the token is found. The adversary reveals the token only when the node is the only possible location of the token. Therefore, to obtain $M(i)$, it takes $2n - 2i$ tests, $1 \le i \le n$, which totals $\Omega(n^2)$ traversals. To prove this holds for all possible *depth–one search* algorithms, we introduce a graph theoretical lemma.

Lemma 1 (Bol) . *Let $F(G)$ be the number of perfect matchings in graph G. Then, $\max\{|E(G)| : |G| = 2n, F(G) = 1\} = n^2$.*

Thus, for depth–one search algorithms, we start with a complete graph K_{2n}, representing all the possible connections between degree two node in our graph. Each test, whether $< 0, i, M >$ is the same as $< 0, j >$ for some i, j, removes one edge from the graph, except if the removal will eliminate all the perfect matchings in the graph. The above lemma states that we cannot determine a unique matching until there are only n^2 edges left. The total number of edges in K_{2n} is $n(2n - 1)$. Therefore, we need at least $n(n - 1)$ tests before we can decide that unique matching in the graph we try to map. $\square$

To verify a matching $i = M(j)$ in the above graph, our robot puts the token at $M(i)$ along $< 0, i >$ and move back to 0 and then to j to check if the token is there. Thus, we traverse six edges for verifying a matched pair. The total verification cost is thus $6n$. The ratio is thus $\Omega(n)$. Therefore, combining this lower bound with the $O(|V| \times |E|)$ algorithm for mapping [DJMW], which is a depth–one strategy, we have

Theorem 1 *The competitive ratio of depth–one strategies for mapping general embedded graphs in the pure Token–Model is $\Theta(n)$.*

A similar construction gives a lower bound for embedded planar graphs.

Corollary 1 *The competitive ratio of depth–one strategies for mapping embedded planar graphs in the pure Token–Model is $\Omega(\log n)$.*

We can, however, design more complicated algorithms. When we move to $< 0, i, M >$ and then to the origin, we denote the origin by $< 0, i, M, 0 >$. In this case, we don't know how to match edges we see now with the ones we saw when we started at the origin. We can then define the edge from which we reach the origin as 0 and order other edges along the clockwise order from this edge: Therefore, $< 0, i, M, 0, j >$ denotes the node we reach along the j-th (*relative to the edge we came back to* 0) clockwise edge from $< 0, i, M, 0 >$. In general, we can use $< 0, i_1, M, 0, i_2, M, 0, i_3, \cdots >$ to denote the node we reached by taking the i_1-th edge (*relative to the last edge we came back to* 0), the matched node and the origin, then taking the i_2-th edge (*relative to the edge we come back to* 0), the matched node and the origin, and so on. Our general test could be in the following form

$$< 0, i_1, M, 0, i_2, M, 0, i_3, \cdots, i_s (or M) > = < 0, j_1, M, 0, j_2, M, 0, j_3, \cdots, j_t, (or M) > .$$

There are $\Omega(n!)$ possible star–like graphs. Whenever we make tests, the adversary can give an answer eliminating no more than half the possible graphs. Thus we have to do at least $\Omega(n \log n)$ tests. However, since we have only one token, each such test needs at least one move. Therefore, we have

Theorem 2 *The competitive ratio for any kind of strategy for the pure Token–Model with a single token is $\Omega(\log n)$.*

However, we conjecture Theorem 1 holds for all possible strategies and with the star–like graph as the example for the lower bound.

Conjecture 1 *The competitive ratio for any kind of strategy for mapping general embedded graphs in the pure Token–Model is $\Theta(n)$.*

Conjecture 2 *The competitive ratio for any kind of strategy for mapping embedded planar graphs in the pure Token–Model is $\Omega(\log n)$.*

However, the star–like planar graphs for Corollary 1 will not give rise to this lower bound since we have an algorithm to map that particular one within linear number of traversals.

3 Exploration with Foot–Prints

As we have seen, the pure token model does not lead to constant competitive algorithms in general graphs. We now restrict our discussion on embedded planar graphs. In this section, we show a competitive algorithm to explore unknown graphs with foot-prints. Then, in the next section, we discuss how this algorithm can be modified to map unknown graphs of n nodes, with n identical tokens.

The backbone of our algorithm for traversing the graph is a DFS tree T. We call the edges on the DSF tree by tree edges, and call other edges by backtrack edges. When we traverse an edge to enter a node for the first time, this edge is made a tree edge. Each node is named the first time it is reached. Thus, for each node already named, say i, we have a tree edge $tree(i) = (j, i)$ in T for some $j < i$. All the tree edges can be identified as edges entering a node without foot-prints on traversing the unknown graph. For backtrack edges, from a node named i we enter an old node u which we may not be able to recognize it. In this case, i is put into a STACK. We will determine the name of node u later. We will also use the map of *partial graph* $P = (V', E')$ similar to that in Section 1. Thus, V' denotes all the nodes already visited. E' denotes all the edges already identified, i.e., names of both its end points are known. For each edges traversed but not yet identified, one of its endpoints is put in $STACK$, we will determine the other endpoint later. Traversals of edges incident to a node i follow the counterclockwise order, starting with the edge $tree(i)$. Thus, our exploration algorithm will find the leftmost DFS tree.

Algorithm 1 *The algorithm with pure foot-prints.*

0. *Initially, name the starting node as 1. Set $COUNT = 2$, $STACK = \emptyset$, draw on the map, $V' = \{1\}$, $E' = \emptyset$. Take an edge e out of node 1 by calling the procedure* explore(e).

1. *On reaching node u from* explore(e) *with $e = (i, u)$, consider two cases:*

   1.1. *If there is no foot-print on u, then name the node as $COUNT$, let $tree(COUNT) = e$, and add e and u into the DFS tree T. We also update the map by adding node $COUNT$ to V' adding the edge e to E'. Then, increment $COUNT$ by one. If there are other edges incident to u, take the edge f counterclockwise first from e at u, and call* explore(f). *Otherwise, call* backtrack(e).

   1.2. *If there are foot-prints on u, push i into $STACK$, and do* backtrack(e).

2. *Upon reaching a node k from operation* backtrack(e) *with $e = (j, k)$. Denote by f the next counterclockwise edge with respect to e if there is one. Do the following until $f = tree(k)$, in which case do* backtrack(f).

   2.1 *If f is untraversed, do* explore(f).

   2.2 *If there is a foot-print on f but f is not on the map, i.e., f is not in E', then add $(pop(STACK), k)$ into E'. Reset f equal to the next counterclockwise edge with respect to f.*

For simplicity, the above algorithm does not deal with the case when we backtrack an edge to the root. In this case, we stop and report the current partial map as the map if f is the edge initially traversed to start our exploration. Figure 1 gives an example of the execution of the algorithm. (In the global and local views, solid edges have foot-prints, dashed edges do not. In the map, dangling edges incident to a node are those whose other end is not yet identified.) The above algorithm traverses each edge no more than twice. We only need to prove that names of two nodes on a backtrack edge are correctly matched.

Theorem 3 *There is an algorithm for a robot to map an unknown embedded planar graph in the pure foot-print model by traversing each edge at most twice.*

4 Exploration with n Identical Tokens

In this section, we show there is a competitive algorithm to explore unknown graphs of n nodes with n identical tokens, one token for one node. However, the above algorithm does not trivially carry over: simply replacing foot-prints by tokens would need $m + n$ tokens. In fact, we achieve this with less tokens by increasing the number of traversals of edges.

Theorem 4 *There is an algorithm for a robot to map an unknown embedded planar graph of n nodes with n identical tokens by traversing each edge at most four times.*

5 Remarks and discussions

We are able to obtain a competitive algorithm for mapping embedded planar graphs with n identical tokens. Can we competitively map embedded planar graphs with a sublinear number of tokens? What is the competitive ratio for mapping general or planar graphs with a single token?

References

[BBFY] E. Bar–Eli, P. Berman, A. Fiat, and P. Yan, "Online Navigation in a Room," SODA 1992, pp. 237–249.

[BCR] R.A. Baeza-Yates, J.C. Culberson, and G.J.E. Rawlins, "Searching in the Plane," To appear in *Information and Computation*.

[BL] K.S.Booth, G.S.Lueker [1976] "Testing the consecutive ones property, interval graphs, and graph planarity using PQ-tree algorithms" J.CSS. 13 (1976), 335-379.

[Bol] B. Bollobas, "Extremal Graph Theory," Academic Press, San Francisco, 1978.

[BRS] A. Blum, P. Raghavan and B. Schieber, "Navigating in Unfamiliar Geometric Terrain," *STOC 1991*, pp.494–504.

[CNA] N.Chiba, T.Nishizeki, S.Abe [1985] "A linear algorithm for embedding planar graphs using PQ-trees" J.CSS. 30 (1985), 54-76.

[DJMW] G.Dudek, M, Jenkin, E. Milios, D. Wilkes, "Using Multiple markers in graph exploration," *SPIE Vol. 1195 Mobile Robots IV(1989)*, pp.77-87.

[DKP] X. Deng, T. Kameda and C. H. Papadimitriou, "How to Learn an Unknown Environment," FOCS 1991, pp.298–303.

[DP] X. Deng, and C. H. Papadimitriou, "Exploring an Unknown Graph," FOCS 1990, pp.355-361.

[FFKRRV] A. Fiat, D.P. Foster, H. Karloff, Y. Rabani, Y. Ravid, and S. Vishwanathan, "Competitive Algorithms for Layered Graph Traversal," FOCS 1991, pp.288-297.

[GMR] L.J. Guibas and R. Motwani and P. Raghavan, " The Robot Localization Problem in Two Dimensions", SODA 1992, pp. 259-268.

[HT] J.E.Hopcroft, R.E.Tarjan [1973] "Efficient planarity testing", JACM 21(4), 1973, 549-568.

[KB] B. Kuipers, and Y. Byun, "A Robot Exploration and Mapping Strategy Based on a Semantic Hierarchy of Spatial Representatinos," *Robotics and Autonomous Systems* 8 (1991) 47–63.

[Kl] R. Klein, "Walking an Unknown Street with Bounded Detour," FOCS 1991, pp. 304–313.

[KP] B. Kalyanasundaram, and K. Pruhs, "A Competitive Analysis of Algorithms for Search Unknown Scenes," manuscripts, 1991.

[KRR] H. Karloff, Y. Rabani, and Y. Ravid, "Lower Bounds for Randomized k-Server and Motion-Planning Algorithms," *STOC 1991*, pp. 278–288.

[LL] T.S. Levitt, and D. T. Lawton, "Qualitative Navigation for Mobile Robots," *Artificial Intelligence* 44 (1990), 305–360.

[PY] C. Papadimitriou and M. Yannakakis, "Shortest paths without a map," *Proc. 16th ICALP*, 1989, pp.610-620.

[RS] R. L. Rivest and R. E. Schapire, "Inference of Finite Automata Using Homing Sequences," STOC 1989, pp.411-420.

[ST] Sleator, D.D. and Tarjan, R.E., Amortized efficiency of list update and paging rules," *Communications of the ACM 28(2)*, 1985, pp.202-208.

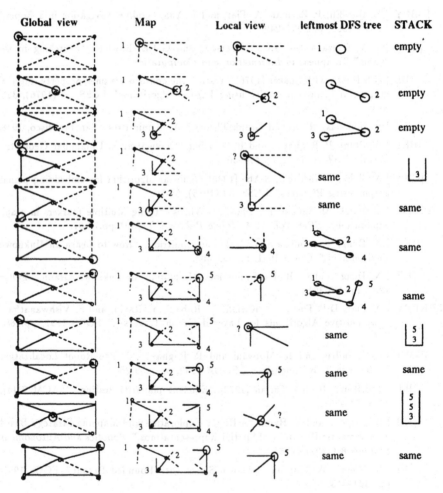

Figure 1.

Global view	Map	Local view	leftmost DFS tree	STACK

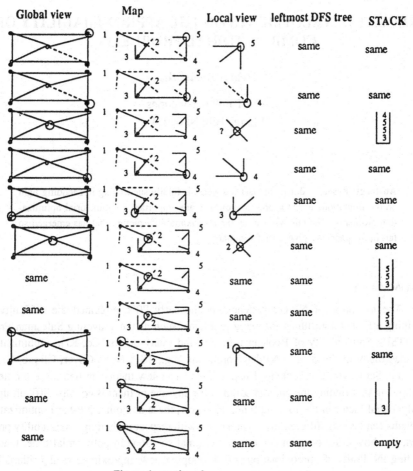

Figure 1: continued.

RECENT DEVELOPMENTS ON THE APPROXIMABILITY OF COMBINATORIAL PROBLEMS

Mihalis Yannakakis

AT&T Bell Laboratories
Murray Hill, NJ 07974

Abstract: Research during the last few years is finally succeeding in shedding light on the limitations in the approximation of many important combinatorial optimization problems. The talk will survey the approximability picture that has emerged so far. In the paper we give a brief summary.

1. SUMMARY

The invention of NP-completeness in the early 70's explained the difficulty in obtaining efficient algorithms for many problems, such as the Traveling Salesman Problem (TSP), Satisfiability of Boolean formulas and many others [Co, Kr]. Optimization provided many of these early NP-hard problems, e.g., TSP, Node Cover, Clique, Graph Coloring, Set Cover, Bin Packing, Knapsack etc. The new theory unified these problems by showing that finding the optimal solution for any one of them is equally difficult up to a polynomial factor in the running time. However, these and other NP-hard optimization problems can be very different from each other with respect to their approximability properties, i.e., how close one can come to the optimal solutions in polynomial time. Understanding the limits of approximation and developing new approximation algorithms has been a very active area of research for a long time, and significant progress has been made in the last few years.

The performance of approximation algorithms is usually measured theoretically by the worst-case ratio between the cost (or value) of the solution that they find and that of the optimal solution. For any given problem, the goal is to determine the "best possible" ratio (closest to 1) that can be achieved by a polynomial-time algorithm, and of course to find an algorithm that achieves it. More modestly, we may classify problems qualitatively into one of the following three categories: (1) Those that can be approximated in polynomial time within *any* constant factor c of optimal, arbitrarily close to 1, i.e., problems that have a *polynomial time approximation scheme* (PTAS); (2) Those that can be approximated with *some* (but not every) constant ratio c; (3) Those that *cannot* be approximated within any constant ratio c, i.e., the best ratio grows with the input size n.

Early work after the introduction of NP-completeness succeeded in classifying some of the problems. For example, it was shown that TSP without triangle inequality cannot be approximated within any constant factor (unless P=NP) [SG], while Knapsack and some other problems involving numbers have polynomial time approximation schemes [Sa] (moreover, they have a stronger type of approximation schemes called full PTAS [IK]). However, the initial optimism subsided at least with regard to many of the more prominent NP-hard combinatorial optimization problems, for which all efforts in classifying them even qualitatively failed. For example, in the case of the Clique problem, it was shown that none of the usual heuristics achieved an approximation ratio of even $n^{1-\varepsilon}$ for any $\varepsilon > 0$ [J1], whereas a PTAS could not be ruled out; in fact Garey and Johnson showed that the problem either had a PTAS or could not be approximated within any constant factor (i.e., Clique cannot be in category (2) above). Similarly, for Graph Coloring, none of the usual heuristics achieved ratio of even $n^{1-\varepsilon}$ [J2], but the best that could be ruled out were ratios less than 2 [GJ1]. For several other problems (including the TSP with triangle inequality, Node Cover, Max Cut, Maximum Satisfiability etc), simple approximation algorithms achieving a constant ratio were found very quickly, but it was then hard to achieve better ratios, or even more ambitiously to obtain a PTAS or alternatively show that the problems do not have a PTAS. As a result of the recent developments we now know how to classify all the above problems.

An obstacle in developing a unifying approximation theory paralleling the NP-completeness theory is that every problem seems to have more or less its own personality in terms of approximability and must be studied individually. However, it turns out that qualitatively many of them are interrelated. In [PY1] we introduced with Papadimitriou a class of optimization problems MAX SNP (and some variants) to capture the type of problems that have simple constant factor approximations but the existence of a PTAS is in doubt. The class was defined using Fagin's logical characterization of NP [Fa], and it was shown that all problems in it have constant approximations. A type of "linear" reduction was also defined between problems that preserves the degree of approximation. A number of well-studied problems have been shown complete or hard for MAX SNP under this reduction, and the list keeps growing. Some complete problem are MAX 3SAT (Maximum Satisfiability of 3CNF Boolean formulas), MAX CUT, many problems on bounded degree graphs such as Node Cover, Independent Set, Dominating Set etc. Many other problems are hard: TSP with triangle inequality [PY2], Steiner Tree [BP], Shortest Superstring [B+], Multiway Cut [D+], 3-dimensional matching [Kn] and others. This means that if any one problem in MAX SNP does not have a PTAS, then neither do all these complete and hard problems.

With the introduction of the class MAXSNP, the important questions involving the boundary of categories (1) and (2) above were unified to the following: Do all problems in MAXSNP have PTAS? Notice that this is a refinement of the P vs. NP problem, in that if P=NP, then the answer is trivially positive. The widespread conjecture was that these two questions are equivalent: All of MAXSNP has PTAS if and only if P=NP. But the proof of this connection seemed as remote as for many other questions in complexity theory.

Help in resolving these questions came from a rather unexpected source. For years researchers working in another front in Complexity Theory, *interactive* and *probabilistically checkable proofs*, were developing a theory of more and more powerful proof systems, whose structure seemed to parallel that of machine-based complexity theory. In 1991, Feige et al. [F+] established a simple but startling connection between this theory and approximability: They showed that the Clique problem cannot be approximated in polynomial time within any constant (and in fact a much larger) factor, unless NP is contained in "quasipolynomial" time (more precisely, unless NP $\subseteq$ DTIME($n^{O(\log\log n)}$). This was tightened in [AS] and [A+] to show that for some $c > 0$ Clique cannot be approximated with ratio n^c unless P=NP (the current value of c is around 1/25 under a somewhat stronger assumption than P≠NP, but still a believable one [BGLR, FK, Zu]). Furthermore, Arora et al. [A+] showed that MAX 3SAT cannot be approximated in polynomial time beyond some constant factor $1 + \varepsilon$ unless P=NP (the current value of ε is around 1% [BGLR]). Thus, MAX 3SAT and all MAX SNP-hard problems do not have a polynomial time approximation scheme.

Several other longstanding problems were shown nonapproximable within any constant (and larger) factors, unless NP is contained in P or in quasipolynomial time. Graph Coloring cannot be approximated within n^c for some c [LY1]. Set Cover cannot be approximated within $c \log n$ for some c (the greedy algorithm achieves ratio $O(\log n)$ [J1]), and the same holds for related problems such as Dominating Set etc [LY1]. The class of maximum induced subgraph problems (find the maximum set of nodes inducing a subgraph that satisfies a given property π) cannot be approximated for any property π that is preserved under node deletion, for example, planar, acyclic, bipartite etc. [LY2]. Quadratic Programming (optimizing a quadratic function over a polytope) cannot be approximated within large factors [BR, FL]. The same holds for the problem of finding the maximum consistent subset of a set of linear equations over the rationals [ABSS] and many other problems [Be, KMR, Zu]. For several problems, especially those that are based on the interactive proof systems with two (or constant number of) provers [LS, FL], the evidence of nonapproximability is often somewhat weaker than P≠NP. As the proof systems are being refined and improved, both the evidence and the degree of nonapproximability get strengthened [BGLR, FK].

Besides this progress on the negative side, there has been in the last few years significant progress also on the positive side, with the development of new methods and good approximation algorithms for several interesting problems; this is a separate large topic we will not cover in this talk.

2. REFERENCES

There is a huge literature on approximation and there are survey articles even for specific problems, for example bin packing [CGJ]. Below is an undoubtedly incomplete bibliography with emphasis on lower bounds and general approximation theory. Chapter 6 of the Garey-Johnson book [GJ2] contains basic definitions and an overview of the work of the 70's, and Chapter 13 of the forthcoming book by Papadimitriou [Pa] covers also the recent results. The Ph.D. thesis of Kann [Kn] contains a detailed enumeration of problems. Johnson's column [J3] and Babai [Ba] survey in different levels of detail the theory of probabilistically checkable (transparent, interactive) proofs and their relation to approximability.

[ABSS]   S. Arora, L. Babai, J. Stern, Z. Sweedyk, "The Hardness of Approximating Problems Defined by Linear Constraints", Proc. 34th IEEE Symp. on Foundations of Computer Science, 1993.

[A+]   S. Arora, C. Lund, R. Motwani, M. Sudan, M. Szegedy, "Proof Verification and Hardness of Approximation Problems", Proc. 33rd IEEE Symp. on Foundations of Computer Science, 14-23, 1992.

[AS]   S. Arora, S. Safra, "Probabilistic Checking of Proofs", Proc. 33rd IEEE Symp. on Foundations of Computer Science, 2-13, 1992.

[AMP]   G. Ausiello, A. Marchetti-Spaccamela, M. Protasi, "Toward a Unified Approach for the Classification of NP-complete Optimization Problems", Theoretical Computer Science 12, 83-96, 1980.

[ADP]   G. Ausiello, A. D'Atri, M. Protasi, "Structure Preserving Reductions Among Convex Optimization Problems", J. Comp. Sys. Sci. 21, 136-153, 1980.

[Ba]   L. Babai, "Transparent Proofs and Limits to Approximation", manuscript, 1993.

[BFL]   L.Babai, L. Fortnow, C. Lund, "Nondeterministic Exponential Time has Two-Prover Interactive Protocols", Computational Complexity 1, 3-40, 1991.

[Be]   M. Bellare, "Interactive Proofs and Approximation: Reductions from Two Provers in One Round", Proc. 2nd Israel Symp. on Theory and Computing Sys., 266-274, 1993.

[BGLR]   M. Bellare, S. Goldwasser, C. Lund, A. Russel, "Efficient Probabilistically Checkable Proofs and Applications to Approximation", Proc. 25th ACM Symp. on Theory of Computing, 294-304, 1993.

[BR]   M. Bellare, P. Rogoway, "The Complexity of Approximating a Nonlinear Program", in Complexity in Numerical Optimization, P. Pardalos ed., World-Scientific, 1993.

[BS]   P. Berman, G. Schnitger, "On the Complexity of Approximating the Independent Set Problem", Information and Computation 96, 77-94, 1992.

[BP]   M. Bern, P. Plassman, "The Steiner Problem with Edge Lengths 1 and 2", Information Processing Letters, 32, 171-176, 1989.

[Bl]   A. Blum, "Some Tools for Approximate 3-coloring", Proc. 30th IEEE Symp. on Foundations of Computer Science, 554-562, 1990.

[B+]   A. Blum, T. Jiang, M. Li, J. Tromp, M. Yannakakis, "Linear Approximation of Shortest Superstrings", Proc. 23rd ACM Symp. on Theory of Computing, 328-336, 1991.

[Ch]   N. Christofides, "Worst-case Analysis of New Heuristics For the Traveling Salesman Problem", Technical Report, GSIA, Carnegie-Mellon, 1976.

[CGJ]   E.G. Coffman Jr., M. R. Garey, D. S. Johnson, "Approximation Algorithms for Bin Packing - An Updated Survey", in Algorithm Design and Computer System Design, G. Ausiello, M. Lucerini, P. Serafini (eds.), 49-106, Springer Verlag, 1984.

[Co]   S. A. Cook, "The Complexity of Theorem Proving Procedures", Proc. 3rd Annual ACM Symp. on Theory of Computing, 151-158, 1971.

[D+]   E. Dahlhaus, D. S. Johnson, C. H. Papadimitriou, P. Seymour, M. Yannakakis, "The Complexity of Multiway Cuts", Proc. 24th Annual ACM Symp. on Theory of Computing, 1992.

[Fa]   R. Fagin, "Generalized First-Order Spectra, and Polynomial-time Recognizable Sets", in R. Karp (ed.), Complexity of Computations, AMS, 1974.

[FK]   U. Feige, J. Kilian, "Two Prover Protocols - Low Error at Affordable Rate", in preparation.

[FL]   U. Feige, L. Lovasz, "Two-prover One-round Proof Systems: Their Power and Their Problems", Proc. 24th Annual ACM Symp. on Theory of Computing, 733-744, 1992.

[F+]   U. Feige, S. Goldwasser, L. Lovasz, S. Safra, M. Szegedy, "Approximating Clique is Almost NP-complete", Proc. 32nd IEEE Symp. on Foundations of Computer Science, 2-12, 1991.

[GJ1]   M. R. Garey, D. S. Johnson, "The Complexity of Near-optimal Graph Coloring", J.ACM 23, 43-49, 1976.

[GJ2]   M. R. Garey, D. S. Johnson, Computers and Intractability: A Guide to the Theory of NP-completeness, Freeman, 1979.

[IK]   O. H. Ibarra, C. E. Kim, "Fast Approximation Algorithms for the Knapsack and Sum of Subset Problems", J.ACM 22, 463-468, 1975.

[Ir]   R. W. Irving, "On Approximating the Minimum Independent Dominating Set", Information Processing Letters 37, 197-200, 1991.

[J1]   D. S. Johnson, "Approximation Algorithms for Combinatorial Problems", J. Comp. Sys. Sc. 9, 256-278, 1974.

[J2]   D. S. Johnson, "Worst Case Behavior of Graph Coloring Algorithms", Proc. 5th Conf. on Combinatorics, Graph Theory and Computing, 513-527, 1974.

[J3]   D. S. Johnson, "The NP-completeness Column: An Ongoing Guide", J. of Algorithms 13, 502-524, 1992.

[Kn]   V. Kann, "On the Approximability of NP-complete Optimization Problems", Ph.D. Thesis, Royal Institute of Technology, Stockholm, 1992.

[Kr]   R. M. Karp, "Reducibility among Combinatorial Problems", in R. E. Miller and J. W. Thatcher (eds.), Complexity of Computer Computations, Plenum Press, 85-103, 1972.

[KLS]   S. Khanna, N. Linial, S. Safra, "On the Hardness of Approximating the Chromatic Number", Proc. 2nd Israel Symp. on Theory and Computing Sys., 250-260, 1993.

[KMR]   D. Karger, R. Motwani, G.D.S. Ramkumar, "On Approximating the Longest Path in a Graph", Proc. Workshop on Discrete Algorithms and Structures, 1993.

[KT]   P. G. Kolaitis, M,. N. Thakur, "Approximation Properties of NP Minimization Classes", Proc. 6th Conf. on Structures in Computer Science, 353-366, 1991.

[LS]   D. Lapidot, A. Shamir, "Fully Parallelized Multiprover Protocols for NEXP-TIME", Proc. 32nd Annual IEEE Symp. on Foundations of Computer Science, 13-18, 1991.

[LY1]   C. Lund, M. Yannakakis, "On the Hardness of Approximating Minimization Problems", Proc. 25th ACM Symp. on Theory of Computing, 286-293, 1993.

[LY2]   C. Lund, M. Yannakakis, "The Approximation of Maximum Subgraph Problems", Proc. 20th Intl. Coll. on Automata, Languages and Programming, 40-51, 1993.

[PR]   A. Panconesi, D. Ranjan, "Quantifiers and Approximation", Proc. 22nd ACM Symp. They of Computing, 446-456, 1990.

[Pa]   C. H. Papadimitriou, Computational Complexity, Addison-Wesley, 1993.

[PY1]   C. H. Papadimitriou, M. Yannakakis, "Optimization, Approximation and Complexity Classes", J. Computer and System Sci. 43, 425-440, 1991.

[PY2]   C. H. Papadimitriou, M. Yannakakis, "The Traveling Salesman Problem with Distances One and Two", Mathematics of Operations Research 18, 1-11, 1993.

[PM]   A. Paz, S. Moran, "Non Deterministic Polynomial Optimization Problems and their Approximation", Theoretical Computer Science 15, 251-277, 1981.

[Sa]   S. Sahni, "Approximate Algorithms for the 0/1 Knapsack Problem", J.ACM 22, 115-124, 1975.

[SG]   S. Sahni, T. Gonzalez, "P-complete Approximation Problems", J. ACM 23, 555-565, 1976.

[Si]   H. U. Simon, "On Approximate Solutions for Combinatorial Optimization Problems", SIAM J. Disc. Math. 3, 294-310, 1990.

[Y1]   M. Yannakakis, "The Effect of a Connectivity Requirement on the Complexity of Maximum Subgraph Problems", J.ACM 26, 618-630, 1979.

[Y2]   M. Yannakakis, "On the Approximation of Maximum Satisfiability", Proc. 3rd ACM-SIAM Symp. on Discrete Algorithms, 1-9, 1992.

[Zu]   D. Zuckerman, "NP-complete Problems Have a Version That's Hard to Approximate", Proc. 8th Conf. on Structure in Complexity Theory, 305-312, 1993.

On the Relationship among Cryptographic Physical Assumptions

Tatsuaki Okamoto

NTT Laboratories
Nippon Telegraph and Telephone Corporation
1-2356 Take, Yokosuka-shi, Kanagawa-ken, 238-03 Japan
Email: okamoto@sucaba.ntt.jp

Abstract. Although the equivalence of some cryptographic *computational* assumptions (e.g., the existence of one-way functions, that of pseudo-random generators, and that of secure signature schemes) has been studied extensively, little study has been done on the equivalence of cryptographic *physical* assumptions.

This paper initiates research in this direction. It shows that three *physical* cryptographic assumptions about channels are equivalent. The three physical assumptions are the existence of anonymous channels, that of direction-indeterminable channels, and that of secure channels.

1 Introduction

One of the most important results in cryptography over the last ten years is that the relationships between and the equivalence of certain *computational* cryptographic assumptions have been clarified. For example, some primitives such as pseudo-random generators, secure bit-commitment, and secure signature schemes have been proven to exist if and only if one-way functions exist [ILL89, Has90, Nao90, NY89, Rom90]. On the other hand, strong evidence that the existence of one-way permutation is not sufficient for the existence of secure secret key exchange has been shown [IR89].

Although the above-mentioned results clarify what cryptographic functions can be realized based only on *computational* assumptions, some *physical* assumptions are known to be essential for their construction. By "*physical assumption*" and "*physical* primitive", we mean ones whose properties depend on no computational condition. (Such assumptions and primitives are often called "*unconditionally secure*"). For example, unconditionally secure multi-party protocols can be constructed using *secure channels* [BGW88, CCD88] (and additionally *broadcast channels* for [BR89]). The *physical separation* between provers is essential in the definition of multi-prover interactive proofs [BGKW88], and a physical bit-commitment primitive, called the *envelope*, is an essential tool for constructing a "perfect" zero-knowledge proof for any language in IP (or PSPACE) [BGGHKMR90]. An *anonymous channel* is often used to construct privacy protecting systems such as electronic secret balloting and electronic cash [Cha81, Cha88, CFN90].

However, little study has been done on the equivalence of cryptographic *physical* assumptions. Although Crépeau and Kilian showed the relationships of some types of oblivious transfer channels, noisy channels and quantum channels [CK88], they gave little attention to the equivalence of these physical assumptions (the existence of these physical primitives), since their major interest was the reduction of an assumption to a trivially weaker assumption. So, to the best of our knowledge, no information about the equivalence of exclusive *physical* cryptographic assumptions has been reported. Here, when we call that two assumptions are "exclusive", there is no trivial inclusion relationship between them.

This paper initiates research in this direction. It shows that three exclusive *physical* assumptions about channels are equivalent. The three physical assumptions are: the existence of anonymous channels, direction-indeterminable channels, and secure channels.

An *anonymous channel* (AC) is a channel in which the originator of a message is anonymous to the receiver, and a third party can know a transmitted message along with the receiver's identity. A *direction-indeterminable channel* (DIC) is a channel in which a third party can know a transmitted message along with a pair of identities, which are the originator's and receiver's identities in random order, but cannot detect the direction of the message (or which is the originator or receiver given the pair of identities). Moreover, the receiver can know the originator's identity. A *secure channel* (SC) is a channel in which a common secret key can be shared between two parties, and an encrypted message (one-time padded by the secret key) is transmitted. Here, an eavesdropper can know both the originator's and receiver's identities of the encrypted message, and the receiver can know the originator's identity. Note that, from the definitions, AC is neither SC nor DIC, DIC is neither AC nor SC, and SC is neither AC nor DIC (i.e., they are exclusive).

Some works informally[1] imply these reductions: the dining cryptographers protocol [Cha88, Wai89, BD89] implies the reduction of AC to SC, and the keyless cryptosystem [AS83] implies the reduction of SC to DIC. This paper, for the first time, formally sheds light on the equivalence problem of the physical assumptions, and proves the equivalence of these exclusive physical assumptions.

In section 3, we define the physical primitives (AC, DIC and SC), as well as what it means for a physical primitive to be reducible to another physical primitive. Section 4 shows that DIC is reducible to AC, that SC is reducible to DIC, and that AC is reducible to SC. That is, we have determined that the three physical assumptions are equivalent.

2 Notations

Let x, y, a_i be binary strings, and n be a non-negative integer. Then $|x|$ denotes the bit length of x, and 0^n denotes "$0 \cdots 0$" ($|0^n| = n$). $x \| y$ denotes the concatenation of x and y. $poly(n)$ denotes a polynomial in n. $x \oplus y$ denotes the bitwise exclusive-or of x and y. $\bigoplus_{i \in [1,\dots,N]} a_i$ denotes $a_1 \oplus a_2 \oplus \cdots \oplus a_N$.

3 Definitions

In this section, we define physical primitives about channels, and the definition for the reduction of a physical primitive to another physical primitive.

3.1 Definitions of Physical Primitives

Definition 1. [Physical primitive] A primitive is *physical*, if no properties of the primitive depend on a computational condition.

---

[1] Here, "informally" means that no formal definitions and reductions are given in their papers.

Definition 2. [Channel]   A physical primitive is called a *"channel"*, if it behaves as follows:

There exist three kinds of parties: an originator, receiver and third party. The originator (A), receiver (B), and third party (C) are interactive Turing machines [GMRa89] with a read-only communication tape and write-only communication tape. When A wants to send a message (one bit) to B, A writes the message along with B's identity on A's write-only communication tape. A "channel," then, reads this message and writes it on B's read-only communication tape. The channel also writes the message on C's read-only tape (if C exists). If C is a passive attacker, C writes a code of "no-change" on the write-only tape. If C is an active attacker, C writes a forged message (one bit) on the write-only tape, and the channel writes the forged message instead of the original message on B's read-only communication tape. The information that A writes on its tape, and that B and C read from their tapes depends on the type of channel (see Definitions 3, 4, and 5). Note that each party can send a message to himself through the channel. If necessary, some additional information, such as the originator's and receiver's identities, and message identifiers, can be transmitted along with the message (see Definitions 3, 4, and 5).

Notes:

- If the above procedure is repeated n times, one can send an n bit message. Sequential n-bit messages from different originators can be distinguished each other using additional information such as the originator's identity. In the case of an anonymous channel, a message identifier can be given to each one bit message to identify bits belonging to the same message.
- Physical channels alone cannot protect against obstruction by an active adversary. To counter obstruction by an active third party, many logical measures such as error detecting/correcting codes, message authentication codes, and digital signatures are often used.

Definition 3. [Anonymous channel (AC)]   A channel is *"anonymous"* if the following conditions are satisfied.

1. **(Transmitting capability:)** The originator writes a message (one bit) along with the receiver's identity on its write-only tape. An anonymous channel writes the message without the originator's identity on the receiver's read-only tape. The anonymous channel also writes the message along with the receiver's identity on the third party's read-only tape.

2. **(Security property:)** Even if the receiver (the third party) is an active attacker and collaborates with parties other than the originator, the probability that the receiver (the third party) guesses the originator correctly is exactly $1/L$, where L is the number of the parties that are not currently collaborating with the receiver (the third party).

Notes:

- When an originator is equivalent to the receiver, condition 2 for a receiver does not apply.

Definition 4. [Direction-indeterminable channel (DIC)]   A channel is *"direction-indeterminable"* if the following conditions are satisfied.

1. **(Transmitting capability:)** The originator writes a message (one bit) along with the receiver's identity on its write-only tape. A direction- indeterminable channel writes the message along with the originator's identity on the receiver's read-only tape. The direction-indeterminable channel also writes the message along with the pair of originator and receiver identities in random order on the third party's read-only tape.
2. **(Security property:)** The third party cannot detect the direction of the message, even if the third party is an active attacker and collaborates with any parties other than the originator and receiver. That is, the third party can guess the correct direction with probability exactly 1/2.

Notes:

− When an originator is equivalent to the receiver, condition 2 does not apply.

Definition 5. [Secure channel (SC)] A channel is *"secure"* if the following conditions are satisfied.

1. **(Transmitting capability:)** First, the originator writes a special code along with the receiver's identity, which implies the "key sharing phase," on the write-only tape. A secure channel writes a random secret key (one bit) on the originator's read-only tape and the same value along with the originator's identity on receiver's read-only tape. (*"key sharing phase"*)
   Next, the originator writes a message (one bit) encrypted (one-time padded) by the common secret key (one bit) along with receiver's identity on its write-only tape (more precisely, this message is started by a specific code which implies the "transmitting phase"). A secure channel writes the encrypted message along with the originator's identity on the receiver's read-only tape. It also writes the message along with the originator's and receiver's identities on the third party's read-only tape. Note that, here, the originator can choose any message whether or not in encrypted or plaintext form. (*"transmitting phase"*)
2. **(Security property:)** A third party cannot determine the secret key for the originator, even if the third party is an active attacker and collaborates with any parties other than the originator and receiver. That is, the third party can guess the correct key with probability exactly 1/2.

Notes:

− When an originator is equivalent to the receiver, condition 2 is trivially satisfied by selecting the secret key randomly within the originator (= receiver).

3.2 Definition of Reduction

Now we provide the definition of the reduction of a channel to another channel.

Definition 6. [Reduction] Let X, Y be two channels. X is reducible to Y if there exists a protocol P such that

1. Each party in P is a probabilistic interactive Turing machine. Party A' in P reads the write-only tape of originator A for X, and another party B' in P writes on the read-only tape of receiver B for X. Hereafter, for simplicity, we suppose the composition of party U' in P and party U for X to be one party, where U' reads (writes) the write-only (read-only) tape of U.

2. Each party in P can use Y as a black-box to communicate with another party.
3. Let n be the total number of steps taken by all parties during P. P realizes the "transmitting capability" of X with probability greater than $1 - \nu(n)$, if all parties in P are honest and third parties are passive. The probability is taken over the coin flips of all parties during P. Here, $f(n) < \nu(n)$ denotes that $\forall$ constants c, $\exists n_0$ such that $\forall n > n_0$, $[f(n) < 1/n^c]$.
4. P satisfies the "security property" of X, even if parties in P and third parties are active attackers.

Notes:

- No computational assumption is used to realize X in the above reduction, P.
- Since P is probabilistic, the probability that P realizes X should be estimated. Condition 3 above guarantees that P realizes X with high probability (or overwhelming probability with respect to the total number of steps during P). This efficiency condition is needed because a physical primitive should be utilized by a "polynomial-time bounded" (w.r.t. a problem size, N) party, and the primitive should work with overwhelming probability in N. The polynomial-time bounded parties can run the reduction P in $n = poly(N)$ steps. Then P can realize X with probability greater than $1 - \nu(N)$. Note that this efficiency condition is needed to make the reduction meaningful for "honest" polynomial-bounded machines who utilize X, but that an "adversary" can be an infinite power machine (so no computational assumption is used for P to realize X).

Definition 7. [Equivalence] Let X, Y be two channels. The assumption of X's existence is equivalent to the assumption of Y's existence, if X is reducible to Y and Y is reducible to X.

Lemma 8. [Transitivity] *Let X, Y, Z be three channels. If X is reducible to Y, and Y is reducible to Z, then X is reducible to Z.*

4 Equivalence of Three Physical Assumptions

This section proves that the three physical assumptions (the existence of AC, DIC, and SC) are equivalent. To prove it, we show three reductions, from DIC to AC, from SC to DIC, and from AC to SC.

4.1 Reduction of DIC to AC

Theorem 9. *A direction-indeterminable channel is reducible to an anonymous channel.*

Sketch of Proof:

If A wants to send a one bit message, m_A, to B through DIC, the protocol $P1$ for the reduction of DIC to AC is as follows:

1. A sends a message M_A to A and B in some random order by using a black-box AC, where M_A is m_A along with the identities of A and B in a random order. When B is equivalent to A, A sends M_A only to himself.
2. A and B receive M_A from AC.

We now show that the above protocol $P1$ realizes DIC with probability 1.
(**Transmitting capability for DIC**) First, B receives A's original message, m_A, from AC correctly. Receiver B also receives A's identity, since M_A, which B receives, includes A's identity. Next, a third party, C, can read message m_A along with the identities of A and B in a random order, since M_A, which C can read, includes the identities of A and B in a random order.

(**Security property for DIC**) A passive third party, C, cannot detect the direction of the message even if C has infinite power and collaborates with any other parties except the pair of originator and receiver, since no infinite power C can determine whether the originator's identity is A or B, given the definition of AC, and A's and B's procedures are independent from the procedures of the other parties.

Even if C and any parties other than A and B are active adversaries, the probability that they can guess the correct direction is no better than that when they are passive, since they can get no additional information after sending the forged message, and the direction is fixed before sending the forged message.
□

4.2 Reduction of SC to DIC

Theorem 10. *A secure channel is reducible to a direction-indeterminable channel.*

Sketch of Proof:
If A wants to send a one bit message, m_A, to B through SC, the protocol $P2$ for the reduction of SC to DIC is as follows:

1. A sends an initiating message to B such as "a message from A to B," which implies the start of key sharing phase, by using a black-box DIC.
2. A selects k random bits $r_1, r_2, \ldots, r_k$, and B selects k random bits $t_1, t_2, \ldots, t_k$. A and B exchange r_i and t_i through DIC repeatedly ($i = 1, 2, \ldots, k$).
3. A and B set $K \leftarrow r_1 \oplus r_2 \oplus \cdots \oplus r_k$. If $r_i = t_i$ for all $i \in \{1, 2, \ldots, k\}$, they give up protocol $P2$.
4. Then, A sends encrypted message $m_A \oplus K$ along with A's identity to B by using DIC.

(**Transmitting capability for SC**) We now show that the above protocol $P2$ realizes SC with probability greater than $1 - \nu(n)$, where n is the total number of steps taken by A and B during the execution of $P2$. First, a common secret key (one bit) K can be shared between two parties with probability $1 - 1/2^k$, since the probability that there is $i \in \{1, 2, \ldots, k\}$ with $r_i \neq t_i$ is $1 - 1/2^k$. Let c_1, c_2 be constants in k such that $c_1 k \geq n \geq c_2 k$. Hence, $c_1' n \geq k \geq c_2' n$, for constants (in n), c_1', c_2'. Therefore, $1 - 1/2^k \geq 1 - 1/2^{c_2' n} > 1 - \nu(n)$. (Key sharing phase of Transmitting capability)

Next, any message can be transmitted through DIC with probability 1. A third party can know the message, the originator, and receiver, and the receiver can know who is the originator. (Transmitting phase of Transmitting capability)
(**Security property for SC**) First, we assume that adversary C is passive. Though C obtains all r_i and t_i, from the security property of DIC, C cannot correctly predict r_i when $r_i \neq t_i$. Moreover, since A's and B's procedures are independent from the procedures of the other parties, no collaboration of parties, except the pair of originator and receiver, can guess the secret key with probability greater than $1/2$.

Next, we assume that adversary C is active. C can get no additional information after sending the forged message, and the value of K is fixed before sending the forged message. Hence, the probability that C can guess the correct value of K is no better than that when C is passive. □

4.3 Reduction of AC to SC

Theorem 11. *An anonymous channel is reducible to a secure channel.*

Sketch of Proof:

Let $A_1, A_2, \ldots, A_N$ be parties in protocol $P3$. W.l.o.g., we assume that any party in $P3$ sends at most one message in a regular interval (one second, one minute, or one hour,..., as fixed in each protocol). If A_l wants to send a one bit message m_l to A_k ($k = l$ is possible) through AC, the protocol $P3$ for the reduction of AC to SC is as follows:

The following procedure is repeated at the regular intervals.

1. A_i and A_j ($i, j \in \{1, 2, \ldots, N\}$, and $i \neq j$) share $2N^2(2N + 1)$ bit secret keys K_{ij} and K_{ji} such that $K_{ij} = K_{ji}$, by $2N^2(2N + 1)$ times using the key sharing phase of SC. A_i keeps K_{ij} and A_j keeps K_{ji}. So, each A_i keeps $N - 1$ secret keys, K_{ij} ($j = 1, \ldots, i - 1, i + 1, \ldots, N$).
2. Here, we explain this protocol when the receiver is A_k.
   (a) If A_l ($l \in \{1, \ldots, N\}$) wants to send a one bit message m_l to A_k, A_l generates an N bit random string r and calculates a $2N + 1$ bit message a_l from m_l and r such that

$$d_l = f(m_l \| r),$$

$$a_l = m_l \| r \| d_l,$$

where f is a non-homomorphic function over operation $\oplus$ such that, for any $x, \ldots, y \in \{0, 1\}^{N+1}$ (the number of $\{x, \ldots, y\}$ is at most N), [2]

$$f(x \oplus \cdots \oplus y) \neq f(x) \oplus \cdots \oplus f(y) \quad (f(x), f(y) \in \{0, 1\}^N).$$

A_l then selects N numbers $b_j \in \{1, \ldots, 2N\}$ ($j = 1, \ldots, N$) randomly. For $i = 1, \ldots, 2N$ and $j = 1, \ldots, N$, A_l generates M_l as follows:

$$M_{i,j} = \begin{cases} 0^{2N+1} & \text{for } i \neq b_j \\ a_l & \text{for } i = b_j \end{cases}$$

$$M_l = M_{1,1} \| \cdots \| M_{2N,1} \| M_{1,2} \| \cdots \| M_{2N,2} \| \cdots \| M_{1,N} \| \cdots \| M_{2N,N}.$$

Here, the size of M_l is $2N^2(2N + 1)$ bits.

Next, A_l generates $2N^2(2N + 1)$ bit message

$$C_l = M_l \oplus K_{l,1} \oplus \cdots \oplus K_{l,l-1} \oplus K_{l,l+1} \oplus \cdots \oplus K_{l,N}.$$

Note that C_k is calculated in the same manner, when $l = k$ (i.e., A_k wants to send a message to himself).

---

[2] the condition "for any $x, \ldots, y$" can be relaxed by "for overwhelming fractions of $x, \ldots, y$".

(b) On the other hand, if A_l wants to send no message to A_k, A_l generates $2N^2(2N+1)$ bit message

$$C_l = K_{l,1} \oplus \cdots \oplus K_{l,l-1} \oplus K_{l,l+1} \oplus \cdots \oplus K_{l,N}.$$

Note that C_k is calculated in the same manner, when $l = k$ (i.e., A_k wants to send no message to himself).

(c) For $l = 1, \ldots, N$, A_l sends C_l to A_k by using the transmitting phase of SC. Note that A_k sends C_k to himself whether A_k wants to send a message or not.

3. When A_k receives C_l ($l = 1, \ldots, N$), A_k calculates

$$M' = C_1 \oplus C_2 \cdots \oplus C_N.$$

For all $i = 1, \ldots, 2N$ and $j = 1, \ldots, N$, A_k checks

$$d'_{ij} = f(m'_{ij} \| r'_{ij}),$$

where $M' = M'_{1,1} \| M'_{2,1} \| \cdots \| M'_{2N,N}$, and $M'_{i,j} = m'_{i,j} \| r'_{i,j} \| d'_{ij}$ ($|m'_{i,j}| = 1, |r'_{ij}| = N$, and $|d'_{ij}| = N$). For each (i, j), if the check fails, A_k discards $M'_{i,j}$. Otherwise A_k accepts one bit message m'_{ij} in $M'_{i,j}$. When $M'_{i,j} = M'_{i',j'}$, A_k considers them to be an identical message.

(Transmitting capability for AC)

We now show that the above protocol $P3$ realizes AC with probability greater than $1 - \nu(n)$, where n is the total number of steps taken by all A_i ($i = 1, \ldots, N$) during the execution of $P3$.

First, we show that A_l's message m_l can be transmitted to A_k with probability greater than $1 - \nu(n)$. Each message is assigned to one slot in each column which consists of $2N$ slots (from $M_{1,j}$ to $M_{2N,j}$). There are N columns ($j = 1, \ldots, N$). In a column, each party selects one slot randomly, since, from the assumption, any party sends at most one message to A_k in an interval. Hence, the probability that A_l selects a slot that collides with somebody else selection is less than $N \times 1/(2N) = 1/2$. Therefore, the probability that there exists at least one column in which A_l's selection collides with none is greater than $(1 - 1/2^N)$. If there is no collision at a slot, A_k can obtain a_l correctly at the slot of M', since

$$\bigoplus_{i \in [1, \ldots, N]} (K_{i,1} \oplus \cdots \oplus K_{i,i-1} \oplus K_{i,i+1} \oplus \cdots \oplus K_{i,N})$$

$$= \bigoplus_{i \in [2, \ldots, N]} \bigoplus_{j \in [1, \ldots, i-1]} (K_{ij} \oplus K_{ji}) = 0^{2N^2(2N+1)}.$$

A_k can check whether a collision has occurred at a slot or not, by checking

$$d'_{ij} = f(m'_{ij} \| r'_{ij}).$$

If the check is passed, the collision must not have occurred from the non-homomorphic property of f. A_k can distinguish a party's message from another's with probability greater than $1 - N/2^N$, since a_l includes a N bit random number and two N bit random numbers can be equivalent with probability $1/2^N$. Therefore, A_l's message can be transmitted to A_k correctly with probability greater than $(1 - 1/2^N)(1 - N/2^N)$. On the other hand, n, the total number of steps during $P3$, is $c_1 N^4 \geq n \geq c_2 N^4$, for constants (in N), c_1, c_2. Therefore, $c_1' n^{1/4} \geq N \geq c_2' n^{1/4}$, for constants (in n), c_1', c_2'. Hence, $(1 - 1/2^N)(1 - N/2^N) \geq 1 - (1 + N)/2^N > 1 - (1 + c_1' n^{1/4})/2^{c_2' n^{1/4}} > 1 - \nu(n)$.

Next, we show that a third party can eavesdrop a message along with the receiver's identity. Since the third party can obtain the transmitted message C_l through the transmitting function of SC, he can obtain exactly the same information as the receiver. So, he gets each message from M' as the receiver does.

(Security property for AC) First, we assume that the adversary is passive. Since K_{ij} ($i \in \{2, \ldots, N\}, j \in \{1, \ldots, i - 1\}$) are randomly and independently selected, for any $l \in \{1, \ldots, N\}$, the distribution of C_l is exactly the same. Therefore, the receiver obtains no information about the originator. Moreover, even if the receiver collaborates with other parties except the originator, for any $l \in \mathcal{L} \subseteq \{1, \ldots, N\}$, the distribution of C_l is exactly the same, since K_{ij} ($i \in \{2, \ldots, N\}, j \in \{1, \ldots, i - 1\}$) are randomly and independently selected, where $\mathcal{L} = \{l \mid A_l$ is not a member of the collaboration$\}$, and the number of parties in $\mathcal{L}$ is L. Therefore, the probability that the collaboration can guess the originator correctly is exactly $1/L$. In addition, since the information that the third party gets is exactly same as the information received by the receiver, the originator is anonymous to the third party as to the receiver, even if he collaborates with any other parties except the originator.

Next, we assume that the adversary is active. The adversary can get no additional information after sending the forged message, and the originator is fixed before sending the forged message. Hence, the probability that the adversary can guess the originator is no better than that when the adversary is passive. $\square$

4.4 Equivalence of Three Assumptions

From the three theorems above and lemma 8, we can immediately obtain the following main theorem.

Theorem 12. *The following three assumptions are equivalent.*

1. *The existence of an anonymous channel,*
2. *The existence of a direction-indeterminable channel,*
3. *The existence of a secure channel.*

5 Conclusion

In this paper, we showed the equivalence of three cryptographic physical assumptions on channels: the existence of anonymous channels, that of direction-indeterminable channels, and that of secure channels.

This initiates a new direction of research about physical assumptions, such as the study of the equivalence class of physical assumptions, and the separation of some equivalence classes of physical assumptions.

Acknowledgments

The author would like to thank Tony Eng, Atsushi Fujioka, Kouichi Sakurai, and Hiroki Shizuya for their many valuable comments on the earlier draft.

378

References

[AS83]   B. Alpern and F.B. Schneider, "Key Exchange Using 'Keyless Cryptography' ", Information Processing Letters, 16, 2, pp.79–81 (1983).

[BD89]   J. Bos, and B. den Boer, "Detection of Disrupters in the DC Protocol", Proc. of EUROCRYPT '89, LNCS 434, Springer–Verlag, pp.320–327 (1990).

[BGGHKMR90] M. Ben-Or, O. Goldreich, S. Goldwasser, J. Håstad, J. Kilian, S. Micali, and P. Rogaway, "Everything Provable is Provable in Zero-knowledge", Proc. of CRYPTO '88, LNCS 403, Springer–Verlag, pp.37–56 (1990).

[BGKW88] M. Ben-Or, S. Goldwasser, J. Kilian, and A. Wigderson, "Multi-Prover Interactive Proofs: How to Remove Intractability Assumptions", Proc. of STOC (1988).

[BGW88]   M. Ben-Or, S. Goldwasser, and A. Wigderson, "Completeness Theorems for Non-Cryptographic Fault-Tolerant Distributed Computation", Proc. of STOC, pp.1–10 (1988).

[BR89]   M. Ben-Or, and T. Rabin, "Verifiable Secret Sharing and Multiparty Protocols with Honest Majority", Proc. of STOC, pp.73–85 (1989)

[CCD88]   D. Chaum, C. Crépeau, and I. Damgård, "Multiparty Unconditionally Secure Protocols", Proc. of STOC, pp.11–19 (1988).

[Cha81]   D. Chaum, "Untraceable Electronic Mail, Return Addresses, and Digital Pseudonyms", Communications of the ACM, Vol.24, No.2, pp.84–88 (1981).

[Cha88]   D. Chaum, "The Dining Cryptographers Problem: Unconditional Sender and Recipient Untraceability", Journal of Cryptology, Vol.1, No.1, pp.65–75 (1988).

[CFN90]   D. Chaum, A. Fiat, and M. Naor, "Untraceable Electronic Cash", Proc. of CRYPTO '88, LNCS 403, Springer–Verlag, pp.319–327 (1990).

[CK88]   C. Crépeau, and J. Kilian, "Achieving Oblivious transfer Using Weakened Security Assumptions", Proc. of FOCS (1988).

[GMRa89] S.Goldwasser, S.Micali and C.Rackoff, "The Knowledge Complexity of Interactive Proof Systems," SIAM J. Comput., 18, 1, pp.186–208 (1989).

[GMRi88] S.Goldwasser, S.Micali and R.Rivest, "A Digital Signature Scheme Secure Against Adaptive Chosen-Message Attacks," SIAM J. Comput., 17, 2, pp.281–308 (1988).

[Has90]   J.Håstad, "Pseudo-Random Generators under Uniform Assumptions," Proc. of STOC (1990)

[ILL89]   R. Impagliazzo, L. Levin and M. Luby, " Pseudo-Random Number Generation from One-Way Functions," Proc. of STOC, pp.12–24 (1989)

[IR89]   R. Impagliazzo, and S. Rudich, " Limits on the Provable Consequence of One-Way Permutations," Proc. of STOC (1989)

[Nao90]   M.Naor, "Bit Commitment Using Pseudo-Randomness," Proc. of Crypto'89, LNCS 435, Springer–Verlag, pp.128–136 (1990).

[NY89]   M.Naor, and M. Yung, "Universal One-Way Hash Functions and Their Cryptographic Applications," Proc. of STOC, pp.33–43 (1989).

[Rom90]   J. Rompel, "One-Way Functions are Necessary and Sufficient for Secure Signature," Proc. of STOC, pp.387–394 (1990).

[Wai89]   M. Waidner, "Unconditional Sender and Recipient Untraceability in spite of Active Attacks", Proc. of EUROCRYPT '89, LNCS 434, Springer–Verlag, pp.302–319 (1990).

Separating Complexity Classes Related to Bounded Alternating ω-Branching Programs

Christoph Meinel[1] and Stephan Waack[2]

[1] FB IV - Informatik, Universität Trier
D-54286 Trier
meinel@uni-trier.de
[2] Inst. für Num. u. Angew. Mathematik, Georg-August-Universität
D-37083 Göttingen
waack@namu01.gwdg.de

Abstract. By means of algebraical rank arguments we derive some exponential lower bounds on the size of bounded alternating ω-branching programs. Together with some polynomial upper bounds these lower bounds make possible the complete separation of the restricted complexity classes $\mathcal{NL}_{ba}$, co-$\mathcal{NL}_{ba}$, $\oplus\mathcal{L}_{ba}$, MOD_p-$\mathcal{L}_{ba}$, p prime.

Introduction

In order to characterize and investigate logarithmic space–bounded complexity classes by means of combinatorial computational devices branching programs [Lee59], Ω-branching programs, $\Omega \subseteq \mathbb{B}_2$ [Mei88, Mei89] and MOD_p-branching programs [DKMW92] were introduced and investigated. Up to now, expect for of some combinatorial cut-and-paste techniques [e. g. Žak84, KMW88], merely considerations of the amount of communication within a given branching program have provided the possibility to derive lower bounds on its size. In this way, exponential lower bounds could be obtained merely for oblivious (nondeterministic, co-nondeterministic, parity) branching programs of linear length [AM86, KMW89, Kra90]. Unfortunately, such oblivious linear length-bounded branching programs seem to be an artificial and rather restricted type of branching programs.

In the following we develop a more adequate theory of communication within branching programs which provides exponential lower bounds on the size of *bounded alternating* (nondeterministic, co-nondeterministic, parity, or MOD_p-) branching programs, a natural and quite general type of branching programs. Such a bounded alternating branching program P is characterized by the property that there are two disjoint subsets $Y, Z \subseteq X$ in the set X of variables tested in P, $\#Y = \#Z = \Theta(\#X)$, such that, on each path of P, the number of alternations between testing variables from Y and testing variables from Z is bounded by $(\#X)^{o(1)}$. Of course, due to a Ramsey theoretic lemma of Alon and Maass [AM86], oblivious branching programs of linear length are bounded alternating. However, we will prove that bounded alternating branching programs are strictly more powerful than oblivious linear length-bounded ones.

Interestingly, the investigation of bounded alternating branching programs also generalizes the settings of classical communication complexity theory in at least two important aspects. On the one hand, it ignores a considerable portion of the input variables (the set $X - (Y \cup Z)$ of size $O(\#X)$) on which no sort of assumption is posed. On the other hand, it does not preassume the sets Y and Z. Hence, the complexity of bounded alternating branching programs reflects in a certain sense the amount of communication with respect to any disjoint subsets (of a partition) of the input.

This paper is structured as follows:

In Section 1 we introduce the algebraically based concept of ω-branching programs, $\omega \colon I\!N \longrightarrow \mathcal{R}$ semiring homomorphism onto a finite semiring $\mathcal{R}$, which provides a general framework for analysing communication complexity aspects in the theory of branching programs. Then, in Section 2, we start the investigation of bounded alternating ω-branching programs.

In Section 3 we present a lower bound technique for bounded alternating ω-branching programs that is mainly based on considerations of certain algebraic invariants of communication matrices. It generalizes ideas of [KMW89, Kra90, DKMW92]. Then, in Section 4, we derive some new exponential lower bounds on the $\mathcal{R}$-ranks of certain communication matrices that seem to be interesting not only in the context of branching programs.

By means of the results of Sections 3 and 4, in Section 5 we derive some exponential lower bounds on the size of bounded alternating ω-branching programs. Finally, in Section 6, we summarize our results in form of separation results concerning complexity classes related to bounded alternating ω-branching programs. In particular, we are able to separate all the corresponding classical complexity classes $\mathcal{NL}_{ba}$, co-$\mathcal{NL}_{ba}$, $\oplus \mathcal{L}_{ba}$, MOD_p-$\mathcal{L}_{ba}$, p prime, from each other.

1 ω-Branching Programs

Let $\mathcal{R} = [\mathcal{R}; +_{\mathcal{R}}, 0_{\mathcal{R}}, \cdot_{\mathcal{R}}, 1_{\mathcal{R}}]$ be a finite semiring with a null element $0_{\mathcal{R}}$ and a one element $1_{\mathcal{R}}$, and let ω, $\omega \colon I\!N \longrightarrow \mathcal{R}$, be a semiring homomorphism from $I\!N = [I\!N; +, 0, \cdot, 1]$ onto $\mathcal{R}$. (Obviously, ω is uniquely determined.)

An ω-*branching program* P is a directed acyclic graph where each node has outdegree $\leq \#\mathcal{R}$. There is a distinguished node v_s, the *source*, which has indegree 0. Exactly two nodes, v_0 and v_1, are of outdegree 0. They are called *0-sink* and *1-sink* of P, respectively. Nodes of outdegree $\#\mathcal{R}$ are labeled with variables x over $\mathcal{R}$, $x \in X = \{x_1, \ldots, x_n\}$, or remain unlabeled as the remaining nodes of P. Each edge starting in a labelled node is labeled with an element r, $r \in \mathcal{R}$, such that no two edges starting in one node have the same label.

Each input $a = a_1 \ldots a_n \in \mathcal{R}^n$ defines some *computational paths* in P: Starting at the source v_s, a node v of P is connected with one of its successor nodes. If v is labeled with the variable x_i, then v is connected with the successor node of v which is reached from v via the edge labeled with a_i. A computational path is said to be *accepting* if it ends up in the 1-sink v_1. The number of accepting computational paths is denoted by $acc_P(a)$, $acc_P(a) \in I\!N$. Now, P is said to *accept* $a \in \mathcal{R}^n$ iff $\omega(acc_p(a)) = 1_{\mathcal{R}}$.

Obviously, ω-branching programs generalize ordinary branching programs where each non-sink node is assumed to be labeled with a Boolean variable. Hence, for each input, there is exactly one computational path which is accepting if it ends up in the 1-sink.

Moreover, ω-branching programs generalize Ω-branching programs introduced in [Mei88], $\Omega \subseteq \mathbb{B}_2$, and MOD_p-branching programs introduced in [DKMW92], $p \in \mathbb{N}$. In order to see this consider the following semirings and semiring homomorphisms

$$\omega_\vee : \mathbb{N} \longrightarrow B = [\{0,1\}; \vee, 0, \wedge, 1], \quad n \longmapsto \begin{cases} 1 & \text{iff } n > 0 \\ 0 & \text{otherwise,} \end{cases}$$

$$\omega_\oplus : \mathbb{N} \longrightarrow \mathbb{F}_2 = [\{0,1\}; \oplus, 0, \wedge, 1], \quad n \longmapsto \begin{cases} 1 & \text{iff } n \text{ is odd} \\ 0 & \text{otherwise,} \end{cases}$$

and

$$\omega_p : \mathbb{N} \longrightarrow \mathbb{F}_p = [\{0, \ldots, p-1\}; +_{(mod\ p)}, 0, \cdot_{(mod\ p)}, 1], \quad n \longmapsto n \ (mod\ p),$$

for each prime $p \in \mathbb{N}$. Obviously,
- disjunctive $\{\vee\}$-branching programs are $\omega_\vee$-branching programs (they accept if at least one computation path is accepting),
- parity $\{\oplus\}$-branching programs are $\omega_\oplus$-branching programs (they accept if the number of accepting computation paths is odd), and
- MOD_p-branching programs, p prime, are ω_p-branching programs (they accept if the number of accepting computation paths equals 1 modulo p).

Due to this correspondence we immediately obtain the following characterization of the well-known nonuniform logarithmic space-bounded complexity classes $\mathcal{L} = L/poly$, $\mathcal{NL} = NL/poly$, $\oplus\mathcal{L} = \oplus L/poly$, $MOD_p\mathcal{L} = (MOD_p L)/poly$ by means of (sequences of) polynomial size ω-branching programs. Remember that the *size* of an ω-branching program P, $size(P)$, is the number of non-sink nodes of P. By $\mathcal{P}_{\omega-BP}$ we denote the complexity class of all languages A, $A \subseteq \mathcal{R}^*$, which are accepted by sequences of polynomial size ω-branching programs.

Proposition 1.
(1) $\mathcal{P}_{BP} = \mathcal{L}$ [PŽ83],
(2) $\mathcal{P}_{\omega_\vee-BP} = \mathcal{NL}$ [Mei88],
(3) $\mathcal{P}_{\omega_\oplus-BP} = \mathcal{P}_{\omega_2-BP} = \oplus\mathcal{L}$ [Mei88], and
(4) $\mathcal{P}_{\omega_p-BP} = MOD_p\text{-}\mathcal{L}$, p prime, [DKMW92]. $\square$

2 Bounded Alternating ω-Branching Programs

Let X be a set of variables over a finite semiring $\mathcal{R}$, and let Y, $Z \subseteq X$, $\#Y = \#Z = \varepsilon n, \varepsilon > 0$, be two disjoint subsets of X. An ω-branching program P, $\omega : \mathbb{N} \longrightarrow \mathcal{R}$ semiring homomorphism onto $\mathcal{R}$, that tests variables of X is said to be of *alternation length* α *with respect to Y and Z* if each path of P can be divided into α segments in which, alternating, variables of Y or variables of Z are *not* tested , and if α is minimal with this property. A sequence of ω-branching programs P_n is called *bounded alternating* if there exist two disjoint subsets $Y, Z \subseteq X$, $\#Y = \#Z = \varepsilon n$, $\varepsilon > 0$, such that P_n is of alternation length $n^{o(1)}$ with respect to Y and Z.

By $\mathcal{P}_{ba\ \omega-BP}$ we denote the set of languages $A, A \subseteq \mathcal{R}^*$ which are accepted by sequences of polynomial size, bounded alternating ω-branching programs. Observe that, since $\{0,1\} \subseteq \mathcal{R}$ for all semirings $\mathcal{R}$, $\{0,1\}$-languages can be considered in any case. For the particular semiring homomorphisms $\omega = \omega_\lor$, $\omega_\oplus$, $\omega_p(p$ prime$)$, we write, due to Proposition 1, $\mathcal{NL}_{ba}$, $\oplus\mathcal{L}_{ba}$, and $MOD_p\mathcal{L}_{ba}$ instead of $\mathcal{P}_{ba\ \omega_\lor-BP}$, $\mathcal{P}_{ba\ \omega_\oplus-BP}$, and $\mathcal{P}_{ba\ \omega_p-BP}$, respectively.

Polynomial size, bounded alternating ω-branching programs seem to be quite powerful computation devices. For instance, the vast majority of arithmetic functions that can be computed by polynomial size branching programs can be computed by bounded alternating ones (in most cases with merely constant alternations). Even problems that are complete in $\mathcal{P}_{\omega-BP}$ belong to $\mathcal{P}_{ba\ \omega-BP}$.

Proposition 2 [MW91].

Let ω-GAP_{mon} denote the GRAPH ACCESSIBILITY PROBLEM that decides for a given acyclic monotone graph $G = (\{1, \ldots, n\}, E)$ whether $\omega(\#[1 \xrightarrow{G} n]) = 1_\mathcal{R}$, where $\#[i \xrightarrow{G} j]$ denotes the number of paths from i to j in G. Then

1. ω-GAP_{mon} is p-projection complete in $\mathcal{P}_{\omega-BP}$, and

2. ω-$GAP_{\mathrm{mon}} \in \mathcal{P}_{ba\ \omega-BP}$. $\square$

Particularly, the monotone GRAPH ACCESSIBILITY PROBLEMs $GAP_{\mathrm{mon}}1$, $GAP_{\mathrm{mon}}2$, ODD-$PATH$-$GAP_{\mathrm{mon}}2$, and MOD_p-$GAP_{\mathrm{mon}}2$ (p prime) which, due to results of [Mei86, Mei89, DKMW92], are known to be p-projection complete in $\mathcal{L}$, $\mathcal{NL}$, $\oplus\mathcal{L}$, and $MOD_p\mathcal{L}$, respectively, belong to the corresponding bounded alternating classes $\mathcal{L}_{ba}$, $\mathcal{NL}_{ba}$, $\oplus\mathcal{L}_{ba}$, and MOD_p-$\mathcal{L}_{ba}$.

How do bounded alternating ω-branching programs fit into the environment of restricted types of branching programs investigated in the past? Beside of read-k times branching programs [BRS93] the most general type of branching programs for which up to now it was possible to establish exponential lower bounds without restricting its length by the number of input variables are oblivious programs of linear length [KMW89]. Remember that, an ω-branching program is said to be *oblivious* if it is leveled (i.e., all paths from the source of the program to any one of its nodes are of the same length) and the property that one node of a level is labeled with a variable x implies that all nodes of that level are labelled by x. A level whose nodes are labeled in this way is called an *input level*. The *length* of an oblivious branching program is the number of its input levels. The classes of all languages which are accepted by sequences of oblivious, polynomial size and linear length ω-branching programs are denoted by $\mathcal{P}_{\mathrm{lin}\ \omega-BP_o}$.

Proposition 3.

Let ω be any of the semiring homomorphisms $\omega_\lor$, $\omega_\oplus$, or ω_p (p prime). Polynomial size bounded alternating ω-branching programs are strictly more powerful than polynomial size oblivious ω-branching programs of linear length. I. e.,

$$\mathcal{P}_{\mathrm{lin}\ \omega-BP_o} \overset{\subsetneq}{\neq} \mathcal{P}_{ba\ \omega-BP} \subseteq \mathcal{P}_{\omega-BP}.$$

Proof. (Omitted) □

Finally let us consider the question whether the classes $\mathcal{P}_{ba\ \omega-BP}$ are closed under complement. Let

$$co\text{-}\mathcal{P}_{ba\ \omega-BP} = \{A \subseteq \mathcal{R}^* \mid (\mathcal{R}^* - A) \in \mathcal{P}_{ba\ \omega-BP}\}$$

be the co-class of the complexity class $\mathcal{P}_{ba\ \omega-PB}$. While we will see later that the class $\mathcal{NL}_{ba} = \mathcal{P}_{ba\ \omega-BP}$ is not closed with respect to complement the classes $MOD_p\mathcal{L}_{ba} = \mathcal{P}_{ba\ \omega_p-BP}$, p prime, are.

Proposition 4.
Let p be prime. Then

$$\mathcal{P}_{ba\ \omega_p-BP} = co\text{-}\mathcal{P}_{ba\ \omega_p-BP}. \quad \square$$

3 The Lower Bound Technique

In the following we generalize ideas of [KMW89, Kra90] and develop a lower bound technique that allows us to prove exponential lower bounds on the size of bounded alternating ω-branching programs. Our lower bound proofs proceed in two steps. In the first step we derive exponential lower bounds on the size of some specially structured bounded alternating ω-branching programs. This is done by estimating certain ranks of communication matrices of a problem. Then, in a second step, we prove that each polynomial size ω-branching program can be transformed into such a special ω-branching program of polynomial size preserving the 'bounded alternating'-property.

Let $\mathcal{R}$ be a finite semiring, and let $\omega : \mathbb{N} \longrightarrow \mathcal{R}$ be the semiring homomorphism onto $\mathcal{R}$. Let P be an ω-branching program with input variables from a set X, and let Y, Z be a partition of X, $Z = X - Y$. Let us consider the set V_X of nodes of P which are labeled with variables from $X = Y \cup Z$. A node v, $v' \in V_X$ is called a *predecessor* in V_X of a node v of P if there is a path in P from v' to v that does not contain any node of V_X. A node v of P is called a *Y-Z-alternation node* if it is labeled with a variable from Z and if all predecessors in V_X of v are labeled with variables from Y. An *alternation node* is a node that is either Y-Z-alternating or Z-Y-alternating. On the other hand, v is said to be *homogeneous* if all its predecessors in V_X are labeled either with variables from Y or Z and if v is not an alternation node. Introducing certain dummy nodes, at the cost of multiplying the size by the constant $\#\mathcal{R}$, P can be converted – without changing the alternation length – to an ω-branching program which consists merely of homogeneous or alternation nodes and accepts the same set. Hence, we assume all our ω-branching programs to consist merely of homogeneous and alternation nodes.

The *alternation depth* of a node v is the maximal number of alternation nodes on a path from the source to v. The alternation depth of the 1-sink v_1 is the *alternation length* of P.

An ω-branching program P is said to be (Y, Z)-*synchronous* if the alternation depth of two consecutive nodes of P differs at most by 1. Let P be a (Y, Z)-synchronous ω-branching program. All nodes of P with the same alternation depth form an (alternation) *segment*. The subsets of all alternation nodes with the same alternation depth are called *alternation levels*. Obviously, each alternation level consists either of Y-Z or of Z-Y-alternation nodes.

Our lower bound technique is mainly based on considerations of invariants of the *communication matrix* $M_{Y,Z}(f)$ of a function f. If $f : \mathcal{R}^Y \times \mathcal{R}^Z \longrightarrow \{0,1\}$ then $M_{Y,Z}(f)$ is a $(\#\mathcal{R})^{\#Y} \times (\#\mathcal{R})^{\#Z}$ matrix which is defined, for $a_Y \in \mathcal{R}^Y$, $a_Z \in \mathcal{R}^Z$, by

$$(M_{Y,Z}(f))_{a_Y, a_Z} = f(a_Y, a_Z).$$

The matrix-invariants we work with are certain matrix-ranks. Remember that, if $M \in I\!M_{\mathcal{R}}(n, m)$ is a $n \times m$ matrix over $\mathcal{R}$, then the *rank* of M over $\mathcal{R}$, $\mathrm{rank}_{\mathcal{R}}(M)$, is the minimum k such that M can be written as $M = A \cdot B$ with $A \in I\!M_{\mathcal{R}}(n, k)$ and $B \in I\!M_{\mathcal{R}}(k, m)$.

For simplicity we denote the addition in $\mathcal{R}$ by $+$, the multiplication by $\cdot$, and the standard inner product by $\langle ., . \rangle_{\mathcal{R}}$ or, even shorter, by $\langle ., . \rangle$.

Lemma 5.

Let Y, Z *be a partition of* X, *and let* P *be a* (Y, Z)-*synchronous* ω-*branching program of alternation length* α *that computes a function* $f : \mathcal{R}^X \longrightarrow \{0,1\}$. *Then*

$$\mathrm{Size}(P) \geq \left(\mathrm{rank}_{\mathcal{R}} M_{Y,Z}(f) \right)^{1/\alpha}.$$

Proof. (Omitted) $\square$

Due to Lemma 5, lower bounds on the $\mathcal{R}$-rank of the communication matrix $M_{Y,Z}(f)$ provide lower bounds on the size of (Y, Z)-synchronous ω-branching programs. The following Lemma proves that such lower bounds also imply lower bounds on the size of arbitrary ω-branching programs.

Lemma 6.

Let (Y, Z) *be a partition of* X. *Each* ω-*branching program* P *of size* σ *can be simulated by a* (Y, Z)-*synchronous* ω-*branching program* P' *of size* σ^2.

Proof. (Omitted) $\square$

Of particular importance for the following is the fact that is suffices to apply the lower bound technique developed in Lemmas 5 and 6 merely to certain subfunctions of a function f in order to derive exponential lower bounds on the size of ω-branching programs for f. If $Y, Z \subseteq X$ are disjoint subsets of X, and if $c \in \mathcal{R}^{X-(Y \cup Z)}$, then we denote by $f^c : \mathcal{R}^{Y \cup Z} \longrightarrow \{0,1\}$ the subfunction $f^c(y, z) := f(c, y, z)$ for all $y \in \mathcal{R}^Y$, $z \in \mathcal{R}^Z$.

Theorem 7.

Let $f : \mathcal{R}^X \longrightarrow \{0,1\}$ *be a function,* $Y, Z \subseteq X$ *two disjoint subsets of* X, *and* P *an*

ω-*branching program of alternation length α with respect to Y and Z that computes f. Then*

$$Size(P) \geq (rank_{\mathcal{R}} M_{Y,Z}(f^c))^{1/2\alpha}$$

for each $c \in \mathcal{R}^{X-(Y \cup Z)}$.

Proof. (Omitted) $\square$

4   Lower Bounds on the Rank of Communication Matrices

Due to Theorem 7, in order to derive exponential lower bounds on the size of bounded alternating branching programs if suffices to derive exponential lower bounds on the rank of certain communication matrices.

Let us start with an easy observation concerning the SEQUENCE EQUALITY TEST SEQ= {SEQ$_{2n}$},

$$\mathrm{SEQ}_{2n} : \mathcal{R}^Y \times \mathcal{R}^Z \longrightarrow \{0, 1\},$$

$\#Y = \#Z = n$, defined by

$$\mathrm{SEQ}_{2n}(y, z) = 1 \text{ iff } y = z.$$

Proposition 8.
Let $\mathcal{R}$ be a finite semiring. Then it holds

$$rank_{\mathcal{R}} M_{Y,Z}(\mathrm{SEQ}_{2n}) = \exp(\Omega(n)). \quad \square$$

Now let $\omega : I\!N \longrightarrow \mathcal{R}$ be a semiring homomorphism and consider the ω-ORTHOGO-NALITY TEST ω-ORT = {ω-ORT$_{2n}$},

$$\omega\text{-ORT}_{2n} : \mathcal{R}^Y \times \mathcal{R}^Z \longrightarrow \{0, 1\},$$

$\#Y = \#Z = n$, defined by

$$\omega\text{-ORT}_{2n}(y, z) = 1 \text{ iff } \langle y, z \rangle_{\mathcal{R}} = O_{\mathcal{R}}. \quad \square$$

Proposition 9.
Let p, q primes with $p \neq q$. Then it holds

$$rank_{I\!\!F_q} M_{Y,Z}(\omega_p\text{-ORT}_{2n}) = \exp(\Omega(n)).$$

Proof. (Omitted) $\square$

Proposition 10.
Let p be prime. Then it holds

$$rank_{I\!B} M_{Y,Z}(\omega_p\text{-ORT}_{2n}) = \exp(\Omega(n)).$$

Proof. (Omitted) $\square$

5 Lower Bounds on the Size of Bounded Alternating ω-Branching Programs

The lower bounds on the $\mathcal{R}$-rank of the communication matrices derived in the last section imply, due to Theorem 7, exponential lower bounds on the size of bounded alternating ω-branching programs.

Let $F = \{f_n\}$ be a sequence of functions,

$$f_n : \mathcal{R}^n \longrightarrow \{0, 1\},$$

and let æ be a symbol not contained in $\mathcal{R}$. Obviously, F is a subproblem of the problem $F^{\text{æ}} = \{f_n^{\text{æ}}\}$,

$$f_n^{\text{æ}} : (\mathcal{R} \cup \{\text{æ}\})^n \longrightarrow \{0, 1\}$$

defined by

$$f_n^{\text{æ}}(a) = 1 \quad \text{iff } red(a) \text{ is of even length } 2m \text{ and } f_{2m}(red(a)) = 1,$$

where $red(a) \in \mathcal{R}^*$ denotes the *reduced word* of a obtained from a by deleting all occurences of æ in a.

Now we consider the æ-extension $SEQ^{\text{æ}} = \{SEQ_{2n}^{\text{æ}}\}$ of the SEQUENCE EQUALITY PROBLEM and the æ-extension $\omega_p\text{-}ORT^{\text{æ}} = \{\omega_p\text{-}ORT_{2n}^{\text{æ}}\}$, p prime, of the ω_p-ORTHOGLONALITY TEST.

Proposition 11.
Let $\mathcal{R}$ be a finite semiring, and let $\omega : I\!N \longrightarrow \mathcal{R}$ be a semiring homomorphism. If P is a bounded alternating ω-branching program that computes $SEQ_{2n}^{\text{æ}}$ then

$$Size(P) = \exp(\Omega(n)). \quad \square$$

Similarly, due to Theorem 7 and Propositions 9 and 10 one obtains the following results.

Proposition 12.
Let p, q primes with $p \neq q$. If P is a bounded alternating ω_q-branching program that computes $\omega_p\text{-}ORT_{2n}^{\text{æ}}$ then

$$Size(P) = \exp(\Omega(n)). \quad \square$$

Proposition 13.
Let p be prime. If P is a disjunctive bounded alternating $\omega_\vee$-branching program that computes $\omega_p\text{-}ORT_{2n}^{\text{æ}}$ then

$$Size(P) = \exp(\Omega(n)). \quad \square$$

387

6 Separation Results

Now, combining the exponential lower bounds obtained in Section 5 with some polynomial upper bounds we are able to establish strong differences in the computational power of bounded alternating ω-branching programs for different ω. These differences imply, in particular, separation results on the corresponding logarithmic space-bounded complexity classes $\mathcal{L}_{ba}$, $\mathcal{NL}_{ba}$, co-$\mathcal{NL}_{ba}$, $\oplus\mathcal{L}_{ba}$, and $MOD_p\text{-}\mathcal{L}_{ba}$, p prime.

We start with a polynomial upper bound on the size of (sequences of) disjunctive $\omega_\vee$-branching programs that compute $\neg\text{SEQ}^{\mathbf{x}}$.

Proposition 14.

$\neg\text{SEQ}^{\mathbf{x}}_{2n}$ can be computed by means of a disjunctive $\omega_\vee$-branching program of size $O(n^4)$ and alternation length 3.

Proof. (Omitted) □

Next we give a polynomial upper bound on the size of (sequences of) ω_p-branching programs that compute $\omega_p\text{-}\text{ORT}^{\mathbf{x}}$.

Proposition 15.

Let p be prime. Then $\omega_p\text{-}\text{ORT}^{\mathbf{x}}_{2n}$ can be computed by means of a ω_p-branching program of size $O(n^4)$ and alternation length 3.

Proof. (Omitted) □

Altogether, the lower bounds of Propositions 11 – 13 and the upper bounds of Propositions 14 – 15 imply the following corollary.

Corollary 16.

(1) $SEQ^{\mathbf{x}} \notin \mathcal{P}_{ba}\,\omega\text{-}BP$ for each ω. Hence, $SEQ^{\mathbf{x}} \notin \mathcal{P}_{ba}\,BP$.

(2) $\neg SEQ^{\mathbf{x}} \in \mathcal{P}_{ba}\,\omega_\vee\text{-}BP$.

(3) $\omega_p\text{-}ORT^{\mathbf{x}} \notin \mathcal{P}_{ba}\,\omega_\vee\text{-}BP$, p prime. Hence, $\omega_p\text{-}ORT^{\mathbf{x}} \notin \mathcal{P}_{ba}\,BP$.

(4) $\omega_p\text{-}ORT^{\mathbf{x}} \notin \mathcal{P}_{ba}\,\omega_q\text{-}BP$, p, q primes, $p \neq q$.

(5) $\omega_p\text{-}ORT^{\mathbf{x}} \in \mathcal{P}_{ba}\,\omega_p\text{-}BP$, p, prime. □

Since the classes $MOD_p\text{-}\mathcal{L}_{ba}$, due to Proposition 4, are closed under complement, and since, due to Propositions 11 and 14, the class $\mathcal{NL}_{ba}$ is not, the bounds of Corollary 16 provide the following separation results concerning the complexity classes $\mathcal{L}_{ba}$, $\mathcal{NL}_{ba}$, co-$\mathcal{NL}_{ba}$, $\oplus\mathcal{L}_{ba}$, and $MOD_p\text{-}\mathcal{L}_{ba}$, $p > 2$ prime, defined by bounded alternating $\omega_\vee$-branching programs and ω_p-branching programs.

Theorem 16.

(1) Any two complexity classes $\mathcal{K}$ and $\mathcal{K}'$

$$\mathcal{K}, \mathcal{K}' \in \{\mathcal{NL}_{ba}, \text{co-}\mathcal{NL}_{ba}, \oplus\mathcal{L}_{ba}, MOD_p\mathcal{L}_{ba}(p > 2\ prime)\}$$

are uncomparable via inclusion, i. e.

$$\mathcal{K} \not\subseteq \mathcal{K}' \text{ and } \mathcal{K}' \not\subseteq \mathcal{K}.$$

(2) Each of these classes properly contain the complexity class $\mathcal{L}_{ba}$.

(3) Finally, the classes $\mathcal{NL}_{ba}$ and co-$\mathcal{NL}_{ba}$ are properly contained in $\mathcal{NL}$,

$$\mathcal{NL}_{ba}, co\text{-}\mathcal{NL}_{ba} \overset{\subsetneq}{\neq} \mathcal{NL},$$

i. e. unbounded alternating disjunctive branching programs are strictly more powerful than bounded alternating ones. $\square$

References

[AM86] N. Alon, W. Maass: Meanders, Ramsey theory and lower bounds, Proc. 27th ACM STOC, 1986, 30–39.

[BRS93] A. Borodin, A. Razborov, R. Smolensky: On lower bounds for read-k times branching programs, Comput. Complexity 3 (1993), 1-18.

[DKMW92] D. Damm, M. Krause, Ch. Meinel, S. Waack: Separating Restricted MOD_p–Branching Program Classes, Proc. STACS'92, LNCS 577,281–292, 1992.

[KMW89] M. Krause, Ch. Meinel, S. Waack: Separating Complexity Classes Related to Certain Input Oblivious Logarithmic Space Bounded Turing Machines, Proc. 4th IEEE Structure in Complexity Theory, 1989, 240–259.

[Kra90] M. Krause: Separating $\oplus\mathcal{L}$ from $\mathcal{L}$, co-$\mathcal{NL}$ and $AL = P$ for Oblivious Turing Machines of Linear Access Time, Proc. MFCS'90, LNCS 452, 385–391.

[Lee59] C. Y. Lee: Representation of Switching Functions by Binary Decision Programs, Bell System Techn. Journal 38 (1959), 985–999.

[Mei88] Ch. Meinel: Polynomial Size Ω-Branching Programs and Their Computational Power, Proc. STACS'88, LNCS 294, 81–90.

[Mei89] Ch. Meinel: Modified Branching Programs and Their Computational Power, LNCS 370, Springer Verlag, 1989.

[MW91] Ch. Meinel, S. Waack: Upper and Lower Bounds for Certain Graph-Accessibility-Problems on Bounded Alternating Branching Programs, MFCS'91, LNCS 520, 337–345.

[PŽ83] P. Pudlák, S. Žak: Space Complexity of Computations, Preprint Univ. of Prague, 1983.

[Žak84] S. Žak: An Exponential Lower Bound for One-Time-Only Branching Programs, Proc. MFCS'84, LNCS 176, 562–566.

The Complexity of the Optimal Variable Ordering Problems of Shared Binary Decision Diagrams

Seiichiro TANI[1], Kiyoharu HAMAGUCHI[2] and Shuzo YAJIMA[2]

[1] Department of Informaiton Science, Faculty of Science, The University of Tokyo,
Tokyo 113, Japan
[2] Department of Information Science, Faculty of Engineering, Kyoto University,
Kyoto 606-01, Japan

Abstract. A binary decision diagram (BDD) is a directed acyclic graph for representing a Boolean function. BDD's are widely used in various areas which require Boolean function manipulation, since BDD's can represent efficiently many of practical Boolean functions and have other desirable properties. However the complexity of constructing BDD's has hardly been researched theoretically. In this paper, we prove that the optimal variable ordering problem of shared BDD's is *NP*-complete, and touch on the hardness of this problem and related problems of BDD's.

1   Introduction

Many problems in digital logic design and testing, artificial intelligence and combinatorics can be expressed as a sequence of operations on Boolean functions. Since the way of representing Boolean functions has a great influence on computation time and storage requirements for operations on Boolean functions, various ways of representation of Boolean functions have been proposed.

A binary decision diagram (BDD) is a directed acyclic graph representation of a Boolean function proposed by Akers [1] and Bryant [2]. BDD's are used as an efficient way of representing a Boolean function, because BDD's have some desirable properties: the size of BDD's is feasible for many practical functions and any Boolean function can be uniquely represented by a BDD for a given variable ordering. A shared binary decision diagram (SBDD) [7] is a powerful extension of BDD's. An SBDD represents multiple Boolean functions simultaneously.

The methods using SBDD's (or BDD's) in various areas of CAD have been demonstrated to be able to handle large scale problems that cannot be handled by conventional methods. For example, state enumeration methods for finite state machines can be used in a wide range of problems such as implementation verification, design verification, and so on. However traditional enumeration methods cannot handle machines with more than a few million states. Coudert et al. [3] first proposed representing a set of states with an SBDD. Owing to the advantage of this technique, machines with more states came to be handled.

It is important to represent Boolean functions by BDD's (an SBDD) with less nodes, because computation time and storage requirements depend on the number of nodes. In general the number of nodes is highly sensitive to the variable ordering. Therefore it is important to find a variable ordering for BDD's (an SBDD) with as few nodes as possible. Some algorithms to compute a minimum

BDD have been proposed, whose input a truth table [5] and a BDD [9]. However the approximation methods [4, 7, 8] such as *dynamic weight assignment* are used when the number of variables is large, because the time complexity of the above minimization algorithms is exponential for the number of variables. From the theoretical point of view, Bryant [2] mentioned without proof that the problem of computing a variable ordering that minimizes the size of the BDD for a given Boolean formula is co *NP*-hard. However it has scarcely been researched theoretically how hard is the problems of finding a optimal variable ordering for a given BDD, and the hardness of this problem has not been shown.

In this paper, we consider the optimal variable ordering problem of SBDD's for a given SBDD (OPTIMALSBDD). We prove that this problem is in NP by constructing the polynomial time nondeterministic algorithm to solve it and prove the NP-hardness of this problem by reducing OPTIMAL LINEAR ARRANGEMENT. The approximation hardness of the minimization problem of BDD's is also touched on.

OPTIMALSBDD contains as a special case the optimal variable ordering problem of BDD's for a given BDD (OPTIMALBDD). Thus OPTIMALBDD is also in NP. However, our result does not carry over for BDD's, and it is left open whether OPTIMALBDD is NP-hard or not. Since the number of nodes of BDD's (an SBDD) for the Boolean formulas of the polynomial size may be exponential for any variable ordering (for example, multiplier functions) and in many of practical cases an SBDD is given as an input, our result settles more practically important problem concerning BDD's.

2 Preliminaries

For Boolean variables x, y, z_i $(i = 1, 2, \cdots, n)$, we denote AND, OR, NOT, EXOR, a tautology function and an inconsistency function by $x \cdot y$, $x + y$, $\overline{x}$, $x \oplus y$, 1 and 0, respectively. For multiple variables we denote AND, OR and EXOR by $\prod_{i=1}^{n} z_i$, $\sum_{i=1}^{n} z_i$ and $\bigoplus_{i=1}^{n} z_i$, respectively. For a function $f(x_1, x_2, \ldots, x_n)$, the resultig function obtained from f by assigning respectively b_i to x_i $(i = 1, 2, \cdots, m, b_i \in \{0, 1\}, m \leq n)$ is denoted $f|_{x_1=b_1, x_2=b_2, \cdots, x_m=b_m}$. The set of the variables which f depends on is denoted $Supp(f)$, that is, $Supp(f) = \{x \mid f|_{x=0} \neq f|_{x=1}\}$.

A binary decision diagram (BDD) is a directed acyclic graph which represents a Boolean function [2]. It has node set N containing two types of nodes, terminal nodes and nonterminal nodes. Each nonterminal node v is labeled with a Boolean variable $var(v)$ and has edges directed toward two children: $low(v) \in N$ and $high(v) \in N$. The edges between v and $low(v)$ and that between v and $high(v)$ are called *0-edge* and *1-edge*, respectively. Each terminal node v has a value $value(v) \in \{0, 1\}$.

There is at most one node labeled the same variable on any path from the rootnode to a terminal node. The ordering of variables on each path is consistent with one another. For a BDD representing $f(x_1, x_2, \ldots, x_n)$, we define a variable ordering of a BDD as $\pi = (\pi[1], \pi[2], \cdots, \pi[n])$, where if $i \neq j$, then $\pi[i] \neq \pi[j]$ and $\pi[i], \pi[j] \in \{x_1, x_2, \cdots, x_n\}$, and if $var(v) = \pi[i]$ and $var(low(v)) = \pi[j]$, then $i > j$ (The same is true of $high(v)$).

A BDD with a rootnode v defines a Boolean function $F(v)$ such that if v is a terminal node with $value(v) =0$ (1), then $F(v) =0$ (1) and that if v is a nonterminal node with $var(v) = x$, then $F(v) = \overline{x}F(low(v)) + xF(high(v))$. In other words, a path from the rootnode to a terminal node corresponds to an assignment to all the variables, that is, if the edge between v and its child on the path is 0-edge (1-edge), then it corresponds to assigning 0 (1) to $var(v)$.

The size of a BDD is the number of nodes of the BDD. We denote the size of a BDD representing a function f on a variable ordering π by $size(f,\pi)$. If there is no confusion, $size(f,\pi)$ is abbreviated as $size(f)$. The size of a BDD can be reduced by applying repeatedly the following transformation rules: If there exists a nonterminal node v where $low(v) = high(v)$, then eliminate v and redirect all incoming edges to $low(v)$. If there exist equivalent nodes u,v, then eliminate v and redirect all edges into u to v. Two nodes are equivalent if one of the following holds:
(1) If both u and v are terminal nodes, the labels of both nodes are equivalent.
(2) If both u and v are nonterminal nodes, $var(u) = var(v)$, $low(u) = low(v)$ and $high(u) = high(v)$.
For a fixed variable ordering, the maximally reduced BDD's is uniquely determined, which is called a canonical BDD for the ordering. In this paper, we call canonical BDD's *BDD's* simply. Examples of BDD's are shown in **Figure1**(a).

Multiple BDD's with the same variable ordering can be joined into a uniquely determined graph by transforming the BDD's under the previously presented rules. This single graph is called a shared binary decision diagram (SBDD) [7]. With the use of SBDD's to represent the set of Boolean functions, not only storage requirement decreases but also the equivalence checking of Boolean functions becomes easy. In practice, SBDD's are used more frequently than BDD's. For example, the SBDD in **Figure1**(b) consists of the four BDD's in **Figure1**(a).

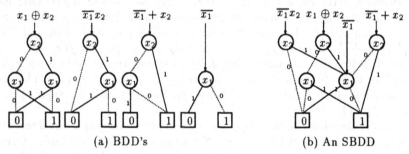

(a) BDD's　　　　　　　　　　(b) An SBDD

Fig. 1. BDD's and an SBDD

3　The Optimal Variable Ordering Problems of SBDD's and Their Complexity

We define the optimal variable ordering problem of SBDD's (OPTIMALSBDD) and prove its NP-completeness in the following.
[OPTIMALSBDD]
INSTANCE: An SBDD with n nodes and a positive integer $k(< n)$

QUESTION: Is there a variable ordering for some SBDD with at most k nodes, which represents the same functions as are represented by the given SBDD ?

Theorem 1 *OPTIMALSBDD is in NP.*

Proof We show a nondeterministic polynomial time algorithm to solve OPTI-MALSBDD. The algorithm chooses a variable ordering π nondeterministically and reconstructs the canonical SBDD with π. If the SBDD with at most k nodes can be reconstructed, the algorithm puts out "yes". However, when the number of the created nodes exceeds k in reconstructing the SBDD, the algorithm stops and puts out "no".

We assume $f_i(i = 1, 2, \ldots, p)$ are functions represented by the given SBDD and *bddtable* is a set of BDD, which is empty initially. Firstly the algorithm creates a new node u and adds the BDD for f_i to bddtable. Note that u will be the rootnode of the BDD for f_i. Secondly the algorithm calls the procedure $RECONST(f, \pi)$. The procedure RECONST reconstructs a canonical BDD in the way of depth first search for a BDD for f and a variable ordering π. RECONST behaves as follows. If f is tautology or inconsistency, then RECONST terminates. If f is neither tautology nor inconsistency, the children of u are created as follows. Unless the BDD for $f|_{\pi[\max\{t|\pi[t]\in Supp(f)\}]=0}$ is in bddtable, $RECONST(f, \pi)$ creates a new node v and 0-edge from u to v, adds the BDD for the function to bddtable and calls $RECONST(f|_{\pi[\max\{t|\pi[t]\in Supp(f)\}]=0}, \pi)$. If the BDD for $f|_{\pi[\max\{t|\pi[t]\in Supp(f)\}]=0}$ is in bddtable, $RECONST(f, \pi)$ creates 0-edge from u to the rootnode of the BDD. Note that the rootnodes of BDD's in bddtable already exists and that the BDD for the restriction of f is smaller than the input SBDD. Next $RECONST(f, \pi)$ executes the same for the BDD for $f|_{\pi[\max\{t|\pi[t]\in Supp(f)\}]=1}$, but 1-edge instead of 0-edge is created.

The procedure RECONST is called less than k times. Since bddtable is a set of BDD, whether the BDD for some function is in bddtable or not can be checked in polynomial time. Thus the time complexity of the deterministic part of the algorithm is $O(n^{O(1)})$. □

Now we consider the optimal variable ordering problem of BDD's (OPTI-MALBDD). OPTIMALBDD is obtained by replacing SBDD with BDD in OP-TIMALSBDD. This problem is a special case of OPTIMALSBDD, that is, a case that the given SBDD in OPTIMALSBDD represents a single Boolean function. Therefore **Theorem1** implies that OPTIMALBDD is in *NP*.

Next we prove the NP hardness of OPTIMALSBDD by reducing OPTIMAL LINEAR ARRANGEMENT defined below to OPTIMALSBDD.

[OPTIMAL LINEAR ARRANGEMENT (OLA)] [6]

INSTANCE: A graph $G = (V, E)$ and a positive integer K

QUESTION: Is there a one-to-one function $\psi : V \longrightarrow \{1, 2, 3, \cdots, |V|\}$ such that $\sum_{(u,v)\in E} |\psi(u) - \psi(v)| \leq K$? $|\psi(u) - \psi(v)|$ is called *cost of edge* (u, v), and $\sum_{(u,v)\in E} |\psi(u) - \psi(v)|$ is called *cost of graph G.*

The basic idea of the reduction is to reflect cost of an edge of G by the size of a BDD and to construct an SBDD whose size reflects cost of G from such BDD's. Now we define a T-phage function as follows. If there is no confusion, we denote a BDD representing a T-phage function simply by a T-phage function.

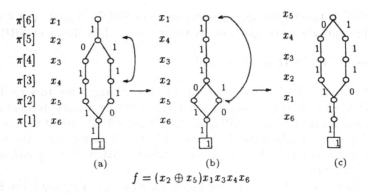

$$f = (x_2 \oplus x_5)x_1x_3x_4x_6$$

Fig. 2. The Property of a T-Phage Function

Definition 1 (T-Phage Function) *A Boolean function f is a T-phage function, if it can be expressed by $f(x_1, x_2, \cdots, x_n) = (x_{i_1} \oplus x_{i_2})\prod_{k \neq i_1, i_2} x_k$, where $(i_1, i_2, \cdots, i_n)$ is a permutation of $(1, 2, \cdots, n)$.* □

An example of a T-phage function is shown in **Figure 2**, where edges into the terminal node labeled 0 are omitted. In the following figures, edges to 0 are abbreviated similarly.

T-phage functions have a convenient property as follows. We assume that f is a T-phage function with n variables and that, without loss of generality, $\pi[l] = x_{i_1}$, $\pi[m] = x_{i_2}$ and $l > m$. We define $\mathcal{F}_i$ as the set of the functions obtained by assigning 0 or 1 to each of the variables $\pi[i], \pi[i+1], \ldots, \pi[n]$ in $Supp(f)$. It is easy to see $size(f) = |\bigcup_{i=1}^{n} \mathcal{F}_i| + 1$. Note that $|\bigcup_{i=1}^{n} \mathcal{F}_i|$ does not contain the number of rootnode. By enumerating the functions in $\bigcup_{i=1}^{n} \mathcal{F}_i$, we obtain $size(f, \pi)$ is $|l - m| + n + 2$.

Thus the size of a T-phage function one-to-one corresponds to cost of an edge (u, v) of G with ψ if we assign a T-phage function $f(x_1, x_2, \ldots, x_n) = (x_u \oplus x_v)\prod_{k \neq u,v} x_k$ to the edge and it is satisfied that $\pi[\psi(v)] = x_v$ for any node v of G. However, if each T-phage function corresponding to each edge depends on the same variables, the SBDD constructed from the T-phage functions does not reflect cost of G, because many nodes of T-phage functions are shared in constructing an SBDD.

To overcome the difficulty that many nodes of T-phage functions on the same set of variables are shared, we introduce a new set of $|V|$ variables $x_{1j}, x_{2j}, \ldots, x_{|V|j}$ for the jth edge (u_j, v_j) of G, and consider for a T-phage function $f_j(x_{1j}, x_{2j}, \ldots, x_{|V|j}) = (x_{u_jj} \oplus x_{v_jj})\prod_{i \neq u_j, v_j} x_{ij}$ $(j = 1, 2, \ldots, |E|)$. Note that there are $|V||E|$ variables. If $x_{v1}, x_{v2}, \ldots, x_{v|E|}$ are arranged consecutively on the variable ordering of the SBDD (the ordering among $x_{v1}, x_{v2}, \ldots, x_{v|E|}$ does not matter) for each node v of G (such ordering will be called well-ordered), the size of the SBDD for the variable ordering one-to-one corresponds to the cost of G with the corresponding variable ordering.

To force such well-ordered property for the SBDD problem, we add a modified T-phage function $h_i = (x_{i1} \oplus x_{i2} \oplus \ldots \oplus x_{i|E|})\prod_{j, k \neq i} \overline{x_{kj}}$ for each i. In minimizing the size of the SBDD, this penalty function makes the variables $x_{i1}, x_{i2}, \ldots, x_{i|E|}$

consecutive on the variable ordering.

The above approach almost works fine, but there still exists a small difficulty caused by shared nodes corresponding to $\pi[1]$. To overcome this completely, we introduce another set of $|V|$ variables for each edge, and as a whole consider the following.

Definition 2 (OLAD) *An OLA-Diagram (OLAD) for graph $G = (V, E)$ given in OLA is an SBDD representing the following functions: for $j = 1, 2, \cdots, |E|$,*

$$
\begin{cases}
f_j = (x_{i_1j} \oplus x_{i_2j})x_{i_3j} \cdots x_{i_{|V|}j} \\
g_j = (y_{i_1j} \oplus y_{i_2j})y_{i_3j} \cdots y_{i_{|V|}j} \\
h_i = (x_{i1} \oplus x_{i2} \oplus \cdots \oplus x_{i|E|} \oplus y_{i1} \oplus y_{i2} \oplus \cdots \oplus y_{i|E|}) \prod_{j,k \neq i} \overline{x_{kj}} \cdot \overline{y_{kj}}
\end{cases}
\tag{1}
$$

where $(i_1, i_2, \cdots, i_{|V|})$ is a permutation of $(1, 2, \cdots, |V|)$. □

In an OLAD, the BDD's for f_j's and g_j's share no node except the terminal nodes because of $Supp(\delta_1) \cap Supp(\delta_2) = \phi$ where δ_1, δ_2 are two functions of f_j's and g_j's. However the BDD's for h_i's share many nodes. Any BDD for f_j (or g_j) and BDD's for h_i's share the terminal nodes and the nonterminal nodes corresponding to the variable $\pi[1]$ (see **Figure3**).

Now we define several terms for the following proof. We denote the set of variables $\{x_{ij}, y_{ij} | 1 \leq j \leq |E|\}$ by B_i for each i and an element of $B = \{B_1, B_2, \ldots, B_{|V|}\}$ is called a variable block. If the variables in each B_i are arranged consecutively on a variable ordering, such a variable ordering and an OLAD with such a variable ordering are called *well-ordered* (*wo*). In the other case, they are called *non-well-ordered* (*nwo*). For a wo-OLAD, we define a variable block ordering $\varphi = (\varphi[1], \varphi[2], \cdots, \varphi[|V|])$ such that, for any variable $\pi[t_1]$ in $\varphi[l_1]$ and any variable $\pi[t_2]$ in $\varphi[l_2]$, it is satisfied that $l_1 < l_2$ if and only if $t_1 < t_2$ where $\varphi[l] \in B$ for each l.

In the following part, firstly we consider influence by exchanging variables in a variable block. Note that, though variables are exchanged in a variable block, the cost of G corresponding to the wo-OLAD must not change since the variable block ordering is not changed. Secondly we show the relation between cost of G and the size of the corresponding wo-OLAD in **Lemma1**, and guarantee that any nwo-OLAD can be transformed into a wo-OLAD with less nodes in **Lemma2**. Finally we finish the proof by using these lemmas.

Since the nonterminal nodes of BDD's for f_j's, g_j's and those of BDD's for h_i's are not shared but the nodes v such that $var(v) = \pi[1]$, we can observe BDD's for f_j's, g_j's and BDD's for h_i's separately. We assume a wo-OLAD. The number of nonterminal nodes of each BDD for h_i's does not vary because, for each i, $Supp(h_i)$ is symmetric in each variable block. If we exchange variables in a variable block but $\varphi[1]$, the number of the nodes of each BDD for f_j's, g_j's does not vary because only one variable in $Supp(f_j)$ (or $Supp(g_j)$) is in each variable block and the BDD's for f_j's, g_j's share no nonterminal node. However the number of nodes of the wo-OLAD may vary when we exchange variables in the variable block $\varphi[1]$, since any nonterminal node v in BDD's for f_j's, g_j's such that $var(v) = \pi[1]$ is shared by BDD's for h_i's. The number of such nodes is one or two. A wo-OLAD in the former case is called *type1* and one in the latter

395

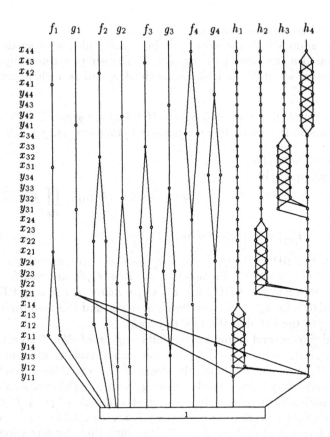

Fig. 3. An example of a wo-OLAD

case is called *type2*. Note that, for some ψ, there are two wo-OLAD with the different number of nodes by a node. However this trouble will be avoided later owing to assigning two T-phage functions to each edge of G.

Then for the two types, we consider the relation between cost of G and the size of the corresponding wo-OLAD . It is trivial that, for the jth edge (u,v) of $G = (V,E)$ given in OLA, $size(f_j) = |\psi(u)-\psi(v)|+|V|+2$ because only one variable in $Supp(f_j)$ is in each variable block. The same is true of $size(g_j)$. Thus cost of the jth edge, that is, $|\psi(u_1)-\psi(u_2)|$ is equal to $\frac{1}{2}(size(f_j)+size(g_j))-|V|-2$.

Lemma 1 *For a graph $G = (V,E)$ with cost K, which is given in OLA, construct an OLAD. If this OLAD is well-ordered, the size of type1 and type2, $k_1(K)$ and $k_2(K)$ are as follows.*

$$k_1(K) = k_2(K) + 1 = 2\{K + 1 - |V| + |E||V|(|V| + 2) - |E|\sum_{i=1}^{|V|-2} i\}$$

Outline of proof We observe BDD's for f_j's and g_j's, and BDD's for h_i's separately. Firstly we count the nodes of the BDD's for f_j's, g_j's and h_i's. Secondly we count the shared nodes in the OLAD. Then we obtain the number of nodes of the OLAD by subtracting the number of shared nodes from $\sum_{j=1}^{|E|}(size(f_j)+size(g_j)) + \sum_{i=1}^{|V|} size(h_i)$. □

$|E|$ attains a maximum value $\frac{|V|(|V|+1)}{2}$ and $K < |E|(|V|-1)$, where K is the cost of G. Hence the upper bound of $k_1(K), k_2(K)$ is $O(|V|^4)$. To finish the proof, we prove the following lemma.

Lemma 2 *An arbitrary nwo-OLAD can be transformed into a functionally equivalent wo-OLAD which has less nodes.*

Proof We show an algorithm which transforms an nwo-OLAD into a wo-OLAD with less nodes. This algorithm behaves as follows. Firstly, assuming that $\pi[1]$ is in B_{u_1}, this algorithm makes the variables in B_{u_1} consecutive without changing the order among them and also without changing the order among variables not in B_{u_1}. After this first stage, we obtain an OLAD with the variable ordering such that $\pi[1], \pi[2], \ldots, \pi[2|E|]$ are all in B_{u_1}. Secondly, assuming that $\pi[2|E|+1]$ is in B_{u_2}, this algorithm makes the variables in B_u consecutive in the same way above. After the second stage, we obtain a OLAD with the variable ordering such that $\pi[1], \pi[2], \ldots, \pi[2|E|]$ are all in B_{u1} and $\pi[2|E|+1], \pi[2|E|+2], \ldots, \pi[4|E|]$ are all in B_{u2}. The algorithm rearranges variables in the same block with $\pi[2|E|(w-1)+1]$ in a similar way at the wth stage for $w = 3, 4, \ldots, |V|$. In this way, the algorithm transforms an nwo-OLAD into a wo-OLAD with less nodes uniquely.

In the following, we prove that the number of nodes does not increase at each stage.

Suppose that after the wth stage, $\pi[2|E|w+1]$ is in B_u and there are q variables $\notin B_u$ between $\pi[2|E|w+1]$ and $\pi[\max\{\, t \mid \pi[t] \in B_u\}]$ on the variable ordering. If $q = 0$, all the variables between $\pi[2|E|w+1]$ and $\pi[\max\{t \mid \pi[t] \in B_u\}]$, that is, $\pi[2|E|(w+1)]$ are already all in B_u after wth stage. Thus the stage does not vary the variable ordering. We assume $q > 0$ in the following part of this proof.

To evaluate the variation of the nodes of the OLAD, we observe the BDD's for f_j's and g_j's and the BDD's for h_i's separately. Without loss of generality, we may assume that the nodes of the BDD for h_i with $B_i = \varphi[|V|]$ is not shared by other BDD's in the OLAD.

A BDD for any h_i can be divided into three parts, that is, subgraphs whose nodes correspond to $(\pi[1] \ldots \pi[t_1-1]), (\pi[t_1], \ldots, \pi[t_2-1])$ and $(\pi[t_2], \ldots, \pi[2mn])$ respectively, where $t_1 = \min\{t \mid \pi[t] \in B_i\}$ and $t_2 = \max\{t \mid \pi[t] \in B_i\}$. These three parts are called *A-part*, *B-part* and *C-part*, respectively. Note that the number of nodes corresponding to each variable in A-part and C-part is only one, and the number of nodes corresponding to each variable in B-part is exactly two. We denote $\min\{t|\pi[t] \in B_u\}$ and $\max\{t|\pi[t] \in B_u\}$ by t_{min} and t_{max} respectively, and the decrease of the number of the nodes of the BDD's for h_i by $-\Delta h_i(-\Delta h_i \geq 0)$.

In the BDD's for h_i's such that $t < t_{min}$ for any $\pi[t] \in B_i (i \neq u)$, only the order of $var(v)$'s for nodes v's in C-part changes at the stage. Thus $-\Delta h_i = 0$.

In the BDD's for h_u, at the stage, the variables $var(v)$'s $\notin B_u$ for the nodes v's in B-part move into the set of the variables $var(v)$'s for the nodes v's in C-part. The number of such variables is q. Therefore the number of the nodes in B-part decreases by $2q$ and the number of the nodes in C-part increases by q. Thus $-\Delta h_u = q$.

In the BDD's for h_i's such that $t > t_{min}$ for any $\pi[t] \in B_i$ $(i \neq u)$, at the stage the variables $var(v)$'s for the nodes v's in B-part and C-part may move into the set of the variables $var(v)$'s for the nodes v's in A-part. Therefore the sum of the number of the nodes in B-part and C-part decreases, say, by δ. Then the number of the nodes in A-part increases by at most δ. Because, in A-part and C-part, the number of the nodes corresponding to each variable is one and, in B-part, the number of the nodes corresponding to each variables is two. Note that the node in A-part of BDD's for h_i's such that $B_i \neq \varphi[|V|]$ is shared (see **Figure3**). In the other case, only the order of the variables $var(v)$'s for the nodes v's in A-part changes at the stage. However in such case the number of the nodes of the BDD for the h_i does not change. Thus $-\Delta h_i \geq 0$. Note that for at least one of such h_i's, $-\Delta h_i > 0$.

Secondly, we observe the variation of nodes of the BDD's for f_j's and g_j's at the stage. We denote the increase of the number of the nodes of each BDD for f_j's and g_j's by Δf_j and Δg_j respectively. In each BDD for f_j's, if we assume that $x_{uj} = \pi[t]$ before the stage and that $x_{uj} = \pi[t - \alpha_j]$ after the stage, it is easy to see $\Delta f_j \leq \alpha_j$. In each BDD for g_j's, we may assume that $\Delta g_j \leq \beta_j$ in the same way as the above. Since the intersection of any two of $Supp(f_j)$ and $Supp(g_j)$ $(j = 1, 2, \ldots, |E|)$ is empty, $\sum_{j=1}^{|E|}(\Delta f_j + \Delta g_j) = \sum_{j=1}^{|E|}(\alpha_j + \beta_j) \leq q$. Thus we obtain $\sum_{j=1}^{|E|}(\Delta f_j + \Delta g_j) + \sum_{i=1}^{|V|} \Delta h_i < 0$.

There are $|V|$ stages until the algorithm stops. At at least one of the $|V|$ stages it is satisfied that $q > 0$, since the OLAD given as an input of the algorithm is non-well-ordered. $\quad\square$

We give a wo-OLAD for some cost K of G given in OLA and $k_1(K)$ as an instance of OPTIMALSBDD. We assume that there exists an OLAD with at most $k_1(K)$ nodes, which represents the same functions as the given wo-OLAD. If this OLAD is well-ordered, there exists function ψ in OLA for the cost at most K because of $k_2(K) < k_1(K) < k_2(K + 1)$ for an arbitrary positive integer K. Even if the OLAD is non-well-ordered, it is guaranteed that there exists a wo-OLAD with at most $k_1(K) - 1$ nodes by **Lemma2**. Thus there exists function ψ for the cost at most K. Therefore OLA is polynomially reducible to OPTIMALSBDD, since the number of nodes of a wo-OLAD is $O(|V|^4)$. Then we obtain the following theorem.

Theorem 2 *OPTIMALSBDD is NP-complete.*

4 Related Problems

The optimal variable ordering problem of BDD's for a given Boolean formula is also an important problem. The co NP-hardness of this problem is shown by reducing the tautology checking problem for a given Boolean formula.

The approximation hardness of the minimization problem of BDD's has not been researched. However the following is easy to show. If $P \neq NP$, then no polynomial time approximation algorithm A for the minimization problem of SBDD's for a given SBDD can guarantee $|A(S) - OPT(S)| \leq K$ for a some fixed constant K, where S is an instance of the problem and $A(S), OPT(S)$ are cost of the solutions by A and an optimization algorithm, respectively. The proof is based on making multiple copies, which is quite standard.

5 Concluding Remarks

It is an unavoidable and important problems to find a variable ordering which minimizes the size of BDD's or SBDD's, since variable ordering has a great influence on the size which computation time and storage requirements for Boolean function manipulation depend on.

In this paper, we consider the optimal variable ordering problem of SBDD's (OPTIMALSBDD) and prove that it is NP-complete. However the NP-hardness of the optimal variable ordering problem of BDD's (OPTIMALBDD) is still an open problem. The NP-hardness of OPTIMALBDD is a stronger conjecture than the NP-hardness of OPTIMALSBDD, because there is a trivial polynomial transformation from OPTIMALBDD to OPTIMALSBDD.

Concerning approximation algorithm, many heuristics for the minimization problem of BDD's have been proposed, but there is no approximation algorithm which guarantees some approximation ratio. Hence it is also important to discuss the approximability of the problem.

References

1. S.B.Akers: Binary Decision Diagrams, IEEE Trans. on Computers, C-27(1978), 509-516.
2. R.E.Bryant: Graph Based Algorithms for Boolean Function Manipulation, IEEE Trans. on Computers, C-35(1986), 677-691.
3. O.Coudert, C.Berthet, and J.C.Madre: Verification of Sequential Machines Based on Symbolic Execution, Proceedings of the Workshop on Automatic Verification Methods for Finite State Systems, Grenoble, France(1989).
4. M.Fujita, H.Fujisawa and N.Kawato: Evaluation and Improvements of Boolean Comparison Method Based on Binary Decision Diagrams, IEEE ICCAD-88(1988), 2-5.
5. S.J.Friedman and K.J.Supowit: Finding the Optimal Variable Ordering for Binary Decision Diagrams, IEEE Trans. on computers, C-39(1986), 710-713.
6. M.R.Garey, D.S.Johnson and L.Stockmeyer: Some Simplified NP-complete Graph Problems, Theoretical Computer Science 1(1976), 237-267.
7. S.Minato, N.Ishiura and S.Yajima: Shared Binary Decision Diagram with Attributed Edges for Efficient Boolean Function Manipulation, Proc. 27th ACM/IEEE DAC(1990), 52-57.
8. M.R.Mercer, R.Kapur and D.E.Ross: Functional Approaches to Generating Ordering for Efficient Symbolic Representations, Proc. 29th ACM/IEEE DAC(1992), 624-629.
9. N.Ishiura, H.Sawada and S.Yajima: Minimization of Binary Decision Diagrams Based on Exchanges of Variables, IEEE ICCAD-91(1991), 472-475.

ON HORN ENVELOPES AND HYPERGRAPH TRANSVERSALS
(Extended Abstract)

Dimitris Kavvadias[1,4]   Christos H. Papadimitriou[2,4,5]   Martha Sideri[3,4]

Abstract: *We study the problem of bounding from above and below a given set of bit vectors by the set of satisfying truth assignments of a Horn formula. We point out a rather unexpected connection between the upper bounding problem and the problem of generating all transversals of a hypergraph, and settle several related complexity questions.*

1. INTRODUCTION

Recently there has been much interest in the *model theory* of Boolean Logic, that is, the relationship between a Boolean formula and the corresponding set of models (in this paper by *model* we shall mean "satisfying truth assignment"). There are at least three distinct motivations for this problem: Identifying plausible Boolean formulae that describe a set of $0-1$ vectors is one form of "discovering structure" in raw data [DP, KKS1]. Besides, simplifications and approximations of a Boolean formula via its models may represent a plausible *vivid form* of the formula, ready to make complicated approximate inferences very rapidly [SK, Pa]. Finally, characterizing the sets of models of various subclasses of Boolean Logic and related formalisms is an important part of understanding the semantic power and usefulness of such formalisms. Therefore, the intricate algorithmic problems involved in going back and forth between a Boolean formula and its set of models have been investigated in the recent literature. This paper is a contribution to this line of research.

Horn formulae (Boolean formulae in conjunctive normal form with at most one positive literal per clause) comprise an important subclass because of their many positive algorithmic properties (for example, their satisfiability problem is P-complete), and they represent a natural style for expressing real-life conditions (logic programming is one manifestation of this aspect of Horn formulae). Accordingly, there has been considerable interest recently in inferring Horn formulae from a given set of models. In [DP] an algorithm is presented for telling whether a given set of vectors in $\{0,1\}^n$ is precisely the set of all models of some Horn formula (in this paper we slightly improve on the performance of this algorithm, Corollary 1 to Proposition 1). If the answer is negative, and there is no Horn formula with precisely the given set of models, then the following natural question arises: What is the Horn formula whose set of models *best approximates* the given set of vectors?

---

[1] Department of Mathematics, Univ. of Patras, Patras, Greece.

[2] University of California at San Diego, La Jolla, California 92093-0114, U.S.A.

[3] Athens University of Economics and Business, Athens 10434 Greece.

[4] Research partially supported by the Esprit project ALCOM.

[5] Research partially supported by the National Sience Foundation.

Set inclusion is perhaps the most natural concept of approximation here; that is, we seek a Horn formula whose set of models includes the given set, and is as small as possible. Such a Horn formula is called the *Horn envelope* of the given set of models. It is easy to see (Corollary 2 to Proposition 1) that this set of models is unique and easy to compute. By the latter we mean that there is an algorithm that outputs this smallest superset in time which is polynomial in both the input *and the output*. This concept of efficiency is the appropriate one here, since the required output may be exponentially larger than the input[6]. *The difficult part is to compute the corresponding Horn formula.* The algorithm should be polynomial in the given data *and the output*, that is, the number of Horn clauses produced.

Several recent papers have addressed this problem. [DP] give an algorithm that does output such a Horn formula in output-polynomial time; the problem is that the Horn formula produced by their algorithm may be *highly redundant*, and therefore have exponentially more clauses than the correct, non-redundant output. [KKS1] present a polynomial algorithm that outputs a Horn formula whose set of models has with high probability a small (as a percentage of 2^n) symmetric difference from the desired minimum enveloping set of models. However, no polynomial algorithm for finding the "Horn envelope" had been known —neither was there any evidence that this algorithmic task is impossible. *One of our main results (Theorem 1) is evidence that computing the Horn envelope is a very hard problem indeed,* since its solution would yield an output-polynomial algorithm for generating all *transversals of a hypergraph*. We introduce this interesting problem next.

A hypergraph is a set H of subsets of $\{1, 2, \ldots, n\}$. A *hitting set* of H is a subset t of $\{1, 2, \ldots, n\}$ such that for all $h \in H$ t intersects h. A *transversal* of H is a minimal hitting set of H. We let $\mathrm{tr}(H)$ be the set of all transversals of H. It turns out that $\mathrm{tr}(\mathrm{tr}(\mathrm{tr}(H))) = \mathrm{tr}(H)$ [Be]. Generating in output-polynomial time $\mathrm{tr}(H)$, given H, is a well-studied algorithmic problem, first proposed in [JYP], which has been open. In a recent paper, [EG] point out that this problem is equivalent to a number of other important problems in Boolean Logic, database theory, switching theory, distributed systems, and artificial intelligence. *In this paper we present a rather unexpected connection between the transversal problem and the problem of generating the Horn envelope of a given set of models* (Lemma 1 and Theorem 1). Our result strongly suggests that there is no output polynomial algorithm for generating the Horn envelope of a set of models.

We also study the opposite problem of computing the Horn *core* (as opposed to envelope) of a given set of models. That is, we are now seeking a Horn formula whose set of models is included in the given set, and is as large as possible. Unlike the case of the Horn envelope, the maximal Horn core is not unique, but *a* maximal core can be computed in polynomial time; however, computing —even approximating—

---

[6] See [JYP, EG] for extensive discussions of the algorithmic issues involved in output-efficient enumeration of combinatorial configurations. Incidentally, in many computational problems considered in this paper the input is a set of models, subsets, or $0 - 1$ vectors; if the size of the input is $2^{\Theta(n)}$, then these problems can be solved trivially in polynomial time by exhaustion. As is commonplace in analyzing the complexity of hypergraph problems, our results are interesting when the input set is considerably smaller than this.

the core with *maximum cardinality* is NP-complete (Theorem 2). Interestingly, for the polynomial upper bound we reduce the problem to a polynomially solvable one about Horn clauses.

In the next section we introduce some basic facts about the model theory of Horn formulae. In Section 3 we introduce the transversal problem and its relationship with Horn envelopes. In Section 4 we discuss the problem of Horn cores, and in Section 5 the problems that are left open by this work.

2. MODELS OF HORN CLAUSES

Let $\{x_1, \ldots, x_n\}$ be a set of Boolean variables. A literal is either a variable or a negation, and a clause is the disjunction (or) of literals. A *Horn clause* is a clause with at most one (unnegated) variable. Examples of Horn clauses are $(\neg x_1 \lor x_3 \lor \neg x_4)$, (x_2), and $(\neg x_1 \lor \neg x_3 \lor \neg x_4)$. Clauses that do have a positive literal are usually denoted in implicational form (for the first two Horn clauses above: $(x_1 \land x_4 \to x_3)$ and $(\to x_2)$)). A Horn formula is a set of Horn clauses.

A *model* t is a vector in $\{0, 1\}^n$ —intuitively, a truth assignment to the Boolean variables. A model satisfies a clause if for some literal of the clause either the literal is x_i and $t_i = 1$, or the literal is $\neg x_i$ and $t_i = 0$. If t and t' are models, we say that $t \le t'$ if $t_i = 1$ implies $t'_i = 1$. We denote by $t \land t'$ the vector which is 1 exactly where both t and t' are 1. If ϕ is a Boolean formula in conjunctive normal form, sat(ϕ) is the set of all models that satisfy all clauses of ϕ.

A basic question is, given a set of models $M \subseteq \{0, 1\}^n$, is there a Horn formula ϕ such that $M = \text{sat}(\phi)$? The answer is quite straightforward, and given below (although researchers in the area have been aware of this characterization, this particular statement and proof have not appeared, to our knowledge, in the literature).

Proposition 1. Let $M \subseteq \{0, 1\}^n$. The following are equivalent:
(a) There is a Horn formula ϕ such that sat(ϕ) = M.
(b) For each $t \notin M$ either there is no $t' \in M$ with $t \le t'$, or there is a unique minimal $t' \in M$ such that $t \le t'$.
(c) If $t, t' \in M$, then also $t \land t' \in M$.

Sketch: That (a) implies (c) is easy. To establish (b) from (c), take t' to be the $\land$ of all $t'' \in M$ such that $t \le t''$. Finally, if we have property (b), we can construct the following set of Horn clauses: For each $t \notin M$ let t' be the model guaranteed by (b); create a Horn clause $((\bigwedge_{t_i=1} x_i) \to x_j)$ for each j such that $t_j = 0$ and $t'_j = 1$. It is easy to see that the set of all these Horn clauses comprise the desired ϕ. □

Corollary 1: Given M, we can test whether there is a Horn formula ϕ with sat(ϕ) = M in $O(m^2 n)$ time, where $m = |M|$. Furthermore, if the answer is positive, then we can generate the clauses of ϕ in polynomial time.

Sketch: We can test part (c) of the Proposition as follows: For any $t, t' \in M$ compute $t \land t'$ and add it to M. Sort by bucket sorting, delete duplicates, and test whether the resulting set contains the same number of elements as M.

To generate the clauses of ϕ, for each $t' \in M$ find the minimal $t \notin M$ such that t' is the model guaranteed by (b), and add the clauses described in the proof of the implication from (b) to (a). $O(m^2n^2)$ such clauses are added, and it is easy to check that all other clauses generated in the proof of the proposition are redundant. $\Box$

Incidentally, the $O(m^2n^2)$ bound in the proof of the corollary is the best possible. Also, notice that the corollary improves slightly on the $O(m^2n \log m)$ algorithm in [DP].

But suppose that M does not satisfy the condition of the proposition; what is the *smallest superset* of M that does? This corresponds to asking, what is the most restrictive Horn formula which implies the Boolean formula with models precisely those in M?

Corollary 2: Given M, there is a unique minimal $\overline{M} \supseteq M$ such that $\text{sat}(\phi) = \overline{M}$ for some Horn formula ϕ. This set can be generated in output polynomial time.

Sketch: The algorithm is this: While the test in Corollary 1 fails, repeat with the new M. This yields the closure of M under $\wedge$, obviously a unique set. $\Box$

Notice that, although we can produce easily the Horn formula ϕ such that $\text{sat}(\phi) = M$, if it exists, we have not said how, if it does not exist, to produce ϕ such that $\text{sat}(\phi) = \overline{M}$. Define the *Horn envelope* of M to be the non-redundant set of Horn clauses ϕ such that $\text{sat}(\phi) = \overline{M}$. The main algorithmic problem studied in this paper is thus, given a set of models M, to generate its Horn envelope. *This is an intriguing and well-known problem,* attacked for example in [DP], [KKS1], as well as the next section.

3. TRANSVERSALS AND ENVELOPES

We defined the transversal problem in the introduction. An immediate but loose connection between this problem and the Horn envelope problem is given by the concept of the *standard clauses,* introduced next.

We can identify a set M of models with a hypergraph H, where each model $t \in M$ contributes a hyperedge in H consisting of all positions at which t is *zero* (notice the departure from the usual correspondence). For each $i \in \{1, 2, \ldots, n\}$ now omit from H all sets in H that do not contain i, and delete i from the remaining sets; call the resulting hyperhraph H_i. Define now the following Horn clauses: First, for each $t \in \text{tr}(H)$ we have the clause $(\bigvee_{i \in t} \neg x_i)$ Finally, for each i and each $t \in \text{tr}(H_i)$ we have the clause $((\bigwedge_{j \in t} x_j) \rightarrow x_i)$. These are called the *standard clauses* associated with M.

The intuition behind the standard clauses is this: The sets in $\overline{M}$ (via the correspondence "set of zeros") are precisely all sets that can be expressed as unions of the hyperedges of H. How can we describe these sets in terms of "constraints?" The standard clauses simply state that, for example, if such a set fails to contain any element in a transversal of H_i, then it cannot contain i. Similarly for the transversals of H. The next lemma (whose technical proof we omit here) states essentially that these constraints capture $\overline{M}$.

Lemma 1: The formula consisting of all standard clauses associated with M is logically equivalent to the Horn envelope of M. $\Box$

Example: As the following examples show, unfortunately the standard clauses may be redundant. Suppose that M consists of the models 01010, 01100, and 00111. The corresponding hyperhraph H_1 has transversals $\{2,3,4\}$ and $\{2,5\}$, while $\{3,4\}$ is one of the transversals of H_5. Thus the following are standard clauses: $(x_3 \wedge x_4 \rightarrow x_5)$, $(x_2 \wedge x_5 \rightarrow x_1)$, and $(x_2 \wedge x_3 \wedge x_4 \rightarrow x_1)$. It is easy to see that the first two logically imply the third. If we omit the third model, then a more benign form of redundancy results: The standard clause $(\neg x_3 \vee \neg x_4)$ logically implies the standard clause $(x_3 \wedge x_4 \rightarrow x_1)$. Also, if in this last example we add the model 10111, then there is no unique Horn envelope, since either one of $(x_3 \wedge x_4 \rightarrow x_1)$ and $(x_3 \wedge x_4 \rightarrow x_5)$ can be omitted (in the light of the standard clauses $(x_1 \rightarrow x_5)$ and $(x_5 \rightarrow x_1)$). $\square$

Despite this loose connection, the standard clauses are useful for establishing a more intriguing relationshp between the two problems:

Theorem 1: If there is an output-polynomial (or polynomial- delay) algorithm for generating the Horn envelope of a given a set of models M, then there is an output-polynomial (respectively, polynomial-delay) algorithm for generating $\text{tr}(H)$, given a hypergraph H.

Sketch: Let H be a hypergraph. Define the following hypergraph $H' = H \cup \{h+j : j \notin h \in H\}$ (where $+$ abbreviates union with a singleton). Define now $\mu(H)$ to be the set of models whose sets of zeros coincide with the hyperedges of H'. We can establish the following:

Lemma 2: The Horn envelope of $\mu(H)$ contains precisely the clauses $(\bigvee_{i \in t} \neg x_i)$ for all $t \in \text{tr}(H)$.

The proof of Lemma 2 establishes that the standard clauses of $\mu(H)$ are precisely $(\bigvee_{i \in t} \neg x_i)$ for all $t \in \text{tr}(H)$; since these clauses can be shown non-redundant, the result would follow from Lemma 1. By the definition of standard clauses, this is tantamount to the following claim, whose proof follows from the definition of H' and completes the proof of Theorem 1.

Claim: $\text{tr}(H) = \text{tr}(H')$. Furthermore, for all j $\text{tr}(H'_j) = \{t \in \text{tr}(H'_j) : j \notin t\}$.

4. HORN CORES

Another natural way of approximating a set M of models in terms of Horn formulae is by finding not the most restrictive Horn formula ϕ with $\text{sat}(\phi) \supset M$ (this would be the Horn envelope we studied so far), but the least restrictive ϕ with $\text{sat}(\phi) \subset M$. Such a formula is called a *Horn core* of M, and its set of models is denoted $\underline{M}$. In contrast to the case of Horn envelopes which are unique, a set of models M may have several Horn cores ϕ, all with maximal $\text{sat}(\phi) = \underline{M} \subset M$. The following result settles the computational complexity of the corresponding algorithmic problems:

Theorem 2. (a) There is a polynomial-time algorithm which, given $M \subseteq \{0,1\}^n$, generates one of the maximal subsets $\underline{M}$ and the corresponding Horn core. In fact, all such maximal sets and formulae can be generated with polynomial delay.

(b) However, it is NP-complete to find the $\underline{M}$ with maximum cardinality.

(c) Furthermore, this latter problem is NP-hard to approximate within any constant factor.

Sketch: The proof of Part (a) is interestingly "incestuous," in that it relies on an algorithmic fact about Horn clauses. Given M, we create a Horn formula with the vectors in M as variables. Intuitively, for $t \in M$ $t =$ true means that $t \in \underline{M}$. For any $t, t' \in M$ there are two cases: If $t \wedge t' \notin M$, then obviously not both models can be in $\underline{M}$; we write the clause $(\neg t \vee \neg t')$. If however $t \wedge t' = t'' \in M$, then we add the clause $((t \wedge t') \rightarrow t'')$. The maximal $\underline{M}$'s then correspond to the maximal models of the resulting Horn formula.

For (b) we reduce the NP-complete problem CLIQUE to the problem of maximum cardinality $\underline{M}$. Given a graph $G = (V, E)$ and an integer k we construct a set of models $M \subseteq \{0,1\}^{|E|}$ (equivalently, subsets of E), as follows: M contains all singletons $\{e\}$, and, for each vertex $v \in V$, the set $\{e \in E : v \in e\}$. It then follows that there is a $\underline{M}$ of size $|E| + k$ if and only if there is a clique of size k in G. For part (c) we amplify the above construction. $\square$

5. DISCUSSION

This paper is a contribution to the study of the algorithmic problems related to knowledge representation. It also adds a new member to the intriguing class of configuration enumeration problems that are related to the transversal problem. It is at present open whether the Horn envelope problem is equivalent to the transversal problem —that is, whether the reduction in Theorem 1 also goes the other way. We conjecture that it does.

There are several interesting related open complexity questions, which can be seen as generalizations of Corollary 1. Given a set of models M, is it the case that $M = \text{sat}(\phi)$ for a ϕ which is (a) Krom (that is, 2-SAT, in conjunctive normal form with at most two literals per clause); (b) 3-SAT; (c) with at most m (a given number) clauses? We know that (a) above is in P. We conjecture that (b) is coNP-complete and that (c) is Σ_2^p-complete.

Another interesting open problem is this: Given a Horn formula, how hard is it to generate its *characteristic models*, that is, a minimal set of models M such that $\text{sat}(\phi) = \overline{M}$. Characteristic models were shown in [KKS2] to be important alternative representations of a Horn formula. As was pointed out by Bart Selman (private communication), our Theorem 1 and the approximation algorithm in [KKS1] imply that generating characteristic models is also related to transversal enumeration.

Finally, it would be interesting if the insights into Horn formulae presented in this paper could lead to improved algorithms for *learning* Horn formulae from equivalence and membership queries [AFP]; such a result would also lead to improved approximations of the Horn envelope using the ideas in [KKS1].

REFERENCES

[AFP] D. Angluin, M. Frazier, L. Pitt "Learning conjunctions of Horn clauses," *1990 FOCS* pp. 186–192.

[Be] C. Berge *Graphes et Hypergraphes*, Dunod, 1980.

[DP] R. Dechter and J. Pearl "Structure identification in relational data," *Artificial Intelligence*, 1993.

[EG] T. Eiter, G. Gottlob "Identifying the minimal transversals of a hypergraph and related problems," *SIAM J. Comp.*, to appear.

405

[JYP] D. S. Johnson, M. Yannakakis, C. H. Papadimitriou "On generating all maximal independent sets," *IPL 27*, 119–123, 1988.

[KKS1] H. A. Kautz, M. J. Kearns, B. Selman "Horn approximations of empirical data," to appear in *Artificial Intelligence*, 1993.

[KKS2] H. A. Kautz, M. J. Kearns, B. Selman "Reasoning with characteristic models," to appear in *AAAI*, 1993.

[Pa] C. H. Papadimitriou "On selecting a satsfying truth assignment," *Proc. 1991 FOCS*.

[SK] B. Selman, H. A. Kautz "Knowledge compilation using Horn approximation," *Proc. AAAI 1991*.

Page Migration Algorithms
Using Work Functions
(Extended Abstract)

Marek Chrobak[1]*   Lawrence L. Larmore[1]*
Nick Reingold[2]**   Jeffery Westbrook[3]***

[1] Department of Computer Science, University of California, Riverside, CA 92521.
[2] AT&T Bell Laboratories, Murray Hill, NJ 07974-0636.
[3] Department of Computer Science, Yale University, New Haven, CT 06520-2158.

Abstract. The *page migration* problem is the management problem for a globally addressed shared memory in a multiprocessor system. Each physical page of memory is located at a given processor, and memory references to that page by other processors incur a cost equal to the network distance. At times the page may migrate between processors, at a cost equal to the distance times the page size factor, D. The problem is to schedule movements on-line so as to minimize the total cost of memory references. We consider the problem under the restriction that movement can only occur after a request has been served and before the next request is known. The major results are: we give randomized $(2 + \frac{1}{2D})$-competitive on-line algorithms for trees (and products of trees, including the hypercube), and for a uniform space when $D = 1, 2$. We show that these algorithms are optimal. We prove a $\frac{85}{27}$ lower bound on the competitiveness constant of a deterministic algorithm (in arbitrary spaces) with $D = 1$, disproving a conjecture by Black and Sleator. We show a deterministic $(2 + \frac{1}{2D})$-competitive algorithm for continuous trees. Our analysis is based on *work functions*, which provide a systematic approach to many competitive analysis problems.

1 Introduction

The *page migration* problem is the management problem for a globally addressed shared memory in a multiprocessor system. Each physical page of memory is located at a given processor, and memory references to that page by other processors incur a cost equal to the network distance. At times the page may migrate between processors, at a cost equal to the distance times the page size factor, D. The problem is to schedule movements on-line so as to minimize the total cost of memory references.

The page migration problem has been considered by the architecture community, where it arises as a memory management problem in a globally addressed

---

* Research partially supported by NSF grant CCR-9112067.
** Research partially supported by NSF grant CCR-8958528.
*** Research partially supported by NSF grant CCR-9009753.

shared memory in a multiprocessor system [3, 4, 16]. A common design for such a system is a network of processors, each of which has its own local memory [8, 14, 18]. In such a design, a programming abstraction of a single global memory address space is supported by a virtual memory system that distributes one or more copies of each physical page of memory among the processors' local memories. If processor p contains a copy of page b, then it can directly read and write memory addresses on b. If p does not have a copy, however, then it must satisfy memory accesses by transmitting a request through the network to some processor q that does have a copy of b and waiting for an answer from q. The communication cost of this request depends upon the network, and may vary depending on the choice of p and q. A very similar situation arises in distributed databases, where multiple copies of a data file may be stored at various local nodes.

Having a given virtual page stored at multiple processors reduces communication overhead during reads, but introduces the problem of maintaining consistency among the multiple copies during writes, something for which most multiprocessors do not provide mechanisms [4]. Therefore, various network designers have proposed restricting each writable page to a single copy [3, 4, 16], and allowing the operating system to move the page between processors in response to changes in access patterns. The *page migration* problem is to decide, on-line, a sequence of page movements that minimizes the cost of communication.

We formalize the page migration problem as follows: An instance of the problem consists of a metric space M, and an integer constant $D \geq 1$ (the page size). Let $s \in M$ denote the current position of the page, which we think of as a "server." A sequence ϱ of requests arrives on line. Each request is a point $r \in M$. Once a request r arrives, it must be immediately served at cost $d(s, r)$, the distance from s to r. This is called the *access cost*. We are then allowed to move the server from s to any other point x at cost $D \cdot d(s, x)$. This is called the *movement cost*. The server is initially placed at point s_0, but it may be moved to an arbitrary point s before the first request, at the usual cost $D \cdot d(s_0, s)$.

Let $\mathcal{A}$ be an on-line algorithm for the page migration problem. We denote the cost of $\mathcal{A}$ on a request sequence ϱ if the initial position is s_0 by $cost_{\mathcal{A}}(s_0, \varrho)$, and we denote the minimum cost of servicing the request sequence ϱ starting from s_0 by $cost_{opt}(s_0, \varrho)$. This minimum cost can be achieved by an off-line algorithm (one which knows the entire sequence ϱ in advance), but, in general, cannot be achieved on-line.

Let $c \geq 1$. Algorithm $\mathcal{A}$ is called *c-competitive* in M if for each $x \in M$ there is a constant a_x

$$cost_{\mathcal{A}}(x, \varrho) \leq c \cdot cost_{opt}(x, \varrho) + a_x, \qquad (1)$$

for all request sequences ϱ. (In our paper all algorithms will have $a_x = 0$.) A randomized algorithm $\mathcal{A}$ is c-competitive if for all sequences ϱ the above inequality holds for the expected cost of $\mathcal{A}$ on ϱ. In this paper we use the notation $cost_{\mathcal{A}}(x, \varrho)$ to denote expected cost of a randomized algorithm $\mathcal{A}$.

In the literature on competitive analysis of randomized algorithms, on-line problems are often viewed as a game played against an adversary that generates requests on-line. An adversary is called *oblivious* if it does not know the location of the page. Our definition of competitiveness corresponds to the oblivious adversary, since such an adversary may as well choose a request sequence in advance and serve it optimally.

Black and Sleator [4] consider two classes of networks: uniform spaces (complete graphs with each edge having length 1), and trees with arbitrary edge lengths. They develop 3-competitive deterministic algorithms for these two cases. In addition, they show that no deterministic algorithm could be better than 3-competitive in any metric space with at least two points. Besides this lower bound, little is known about deterministic algorithms that work for arbitrary metric spaces; the problem seems quite hard. Black and Sleator conjecture that a 3-competitive algorithm exists for all metric spaces. The best currently known upper bound of 7 was given recently by Awerbuch *et al.* [1]. Westbrook [17] gives a randomized algorithm for the uniform network whose competitiveness constant approaches $(5 + \sqrt{17})/4 \approx 2.28$ as D approaches infinity, a $(1 + \phi)$-competitive randomized algorithm[4] for general networks, and a randomized algorithm that is 3-competitive against an *on-line adaptive adversary*.

Similar memory management problems are examined in [2, 5, 10, 11, 12, 13, 15]. Practical issues and applications of page migration are discussed more fully in [3, 4, 16]. The closely related *file assignment* problem, an important problem in distributed databases, involves the static allocation of multiple copies of data files to the nodes of a distributed system, see [9, 19, 1].

A major difficulty in traditional competitive analysis has been to understand optimal off-line algorithms. In particular, it is almost always impossible to know when, or to where, an optimal algorithm will move its server (the page). The approach in this paper is to use *work functions*, which completely describe the actions of an optimal off-line algorithm, to the extent that an on-line algorithm can know it. Work functions were defined in [7] and are similar to the residues of [13]. Work functions provide a systematic approach to problems of competitive analysis.

Overview of the paper. In Sect. 2 we define work functions and our approach to competitive analysis. In Sect. 3 we consider the simple case where the metric space consists of only two points. We present a $(2 + \frac{1}{2D})$-competitive randomized algorithm, EDGE , and prove the matching lower bound. In Sect. 4 we consider uniform spaces of arbitrary size. Quite unexpectedly, this problem appears to be very hard. We give an extension of EDGE to uniform spaces, prove that this extension is optimal for $D = 1, 2$, and conjecture that it is optimal for any $D \geq 3$. In Sect. 5 we present a randomized algorithm for trees called FACTOR. FACTOR can be viewed as a "product" of a number of instances of RAND (one instance for each edge). FACTOR is $(2 + \frac{1}{2D})$-competitive, *i.e.*, optimal. FACTOR can also be applied, with the same competitive constant, to metric spaces that are Cartesian products of trees with the $\mathcal{L}^1$ metric, including hypercubes and meshes.

---

[4] $\phi = 1.618\ldots$, the Golden Ratio.

We then present a number of results about deterministic algorithms. In Sect. 6 we consider continuous trees, and we show how to use continuity to turn a randomized algorithm into a deterministic one with the same constant, $2 + \frac{1}{2D}$. In Sect. 7 we disprove the conjecture of Black and Sleator by demonstrating a simple metric space for which there is no algorithm with competitiveness constant better than $\frac{85}{27} \approx 3.148$ for $D = 1$. In Sect. 8, we also give an optimal 3-competitive algorithm for the space consisting of a single triangle and $D = 1$. This algorithm uses a technique called *forgiveness*, that has also been recently applied to define an 11-competitive algorithm for the 3-server problem [6].

2 Work Functions, Algorithms, and Competitiveness

Work Functions. In their original paper on the k-server problem, Manasse *et al.* [13] described a dynamic program for computing the optimum cost in an instance of the k-server algorithm. They discovered that the information carried in the matrix of this dynamic program can be useful both in executing an on-line algorithm and in analyzing its performance.

Work functions (see [7]) extend and generalize this dynamic programming approach. Fix D. We say that a function $\omega : M \to \mathbb{R}^+$ is a *work function* if it satisfies the following two conditions:

(wf1) $\omega(x) - \omega(y) \le D \cdot d(x, y)$ for all $x, y \in M$.

(wf2) For some point $x \in M$, $\sup_y \{D \cdot d(x, y) - \omega(y)\} < \infty$.

Let W_M be the set of all work functions on M. The *update operator*, $\wedge$, is defined by $(\omega \wedge r)(x) = \inf_z \{\omega(z) + d(z, r) + D \cdot d(z, x)\}$.

It is quite easy to show that $\wedge$ is well-defined, that is, that $\omega \wedge r$ is a work function. In general, the infimum in the definition of $\wedge$ may not be achieved. However, all work functions described in this paper will have a well-defined minimum, and hence we may replace the infimum by a minimum in the above statements. Inductively, we can extend the update operator to sequences of requests: $\omega \wedge \varepsilon = \omega$ and $\omega \wedge \varrho r = (\omega \wedge \varrho) \wedge r$. The *characteristic function* of a point, χ_x, is defined by $\chi_x(y) = D \cdot d(x, y)$.

Lemma 1. *Let ϱ be a request sequence. Then the value $\omega(y) = (\chi_s \wedge \varrho)(y)$ is the minimum cost to service request sequence ϱ starting with the server at point s and ending with the server at point y.*

A work function ω is called *reachable* if $\omega = \chi_x \wedge \varrho$ for some x, ϱ. Note that our definition allows work functions that are not reachable. (This will be useful in our algorithm for the triangle.) When analyzing the behavior of an on-line algorithm after processing request sequence ϱ starting at position s, we will refer to the function $\omega = \chi_s \wedge \varrho$ as the *current* work function. Suppose the next request is r. The new current work is $\omega' \wedge r = \chi_s \wedge \varrho r$. We call the quantity $\inf(\omega') - \inf(\omega)$ the *amortized optimal cost* of request r, denoted $\Delta cost_{opt}$.

An *offset function* is a work function whose infimum value is zero. One special example of a work function which is also an offset function is χ_x. For a work function ω define offset(ω) to be the offset function given by $\omega - \inf(\omega)$.

Our on-line algorithms are based on offset functions. At any time, the algorithm maintains a current offset function that it uses to make decisions about the server movement. The current offset function will be exactly the one defined by offset($\chi_s \wedge \varrho$), where s is the initial position and ϱ is the request sequence up to the current time. Since we will be dealing primarily with offsets, we define another update operator Δ by $\omega\Delta r = \omega \wedge r - \inf(\omega \wedge r)$. Note that $\omega\Delta r$ is always an offset function. The operator Δ can be extended to sequences of requests in an analogous way to $\wedge$. We say an offset function is *reachable* if it is of the form $\chi_x \Delta \varrho$ for some $x \in M$ and some sequence of requests $\varrho \in M^*$.

Randomized Algorithms. A probability distribution μ on a metric space M is said to be *finite* if there exists a finite set $X \subseteq M$. such that $\mu(X) = 1$. Let Π_M denote the set of all finite probability distributions on M. Throughout this paper we say "distribution" for short when we mean finite probability distribution. If $\mu \in \Pi_M$ and $\mu(X) = 1$, we will say that X *supports* μ. The smallest set that supports μ is called the *support of* μ. If the support of μ is a singleton, $\{x\}$, by an abuse of notation, we will identify x and μ, *e.g.*, we will write $\mu = x$.

In this paper, a randomized algorithm $\mathcal{A}$ will be viewed as a *deterministic* algorithm whose states, at each step, will be server distributions on M.

We say that a distribution α on $M \times M$ is a *transport* between distributions μ and ν on M, if $\mu(x) = \sum_u \alpha(x, u)$ and $\nu(x) = \sum_u \alpha(u, x)$ for all $x \in M$. A transport describes a way for a randomized algorithm $\mathcal{A}$ to change the distribution of its server from μ to ν, by specifying for each pair of points the probability that the server starts at x and ends at y. If the server position is x, then $\mathcal{A}$ moves its server to position y with probability $\alpha(x, y)/\mu(x)$.

The cost of transport α is defined by $\|\alpha\| = \sum_x \sum_y d(x, y) \cdot \alpha(x, y)$, and the transport distance between μ and ν is $\delta(\mu, \nu) = \min_\alpha \|\alpha\|$, where the minimum is over all transports α between μ and μ. We call an α for which $\|\alpha\| = \delta(\mu, \nu)$ a *minimal transport* from μ to ν.

Note that for $x, y \in M$, $d(x, y)$ (the distance from x to y) and $\delta(x, y)$ (the transport distance between distributions supported by $\{x\}$ and $\{y\}$) yield the same value. The access cost incurred by $\mathcal{A}$ is simply the expected value, given μ, of the distance between the server and r, which is equal to the transport distance $\delta(\mu, r) = \sum_x d(x, r) \cdot \mu(x)$. The movement cost is given by $D \cdot \|\alpha\|$, where α is the transport used by $\mathcal{A}$. We assume that any algorithm uses a minimum transport to move between distributions.

Let $\mathcal{A}(x_0, \varrho)$ be the distribution of the server of $\mathcal{A}$ after processing ϱ. Write $\varrho = r_1 \ldots r_k$, and let $\mu_i = \mathcal{A}(x_0, r_1 \ldots r_{i-1})$ Then

$$cost_{\mathcal{A}}(x_0, \varrho) = \sum_{i=0}^{k-1} \left(\delta(\mu_i, r_{i+1}) + D \cdot \delta(\mu_i, \mu_{i+1}) \right). \tag{2}$$

We call a randomized algorithm *stable* if it has the property that the current distribution depends only on the current offset function. In this case an algorithm can be defined simply as a function $\mathcal{A} : W_M^0 \to \Pi_M$. All algorithms described in this paper are stable.

Competitive Analysis with Offset Functions. Our proofs of competitiveness are based on the following method. Let $\mathcal{A}$ be a stable randomized on-line algorithm. (The method can be easily generalized to other types of on-line algorithms). We say that $\Phi : W_M^0 \longrightarrow \mathbb{R}^+$ is a *c-potential for* $\mathcal{A}$ (or simply *potential*) if it satisfies the following condition: for any $r \in M$ and any reachable offset $\omega \in W_M^0$

$$\delta(\mu, r) + D \cdot \delta(\mu, \mu') + \Phi(\omega') - \Phi(\omega) \leq c \cdot \min_z(\omega \wedge r)(z). \qquad (3)$$

where $\mu = \mathcal{A}(\omega)$ and $\mu' = \mathcal{A}(\omega')$. The quantity $\inf(\omega \wedge r)$ is called the amortized optimum cost. It is well-known that the existence of a c-potential implies that $\mathcal{A}$ is c-competitive.

We will often write the inequality (3) symbolically as

$$\Delta cost_{\mathcal{A}} + \Delta \Phi \leq c \cdot \Delta cost_{opt},$$

where $\Delta cost_{\mathcal{A}} = \delta(\mu, r) + D \cdot d(\mu, \mu')$, $\Delta \Phi = \Phi(\omega') - \Phi(\omega)$ and $\Delta cost_{opt} = \inf(\omega \wedge r)$.

3 Migration on Two Points

In this section we consider the page migration problem when the metric space M has only two points, x and y. Without loss of generality we can assume that the distance between these points is 1. In this case, an offset function is a pair of numbers: a value at x and a value at y. We will write $(\omega(x), \omega(y))$ for the offset function ω. The reachable offset functions are those in which one value is 0 and the other is an integer from 0 to D.

Algorithm RAND. Suppose the current offset function is $(0, b)$, where $0 \leq b \leq D$. The algorithm EDGE will use the probability distribution that places mass $p_b = \frac{D+b}{2D}$ on x and $1 - p_b = \frac{D-b}{2D}$ on y. When the offset function is $(b, 0)$, EDGE is defined in the obvious, symmetric manner.

Theorem 2. EDGE *is* C_D-*competitive for* $C_D = 2 + \frac{1}{2D}$. *Furthermore, no on-line algorithm for the page migration problem on two points can be less than* C_D-*competitive.*

4 Uniform Spaces

Let $C_D = 2 + \frac{1}{2D}$. We will define a C_D-competitive randomized algorithm for a uniform space, *i.e.*, a metric space M where $d(x, y) = 1$ for all distinct $x, y \in M$. The algorithm is called UNIFORM. As in the previous section, UNIFORM is stable, and thus it can be described as a function that assigns a probability distribution on M to each offset function. The algorithm is an extension of the edge algorithm.

Let ω be the current offset function. For each $x \in M$ we define its *weight* as $k_\omega(x) = (D - \omega(x))/(D + \omega(x))$. We let UNIFORM have the distribution

$$\mu(x) \;=\; \frac{k_\omega(x)}{\sum_{y \in M} k_\omega(y)}.$$

If M has only two points then UNIFORM is identical to algorithm EDGE .

Theorem 3. *In a uniform space,* UNIFORM *is* C_D*-competitive for* $D = 1, 2$.

Conjecture: UNIFORM is C_D-competitive for each $D \geq 3$.

Subsequent to this work, Lund and Reingold have constructed a randomized algorithm that is C_D-competitive in any uniform space. Their algorithm is a different generalization of EDGE .

5 Trees

In this section we present a $(2 + \frac{1}{2D})$-competitive algorithm, FACTOR, for trees. This algorithm uses a technique called "factoring" that reduces one instance of the page migration problem on a tree to a number of instances of the problem on two-point spaces, one for each edge of the tree.

We define a finite metric space M to be a *tree* if its points are the vertices of an acyclic connected graph whose edges have nonnegative weights, and its distances are the weighted path lengths in that graph. We will denote an edge of M between vertices $x, y \in M$ by $\{x, y\}$, and the set of all edges of M by E. For $x, y \in M$, we denote the set of vertices that are on the unique path in M from x to y by $[x, y]$.

Let M be a tree. We refer to the page migration problem on M as PMP^M. We will show that PMP^M "induces" in a natural way an on-line problem PMP^e on each edge $e = \{x, y\}$, which we view as a space of two points at distance $d(x, y)$. The removal of e disconnects M into two subtrees M_x and M_y, one containing x and the other containing y. For each $z \in M$, we define $z^e = x$ if $z \in M_x$ and $z^e = y$ if $z \in M_y$. This defines a map from request sequences in M to request sequences in each $e = \{x, y\}$. Any algorithm $\mathcal{A}$ on M induces an algorithm $\mathcal{A}^e$ on e as follows: Whenever $\mathcal{A}$ moves its server from u to v, $\mathcal{A}^e$ moves a server from u^e to v^e (u^e may be equal to v^e). Notice that for this move $\mathcal{A}$ pays $D \cdot d(u, v)$ while $\mathcal{A}^e$ pays zero unless e is on the path from u to v, when it pays D times the length of e. Thus for any movement of the server, the cost incurred by $\mathcal{A}$ is the sum of all the movement costs incurred by the $\mathcal{A}^e$. Similary, any access cost incurred by $\mathcal{A}$ is the sum of all the remote access costs incurred by the $\mathcal{A}^e$. Thus for any request sequence, ϱ, $cost_{\mathcal{A}}(x, \varrho) = \sum_e cost_{\mathcal{A}^e}(x^e, \varrho^e)$. Clearly, if each $\mathcal{A}^e$ is C_D-competitive then $\mathcal{A}$ must be, as well.

We will construct an algorithm, FACTOR, such that each FACTOR^e is equivalent to EDGE on e.

For a given work function ω, and an edge $e = \{x, y\}$ we define the *slope of ω between x and y* as

$$S_\omega(x, y) = \frac{\omega(y) - \omega(x)}{d(x, y)}.$$

Note that $S_\omega(x, y) = -S_\omega(y, x)$. For simplicity, we will write $S(x, y)$ wherever ω is understood.

Algorithm FACTOR. We now define algorithm FACTOR for tree-like networks. Let ω be the current offset function. Then the probability that the server is on $x \in M$ is

$$\Pr_\omega[x] = 1 - \sum_{y \in N(x)} \frac{D - S_\omega(x, y)}{2D}.$$

Notice that if $\{x, y\}$ were the only edge in M, and $S(x, y) \geq 0$, then the distribution given above is the distribution given in Sect. 3 for an offset function which takes on value 0 on x and $S(x, y) \cdot d(x, y)$ on y. This ensures that FACTOR handles each single edge correctly.

Theorem 4. *If M is a tree then* FACTOR *is C_D-competitive on M.*

FACTOR can be extended to infinite trees, or to dynamic trees, that is trees that can grow branches during our algorithm's computation. It can also be used to design an optimally competitive randomized algorithm for any product of trees, by using one instance of FACTOR in each dimension [4]. The most common examples of such product topologies are hypercubes and meshes.

6 An Optimal Deterministic Algorithm for Continuous Trees

Let T be tree in which each edge e has some length $0 \leq l(e) \leq \infty$. Consider each edge e of T to be an interval of length $l(e)$ and consider the union of these edges, attached according to T. We call this space the *continuous tree based on T*. Any space which is obtained from some (weighted) tree in this fashion is called a continuous tree.

Suppose M is a continuous tree, and μ is any distribution on M. The *second moment* of μ is a function $I^\mu : M \to \mathbb{R}$ defined by

$$I^\mu(z) = \sum_x \frac{1}{2} d(x, z)^2 \cdot \mu(x).$$

We define $\bar{\mu}$, a *barycenter* of μ, to be any point where I^μ is minimized.

Algorithm BARY for continuous trees. If M is a continuous tree, and if ϱ is any sequence of requests, let μ be the distribution of our server on M given by the randomized algorithm FCTOR, after serving the sequence ϱ. After BARY serves ϱ, the server is deterministically at $\bar{\mu}$.

Theorem 5. *BARY is $(2 + \frac{1}{2D})$-competitive on any continuous tree.*

7 A Lower Bound

Black and Sleator [4] conjectured the existence of a 3-competitive deterministic page migration algorithm for all metric spaces and for all values of D. In this section we disprove this conjecture by demonstrating that for $D = 1$ no deterministic page migration algorithm for general metric spaces is c-competitive for any $c < \frac{85}{27}$.

Theorem 6. *If $c < \frac{85}{27}$, then there is no deterministic c-competitive algorithm for the page migration problem.*

8 An Algorithm for Three Points

The results of Black and Sleator [4] imply that there is a 3-competitive algorithm for each D and each metric space with two points. For $D = 1$, we extend it to three points in the theorem below. Our algorithm is based on a new technique called *forgiveness* that has been recently used in an 11-competitive algorithm for 3 servers [6].

Theorem 7. *Let M be a triangle, that is a metric space with only three points. Then there is a 3-competitive algorithm in M for $D = 1$.*

Four point spaces. One might wonder whether Black and Sleator's conjecture holds for spaces of fixed size. According to a computer program that we have written, there exists a 4-point metric space M such that no on-line algorithm for the page migration problem for $D = 1$ on M is 3-competitive. In fact, it is the same space used for the proof of Theorem 6.

References

1. B. Awerbuch, Y. Bartal and A. Fiat. Competitive distributed file allocation. *Proc. of 25th Symposium on Theory of Computation*, 1993, pages 164–173.
2. P. Berman, H. J. Karloff, and G. Tardos. A competitive three-server algorithm. In *Proc. 1st ACM-SIAM Symp. on Discrete Algorithms*, pages 280–290, 1990.
3. D. Black, A. Gupta, and W. Weber. Competitive management of distributed shared memory. In *Proceedings, Spring Compcon 1989*, pages 184–190. IEEE Computer Society, San Francisco, CA., 1989.

4. D. L. Black and D. D. Sleator. Competitive algorithms for replication and migration problems. Technical Report CMU-CS-89-201, Department of Computer Science, Carnegie-Mellon University, 1989.

5. M. Chrobak, H. Karloff, T. Payne, and S. Vishwanathan. New results on server problems. *SIAM Journal on Discrete Mathematics*, 4: 172–181, 1991. Also in *Proc. 1st ACM-SIAM Symp. on Discrete Algorithms*, pages 291–300, 1990.

6. M. Chrobak and L. L. Larmore, Generosity helps, or an 11-competitive algorithm for three servers. In *Proc. 3rd ACM-SIAM Symp. on Discrete Algorithms*, pages 196–202, 1992.

7. M. Chrobak and L. L. Larmore, The server problem and on-line games, In *Proceedings of the DIMACS Workshop on On-Line Algorithms, American Mathematical Society*, vol. 7, pages 11–64, 1992.

8. W. Crowther, J. Goodhue, E. Starr, R. Thomas, W. Milliken, and T. Blackadar. Performance measurements on a 128-node butterfly parallel processor. In *Proc. International Conf. on Parallel Processing*, IEEE Computer Society, pages 531–540, 1985.

9. L. W. Dowdy and D. V. Foster. Comparative models of the file assignment problem. *ACM Computing Surveys*, 14(2):287–313, 1982.

10. Amos Fiat, Richard Karp, Michael Luby, Lyle McGeoch, Daniel D. Sleator, and Neil Young. On competitive algorithms for paging problems. *Journal of Algorithms*, 12(4):685–699, 1991.

11. A. Karlin, M. Manasse, L. Rudolph, and D. Sleator. Competitive snoopy caching. *Algorithmica*, 3(1):79–119, 1988.

12. A. R. Karlin, M. S. Manasse, L. A. McGeoch, and S. Owicki. Competitive randomized algorithms for non-uniform problems. In *Proc. 1st ACM-SIAM Symp. on Discrete Algorithms*, pages 301–309, 1990.

13. M. Manasse, L. A. McGeoch, and D. Sleator. Competitive algorithms for on-line problems. *Journal of Algorithms*, 11: 208–230, 1990. Also in *Proc. 20th ACM Symposium on Theory of Computing*, pages 322–333, 1988.

14. G. Pfister et. al. The IBM research parallel processor prototype: Introduction and architecture. In *Proc. International Conf. on Parallel Processing*, IEEE Computer Society, pages 764–771, 1985.

15. P. Raghavan and M. Snir. Memory versus randomization in on-line algorithms. Research Report RC 15622 (No. 69444), IBM T. J. Watson Research Center, 1990. Also in *Proc. 16th International Colloquium on Automata, Languages, and Programming*, Lecture Notes in Computer Science vol. 372, Springer-Verlag, pages 687–703, 1989.

16. C. Scheurich and M. Dubois. Dynamic page migration in multiprocessors with distributed global memory. *IEEE Transactions on Computers*, 38(8):1154–1163, August 1989.

17. J. Westbrook. Randomized algorithms for multiprocessor page migration. In *Proceedings of the DIMACS Workshop on On-Line Algorithms, American Mathematical Society*, vol. 7, pages 135–150, 1992.

18. A. Wilson. Hierarchical cache/bus architecture for shared memory multiprocessors. In *Proc. 14th International Symp. on Computer Architecture*, pages 244–252. ACM SIGARCH/IEEE Computer Society, 1987.

19. O. Wolfson. A distributed algorithm for adaptive replication data. Technical Report CUCS-057-90, Department of Computer Science, Columbia University, 1990.

Memory Paging for Connectivity and Path Problems in Graphs *
(Extended abstract)

Esteban Feuerstein** and Alberto Marchetti-Spaccamela

Dipartimento di Informatica e Sistemistica, Università di Roma "La Sapienza",
via Salaria 113, I-00198 Roma, Italia.

1 Introduction

On-line problems require to serve a sequence of requests each of which is to be served without knowing future requests. Such problems arise for example in on-line memory management where an algorithm has to decide which page of memory must be ejected from fast memory when a page fault occurs. In the Paging Problem [15] we assume that there is a fast memory (the *cache*) that can contain at most k pages; if there is a request to a page that is present in the cache then the request can be answered with zero cost; otherwise a page fault occurs and it is necessary to move that page from secondary memory to the cache paying a unit cost. The decision on which page to eject must be taken on-line, without knowing future requests.

In this paper we study an extension of the paging problem to graph problems. The problems that we will consider extend the paging problem to the case in which each piece of information, eventually combined with other pieces, can be used to infer information not directly present in the cache. As an example, suppose you are given a graph and that the edges of its transitive closure are stored in memory. If you want to answer connectivity queries of the type "is i connected with j?" then if the cache contains positive information about the connectivity between i and a vertex h and between h and j it is possible to answer the query even if the cache does not contain the edge (i, j) .

Another extension with respect to the paging problem that we will consider is to the case in which the answer to a query is not a single piece of information stored in memory. As an example assume that the memory contains edges of the graph and one is interested in answering queries of the type "report a path from i to j" then it is necessary to have in memory all the edges of a path from i to j. Note that in general there could be many possible paths and, therefore, many possible answers.

During the last years considerable attention has been devoted to competitive analysis of on-line algorithms, particularly to extend and generalize the Paging Problem. This includes the weighted version of the paging problem [14], the problem of maintaining caches in a multiprocessor system [11], the k-server problem [5, 6, 8, 9, 12]. All above extensions to the paging problem are based in considering increasingly

---
* This work was partly supported by ESPRIT BRA Alcom II under contract No.7141, and by Italian Ministry of Scientific Research Project 40% "Algoritmi, Modelli di Calcolo e Strutture Informative".

** Supported by an ICSC-World Laboratory grant.

more complex models, but share essentially the same fundamental "individuality" property: in the case of paging, each request to a page of memory requires *that* page to be present in the cache; in the k-server problem *one* server must be present in the required location to serve the request.

In this paper we will consider problems in which the information present in the cache can be combined following certain rules to *infer* or *deduce* new information; or to cases in which the answer to a query is given by a set of items that must be present in the cache. The problems that we consider are a special case of the more general notion of Metrical Task Systems [4] that allows to model a great variety of on-line problems. However the generality of that approach implies that the results that have been proved are rather negative results. In fact, while for the different versions of the paging and the k-server problem competitive algorithms with small constant competitiveness coefficients have been found or conjectured to exist [5, 6, 8, 9, 11, 12, 14, 15], for Metrical Task Systems a lower-bound on the competitivity of any on-line algorithm has been proved that is linear in the number of different states [4]. In our case, that number is exponential in the size of the cache. However, the competitivity coefficients we obtain depend linearly or at most quadratically in the size of the cache and are either optimal or a constant factor away from optimality.

Besides their theoretical interest, the problems that we consider are significant for their application to the memory management problem of data structures for graphs. As an example, assume you have to answer a sequence of connectivity queries on a graph G with a huge number of vertices and edges, such that it may not be stored in main memory, or such that its information is distributed among the processors of a network. The strategy we propose consists in maintaining part of the information in a small and fast memory (the cache), with the objective of using that information to answer queries without accessing secondary memory (or at least reducing the number of such accesses), reducing the number of messages sent to other nodes in the network or reducing the time spent for the computation of the answer. Another application concerning communication in a network will be discussed in section 5.

To the best of our knowledge, there is not previous work on this subject, but a considerable amount of work on related matters can be found in the literature. Borodin et al. ([3]) considered the paging problem in the case in which the sequence of requests follows a pattern given by a previously known access graph. Nodine et al. ([10]) studied the speed-up that may be obtained using redundancy and efficient blocking in the case of searching in particular classes of graphs that are too large to fit in internal memory. The difference is that in our problem all queries *refer* to an underlying graph, but the sequence of requests does not follow any specified pattern. Attention has been also devoted to efficiently storing particular data structures in secondary memory in a dynamic environment ([13, 16]) and to hierarchichal memory models ([1, 2]). Our problem is different from the cited works in the sense that we do not have to minimize the number of accesses to secondary memory necessary to solve an off-line problem but to answer in an on-line fashion an arbitrarily long sequence of queries.

In the reminder of the paper we present a general framework in which different problems can be defined depending on the type of query we are interested in, the kind of information stored in memory, the cost function and the class of graphs that is considered (section 2). Section 3 is devoted to the case in which queries consist in requests of a path joining two vertices of an arbitrary graph that is stored in secondary memory. We show a $k(d+1)/2$ lower bound on the size of the cache for any algorithm to be competitive, where d is the diameter of the graph and k is the size of the adversary's cache. We also study this problem in the particular case of trees, giving lower and upper bounds for the competitivity of deterministic and randomized algorithms for different cost models. In section 4 we consider the particular case in which connectivity queries are asked. We establish a quadratic lower bound for the size of the cache, and a lower bound of k for the competitivity of any on line algorithm *with any cache size*. We also show a strategy for this problem that is a constant factor away from optimal. Section 5 presents optimal solutions to the connectivity problem in the particular case of connected graphs, that find direct application in a communication problem for high-speed computer networks. Section 6 depicts conclusions and open problems. For space reasons many proofs are not reported in this abstract. A full version of this article is available from authors.

2 Preliminaries

Given a graph G we want to answer sequences of queries $Connected(a,b)$, meaning "is there a path from a to b?", or $Path(a,b)$, meaning "give a path joining vertices a and b". Queries must be answered in an on-line fashion: the i-th query must be answered before the following queries are posed.

We are interested in strategies that minimize the number of accesses to secondary memory done to serve a sequence of requests. With this aim, we will analyze the behavior of algorithms that use two levels of memory: a small and fast level (the *cache*) whose size will be denoted k, and a secondary level of memory that allows to store the whole graph or its transitive closure, depending on the particular characteristics of the problem at hand. Each query can be answered in two possible ways that depend on whether it is possible to answer using only the information stored in cache (nothing is charged) or not (a cost is charged).

We can limit our attention to algorithms that store the same kind of information in cache and in secondary memory. In particular we will consider the following cases:

1. we store in secondary memory and in the cache *edges of the graph*,
2. we store in secondary memory and in the cache *edges of the transitive closure of the graph*. This means that each pair of nodes has a *yes-edge* if they are in the same connected component of the graph or a *no-edge* if they are in different connected components.

As for the cost charged to an algorithm in the case it cannot answer a query with the information present in its fast memory, there are different possibilities, each one of which may be more adapted to different situations. We will consider three different cost models:

1. *Full-cost model*: when a path joining a and b is not present in cache, charge the algorithm the length of the path that must be looked for in secondary memory.
2. *Partial-cost model*: when a path joining a and b is not present in cache, charge the algorithm the number of edges it brings into the cache to have a path between the query nodes.

3. *0/1-cost model*: when a path joining a and b is not present in cache, charge the algorithm a unit cost.

The three cost models may be suitable in the case where edges of the graph are stored, while the 0/1-cost model is particularly adapted to the case of *Connected* queries and where the cache contains information about the graph's transitive closure. From the above framework, numerous different problems arise, depending on all the combinations of the parameters we have defined, applied to different classes of graphs.

An on-line algorithm for a certain problem is c-competitive [11] if the cost afforded by it to serve *any* sequence of requests is less than c times the cost charged to the optimal algorithm for that sequence plus a constant.

For the paging problem it has been shown [15] that a simple First In First Out (FIFO) rule is optimal from a competitive point of view. In fact if both the algorithm and the adversary have the same memory of size k then FIFO achieves a competitive ratio of k and this ratio is optimal since k is also a lower bound.

Note that the lower-bounds for the competitivity of on-line strategies for the Paging problem hold for all the problems we introduce, independently from the cost model, the kind of information stored or the particular class of graphs chosen. This implies that, for all the considered problems, k is a lower bound for deterministic algorithms and H_k (the k-th harmonic number) is a lower bound for randomized algorithms against an oblivious adversary [7] (see [14] for the definitions of the different kinds of adversaries).

In the proofs of competitiveness we will often consider an adversary that serves the same sequence of requests as the on-line algorithm. We will use the following lemma, whose standard proof is left to the reader. This or similar lemmas have been used in previous works on on-line algorithms (see for example [6]).

Lemma 2.1 *Lets call C_{ADV} and C_{ALG} denote the total costs charged respectively to an adversary (ADV) and on-line algorithm ALG to serve a sequence of requests. Suppose that Φ is a nonnegative potential function with initial value Φ_0 such that*

1. *when the adversary makes a move, the increment in the potential is not more than α times the cost paid by the adversary, and*
2. *when the on-line algorithm serves a request, Φ decreases by at least β times the cost paid by the algorithm to serve the request*

Then $C_{ALG} \leq (\alpha/\beta)C_{ADV} + \Phi_0$, and hence ALG is (α/β)-competitive.

3 The Path Paging Problem

In this section we consider the following problem: suppose to have a connected graph, and to have to answer queries of the form $Path(a, b)$, meaning "give a path joining vertices a and b". We call this problem the *Path Paging problem*. An on-line algorithm for this problem maintains a cache consisting of k edges of the graph, and serves a query $Path(a, b)$ without cost if its cache already contains edges of a path joining a and b. If such edges are not present, then a set of edges forming a shortest-path from a to b is provided to the on-line algorithm, a cost been charged to it that depends on the chosen cost model.

With this definition, the problem makes sense only if the diameter of the considered graph is bounded by k, the size of the cache. In the reminder of this section we will study the Path Paging problem. The first result we give is a lower bound for the size of the cache of an on-line algorithm for it to be competitive in the case of arbitrary connected graphs, for all cost models. Then we analyze lower and upper bounds for the the Path Paging problem under the different cost models, both for deterministic and randomized algorithms in the particular case of trees.

We will show that no strategy can achieve a bounded competitiveness coefficient against an adversary with a cache of size k if its own cache is of size $K < k(d+1)/2$, where d is the diameter of the graph.

Theorem 3.1 *No deterministic strategy for the Path Paging problem for arbitrary graphs with a cache of size $K < k(d+1)/2$ can achieve a bounded competitive ratio against an adversary with cache size k.*

3.1 The Path Paging problem for trees

For the Full-cost model the first result we will prove is a somehow negative result, that is, high lower bounds both for deterministic and randomized algorithms. However, we also prove that the FIFO strategy is almost optimal in this model. FIFO consists in evicting the edges that have been present in cache for the longest time each time a new edge has to be brought into the cache. Next theorem shows that FIFO is at most a factor of 2 away from optimal for this problem.

Theorem 3.2 *The FIFO algorithm is $(k^2 - k + 4)/2$-competitive for the path-paging problem for trees under the full-cost model.*

Proof: Consider the sequence of queries divided in phases, such that the first phase starts with the first query that implies a fault of FIFO, and a phase starting with query σ_i ends with query σ_j such that j is the minimum integer such that the paths referred in the queries $\{\sigma_i, \sigma_{i+1}, \ldots, \sigma_{j+1}\}$ require more than k edges to be answered. During a phase the adversary faults at least once, and hence its cost is at least 1. The paths requested during a phase involve at most k edges, and it is possible to see that the maximum that FIFO may pay in such a phase is exactly $(k^2 - k + 4)/2$. $\square$

If we consider randomized algorithms, it is possible to prove the following

Theorem 3.3 *No randomized on-line strategy for the path-paging problem under the full-cost model is c competitive with $c < (k+1)/2$, even if oblivious adversaries are considered.*

Proof(sketch): We shall consider a randomized oblivious adversary (that is an adversary that generates a whole random sequence of queries before starting to serve it). Let C_A and C_{ADV} be the random variables that denote the costs of algorithm A and ADV respectively. Consider an initial configuration in which ADV and A have in cache $k - 2$ edges forming a path between vertex o and vertex x plus two edges (o, y) and (o, z). Consider a random sequence of queries made of an arbitrarily large number of phases, each phase done in this way: the adversary tosses a fair coin, if the coin is HEAD the adversary asks for edge (o, w), evicts edge (o, z) and requests $q(x, y)$; if the coin is TAIL the adversary asks for edge (o, w), evicts edge (o, y) and requests $q(x, z)$. After the phase vertex w is "renamed" z or y so as to always have again the initial configuration. The expected cost charged to the adversary in each phase is 1 (in fact 1 is the exact cost).

Let C_i be the expected cost charged to algorithm A during phase i. Then

$$E[C_i] = E[C_i|HEAD]*1/2 + E[C_i|TAIL]*1/2 = (E[C_i|HEAD]+E[C_i|TAIL])*1/2$$

Let p_y, p_z be respectively the probabilities that A evicts edge $(o, y), (o, z)$ when the fault on (o, w) occurs. Then, we have that $E[C_i|HEAD] = 1 + (1 - p_z)(k - 1)$ and $E[C_i|TAIL] = 1 + (1 - p_y)(k - 1)$. Since $p_y + p_z \leq 1$ we have

$$E[C_i] = (2 + (2 - p_z - p_y)(k - 1)) * 1/2 \geq (2 + k - 1) * 1/2 = (k + 1)/2$$

The above inequality implies that the competitiveness coefficient of any algorithm, must necessarily be greater or equal than $(k + 1)/2$. □

The somehow disappointing results for the full-cost model are overcome when we consider the partial-cost model, in which the cost charged to an algorithm for a query that is not possible to answer with the contents of the cache is the number of missing edges that complete the requested path.

In the following we propose a strongly competitive algorithm, namely a variation of the *Least Recently Used* (LRU) algorithm for the Paging Problem ([15]). LRU evicts the edges that have not been used for the longest time when space for a new edge is needed in the cache. The strategy can be implemented in the following way: LRU maintains for every edge of the tree an integer valued variable $a[e]$, in the range $1..k$ that is initialized to 0 for every edge not present in the cache, and to the values $1..k$ for all the edges present in the cache. The intuition behind this implementation is that higher values of variable $a[e]$ correspond to more recently used edges.

Theorem 3.4 *LRU is k-competitive under the partial-cost model.*

Proof(sketch): We use Lemma 2.1 and the potential function $\Phi = \sum_{e \in ADV} k - a[e]$. We suppose that the adversary serves each request first, and then it poses it to LRU, and we analize the variation of the potential in the different situations that may arise. □

In the reminder of this section we will show that a randomized strategy similar to the marking algorithm developed in [7] for the paging problem achieves the same competitivity factor, namely $2H_k$, where H_k denotes the kth harmonic number. The strategy can be described as follows: The algorithm maintains a set of marked edges. Initially the marked edges are exactly those that are present in the cache. After each request, the marks are updated, then some edges are evicted to make place to other edges if necessary, as follows: Each time a query $Path(a, b)$ is requested, all the edges in the path are marked. When more that k edges are marked, all the marks except those in the most recently marked edges are erased. If the requested path is completely present in cache, then nothing is done. If there are $x > 0$ edges missing, then randomly choose among the unmarked edges x edges to evict, to make place to the x missing edges.

Theorem 3.5 *M is $2H_k$ competitive under the partial-cost model.*

We recall that in [7] it has been proved that H_k is a lower bound for the competitivity of any algorithm for the Paging problem, and hence it is also a lower bound for the Path paging problem for trees. Therefore the proposed algorithm is at most a factor of 2 from optimality.

If the 0/1-cost model is used, we can prove that

Theorem 3.6 *LRU and FIFO are k-competitive for the Path Paging problem under the 0/1 cost model.*

Proof*(sketch)***:** The proof uses Lemma 2.1 and the following potential function: $\Phi = \max_{e \in LRU, e \notin ADV} a[e]$. Without loss of generality, we suppose that the adversary pays for the evicted edges instead of paying for the incoming ones. □

We recall that k is a lower bound for the competitiveness of any on-line algorithm, and hence LRU and FIFO are optimal.

4 Connectivity Paging

In this section we will consider the case in which we have to answer *Connected* queries about an arbitrary graph, from now on this problem will be referred to as the *Connectivity Paging problem*. We will analyze the case in which the cache contains a subset of the pairs of nodes, together with the information whether the two nodes lie in the same connected component of the graph or not. In the first case a pair (i, j) is called a *yes-edge*, and in the latter it is called a *no-edge*. Both kinds of pairs will be referred to as *edges* when no confusion arises.

A query *Connected*(i, j) may be answered with no cost if there is a path in the cache containing at most one no-edge between i and j. Obviously, if there is a path containing no no-edges the answer to *Connected*(i, j) will be yes (pair (i, j) is called a *deduction*), conversely if there is a path containing one no-edge, the answer to the query will be negative (in this case we call pair (i, j) a *no-deduction*). Any other kind of path (that is, paths with more than one no-edge) does not provide information to answer the connectivity query between i and j. If the cache does not contain a path between i and j that allows to *deduce* the query in the previous ways, then an access to secondary memory must be done, charging a cost to the algorithm. We require that after each query *Connected*(i, j), the information about the connectivity between i and j is present in the cache. That is, if i and j are connected the cache contains the pair (i, j) or a path between them, and if i and j are not connected, the cache contains either the negative pair (i, j) or two paths P_{iv}, P_{wj} and the negative pair (v, w). We also assume that if a fault occurs on serving query *Connected*(i, j), then the query is served by bringing into the cache the edge (i, j).

A natural consequence of our approach is that the 0/1-cost model is the most reasonable to apply in this case. First of all, we show that no strategy can achieve a bounded competitiveness coefficient against an adversary with a cache of size k if its own cache is of size $K < f(k)$, where $f(k)$ is the maximum number of no-deductions possible with k edges. We also prove that $f(k) \simeq k^2/2$.

Theorem 4.1 *No deterministic strategy with a cache of size $K < f(k)$ can be competitive against an adversary with cache size k.*

The maximum number of NO-deductions possible with a cache of size k, namely $f(k)$, appears to be crucial to our analysis, as its value appears not only in the expression denoting the lower bound on the size of the cache of an algorithm for it to be competitive but also in the competitivity coefficient achieved by FIFO.

Lemma 4.2 *Configurations that maximize the number of NO-deductions consist in a certain number of spanning subtrees of the graph (connected components or CC for short), whose sizes differ at most in 1, and all the no-edges "connecting" all possible pairs of connected components, i.e. one no-edge between a vertex in CC a and one vertex in CC b for every pair of CCs a and b.*

Proof(sketch): The proof proceeds by showing a transformation between configurations that is monotonous in the number of possible no-deductions and that has fixpoints precisely in configurations of the claimed type. □

It remains to show which is the optimal size of the CCs as a function of k. To do that, we maximize the number of no-deductions possible in a situation as the one described above, as the number l of connected components varies. In the following proposition we give an upper and a lower bound for $f(k)$.

Proposition 4.3 $\frac{k^2}{2} - \frac{3}{4}k^{5/3} < f(k) < \frac{k^2}{2}$. *Moreover, the value of l that maximizes the number of CCs is $O(k^{1/3})$.*

The second negative result we prove in this section is a lower-bound on the competitivity of any on-line algorithm, with a cache of any size.

Theorem 4.4 *No on-line algorithm for the Connectivity paging problem for arbitrary graphs can be c-competitive against an adversary with cache size k with $c \leq k$.*

Proof: Consider a configuration in which the adversary has $k-1$ yes-edges forming a tree T of k nodes and that the on-line algorithm has no edges joining nodes of T. If the adversary asks for a no-edge (i, j) with j in the tree, it will automatically be able to answer with no further cost k queries of the form $Connected(i, t), t \in T$. As the on-line algorithm has no yes-edges connecting nodes in T, it will have a fault for each one of those queries. A similar sequence of requests can start after all k queries have been posed to the on-line algorithm, and this can be repeated forever. □

In the following, we will prove an upper bound for the competitivity of FIFO for the Connectivity Paging problem. For this, we will consider the following implementation of FIFO: let $f'(k) = f(k)+$maximum number of positive edges in an optimal configuration of ADV. Note that $f'(k) \leq f(k) + k - 1$. An array a is maintained, with values from 1 to $f'(k)$, such that, whenever FIFO has a fault, the values of a are decremented for every e such that $a[e] > 0$. The edge whose a becomes exactly $f'(k) - f(k)$ is the one evicted.

Theorem 4.5 *FIFO is $f'(k)$-competitive against an adversary with cache size k.*

Proof*(sketch)*: We use Lemma 2.1 and the potential function $\Phi = \max_{e \notin D(ADV)} a[e]$, where $D(A)$ denotes the set of pairs of nodes such that the adversary may answer without paying queries about their connectivity in a certain configuration of its cache. Note that whenever FIFO has a fault, there must be at least one edge e with $a[e] \geq 0$ such that $e \notin D(ADV)$. □

5 Link paging

An important feature due to the use of high speed network is that the cost associated to the communication does not significantly depend on the length of the path but only on the eventual reconfiguration of the network. Suppose to have a computer network in which all connections between pairs of nodes are potentially possible, but actually there are at most k pairs of nodes connected at any moment. The requests are pairs of nodes to be connected, and one request is served with no cost if there is a path between the nodes in the current configuration of the k links, otherwise both nodes are linked by switching one of the k physical links to connect the requested pair, and the cost is 1. We call this problem the *Link Paging* problem. An algorithm for the Connectivity Paging problem in the particular case in which the graph is complete is also an algorithm for this problem.

Theorem 5.1 *FIFO is $\frac{K}{K-k+1}$-competitive against adversaries with cache size $k \leq K$, and this is optimal.*

Proof*(sketch)*: Let $G_1 = (V_1, E_1)$ and $G_2 = (V_2, E_2)$ be the graphs induced by the set of edges of the algorithm and of the adversary, respectively. Let $H = (V_1 \cup V_2, E_1 \cup E_2)$. The potential function in this case is defined as follows: $\Phi = kK - \sum_{e \in S} a[e]$, where the set S is defined by the following procedure:

```
S := ∅
While there is a cycle in H do
    e := edge in FIFO's cache belonging to a cycle s.t. a[e] is minimum
    S := S ∪ {e}; H := H − {e}
```

Note that $|S| \leq k$ and $(|S| = k) \Rightarrow (D(A) \subseteq D(F))$, where $D(A)$ and $D(F)$ denote the set of pairs that may be connected by the adversary and FIFO respectively. We show that when the adversary has a fault, $\Delta\Phi \leq K$ and when FIFO has a fault $\Delta\Phi \leq -(K - k + 1)$. □

6 Conclusions and open problems

In this paper we have proposed a framework for studying memory management problems for maintaining connectivity information and paths on graphs. We have proved upper and lower bounds for numerous problems within the framework. Many open problems require further investigation.

One of the open problems is that of giving upper and lower bounds on the competitivity of on-line algorithms for the Path paging problem on arbitrary connected graphs. An interesting matter currently under study is that the lower bound of theorem 3.1 can be beaten if we consider a different model in which at every fault on query $Path(a, b)$ an on-line algorithm may request any particular set of edges containing a path from a to b together with other paths.

As far as the Connectivity Paging problem, it remains an open question whether FIFO achieves a better competitive ratio that the one we have proved. In fact, we conjecture that the competitive ratio of FIFO is $f(k)/2$, and that that value is a lower bound. An interesting extension of the Connectivity Paging problem consists in considering algorithms that in front of a fault on query $Connected(a, b)$ are not forced to bring to their cache the edge (a, b) of the transitive closure but may look for other edges that allow to answer the query. Note that the proof of the lower bound on the size of the cache in theorem 4.1 does not hold in this model.

Acknowledgments: We would like to thank Giorgio Ausiello, Xavier Messeguer and Nicola Santoro for useful discussions about this work.

References

1. A. Aggarwal, B. Alpern, A.K. Chandra and M. Snir, A model for hierarchical memory, *Proc. 19th Annual ACM Symposium on Theory of Computing* 305-314 (1987).
2. A. Aggarwal and A.K. Chandra, Virtual memory algorithms, *Proc. 20th Annual ACM Symposium on Theory of Computing* 173-185 (1988).
3. A. Borodin, S. Irani, P. Raghavan and B. Schieber, Competitive paging with locality of reference, *Proc. 23rd Annual ACM Symp. on Theory of Computing* 249-259 (1991).
4. A. Borodin, N. Linial, and M. Saks, An optimal online algorithm for metrical task systems, *Proc. 19th Annual ACM Symposium on Theory of Computing* 373-382 (1987).
5. M. Chrobak, H. Karloff, T. Payne, and S. Vishwanathan, New Results on server problems, *Proc. 1st. Annual ACM-SIAM Symp. on Discrete Algorithms* 291-300 (1990).
6. M. Chrobak and L. Larmore, An optimal online algorithm for k servers on trees, *SIAM Journal on Computing* 20 144- 148 (1991).
7. A. Fiat, R.M. Karp, M. Luby, L.A. McGeoch, D.D. Sleator and N.E. Young, Competitive paging algorithms, *Journal of Algorithms* 12 685-699 (1991).
8. A. Fiat, Y. Rabani and Y. Ravid, Competitive k-server algorithms, *Proc. 31th. Annual Symposium on Foundations of Computer Science* 454-463 (1990).
9. E.F. Grove, The harmonic online k-server algorithm is competitive, *Proc. 21th Annual ACM Symposium on Theory of Computing*, 260-266, (1991).
10. M. Nodine, M. Goodrich, J. S. Vitter, Blocking for external graph searching, *Tech. Rep. CS-92-44* (1992).
11. A. Karlin, M. Manasse, L. Rudolph and D. Sleator, Competitive snoopy caching, *Algorithmica* 3 79-119 (1988).
12. M.S. Manasse, L.A. McGeoch and D. Sleator, Competitive algorithms for on-line problems, *Proc. 20th ACM Symp. on Theory of Computing* 322-333 (1988).
13. M. Overmars, M. Smid, M. de Berg and M. van Kreveld, Maintaining range trees in secondary memory, part I: partitions, *Acta Informatica* 27 423-452 (1990).
14. P. Raghavan and M. Snir, Memory versus randomization in on-line algorithms, *IBM Research Report* RC 15622 (1990).
15. D. Sleator and R.E. Tarjan, Amortized efficiency of list update and paging algorithms, *Comm. ACM* 28 202-208 (1985).
16. M. Smid and P. van Emde Boas, Dynamic data structures on multiple storage media, a tutorial, *Technical Report Universitat des Saarlandes* a 15/90, (1990).

Randomized Competitive Algorithms for Successful and Unsuccessful Search on Self-adjusting Linear Lists*

Lucas Chi Kwong Hui[1] and Charles U. Martel[2]

[1] Department of Information Systems & Computer Science,
National University of Singapore, Singapore 0511,
REPUBLIC OF SINGAPORE (lhui@iscs.nus.sg)
[2] Department of Computer Science, University of California, Davis,
Davis, CA 95616, USA (martel@cs.ucdavis.edu)

Abstract. This paper studies the classical Dictionary problem using a self-adjusting linear list. We design and analyze randomized, on-line algorithms for a sequence of successful and unsuccessful searches which are competitive with off-line algorithms. Our algorithms combine our ps bit technique which speeds up unsuccessful search with the randomized move-to-front scheme of Reingold, Westbrook, and Sleator, which they used to speed up successful search.

1 Introduction and Summary

The operations performed by the abstract data type *Dictionary*, insert, delete and find, are among the most fundamental for good algorithm design. One important focus has been to design dictionaries which have faster lookup times for more frequently accessed elements. Since the access frequencies are usually unknown and may vary over time, it is attractive to use *self-adjusting* data structures which dynamically change to reflect the observed access pattern. There has been a great deal of work on self-adjusting dictionaries [1, 2, 3, 4, 5, 6, 7, 8].

Probably the most classical self-adjusting dictionary implementation is an unsorted linear list. Arranging the items in order of *decreasing access probability* gives an optimal static linear list. For self-adjusting linear lists, Sleator and Tarjan [7] have shown that considering successful searches, insertions and deletions only, the cost of the Move-To-Front (MTF) in a linear list is within a factor of 2 of a large class of self-adjusting schemes, even offline strategies which know all future requests. However, their setting (like most of the other analyses of self-adjusting linear lists) considers only successful search. Unsuccessful searches require a complete scan of the list and are thus very expensive. Considering unsuccessful searches is particularly important for dynamic lists which allow insertions, as an insertion is usually preceded by an unsuccessful search to prevent duplicate list elements.

---

* This work was supported by US National Science Foundation Grant CCR 90-23727

In [3], an online self-adjusting linear list algorithm, called *MP* is presented. Using a modified version of the move-to-front strategy, and uses only two extra bits per key as auxiliary data structures (the *ps bits*), MP supports fast successful and unsuccessful searches. In particular, if there are many $find(x)$ operations where x is not in the list, then the average cost for processing each $find(x)$ is low. The cost of MP is within a factor of 3 of a large class of self-adjusting offline strategies [3].

For successful searches only, randomization can improve the result in [7]. The expected cost of a randomized version of MTF, BIT [4, 6], is at most 1.75 times the cost of a class of offline (oblivious) self-adjusting adversaries. A refined version of BIT has expected cost no greater than $\sqrt{3} \approx 1.732$ times the cost of an offline (oblivious) optimal self-adjusting adversary.

In this paper we adopt the same randomization technique of BIT to MP, and we show that the expected cost of the new algorithm, BP, is at most 2.5 times the cost of an offline (oblivious) adversary, considering successful and unsuccessful searches as well as insertions. However, we do not consider deletions, since we cannot yet analyze our algorithm's performance for deletions.

2 Setting

We represent our dictionary by a linear list. We process a $find(x)$ request by scanning the list until we hit x or until we determine that x is not in the list. For any key x, let x_p denote the *predecessor* of x, the largest key in the list smaller than x, and let x_s denote the *successor* of x, the smallest key in the list greater than x. The predecessor of the smallest key in the list, and any smaller value is $-\infty$. Similarly the successor of the largest key in the list, and any larger value is $+\infty$.

We only consider algorithms that have the following behavior: During a $find(x)$ operation, if x is *not* in the list, then the algorithm will stop after finding both x_p and x_s (conceptually we treat $+\infty$ and $-\infty$ as if they are at the front of the list, so they are "found" when we start the search). After x_p and x_s are found the algorithm is free to move x_p and x_s forward to any desired position. The action of moving x_p (or x_s) one position forward is called a *free exchange*. If x is in the list, the algorithm will stop after finding x, and can move x forward by free exchanges. The algorithm can also pay one unit cost to swap two neighboring keys. This is called a *paid exchange*.

To insert key x, we first do a find(x) and then add x to the head of the list if the find is unsuccessful. Otherwise it is treated as a successful $find(x)$ operation. A deletion of x also starts with a find, and key x is deleted if the find is successful. Otherwise it is treated as an unsuccessful find operation.

Our cost measure is as follows. The cost of a find is the number of elements in the list traversed. The cost of an insertion is the cost of the associated find plus one (for adding the new item). The cost of a deletion is just that of the associated find. Finally, the cost of a request is the sum of all costs due to paid exchanges, and traversal of the list.

To simplify the notation we write $find([x_p, x_s])$ as a short form for finding a key value x not in the list. Note that we do not assume the algorithm is given x_p, x_s. We write $find([x_p, x_s])$ for $find(x)$ when x is not in the list for descriptive purposes only.

The MP algorithm MP [3] is a deterministic linear list algorithm which uses *ps bits*. Each list item x has two extra bits, the p bit and the s bit. If x_p is ahead of x in the list then the p bit of x is 0; otherwise the p bit of x is 1. The s bit of x is defined similarly with respect to x_s. We use the following result in [3]:

Lemma 1. *With the ps bits, for an unsuccessful search $find([x_p, x_s])$ we can conclude x is not in the list once we have found both x_p and x_s.*

After a successful search $find(x)$, MP moves x to the front of its list. After an unsuccess search $find([x_p, x_s])$, MP moves both x_p and x_s to the front, preserving the order of x_p and x_s. For an insertion $insert(x)$, MP first searches the list to make sure that x is not in the list (like an unsuccessful search). If x is not in the list, then MP inserts x in the front of the list. MP is 3-competitive to offline adversaries which finish an unsuccessful search $find([x_p, x_s])$ by finding both x_p and x_s.

Competitive analysis of randomized algorithms Let $A(\sigma)$ denote the cost of a list update algorithm A in serving all requests in an access sequence σ, and $OPT(\sigma)$ be the minimum cost to serve σ. A deterministic algorithm, A (like MTF or MP), is *c-competitive* if there exists a b such that for all size lists and all request sequences σ, $A(\sigma) \leq c \cdot OPT(\sigma) + b$.

The competitiveness of a randomized list update algorithm can be defined with respect to three kinds of adversaries: *oblivious, adaptive online,* and *adaptive offline* [6]. The *oblivious* adversary generates a complete request sequence before the online algorithm begins to process it. A randomized online algorithm, A, is *c*-competitive against an oblivious adversary if there is a constant b such that for every finite request sequence σ, $E[A(\sigma)] \leq c \cdot OPT(\sigma) + b$. The expectation is over the random choices made by A. Throughout this paper, we compare our algorithm to oblivious adversaries. We use "*c*-competitive" as an abbreviation for "*c*-competitive against an oblivious adversary."

The BIT randomized algorithm The BIT algorithm in [4, 6] is a randomized version of the MTF algorithm in [7]. BIT is 1.75-competitive to oblivious offline adversaries. For each item x in the list, BIT use exactly one random bit, $b(x)$. The initial value of $b(x)$ is set randomly. From then on BIT runs completely deterministically: after finding x, BIT first complements $b(x)$, and then moves x to the front if $b(x) = 1$. BIT shares the same drawback as MTF: it does not handle unsuccessful searches efficiently. Our BP algorithm is an extension of BIT to handle unsuccessful searches.

3 Competitive analysis of the BP algorithm

We associate three bits with each key x in BP's list. Two of them are the *ps bits* as described in algorithm MP. By lemma 1, we can finish an unsuccessful search $find([x_p, x_s])$ by finding both x_p and x_s. The third bit, $b(x)$, is used as the random bit. Initially $b(x)$ is set to 0 or 1 with equal probability. After a successful $find(x)$, we toggle $b(x)$. If $b(x)$ changes to 1 we move x to the front, otherwise the list remains unchanged. For an unsuccessful search, for each of the two boundary keys $(x_p$ and $x_s)$, we toggle its random bit; if a key's random bit changes to 1 we move it to the front, otherwise it remains in its position. If both boundary keys have to be moved to front, BP preserves their relative order. The ps bits are maintained with constant extra time using the techniques described in [3].

We are going to prove that BP is 2.5-competitive to OPT, the optimal offline adversary which finishes an unsuccessful search by finding the two boundary keys. In §3.1 and §3.2 we give the proof for successful and unsuccessful searches only. In §3.3 we extend the analysis to include insertions. The analysis of deletion is still an open issue and is not discussed in this paper.

Given an access sequence σ, we compare $E[BP(\sigma)]$ to $OPT(\sigma)$. We assume that initially BP and OPT have the same list. Define a pair of keys (y, x) to be an *inversion* if y occurs before x in BP's list while y occurs after x in OPT's list. An inversion (y, x) is called a *type I* inversion if $b(x) = 0$ and is called a *type II* inversion if $b(x) = 1$. Let ϕ_1 denote the number of type I inversions and ϕ_2 denote the number of type II inversions. Define the *potential function* $\Phi = 2\phi_2 + \phi_1$. Φ is initially zero and Φ is always non-negative. For accounting purpose, we assume that for each request, (Step 1) first BP serves the request (and performs the move-to-front actions), then (Step 2) OPT serves the request and performs its free exchanges, then (Step 3) OPT performs any paid exchanges.

Let BP_j be the expected amortized cost of BP for the jth access, and OPT_j be the cost of OPT. Note that given σ and the initial list, OPT_j is a fixed value which does not depend on BP's random bits. BP_j consists of two parts: (1) the expected cost of BP in accessing the list, denoted as $\mathcal{A}_j^{bp}$, and (2) the expected change of potential function, denoted as $\overline{\Delta\Phi_j}$.

$\overline{\Delta\Phi_j}$ is divided into 3 parts, corresponding to the expected change in Φ (i) from BP's move-to-front self-adjusting actions, denoted as $\overline{\Delta\Phi_j}^{mtf}$, (ii) from OPT's free exchanges, denoted as $\overline{\Delta\Phi_j}^{fe}$, and (iii) from OPT's paid exchanges, denoted as $\overline{\Delta\Phi_j}^{pe}$. Note that BP_j, $\mathcal{A}_j^{bp}$, $\overline{\Delta\Phi_j}^{mtf}$, $\overline{\Delta\Phi_j}^{fe}$, and $\overline{\Delta\Phi_j}^{pe}$ are expected values. Similarly, OPT_j consists of two parts: (1) OPT' cost in accessing the list, denoted as $\mathcal{A}_j^{opt}$, and (2) the cost of OPT's paid exchanges, denoted as $\mathcal{P}_j$. Both $\mathcal{A}_j^{opt}$ and $\mathcal{P}_j$ are constants.

We prove that BP is 2.5-competitive to OPT for successful and unsuccessful searches only (theorem 7) in §3.1 and §3.2. This proof adopts the proof technique in [4, 6].

3.1 Successful search $find(x)$

For any integer j and item x, let $b_j(x)$ be the value of $b(x)$ after the jth access ($b_0(x)$ is the initial random value of $b(x)$). Assume access j is a successful search

$find(x)$. Let k be the position of x in OPT's list, then $k = A_j^{opt}$. Let R be the expected number of inversions (w, x) at the access time.

(Step 1: BP serves the request) We want to prove

$$A_j^{bp} + \overline{\Delta\Phi_j}^{mtf} \leq 1.75 A_j^{opt} - 0.75. \tag{1}$$

We have $A_j^{bp} \leq k + R$. Let $\overline{\Delta\Phi_j}^{mtf} = A + B + C$ where A is the expected change of Φ due to new inversions created during the access, B is the expected change of Φ due to old inversions removed, and C is the expected change of Φ due to old inversions that change type, thus $A_j^{bp} + \overline{\Delta\Phi_j}^{mtf} \leq k + R + A + B + C$. We first show $B + C = -R$. If $b_{j-1}(x) = 1$ then x stays in place and $b_j(x)$ is 0; $B = 0$ and $C = -R$, since each inversion (w, x) goes from type II to type I. If $b_{j-1}(x) = 0$ then x moves to front and removes all inversions (w, x); $B = -R$ since each inversion is of type I and $C = 0$. In both cases $B + C = -R$ and hence $A_j^{bp} + \overline{\Delta\Phi_j}^{mtf} \leq k + A$.

To consider A we need the following lemmas, analogous to the treatment in [4].

Lemma 2. *For any item x and any j:*
(i) $b_j(x)$ is equally likely to be 0 or 1,
(ii) $b_j(x)$ is independent of the position of x in OPT's list, and
(iii) $b_j(x)$ and $b_j(y)$ are independent for $x \neq y$.

Proof. Since OPT is an oblivious adversary, its choice of the sequence, and its algorithm for satisfying the requests (the optimal one) are all chosen before the initial random bits of BP are chosen.

The $b_0()$ values are chosen uniformly at random. If access j is an access to the key x, the gap after x, or the gap before x, then $b_j(x)$ is the complement of $b_{j-1}(x)$, otherwise $b_j(x) = b_{j-1}(x)$. Let $N_{x,j}$ be the total number of times that x, the gap before x, or the gap after x was accessed in accesses 1 to j, then

$$b_j(x) = (b_0(x) + N_{x,j}) \bmod 2.$$

Given x and j, $b_0(x)$ is uniformly distributed and $N_{x,j}$ is a constant, thus the distribution of $b_j(x)$ remains uniform. All three parts of the lemma follow. □

Note that lemma 2 only says that $b_j(x)$ is independent of the position of x in OPT's list, but $b_j(x)$ is *not* independent of x's position in BP's list.

Lemma 3. *Let H be the expected increase in Φ due to new inversions created by moving a key x to the front of BP's list; k' be the number of keys ahead of x in OPT's list; and Z be the expected increase of Φ due to a newly created inversion (x, z_i) where $b(z_i)$ is unknown, then $H \leq k' \cdot Z$.*

Proof. Let x be moved to the front of BP's list. In the worst case a new inversion (x, z_i) is created for each of the k' items z_i preceding x in OPT's list, so $H \leq k' \cdot Z$. □

Lemma 4. *Let Z be defined as in lemma 3, then $Z = 1.5$.*

Proof. Z is 1 if $b(z_i) = 0$, and is 2 if $b(z_i) = 1$. By lemma 2 $Z = (\frac{1}{2}\cdot 2)+(\frac{1}{2}\cdot 1) = 1.5$. $\square$

Now we compute A. New inversions can only be created when $b_{j-1}(x) = 0$, so x moves to front. By lemmas 2, 3, and 4, $A = \frac{1}{2}H \le \frac{1}{2}k'1.5$ where $k' = k-1$, so $A \le \frac{1}{2}(k-1)1.5 = 0.75k - 0.75$. Thus

$$A_j^{bp} + \overline{\Delta\Phi}^{mtf} \le k + A$$
$$\le 1.75k - 0.75$$
$$= 1.75A_j^{opt} - 0.75,$$

eq. (1) is proved.

(Step 2) We want to show $\overline{\Delta\Phi}_j^{fe} \le 0$. A free exchange can only occur immediately after an access to x by OPT, and must move x forward. By lemma 2, with $\frac{1}{2}$ probability $b_j(x)$ is 1, so x is moved to the front of BP's list. In this case a free exchange always removes an inversion. The expected change in Φ is $\frac{1}{2}(-1+-2) = -\frac{3}{2}$. With probability $\frac{1}{2}$, $b_j(x)$ is 0 so x stays fixed in BP's list. In the worst case each free exchange will create one type I inversion. Therefore the expected change in Φ is ≤ 1. In summary, the expected $\Delta\Phi$ for a free exchange is $\le \frac{1}{2}(-\frac{3}{2}+1) = -\frac{1}{4}$. This implys $\overline{\Delta\Phi}_j^{fe} \le 0$.

(Step 3) OPT pays 1 for the exchange. In the worst case the exchange creates an inversion which increases Φ by 2 or by 1 with $\frac{1}{2}$ probability (by lemma 2), thus we have $\overline{\Delta\Phi}_j^{pe} \le 1.5P_j$.

Lemma 5. *If access j is a successful search, then $BP_j \le 1.75OPT_j - 0.75$.*

Proof.

$$BP_j = A_j^{bp} + \overline{\Delta\Phi}_j^{mtf} + \overline{\Delta\Phi}_j^{fe} + \overline{\Delta\Phi}_j^{pe}$$
$$\le 1.75A_j^{opt} - 0.75 + 1.5P_j$$
$$\le 1.75(A_j^{opt} + P_j) - 0.75$$
$$= 1.75OPT_j - 0.75.$$

$\square$

3.2 Unsuccessful search $find([x_p, x_s])$

Let access j be an unsuccessful search $find([x_p, x_s])$, and k_p, k_s be the position of x_p, x_s in OPT's list, respectively. Without loss of generality, assume x_s is in front of x_p in OPT's list, so $A_j^{opt} = k_p$. Let R_p be the expected value of the total number of inversions (w, x_p) for all $w \ne x_s$, and R_s be the expected value of the total number of inversions (w, x_s) for all $w \ne x_p$ at the time of the access.

Let $\overline{\Delta\Phi}_j^{mtf} = A_p + B_p + C_p + A_s + B_s + C_s + D$, where A_p is the expected change in Φ due to new inversions (x_p, w) (for $w \neq x_s$) created during the access, B_p is the expected change in Φ due to old inversions (w, x_p) (for $w \neq x_s$) removed during the access, and C_p is the expected change in Φ due to old inversions (w, x_p) (for $w \neq x_s$) that change type during the access. Let A_s, B_s, and C_s be defined similarly with respect to inversions (w, x_s), for $w \neq x_p$. Finally, let D be the expected change of Φ due to the creation/removal/type-changing of the inversion (x_p, x_s) in (Step 1).

Also, let $\overline{\Delta\Phi}_j^{fe} = \mathcal{F}_j' + \mathcal{F}_j$ where $\mathcal{F}_j'$ is the expected change of Φ by swapping x_p and x_s in (Step 2) using a free exchange, if this free exchange happens; and $\mathcal{F}_j$ is the expected change of Φ due to other free exchanges (a swap between x_p and $y \neq x_s$, or a swap between x_s and $z \neq x_p$).

By lemma 2, the following four cases happen with equal $(\frac{1}{4})$ probabilities:
   (**Case 1**) $b_{j-1}(x_s) = b_{j-1}(x_p) = 1$.
   (**Case 2**) $b_{j-1}(x_s) = b_{j-1}(x_p) = 0$.
   (**Case 3**) $b_{j-1}(x_s) = 1, b_{j-1}(x_p) = 0$.
   (**Case 4**) $b_{j-1}(x_s) = 0, b_{j-1}(x_p) = 1$.
Each case is divided into two sub-cases: (a) x_s precedes x_p in BP's list, and (b) x_p precedes x_s in BP's list. In the eight different sub-cases, we compute an upper bound on A_j^{bp}, A_s, A_p, B_s, B_p, C_s, C_p, D, and $\mathcal{F}_j'$. Our first goal is to show that

$$A_j^{bp} + \overline{\Delta\Phi}_j^{mtf} + \mathcal{F}_j' \leq 2.5A_j^{opt} - 1.75 \qquad (2)$$

holds for all cases. To simplify the presentation, we only give the method of how to compute different variables, and give (**Case 2**) as an example. Also we skip the condition part of conditional expected values. For example when we write "$A_p \leq 1.5(k_p - 2)$ for sub-case 2.a", we actually mean that "the conditional expected value of A_p given that sub-case 2.a happens is $\leq 1.5(k_p - 2)$".

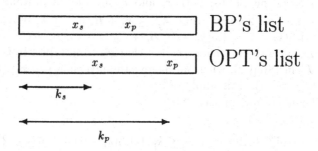

Fig. 1. Unsuccessful search – (x_s precedes x_p in BP's list)

To compute A_j^{bp}: For the four sub-cases where x_s precedes x_p before the access in BP's list (sub-cases 1.a, 2.a, 3.a, and 4.a), $A_j^{bp} \leq k_p + R_p$. For the other four sub-cases (sub-cases 1.b, 2.b, 3.b, and 4.b), $\mathcal{A}_j^{bp} \leq k_s + R_s + 1$ (the extra 1 is for searching x_p in BP's list).

To compute A_p, similar arguments as in a successful search are used. For sub-cases 1.a, 1.b, 4.a, and 4.b, $b_{j-1}(x_p) = 1$, no inversions are created, thus $A_p = 0$. For sub-cases 2.a, 2.b, 3.a, and 3.b, $b_{j-1}(x_p) = 0$, so x_p moves to front. In the worst case a new inversion (x_p, y_i) is created for each of the $k_p - 2$ items y_i preceding x_p in OPT's list, where $y_i \neq x_s$. By similar arguments as that in lemmas 3 and 4, $A_p \leq (k_p - 2)1.5$, where 1.5 is the expected increase in Φ of a new inversion (x_p, y_i).

Computing A_s is similar to computing A_p, for sub-cases 2.a, 2.b, 4.a, and 4.b, where x_s moves to front, $A_s \leq 1.5(k_s - 1)$; for sub-cases 1.a, 1.b, 3.a, and 3.b, where x_s stays, $A_s = 0$.

Like the successful find case, we have $B_p + C_p = -R_p$ and $B_s + C_s = -R_s$ for all sub-cases.

To compute D, and $\mathcal{F}_j'$, we have to consider each sub-case individually.

Now we analyse **Sub-case 2.a:** $b_{j-1}(x_s) = b_{j-1}(x_p) = 0$ and x_s precedes x_p in BP's list. See figure 1. Both x_s and x_p are moved to front. We have:
$A_j^{bp} \leq k_p + R_p$; $A_p = 1.5(k_p - 2)$; $A_s = 1.5(k_s - 1) \leq 1.5(k_p - 2)$;
$B_p + C_p = -R_p$; $B_s + C_s = -R_s$.
$D = 0$ as the inversion (x_p, x_s) does not exist.
For $\mathcal{F}_j'$ in (Step 2), if OPT moves x_p ahead of x_s it creates the inversion (x_s, x_p) in BP's list, Φ is increased by 2 (as $b_j(x_p) = 1$), so $\mathcal{F}_j' \leq 2$. In summary,

$$
\begin{aligned}
\mathcal{A}_j^{bp} + \overline{\Delta\Phi}_j^{mtf} + \mathcal{F}_j' &= \mathcal{A}_j^{bp} + (A_s + B_s + C_s + A_p + B_p + C_p + D) + \mathcal{F}_j' \\
&\leq k_p + 1.5(k_p - 2) + 1.5(k_p - 2) - R_s + 2 \\
&\leq 4k_p - 4.
\end{aligned}
$$

For **sub-case 2.b:** $b_{j-1}(x_s) = b_{j-1}(x_p) = 0$ and x_p precedes x_s in BP's list. After accessing the list both x_s and x_p are moved to front. Similar arguments as in sub-case 2.a show that $\mathcal{A}_j^{bp} \leq k_s + R_s + 1 \leq k_p + R_s$; $A_p = 1.5(k_p - 2)$; $A_s = 1.5(k_s - 1) \leq 1.5(k_p - 2)$; $B_p + C_p = -R_p$; $B_s + C_s = -R_s$.
For D, since the inversion (x_p, x_s) changes from type I to type II, $D = 1$.
For $\mathcal{F}_j'$ in (Step 2), when we remove the inversion (x_p, x_s), Φ is decreased by 2 (as $b_j(x_s) = 1$). This implys $\mathcal{F}_j' \leq 0$.
Therefore for sub-case 2.b,

$$
\begin{aligned}
\mathcal{A}_j^{bp} + \overline{\Delta\Phi}_j^{mtf} + \mathcal{F}_j' &\leq k_p + 1.5(k_p - 2) + 1.5(k_p - 2) - R_p + 1 \\
&\leq 4k_p - 5.
\end{aligned}
$$

So for **(Case 2)**,

$$
\mathcal{A}_j^{bp} + \overline{\Delta\Phi}_j^{mtf} + \mathcal{F}_j' \leq \max \left\{ \begin{matrix} 4k_p - 4 \\ 4k_p - 5 \end{matrix} \right\} = 4k_p - 4.
$$

Using similar technique, we show that

$$\text{for (Case 1)}, \quad A_j^{bp} + \overline{\Delta\Phi}_j^{mtf} + \mathcal{F}_j' \leq k_p + 1,$$

$$\text{for (Case 3)}, \quad A_j^{bp} + \overline{\Delta\Phi}_j^{mtf} + \mathcal{F}_j' \leq 2.5k_p - 2,$$

$$\text{and for (Case 4)}, \quad A_j^{bp} + \overline{\Delta\Phi}_j^{mtf} + \mathcal{F}_j' \leq 2.5k_p - 2.$$

Since the above four cases appear with equal probability $(1/4)$, the overall expected value of $A_j^{bp} + \overline{\Delta\Phi}_j^{mtf} + \mathcal{F}_j'$ is as follows.

$$
\begin{aligned}
A_j^{bp} + \overline{\Delta\Phi}_j^{mtf} + \mathcal{F}_j' &\leq \frac{1}{4}\{(k_p + 1) + (4k_p - 4) + (2.5k_p - 2) + (2.5k_p - 2)\} \\
&= 2.5k_p - 1.75 \\
&= 2.5A_j^{opt} - 1.75.
\end{aligned}
$$

Now eq. (2) is proved. Also, similar arguments as that in successful search case imply that $\mathcal{F}_j \leq 0$, and $\overline{\Delta\Phi}_j^{pe} \leq 1.5P_j$. These three facts together lead to the following lemma.

Lemma 6. *For an unsuccessful search access j, $BP_j \leq 2.5OPT_j - 1.75$.*

Proof.

$$
\begin{aligned}
BP_j &= A_j^{bp} + \overline{\Delta\Phi}_j^{mtf} + \overline{\Delta\Phi}_j^{fe} + \overline{\Delta\Phi}_j^{pe} \\
&= (A_j^{bp} + \overline{\Delta\Phi}_j^{mtf} + \mathcal{F}_j') + \mathcal{F}_j + \overline{\Delta\Phi}_j^{pe} \\
&\leq 2.5A_j^{opt} - 1.75 + 1.5P_j \\
&\leq 2.5(A_j^{opt} + P_j) - 1.75 \\
&= 2.5OPT_j - 1.75. \quad \square
\end{aligned}
$$

Theorem 7. *For any access sequence σ consisting of successful and unsuccessful searches only, $E[BP(\sigma)] \leq 2.5 \cdot OPT(\sigma)$.*

Proof. Combining lemma 5 with lemma 6, summing up all successful and unsuccessful searches proves the theorem. $\square$

3.3 Insertions

The extension of BP to handle insertions is straightforward. On an insertion of the key x, after performing a search to make sure that x is not in the list, BP inserts x at the front of the list and randomly initializes $b(x)$. As both BP and OPT insert in the front of the list, we decompose an insertion into an unsuccessful find and an *add-item* operation which puts the new item at the list head. For each add-item operation, cost of BP = cost of OPT = 1, and $\Delta\Phi$ is 0. Thus BP is 2.5-competitive to OPT for a request sequence of (successful and unsuccessful) searches and insertions.

4 Conclusions and discussions

1. We have shown that randomization can be used to improve the performance of MP (move-to-front with ps bits) against oblivious adversaries, by reducing the competitive coefficient from 3 to 2.5. We can further improve the competitive coefficient to $2\sqrt{3} - 1 \approx 2.464$ by refining BP using similar technique as in [6].

2. Analyzing deletion is an unsolved issue in the application of ps bits [3], therefore in our analysis of BP, we can only analyze request sequences consist of successful searches, unsuccessful searches, and insertions. To include deletion in our analysis we first have to solve the problem of analyzing deletions (in algorithms using ps bits) in the deterministic case.

3. We only prove an upper bound for BP against an oblivious adversary. One interesting problem is to find a (hopefully matching) lower bound in this setting.

4. In this paper we use adversaries that can finish an unsuccessful search $find([x_p, x_s])$ by finding both x_p and x_s. What if we are comparing BP to stronger adversaries that can finish a $find([x_p, x_s])$ by finding either x_p and x_s? We leave this as an open problem.

References

1. J. Bentley and C. McGeoch. Worst-case analysis of self-organizing sequential search heuristics. In *Proc. of 20th Allerton Conf. on Communication, Control, and Computing*, pages 452–461, 1983.

2. G. Frederickson. Self-organizing heuristics for implicit data structures. *SIAM Journal on Computing*, 13(2):277–291, 1984.

3. L. Hui and C. Martel. On efficient unsuccessful search. In *Proc. of the 3rd ACM-SIAM Symposium on Discrete Algorithms*, pages 217–227, 1992.

4. S. Irani, N. Reingold, J. Westbrook, and D. Sleator. Randomized competitive algorithms for the list update problem. In *Proc. of the 2nd ACM-SIAM Symposium on Discrete Algorithms*, pages 251–260, 1991.

5. J. McCabe. On serial files with relocatable records. *Oper. Res.*, 12:609–618, 1965.

6. N. Reingold, J. Westbrook, and D. Sleator. Randomized competitive algorithms for the list update problem. *Algorithmica*, to appear.

7. D. Sleator and R. Tarjan. Amortized efficiency of list update and paging rules. *Commun. of the A.C.M.*, 28(2):202–208, 1985.

8. D. Sleator and R. Tarjan. Self-adjusting binary search trees. *J.A.C.M.*, 32(3):652–686, 1985.

Randomized On-line Algorithms for the Page Replication Problem

Hisashi Koga

Department of Information Science, Faculty of Science,
The University of Tokyo
7-3-1 Hongo, Bunkyo-ku, Tokyo-to 113, Japan
e-mail: nwa@is.s.u-tokyo.ac.jp

Abstract. In a distributed shared memory system, each read-only page needs to be located at appropriate processors by replication to make the total access cost lower. In this paper, on-line algorithms to implement this low-cost locating are considered in terms of competitiveness. Specifically, we first show that the application of algorithms based on randomized algorithms for the page migration problem to the page replication problem can beat the deterministic lower bound for tree networks. Coin-flipping algorithms are also investigated in this paper. We prove that a coin-flipping algorithm is 2-competitive for trees. For circles we propose the first 4-competitive coin-flipping algorithm which may be generalized for general networks.

1 Introduction

A common design for a shared memory multiprocessor system is a network of processors, each of which has its own local memory. In such a design, a programming abstraction of a simple global memory is supported by a virtual memory system that distributes the physical pages among the local memories.

In a distributed shared memory system, when processor q wishes to access memory address a of page b, first q examines whether the page is contained in its local memory. If so, the page access is done at 0 cost. If not, q searches the processor p having the page b and sends a access request to p. Then the processor p responses to the request and the value of the location a is transmitted back to q. The cost of this action is proportional to the distance between p and q. However, if q requires the page access to b frequently, the migration/replication of a full page of b to q may result in spending lower cost in total, because once q has the page copy, q accesses the page at 0 cost. On the other hand, moving a full page incurs a large amount of communication cost proportional to the distance.

For writable pages, only one page copy must exist to avoid the difficulty of maintaining consistency among multiple copies. In such a situation, it is important to consider the *page migration problem* whose purpose is to devise residency strategies that decide which local memory should have the only copy of a writable page to reduce the cost in processing a sequence of page-access requests [7].

For read-only pages, many copies may exist at the same time, because the consistency cannot be broken. Therefore, to find residency strategies that decide which subset of the local memories should contain the page copy is essential. This problem is called the *page replication problem*.

In this paper, we focus on on-line algorithms for the page replication problem. An algorithm is said to be on-line if it processes a request based only on that request and past requests. To evaluate on-line algorithms, we use competitiveness, introduced in [8].

As for deterministic algorithms, Black and Sleator [2] show that no on-line algorithm can be better than 3-competitive in the page migration problem, and that no on-line algorithm can be better than 2-competitive in the page replication problem. They actually devised a 2-competitive algorithm for trees in the page replication problem.

On the other hand, in the case of randomized algorithms in the migration problem, more powerful results are obtained by Westbrook [9]. He gives a $\frac{5+\sqrt{17}}{4} \simeq 2.28$-competitive randomized algorithm for uniform networks. He also proves a coin-flipping algorithm is 3-competitive against any topologies of networks. However, for the page replication problem, no further result is known.

In this paper, by extending randomized algorithms for the migration problem to the replication problem, we show the following.
(1) We give a $1 + \frac{\sqrt{2}}{2} \simeq 1.71$-competitive algorithm for trees. This beats the deterministic lower bound of 2-competitiveness.
(2) We show that a coin-flipping algorithm is 2-competitive for trees. However, this competitive ratio cannot be extended to circles of processors.
(3) We introduce a newly developed coin-flipping algorithm which is 4-competitive for any circle. For circles, there has been no competitive algorithm, and our algorithm is the first competitive one.

The page replication problem is a fundamental on-line problem. For example, the simplest case corresponds to the ski rental problem [6]. The results and techniques developed in this paper would have more applications for on-line problems.

2 Competitive Analysis

An on-line algorithm is an algorithm which must satisfy a request without knowing the future requests. In this paper, we focus on the competitiveness of on-line algorithms introduced by Sleator and Tarjan [8].

The definition of competitiveness is as follows. The cost of an algorithm A on request sequence σ is denoted by $C_A(\sigma)$. Generally, a deterministic on-line algorithm A is c-competitive if, for all request sequences σ, there is a constant b such that $C_A(\sigma) \leq c \cdot C_{OPT}(\sigma) + b$. Here OPT is the off-line algorithm which achieves the optimum cost on σ knowing the entire sequence in advance.

However, in the page replication problem, the trivial algorithm that initially copies the page to all nodes becomes 0-competitive by setting b to the sum of the costs of the initial page replications. To give meaningful results, we must redefine the competitiveness according to [2]. A deterministic on-line algorithm A is c-competitive if, for all request sequences σ, $C_A(\sigma) \leq c \cdot C_{OPT}(\sigma)$.

A randomized on-line algorithm B is c-competitive against an oblivious adversary (see [1] for detail) if, for all request sequences σ, $E[C_B(\sigma)] \leq c \cdot C_{OPT}(\sigma)$. The expectation is taken over the random choice made by the on-line algorithm.

3 Formal Definition of the Page Replication Problem

The component of this problem is an undirected graph G which represents the network. The vertices correspond to processors. The edges represent the links between two adjacent processors, and their length denotes the distance between them. Let δ_{ij} be the length of the shortest path between node i and node j.

In the page replication problem, we concentrate on a particular page b. Satisfying a request from q costs the distance from q to the nearest node p with the page copy (i.e. δ_{pq}) if q does not hold the copy of b yet. If q holds the copy, the request is satisfied at 0 cost. When q does not have the page copy, p can replicate the page to q at the cost of $r\delta_{pq}$ after satisfying the request, where r is a constant bigger than 1 proportional to the page size.

However, in Section 5.2, because an algorithm in which a request from q is not always satisfied by the nearest node with the page along the shortest path is considered, the above definition of costs cannot be applied. Thus a new distance function D must be introduced.

The replication problem is to decide (in an on-line fashion) which nodes should have the page to process the request sequence at low cost provided initially only one particular node, (which we call s in this paper) has the page.

In this paper three assumptions, which are originally used for the deterministic page replication problem in [2], are made to simplify the problem.
1. Once a node has the copy of the page, it never drops it.
2. A node can replicate the page copy only to its adjacent nodes.
3. Every local memory has infinite capacity.

From these assumptions the set of nodes with a page copy shall always be a connected component of the graph.

4 Randomized Algorithms Using Counters

In this section, we describe randomized algorithms using counters based on the algorithm UNIFORM for the migration problem in [9]. We first analyze these algorithms when the graph is a single edge between two nodes and then show that these algorithms can be easily extended to arbitrary trees. As UNIFORM beats the deterministic lower bound for the migration problem, these algorithms break the deterministic lower bound of 2-competitiveness.

4.1 2 Nodes

Let s and t be 2 nodes. W.l.g. we may assume that $\delta_{st} = 1$. Let k be the number of requests from t. Since requests from s incur no cost, it suffices to consider k. Our randomized algorithm RANDOM COUNT for 2 nodes can break the deterministic lower bound of 2. This algorithm can be used for the ski rental problem [6] by considering k as the frequency of going skiing.

> **Algorithm RANDOM COUNT for 2 nodes:** This algorithm maintains a counter c_t on node t. To begin the algorithm, c_t is set to a value between 0 and $l - 1$ chosen at random where l is a constant bigger than r. Each time a request at node t is generated, c_t is incremented. When c_t reaches l, the page replication occurs.

Table 1. The cost ratio of RANDOM COUNT to OPT

	OPT	RANDOM COUNT	$\dfrac{\text{RANDOM COUNT}}{\text{OPT}}$
$k < r$	k	$\dfrac{-k^2+2k(l+r)+k}{2l}$	$\dfrac{-k+2(l+r)+1}{2l}$
$r \le k < l$	r	$\dfrac{-k^2+2k(l+r)+k}{2l}$	$\dfrac{-k^2+2(l+r)k+k}{2lr}$
$k \ge l$	r	$r + \dfrac{l+1}{2}$	$1 + \dfrac{l+1}{2r}$

Theorem 1. *Algorithm RANDOM COUNT for 2 nodes is μ-competitive where* $\mu = \max\left\{1+\frac{r}{l}, 1+\frac{l+1}{2r}\right\}$.

Proof. There are two cases to consider about the value of k. Let m be the initial value of c_t selected at random $(0 \le m \le l-1)$.

$k \ge l$: In this case, whatever value m should be, the replication to t occurs. Thus the expected cost of algorithm RANDOM COUNT is :

$$\frac{1}{l}\sum_{m=0}^{l-1}(l-m+r) = r + \frac{l+1}{2}$$

$k < l$: In this case only when $m \ge l-k$, the replication occurs. Thus the expected cost is :

$$\frac{1}{l}\sum_{m=0}^{l-k-1}k + \frac{1}{l}\sum_{m=l-k}^{l-1}(l-m+r) = \frac{-k^2+2k(l+r)+k}{2l}$$

On the other hand, the optimal algorithm replicates the page to t immediately before responding any request and incurs a cost of r when $k \ge r$. When $k < r$, it never replicates the page and incurs a cost of k.

The cost ratio of RANDOM COUNT to the optimum off-line algorithm is summarized in Table 1. Because (a) when $k < r$, $\frac{-k+2(l+r)+1}{2l}$ is always less than $1+\frac{r}{l}$, (b) when $r \le k < l$, $\frac{-k^2+2(l+r)k+k}{2lr}$ is always less than $1+\frac{l+1}{2r}$, and (c) when $k \ge l$, $1+\frac{l+1}{2r}$ is a constant, the ratio of $E[C_{\text{RANDOM}}]$ to C_{OPT} is always less than or equal to μ and the proof is completed. □

Especially if $l = \dfrac{\sqrt{1+8r^2}-1}{2}$, then $\mu = 1 + \dfrac{\sqrt{8r^2+1}+1}{4r}$ which becomes $1 + \frac{\sqrt{2}}{2} \simeq 1.71$ when $r \to \infty$ and breaks the deterministic lower bound of 2.

4.2 Trees

An easy extension of algorithm RANDOM COUNT is to the case in which the graph is a tree.

Algorithm RANDOM COUNT for trees: Each node v has a counter c_v. To begin the algorithm, all counters are set to the same value between 0 and $l - 1$ selected at uniformly random. When a node i without the page requests, the counter of every node along the path from i to the closest node with the page is incremented. When a counter reaches l, the page is replicated to the corresponding node.

The description of RANDOM COUNT implies that the longer the distance between some node and s is, the smaller the counter value for the node becomes. Thus, the nodes holding the page form a connected component.

Theorem 2. *Algorithm RANDOM COUNT for trees is μ-competitive where $\mu = \max \left\{ 1 + \frac{r}{l}, 1 + \frac{l+1}{2r} \right\}$.*

Proof sketch. The definition of the cost allows us to analyze it by partitioning the total cost into parts incurred by each edge (equal to the length of the edge). Note that RANDOM COUNT for trees acts the same as RANDOM COUNT for 2 nodes with respect to each edge. Since the sum of the costs incurred by all the edges is the total cost and, for each edge, RANDOM COUNT is μ-competitive, RANDOM COUNT for trees is μ-competitive totally. $\square$

5 Coin-Flipping Algorithms

It is natural to consider the simple replication strategy which decides when to replicate the page to the requesting node judging from the result of coin-flipping at each request. We call this strategy a coin-flipping algorithm.

Algorithm Coin-Flipping (CF): A coin-flipping algorithm replicates the page to the requesting node with the probability $p > 0$ at each request if the requesting node does not have the page. During this replication, the page must be also replicated to all the nodes on the page-access path from the assumptions in Section 3. Note that this additional replication can be done with no extra cost.

CF algorithms have the advantage of not using counters. Westbrook [9] proves that a CF algorithm can be 3-competitive for any networks with proper choice of p in the migration problem. In this section we investigate the competitiveness of CF algorithms in the page replication problem.

5.1 Trees

First the competitive ratio of the CF algorithm for tree networks is investigated. Before describing Theorem 4, we give the definition of a *boundary node* used in the proof of Theorem 4.

Definition 3. For algorithm B, B's *boundary node* of some node (say d) which does not have the page yet is the node with the page in algorithm B on the path from s to d which is the furthest from s. Since trees are considered, the number of boundary nodes for a node without the page is exactly 1. Note that when d requests, the request is satisfied by d's boundary node.

Theorem 4. *The coin-flipping algorithm with $p = \frac{1}{r}$ is 2-competitive for any tree.*

Proof. We shall show that, for any algorithm A, $E[C_{CF}(\sigma)] \leq 2C_A(\sigma)$ for any request sequence σ. Our proof uses the standard technique comparing simultaneous runs of CF and A on σ by merging the actions generated by CF and A into a single sequence of events. This sequence contains two types of events: (Type I) A replicates the page, and (Type II) a request is satisfied by both A and CF, which may accompany CF replicating the page. We shall give a non-negative potential function Φ (initially 0) such that the following inequality holds for all kinds of events.

$$2\Delta C_A - E[\Delta C_{CF}] \geq E[\Delta\Phi] \tag{1}$$

where the Δ indicates the change in the value as a result of the event. Summing this inequality for all events results in

$$2C_A - E[C_{CF}] \geq E[\Phi_{end}] - E[\Phi_{start}] . \tag{2}$$

Since $\Phi_{start} = 0$ and $\Phi_{end} \geq 0$, the right side of (2) is non-negative. Thus the proof ends. It remains to specify the potential and verify (1) for all events.

Let S be the set of nodes where only A has the page. The potential function Φ is defined as:

$$\Phi = 2r \sum_{j \in S} \delta_{jp(j)}$$

where $p(j)$ is the parent node of j when considering node s as the root.

From now on (1) is proved for all events.

(Type I) The event is A replicating the page. Let the source node be k and the destination node be j. In this case, $\Delta C_A = r\delta_{kj}$ and $\Delta C_{CF} = 0$, so we must show $E[\Delta\Phi] \leq 2r\delta_{kj}$. There are two cases to consider based on whether j belongs to S after the replication.
$j \notin S : \Delta\Phi = 0$
$j \in S :$ if $k \in S$ before the replication $\Delta\Phi = 2r\delta_{kj}$ else $\Delta\Phi \leq 2r\delta_{kj}$

(Type II) The event is the servicing of a request by both algorithms. Let d be the requesting node. There are two cases:
(a) node d already had the page in CF.
$\Delta C_{CF} = 0$. ΔC_A is at least 0. $E[\Delta\Phi] = 0$, because no replication occurs in CF. Thus (1) is satisfied.
(b) node d did not have the page in CF yet.
Let i be CF's boundary node of d. The servicing of the request by CF costs δ_{id} and with the probability of $\frac{1}{r}$, the replication occurs. Thus $E[\Delta C_{CF}] = \delta_{id} + \frac{1}{r}r\delta_{id} = 2\delta_{id}$. About ΔC_A, there are three cases depending on A's configuration of nodes with the page. Let j be A's boundary node of d in the first two cases.

A has not replicated the page beyond i when the request occurs. $\Delta C_A = \delta_{jd}$, and we must show that $E[\Delta\Phi] \leq 2\delta_{jd} - 2\delta_{id} = 2\delta_{ij}$. However, $E[\Delta\Phi] = 0$, because Φ does not change whether CF replicates the page or not.

A has replicated beyond i, but not beyond d. $\Delta C_A = \delta_{jd}$, and we must show that $E[\Delta\Phi] \leq 2\delta_{jd} - 2\delta_{id} = -2\delta_{ij}$ (Note that $\delta_{id} > \delta_{jd}$). $\Delta\Phi = 0$ if no replication occurs in CF. Otherwise $\Delta\Phi = -2r\delta_{ij}$ Thus, $E[\Delta\Phi] = -\frac{1}{r} \cdot 2r\delta_{ij} = -2\delta_{ij}$.
A has replicated beyond d. In this case $\Delta C_A = 0$. Therefore we must show $E[\Delta\Phi] \leq 0 - 2\delta_{id} = -2\delta_{id}$. If the replication is not done in CF, $\Delta\Phi = 0$, else $\Delta\Phi = -2r\delta_{id}$. Thus $E[\Delta\Phi] = \frac{1}{r}(-2r\delta_{id}) = -2\delta_{id}$.

Now that (1) is proved for all events, The entire proof completes. ☐

In this section, we have proved that, for trees, the coin-flipping algorithm with $p = \frac{1}{r}$ is 2-competitive, the deterministic lower bound in the page replication problem. On the other hand, in the migration problem, Westbrook [9] shows that the coin-flipping algorithm with $p = \frac{1}{2d}$ is 3-competitive against any topologies, where d is the page size factor. This "3-competitive" is also the deterministic lower bound. Here, the natural question whether this ratio of the CF algorithm can be extended to any topologies in the page replication problem occurs. This question is answered negatively. That is,

Theorem 5. *A request sequence and a graph exist such that, for the request sequence on the graph, the cost incurred by the coin-flipping algorithm with $p = \frac{1}{r}$ is more than two times the optimal minimum cost.*

5.2 A Generalized Coin-Flipping Algorithm for Circles

For circles, the page replication problem is more complicated than for trees. The reason is that since there are two paths between any two nodes in circles, the on-line algorithms must determine which to use for satisfying requests and replicating the page. In this section a new generalized coin-flipping algorithm for circles in which response to a request and replication of the page is not necessarily from the nearest node along the shortest path is introduced. Although accessing the farther node may seem strange, this algorithm is 4-competitive.

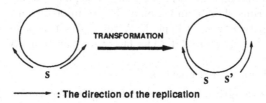

: The direction of the replication

Fig. 1. Transformation of G into G'

Let G be the underlying circle. Even if we replace s by two nodes s and s' such that both of them initially has the page, and the length of the shortest path between them (which is called $\overline{ss'}$) is 0, the situation does not change. Thus from now on such modified circle G' is considered (see Fig. 1). As an algorithm runs, the page is replicated clockwise from s and counterclockwise from s'. First a new distance function D is defined. D_{ij} is defined as follows :

$$D_{ij} = \begin{cases} \text{if } i = j & 0 \\ \text{if } i \neq j & \text{the length of the path between node } i \\ & \text{and node } j \text{ in which } \overline{ss'} \text{ is not contained} \end{cases}$$

This definition is natural since paths containing $\overline{ss'}$ is not used for the page access operations.

Before describing the algorithm, we introduce the following two concepts.

Definition 6. The *left (right) boundary node* of algorithm A at some time is the endmost node to which the page has been replicated clockwise (counterclockwise) in A before the time. For any algorithm, initially the left (right) boundary node is node s (s'). We abbreviate left (right) boundary node as lbn (rbn).

Hereinafter, let a be the lbn of GCF, and b be the rbn of GCF when a request occurs.

Definition 7. The *nodecover* of an algorithm A at some time is the arc $L_A s R_A$, where L_A is the lbn of A and R_A is the rbn of A at the time.

> **Algorithm Generalized Coin-Flipping (GCF) for circles**: When a request occurs at some node d, if d holds the page, the request is satisfied at 0 cost and nothing else occurs. Otherwise with the probability of $\frac{D_{ad}}{D_{ab}}$, the request is satisfied by b and with the probability of $\frac{D_{bd}}{D_{ab}}$, the request is satisfied by a. After satisfying the request, GCF replicates the page from the node satisfying the request to node d with the probability of $\frac{1}{r}$. Like a normal coin-flipping algorithm the page is also replicated to all the nodes on the page access path.

Theorem 8. *GCF is 4-competitive for any circles.*

Proof. As in Section 5.1, we shall give a non-negative potential function Φ (initially 0) and prove that, for any algorithm A and for any request sequence σ,

$$4\Delta C_A - E[\Delta C_{GCF}] \geq E[\Delta\Phi] \qquad (3)$$

for all kinds of events contained in the merged sequence of actions generated in the simultaneous runs of GCF and A on σ. The components of the event sequence are classified into two types: (Type 1) A replicates the page, and (Type 2) a request is satisfied by A and GCF. The latter may accompany GCF replicating the page.

The potential function is given as:

$$\Phi = 4r \times \text{length of (nodecover of A} \cap \text{(nodecover of GCF)}^c.)$$

From now on (3) is proved for all kinds of events.

(Type 1) The event is A replicating its page. Let the source node be k and the destination node be j. In this case, $\Delta C_A = r\delta_{kj}$ and $\Delta C_{GCF} = 0$, so we must show $E[\Delta\Phi] \leq 4r\delta_{kj}$. There are four cases depending on whether k and j belong to the nodecover of GCF before the replication. If $k, j \notin$ nodecover of GCF, $E[\Delta\Phi] = 4r\delta_{kj}$. Else $E[\Delta\Phi] \leq 4r\delta_{kj}$. Thus (3) holds.

(Type 2) The event is the servicing of a request by both algorithms. Let d be the requesting node. Let x be the lbn of A and y be the rbn of A.
(i) If $d \in$ nodecover of GCF before the request, $\Delta C_{GCF} = 0$ and ΔC_A is at least 0. $E[\Delta\Phi] = 0$ because no replication occurs in GCF. Thus (3) holds.

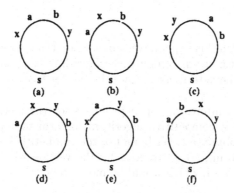

Fig. 2. Six cases about the locations of $a,b,x,$ and y

(ii) If $d \notin$ nodecover of GCF before the request:

First we define the ordering of nodes such that for two nodes m, n:

$$m < n \quad \text{if} \quad D_{sm} < D_{sn} .$$

Obviously always $a < b$ and $x < y$. Note that since $d \notin$ nodecover of GCF, $a < d < b$. There are six cases to consider about the ordering of a, b, x and y under the condition that $a < b$ and $x < y$ as illustrated in Fig. 2. However, in this paper we show the proof of (3) only for case (a) and case (b) due to the space limitation. In both cases, $E[\Delta C_{\text{GCF}}] = \frac{D_{bd}}{D_{ab}}(D_{ad} + \frac{1}{r} r D_{ad}) + \frac{D_{ad}}{D_{ab}}(D_{bd} + \frac{1}{r} r D_{bd}) = 4\frac{D_{ad}D_{bd}}{D_{ab}}.$

In case (a), $\Delta C_A \geq \min(D_{xd}, D_{yd})$. $E[\Delta\Phi] = 0$. Moreover,

$$E[\Delta C_{\text{GCF}}] = 4\frac{D_{ad}D_{bd}}{D_{ab}} \leq 4D_{ad} \leq 4D_{xd}.$$

$$E[\Delta C_{\text{GCF}}] = 4\frac{D_{ad}D_{bd}}{D_{ab}} \leq 4D_{bd} \leq 4D_{yd}.$$

Thus, $4\min(D_{xd}, D_{yd}) - 4\frac{D_{ad}D_{bd}}{D_{ab}} \geq 0$ and (3) holds for this case.

In case (b), there are two cases about the location of d.

if $a < d \leq x$: $\Delta C_A = 0$. The expected potential change is $\frac{1}{r}(-4r D_{ad})$ if the request is satisfied by a in GCF and $\frac{1}{r}(-4r D_{xd})$ if satisfied by b. Thus $E[\Delta\Phi] = -4\frac{D_{ad}D_{bd}}{D_{ab}} - 4\frac{D_{xd}D_{ad}}{D_{ab}}$. Therefore

$$4\Delta C_A - E[\Delta C_{\text{GCF}}] = -4\frac{D_{ad}D_{bd}}{D_{ab}} \geq -4\frac{D_{ad}D_{bd}}{D_{ab}} - 4\frac{D_{xd}D_{ad}}{D_{ab}} = E[\Delta\Phi].$$

if $x < d < b$: $\Delta C_A \geq \min(D_{xd}, D_{yd})$. The potential changes only when the request is satisfied by a. Thus the expected potential change is $-4\frac{D_{bd}D_{ax}}{D_{ab}}$. Therefore, $E[\Delta C_{\text{GCF}}] + E[\Delta\Phi] = 4\frac{D_{bd}D_{xd}}{D_{ab}}$. Here,

$$4D_{xd} - 4\frac{D_{bd}D_{xd}}{D_{ab}} = 4D_{xd}(1 - \frac{D_{bd}}{D_{ab}}) \geq 0$$

$$4D_{yd} - 4\frac{D_{bd}D_{xd}}{D_{ab}} \geq 4D_{yd} - 4D_{bd} = 4D_{by} \geq 0$$

The above inequalities show that $4\Delta C_A \geq E[\Delta C_{\text{GCF}}] + E[\Delta\Phi]$. Since (3) is proved for all events, Theorem 8 is proved. $\qquad\square$

6   Remarks

Our results obtained in this paper have demonstrated the power of randomization since, in deterministic cases, competitive algorithms are not found except for tree networks, and the deterministic lower bound for tree networks is beaten by randomization.

There are several interesting open problems for the page replication problem. One is to find a competitive algorithm for arbitrary networks. Another open problem is to obtain a lower bound of the competitiveness for randomized algorithms. It is also important to consider the version where the assumptions of Section 3 are removed, which would be more practical.

7   Acknowledgement

The author would like to thank Dr. Susanne Albers for helpful advice for this research.

References

1. S. Ben-David, A. Borodin, R. Karp, G. Tardos, and A. Widgerson. "On the power of randomization in on-line algorithms". In *Proceedings of the 22th ACM Symposium on Theory of Computing*, pages 379–386, 1990.
2. D.L. Black and D.D. Sleator. "Comptitive algorithms for replication and migration problem". Technical Report CMU-CS-89-201, Department of Computer Science, Carnegie Mellon University, 1989.
3. A. Fiat, R.M. Karp, M. Luby, L.A. McGeoch, D.D. Sleator, and N.E. Young. "Competitive pageing algorithms". *Journal of Algorithms*, pages 685–689, 1991.
4. A. Karlin, M. Manesse, and L. Rudolph. "Competitive snoopy caching". *Algorithmica*, 28:79–119, 1988.
5. S. Irani, N. Reinold, and J. Westbrook. "Randomized algorithms for the list update problem". In *Proceedings of the 2nd ACM-SIAM Symposium on Discrete Algorithms*, pages 251–260, 1988.
6. P. Raghavan. "Lecture notes on randomized algorithms". Reserach Report RC 15340, IBM T.J. Watson Research Center, 1989.
7. C. Scheurich and M. Dubois. "Dynamic page migraion in multiprocessors with distributed shared memory". *IEEE Transactions on Computers*, 38(no.8):1154–1163, 1989.
8. D.D. Sleator and R.E. Tarjan. "Amortized efficiency of list update and paging rules". *Communications of the ACM*, 28:202–208, 1985.
9. J. Westbrook. "Randomized algorithm for multiprocessor page migration". In *Proceedings of a DIMACS Workshop*, pages 135–149, 1991.

New Algorithms for Minimizing the Longest Wire Length during Circuit Compaction [*]

(Extended Abstract)

Susanne E. Hambrusch[1] and Hung-Yi Tony Tu[2]

[1] Department of Computer Sciences, Purdue University, West Lafayette, IN 47907, USA
[2] Department of Information Sciences, Providence University, Taichung, Taiwan, ROC

1 Introduction

Circuit compaction is the process of converting a symbolic layout into an actual layout that satisfies the design rules and minimizes a set of objective functions [5, 8]. One-dimensional (1-D) compaction allows layout elements to slide in one direction only and is generally preferred over the computationally intractable 2-dimensional compaction. Consider performing 1-D compaction along the horizontal direction. Layout elements are now allowed to slide horizontally as long as no distance constraint is violated and the relative order of the layout elements is preserved. Horizontal wires can change their lengths. A layout generated by most conventional width-minimizing compaction algorithm contains unnecessarily long horizontal wires, and controlling the wire length is crucial in circuit design [8, 9, 11]. In this paper we present new and efficient algorithms for minimizing the length of the longest horizontal wire during 1-D compaction. Assume the given layout contains n_h horizontal wires, n_v vertical wires, and rectilinear polygonal layout components composed of n_r vertical edges. We present an $O(n_h \cdot n \log n)$ time algorithm for generating a new layout which minimizes the length of the longest wire and for which, among all layouts having the same minimum longest wire length, one of minimum width is generated, $n \leq 2n_h + n_v + n_r$. We also extend our approach to minimize a specified tradeoff between the longest wire length and layout width. More precisely, given a tradeoff function $\alpha \cdot W + \beta \cdot L$ between the width W and the longest wire length L, and constants α, β, we present an $O(n_h \cdot n \log n)$ time algorithm for minimizing $\alpha \cdot W + \beta \cdot L$.

The best previously known algorithm for minimizing the longest wire length follows from Laio and Wong's result in [6]. From a given layout, Laio and Wong's algorithm generates a layout of minimum width in which the length of each horizontal wire lies between a specified upper and lower bound. Translating their running time into our framework gives an $O((n_h + n_r) \cdot n)$ time algorithm. Using this algorithm for performing a binary search on the longest wire length, a layout minimizing the longest wire length can be generated in $O(\log L \cdot (n_h + n_r) \cdot n)$

[*] Research supported in part by DARPA under contract DABT63-92-C-0022.

time, where L is the longest wire length in the initial layout. Our algorithm is based on a different approach and its $O(n_h \cdot n \log n)$ running time is independent of L and asymptotically faster for $\log n < (1 + n_r/n_h) \cdot \log L$. More importantly, our approach can be generalized to generate a layout minimizing a given tradeoff function between the longest wire length and the layout width. Other compaction algorithms minimizing the layout width, the longest wire length, or the total wire length are described in [2, 3, 5, 6, 7, 10, 11, 12]. None of these algorithms can trade layout width for the longest wire length.

We briefly sketch the approach underlying our algorithms. The layout elements are either layout components of rectilinear polygonal shape or vertical or horizontal wires. A *configuration* assigns to the leftmost edge of every layout element an x-position in the layout area. The input is a feasible configuration (i.e., the x-positions associated with the layout elements result in a layout satisfying the constraints). From this initial configuration we generate a configuration minimizing the longest wire length or the tradeoff function over a number of iterations. Each iteration produces a configuration with smaller longest wire length or a smaller tradeoff function value, respectively. The relevant constraints and distances are represented by graphs and, within the area of compaction methods, our algorithm is a graph-based solution [5]. A new configuration is generated from a previous one by assigning to each layout element a speed according to which it moves to the right. The concept of assigning speeds to layout element is novel in the area of compaction algorithms. The speeds are computed by a longest path computation in the longest wire minimizing algorithm and by a shortest path computation in the tradeoff minimizing algorithm.

In the longest wire minimizing algorithm the movement to the right reduces the length of longest wires. At the same time, it can increase the length of wires that are originally not the longest ones and it changes distances between layout elements. One iteration stops when a non-longest wire turns into a longest one or any further movement would violate the distance constraints. If a further reduction in the longest wire length is possible, the next iteration continues the movement with updated speeds. The scenario for the tradeoff minimizing algorithm is similar. We show that a final configuration is found after at most $n_h \cdot n$ iterations. The $O(n_h \cdot n \log n)$ time bound is achieved by performing updates, rather than re-computations on the data structures.

2 Preliminaries

Throughout, a horizontal wire is a horizontal segment of maximum length. Every wire segment not belonging to a horizontal wire forms a vertical wire. The layout shown in Figure 1(a) contains five horizontal and seven vertical wires. Layout elements are grouped into *cells* as follows. Partition the layout components and vertical wires into maximal sets, so that each set represents one rigid object that can only move as one entity. One such rigid object induces one cell of the layout. In addition, if an endpoint of a horizontal wire is not connected to a layout component, then this endpoint induces a rectangular cell of width 0. Figure 1(b) shows the cells induced by the layout of Figure 1(a).

A cell U_j is *visible* from a cell U_i if and only if one can draw a horizontal line connecting U_i and U_j starting at a position x_1 and ending at position x_2, $x_1 \leq x_2$, and no position between x_1 and x_2 on this horizontal line is occupied by a cell. Every such horizontal line between two cells has a length of $x_2 - x_1$. The *distance* between two cells is the minimum over all lengths associated with horizontal lines connecting cells U_i and U_j and inducing visibility between them.

Let n be the number of cells in configuration $\mathcal{C}$, $n \leq 2n_h + n_v + n_r$. In our algorithm the information about configuration $\mathcal{C}$ is represented by two n-vertex graphs, the *distance graph* $G_d = (V_d, A_d)$ and the *wire graph* $G_w = (V_w, A_w)$. In both graphs the vertices correspond to the cells of configuration $\mathcal{C}$ and vertex u_i represents cell U_i, $1 \leq i \leq n$.

The distance graph models visibility and the distances and thus the constraints between the cells (in related literature this graph is also called the constraint graph). When cell U_j is visible from cell U_i, the distance graph contains the arc (u_i, u_j) and its weight, $d(u_i, u_j)$, is set to the distance between U_i and U_j in configuration $\mathcal{C}$. Setting the weights this way guarantees that cells will not overlap. Other constraints (e.g., requiring a certain minimum separation between cells) can also be incorporated into the distance graph. For every cell induced by an endpoint of a horizontal wire the following arcs are included. Let H be a horizontal wire and let $V_{up,l}$ and $V_{up,r}$ be the leftmost and rightmost vertical wire incident to H, respectively, and lying above H. Vertical wires $V_{dw,l}$ and $V_{dw,r}$ lying below H are defined analogously. Assume U_i is the cell corresponding to the left endpoint of horizontal wire H and that $V_{up,l}$ (resp. $V_{dw,l}$) exists. If cell U_j contains vertical wire $V_{up,l}$ (resp. $V_{dw,l}$), we include the arc (u_i, u_j) with its weight $d(u_i, u_j)$ set to the distance between the left endpoint of H and vertical wire $V_{up,l}$ (resp. $V_{dw,l}$). Arcs (u_{12}, u_1), (u_{12}, u_2), (u_{14}, u_{10}), and (u_{14}, u_{11}) of Figure 1(c) are created according to these rules. When cell U_i corresponds to the right endpoint of horizontal wire H and $V_{up,r}$ (resp. $V_{dw,r}$) is contained in cell U_j, we include the arc (u_j, u_i) and set its weight accordingly.

The wire graph G_w records the lengths of the longest wires between cells. Arc (u_i, u_j) is in G_w if cell U_j is visible from cell U_i and there exists at least one horizontal wire having its left endpoint incident to U_i and its right endpoint incident to U_j. The weight of this arc, $w(u_i, u_j)$, is set to the length of the longest such wire between U_i and U_j. Throughout, let $H_{i,j}$ represent the longest wire connecting cells U_i and U_j so that the left endpoint of this horizontal wire is incident to U_i and its right endpoint is incident to U_j. From now on we assume that among all wires connecting two cells U_i and U_j all but the longest one have been removed.

3 Overall Approach

In this section we describe the overall approach of our algorithm minimizing the length of the longest wire. The algorithm for minimizing a tradeoff function uses a similar approach. As stated in the introduction, we generate a final configuration by moving cells to the right with certain speeds. A *speed assignment* assigns to every cell U_i a non-negative real number $speed(U_i)$, $1 \leq i \leq n$. A speed assignment is *legal* for configuration $\mathcal{C}$ when: (1) If $d(u_i, u_j) = 0$, then $speed(U_i) \leq speed(U_j)$. This guarantees that, when the distance between U_i and U_j is zero, moving cell U_i according to its speed does not cause an overlap with the slower moving cell U_j. (2) If $H_{i,j}$ is a wire of maximum length, then $speed(U_i) > speed(U_j)$; i.e., the length of the longest wire $H_{i,j}$ decreases.

We represent the arcs of the distance graph having weight 0 and arcs of the wire graph corresponding to longest wires in the *speed graph* $G_s = (V_s, A_s)$ with $V_s = V_d \cup \{u_s\}$. For every vertex u_i in V_d we include the arc (u_s, u_i) of cost 0. For every arc (u_i, u_j) of weight 0 in the distance graph, the speed graph contains the arc (u_i, u_j) of cost 0. If (u_i, u_j) is an arc in G_w, speed graph G_s contains the arc (u_j, u_i) having cost 1. Observe that the direction is now from u_j to u_i. See Figure 1(c) for an example.

Lemma 3.1 *Let $\mathcal{C}$ be a configuration. The longest wire length is not minimized if and only if there exists a legal speed assignment for $\mathcal{C}$.*

Proof. (Sketch) If there exists a legal speed assignment, the longest wire length can be reduced by moving the cells to the right according to the associated speeds. Assume now that the longest wire length is not minimized in $\mathcal{C}$. Let $x_{\mathcal{C}}(U_i)$ be the x-position of the leftmost vertical edge of cell U_i in configuration $\mathcal{C}$. A re-positioning argument shows that there exists a configuration $\mathcal{C}^*$ in which the longest wire length is a minimum and $x_{\mathcal{C}}(U_i) \leq x_{\mathcal{C}^*}(U_i)$ holds for every i. Setting $speed(U_i) = x_{\mathcal{C}^*}(U_i) - x_{\mathcal{C}}(U_i)$, $1 \leq i \leq n$, results in a legal speed assignment for configuration $\mathcal{C}$. □

There can exist many legal speed assignments for configuration $\mathcal{C}$. Our algorithm determines a legal speed assignment by performing a single-source longest path computation in the speed graph. Assume that G_s contains no positive cycle. Let $\mathcal{L}(u_i)$ be the length of the longest path from source u_s to vertex u_i in G_s and let $P = < u_s, u_{k_1}, \ldots, u_{k_l} = u_i >$ be the associated path. A simple argument shows that setting speeds equal to the longest path entries results a legal speed assignment. Intuitively, the existence of path P implies that one can travel from cell U_{k_1} to cell U_i using $\mathcal{L}(u_i)$ longest wires. By travel we mean that there exists a path from u_{k_1} to u_i in the speed graph that corresponds to moving from U_{k_1} to U_i along horizontal wires of maximum length, through the interior of cells, and from one cell to another cell as long as the distance between these two cells is 0. Moving every cell U_{k_r} on path P to the right with a speed of $\mathcal{L}(u_{k_r})$, $1 \leq r \leq l$, reduces the length of the longest horizontal wires associated with path P. We summarize the discussion in the following theorem and refer to [4] for details.

Theorem 3.1 *There exists a legal speed assignment for $\mathcal{C}$ if and only if the speed graph of $\mathcal{C}$ contains no cycle of positive cost. If no positive cycle exists, setting $speed(U_i) = \mathcal{L}(u_i)$, $1 \leq i \leq n$, gives a legal speed assignment.*

The entries $\mathcal{L}(u_i)$ are computed similar to the Bellman-Ford algorithm [1] for solving a single-source shortest path problem on a graph with negative weights using of relaxation. The speed graph G_s contains at most n_h arcs having cost 1. Hence, if the length of the longest path from source u_s to a vertex exceeds n_h, G_s contains a positive cycle and no legal speed assignment exists. Since the longest path from u_s to each vertex will not exceed n_h, the longest path algorithm has an $O(n \cdot n_h)$ running time.

Moving the cells to the right according to the speeds specified in the legal speed assignment reduces the longest wire length and changes the distances between cells. After a certain amount of movement, the speed assignment may no longer be legal. We need to re-compute legal speeds when one (or both) of the following occurs. An arc (u_i, u_j) having a positive weight in distance graph turns into an arc of weight 0, or a wire $H_{i,j}$ which is not a longest wire before the movement to the right turns into a longest wire. We refer to the first event as a *distance event* and to the second one as a *wire event*. The earliest time at which either event occurs is called the *event time* for configuration C. The remainder of this section describes how to determine the event time.

Assume a legal speed assignment has been determined for configuration C. Let t_d (resp. t_w) be the earliest time at which a distance (resp. wire) event occurs. The event time is $\min\{t_d, t_w\}$. Consider first the computation of t_d. Let (u_i, u_j) be an arc of the distance graph with $d(u_i, u_j) > 0$ and $speed(U_i) > speed(U_j)$. The *zero time* $zt_{i,j}$ of (u_i, u_j) is the time at which $d(u_i, u_j)$ turns 0 when U_i and U_j move to the right with their respective speeds; i.e., $zt_{i,j} = \frac{d(u_i, u_j)}{speed(U_i) - speed(U_j)}$. Time t_d is determined in $O(n)$ time by selecting the minimum among these zero times.

Consider now the computation of t_w. Let (u_i, u_j) be an arc of G_w representing horizontal wire $H_{i,j}$ in C and let $length_{i,j}(t) = w(u_i, u_j) - t \cdot (speed(U_i) - speed(U_j))$. The value of $length_{i,j}(t)$ represents the length of wire $H_{i,j}$ at time t when cells U_i and U_j move to the right with their respective speeds. The slope of $length_{i,j}(t)$ is negative (resp. positive) when the length of $H_{i,j}$ reduces (resp. increases). Let $ENV(t)$ be the upper envelope of the *length* functions. Then, t_w is the minimum of $ENV(t)$. For every wire $H_{i,j}$ whose length is reduced, the slope of $length_{i,j}(t)$ is not larger than -1. There exists at least one longest wire, say $H_{i,j}$, with $speed(U_i) = speed(U_j) + 1$, and thus the slope of $length_{i,j}(t)$ is -1. This implies that $ENV(t)$ contains only one line segment of negative slope and it has slope -1. Let $L(C)$ be the length of the longest wire in C and let $length_L(t) = L(C) - t$. The minimum of the upper envelope occurs at the intersection of $length_L(t)$ and $length_{p,q}(t)$ having a nonnegative slope for some horizontal wire $H_{p,q}$. We thus determine for each wire $H_{p,q}$ with $speed(U_p) \leq speed(U_q)$ the time at which $length_L(t)$ intersects $length_{p,q}(t)$. We call this the *intersection time* of wire $H_{p,q}$. Clearly, the minimum of $ENV(t)$ can be determined in $O(n_h)$ time.

4 Longest Wire Minimizing Algorithm

In this section we describe an $O(n_h \cdot n \log n)$ time algorithm for generating a configuration minimizing the length of the longest horizontal wire. The algorithm performs at most $n_h \cdot n$ iterations, with each iteration generating a configuration

having smaller longest wire length. Let $\mathcal{C}_i$ be the configuration at the beginning
of the i-th iteration, $i \geq 1$. Also, let G_d^i, G_w^i and G_s^i be the distance graph, wire
graph and speed graph of $\mathcal{C}_i$, respectively. Accordingly, $d_i(u_j, u_k)$ and $w_i(u_j, u_k)$
are the weight of arc (u_j, u_k) in distance graph G_d^i and wire graph G_w^i, respectively. The i-th iteration determines whether a legal speed assignment for
configuration $\mathcal{C}_i$ exists. If none exists, configuration $\mathcal{C}_i$ has minimum longest wire
length. Otherwise, let $speed_i(\cdot)$ be the computed entries. We then compute the
event time $t_{e,i}$ for configuration $\mathcal{C}_i$ and move every cell U_j distance $t_{e,i} \cdot speed_i(U_j)$
to the right. The movement to the right gives configuration $\mathcal{C}_{i+1}$.

We next sketch the ideas for showing that this process terminates after at
most $n_h \cdot n$ iterations. We refer to [4] for details. First, the speed of a cell cannot
decrease from one iteration to the next. An intuitive argument is as follows. Let
$\mathcal{L}^i(u_j)$ be the longest path length from u_s to u_j in G_s^i. The entry $\mathcal{L}^i(u_j)$ implies
that one can travel from a cell U_{k_1} to U_j using $\mathcal{L}^i(u_j)$ longest horizontal wires.
After the movement to the right these longest wires remain longest wires and the
number of longest horizontal wires available for traveling from U_{k_1} to U_j could
only have increased. Observe that longest horizontal wires that are not on any
longest path could have reduced faster and may no longer be longest wires after
the movement to the right stopped. Secondly, in every iteration there exists at
least one cell U_j with $speed_i(U_j) < speed_{i+1}(U_j)$. This holds since there exists at
least one arc (u_p, u_q) that is not in speed graph G_s^i, but is in G_s^{i+1}. The addition
of this arc causes an increase in either $speed_i(U_p)$ or $speed_i(U_q)$. Since the speed
of a cell is bounded by n_h, the claimed bound of $n_h \cdot n$ follows immediately.

From the previous section it follows that one iteration can be completed in
$O(n_h \cdot n)$ time, resulting in an $O(n_h^2 \cdot n^2)$ time algorithm. Our $O(n_h \cdot n \log n)$ time
algorithm updates, rather than recomputes, the data structures and information
used in the i-th iteration from that used in the $(i-1)$-st iteration. The rest of
this section sketches the details of one iteration.

Assume that the first iteration starts at time 0 and that the i-th iteration
starts at time $t_{e,i-1}$ and ends at time $t_{e,i}$. Let speed graph G_s^i be obtained from
G_s^{i-1} by deleting the arcs in set D_{i-1} and adding the arcs in set A_{i-1}. At the
beginning of the i-th iteration, $i \geq 2$, we have the following information:

(i) a legal speed assignment for G_s^{i-1},
(ii) heap $\mathcal{H}_A$ (resp. heap $\mathcal{H}_B$) storing the zero times (resp. intersection times) and
their corresponding arcs (resp. wires) induced by the legal speed assignment
of G_s^{i-1}, and
(iii) event time $t_{e,i-1}$ and arc sets A_{i-1} and D_{i-1}.

The first iteration sets up the necessary graphs and data structures and determines event time $t_{e,1}$ and arc sets A_1 and D_1. This computations can easily be
done within $O(n_h \cdot n)$ time. Assume $i \geq 2$. The first step of the i-th iteration
creates speed graph G_s^i from G_s^{i-1}. We then determine a legal speed assignment
for G_s^i from the one computed for G_s^{i-1}. Since the speed of a cell cannot decrease, only the arcs in A_{i-1} can cause a change in the $speed$-entries. Every arc
in A_{i-1} is considered and its effect on the $speed_i$-entries is determined (i.e., we

determine which *speed*-entries increase). This is done by performing relaxation on arcs of the speed graph. Should a $speed_i$-entry get a value larger than n_h, the algorithm terminates and $\mathcal{C}_i$ is a final configuration.

Once the legal speed assignment for speed graph G_s^i has been determined, heaps $\mathcal{H}_A$ and $\mathcal{H}_B$ are updated. Consider first heap $\mathcal{H}_A$. Every element of $\mathcal{H}_A$ corresponds to an arc of the distance graph inducing a zero time. Let (u_j, u_k) be an arc of the distance graph. If the speed of cell U_j or that of cell U_k is increases in the i-th iteration, the zero time associated with arc (u_j, u_k) may need to be updated. This involves the deletion and the re-insertion of an heap entry. In order to compute the new zero time $zt_{j,k}^i$, we need the value of $d_i(u_j, u_k)$. Since it is not possible to compute all d_i- and w_i- entries (doing so would result in an $O(n_h \cdot n^2)$ running time), we do following. Assume that iteration p, $p < i$, was the last iteration in which either the the speed of cell U_j or U_k was increased. Whenever the speed of a cell increases, we update the d-entries of all the arcs incident to this cell. This implies that during the p-th iteration $d_p(u_j, u_k)$ was computed. During iterations $p + 1, \ldots, i - 1$ the speed of neither U_j or U_k was increased. Then, $d_i(u_j, u_k) = d_p(u_j, u_k) + (speed_p(U_k) - speed_p(U_j)) \times (t_{e,i-1} - t_{e,p})$. The w_i-entries are handled in an analogous way and thus, whenever a d_i- or w_i-entry is needed, it can be computed in $O(1)$ time. The updating of heap $\mathcal{H}_B$ is similar since the intersection time of a wire can only change when the speed of a cell incident to this wire increases.

The last step of the i-th iteration determines the event time $t_{e,i}$ and arc sets A_i and D_i. Event time $t_{e,i}$ is computed by determining the minimum entry in each heap. Assume (u_j, u_k) is an arc to be included in A_i. If (u_j, u_k) has a positive weight in the distance graph G_d^i, but weight 0 in G_d^{i+1}, then arc (u_j, u_k) induces a distance event. This implies that the zero time associated with arc (u_j, u_k) is a minimum in heap $\mathcal{H}_A$. On the other hand, if (u_j, u_k) represents a wire $H_{k,j}$ which is not a longest wire in the i-th iteration, but is a longest wire in the $(i+1)$-st iteration, then (u_j, u_k) induced a wire event. Hence, the intersection time associated with (u_j, u_k) is a minimum in heap $\mathcal{H}_B$. Set A_i is thus formed by the arcs in heaps $\mathcal{H}_A$ and $\mathcal{H}_B$ causing event time $t_{e,i}$. Arc set D_i contains the arcs to be deleted from G_s^i. If arc (u_j, u_k) is in D_i, then the speed of one of the associated cells must have been increased in the i-th iteration. We refer to [4] for details. Hence, set D_i is created while the $speed_i$-entries are updated.

Consider the time complexity of this algorithm. Assume we perform I iterations. If an arc (u_j, u_k) is added into D_i, the speed of at least one of U_j and U_k increased. Since the speed of a cell cannot exceed n_h, $\sum_{i=1}^{I} |D_i| \le n_h \cdot n$. If an arc is included into A_i, it was either not in G_s^1 or it was deleted in some earlier iteration. Hence, $\sum_{i=1}^{I} (|A_i| + |D_i|) = O(n_h \cdot n)$. This bounds all computations, excluding the updating of the heaps, to sum up to $O(n_h \cdot n)$ over all I iterations. The total time spent in maintaining heaps $\mathcal{H}_A$ and/or $\mathcal{H}_B$ is $O(n_h \cdot n \log n)$. Combining this discussion with the fact that the algorithm performs at most $n_h \cdot n$ iterations gives the claimed $O(n_h \cdot n \log n)$ time for determining a configuration minimizing the longest wire length.

Let $\mathcal{C}_f$ be the configuration of minimum longest wire length generated by our algorithm, and let $L(\mathcal{C}_f)$ be the length of the longest wire in $\mathcal{C}_f$. $\mathcal{C}_f$ may not have minimum width among all configurations having wire length $L(\mathcal{C}_f)$. To generate a configuration of minimum width, we use Liao and Wong's [6] algorithm once using $\mathcal{C}_f$ as the input and setting the upper bound of wire length to $L(\mathcal{C}_f)$. This step takes $O(n_h \cdot n \log n)$ time.

5 Tradeoff Function Minimizing Algorithm

In this abstract we only sketch the ideas underlying our algorithm minimizing a tradeoff function between the layout width and longest wire length. Let $W(\mathcal{C})$ and $L(\mathcal{C})$ be the layout width and longest wire length of configuration $\mathcal{C}$, respectively. In order to control the width of the configurations we add two cells U_l and U_r to the initial layout. Each has a height equal to that of the layout and a width of 0. Cell U_l (resp. U_r) is positioned immediately to the left (resp. right) of the leftmost (resp. rightmost) cell of the layout. We call a configuration $\mathcal{C}$ a *min-width configuration* if width $W(\mathcal{C})$ is a minimum among all configurations having longest wire length not larger than $L(\mathcal{C})$. Given a min-width configuration $\mathcal{C}$, if we want to reduce $\alpha \cdot W(\mathcal{C}) + \beta \cdot L(\mathcal{C})$ by decreasing the layout width, the length of longest wire must increase.

We first generate the min-width configuration minimizing the longest wire length. Let $\mathcal{C}_f$ be this configuration. Starting with $\mathcal{C}_f$, we generate $\mathcal{C}^*$ over a number of iterations, with each iteration generating a new min-width configuration. New configurations are generated by moving cells to the right. The movement is controlled by a value-reducing speed assignment. A *value-reducing speed assignment* assigns to every cell in a configuration $\mathcal{C}$ a non-negative real number $speed(\cdot)$. It is legal when: (1) If $d(u_i, u_j) = 0$ in the distance graph, then $speed(U_i) \le speed(U_j)$. (2) $speed(U_l) > speed(U_r)$; i.e., the layout width decreases. (3) There exists a time t such that when every cell U_i moves to the right a distance of $speed(U_i) \cdot t$ and $\mathcal{C}^t$ represents the resulting configuration, we have $\alpha \cdot W(\mathcal{C}) + \beta \cdot L(\mathcal{C}) > \alpha \cdot W(\mathcal{C}^t) + \beta \cdot L(\mathcal{C}^t)$. If $\mathcal{C}$ is a min-width configuration, a legal value-reducing speed assignment implies that one can reduce the tradeoff function value by moving the cells to the right according to the associated speeds.

We determine a legal value-reducing speed assignment by performing a single-source shortest path computation in the speed graph. Let $\mathcal{S}(u_i)$ be the length of the shortest path from u_l to u_i in the speed graph. If no path exists, we set $\mathcal{S}(u_i) = n_h$. Intuitively, the existence of a shortest path from u_l to u_i implies that in order to travel from U_l to U_i, one has to go through at least $\mathcal{S}(u_i)$ longest wires. When decreasing the layout width, the increase in the length of the longest wire needs to be as small as possible. We thus distribute the increase evenly among the $\mathcal{S}(u_i)$ longest wires. It can be shown that setting $speed(U_j) = n_h - \mathcal{S}(u_j)$, for every cell U_j, results in a legal value-reducing speed assignment. For example, Figure 2(a) shows a min-width configuration in which H_1 and H_2 are the longest wires. If we want to decrease the layout width of this configuration, the length of either H_1 or H_2 or both must increase. Rather than increasing the length of either H_1 or H_2, evenly increasing the length of both wires minimizes the increase of longest wire length. Figure 2(c) shows the min-width configuration generated by evenly distributing the increase of longest wire length between H_1 and H_2.

Assume we are given a min-width configuration $\mathcal{C}$. Let $\gamma = \frac{\alpha}{\beta}$. If $\gamma \cdot \mathcal{S}(u_r) > 1$, then, setting $speed(U_j) = n_h - \mathcal{S}(u_j)$, for every cell U_j, results in a legal value-reducing speed assignment for $\mathcal{C}$. Moving the cells to the right reduces the value of the tradeoff function and generates a new min-width configuration. If $\gamma \cdot \mathcal{S}(u_r) \leq 1$, no value-reducing speed assignment exists and we found $\mathcal{C}^*$. We perform at most $n_h \cdot n$ iterations and by implementing each iteration so that re-computations are avoided, the $O(n_h \cdot n \log n)$ time is achieved.

References

1. T. H. Cormen, C. E. Leiserson, and R. L. Rivest. *Introduction to Algorithms*. The MIT Press, 1990.
2. S. Gao, M. Kaufmann, and F. M. Maley. Advances in homotopic layout compaction. In *Proceedings of 1989 ACM SPAA*, pages 273–282, 1989.
3. S. E. Hambrusch and H. Y. Tu. Minimizing total wire length during 1-dimensional compaction. *INTEGRATION, the VLSI journal*, 14(2), 1992.
4. S. E. Hambrusch and H. Y. Tu. *New Algorithms for Minimizing the Longest Wire Length during Circuit Compaction*. Technical Report, Purdue University, 1993.
5. T. Lengauer. *Combinatorial Algorithms for Integrated Circuit Layout*. J. Wiley & Sons, 1990.
6. Y. Z. Liao and C. K. Wong. An algorithm to compact a VLSI symbolic layout with mixed constraints. *IEEE Transactions on CAD of Integrated Circuits and Systems*, 2(2):62–69, April 1983.
7. D. Marple. A hierarchy preserving hierarchical compactor. In *Proceedings of 27th ACM/IEEE Design Automation Conference*, pages 375–381, 1990.
8. D. A. Mlynski and C. H. Sung. Layout compaction. In T. Ohtsuki, editor, *Layout Design and Verification*, pages 199–235. Elsevier Science Publ., 1986.
9. A. R. Newton. Symbolic layout and procedural design. In G. DeMicheli, A. Sangiovanni-Vincentelli, and P. Antognetti, editors, *Design Systems for VLSI Circuits*, pages 65–112, Boston, 1987. Martinus Nijhoff Publishers.
10. A. Onozawa. Layout compaction with attractive and repulsive constraints. In *Proceedings of 27th ACM/IEEE Design Automation Conference*, pages 369–374, 1990.
11. W. L. Schiele. Improved compaction by minimized length of wires. In *Proceedings of 20th ACM/IEEE Design Automation Conference*, pages 121–121, 1983.
12. B. X. Weis and D. A. Mlynski. A graphtheoretic approach to the relative placement problem. *IEEE Transactions on Circuits and Systems*, 35(3):286–293, 1988.

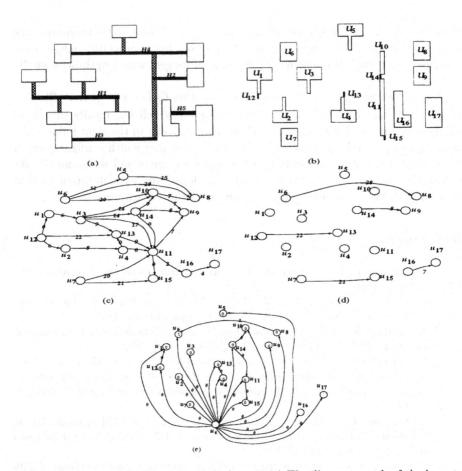

Fig. 1. (a) A layout. (b) The cells of the layout. (c) The distance graph of the layout. (d) The wire graph of the layout. (e) The speed graph of the layout.

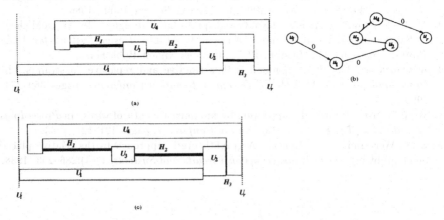

Fig. 2. (a) A min-width configuration. (b) The speed graph of the layout shown in (a). (c) The min-width configuration generated by evenly distributing the increase of longest wire length between H_1 and H_2.

We consider these problems in all contexts for which linear-time sequential algorithms are known instead of considering only river routing, where each net is restricted to have exactly one terminal on each side of the channel. The input we assume is two arrays of terminals sorted by x-coordinate. The top terminals (and, in arithmetic contexts, their x-coordinates) are denoted $t_0, t_1, \ldots, t_{m-1}$, and the bottom terminals are denoted $b_0, b_1, \ldots, b_{n-1}$. We use N to denote $m+n$. Associated with each terminal is a net number. Terminals belonging to the same net are to be connected together. We also assume that each terminal has a pointer to the next terminal of the same net in a clockwise ordering of the terminals. (It may be possible to eliminate this assumption under some circumstances, e.g., constant number of terminals per net, allowing randomization, and/or allowing a modest increase in time or work, by applying results on sorting of small integers (e.g, [10, 15]).) For simplicity, we use a rectilinear, grid-based model in which terminals lie on gridpoints and wires are disjoint paths through grid edges. Also, for convenience, we allow routing on channel boundaries.

We henceforth assume that each net has two terminals. Multiterminal nets can be handled by a transformation described in [7], that might be considered "folklore". Then a *single-sided net* has its two terminals on the same side of the channel, whereas a *two-sided net* is the type of net allowed in river routing.

We also assume henceforth that the channel is routable in one layer, i.e., no two nets are topologically forced to cross. (This condition can be verified without increasing the running time of our parallel algorithm by doing parentheses matching [2].)

The results obtained in this paper are summarized in Table 1, where the river routing model is as described above, the general model includes any single-layer channel routing problem, and the intermediate model is one in which all single-sided nets are on one side of the channel.

Table 1. Running time and work (processor-time product) for the algorithms presented in this paper.

problem	model	CREW		CRCW	
		time	work	time	work
min. sep.	general	$O(\lg N)$	$O(N)$	$O(\lg N/\lg\lg N)$	$O(N)$
offset range	intermediate	$O(\lg N)$	$O(N)$	$O(\lg N/\lg\lg N)$	$O(N)$
optimal offset	intermediate	$O(\lg N \lg\lg N)$	$O(N \lg\lg N)$	$O(\lg N)$	$O(N \lg\lg N)$
optimal offset	river	$O(\lg N \lg\lg N)$	$O(N \lg\lg N)$	$O((\lg\lg N)^2)$	$O(N \lg\lg N)$

Most prior work on single-layer routing has been limited to sequential models of computation; linear-time sequential algorithms for the problems considered in this paper can be found in [7], [9], and [16]. For the river routing model only, parallel algorithms for the problems in this paper are given by Aggarwal and

Park [1]. For optimal offset, their results are $O(\lg^2 N)$ time on the CREW and $O(\lg N \lg \lg N)$ time on the CRCW, both with $O(N \lg N)$ work. We obtain superior bounds even for the more general problem labeled as the "intermediate" model in Table 1. We improve these bounds even further for river routing on the CRCW by using the offset range results of Aggarwal and Park quoted in Sect. 3.3. Chang, JáJá, and Ryu [4], independently obtain the same bounds as Aggarwal and Park for minimum separation in the river routing model (matching our CREW bounds for the general model and improving the CRCW bounds of the general model) and also give an optimal (work) algorithm for routability testing for switchboxes.

The remainder of this paper is organized as follows. In Sect. 2, we explain the parallel operations used in this paper. We also indicate how to conveniently express the routability conditions for *single-layer channel routing*. These conditions are then used in Sect. 3 to solve the minimum separation, offset range and optimal offset problems. Section 4 gives some concluding remarks.

2 Preliminaries

2.1 Basic Parallel Operations

Given a sequence of N elements $\{x_1, x_2, \ldots, x_N\}$ with a binary associative operator $*$, the *prefix sums* are all the partial sums defined by:

$$p_i = x_1 * x_2 * \ldots * x_i, 1 \le i \le N$$

For a given array $A(i)$ with $1 \le i \le N$, the *range maxima problem* is to find the element with maximum value between two given positions i and j. A query can be answered in $O(1)$ time after the preprocessing described in [11].

The range maxima preprocessing can be implemented in $O(\lg N)$ (respectively $O(\frac{\lg N}{\lg \lg N})$) time with $O(N)$ work on a CREW (CRCW) PRAM [3]. The prefix sums computation can be performed with the same time and processor bounds on the CREW [13], and on the CRCW as long as the input elements are integers in the interval $[1, N]$ [5].

2.2 Cut Conditions

We need a few definitions in order to use a general theory of single-layer routing developed by Maley. Define a *critical cut* to be a line segment that connects a top and bottom terminal or runs from a terminal straight across to the opposite side of the channel. Define a *pivotal cut* to be a line segment that connects a top and bottom terminal or runs at 45° from a terminal to the opposite side of the channel. Also let the *flow* across a cut χ be the number of nets that must cross χ, namely those nets having terminals on both sides of χ and those having an endpoint of χ as a terminal. The *capacity* of χ is one greater than the maximum

of the horizontal and vertical separations of its endpoints; if χ is the line segment from (x_1, y_1) to (x_2, y_2), then

$$capacity(\chi) = max\{|x_1 - x_2|, |y_1 - y_2|\} + 1 \ .$$

The cut χ is *safe* if $flow(\chi) \leq capacity(\chi)$, which means that there is enough space along χ for the wires to get through.

Lemma 1. *A channel is routable if and only if every critical cut or every pivotal cut is safe.* □

This lemma follows from the corresponding results in [14, §2.1,2.3,2.6.5]. (Our slightly different definitions of flow and capacity allow Maley's formulation in terms of cuts emanating from "feature" endpoints to correspond to cuts emanating from terminals. Since we allow routing on the channel boundaries, the only "features" are terminals and two routing obstacles (horizontal lines) located one unit outside of what we have been referring to as the channel boundaries.)

We can further strengthen the result for critical cuts as follows. Define the *span* of a cut χ to be the horizontal distance between its endpoints. Call χ *sparse* if χ is not vertical and $flow(\chi) \leq span(\chi) + 1$, and *dense* otherwise. A sparse cut is safe regardless of the separation, but a dense cut χ is safe if and only if the separation is at least $flow(\chi) - 1$.

Lemma 2. *The minimum channel separation is the maximum of $flow(\chi) - 1$ over dense critical cuts χ.* □

When all single-sided nets are on the bottom, we can strengthen the result for pivotal cuts, but first we must review results regarding *contours* of single-sided nets. Define the *contour* of single-sided nets on the bottom to be the upper boundary of the routing region consumed in the routing of these nets that minimizes total wire length. That is, when the nets are routed as tightly as possible against the boundary of the channel, the contour is formed by the uppermost nets and portions of the channel boundary.

The following Lemma from [4] shows that a contour of single-sided nets can be found efficiently.

Lemma 3. *The bendpoints in the contour of a set of N single-sided nets can be found in $O(\lg N)$ (respectively $O(\frac{\lg N}{\lg \lg N})$) time using $O(\frac{N}{\lg N})$ (respectively $O(\frac{N \lg \lg N}{\lg N})$) processors on a CREW (CRCW) PRAM.* □

Now, we are ready to state a result of [9] relating to pivotal cuts:

Lemma 4. *A channel with all single-sided nets on the bottom is routable if and only if all 45° cuts from bottom terminals of two-sided nets and all 45° cuts crossing the contour of single-sided nets at a convex corner are safe.* □

3   The Algorithms

3.1   The Minimum Separation Problem

Our algorithm for this problem is based on Lemma 2. To ensure that vertical cuts are captured, first add a dummy terminal across from each real terminal. Then we find the minimum separation that makes all dense (critical) cuts emanating from bottom terminals safe. To find all dense cuts emanating from b_j, we search for the two farthest dense cuts, one going to the right and one going to the left from b_j; these two cuts form a "cone" such that cuts emanating from b_j are dense if and only if they lie inside the cone.

We now provide some further definitions and notations used in this subsection. First, we say that a terminal is covered by a single-sided net on its side of the channel if it lies in the closed interval defined by the endpoints of the net. Two-sided nets are said to lie to the left or right of a terminal on the top (or bottom) according to the location of the net's top (respectively bottom) terminal. Also, a two-sided net is a right net if its top terminal is to the right of its bottom terminal; it is a left net if its top terminal is to the left of its bottom terminal. Define $R(\tau)$ to be the number of right nets to the left of terminal τ, and $L(\tau)$ to be the number of left nets to the left of τ. Also define $S(\tau)$ to be the number of single-sided nets covering τ. Define IL(τ) (and IR(τ)) to be 1 if τ is a terminal of a left (respectively right) net, and zero otherwise.

The heart of this algorithm is to form the cone for each terminal b_j. According to the definition, a nonvertical cut $\overline{t_i b_j}$ is dense if $|t_i - b_j| + 1 < flow(\overline{t_i b_j})$. We now show how to find the farthest cuts emanating from each terminal b_j on the bottom that are dense. Note that for any dense cut $\overline{t_i b_j}$, $R(t_i) \le R(b_j)$, and $L(t_i) \ge L(b_j)$. Thus, for dense cuts, $flow(\overline{t_i b_j}) = L(t_i) - L(b_j) + \text{IL}(t_i) + R(b_j) - R(t_i) + \text{IR}(b_j) + S(t_i) + S(b_j)$. Defining, bl($j$) to be $b_j + L(b_j) - R(b_j) - S(b_j) - \text{IR}(b_j) + 1$ and tl(i) to be $t_i + L(t_i) - R(t_i) + S(t_i) + \text{IL}(t_i)$, we need to find the smallest t_i such that bl(j) < tl(i); for similar definitions of tr(i) and br(j), we also find the largest t_i such that tr(i) < br(j). It can be shown that the four functions bl, tl, br, and tr are non-decreasing. Now, we can give an algorithm for forming the cone for each bottom terminal.

procedure FIND-CONES

1. Compute tl(i), bl(j), tr(i) and br(j) for $0 \le i \le m - 1$ and $0 \le j \le n - 1$.
2. Merge tl(i) with bl(j) and tr(i) with br(j) in order of nondecreasing values. If a tie occurs, put br(j) before tr(i) and put bl(j) after tl(i).
3. For each bl(j), find the nearest tl(i) to the right in the merged sequence. If we do not find such a tl(i) corresponding to a t_i with lesser x-coordinate than b_j, then the farthest dense cut to the left from b_j is vertical. Similarly, for each br(j), find the nearest tr(i) to the left in the merged sequence, and select a vertical cut if necessary.

The merging can be done using the approach of Kruskal [12] in $O(\lg\lg N)$ time with $O(N)$ work on a CREW PRAM. Also steps 1 and 3 can be implemented by using prefix-sums. Therefore, algorithm FIND-CONES can be implemented with optimal work in $O(\lg N)$ (respectively $O(\frac{\lg N}{\lg\lg N})$) time on a CREW (CRCW) PRAM.

Once we find the cone for each bottom terminal, we can use the information to find the minimum separation for the single-layer channel routing problem:

procedure MINIMUM-SEPARATION

1. Apply algorithm FIND-CONES to find the farthest dense cuts to form a cone for every terminal on the bottom.
2. Find the maximum flow $F(b_j)$ among the cuts inside the cone for every terminal b_j.
3. The minimum separation is $-1 + \max\{F(b_1), F(b_2), \ldots, F(b_n)\}$.

Theorem 5. *Algorithm* MINIMUM-SEPARATION *finds minimum separation for single-layer channel routing with $O(N)$ work in time $O(\lg N)$ on a CREW PRAM and in time $O(\frac{\lg N}{\lg\lg N})$ on a CRCW PRAM.*

Proof. We have already explained how step 1 can be performed within the specified time and processor bounds, and step 3 simply involves a minimum that can be computed in the same bounds. To find the maximum flow in each cone in step 2, we use the range maxima technique. Since, the flow for a dense cut $\overline{t_i b_j}$ is $L(t_i) - L(b_j) + \mathrm{IL}(t_i) + R(b_j) - R(t_i) + \mathrm{IR}(b_j) + S(t_i) + S(b_j)$, and the terms dependent on j are fixed for any given cone, the task is to find the maximum of $\mathrm{tl}(i) - t_i$ over each cone. The preprocessing for range maxima and the single query per b_j can also be implemented within the stated bounds. $\Box$

3.2   The Offset Range Problem

In this subsection, we consider the offset range problem for single-layer channel routing with single-sided nets on one side. Without loss of generality, assume that all single-sided nets are on the bottom. Additional notation used in this subsection is as follows. Let s be the separation and d the offset (the positive or negative distance by which the upper block of terminals is moved right from its original position) . Define $T(\tau)$ to be the number of two-sided nets to the left of the terminal τ. Also, define $\mathrm{IT}(\tau)$ to be one if τ belongs to a two-sided net, and zero otherwise.

According to Lemma 4, we only need to ensure that all $45°$ cuts from bottom terminals of two-sided nets and all $45°$ cuts crossing the contour of single-sided nets at a convex corner are safe. In this subsection, we achieve that task by checking all $45°$ cuts from bottom terminals $b_0, b_1, \ldots, b_{n-1}$. For a cut χ emanating from b_j, the flow contributed by single-sided nets is $S(b_j)$; to be safe,

χ can accommodate at most $s + 1 - S(b_j)$ more flow. For simplicity, we denote $s + 1 - S(b_j)$ as e_j. Then, for the left-going $45°$ cut from b_j to be safe, we need that $b_j - s > t_i + d$ when the number of two-sided nets crossing cut $\overline{t_i b_j}$ exceeds e_j. Finding the i that gives the tightest constraint, we have $b_j - s > t_{T(b_j)-e_j+IT(b_j)} + d$. A similar argument for the right-going cut yields $b_j + s < t_{T(b_j)+e_j+1} + d$. So the feasible offsets, if any, are given by

$$\max_{0 \leq j \leq n-1} \{b_j + s - t_{T(b_j)+e_j+1}\} < d < \min_{0 \leq j \leq n-1} \{b_j - s - t_{T(b_j)-e_j+IT(b_j)}\} .$$

The most difficult operation involved in the computation just specified is finding the $S(b_j)$ and $T(b_j)$ values, for which performing prefix sums suffices, so we have the following theorem:

Theorem 6. *The offset range for single-layer channel routing with single-sided nets on one side can be found with $O(N)$ work in $O(\lg N)$ (respectively $O(\frac{\lg N}{\lg \lg N})$) time on a CREW (CRCW) PRAM.* $\square$

3.3 The Optimal Offset Problem

This subsection considers the optimal offset problem for river routing and channels with single-sided nets on one side. In both cases, the optimal offset problem is solved by using an algorithm for offset range as a subroutine. For channels with single-sided nets on one side, we use the results of Sect. 3.2 to obtain optimal offset results better than those of Aggarwal and Park, even though their results apply only to river routing. For river routing on the CRCW, we further improve their optimal offset bounds by using their results for offset range, $O(\lg \lg N)$ time and $O(N)$ work. (On the CREW, their offset range results offer no improvement over the results of Sect. 3.2.)

We first explain how to solve the problem for river routing, and then we extend the algorithm to channels with single-sided nets on one side. For ease of presentation, we assume the number of nets N is 2^p for some integer p.

Our algorithm adapts Mirzaian's halving technique [16] for relating optimal offset to offset range. Let T^0 be the original set of two-sided nets, and define T^k to be the set of even-numbered nets of T^{k-1}, for $1 \leq k \leq p$. Also define optsep(A) to be the minimum separation attainable with an optimal offset for a channel with the set A of nets. (Once optsep(A) is determined, the solution of the offset range problem can be used to determine the optimal offsets.) The following lemma adapted from [16][2] states the relationship between optsep(T^k) and optsep(T^{k+1}):

Lemma 7. *Let $s^k = $ optsep(T^k) and $s^{k+1} = $ optsep(T^{k+1}), then $0 \leq s^k - 2s^{k+1} \leq 2$.* $\square$

---

[2] This lemma differs slightly from [16] because we are allowing routing on both boundaries of the channel.

From the above lemma, once we know s^{k+1}, then s^k can be solved by checking only three possible separations to achieve the optimal offset. Now, s^{k-1} can also be solved by checking all the possible separations derived from the three separations. (Again each possible separation of s^k induces 3 possible separations for s^{k-1}.) By considering the union of these separations, we only have to check 7 possible separations to solve s^{k-1}. Continuing this line of reasoning, we have the following corollary:

Corollary 8. *If s^k is known, then s^{k-l} can be solved by checking only $2^{l+1} - 1$ possible separations.* □

Our algorithm finds the optimal separations $s^{p/2^i}$, for $1 \leq i \leq \lg p$. Each separation is determined from previously computed separations by using Corollary 8. Figure 1 shows the algorithm to solve the optimal offset problem.

```
    procedure OPTIMAL-SEPARATION
1       for i ← lg p to 1 do
2           find T^{p/2^i}
3       endfor
4       for i ← 1 to lg p do
5           find s^{p/2^i}
6       endfor
7       find s^0
```

Fig. 1. This algorithm finds the minimum separation for the optimal offset problem.

Theorem 9. *The optimal offset for river routing can be found in $O(\lg N \lg \lg N)$ (respectively $O((\lg \lg N)^2)$) time using $O(\frac{N}{\lg N})$ (respectively $O(\frac{N}{\lg \lg N})$) processors on a CREW (CRCW) PRAM.*

Proof. First we explain the CREW result. Lines 1–3 can be executed in time $O(\lg N)$ time using $O(\frac{N}{\lg N})$ processors. (To see this, it suffices to consider repeated "halving" of the net set, with the time at each stage decreasing geometrically to a constant as the number of nets decreases.) In lines 4–6, each $s^{p/2^i}$ can be found in $O(p) = O(\lg N)$ time as follows. First, $s^{p/2}$ has at most $2^{p/2}$ possible values because there are only $2^{p/2}$ nets in $T^{p/2}$. The feasibility of each separation can be checked in $O(p)$ time using $O(\frac{2^{p/2}}{p})$ processors by the offset range algorithm of Sect. 3.2 or [1]. There is a total of $O(\frac{2^p}{p})$ processors, so all possible separations can be checked at the same time. The minimum separation among the feasible separations is the optimal separation. Now, suppose $s^{p/2^i}$ is known; then at most $O(2^{p/2^{i+1}})$ possible values need to be checked for $s^{p/2^{i+1}}$ by Corollary 8. By a similar argument as before, $s^{p/2^{i+1}}$ can be found in $O(p)$ time. Finally, Line 7 is again an $O(p)$ time computation. The total time, including all the passes through the loop in lines 4–6, is $O(p \lg p)$.

On the CRCW, we can do each pass through the loop in lines 1–3 in time $O(\lg \lg N)$ with $O(N/\lg \lg N)$ processors, and there are $O(\lg \lg N)$ passes. The remaining analysis is the same as before except that we use the CRCW result of Aggarwal and Park [1] for offset range, yielding $O(\lg p)$ time to compute each $s^{p/2^i}$ (and s^0). □

When there are single-sided nets, we first use Lemma 3 to find the *contour* of the single-sided nets. At each column, we define the *extension* of the contour to be the distance that the contour extends into the channel at that column. Let T^0 be the original set of two-sided nets and let L^0 be the original contour of single-sided nets. We define T^k as before, and recursively define L^k to be a contour with the extensions of L^{k-1} divided by two (and rounded down to integral values). (The new problem $T^k \cup L^k$ can be viewed as an ordinary problem instance with at most half as many nets as $T^{k-1} \cup L^{k-1}$ by using Lemma 4. The bottom endpoints of the cuts going through convex corners of the new contour can be viewed as the new bottom terminals of single-sided nets that are of interest.) The following lemma from [9] states the relationship between $\text{optsep}(T^k \cup L^k)$ and $\text{optsep}(T^{k+1} \cup L^{k+1})$:

Lemma 10. *Let* $s^k = \text{optsep}(T^k \cup L^k)$ *and* $s^{k+1} = \text{optsep}(T^{k+1} \cup L^{k+1})$. *Then* $s^k \geq 2s^{k+1} - 1$ *and* $s^k \leq 2s^{k+1} + 2$. □

Now we can use the halving technique as before; the only difference is that we have to check 4 possible separations for s^k once s^{k+1} is known. We follow a similar analysis as in Theorem 9, but the CRCW details must be altered to avoid using the result of Aggarwal and Park that only applies to river routing. Instead, we plug in the offset range results from Sect. 3.2.

Corollary 11. *The optimal offset problem for channels with single-sided nets on one side can be solved in* $O(\lg N \lg \lg N)$ *(respectively* $O(\lg N)$*) time using* $O(\frac{N}{\lg N})$ *(respectively* $O(\frac{N \lg \lg N}{\lg N})$*)) processors on a CREW (CRCW) PRAM.* □

4 Conclusion

This paper has provided efficient parallel algorithms for (1) the minimum separation problem for general single-layer channels and (2) offset problems for single-layer channels in which only one side of the channel has multiple connections to a single net. In addition, we have improved previous results for optimal offset in river routing problems, where each net has exactly one terminal on each side of the channel. An obvious open question is whether any of the bounds on time or work can be improved. In particular, the algorithms for optimal offset use $N \lg \lg N$ work rather than the $O(N)$ work that can be achived sequentially. An additional open question is whether offset problems can be efficiently solved in parallel when both sides of the channel contain single-sided nets; for sequential computation, this problem is considered in [8].

5 Acknowledgements

Thanks to Uzi Vishkin and Joseph JáJá of the University of Maryland, Omer Berkman of King's College, and Yossi Matias of AT&T Bell Labs for helpful discussions.

References

1. Alok Aggarwal and James Park. Notes on searching in multidimensional monotone arrays. In *29th Annual Symposium on Foundations of Computer Science*, pages 497–512. IEEE Computer Society Press, 1988.
2. O. Berkman, Baruch Schieber, and U. Vishkin. Some doubly logarithmic optimal parallel algorithms based on finding all nearest smaller values. Technical Report UMIACS-TR-88-79, University of Maryland Institute for Advanced Computer Studies, October 1988. To appear in *J. Algorithms*.
3. O. Berkman and U. Vishkin. Recursive star-tree parallel data-structure. Technical Report UMIACS-TR-90-40, University of Maryland Institute for Advanced Computer Studies, March 1990.
4. Shing-Chong Chang, Joseph JáJá, and Kwan Woo Ryu. Optimal parallel algorithms for one-layer routing. Technical Report UMIACS-TR-89-46, University of Maryland Institute for Advanced Computer Studies, April 1989.
5. R. Cole and U. Vishkin. Faster optimal prefix sums and list ranking. *Information and Control*, 81:334–352, 1989.
6. Danny Dolev, Kevin Karplus, Alan Siegel, Alex Strong, and Jeffrey D. Ullman. Optimal algorithms for structural assembly. *VLSI Design*, pages 38–43, 1982. Earlier version in *Proceedings of the 13th ACM Symposium on Theory of Computing*.
7. Ronald I. Greenberg and F. Miller Maley. Minimum separation for single-layer channel routing. *Information Processing Letters*, 43(4):201–205, September 1992.
8. Ronald I. Greenberg and Jau-Der Shih. Feasible offset and optimal offset for single-layer channel routing. In *Proceedings of 2nd Annual Israel Symposium on Theory of Computing and Systems*, pages 193–201. IEEE Computer Society Press, June 1993. Revised version submitted to *SIAM Journal on Discrete Mathematics*.
9. Ronald I. Greenberg and Jau-Der Shih. Single-layer channel routing and placement with single-sided nets. Submitted to *Discrete Applied Mathematics*, 1993.
10. Torben Hagerup. Constant-time parallel integer sorting. In *Proceedings of the 23rd ACM Symposium on Theory of Computing*, pages 299–306. ACM Press, 1991.
11. Joseph JáJá. *An Introduction to Parallel Algorithms*. Addison-Wesley, 1992.
12. D. Kruskal. Searching, merging and sorting in parallel computation. *IEEE Trans. Computers*, C-32(10):942–946, October 1983.
13. R. E. Ladner and M. J. Fischer. Parallel prefix computation. *Journal of the ACM*, 27(4):831–838, October 1980.
14. F. Miller Maley. *Single-Layer Wire Routing and Compaction*. MIT Press, 1990.
15. Yossi Matias and Uzi Vishkin. Converting high probability into nearly-constant time — with applications to parallel hashing. In *Proceedings of the 23rd ACM Symposium on Theory of Computing*, pages 307–316. ACM Press, 1991.
16. Andranik Mirzaian. River routing in VLSI. *Journal of Computer and System Sciences*, 34:43–54, 1987.

Consecutive Interval Query and Dynamic Programming on Intervals

Alok Aggarwal[1] and Takeshi Tokuyama[2]

[1] IBM Research Division, T. J. Watson Research Center, P.O.Box 218, Yorktown Heights, NY 10598
[2] IBM Research Division, Tokyo Research Laboratory, 1623-14, Shimotsuruma, Yamato-shi, Kanagawa, 242 Japan

Abstract. Given a set of n points on a line and a set of m weighted intervals defined on these points, we consider a particular dynamic programming problem on these intervals. If the weights are all nonnegative or all nonpositive, we solve this dynamic programming problem efficiently by using matrix searching in Monge arrays, together with a new query data structure which we call the *consecutive interval query* structure. We invoke our algorithm to obtain fast algorithms for the sequential partition of a graph and for the partial clique covering of an interval graph.

1   Introduction

Let U be a set of n integers $\{1, 2, .., n\}$ that can be regarded as a set of points on a line, and let Z be a set of m intervals which have their endpoints in U. We associate a weight $w(I)$ with each interval. Let $W(i,j) = \sum_{I \subset (i,j]} w(I)$. Assume that there exists a non-decreasing sequence $f(i); i = 1, 2, .., n - 1$ such that $i \leq f(i) \leq n$. We define $\tilde{W}(i,j) = W(i,j)$ if $i \leq j \leq f(i)$ and ∞, otherwise. The problem is to compute two functions $D(i)$ and $E(i)$ on U, where it is assumed that $D(i)$ can be computed in $O(1)$ time from $E(i)$, $E(1) = 0$, and $E(i)$ is defined by the following recurrence:

$$E(i) = \min_{j>i}\{D(j) + \tilde{W}(i,j)\}. \tag{1}$$

In [1], Asano showed several problems in computational geometry and graph algorithms can be solved by using the recurrence given above and he also gave an $O(m \log n)$ time algorithm for solving this recurrence. (Note that his algorithm is worse than the simple dynamic programming algorithm that takes $O(n^2)$ time, when $m > n^2/\log n$.) His algorithm needs a solution of the *interval query* problem that can be defined as follows: "Given a set of m weighted intervals, preprocess the set so that the total weight of the intervals contained in a query interval I can be computed efficiently". A well-known data structure called the *interval tree* answers the query in $O(\log n)$ time after using $O(m \log n)$ time in preprocessing. Thus, the preprocessing time dominates the total time complexity of Asano's dynamic programming algorithm.

Parallel Algorithms for Single-Layer Channel Routing*

Ronald I. Greenberg, Shih-Chuan Hung, and Jau-Der Shih

Electrical Engineering Department
University of Maryland
College Park, MD 20742

Abstract. We provide efficient parallel algorithms for the minimum separation, offset range, and optimal offset problems for single-layer channel routing. We consider all the variations of these problems that have linear-time sequential solutions rather than limiting attention to the "river-routing" context, where single-sided connections are disallowed. For the minimum separation problem, we obtain $O(\lg N)$ time on a CREW PRAM or $O(\frac{\lg N}{\lg \lg N})$ time on a CRCW PRAM, both with optimal work (processor-time product) of $O(N)$, where N is the number of terminals. For the offset range problem, we obtain the same time and processor bounds as long as only one side of the channel contains single-sided nets. For the optimal offset problem with single-sided nets on one side of the channel, we obtain time $O(\lg N \lg \lg N)$ on a CREW PRAM or $O(\lg N)$ time on a CRCW PRAM with $O(N \lg \lg N)$ work. Not only does this improve on previous results for river routing, but we can obtain an even better time of $O((\lg \lg N)^2)$ on the CRCW PRAM in the river routing context.

1 Introduction

Much attention has been given to single-layer routing for VLSI. Most popular has been river routing [6], the connection of two (horizontal) rows of corresponding terminals using the channel region between the rows of terminals; see also [16] and the references therein. More general arrangements of modules and nets have been considered for testing routability of terminals in fixed positions, but it is also desirable to answer more sophisticated questions. For example, the *minimum separation* problem involves finding the minimum vertical separation between two rows of terminals that is required for routability (given that the horizontal positions of the terminals are completely fixed). In other problems, we are allowed to offset the upper row of terminals as a block to the left or the right, though the individual terminals do not shift position relative to one another. In particular, the *optimal offset problem* involves finding the offset that minimizes the amount of separation necessary to route the channel. The *offset range problem* involves finding all offsets that give enough room to route at a given separation.

* Supported in part by NSF grant CCR-9109550.

In this paper, we solve the above dynamic programming problem by using a different approach. Our algorithm runs in $O(m + n \log \log n)$ time for the concave version of this problem and in $O(m + n \log n)$ time for the convex version; the problem is called concave (convex) if all weights are non-negative (resp. non-positive). Our algorithm uses a data structure for solving a different query problem (*the consecutive query problem*) in conjuction with searching in Monge arrays [2, 3, 10]; this data structure is sufficient for solving our dynamic programming problem but not for solving the general interval query problem. Finally, we use our algorithms to obtain efficient algorithms for the following problems:

Sequential partition of graph (Figure 1). Let G be an undirected weighted graph on n nodes and m edges, where m may be much larger than n. Let the nodes be numbered according to a specified order. The problem is to partition these nodes into subsets such that each subset consists of contiguous nodes, each subset contains at most K nodes, and the total weight of edges connecting nodes in different subsets is minimized.

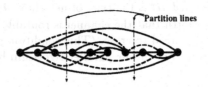

Fig. 1. Sequential partition of graph, n=10, K=4. Solid edges have weight 2, and broken edges have weight 1.

Partial clique covering of interval graph (Figure 2). Given a set of n points on a line and a set of m weighted intervals on these points, let G be the associated *interval graph*. Given a number K, compute K subgraphs of G so that the total weight of the intervals associated with the nodes in the union of the subgraphs is maximized.

Fig. 2. Partial clique cover of interval graph, K=2

2 Consecutive Interval query

The interval query problem is defined as follows:

Given two integers i and j such that $1 \le i < j \le n$, compute the total weight $W(i,j)$ of intervals of Z that are covered by the closed interval $[i,j]$.

Many $O(\log n)$ query-time and $O(m \log n)$ preprocessing-time data structures for the interval query problem are known in the literature, e.g., the *segment tree* and the *orthogonal range tree* [12]. However, for these data structures, the $O(m \log n)$ preprocessing time dominates the time complexity when $m \gg n$. We can, indeed, improve the preprocessing time by providing a data structure that answers the query in $O(n^\epsilon)$ query time and $O(m)$ preprocessing time. However, such a data structure will improve the time bound of Asano's algorithm only when $m \ge n^{1+\epsilon}$. Similarly, we can provide a data structure that answers a query in $O(\log^2 n/\log \log n)$ query time and $O(m \log n/\log \log n)$ preprocessing time but this also improves Asano's algorithm for only certain values of m and n. Keeping this in view, we restrict our attention to the dynamic programming problem at hand, and show in section 3 that the sequence of queries that we need, have some nice properties. In particular, this sequence of queries can be answered by solving the consecutive query problem given below. Consequently, in this section, we describe a data structure that can be constructed using $O(m)$ preprocessing time and that answers the *consecutive query* problem efficiently; more precisely, we answer the query $W(i+i_0, j+j_0)$ in $O(\frac{i_0+j_0}{\log^c n} + c\log n)$ time (where c is a non-negative constant, and can be 0) as long as we have already queried $W(i,j)$ before. This data structure is constructed using *fractional cascading* [5].

The consecutive query problem: Query $W(i+i_0, j+j_0)$ efficiently after we have queried $W(i,j)$.

We first describe a data structure for consecutive query in $O(i_0+j_0)$ time with $O(m)$ preprocessing time. We map an interval $I = [(i,j)]$ to a point $p(I) = (i,j)$ in the $n \times n$ planar grid G. Then, we obtain a set $S(Z)$ of m points in the grid G. An interval $J = [a,b]$ is contained the interval $I = [i,j]$ if and only if $a \ge i$ and $b \le j$; in other words, $p(J)$ is located in the South-East of $p(I)$. Thus, the problem is reduced to a South-East rectilinear range search problem in a grid (usually, the problem is referred as the North-West search, after a reflection transformation). The value $W(i+i_0, j) - W(i,j)$ is simply the total weight of the points located in the rectangle whose corners are (i,n), (i,j), $(i+i_0,n)$, and $(i+i_0,j)$. Let R_k be the set of points located on the k-th row of the grid. We equip the sorted list of R_k with respect to the ordinate for each $k = 1, 2, .., n$. This can be obtained in $O(n+m)$ time as follows:

We compute the sorted list of m points with respect to the ordinate; this can be done in $O(m+n)$ time by using bucket sorting. Then, we distribute the list into $R_1, .., R_k$, to obtain the sorted list of each subset. For each element x of a list, we store the sum of the weight of its elements that are greater than x with respect to the sorted order. Now, it is easy to compute $W(i+i_0, j) - W(i,j)$ in $i_0 \log n$ time by locating j in each of $R_{i+1}, .., R_{i+i_0}$. To avoid consuming $\log n$

time for searching in each list, we adapt the *fractional cascading* method of [5], which attains $O(\log n + i_0)$ query-time with $O(m + n)$ preprocessing-time.

Theorem 1. *[5] Let G be an undirected graph with maximum node degree d. Suppose a sorted list of elements in an totally ordered set A is associated with each node of G. Let s be the total size of these lists. Then, with $O(s)$ preprocessing time we can construct a data structure such that we can locate any given element a of A in all lists on a path p of length l of G in $O(l \log d + \log s)$ time.*

In our case, the underlying graph G is the path $v_1, v_2, .., v_n$, and the list R_k is the one associated with v_k. Next, an augmented list $\tilde{R}_k$ is constructed for each list R_k, and the location of j in R_k can be computed from that in $\tilde{R}_k$ in $O(1)$ time. Further, the location of j in $\tilde{R}_{k+1}$ is known from that in $\tilde{R}_k$ in $O(1)$ time by using the *bridge* between $\tilde{R}_k$ and $\tilde{R}_{k+1}$. Since the method is well-known, we omit the details and refer the readers to Cole [6] and Chazelle-Guibas [5]. We call above structure the *row-structure*. Note that in such a data structure, when we compute $W(i + i_0, j) - W(i, j)$, we spend $O(\log n)$ time for locating j in the initial list $\tilde{R}_{i+1}$, and the remaining computation is done in $O(i_0)$ time. We give a similar structure (called *column-structure*) for computing $W(i + i_0, j + j_0) - W(i + i_0, j)$. The sorted list of the points in the k-th column is denoted by C_k, and the associated augmented list is denoted by $\tilde{C}_k$. Using the above arguments, it is clear that we can compute $W(i + i_0, j + j_0) - W(i, j)$, and hence $W(i + i_0, j + j_0)$, in $O(i_0 + j_0 + \log n)$ time. We show below that we can remove the $\log n$ term for the consecutive search problem.

When we query $W(i, j)$, we remember not only the value $W(i, j)$ but also both the position of j in $\tilde{R}_i$ and the position of i in $\tilde{C}_j$. When we compute $W(i + i_0, j + i_0)$, we can find the location of j in $\tilde{R}_{i+1}$ from that in $\tilde{R}_i$ in $O(1)$ time. Thus, it takes $O(i_0)$ time for computing $W(i + i_0, j) - W(i, j)$. Besides, we can find the location of $i + i_0$ in $\tilde{C}_j$ in $O(i_0)$ time since it suffices to move up the list $\tilde{C}_j$ from the location of i, at most $O(i_0)$ steps. Therefore, we can compute $W(i + i_0, j + j_0) - W(i + i_0, j)$ in $O(i_0 + j_0)$ time. Furthermore, we can simultaneously compute the location of $i + i_0$ in $\tilde{C}_{j+j_0}$ and that of $j + j_0$ in $\tilde{R}_{i+i_0}$. Hence, we obtain the following:

Theorem 2. *A data structure can be built that answers the consecutive query $W(i + i_0, j + j_0)$ after querying $W(i, j)$ in $O(i_0 + j_0)$ time; this structure can be constructed in $O(m)$ preprocessing time. If we only know the value $W(i, j)$, then this consecutive query can be answered in $O(i_0 + j_0 + \log n)$ time by using the same data structure.*

We construct an $O(\frac{i_0 + j_0}{(\log n)^c} + \log n)$ query time data structure for any given constant c. Let $L = \lfloor \log n \rfloor$, and $h(s) = \lfloor \frac{n}{L^s} \rfloor$ for $s = 1, 2, .., c$. For each $k = 1, 2, .., h(s)$, we consider the set $R(s)_k$ of the points of $S(Z)$ whose abscissa is in the interval $[(k-1)L^s, kL^s)$ if $s \neq 0$, where $R(0)_k = R_k$. For each s and k, we sort $R(s)_k$ with respect to the ordinate values, and store the sorted list. If more than two points has the same ordinate value, we sort them with respect to the x-coordinate value; it is easy to see the sorting can be done in $O(m + n)$ time. We construct a fractional cascading structure for the set of lists $\{R(s)_0, .., R(s)_{h(s)} | s = 0, 1, .., c\}$. The underlying graph has nodes $\{v(s)_1, v(s)_2, .., v(s)_{h(s)} | s = 0, 1, .., c\}$. The edges are given between $(v(s)_i, v(s)_{i+1}))$ for $i = 1, 2, .., h(s) - 1$ and $s = 0, 1, .., c$, and $(v(s)_k, v(s-1)_{kL})$ for $k = 1, 2, .., h(s) - 1$ and $s = 1, 2, ..c$, where the list $R(s)_j$ is associated with $v(s)_j$. We denote $\tilde{R}(s)_i$ for the augmented list of $R(s)_j$.

Lemma 3. *By using the above data structure, $W(i + i_0, j) - W(i, j)$ can be computed in $O(\frac{i_0}{(\log n)^c} + \log n)$ time.*

Proof. Since there is a path of length at most $\frac{c i_0}{(\log n)^c} + c \log n$ from $v(i)$ to $v(i+i_0)$ in the underlying graph, the above lemma follows from Theorem 1. □

By constructing a similar structure for the columns, we can similarly compute $W(i + i_0, j + j_0) - W(i + i_0, j)$. Hence, we have the following theorem:

Theorem 4. *A data structure can be constructed such that it answers the query $W(i + i_0, j + j_0)$ in $O(\frac{i_0 + j_0}{(\log n)^c} + c \log n)$ time as long as the query $W(i, j)$ has been answered before; this data structure can be constructed in $O(m + n)$ time; where the constant c is non-negative but can be zero.*

3 Dynamic programming on intervals

3.1 Dynamic programming and Matrix searching

In this section, we show how to solve the dynamic programming on intervals by using above data structure. Let us give a brief summary of dynamic programming on concave (or convex) functions.

A matrix $A = (A(i, j))_{i,j=1,2,..,n}$ is called a convex Monge (concave Monge) matrix if $A(i, j) + A(i + 1, j + 1) \geq A(i, j + 1) + A(i + 1, j)$ (resp. $A(i, j) + A(i + 1, j+1) \leq A(i, j+1) + A(i+1, j)$) for $1 \leq i, j \leq n$. A matrix A is called a staircase matrix if $A(i, j)$ is ∞ unless $i < j < f(i)$ for some non-decreasing sequence $f(i)$. A staircase matrix is called a convex staircase Monge matrix (or concave staircase Monge matrix) if the Monge relation holds within the staircase. Note that if a matrix is concave staircase, then it is a concave Monge matrix, although the same statement is not true for a convex staircase matrix. Let $U = \{1, 2, .., n\}$. A function $F(i, j)$ on $U \times U$ is called *convex* (resp.*concave*) if the associated matrix is convex staircase Monge (resp. concave staircase Monge). The convex (resp. concave) dynamic programming problem is to compute a function $E(i)$ by using the inductive formula $E(i) = \min_{j<i}\{D(j) + F(i, j)\}$, where $D(j)$ can be computed in $O(1)$ time from $E(j)$, and $F(i, j)$ is convex (or concave).

It is known that such kind of dynamic programming problems can be solved efficiently by using the matrix searching technique [10, 8, 7]. In particular, if it takes $O(q)$ time to compute $F(i, j)$ for an arbitrary (i, j), it is known the concave problem can be solved in $O(nq)$ time [10], and convex problem can be solved in in $O(nq\alpha(n))$ time [11], where $\alpha(n)$ is the inverse Ackermann function.

3.2   Dynamic programming for concave problems

Let S be a set of m intervals whose endpoints are among $\{1, 2, ..., n\}$. Each interval e has a nonnegative weight $w(e)$. We define $W(i, j)$, $\tilde{W}(i, j)$, $D(i)$, and $E(i)$ as in the introduction, where the function W is the weight function. The following lemma is a key observation, which is easy to verify:

Lemma 5. *If $w(e)$ are nonnegative, $\tilde{W}(i, j)$ is a concave function.*

We define the matrix A by $A(i, j) = D(i) + \tilde{W}(i, j)$. By definition, $E(j)$ is the minimum element in the j-th column. The following lemma is easy to show:

Lemma 6. *A is concave staircase Monge.*

If it takes $O(q)$ time to query the value of the weight function W for an arbitrary (i, j), the problem is solved in $O(nq)$ time [10]. Hence, if we use segment tree as the query structure, the overall time complexity becomes $O(m \log n)$, which matches that of Asano [1]; we improve it to $O(m + n \log \log n)$ by using the consecutive interval query data structure. We investigate the ordering of querying W in the algorithm of Klawe [10], and show that consecutive query structure is an efficient amortized query time structure.

The algorithm of Klawe [10] consists of a series of off-line matrix searching in rectangular submatrices; here, off-line means that $D(i)$ is known for each row index i of the matrix even before the algorithm begins. Each rectangular submatrix has contiguous column indices and row indices. The total sum of the column size and row size of the submatrices is $O(n)$. Therefore, we first consider the problem of searching all column minima of a (rectangular) Monge matrix M of size $n \times l$ where the value of $D(i)$ is known for the i-th row of M. We use the consecutive query structure, so that $M(i + i_0, j + j_0)$ can be computed in $O(i_0 + j_0)$ time from $M(i, j)$.

Lemma 7. *All column minima of M can be computed in $O((n + l) \log l)$ time.*

Proof. Let us denote the row index of the minimum of i-th column by $r(i)$. We first compute $r(l/2)$ (we can assume that l is even). This takes $O(n + l)$ time since we can find all entries $M(i, l/2)$ in $O(n + l)$ time by using the consecutive search. Because of the concave Monge property, the row indices of the column minima is a non-increasing sequence. Thus, $r(j) \geq r(l/2)$ (resp. $r(j) \leq r(l/2)$) if $j > l/2$ (resp. $j < l/2$). So, it suffices to compute in the shaded regions in Figure 3. Lemma 7 then easily follows from the corresponding recursion. □

We improve the time complexity of Lemma 7 by using the data structure for computing $W(i + i_0, j + j_0)$ from $W(i, j)$ in $O(\frac{i_0 + j_0}{\log^2 n} + \log n)$ time. Let A be our $n \times n$ staircase Monge matrix, and let M be its submatrix corresponding to contiguous k rows and l columns of A.

Lemma 8. *All column minima of M can be computed in $O((k + l) \log \log n)$ time.*

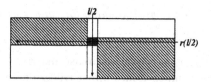

Fig. 3. Possible location of minima

Proof. If both l is smaller than $\log^3 n$, then the lemma follows from Lemma 7, since $\log l = O(\log\log n)$. Consequently, assume that l is larger than $\log^3 n$.

We consider the columns which has a column index of an integer multiple of $L = \lfloor \log^2 n \rfloor$. Assume that we have already computed the minimum of those columns. Given i, for every j such that $iL \geq j \geq (i+1)L$, $r(iL) \geq r(j) \geq r((i+1)L)$ because of Monge property. Therefore, it suffices to search in submatrices $M_1, .., M_s$, where $s = \lceil l/L \rceil$, such that M_i is a $(L-1) \times k_i$ matrix and $\sum_{i=1}^{s} k_i = k$. These sumbatrices can be computed in $O((k+l)\log\log n)$ time in total, since each submatrix has $\log^2 n$ contiguous columns. Thus, it suffices to compute $r(iL)$ for $i = 1, 2, .., s$. We first use the REDUCE subroutine of the SMAWK algorithm of [2]. Since REDUCE computes the entry in a consecutive manner, we can remove $k - \frac{l}{L}$ rows from the searched matrix in $O(k + l)$ time by using the consecutive query structure with $c = 0$. Thus we have a matrix M' which has $\frac{l}{L}$ rows and $\frac{l}{L}$ columns. However, neither the column indices nor the row indices are contiguous. We process this matrix by using the algorithm implicit in Lemma 7. By using the consecutive query structure, we can find all element on a column (and a row) of M' in $O(\frac{l}{\log^2 n} + \frac{l\log n}{L}) = O(\frac{l}{\log n})$ time. Thus, the algorithm of Lemma 7 computes all column minimum of M' in $O(l)$ time. □

Using the algorithm of Klawe [10] together with the matrix searching algorithm of Lemma 8, we obtain the following:

Theorem 9. *The concave dynamic programming problem on intervals can be solved in $O(m + n\log\log n)$ time.*

3.3 Dynamic programming for convex problems

This section uses the same notation as the concave case except that each interval e has a nonpositive weight.

Lemma 10. *If $w(e)$ are nonpositive for all e, $\tilde{W}(i,j)$ is a concave function.*

The algorithms of Klawe-Kleitman [11] and Aggarwal-Klawe [2] are based on matrix searching in rectangular submatrices; the difference from the convex case is that these submatrices do not have contiguous column indices. We first give a relatively naive $O(m + n\log n\log\log n)$ time algorithm, and then improve it to $O(m + n\log n)$ time.

Proposition 11. *The dynamic programming for convex weights can be solved in* $O(m + n \log n \log \log n)$ *time.*

Proof. The matrix $A(i, j)$ is a convex staircase matrix. As shown in [2], a s-taircase matrix can be decomposed into upper triangular submatrices and lower triangular submatrices. The total sum of the row size and column size of these submatrices is $O(n)$, and each submatrices has contiguous row indices and column indices. Thus, we can solve our problem as a series of matrix searching problems in triangular matrices. When we compute a lower triangular matrix, the value $D(i)$ for each row index of it has been already computed, thus we can process it by using Lemma 8. Hence, we can assume that $A(i, j)$ is an upper triangular matrix. Let $T(n)$ be the computing time. We apply a simple divide and conquer method. We divide the matrix $A(i, j)$ into four $n/2 \times n/2$ submatrices which are upper-left, upper-right, lower-left, and lower-right with respect to the $n/2$-th row and the $n/2$-th column. All entries in the lower-left submatrix are infinity, hence we need not compute it. We first compute, in $T(n/2)$ time, the column minimums of the upper-left submatrix, which is a triangular matrix of size $n/2 \times n/2$. We next compute the column minima of the right upper submatrix; this is a rectangular Monge matrix, and we know the value of $D(i)$ for each row index i of it. Thus, we can compute in $O(n \log \log n)$ time. Then, we search in the lower-right submatrix, which is upper triangular, in $T(n/2)$ time. Finally, we compare the column maxima of the upper-right submatrix and those of the lower-right submatrix in $O(n)$ time. Hence, $T(n) = 2T(n/2) + O(n \log \log n) = O(n \log n \log \log n)$. □

We can improve the time complexity of the above algorithm by a factor of $O(\log \log n)$; the new algorithm is based on the algorithm of Aggarwal-Klawe [2]. Because of space limitation, we omit it in this version.

Theorem 12. *The dynamic programming for a convex problem can be solved in* $O(m + n \log n)$ *time.*

4 Applications

4.1 Sequential partition of a graph

The optimal partition of a weighted graph is defined as follows: Given a graph $G = (V, E)$ with n nodes and m edges, an edge-weight function $w : E \to R^+$, a vertex-weight function $\phi : V \to R^+$, and a positive number K. A feasible partition is the partition of V into subsets $V_1, ..., V_s$ such that the sum of weights of the vertices in each cluster is no more than K. The optimal partition is the feasible partition that minimizes the total weight of the edges connecting different subsets.

The problem of computing an optimal partition of a graph occurs in several areas in computer science. Unfortunately, the problem of finding an optimal partition of a weighted graph is NP hard. Consequently, a popular heuristic for finding an approximation uses the sequential partition technique [9]. In this technique, there is a linear ordering of the vertices, say, labeled from 1 to n. A sequential partition of G is a partition such that each cluster V_i consists of consecutive vertices with respect to this order. An optimal sequential partition is a sequential partition which minimizes the total cost of the edges connecting different clusters. We can consider a vertex of G as an integral point of the interval $I = [1, n]$, the edge between i and j a sub-interval of I, and a cluster of a sequential partition can also be regarded as an interval. Consequently, the problem becomes that of computing the functions $E(i)$ on U, such that $E(0) = 0$ and

$$E(i) = \max_{i-f(i) \leq j \leq i} \{E(j) + W(i,j)\} \tag{2}$$

for a positive integer valued function f that is defined to be the smallest integer such that the summation of vertex weights of $\{v_{i-f(i)}, .., v_i\}$ does not exceed K.

Kernighan[9] first gave an algorithm for finding an optimal sequential partition. His algorithm is based on dynamic programming and runs in $O(n^2)$ time. Asano[1] improved the time complexity to $O(m \log n)$ time; we obtain the following:

Theorem 13. *The optimal sequential partition of G is obtained in $O(m+n \log n)$ time.*

Proof. It is easy to see that the function f can be computed in $O(m)$ time. The rest of the computation can be formulated as a convex dynamic programming on intervals by replacing each weight by its negative; this can be solved in $O(m + n \log n)$ time. □

4.2 Partial clique covering of interval graph

Given an interval graph G with m intervals on n terminals. Each interval has a positive weight. It is well-known that its maximal clique can be computed in $O(m)$ time, and the minimal clique covering can be computed in $O(m)$ time if the terminals are sorted. Consider the following two problems:

K-large clique problem: Given a number K, find K cliques $C_1, .., C_K$ of G such that the total weight in $\cup_{i=1}^{K} C_i$ is maximized.

Parametric partial clique covering problem: Given a nonpositive parameter t, find cliques $C_1, .., C_s$ such that the sum of ts and the total weight in $\cup_{i=1}^{s} C_i$ is maximized.

It can be easily seen that the partial parametric clique covering problem is a concave dynamic programming on intervals. Thus, we have the following:

Theorem 14. *The parametric partial clique covering problem can be solved in $O(m + n \log \log n)$ time.*

For K-large clique problem, we can easily give an $O(m+Kn \log \log n)$ time by applying matrix searching K times. Furthermore, by using a parametric searching algorithm, we can solve the problem by solving parametric partial clique covering problem on $\sqrt{K \log n}$ different parameters (for details, see [4]).

475

Theorem 15. *The K-large clique problem can be solved in* $O((\min\{K, \sqrt{K\log n}\}n\log\log n) + n)$ *time.*

5 Concluding remarks

In this paper, we only dealt with the convex case and the concave case of the dynamic programming problem of intervals although there are several problems which can be formulated as dynamic programming on intervals but with mixed weights (i.e., $w(e)$ may have arbitrary real value). It remains open whether the general case is solved in $o(m\log n)$ time.

References

1. T. Asano, "Dynamic Programming on Intervals," Proc. of ISA, LNCS 557, Springer Verlag (1991), 199-207.
2. A. Aggarwal and M. Klawe, "Applications of Generalized Matrix Searching to Geometric Algorithms," Discrete Appl. Math. 27(1990), 2-23
3. A. Aggarwal, M. Klawe, S. Moran, P. Shor, and R. Wilber, "Geometric Applications of a Matrix-Searching Algorithm," Algorithmica 2 (1987), 209-233.
4. A. Aggarwal, B. Schieber, and T. Tokuyama, "Finding a Minimum Weight K-Link Path in Graphs with Monge Property and Applications", Proc. 9th ACM Symp. on Computational Geometry (1993), 189-197.
5. B. Chazelle and L. Guibas, "Fractional Cascading: I. A Data Structure Technique," Algorithmica 1(1986), 133-162
6. R. Cole, "Searching and Storing Similar Lists," J. of Algorithms 7 (1986), 202-220.
7. D. Epstein, Z. Galil, R. Giancarlo, and G. Italiano, "Sparse Dynamic Programming," Proc. of the First ACM-SIAM Symp. on Discrete Algorithms, (1990), 513-522.
8. L. Larmore and B. Schieber, "On-line Dynamic Programming with Applications to the Prediction of RNA Secondary Structure," J. of Algorithms 12 (1991), 490-515.
9. B. Kernighan, "Optimal Sequential Partitions of Graphs," J. ACM 18(1971) 34-40.
10. M. Klawe, "A Simple Linear Time Algorithm for Concave One-Dimensional Dynamic Programming," Technical Report, University of British Columbia, Vancouver, 1989.
11. M. Klawe and D. Kleitman, "An Almost Linear Time Algorithm for Generalized Matrix Searching," Technical Report, IBM Almaden R.C., 1988.
12. F. Preparata and M. Shamos, Computational Geometry – An Introduction, 1988 (2nd edition), Springer-Verlag.

An Improved Algorithm for the Traveler's Problem

Alok Aggarwal[1] and Takeshi Tokuyama[2]

[1] IBM Research Division, T. J. Watson Research Center, P.O. Box 218, Yorktown
Heights, NY 10598
[2] IBM Research Division, Tokyo Research Laboratory, 1623-14, Shimotsuruma,
Yamato-shi, Kanagawa, 242 Japan

Abstract. In a paper in Journal of Algorithms, Vol.13, (1992), p.148-
160, Hirschberg and Larmore introduced the *traveler's problem* as a subroutine for constructing the B-tree. They gave an $O(n^{5/3} \log^{1/3} n)$ time
algorithm for solving traveler's problem of size n. In this paper, we improve their time bound to $O(n^{3/2} \log n)$. As a consequence, we build a
B-tree in $O(n^{3/2} \log^2 n)$ time as compared to the $O(n^{5/3} \log^{4/3} n)$ time
algorithm of Hirschberg and Larmore for the same problem.

1 Traveller's problem

In a recent paper in Journal of Algorithms, Hirschberg and Larmore [6] introduced the *traveler's problem*, which can be roughly defined as follows: Consider
some route on which a traveler travels, and he has B dollars as his travel budget. There are n hotels along this route. For any hotel, v, $cost(v)$ is the cost of
staying at that hotel, and $Farthest(v)$ is the farthest hotel to which the traveler
can travel to in a day, from v. Then, the problem is, starting from each hotel,
what is the farthest hotel he can reach within the given budget? (The number
of days for travelling is immaterial as far as this problem is concerned.) In their
paper, Hirschberg and Larmore [6] showed that this problem can be solved for
all $v = 0, 1, 2, .., n$ in $O(n^{5/3} \log^{1/3} n)$ time. As a corollary, they also solved the
following problem:

B-tree for dictionary: Suppose we have a dictionary with n words $\{w_1, ..., w_n\}$.
Each word has a given length $c(1), .., c(n)$. We construct a search tree such that
the words are stored in both internal and external nodes. Each node of this tree
has a fixed capacity, enabling it to store varying numbers of words depending
on the length of these words, and the internal nodes contain enough information
to guide the search for a desired word. The problem is to find a tree with the
minimum height.

Hirschberg and Larmore [6] showed that such a B-tree can be constructed by solving the traveler's problem $O(\log n)$ times; thus, it can be solved in
$O(n^{5/3} \log^{4/3} n)$ time. In this paper, we give the following improvement:

Theorem 1. *The traveler's problem can be solved in $O(n^{3/2} \log n)$ time. As a
corollary, the B-tree problem can be solved in $O(n^{3/2} \log^2 n)$ time.*

This paper has 4 sections. Section 2 defines the traveler's problem precisely and reviews some important lemmas from [6]. Section 3 discusses the notion of a length limited path and its importance to this problem, whereas, section 4 concludes with a few open problems.

2 Preliminaries

Let $V = \{0, .., n\}$, let the distance between vertices u and v of V is $|u - v|$, and let $cost(v)$ be a positive number assigned to $v \in V$, which is called the cost of v. We define a function $Farthest$ from V to V. This function is monotone, i.e., if $v > u$, $Farthest(v) \geq Farthest(u)$. Also, $Farthest(v) > v$. A weighted direct acyclic graph (DAG) is a directed graph on node set $\{1, 2, .., n\}$, with edge set $\{(u, v) : 1 \leq u < v \leq n\}$. Each directed edge (u, v) has a positive (possibly infinite) weight $weight(u, v)$. Given an instance of traveler's problem, we define a weighted DAG G on the vertex set V with directed edge set $E = \{(u, v) : 0 \leq u < v \leq n\}$. The weight is defined by $weight(u, v) = cost(v)$ if $v \leq Farthest(u)$, otherwise, it is ∞. We fix a number B, called the budget. For a path p on G, $Expense(p)$ is the total sum of weights of edges on it, and this is called the expense of p. We often use a *link* of a path and an edge on a path interchangeably. We often call a vertex on a path a *stop*. A path is *feasible* if its expense is less than B. The *length* of a path is the number of links between its starting vertex and its terminus. The traveler's problem (Figure 1) is to find the longest feasible path $Opt(v)$ starting from each vertex $v = 0, 1, .., n$ of G, and if there are several feasible paths of the same length, to choose cheapest one. (Note: In [6], the weight of last link is not counted. However, it is clear and shown in their paper that the above definition gives an equivalent problem).

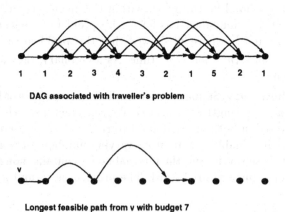

1   1   2   3   4   3   2   1   5   2   1

DAG associated with traveller's problem

v

Longest feasible path from v with budget 7

Fig. 1. Traveller's problem

We denote the starting vertex and the terminus of a path p by $Start(p)$ and $Term(q)$, respectively. Without loss of generality, we can assume that there is a unique optimal path starting from a given vertex. More precisely, we assume that there is a total ordering on the set of feasible paths starting from a given vertex, such that for any pair of such paths, p and q, p is better than q if either (1) $Term(p) > Term(q)$, (2) $Term(p) = Term(q)$ and the expense of p is cheaper than that of q, or (3) $Term(p) = Term(q)$, both have the same expense, and p is lexicographically greater than q. To obtain the improved algorithm, we use techniques similar to those given in [6]; we classify the paths into *bottleneck paths* and *length limited paths*. Our improvement is obtained by improving the time for computing the length limited paths, which, in turn, is achieved by introducing the notion of *milestones*. The following is a key lemma from [6]:

Lemma 2. *For each starting vertex, the optimal path starting from it can be computed in $O(n)$ time.*

Proof. The SSA algorithm of Hirschberg-Larmore [6] is a linear time algorithm for computing such an optimal path. Consequently, the traveler's problem can be solved in $O(n^2)$ time. □

Hirschberg and Larmore [6] improved this time complexity to $O(n^{5/3} \log^{1/3} n)$ by classifying the paths into *bottleneck paths* and *length limited paths*. Note that for each path starting before v and ending after $Farthest(v)$, it must use at least one vertex in the closed interval $[v, Farthest(v)]$. Consequently, any such path is *bottlenecked* by $[v, Farthest(v)]$. A path is called l-bottlenected if it is bottlenecked by an interval of length less than l. Hirschberg and Larmore [6] showed the following:

Proposition 3. *All optimal paths that are l-bottlenecked can be found in $O(nl \log n)$ time.*

Finally, we need some more notions from previous papers. Any weighted DAG has convex *Monge Property* if for any $a < b < c < d$, $weight(a,c) + weight(b,d) \le weight(a,d) + weight(b,c)$; it has concave Monge property if the reverse-inequality holds. There are several results on DAGs with Monge property [1, 3, 7] and there are many applications of such results in the literature. Given an instance of traveler's problem, we construct a DAG, G, on $\{1, 2, .., n\}$, where $weight(i,j) = cost(j)$. It can be easily seen that this graph has convex Monge property; moreover, the equality holds unless $weight(a,d) = \infty$. Consequently, we say that this DAG obeys the *strong Monge property*.

3 Length limited paths

In this section, we compute optimal paths which are not l-bottlenecked. We say G is l-dense if $[v, Farthest(v)]$ has length more than l except the case $v = 0$ or $Farthest(v) = n$. Given G, if G is not l-dense, then there exists an interval $I = [v, Farthest(v)]$ of length less than or equal to l and neither $v = 0$

nor $Farthest(v) = n$. We choose smallest such v satisfying that $v > l$ and $Farthest(v) < n - l$. If no such v exists, we choose the v nearest to the center vertex $\lfloor n/2 \rfloor$. We need not consider the paths starting before I and ending after I since they must be l-bottlenecked paths. If $n \geq 2l$, we cut the graph G to make a pair of graphs on $0, ..., Farthest(v) - 1, Farthest(v)$ and $v, v+1, ..., n$. Cutting recursively, we obtain a set of subgraphs which are either l-dense or have size less than $2l$. Observe that this set of subgraphs can be created so that the sum of the lengths does not exceed $2n$. The optimal path on a graph whose size is smaller than $2l$ can be easily computed since it has at most two links if it is not l-bottlenecked. Therefore, we can assume G is l-dense. Let $K = \lceil n/l \rceil$.

Lemma 4. *An optimal path has at most two stops in $[v, Farthest(v)]$ for each* v.

Proof. Suppose an optimal path has three (or more) stops in the interval $[v, Farthest(v)]$. Then, a cheaper feasible path can be obtained by skipping the middle stop, which is contradiction. □

As a consequence of Lemma 4, an optimal path on an l-dense graph contains at most $2K$ links. The *weak interleaving property* [2, 5] of the paths on a DAG with Monge property plays a key role in the algorithm design; to define this property we need the following notation. Let $h(v)$ be the number of links in the optimal path starting from v. Because the optimal path is unique, $h(v)$ is well-defined. Let $First(v)$ be the first stop on the optimal path starting from v.

Lemma 5. (Weak Interleaving Lemma). *For any $u \in [v, First(v)]$, Let $p :$ $u = u_0, u_1, .., u_L$ and $q : v = v_0, v_1, ..., v_M$ be optimal paths starting from u and v respectively. Then, $u_L \geq v_M$, and $v_{i-1} < u_i < v_{i+2}$ for each $i = 1, 2, .., M - 2$.*

Proof. From definition, $u_L \geq v_M$ is trivial. Assume $u_i \geq v_{i+2}$ for an index i. (For example, $u_4 \geq v_6$ in Figure 2 (a).) There is no link e of p such that more than two stops of q are located between the endpoints of e; otherwise we can delete an endpoint of q and obtain a cheaper path. Hence, there are at least two links of p such that each of these contains two stops of q in the corresponding left-open interval (i.e., the interval $(a, b]$ corresponds to an edge (a, b)). Let (u_i, u_{i+1}) and (u_j, u_{j+1}) be such links ($i < j$). Let v_s be the higher index stop of q in the interval $(u_i, u_{i+1}]$, and v_t be the lower index stop in the interval $(u_j, u_{j+1}]$. Then, exchanging the path $u_{i+1}, ..., u_j$ with $v_s, ..., v_t$, we obtain a pair $\{\check{p}, \check{q}\}$ of paths. It can be clearly seen that these paths satisfy: (1) $\check{p} \neq p$, (2) The union of stops of $\check{p}$ and $\check{q}$ is the same as that of p and q, (3) $Start(\check{p}) = Start(p)$, $Term(\check{p}) = Term(p)$, $Start(\check{q}) = Start(q)$, and $Term(\check{q}) = Term(q)$, and (4) $Expense(\check{p}) + Expense(\check{q}) = Expense(p) + Expense(q)$. The new pair of paths is shown in the Figure 2 (b). From (4), either (a) $Expense(\check{p}) < Expense(p)$ or $Expense(\check{q}) < Expense(q)$, or (b) $Expense(\check{p}) = Expense(p)$ and $Expense(\check{q}) = Expense(q)$. In case of (b), because of (1) and (2), either $\check{p}$ is lexicographically larger than p, or $\check{q}$ is lexicographically larger than q. Thus, either $\check{p}$ is a better path than p or $\check{q}$ is a better path than q. This contradicts the optimality of p

and q. If $v_{i-1} \geq u_i$, there are two links of q each of which contains two stops of p, and a contradiction can be obtained similarly. □

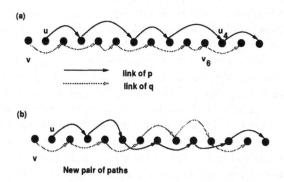

Fig. 2. Proof of weak interleaving property

Corollary 6. *For any* $u \in [v, First(v)]$, $h(u) \geq h(v) - 1$.

Lemma 7. *Define the "stage vertices"* s_i *by* $s_0 = 0$, *and* $s_i = First(s_{i-1})$ *for* $i \geq 1$. *Then, the number of stage vertices are at most* $2K$, *and they can be computed in* $O(Kn)$ *time.*

Proof. Assume that there are three stage vertices v_1, v_2, v_3 in an interval $[u, Farthest(u)]$. Without loss of generality, we can assume that v_2 (resp. v_3) is the first stop of the optimal path p (resp. q) from v_1 (resp. v_2). It is clear that the terminus of q is larger or equal to the terminus of p. Because of the monotonicity of the function $Farthest$, v_3 is smaller than $Farthest(v_1)$. Hence, the path that starts from u and that uses the same stops as q is a better path than p if either $Term(q) > Term(p)$ or $Expense(q) \leq Expense(p)$; this is a contradiction. Thus, we can assume that $Term(q) = Term(p)$ and $Expense(q) > Expense(p)$. Now, the path starting from v_2 using the stops of p (except v_1) is a better path than q. This is again a contradiction. Hence, the number of stage vertices are at most $2K$, and the time required for computing these is obtained immediately from Lemma 2. □

Let S be the set of stage vertices. A sorted superset $T = \{t_1, t_2, ..., t_N\}$ of S is called a set of *milestones* if either the length of $[t_i, t_{i+1}]$ is less than $2 \log n$ or $h(t_i) - 1 \leq h(t_{i+1}) \leq h(t_i) + 1$ for $i = 1, 2, .., N - 1$.

Lemma 8. *We can find a set of milestones of size* $O(K)$ *in* $O(nK \log n)$ *time.*

Proof. Let $k_i = h(s_i) - h(s_{i-1})$. Let m be the number of stage vertices. Since $h(s_i) - 1 \leq h(s_{i+1})$ and $m \leq 2K$, $\sum_{i=1}^{m} |-k_i|^+ \leq 2K$, where $|a|^+ = \max(0, a)$.

Since $\sum_{i=1}^{K} k_i = h(s_K) - h(s_1) \leq 2K$, it follows that $\sum_{i=1}^{K} |k_i| \leq 6K$, where $|a|$ means the absolute value of a. Hence, it suffices to find a set of $O(k_i)$ milestones between each s_{i-1} and s_i in $O(nk_i \log n)$ time.

Suppose we have a set $\tilde{T} = \{t_1, t_2, .., t_N\}$ of milestones of size N between $s = s_{i-1}$ and $s' = s_i$. Without loss of generality, we assume $k = k_i = h(s') - h(s) \geq 2$. We show below how to obtain a set of milestones of size $k - 1$ (excluding s and s') from it. From Lemma 6, $h(v) - 1 \leq h(w)$ for any pair (v, w) of vertices satisfying $s \leq v < w \leq s'$ since it is easy to see that $w \in [v, First(v)]$. Hence, there is no decreasing subsequence of length 2 in $\tilde{T}$; otherwise there is a pair $t_i < t_j$ satisfying $h(t_i) \geq h(t_j) + 2$, thereby, yielding a contradiction. Suppose $h(t_i) \geq h(t_{i+1})$ for an index i. Then, because of above observation, the sequence obtained by deleting t_{i+1} from $\tilde{T}$ is also a set of milestones. Continuing this deletion, we obtain a set of milestones on which the function $h()$ is strictly increasing in the interval $[s, s']$. This set has $k - 1$ milestones. Thus, if we obtain a set of milestones of any size, we can obtain a set of milestones of size $k - 1$ in $O(k)$ time. Below, we give an algorithm which compute a set of milestone of size $O(k \log n)$ in $O(kn \log n)$ time; hence we have a set of milestones of size $k - 1$ in the same time complexity.

An interval $I = [u, v]$ is called a *completed* interval if $h(u) - 1 \leq h(v) \leq h(u) + 1$. Our algorithm computes the optimal path starting from the midpoint m of the interval I, which is initially $[s, s']$, and continue the process to each sub-interval (obtained by cutting I by m) unless it is a completed interval. It is clear that the set of midpoints computed by the algorithm is a set of milestones. A sub-interval $I = [u, v]$ is called *restricted* interval if $h(u) < h(v)$ and there is a milestone $w < u$ that has been computed and that satisfies $h(w) > h(u)$; otherwise I is called *free* interval. Let $Num(I)$ be the upper bound on the number of milestones inserted by the algorithm into a sub-interval I of $[s, s']$. We denote $f(I)$ (resp. $g(I)$) for $Num(I)$ if I is free (resp. restricted). The upper bound of $f(I)$ (resp. $g(I)$) among free (resp. restricted) sub-intervals $I = [u, v]$ of length a and $h(v) - h(u) = y$ is denoted by $f(y, a)$ (resp. $g(y, a)$). We show that $f(y, a) \leq y \log a$ and $g(y, a) \leq (y - 1) \log a$. Without loss of generality we can assume $y \geq 2$.

We choose the mid-point m of an interval $I = [u, v]$. First, assume I is a free interval. If $h(u) \leq h(m) \leq h(v)$, $f(I) \leq f([u, m]) + f([m, v]) + 1 \leq y \log(a/2) + 1 = y \log a - y + 1 < y \log a$. If $h(m) = h(u) - 1$, then we need not insert a milestone in $[u, m]$ since it is a completed interval. The interval $[m, v]$ is a restricted interval, it follows that $f(I) \leq g(y + 1, a/2) + 1 = y \log(a/2) + 1 < y \log a$.

Next, assume I is a restricted interval. Then, clearly, $h(u) \leq h(m) \leq h(v)$. If $h(u) < h(m)$, then the interval $[u, m]$ is a restricted interval. Thus, $g(I) \leq g([u, m]) + f([m, v]) + 1 \leq (y - 1) \log(a/2) + 1 \leq (y - 1) \log a$. If $h(u) = h(m)$, then the interval $[m, v]$ is a restricted interval. Thus, $g(I) \leq g([m, v]) + 1 \leq (y - 1) \log(a/2) + 1 \leq (y - 1) \log a$. This proves the cardinality of the set of milestones computed by the algorithm is $O(k \log n)$. The time complexity for computing these follows from Lemma 2. □

In contrast to Lemma 5, the terminus of the best L link path starting from v may be less than the terminus of the best L link path starting from $u < v$. In fact, it is possible that no feasible L link path starting from v exists, even if one starting from u may exist. For mitigating the above inconvenience, we modify the traveler's problem slightly so that **traveler can go backward** in the first link. There is no limit on the distance of the backward link, and it can be a loop. The weight of a backward link is the cost of its left vertex. The best L-link path is the L-link path which reaches the terminus with largest index using the given budget. This modification does not affect the optimal path (which does not have the restriction on link number), since it is always better to skip a link than go backward. Note that Hirschberg and Larmore [6] use a different modification which gives the same effect in principle. The following lemma is given in [6]:

Lemma 9. (Strong Order Lemma) *Under the modification that the first link may goes backward, let $p : u = u_1, u_2, ..., u_L$ and $q : v = v_1, v_2, .., v_L$ be the best L-link paths respectively, from u and v, where $v > u$. Then $v_i \geq u_i$.*

Proof. Since the traveler starting from v can choose exactly the same stops as that of p (because travelers can go backward), $v_L \geq u_L$. If $v_i < u_i$ for some index i, there exists an index $j < i$ such that the link (u_j, u_{j+1}) of p containing two stops of q. On the other hand, there exist an index $k > i$ so that the link (v_k, v_{k+1}) of q containing two stops of p. Thus, we can construct a pair of L link paths from u to u_L and v to v_L with same total expense similarly to Lemma 5; this yields a contradiction. $\square$

Lemma 10. (Weak Order Lemma)
Let $p : u = u_1, u_2, .., u_M$ be the optimal path from u, $q : v = v_1, v_2, .., v_L$ be the best L-link path from v.
(A) If $u < v < u_1$ and $v_L > u_M$. Then,
(A1): If $L \leq M$, then $v_i \geq u_i$ for $i = 1, 2, .., L$.
(A2): If $L > M$, $v_{i+L-M} \geq u_i$. for $i = 1, 2, .., M$
(B): If $v < u$. Then,
(B1): If $L \geq M$, then $v_i \leq u_i$ for $i = 1, 2, .., M$ and $v_i \leq u_M$ for $i = M + 1, .., L$.
(B2): If $L < M$, then $v_i \leq u_{i+M-L}$ for $i = 1, 2, .., L$.

Proof. We consider the case where $u < v < u_1$, $v_L > u_M$, and $L > M$. The other cases can be proved in the same way.
Assume that (A2) fails. Then, there is an index $s \leq M$ such that $v_{s+L-M} < u_s$. Since there are $M - s - 1$ links of q and $M - s$ links p on the right of u_s, and $Term(q) = v_L > u_M = Term(p)$, there exists a link of q on the right of v_{s+L-M} containing at least two stops of p. Also, there exists a link of p on the left of u_s containing at least two stops of q. Thus, there are indices i and j such that there are more than one stop of q between u_i and u_{i+1} and there are more than one stops of p between v_j and v_{j+1}. We consider the rightmost stop $v(cut)$ of q between u_i and u_{i+1} by $v(cut)$ and the leftmost stop $u(cut)$ of p between v_j and v_{j+1}. Furthermore, we select the pair of i, j above suitably so that the

number of links of p between u_i and $u(cut)$ equal to that of links of q between $v(cut)$ and v_j. Thus, we can make a pair of M link path from u to u_M and L link path from v to v_L with the same total expense by exchanging the sub-path of p between u_{i+1} and $u(cut)$ (including the two stops), with that of q between $v(cut)$ and v_j. Similar to the proof of Lemma 5, we have now created either a better M-link path than p, or a better L-link path than q, this contradicts the optimality. □

The following lemma is given by Hirschberg-Larmore [6] (as Lemma 7):

Lemma 11. *Given sequences $a_1 \leq .., \leq a_L$ and $b_1 \leq, ... \leq b_L$ satisfying $a_i \leq b_i$ for $i = 1, 2, .., L$, we can compute the best L-link path starting from a vertex w such that its i-th stop w_i satisfies $a_i \leq w_i \leq b_i$ for $i = 1, 2, .., L$ in $O(\sum_{i=1}^{L}(b_i - a_i + 1))$ time.*

By using Lemma 11 and weak order lemma, we can show the following three lemmas (we omit the proofs in this vertion):

Lemma 12. *Suppose we have the best L-link paths $u = u_0, u_1, .., u_L$ and $v = v_0, v_1, .., v_L$. Then, we can compute the best L-link path starting from a vertex $w \in [u, v]$ in $O(\sum_{i=0}^{L}(v_i - u_i))$ time. Furthermore, we can compute the best L-link paths starting from all vertices $w \in [u, v]$ in $O(\sum_{i=0}^{L}(v_i - u_i) \log n)$ time.*

Lemma 13. *Suppose we have an optimal path $u = u_0, u_1, .., u_M$ and a best L-link path $v = v_0, v_1, .., v_L$.*
(1) If $v < u$, we can compute the best L-link path starting from a vertex $w \in [v, u]$ in $O(|L - M|n + \sum_{i=0}^{\min(L,M)}(u_i - v_i))$ time.
(2) If $u < v < u_1$, we can compute the best L-link path starting from a vertex $w \in [u, v]$ in $O(|L - M|n + \sum_{i=0}^{\min(L,M)}(v_i - u_i))$ time, if its terminus is larger than u_M; otherwise, we can determine that the terminus is smaller than or equal to u_M in the same time complexity.

Lemma 14. *Suppose we have an optimal path $u = u_0, u_1, .., u_L$ and an optimal path $v = v_0, v_1, .., v_M$. Further, we assume $u < v < u_1$. Then, we can compute the best K-link path starting from a vertex $w \in [u, v]$ in $O((|K - L| + |K - M|)n + \sum_{i=0}^{\min(L,M)}(v_i - u_i))$ time, if its terminus is larger than u_L; otherwise, we can determine that the terminus is smaller than or equal to u_L in the same time complexity.*

Let s and t be two consecutive milestones. Assume $|h(t) - h(s)| \leq 1$, and let $h(s) = M$. Because of Lemma 6, $h(s) - 1 \leq h(u) \leq h(t) + 1$ for each $u \in [s, t]$; hence, $h(u)$ is one of $M - 1, M, M + 1, M + 2$. Let L be one of $M - 1, M, M + 1$, or $M + 2$. An L-link path p starting from $u \in [s, t]$ is called a *candidate path* if it is the best L-link path starting from u, and its terminus is larger than or equal to the terminus $Term(Opt(s))$ of the optimal path starting from s. Obviously, the optimal path starting from u is one of those candidate paths.

Lemma 15. *If $|h(t) - h(s)| \leq 1$, we can compute the candidate paths for all $u \in I = [s,t]$ in $O(n \log n)$ time.*

Proof. It suffices to consider the candidate L-link paths for each L of $\{M - 1, M, M + 1, M + 2\}$. We first decide whether the terminus of the best L-link path starting from the center vertex μ of I is larger than $Term(Opt(s))$, by using Lemma 14. If the terminus is not larger than $Term(Opt(s))$, we cut off the interval $[s, \mu]$ as *inactive* (we need not compute the paths starting from this interval) and continue for interval $[\mu, t]$ (called *active interval*) similarly by using Lemma 13. If the terminus is larger than $Term(Opt(s))$, we compute the best L-link path, and subdivide the interval $[s,t]$ into two intervals $[s, \mu]$ and $[\mu, t]$. We process each interval recursively using Lemma 13, until the best L link paths for both end-points of the interval have been computed or the interval contains no internal vertex. Note that, after the first L-link path has been found, only either the leftmost or the rightmost sub-interval of the active interval can be further subdivided. Thus, after $O(\log n)$ path-finding operations, we can subdivide the interval such that the condition of Lemma 12 holds.

We define the potential $\Phi(y)$ by $\Phi(y) = \sum_{i=0}^{h} y_i$, where $y_0, .., y_h$ is the path y. We need $O(\{|M - L|n + \Phi(opt(t)) - \Phi(opt(s))\} \log n)$ time for the first $O(\log n)$ path finding because of Lemmas 13 and 14. Let $p^1, p^2, .., p^m$ be the L-link paths computed so far (hence $m = O(\log n)$) ordered with respect to their starting vertices. We can find all L-link paths in the active interval in $O(\sum_{i=1}^{m-1}(\Phi(p^{i+1}) - \Phi(p^i) + 1) \log n)$ time because of Lemma 12; hence in $O((\Phi(opt(s)) - \Phi(opt(t)) + m) \log n)$ time. The total time complexity is $O(n \log n)$ since $m = O(\log n)$ and $|\Phi(opt(s)) - \Phi(opt(t))| = O(n)$ because of Lemma 5. $\qquad\Box$

Proposition 16. *The optimal paths can be computed in $O(Kn \log n)$ time for all vertices on an l-dense graph.*

Proof. We can compute all optimal paths starting from milestones in $O(Kn \log n)$ time. From Lemma 15, we can compute all optimal paths starting from the interval between each pair of consecutive milestones in $O(n \log n)$ time. Since there are $O(K)$ milestones, the total time complexity is $O(Kn \log n)$. $\qquad\Box$

Theorem 17. *The traveler's problem can be solved in $O(n^{3/2} \log n)$ time. Consequently, B-tree problem can be solved in $O(n^{3/2} \log^2 n)$ time.*

Proof. We equate $O(nl \log n)$ and $O(nK \log n)$ under the constraint $K = O(n/l)$. $\qquad\Box$

4 Concluding remarks

We have given an $O(n^{3/2} \log n)$ time algorithm for the traveler's problem, which improves the known bound by a factor of $n^{1/6}/\log^{2/3} n$. No nontrivial lower bound for this problem is known, and it is an open problem to improve the time complexity, hopefully to $O(n \log^c n)$ for some constant c. The solution to the traveler's problem can be applied to the efficient construction of the optimal B-tree for a dictionary; finding more applications of the traveler's problem and obtaining fundamental properties of paths on a DAG with Monge property remain open problems. Computing the shortest path in a Monge DAG has several applications; this paper provided one such application, and we refer the reader to [4] for more. Using Larmore and Schieber [7], the shortest path can be computed in linear time sequentially; however, the only polylogarithmic parallel-time algorithms for computing such paths take $O(n^2)$ processors [3].

Acknowledgement. The authors thank Baruch Schieber for helpful discussions.

References

1. A. Aggarwal, M. Klawe, S. Moran, P. Shor, and R. Wilbur, Geometric Applications of a Matrix-Searching Algorithm, *Algorithmica* **2** (1987), 209-233.
2. A. Aggarwal, J. Chang and C. Yap, Minimum Area Circumscribing Polygons, *The Visual Computer* **1** (1985), 112-117.
3. A. Aggarwal and J. Park, Notes on Searching in Multidimensional Monotone Arrays, *Proc. of the 29th Annual IEEE Symposium on the Foundations of Computing* (1988), 497-512.
4. A. Aggarwal, B. Schieber, and T. Tokuyama, Finding a Minimum Weight K-Link Path in Graphs with Monge Property and Applications, Proc. of *The 9-th ACM Symp. on Computational Geometry*, (1993), 189-197.
5. J. Boyce, D. Dobkin, R. Drysdale, and L. Guibas, Finding Extremal Polygons, *SIAM J. on Computing* **14** (1985), 134-147.
6. D. Hirschberg and L. Larmore, The Traveler's Problem, *J. of Algorithms* **13** (1992), 148-160.
7. L. Larmore and B. Schieber, On-line Dynamic Programming with Applications to the Prediction of RNA Secondary Structure, *J. of Algorithms* **12** (1991), 490-515.

Vehicle Scheduling on a Tree with Release and Handling Times

Yoshiyuki KARUNO †   Hiroshi NAGAMOCHI ‡   Toshihide IBARAKI ‡

† *Department of Mechanical and System Engineering*
Kyoto Institute of Technology

‡ *Department of Applied Mathematics and Physics*
Kyoto University

Abstract

Let $T = (V, E, v_0)$ be a rooted tree, where V is a set of n vertices, E is a set of edges and $v_0 \in V$ is the root. The travel times $d(v_i, v_j)$ and $d(v_j, v_i)$ are associated with each edge $(v_i, v_j) \in E$, and a job, which is also denoted as v_i, is located at each vertex v_i. Each job v_i has release time $r(v_i)$ and handling time $h(v_i)$. The TREE-VSP (Vehicle Scheduling Problem on a Tree) asks to find a routing schedule of the vehicle such that it starts from root v_0, visits all jobs $v_i \in V$ for processing, and returns to v_0. The processing of a job v_i cannot be started before its release time $t = r(v_i)$ (hence the vehicle may have to wait if it arrives at v_i too early) and takes $h(v_i)$ time units once its processing has been started (no interruption is allowed). The objective is to find a schedule that minimizes the completion time (i.e., the time to return to v_0 after processing all jobs). We first prove that TREE-VSP is NP-hard. Then we show that TREE-VSP with depth-first routing constraint can be exactly solved in $\Theta(n \log n)$ time. Finally we show that, if we regard this exact algorithm as an approximate algorithm for TREE-VSP without such routing constraint, its worst-case ratio is at most two, and that this bound is tight.

1   Introduction

The problem of routing and scheduling a vehicle, such as automated guided vehicle, handling robot, bus, truck and so forth, on a given road network is an important problem encountered in various applications. In this paper, we consider a single-vehicle scheduling problem, in which the vehicle is assumed to process all jobs at different locations by following a tree-shaped road network that connects such locations (represented as vertices). Each job has its own release time and handling time. The tree-shaped road network can be found, for example, in a building with simple structure of corridors (each floor corresponds to a subtree and each room a leaf vertex).

The problem is mathematically defined as follows. Let $T = (V, E, v_0)$ be a rooted tree that represents the road network, where V is a set of n vertices, E is a set of edges and $v_0 \in V$ is a vertex called the root. We assume that the initial location of the vehicle (at time $t = 0$) is root v_0. The travel time of the vehicle is $d(v_i, v_j) \geq 0$ to traverse $(v_i, v_j) \in E$ in this direction, and is $d(v_j, v_i) \geq 0$ to traverse it in the opposite direction. There is a job at each vertex $v_i \in V$, which must be processed by the vehicle (such as picking up an item). The job at vertex v_i is also denoted as v_i. Each job v_i has its release time $r(v_i)$ and handling time $h(v_i)$. That is, the vehicle cannot start processing job v_i before $t = r(v_i)$, and needs $h(v_i)$ time units to finish its processing. The vehicle at a vertex v_i is allowed to wait until $r(v_i)$ if necessary, and is also allowed to move to other vertices without processing v_i if it is more advantageous (in this case, the vehicle has to come back to v_i later to process job v_i). A routing schedule of the vehicle is completely specified by a sequence $\pi = (v_{j_1}, v_{j_2}, ..., v_{j_n})$ of jobs to be processed, i.e., the vehicle first moves to v_{j_1} (which may possibly be v_0) along the unique path from v_0 to v_{j_1} in T, taking the travel time of the length of the path, waits until $r(v_{j_1})$ if the arrival time is before $r(v_{j_1})$ and processes job v_{j_1}, then it moves to v_{j_2} along the unique path from v_{j_1} to v_{j_2} in T, waits until $r(v_{j_2})$ if necessary and processes v_{j_2}, and so on, until it returns to v_0 after finishing v_{j_n}. Our objective to find an optimal schedule π^* that minimizes the completion time $C(\pi)$, i.e., the time to return to v_0 after finishing all jobs. We call this problem TREE-VSP (Vehicle Scheduling Problem on a Tree).

This type of problems is studied under the name of the vehicle routing and scheduling problem with time window constraints (e.g., see [10],[11]), which is more general in the sense that it usually allows multiple vehicles and a time window (i.e., not only release time, but also deadline) at each v_i, and that the road network can be a general graph. Even with our setting, i.e., only with release times, the problem is NP-hard if the network is general, since it contains the Traveling Salesman Problem (TSP) as a special case. However, Psaraftis et al. [9] proved that a special case in which the road network is represented as a straight-line can be solved in polynomial time if all handling times are zero. But they conjectured that it is NP-hard if the jobs have positive handling times. Although this conjecture is still open, we show in this paper that the problem becomes NP-hard if the underlying road network is a tree, i.e., TREE-VSP, even if no job has handling time. We also show that TREE-VSP can be exactly solved in $\Theta(n \log n)$ time if we add a constraint that the vehicle has to process all jobs in depth-first manner. Note that this exact algorithm can be viewed as an approximate algorithm for the TREE-VSP without such routing constraint. We therefore analyze its worst-case ratio, and show that the ratio is at most two, and that the bound is tight.

Another interesting single-vehicle problem on tree-shaped network is provided in Frederickson et al. [6]. The problem is to find a minimum cost tour to deliver a set of objects to their corresponding destinations by a vehicle. The vehicle can carry only one object at a time. They presented two polynomial time algorithms for the problem, allowing in addition that the objects can be temporarily kept on intermediate vertices (i.e., preemptive). Atallah et al. [1] also contains polynomial time algorithms for the cases of a simple path or a simple cycle in both preemptive and nonpreemptive versions, and one of their algorithm (for the case of a simple cycle and nonpreemptive) is improved in [5]. If the road network is a general graph, the nonpreemptive version is known to be NP-hard [7].

2 NP-hardness of TREE-VSP

In this section, we show that TREE-VSP is NP-hard by a reduction from PARTITION, which is known to be NP-complete (e.g., see [8]):

PARTITION

INSTANCE: A finite set $I = \{1, 2, ..., n\}$, a "size" $a_i \in Z^+$ for each $i \in I$, and a bound $B \in Z^+$ such that $\sum_{i=1}^{n} a_i = 2B$, where Z^+ denotes the set of positive integers.

QUESTION: Is there a subset $I' \subseteq I$ such that

$$\sum_{i \in I'} a_i = \sum_{i \in I - I'} a_i = B \ ? \tag{1}$$

488

Theorem 1. TREE-VSP is *NP*-hard.

Proof. Given the above instance of PARTITION, we transform it into the following instance of TREE-VSP. Let $V = \{v_0, v_1, v_2, ..., v_{2n}\}$, where v_0 is the root vertex. Define edges (v_0, v_i) and (v_i, v_{n+i}) with travel times $d(v_0, v_i) = d(v_i, v_0) = d(v_i, v_{n+i}) = d(v_{n+i}, v_i) = a_i$ for $i = 1, 2, ..., n$ (see Figure.1). Furthermore, the release and handling times are given as follows:

$$r(v_i) = \begin{cases} 4B + 3a_i, & i = 1, 2, ..., n \\ 0, & i = 0, n+1, n+2, ..., 2n, \end{cases} \quad (2)$$

and

$$h(v_i) = 0, \quad i = 0, 1, 2, ..., 2n. \quad (3)$$

We show that, for this instance, there is an optimal schedule π^* with $C(\pi^*) = 10B$ if and if only the instance of PARTITION has a solution; otherwise, $C(\pi^*) > 10B$.

For a given subset $I' \subseteq I$, let $V'(I')$ and $V''(I')$ denote the set of vertices defined by

$$V'(I') = \{v_i \mid i \in I'\}, \quad (4)$$

$$V''(I') = \{v_i, v_{n+i} \mid i \in I - I'\}. \quad (5)$$

(I) The instance of PARTITION has a solution:

Let I' be the solution to PARTITION. The vehicle first processes all jobs $v_{n+i}, i \in I'$. This requires $4B$ time to traverse the paths from v_0 to v_{n+i} and from v_{n+i} to v_0 (since $\sum_{i \in I'} a_i = B$). At time $t = 4B$, the vehicle is at root v_0. Then the vehicle visits the remaining jobs in $V''(I')$ (for each $k \in I - I'$, it first processes v_{n+k} and then v_k) and $V'(I')$ and returns to root v_0. This requires $4B$ and $2B$ time, respectively. Thus,

$$C(\pi^*) = 4B + 4B + 2B = 10B.$$

(II) The instance of PARTITION does not have a solution:

At time $t = 4B$, assume that the vehicle has already processed the jobs v_{n+i} (since no jobs v_i can be processed by the release time constraint) such that $i \in J(\subseteq I)$, and let

$$B' = \sum_{i \in J} a_i, \quad B'' = \sum_{i \in I-J} a_i. \quad (6)$$

Then, since PARTITION has no solution, $B' \neq B$ (and hence $B'' \neq B$ by the property $B' + B'' = 2B$). We can assume without loss of generality that the vehicle did not wait at any vertex before $t = 4B$.

1. $B' < B$: Then, at time $t = 4B$, the vehicle is located in a certain path (v_0, v_k, v_{n+k}) with $k \in I - J$, at distance ε from v_0 (see Figure.2). This ε satisfies $0 < \varepsilon < 2a_k$ and $4B' + \varepsilon = 4B$, i.e.,

$$B' = B - \varepsilon/4, \quad B'' = B + \varepsilon/4. \quad (7)$$

(a) Assume that, after visiting v_{n+k}, the vehicle waits at v_k until its release time $r(v_k) = 4B + 3a_k$, and then returns to v_0 by taking a_k travel time. To visit the remaining vertices in $V''(J) - \{v_k, v_{n+k}\}$ and $V'(J)$ (and to return to v_0), the vehicle needs additional $4(B'' - a_k)$ and $2B'$ time units, respectively. Thus,

$$C(\pi^*) = (4B + 3a_k) + a_k + 4(B'' - a_k) + 2B' = 10B + 2(B - B') > 10B.$$

(b) Assume that, after visiting v_{n+k}, the vehicle directly returns to root v_0 without waiting at v_k. To process all jobs in $V''(J) - \{v_k, v_{n+k}\}$ and $\{v_k\} \bigcup V'(J)$ (and to return to v_0), the vehicle needs $4(B'' - a_k)$ and $2a_k + 2B'$ time units, respectively. Thus,

$$C(\pi^*) = 4B + (4a_k - \varepsilon) + 4(B'' - a_k) + 2a_k + 2B' = 10B + (4a_k - \varepsilon)/2 > 10B.$$

2. $B' > B$: Then, at time $t = 4B$, the vehicle is located in a certain path (v_0, v_k, v_{n+k}) with $k \in J$, at distance ε from v_0. This ε satisfies $0 < \varepsilon \leq 2a_k$ and $4B' - \varepsilon = 4B$, i.e.,

$$B' = B + \varepsilon/4, \quad B'' = B - \varepsilon/4. \tag{8}$$

In this case the vehicle has already processed job v_{n+k}. Then it either waits at v_k until its release time or returns to v_0 without processing v_k at this moment. The rest of analysis goes in a manner similar to case 1. In either case it is easy to prove that $C(\pi^*) > 10B$.

Therefore, we conclude that $C(\pi^*) > 10B$ if the instance of PARTITION has no solution. □

3 TREE-VSP with Depth-First Routing Constraint

Introduce the following depth-first routing constraint to problem TREE-VSP: once the vehicle reaches a vertex v from its parent in T, it cannot return to the parent unless it has completed all jobs in the subtree rooted at vertex v. Thus each edge $(v_i, v_j) \in E$ is traversed exactly two times (that is, one from v_i to v_j and another from v_j to v_i), in order to process all jobs.

To describe a schedule under depth-first routing constraint, we introduce some notations. Let $v_{p,i}$ denote the i-th child of vertex $v_p \in V, i = 1, 2, ..., k_p$, and $T(v_{p,i})$ the subtree of T rooted at $v_{p,i}$ (see Figure.3). Note that $k_p = 0$ if v_p is a leaf vertex. For convenience, we also denote v_p by $v_{p,0}$, and the graph consisting only of $v_{p,0}$ (and no edge) by $T(v_{p,0})$. A schedule $\pi = \pi(v_0)$ under depth-first routing constraint is recursively defined by

$$\pi(v_p) = (\pi(v_{p,\psi(0)}), \psi(v_{p,\psi(1)}), ..., \psi(v_{p,\psi(k_p)})), \quad v_p \in V \tag{9}$$

where $\psi = (\psi(0), \psi(1), ..., \psi(k_p))$ is a permutation on $\{0, 1, ..., k_p\}$ and $\pi(v_{p,i}) = v_{p,i}$ if $T(v_{p,i})$ consists only of one vertex (i.e., $i = 0$ or $T(v_{p,i})$ is a leaf vertex).

In Figure.4, for example, we consider that $v_{01} = v_1$, $v_{02} = v_2$, $v_{03} = v_3$, $v_{11} = v_4$, $v_{12} = v_5$, $v_{31} = v_6$, $v_{32} = v_7$ and $v_{i,0} = v_i$ for all i. A schedule π defined by

$$\pi(v_0) = (\pi(v_{01}), \pi(v_{00}), \pi(v_{03}), \pi(v_{02})),$$

$$\pi(v_{01}) = (\pi(v_{10}), \pi(v_{11}), \pi(v_{12})), \quad \pi(v_{03}) = (\pi(v_{32}), \pi(v_{31}), \pi(v_{30}))$$

gives the sequence of vertices $\pi = (v_1, v_4, v_5, v_0, v_7, v_6, v_3, v_2)$.

3.1 Basic properties

The completion time $C(\pi)$ of a schedule π is equal to the sum of travel times on edges, handling times at vertices and waiting times at vertices. However, since the sum of travel times and the sum of handling times are constants, minimizing $C(\pi)$ is equivalent to minimizing the sum of the waiting times at vertices.

Let $W_t(v_{p,i})$ denote the sum of waiting times when the vehicle starting from the parent v_p at time t visits $v_{p,i}$ and then processes all jobs in $T(v_{p,i})$ (including also $v_{p,i}$ itself) according to an optimal schedule under depth-first routing constraint, and returns to v_p. Note that, for $v_{p,0} = v_p$, this definition states that, starting at time t, the vehicle processes $T(v_{p,0})$ (i.e., v_p itself) immediately at time $\max\{r(v_p),\ t\}$; hence $W_t(v_{p,0}) = \max\{r(v_p) - t,\ 0\}$.

Lemma 1. For $t > t'$, the above W satisfies

$$W_t(v_{p,i}) = \max\{W_{t'}(v_{p,i}) - t + t',\ 0\} \qquad (10)$$

for any $v_{p,i} \in V$.

Proof. By changing the start time from t' to t, the completion time of $T(v_{p,i})$ can be decreased at most $t - t'(> 0)$, and hence

$$W_t(v_{p,i}) \geq \max\{W_{t'}(v_{p,i}) - (t - t'),\ 0\}.$$

However, if we start from v_p at time t and process $T(v_{p,i})$ by the same schedule as an optimal schedule $\pi^*(v_{p,i})$ of start time t', then

$$W_t(v_{p,i}) \leq W_t^*(v_{p,i}) = \max\{W_{t'}(v_{p,i}) - (t - t'),\ 0\},$$

where $W_t^*(v_{p,i})$ is the sum of waiting times by $\pi^*(v_{p,i})$. This proves (10). $\square$

Lemma 2. For TREE-VSP with depth-first routing constraint, there is an optimal schedule π^* such that

$$\pi^*(v_0) = (\pi^*(v_{0,\psi(0)}), \pi^*(v_{0,\psi(1)}), ..., \pi^*(v_{0,\psi(k_0)})), \qquad (11)$$

where permutation $\psi = (\psi(0), \psi(1), ..., \psi(k_0))$ on $I_0 = \{0, 1, ..., k_0\}$ satisfies

$$W_0(v_{0,\psi(0)}) \leq W_0(v_{0,\psi(1)}) \leq \cdots \leq W_0(v_{0,\psi(k_0)}), \qquad (12)$$

and $\pi^*(v_{0,i})$ are optimal schedules for subtrees $T(v_{0,i})$ of start time 0, $i = 0, 1, ..., k_0$.

Proof. Let $\pi(v_0) = (\pi(v_{0,\sigma(0)}), \pi(v_{0,\sigma(1)}), ..., \pi(v_{0,\sigma(k_0)}))$ be any schedule, where $\sigma = (\sigma(0), \sigma(1), ..., \sigma(k_0))$ is a permutation on I_0. If $\psi(0) = \sigma(j)$ and $j \neq 0$, we construct schedule π' from π by

$$\pi'(v_0) = (\pi(v_{0,\sigma(j)}), \pi(v_{0,\sigma(0)}), ..., \pi(v_{0,\sigma(j-1)}), \pi(v_{0,\sigma(j+1)}), ..., \pi(v_{0,\sigma(k_0)})).$$

First, compare the sums of waiting times by π' and π, when the vehicle has returned to v_0 after completing subtree $T(v_{0,\sigma(0)})$. The sum of waiting times by π is exactly $W_0(v_{0,\sigma(0)})$, while the sum of waiting times by π' is

$$
\begin{aligned}
W_0(v_{0,\sigma(j)}) + W_{s_0'}(v_{0,\sigma(0)}) &\leq W_0(v_{0,\sigma(j)}) + W_w(v_{0,\sigma(0)}) \\
&= W_0(v_{0,\sigma(j)}) + (W_0(v_{0,\sigma(0)}) - W_0(v_{0,\sigma(j)})) \\
&= W_0(v_{0,\sigma(0)}),
\end{aligned}
$$

where $w = W_0(v_{0,\sigma(j)})$ and $s_0'(\geq w)$ is the start time of processing $T(v_{0,\sigma(0)})$ in $\pi'(v_0)$. Note that Lemma 1 and $W_0(v_{0,\sigma(0)}) \geq W_0(v_{0,\sigma(j)})$ are used to derive the first and second lines.

The start time s_i of subtree $T(v_{0,\sigma(i)})$ by π is not later than the start time s_i' of the same subtree by π', for $i \neq 0$, j. This implies

$$W_{s_i'}(v_{0,\sigma(i)}) \leq W_{s_i}(v_{0,\sigma(i)}), \quad i \neq 0, \, j$$

and hence

$$\sum_{i=0}^{k_0} W_{s_i'}(v_{0,\sigma(i)}) \leq \sum_{i=0}^{k_0} W_{s_i}(v_{0,\sigma(i)}).$$

Therefore we can assume without loss of generality that $T(v_{0,\psi(0)})$ is the first subtree processed in an optimal schedule π^*.

Now, let t_0 be the completion time of $T(v_{0,\psi(0)})$ by an optimal schedule. By applying the above argument to the rest of subtrees, we see that the second subtree in the optimal schedule π^* minimizes $W_{t_0}(v_{0,i})$ among all $i \neq \psi(0)$. However, by Lemma 1, the one that minimizes $W_0(v_{0,i})$ also minimizes $W_{t_0}(v_{0,i})$. This shows that $T(v_{0,\psi(1)})$ can be considered as the second subtree in π^*. By repeating this argument, the lemma statement will be eventually proved. $\square$

Theorem 2. TREE-VSP with depth-first routing constraint has an optimal schedule π^*, which is recursively defined for every $v_p \in V$ by a permutation $\psi = (\psi(0), \psi(1), ..., \psi(k_p))$ on $\{0, 1, ..., k_p\}$ satisfying

$$W_0(v_{p,\psi(0)}) \leq W_0(v_{p,\psi(1)}) \leq \cdots \leq W_0(v_{p,\psi(k_p)}). \tag{13}$$

Proof. By generalizing the argument of Lemma 2, it is not difficult to see that, in an optimal schedule π^*, the order of subtrees in $T(v_p)$, starting at time s_p, is given by a permutation ψ with

$$W_{s_p}(v_{p,\psi(0)}) \leq W_{s_p}(v_{p,\psi(1)}) \leq \cdots \leq W_{s_p}(v_{p,\psi(k_p)}).$$

However, by Lemma 2, property (13) implies this ordering. $\square$

3.2   Polynomial time algorithm

Based on Theorem 2, we present a polynomial time algorithm, called DF-Routing, to solve TREE-VSP with depth-first routing constraint. Now introduce a new notation $W_t'(v_p)$ as follows:

$W_t'(v_p)$: the sum of waiting times when the vehicle starting from v_p at time t processes all jobs in $T(v_p)$ according to an optimal schedule under depth-first routing constraint, and returns to v_p.

Note that this W_t' and the W_t of Section 3.1 are related as follows:

$$W_t(v_{p,i}) = W_{t+d(v_p,v_{p,i})}'(v_{p,i}) = \max\{W_t'(v_{p,i}) - d(v_p, v_{p,i}), 0\}, \tag{14}$$

where $d(v_p, v_{p,0}) = d(v_{p,0}, v_p) = 0$ are assumed for convenience. In addition, $D(v_p)$ and $H(v_p)$ denote the sum of travel times and the sum of handling times, respectively, which are incurred when the vehicle has processed $T(v_p)$. The subschedule $\pi(v_p)$ is computed in SEARCH$(v_p; W_0'(v_p), D(v_0), H(v_p), \pi(v_p))$, by recursively calling this procedure for its children $v_{p,i}, i = 1, 2, ..., k_p$. $WAIT, DIST$ and $HAND$ are temporary storages used to calculate $W_0'(v_p)$, $D(v_p)$ and $H(v_p)$, respectively.

Algorithm DF-Routing

begin

    $\text{SEARCH}(v_0; W_0'(v_0), D(v_0), H(v_0), \pi(v_0))$

end.

procedure $\text{SEARCH}(v_p; W_0'(v_p), D(v_p), H(v_p), \pi(v_p))$

begin

    $W_0'(v_p) := r(v_p); \ D(v_p) := 0; \ H(v_p) := h(v_p); \ \pi(v_p) := (v_p);$

    $k_p :=$(the number of children of v_p in T);

    if $k_p = 0$ **then** return **else**

        (Computation of optimal schedule at v_p)

        $W_0(v_{p,0}) := r(v_p);$

        for $i := 1$ **to** k_p

            begin

                $\text{SEARCH}(v_{p,i}; \ W_0'(v_{p,i}), D(v_{p,i}), H(v_{p,i}), \pi(v_{p,i}));$

                $W_0(v_{p,i}) := \max\{W_0'(v_{p,i}) - d(v_p, v_{p,i}), \ 0\}$

            end;

        Compute a permutation ψ on $\{0, 1, .., k_p\}$ such that

            $W_0(v_{p,\psi(0)}) \leq W_0(v_{p,\psi(1)}) \leq \cdots \leq W_0(v_{p,\psi(k_p)});$

        $\pi(v_p) := (\pi(v_{p,\psi(0)}), \pi(v_{p,\psi(1)}), ..., \pi(v_{p,\psi(k_p)}));$

        (Computation of $W_0'(v_p), D(v_p)$ and $H(v_p)$)

        $WAIT := 0; \ DIST := 0; \ HAND := 0;$

        for $i := 0$ **to** k_p

            begin

                $DIST := DIST + d(v_p, v_{p,\psi(i)});$

                $WAIT := WAIT + \max\{W_0'(v_{p,\psi(i)}) - \{WAIT + DIST + HAND\}, \ 0\};$

                $HAND := HAND + H(v_{p,\psi(i)});$

                $DIST := DIST + D(v_{p,\psi(i)}) + d(v_{p,\psi(i)}, v_p)$

            end;

        $W_0'(v_p) := WAIT; \ D(v_p) := DIST; \ H(v_p) := HAND;$

        return

end.

Lemma 3. The algorithm DF-Routing solves TREE-VSP with depth-first routing constraint in $O(n \log n)$ time.

Proof. The DF-Routing constructs π^* of Theorem 2 by searching T in the depth-first manner. Therefore its correctness follows from Theorem 2. To analyze its time complexity, note that traversing all edges in T requires $O(n)$ time. At each vertex v_p, it sorts $(k_p + 1)$ elements (where k_p is the number of children of v_p). Since $\sum_{p=0}^{n-1} k_p = n - 1$, the total time for sorting is $\sum_{p=0}^{n-1} O((k_p + 1)\log(k_p + 1)) = O(n\log n)$. Hence the time complexity of DF-Routing is $O(n\log n)$. $\square$

Lemma 4. TREE-VSP with depth-first routing constraint requires $\Omega(n\log n)$ time.

Proof. Given n distinct positive integers $a_1, a_2, ..., a_n$ such that $\min_{1 \leq i < j \leq n} |a_i - a_j| \geq 2$, the problem instance shown in Figure.5 has the unique depth-first optimal schedule $\pi^* = (v_0, v_{i_1}, v_{i_2}, ..., v_{i_n})$ (without loss of correctness, we can assume that v_0 is the first processed job) such that $a_{i_1} < a_{i_2} < ... < a_{i_n}$. Therefore, any algorithm for TREE-VSP with depth-first routing constraint must contain sorting of n elements, which is known to have $\Omega(n\log n)$ time lower bound. $\square$

Combining these, we have the following theorem.

Theorem 3. TREE-VSP with depth-first routing constraint has $\Theta(n\log n)$ time complexity, which is realized by algorithm DF-Routing. $\square$

3.3 Worst-case analysis as an approximate algorithm

The algorithm DF-Routing can also be used as an approximate algorithm for the original TREE-VSP without depth-first routing constraint. In order to examine its worst-case performance, we introduce the following notations.

$C(\pi^*)$: The optimal completion time for the original TREE-VSP.

$C(\pi^*_{DFR})$: The completion time obtained by algorithm DF-Routing.

$D = \sum_{(v_i, v_j) \in E} (d(v_i, v_j) + d(v_j, v_i))$: The sum of travel times for all edges.

$H = \sum_{v_i \in V} h(v_i)$: The sum of all handling times.

$r_{max} = \max_{v_i \in V} r(v_i)$: The maximum release time.

Theorem 4. The approximate schedule obtained by algorithm DF-Routing satisfies

$$C(\pi^*_{DFR}) \leq 2 \cdot C(\pi^*), \tag{15}$$

and this bound is tight.

Outline of the proof: Consider the following two cases: (a) $r_{max} \geq D + H$: Then it is clear that $C(\pi^*) \geq r_{max}$ and $C(\pi^*_{DFR}) \leq r_{max} + D + H$. (b) $r_{max} < D + H$: Then it can also be seen that $C(\pi^*) \geq D + H$ and $C(\pi^*_{DFR}) \leq r_{max} + D + H$. In both cases, we have (15).

To show that the bound 2 is tight, we provide the instance of Figure.6. It holds that $C(\pi^*_{DFR}) = 2(\varepsilon + d)n + 2\varepsilon + 2(\varepsilon + d)(n - 1)$, $C(\pi^*) = 2(\varepsilon + d)n + 2\varepsilon n$ (since an optimal schedule $\pi^* = (v_{n+1}, ..., v_{2n}, v_1, ..., v_n, v_0)$). Hence we have $C(\pi^*_{DFR})/C(\pi^*) \to 2$ when $\varepsilon \to 0$ and $n \to \infty$. $\square$

4 Conclusion

In this paper we considered the single-vehicle routing and scheduling problem on a tree, called TREE-VSP. After showing its *NP*-hardness by reducing PARTITION to TREE-VSP, we examined TREE-VSP under depth-first routing constraint, and presented an $\Theta(n \log n)$ time exact algorithm (DF-Routing). We also derived its worst-case performance ratio when this algorithm was used as an approximate algorithm for the original TREE-VSP.

However, since PARTITON is only weakly *NP*-complete, it remains to examine whether TREE-VSP is strongly *NP*-hard or not. Among other interesting topics associated with TREE-VSP are different types of road networks such as circular paths and planar networks, and different constraints such as deadline of jobs. It is known that our proof of Theorem 1 can be modified to show TREE-VSP with deadlines, instead of release times, is also *NP*-hard.

References

1 Atallah , M. and S. Kosaraju, Efficient solutions to some transportation problems with application to minimizing robot arm travel, *SIAM J. Comput.*, 17(1988), 849-869.

2 Baker, E., An Exact Algorithm for the Time-Constraint Traveling Salesman Problem, *Operations Research*, 31-5(1983), 938-945.

3 Bramel, J., E. Coffman, Jr., P. Shor and D. Simchi-Levi, Probabilistic Analysis of the Capacitated Vehicle Routing Problem with Unsplit Demands, *Operations Research*, 40-6(1992), 1095-1106.

4 Dell'Amico, M., M. Fischetti and P. Toth, Heuristic Algorithms for the Multiple Depot Vehicle Scheduling Problem, *Management Science*, 39-1(1993), 115-125.

5 Frederickson, G., A Note on the Complexity of a Simple Transportation Problem, *SIAM J. Comput.*, 22-1(1993), 57-61.

6 Frederickson, G. and D. Guan, Preemptive Ensemble Motion Planning on a Tree, *SIAM J. Comput.*, 21-6(1992), 1130-1152.

7 Frederickson, G., M. Hecht and C. Kim, Approximation Algorithms for Some Routing Problems, *SIAM J. Comput.*, 7-2(1978), 178-193.

8 Garey, M. and D. Johnson, *Computers and Intractability: A Guide to the Theory of NP-Completeness*, W. H. Freeman and Company, San Francisco, 1979.

9 Psaraftis, H., M. Solomon, T. Magnanti and T. Kim, Routing and Scheduling on a Shoreline with Release Times, *Management Science*, 36-2(1990), 212-223.

10 Solomon, M., Algorithms for the Vehicle Routing and Scheduling Problem with Time Window Constraints, *Operations Research*, 35-2(1987), 254-265.

11 Solomon, M. and J. Desrosiers, Time Window Constrained Routing and Scheduling Problems: A Survey, *Transportation Sci.*, 22(1988), 1-13.

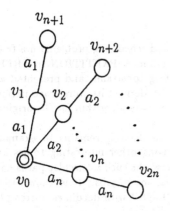

Figure.1   The instance of TREE-VSP corresponding to PARTITION.

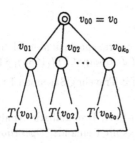

Figure.3   The subtrees rooted at v_0.

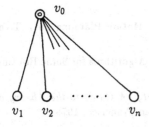

$d(v_0, v_i) = d(v_i, v_0) = 1, \ i = 1, 2, ..., n$

$r(v_0) = 0, \ r(v_i) = a_i, \ i = 1, 2, ..., n$

$h(v_i) = 0, \qquad\qquad i = 0, 1, ..., n$

Figure.5   Proof of Lemma 4.

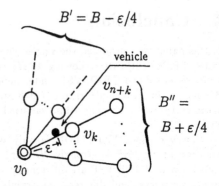

Figure.2   Location of the vehicle and the remaining jobs at time $t = 4B$ when PARTITION does not have a solution.

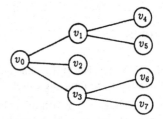

Figure.4   An example tree.

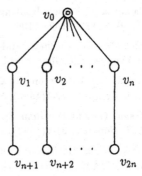

$d(v_0, v_i) = d(v_i, v_0) = \varepsilon (> 0), \qquad i = 1, 2, ..., n$

$d(v_i, v_{n+i}) = d(v_{n+i}, v_i) = d(> 0), \qquad i = 1, 2, ..., n$

$r(v_i) = \begin{cases} 2(\varepsilon + d)n + 2\varepsilon n, & i = 0 \\ 2(\varepsilon + d)n + \varepsilon, & i = 1, 2, ..., n \\ 0, & i = n+1, n+2, ..., 2n \end{cases}$

$h(v_i) = 0, \qquad\qquad i = 0, 1, ..., 2n$

Figure.6   An example with $C(\pi^*_{DFR})/C(\pi^*) \to 2$ in Theorem 4.

Scheduling Algorithms for a Chain-like Task System

Chan, Chi-Lok*,Gilbert, Young**

The Chinese University of Hong Kong

Abstract. The optimal allocation of a chain-like task system on the chain-like network computers was first presented by Bokhari with time complexity $O(m^3 n)$[1], where m and n denote the number of modules and the number of processors respectively. Sheu and Chiang improved it and gave an $O(\min\{m, n\}m^2)$ algorithm[2]. Hsu had further developed a two phase approach with the worst case time complexity of $O(m + (m' - n)^2 n)$[3] where m' denotes the number of the remaining modules after the merge phase. For all of these papers, none of them answers the decision version of this problem, that is whether there exists a schedule with schedule length less than a given deadline. In this paper, two algorithms are presented. The first one answers the decision problem and gives a feasible schedule if there exists one. It is an optimal algorithm with time complexity of $O(m)$. The second one finds an optimal schedule in $O(m + m' \log m' + m'^2 - n^2)$ time using the first algorithm. It is shown that our approach is more efficient than Hsu's one if $m' - n = \omega(\sqrt{\log n})$.

Keywords: bottleneck processor, chain-like task system, completion time, contiguity constraint, feasible length-$\mathcal{K}$ schedule, layered graph, merged module, optimal schedule, schedule length, un-mergeable modules.

1 Introduction

The problem we investigate in this paper is the allocation of a chain-like task system on the chain-like network computers which was first presented by Bokhari [1]. A chain-like task system consists of m modules scheduled on n processors. Each module is associated with an execution time and each module communicates with its neighboring modules with a communication cost (Fig. 1).

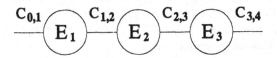

Fig. 1: A chain-like task system

* E-mail: clchan@cs.cuhk.hk
** E-mail: young@cs.cuhk.hk

A typical application of chain-like task system is pipelining system with an essentially unending series of data. The maximum rate of processing is now determined by the processor that takes the longest amount of time (schedule length) to perform its tasks (the bottleneck processor). The optimal allocation problem is to find a schedule in which the bottleneck processor has the minimal execution time. Bokhari had presented an algorithm with time complexity $O(m^3n)$ [1]. Sheu and Chiang showed that a more general assignment problem can be solved with time complexity $O(\min\{m,n\}m^2)$ [2]. Hsu proposed an approach which is composed of two phases: merge phase and assignment phase[3]. The time complexity of the merge phase is $O(m)$ and the overall complexity is $O(m+(m'-n)^2n)$ where m' denotes the number of the remaining modules after the merge phase ($m' \leq m$). These three algorithms use a similar approach, a layered graph method. In this paper, we propose an $O(m)$ algorithm for answering whether there exists a schedule with schedule length no greater than a constant K. Furthermore, we propose another approach for finding an optimal schedule in $O(m + m'\log m' + m'^2 - n^2)$ time.

The remainder of this paper is organized as follows: section 2 states the problem assumptions and notations used in this paper. In section 3, an optimal greedy algorithm for the decision problem is presented. Section 4 proposes another approach for finding an optimal schedule. Finally, we draw some concluding remarks in the last section.

2 Problem Assumptions and Notations Definition

1. A chain-like system consists of m modules, $t_1, t_2, \ldots, t_m$, scheduled on n processors. Execution time of module i is denoted as E_i and communication time of adjacent modules i and $i+1$ is denoted as $C_{i,i+1}$ ($0 \leq i \leq m$). All of execution times and communication times are positive except that $C_{0,1}$ and $C_{m,m+1}$ equal zero.
2. Adjacent modules must be assigned to adjacent processors or the same processor, which is called contiguity constraint [1].
3. If adjacent modules are assigned to the same processor, their communication (intra-communication) time is assumed to be zero.
4. The elapsed time required to complete the execution and communication for a module (may be a sequence of adjacent modules) on a processor is called completion time. The completion time of a module i,denoted as CMP_i, is equal to $C_{i-1,i} + E_i + C_{i,i+1}$. The completion time of modules $i, i+1, \ldots, j$, denoted as $CMP_{i\oplus i+1\oplus\cdots\oplus j}$, is equal to $C_{i-1,i} + \sum_{k=i}^{j} E_k + C_{j,j+1}$.
5. For a schedule S of m modules on n processors, if $t_j, t_{j+1}, \ldots, t_k$ are modules scheduled on processor i, $1 \leq j \leq k \leq m$ and $1 \leq i \leq n$, $CP_i(S)$ denotes the completion time of processor i, ie.

$$CP_i(S) = CMP_{j\oplus\cdots\oplus k}$$

6. The schedule length of a schedule, $\mathcal{L}(S)$, is defined as the maximum completion time of all processors in S, ie.

$$\mathcal{L}(S) = \max_i \{CP_i(S)\}$$

7. A schedule is said feasible if it satisfies the contiguity constraint. An optimal schedule, S°, is a feasible schedule with minimal schedule length, ie.

$$\mathcal{L}(S^\circ) \leq \mathcal{L}(S')$$

where S' is any possible feasible schedule.

8. Two modules, t_i and t_{i+1} are mergeable if the completion time of the merged module is no greater than the maximum completion time of t_i and t_{i+1} (Fig. 2),

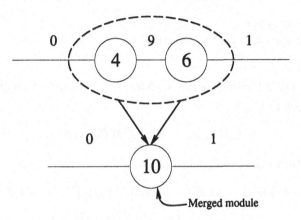

Fig. 2: Merge two modules.

ie.

$$CMP_{i\oplus i+1} \leq \max\{CMP_i, CMP_{i+1}\}$$

9. Two modules, t_i and t_{i+1} are un-mergeable if they are not mergeable, ie.

$$C_{i,i+1} < E_i + C_{i-1,i} \quad \text{and} \quad C_{i,i+1} < E_{i+1} + C_{i+1,i+2}$$

3 An Algorithm for the Decision Problem

In this section, we present an algorithm for answering whether there exists a schedule S such that the schedule length of S, $\mathcal{L}(S)$, is less than or equal to $\mathcal{K}$, a given constant (called a decision problem).

The following lemmas are useful for the proof of correctness of our algorithm.

Lemma 1. *For m' un-mergeable modules, $t_1, \ldots, t_{m'}$, in a chain-like task system, $1 \leq i \leq p \leq j \leq m'$.*

$$CMP_{i\oplus i+1\oplus\cdots\oplus j} \geq CMP_{p\oplus p+1\oplus\cdots\oplus j}$$

Proof. By induction on p, for any $1 \leq i \leq p \leq j \leq m'$,

For $p = i$,

$$CMP_{p \oplus p+1 \oplus \cdots \oplus j} = CMP_{i \oplus i+1 \oplus \cdots \oplus j}$$

ie. it is true for $p = i$

Assume it is true for $p = k < j$

ie. $CMP_{i \oplus i+1 \oplus \cdots \oplus j} \geq CMP_{k \oplus k+1 \oplus \cdots \oplus j}$

For $p = k + 1$,

$$\begin{aligned}
CMP_{k+1 \oplus \cdots \oplus j} &= CMP_{k \oplus \cdots \oplus j} - C_{k-1,k} - E_k + C_{k,k+1} \\
&< CMP_{k \oplus \cdots \oplus j} \qquad (C_{k,k+1} < C_{k-1,k} + E_k) \\
&\leq CMP_{i \oplus \cdots \oplus j} \qquad \text{(induction hypothesis)}
\end{aligned}$$

ie. It is true for $p = k + 1$

So, it is true for $i \leq p \leq j$ by induction.

$\square$

Lemma 2. *For m' un-mergeable modules, $t_1, \ldots, t_{m'}$, in a chain-like task system, $1 \leq i \leq p \leq j \leq m'$.*

$$CMP_{i \oplus i+1 \oplus \cdots \oplus j} \geq CMP_{i \oplus i+1 \oplus \cdots \oplus p}$$

Proof. Similar to the proof of Lemma 1. $\square$

Definition. A schedule is called a feasible length-$\mathcal{K}$ schedule if

$$\mathcal{L}(S) = max_{1 \leq i \leq n}\{CP_i(S)\} \leq \mathcal{K}$$

where $\mathcal{K}$ is an integer constant.

The following algorithm decides whether there exists a feasible length-$\mathcal{K}$ schedule. The input modules are assumed un-mergeable. It schedules as many modules on a processor as possible successively and the completion time of each processor is no greater than $\mathcal{K}$. If all modules can be scheduled properly, it returns 'Yes' and the corresponding schedule; otherwise it returns 'No'.

Algorithm 1: (Greedy Algorithm)

Input: m' un-mergeable modules in a chain-like task system with their execution times and communication times, n processors and $\mathcal{K} \in \mathcal{Z}^+$

Output: 'Yes' and gives a feasible length-$\mathcal{K}$ schedule S if there exists one; otherwise, 'No'.

(1) $j = 1, CP = 0$ // j denotes the processor no.
(2)   for $i = 1$ to m' // for each module i,
(3)   if $CP + E_i + C_{i,i+1} - C_{i-1,i} > \mathcal{K}$ //if we cannot schedule t_i on p_j
(4)   if $j = n$ // no processor left
(5)   return 'No'
(6)   else

(7) $\qquad CP = C_{i-1,i} + E_i + C_{i,i+1} \qquad // \ CP = CMP_i$
(8) $\qquad$ if $(CP > \mathcal{K}) \qquad //$ schedule length $> \mathcal{K}$
(9) $\qquad$ return 'No'
(10) $\qquad j = j + 1$
(11) $\quad$ else
(12) $\qquad CP = CP + E_i + C_{i,i+1} - C_{i-1,i} \qquad //$ schedule t_i on p_j
(13) $\quad S[i] = j$
(14) return 'Yes' and S

For the above algorithm, CP is the current completion time of processor j and S is an array such that $S[i] = j$ if t_i is scheduled on processor j.

Since each module is examined only once and there are m' modules (line 2), the time complexity of Algorithm 1 is $O(m')$. For a mergeable chain-like task system with m modules, $O(m)$ time is needed for merging [3] and $O(m')(m' \le m)$ time is used for Algorithm 1, the overall time complexity is $O(m)$.

Theorem 1 *For m' un-mergeable modules scheduled on n processors, Algorithm 1 returns a feasible length-$\mathcal{K}$ schedule S_n iff there exists a feasible length-$\mathcal{K}$ schedule S'_n for the m' modules on n processors.*

Proof.
Part 1: "$\Rightarrow$" part.

If Algorithm 1 answers 'Yes' and returns S_n, it is obvious that S_n is a feasible length-$\mathcal{K}$ schedule because we schedule t_j on processor i iff the completion time of processor i is less than or equal to $\mathcal{K}$, $1 \le j \le m'$, ie.

$$\max_{1 \le i \le n} \{CP_i(S_n)\} \le \mathcal{K}$$

we can let

$$S'_n = S_n \qquad \text{and} \qquad \max_{1 \le i \le n}\{CP_i(S'_n)\} = \max_{1 \le i \le n}\{CP_i(S_n)\} \le \mathcal{K}$$

Part 2: "$\Leftarrow$" part.

Suppose there exists a feasible length-$\mathcal{K}$ schedule S'_n and S_n is the schedule obtained from Algorithm 1 (suppose we do not know whether Algorithm 1 answers 'Yes' or 'No' and S_n may not be a feasible length-$\mathcal{K}$ schedule) We have to prove that

$$\max_{1 \le i \le n}\{CP_i(S'_n)\} \le \mathcal{K} \Rightarrow \max_{1 \le i \le n}\{CP_i(S_n)\} \le \mathcal{K}$$

by induction on n, the number of processor.

For $n = 1$, there is only one processor so that all m' modules are scheduled on processor 1 in S'_1. By Algorithm 1, we always schedule as many modules as possible on each processor so that $S_1 = S'_1$ and

$$\max_{i=1}\{CP_i(S_1)\} = \max_{i=1}\{CP_i(S'_1)\} \le \mathcal{K}$$

So, it is true for $n = 1$.

Assume it is true for $n = k$, ie. if there exists a feasible length-$\mathcal{K}$ schedule S'_k, then Algorithm 1 returns a feasible length-$\mathcal{K}$ schedule S_k,

$$\max_{1 \le i \le k}\{CP_i(S'_k)\} \le \mathcal{K} \Rightarrow \max_{1 \le i \le k}\{CP_i(S_k)\} \le \mathcal{K}$$

For $n = k + 1$, let S'_{k+1} is any feasible length-$\mathcal{K}$ schedule and S_{k+1} is the schedule returned by Algorithm 1. Firstly, for S'_{k+1}

$$\max_{1 \le i \le k+1}\{CP_i(S'_{k+1})\} \le \mathcal{K}$$

We compare the modules scheduled on processor 1 in S'_{k+1} and S_{k+1}. Let n'_1 and n_1 be the number of modules scheduled on processor 1 in S'_{k+1} and S_{k+1} respectively,

Case 1: $n'_1 = n_1$

For processor 1, the schedules in S'_{k+1} and S_{k+1} are the same, ie.

$$CP_1(S_{k+1}) = CP_1(S'_{k+1}) \le \mathcal{K} \qquad \text{... (i)}$$

Consider the modules scheduled on processor 2 to k+1 for both schedules, they schedule the same modules on k processors. By induction hypothesis,

$$\max_{2 \le i \le k+1}\{CP_i(S'_{k+1})\} \le \mathcal{K}$$
$$\Rightarrow \max_{2 \le i \le k+1}\{CP_i(S_{k+1})\} \le \mathcal{K} \qquad \text{... (ii)}$$

By (i) and (ii),

$$\max_{1 \le i \le k+1}\{CP_i(S_{k+1})\} \le \mathcal{K}$$

Case 2: $n'_1 < n_1$

In this case, for S'_{k+1}, we can re-schedule some modules $(t_{n'_1+1}, t_{n'_1+2}, \ldots, t_{n_1})$ on processor 1 and remove these modules from some other processors. If there are some processors becoming no module scheduled on them after re-scheduling, we can relabel the processor so that the contiguity constraint is not violated. We call the resultant schedule S''_{k+1}. The completion time of processor 1 in S''_{k+1} is less than or equal to $\mathcal{K}$ because we can schedule n_1 modules on processor 1 by Algorithm 1 only if $CMP_{1 \oplus 2 \oplus \cdots \oplus n_1} \le \mathcal{K}$. For S''_{k+1}, compared with S'_{k+1}, the number of module remains in processor j, $2 \le j \le k+1$, is reduced (because some modules or none are removed). By Lemma 1, the completion time of processor j in S''_{k+1} is less than or equal to that in S'_{k+1}, ie.

$$CP_j(S''_{k+1}) \le CP_j(S'_{k+1}), 2 \le j \le k+1 \qquad \text{and}$$

$$\max_{2 \le i \le k+1}\{CP_i(S''_{k+1})\} \le \max_{2 \le i \le k+1}\{CP_i(S'_{k+1})\} \le \mathcal{K}$$

So, S''_{k+1} is also a feasible length-$\mathcal{K}$ schedule and the number of modules scheduled on processor 1 is equal to n_1. By case 1, S_{k+1} is a feasible length-$\mathcal{K}$ schedule as well.

Case 3: $n_1' > n_1$

In this case, t_1 to t_{n_1} are scheduled on processor 1 in S_{k+1} and t_1 to $t_{n_1'}$ are scheduled on processor 1 in S_{k+1}' where $n_1' > n_1$.

However, t_{n_1+1} cannot be scheduled on processor 1 by Algorithm 1 (line 3), so

$$CMP_{1\oplus\cdots\oplus n_1 \oplus n_1+1} > \mathcal{K}$$

By Lemma 2, $n' \geq n_1 + 1$

$$CP_1(S_{k+1}') = CMP_{1\oplus\cdots\oplus n_1'} \geq CMP_{1\oplus\cdots\oplus n_1 \oplus n_1+1} > \mathcal{K}$$

which is contradictory to the fact that S_{k+1}' is a feasible length-$\mathcal{K}$ schedule. The case $n_1' > n_1$ never occurs.

By case 1, 2 and 3, if S_{k+1}' is a feasible length-$\mathcal{K}$ schedule, S_{k+1} is a feasible length-$\mathcal{K}$ schedule as well, ie. it is true for $n = k + 1$.

By induction on n, Theorem 1 is true for any number of processor. $\qquad\square$

4 An Algorithm for Optimal Schedule

In the previous section, we present an $O(m)$ algorithm for deciding the existence of a feasible length-$\mathcal{K}$ schedule. Now we can apply this algorithm to find an optimal schedule.

4.1 Algorithm

It is obvious that the schedule length of any schedule is equal to the longest completion time of the n processors, ie. the completion time of the bottleneck processor. We can construct a set, Λ, which contains all possible values of schedule length, By Hsu's result, we can only consider a chain-like task system with un-mergeable modules only [3] and

$$\Lambda = \{CMP_{i\oplus\cdots\oplus j} : 1 \leq i \leq j \leq m'\}$$

By Hsu's result, for m' un-mergeable modules scheduled on n processors, if $m' \leq n$, we can obtain an optimal schedule by assigning one merged module to one processor. On the other hand, if $m' > n$, all processors must be utilized and at least one module is scheduled on each processor [3]. As a result, for $m' > n$, the maximum number of modules scheduled on any processor is $m' - n + 1$, so

$$\Lambda = \{CMP_{i\oplus\cdots\oplus j} : 1 \leq i \leq j \leq m' \text{ and } j - i \leq m' - n\}$$

where Λ covers all possible minimal schedule length and

$$|\Lambda| = \sum_{i=n}^{m'} i = \sum_{i=1}^{m'} i - \sum_{i=1}^{n-1} i = O(m'^2 - n^2)$$

Now, we can find the median of the elements in Λ and apply Algorithm 1 as a feasibility test. Then we can eliminate half of the elements in Λ. Repeat the steps until there is only one element in Λ and this is the optimal schedule length. The following algorithm summarize the above method.

Algorithm 2:

Input: m modules in a chain-like task system with their execution times
            and communication times, n processors
Output: An optimal schedule S.

(1)   merge the m modules into m' un-mergeable modules by Hsu's algorithm
(2)   if $m' \leq n$
(3)   return S where $S[i] = i, 1 \leq i \leq m'$
(4)   construct a set Λ where
            $\Lambda = \{CMP_{i\oplus\cdots\oplus j} : 1 \leq i \leq j \leq m'$ and $j - i \leq m' - n\}$
(5)   find the median of the elements in Λ
(6)   apply Algorithm 1 with the median as the deadline, $\mathcal{K}$
(7)   if Algorithm 1 returns 'Yes'
(8)   remove all elements which are larger than $\mathcal{K}$ in Λ
(9)   otherwise
(10)   remove all elements which are no greater than $\mathcal{K}$ in Λ
(11) if $|\lambda| \neq 1$
(12)   repeat step 5
(13) return the corresponding schedule S

4.2   Complexity Analysis

The time complexity of Algorithm 2 consists of several parts. Firstly, $O(m)$ time for merging the modules(line 1). For the construction of Λ (line 4), we need $O(m'^2 - n^2)$ time. For step 5 to 12, we can find the median of N elements in $O(N)$ [4][5][6]. The feasibility test, Algorithm 1, takes $O(m')$ time and we can remove the elements (line 8, 10) in $O(N)$ time. Let $X(N)$ be the time taken for step 5 to 12,

$$X(N) = X(\frac{N}{2}) + N + m'$$

$$X(m'^2 - n^2) = O(m' \log m' + m'^2 - n^2)$$

As a result, the total time complexity of Algorithm 2 is

$$T(m,n) = O(m + m' \log m' + m'^2 - n^2)$$

The time complexity of Hsu's algorithm is [3]

$$T'(m,n) = O(m + (m' - n)^2 n)$$

Comparison between the two algorithms

If $m' \leq n$, Algorithm 2 is identical to Hsu's algorithm and the time complexity is $O(m)$.

If $m' > n$, let $m' - n = f(n)$, $f(n) > 0$ for all n,

$$T(m,n) = O(m + m' \log m' + m' f(n))$$

$$T'(m,n) = O(m + n f^2(n))$$

Case 1: $f(n) = o(\sqrt{\log n}) \Rightarrow m' = O(n)$

$$T(m,n) = O(m + n \log n + n f(n)) = O(m + n \log n)$$
$$T'(m,n) = O(m + n f^2(n)) \leq O(m + n \log n) = T(m,n)$$

Case 2: $O(\log n) = f(n) = \omega(\sqrt{\log n}) \Rightarrow m' = O(n)$

$$T(m,n) = O(m + n \log n + n f(n)) = O(m + n \log n)$$
$$T'(m,n) = O(m + n f^2(n)) \geq O(m + n \log n) = T(m,n)$$

Case 3: $O(n) = f(n) = \Omega(\log n) \Rightarrow m' = O(n)$

$$T(m,n) = O(m + n \log n + n f(n)) = O(m + n f(n))$$
$$T'(m,n) = O(m + n f^2(n)) \geq O(m + n f(n)) = T(m,n)$$

Case 4: $f(n) = \Omega(n) \Rightarrow m' = O(f(n))$

$$T(m,n) = O(m + f(n) \log f(n) + f^2(n)) = O(m + f^2(n))$$
$$T'(m,n) = O(m + n f^2(n)) \geq O(m + f^2(n)) = T(m,n)$$

For case 2, 3 and 4, $T'(m,n) \geq T(m,n)$. Combining these cases, $T(m,n) \leq T'(m,n)$ if $f(n) = \omega(\sqrt{\log n})$. In other words, our algorithm is asymptotically better than Hsu's algorithm if $m - n = \omega(\sqrt{\log n})$.

5 Conclusion

In this paper, two algorithms are presented for answering the decision problem, and optimally assigning a chain-like task system on a chain-like network computers. The time complexity of the algorithm for the decision problem is $O(m)$ where m is the number of modules in the system. This is an optimal algorithm which returns a feasible schedule if there exists one. For optimal schedule, Hsu has improved the Sheu and Chiang's algorithm [2] and solved this problem by a two phase algorithm with time complexity $O(m + (m' - n)^2)$ if $m' > n$ where m' is the number of modules after the merge phase [3]. If $m' \leq n$, the time complexity is $O(m)$. All of the existing algorithms use a layered graph approach.

In this paper, we use a new approach which is based on a searching strategy to find an optimal schedule. The algorithm (Algorithm 2) presented in this paper is identical to Hsu's algorithm if $m' \leq n$. For $m' > n$, the time complexity is $O(m + m' \log m' + m'^2 - n^2)$. In the case of $m' - n = \omega(\sqrt{\log n})$, our algorithm is more efficient than Hsu's algorithm.

References

1. Shahid H. Bokhari. Partitioning problems in parallel, pipelined, and distributed computing. *IEEE Transactions on Computers*, 37:48–57, January 1988.
2. J.P. Sheu and Z.F. Chaing. Efficient allocation of chain-like task on chain-like network computers. *Information Processing Letters*, 36:241–245, 1990.
3. C.C. Hsu. A two-phase approach for the optimal assignment of a chain-like task on a chain-like network computer. Technical report, National Taiwan Institute of Technology, 1993.
4. Manuel Blum, Robert W. Floyd, Vaughan Pratt, Rivest Ronald L., and Robert E. Tarjan. Time bounds for selection. *Journal of Computer and System Science*, 7(4):448–461, 1973.
5. C.A.R. Hoare. Algorithm 63(partition) and algorithm 65(find). *Communication of ACM*, 4(7):321–322, 1961.
6. Robert W. Floyd and Ronald L. Rivest. Expected time bounds of selection. *Communication of ACM*, 18(3):165–172, 1975.

Weighted Independent Perfect Domination on Cocomparability Graphs

Gerard J. Chang [*†]
Department of Applied Mathematics,
National Chiao Tung University
Hsinchu 30050, Taiwan, Republic of China
Email: gjchang@cc.nctu.edu.tw

C. Pandu Rangan and Satyan R. Coorg [‡]
Department of Computer Science and Engineering,
Indian Institute of Technology
Madras 600 036, India

Abstract

Suppose $G = (V, E)$ is a graph in which every vertex $v \in V$ is associated with a cost $c(v)$. This paper studies the weighted independent perfect domination problem on G, i.e., the problem of finding a subset D of V such that every vertex in V is equal or adjacent to exactly one vertex in D and $\sum \{c(v) : v \in D\}$ is minimum. We give an $O(|V||E|)$ algorithm for the problem on cocomparability graphs. The algorithm can be implemented to run in $O(|V|^{2.37})$ time. With some modifications, the algorithm yields an $O(|V| + |E|)$ algorithm on interval graphs, which are special cocomparability graphs.

Key words. independent perfect domination, cocomparability graph, interval graph

[*]Supported partly by the National Science Council of the Republic of China under grant NSC82-0208-M009-050.

[†]DIMACS permanent member.

[‡]Current address: Laboratory for Computer Science, MIT, Cambridge, MA 02139. Email: satyan@abp.lcs.mit.edu.

1 Introduction

A *dominating set* of a graph $G = (V, E)$ is a subset D of V such that every vertex not in D is adjacent to some vertex in D. The concept of domination in graph theory arises naturally from the facility location problem in operations research. Depending on the different requirements of various location problems, domination has many variants, e.g., independent domination, connected domination, total domination, edge domination, k-domination, and perfect domination.

A *perfect dominating set* of a graph $G = (V, E)$ is a subset D of V such that every vertex not in D is adjacent to *exactly one* vertex in D. The *perfect domination problem* is to find a minimum-sized perfect dominating set. Suppose that every vertex $v \in V$ is associated with a cost $c(v)$ and every edge $e \in E$ has a cost $c(e)$. The *weighted perfect domination problem* is to find a perfect dominating set D such that its total cost

$$c(D) = \sum \{c(v) : v \in D\} + \sum \{c(u, v) : u \notin D, v \in D, \text{ and } (u, v) \in E\}$$

is minimum. Note that the perfect domination problem is just the weighted perfect domination problem with $c(v) = 1$ for each vertex v and $c(e) = 0$ for each edge e. Yen and Lee [20] proved that the perfect domination problem is NP-complete for bipartite graphs and chordal graphs. They also gave linear time algorithms for the weighted perfect domination problem on trees [20] and series-parallel graphs [21].

Yen [19] and Yen and Lee [22] also considered the following variants of perfect domination. A perfect dominating set D is *independent, connected,* or *total* if the subgraph $G[D]$ induced by D has no edge, is connected, or has no isolated vertex, respectively. They gave NP-complete results of these variants in bipartite graphs and chordal graphs, except for connected domination in chordal graphs. They also gave linear time algorithms for these variants in trees, block graphs, and series-parallel graphs. On the other hand, Chang and Liu [7] gave a linear time algorithm for the weighted connected perfect domination problem in chordal graphs by using clique-tree structures of chordal graphs. They [8] also use the same technique to give linear time algorithms for the weighted perfect domination problem and its three variants in interval graphs. Independent perfect domination was called *efficient domination* by Bange, Barkauskas, and Slater [2], who proved that the efficient domination problem is NP-complete for general graphs and gave a linear time algorithm for the problem on trees. If we consider the application of perfect domination to coding theory, Biggs [6] studied perfect d-codes. A *perfect d-code* of a graph $G = (V, E)$ is a vertex set C such that every vertex $v \in V$ is within distance d of exactly one vertex in D. In conjunction with the study of the interconnection networks used in parallel computers, Livingston and Stout [13] studied *perfect d-dominating sets*, which are precisely the perfect d-codes. The concept of independent perfect domination in this paper is the same concept as their perfect 1-domination.

The main results of this paper are an $O(|V||E|)$ algorithm for the weighted independent perfect domination problem on a cocomparability graph $G = (V, E)$ and an $O(|V|+|E|)$ time algorithm for the same problem on an interval graph. The algorithm for cocomparability graphs can also be implemented to run in $O(|V|^{2.37})$ time. Note that cocomparability graphs include interval graphs and permutation graphs.

For any independent perfect domination set D, $c(D) = \bar{c}(D)$ if $\bar{c}(v) = c(v) + \sum\{c(u,v) : (u,v) \in E\}$ for every $v \in V$ and $\bar{c}(e) = 0$ for every $e \in E$. So, for the solution to the weighted independent perfect domination in this paper, without loss of generality, we may assume that $c(e) = 0$ for all $e \in E$. In this case, $c(D) = \sum\{c(v) : v \in D\}$.

2   Cocomparability Graphs

In this section we give an $O(|V||E|)$ algorithm for the weighted independent domination problem on a cocomparability graph $G = (V, E)$ in which each vertex v is associated with a cost $c(v)$. Domination and its three variants on cocomparability graphs have been studied by Kratsch and Stewart [12] and Arvind and Pandu Rangan [1].

A *comparability graph* is a graph $G = (V, E)$ whose vertex set has a *transitive ordering*, i.e., an ordering of V into $1, 2, \ldots, n$ such that

$$i < j < k, (i,j) \in E, \text{ and } (j,k) \in E \text{ imply } (i,k) \in E. \quad \text{(TO)}$$

There is an $O(|V|^2)$ algorithm [17] to test if a graph is a comparability graph. In the case of a positive answer, the algorithm produces a transitive ordering. A *cocomparability graph* is the complement of a comparability graph, or, equivalently, if its vertex set has a *cocomparability ordering*, which is an ordering of V into $1, 2, \ldots, n$ such that

$$i < j < k \text{ and } (i,k) \in E \text{ imply } (i,j) \in E \text{ or } (j,k) \in E. \quad \text{(CCO)}$$

In this section we assume that $G = (V, E)$ is a cocomparability graph with a given cocomparability ordering. For technical reasons, we add two isolated vertices 0 and $n+1$ with $c(0) = c(n+1) = 0$ to G to obtain a new cocomparability graph, which for simplicity, we again call as G with a cocomparability ordering $0, 1, 2, \ldots, n, n+1$. Note that D is an independent perfect dominating set of the original graph if and only if $D \cup \{0, n+1\}$ is an independent perfect dominating set of the new graph. For convenience we need the following notation, where v is a vertex:

$N(v) = \{u \in V : (u, v) \in E\}$.
$N[v] = \{v\} \cup N(v)$.
$\text{high}(v) = \max N[v]$.
$\text{low}(v) = \min N[v]$.
$N^+[v] = \{u \in N[v] : u \geq v\}$.
$d^+(v) = |N^+[v]|$.
$N^-[v] = \{u \in N[v] : u \leq v\}$.
$d^-(v) = |N^-[v]|$.

Theorem 1 $D = \{0 \equiv v_0 < v_1 < v_2 < \ldots < v_r < v_{r+1} \equiv n+1\}$ *is an independent perfect dominating set of a cocomparability graph G if and only if the following three conditions hold for all $1 \leq i \leq n+1$.*

(1) $\text{high}(v_{i-1}) < v_i$.
(2) $v_{i-1} < \text{low}(v_i)$.
(3) $\{x \in V : v_{i-1} \leq x \leq v_i\}$ is the disjoint union of $N^+[v_{i-1}]$ and $N^-[v_i]$.

Proof. Suppose D is an independent perfect dominating set of G.

(1) Suppose $\text{high}(v_{i-1}) \geq v_i$. Note that $v_i \neq \text{high}(v_{i-1})$, since D is independent. Then $v_{i-1} < v_i < \text{high}(v_{i-1})$. By (CCO), either $(v_{i-1}, v_i) \in E$ or $(v_i, \text{high}(v_{i-1})) \in E$. The former case contradicts the assumption that D is an independent set. The latter case contradicts the fact that a vertex not in D is adjacent to exactly one vertex in D. So $\text{high}(v_{i-1}) < v_i$.

(2) Suppose $v_{i-1} \geq \text{low}(v_i)$. Note that $\text{low}(v_i) \neq v_{i-1}$, since D is independent. Then $\text{low}(v_i) < v_{i-1} < v_i$. By (CCO), either $(\text{low}(v_i), v_{i-1}) \in E$ or $(v_{i-1}, v_i) \in E$. The former case contradicts the fact that a vertex not in D is adjacent to exactly one vertex in D. The latter case contradicts the assumption that D is an independent set. So $v_{i-1} < \text{low}(v_i)$.

(3) For any vertex x such that $v_{i-1} < x < v_i$, we claim that x is adjacent to v_{i-1} or v_i. By the definition of perfect domination, x is adjacent to exactly one v_j. If $j > i$, then $x < v_i < v_j$ and $(x, v_j) \in E$ and $(x, v_i) \notin E$. By (CCO), $(v_i, v_j) \in E$, which contradicts the assumption that D is an independent set. If $j < i - 1$, then $v_j < v_{i-1} < x$, and $(v_j, x) \in E$ and $(v_{i-1}, x) \notin E$. By (CCO), $(v_j, v_{i-1}) \in E$, a contradiction again. So x is adjacent to v_{i-1} or v_i. This, together with (1) and (2), implies (3).

Conversely, suppose conditions (1) to (3) hold. For any vertex x not in D, assume $v_{i-1} < x < v_i$. First, by (3), x is adjacent to exactly one of v_{i-1} and v_i. Suppose x is adjacent to some other v_j with $j-1 \geq i$ or $j+1 \leq i-1$. If $j-1 \geq i$, then $\text{low}(v_j) \leq x < v_i \leq v_{j-1}$, which contradicts (2). If $j+1 \leq i-1$, then $\text{high}(v_j) \geq x > v_{i-1} \geq v_{j+1}$, which contradicts (1). Q.E.D.

Theorem 1 can be rewritten in the following form, which is more useful in designing an efficient algorithm to solve the weighted independent perfect domination problem on cocomparability graphs.

Theorem 2 $D = \{0 \equiv v_0 < v_1 < v_2 < \ldots < v_r < v_{r+1} \equiv n+1\}$ *is an independent perfect dominating set of a cocomparability graph* G *if and only if the following four conditions hold for all* $1 \leq i \leq n+1$.

(1) $\text{high}(v_{i-1}) = \max(\{0, 1, 2, \ldots, v_i\} - N[v_i])$.
(2) $v_{i-1} < \text{low}(v_i)$.
(3) $N[v_{i-1}] \cap N[v_i] = \emptyset$.
(4) $d^+(v_{i-1}) + d^-(v_i) = v_i - v_{i-1} + 1$.

Working from Theorem 2, we can derive the following algorithm for finding a weighted independent dominating set of a cocomparability graph.

Algorithm WIPD-CC. Find a weighted independent perfect dominating set of a cocomparability graph.
Input. A cocomparability graph $G = (V, E)$ with a cocomparability ordering $0, 1, \cdots, n, n+1$, in which each vertex v is associated with a weight $c(v)$.
Output. A minimum weighted independent perfect dominating set D.
Method.

```
1.    cost(0) ← 0;
2.    for v = 1 to n + 1 do
3.        cost(v) ← ∞;
4.        h ← max({0, 1, . . . , v} − N[v]);
5.        for all u ∈ N⁻[h] satisfying
          (C1) high(u) = h,
          (C2) u < low(v),
          (C3) N[u] ∩ N[v] = ∅,
          (C4) d⁺(u) + d⁻(v) = v − u + 1 do
6.                if ( cost(u) + c(v) < cost(v) )
7.                then { cost(v) ← cost(u) + c(v); previous(v) ← u; }
          end do;
      end do;
8.    D ← ∅;
9.    v ← previous(n + 1);
10.   while ( v ≠ 0 ) do {D ← D ∪ {v};  v ← previous(v); }
```

Theorem 3 *Algorithm WIPD-CC gives a minimum weighted independent perfect dominating set D of a cocomparability graph $G = (V, E)$ in $O(|V||E|)$ time.*

Proof. The correctness of Algorithm WIPD-CC follows from Theorem 2.

Note that the functions high, low, d^+, and d^- can be calculated in $O(|V| + |E|)$ time. Steps 2 to 7 of the algorithm take $n + 1$ iterations. To implement step 4 and (C3) of step 5 efficiently, we use an array mark$[0..n + 1]$, which is set to 0 initially. At iteration v we reset mark(x) to be v for all $x \in N[v]$. A simple backward do loop for x from v down to 0 will find the first x with mark$(x) < v$ in $O(|N[v]|)$ time. This is the desired h. For condition (C3) of step 5, we only need to check if mark$(x) < v$ for all $x \in N[u]$. This takes $O(|N[u]|)$ time. So the do loop from steps 5 to 7 takes $O(|E|)$ time. Altogether, steps 2 to 7 take $O(|V||E|)$ time. Steps 8 to 10 clearly take $O(|V| + |E|)$ time. Q.E.D.

There is an alternative way to check condition (C3) of step 5, by using matrix multiplication. Consider the *closed neighborhood matrix* $A = (a_{uv})$ of the graph G defined by $a_{uv} = 1$ if and only if $u \in N[v]$. Let $AA = (b_{uv})$. Then $N[u] \cap N[v] = \emptyset$ if and only if $b_{ij} = 0$. If we pre-calculate AA, then condition (C3) of step 5 can be checked in a constant time. Thus, the fastest known algorithm for matrix multiplication to date [9] gives us the following theorem.

Theorem 4 *Algorithm WIPD-CC can be implemented in $O(|V|^{2.37})$ time.*

3   Interval Graphs

An *interval graph* is a graph each of whose vertices can be associated with an interval in the real line so that two vertices are adjacent in the graph if and only if the corresponding intervals intersect. Interval graphs can be recognized in linear time. If the graph is an interval graph we can get the set of intervals also within the same bound [10].

Domination and its variants in interval graphs have been studied extensively [3, 4, 5, 11, 14, 15, 16, 18]. The following vertex ordering methodology is of particular interest from our point of view.

Theorem 5 ([16]) *A graph $G = (V, E)$ is an interval graph if and only if it has an interval ordering, i.e., its vertex set V can be ordered into $[1, 2, \ldots, n]$ such that*

$$i < j < k \text{ and } (i, k) \in E \text{ imply } (j, k) \in E. \tag{IO}$$

Note that an interval ordering is a cocomparability ordering. A vertex v is said to have a *consecutive neighborhood* if $N[v]$ contains precisely all vertices from low(v) to high(v). The property (IO) implies that $N^-[v]$ contains precisely all vertices from low$[v]$ to v. Therefore Theorem 1 can be rewritten for interval graphs as follows:

Theorem 6 $D = \{0 \equiv v_0 < v_1 < v_2 < \ldots < v_r < v_{r+1} \equiv n+1\}$ *is an independent perfect dominating set of an interval graph G if and only if all v_i have consecutive neighborhoods and V is the disjoint partition of all $N[v_i]$ where $0 \leq i \leq r+1$.*

Theorem 6 suggests the following algorithm for the weighted independent perfect domination problem on interval graphs.

Algorithm WIPD-I. Find a minimum weighted independent perfect dominating set of an interval graph.
Input. An interval graph $G = (V, E)$ with an interval ordering $0, 1, \cdots, n, n+1$, in which each vertex v is associated with a weight $c(v)$.
Output. A minimum weighted independent perfect dominating set D of G.
Method.
1. find the set C of all vertices having consecutive neighborhoods;
2. cost$(0) \leftarrow 0$;
3. **for** $v = 1$ **to** $n+1$ **do**
4.   cost$(v) \leftarrow \infty$;
5.   **for all** $u \in N^-[v] \cap C$ with $v = $ high(u) **do**
6.   **if** (cost(low$(u) - 1$) $+ c(v) < $ cost(v))
7.   **then** { cost$(v) \leftarrow$ cost(low$(u) - 1$) $+ c(v)$;
                    previous$(v) \leftarrow$ low$(u) - 1$;
                    center$(v) \leftarrow u;$ }
            end do;
        end do;
8. $D \leftarrow \emptyset$;
9. $v \leftarrow$ previous$(n+1)$;
10. **while** ($v \neq 0$) **do** $\{D \leftarrow D \cup \{$center$(v)\}$; $\quad v \leftarrow$ previous$(v); \}$

Theorem 7 *Algorithm WIPD-I gives a minimum weighted independent perfect dominating set D of an interval graph $G = (V, E)$ in $O(|V| + |E|)$ time.*

Proof. The correctness of Algorithm WIPD-I follows from Theorem 6 and the fact that each cost(v) is precisely the minimum weighted sum $\sum_{i=0}^{j} c(v_i)$, where $0 \equiv v_0 < v_1 < \cdots < v_j < v$ are vertices having consecutive neighborhoods and $\{0, 1, \cdots, v\}$ is the disjoint union of their closed neighborhoods. The claim about the running time follows from the fact that each iteration v takes $O(|N[v]|)$ time. Q.E.D.

References

1. K. Arvind and C. Pandu Rangan, Efficient algorithms for domination problems on cocomparability graphs, Technical Report TR-TCS-90-18 (1990), Department of Computer Science and Engineering, Indian Institute of Technology, Madras.

2. D. W. Bange, A. E. Barkauskas, and P. T. Slater, Efficient dominating sets in graphs, *Applications of Discrete Mathematics*, R. D. Ringeisen and F. S. Roberts. eds., SIAM, Philad. (1988)189-199.

3. A. A. Bertossi, Total domination in interval graphs, *Inform. Processing Letters* 23 (1986) 131-134.

4. A. A. Bertossi, On the domatic number of interval graph, *Inform. Processing Letters* 28 (1988) 275-280.

5. A. A. Bertossi and A. Gori, Total domination and irredundance in weighted interval graphs, *SIAM J. Disc. Math.* 1 (1988) 317-327.

6. N. Biggs, Perfect codes in graphs, *J. Comb. Theory, Series B* 15 (1973) 289-296.

7. M. S. Chang and Y. C. Liu, Polynomial algorithms for the weighted perfect domination problems on chordal graphs and split graphs, (1993) manuscript.

8. M. S. Chang and Y. C. Liu, Polynomial algorithms for the weighted perfect domination problems on interval graphs, (1993) manuscript.

9. D. Coppersmith and S. Winograd, Matrix multiplication via arithmetic progression, *Proceedings 19th Annual ACM Sysposium on Theory of Computing* (1987) 1-6.

10. M. C. Golumbic, *Algorithmic Graph Theory and Perfect Graphs*, Academic Press, New York (1980).

11. J. K. Keil, Total domination in interval graphs, *Inform. Processing Letters* 22 (1986) 171-174.

12. D. Kratsch and L. Stewart, Domination on cocomparability graphs, (1989), preprint.

13. M. Livingston and Q. F. Stout, Perfect dominating sets, *Congressus Numerantium* 79 (1990) 187-203.

14. T. L. Lu, P. H. Ho and G. J. Chang, The domatic number problem in interval graphs, *SIAM J. Disc. Math.* 3 (1990) 531-536.

15. G. Ramalingam and C. Pandu Rangan, Total domination in interval graphs revisited, *Inform. Processing Letters* 27 (1988) 17-21.

16. G. Ramalingam and C. Pandu Rangan, A unified approach to domination problems on interval graphs, *Inform. Processing Letters* 27 (1988) 271-274.

17. J. Spinrad, Transitive orientation in (n^2) time, *15th STOC Proceedings* (1983) 457-466.

18. A. Srinivasa Rao and C. Pandu Rangan, Linear algorithm for domatic number problem on interval graphs, *Inform. Processing Letters* 33 (1989/90) 29-33.

19. C. C. Yen, *Algorithmic Aspects of Perfect Domination*, Ph. D thesis, Institute of Information Science, National Tsing Hua University, Taiwan (1992).

20. C. C. Yen and R. C. T. Lee, The weighted perfect domination problem, *Inform. Processing Letters* 35 (1990) 295-299.

21. C. C. Yen and R. C. T. Lee, A linear time algorithm to solve the weighted perfect domination problem in series-parallel graphs, *European J. Operational Research* (1992), to appear.

22. C. C. Yen and R. C. T. Lee, The weighted perfect domination problem and its variants, manuscript.

Plane Sweep Algorithms for the Polygonal Approximation Problems with Applications

D.P. Wang, N.F. Huang, H.S. Chao and R.C.T. Lee

Department of Computer Science

National Tsing—Hwa University

Hsinchu Taiwan 30043

Republic of China

E—mail: rctlee@nthu.edu.tw

Abstract

In this paper, we consider the minimum Chebyshev ϵ polygonal approximation problem. For this problem, we can use the plane sweep strategy to solve it in $O(n^2)$ time.

1 Introduction

In this paper, we consider the the minimum Chebyshev ϵ polygonal approximation problem [HS 91], which is defined as follows: We are given a set of points $S=\{p_1,p_2,...,p_n\}$ in R^2 with $p_i=(x_i,y_i)$ and $x_1<x_2<...<x_n$. Our job is to construct a polygonal approximation function f which fits S well. We shall use the following Chebyshev error function:

$$E_c(f,S) = \max_{(x_i,y_i)\in S} |y_i-f(x_i)|$$

In addition to the above error which we like to minimize, we also have to consider another parameter, the number of turns of the polygonal approximation function. Therefore, the problem can be formally defined as follows: Given a set of points S and an integer t, $t\geq0$, find a polygonal approximation function f fitting S such that the number of turns of f is not larger than t and the Chebyshev error is minimized. The presently best algorithm for solving this problem is $O(n^2 log n)$ time [HS 91].

Fig. 1 shows a solution of the above problem where $n=6$ and $t=2$.

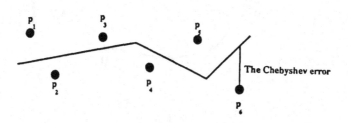

The Chebyshev error

Fig. 1

We will use the plane sweep technique to solve this problem. The organization of this paper is as follows: In Section 2, we will introduce some basic idea of our algorithm. In Section 3, we will present our algorithm and prove its correctness. Finally, we will give concluding remarks in Section 4.

2. Characteristics

For this problem, we shall solve it by using a concept, called ϵ_{ij}, which was proposed by Hakimi and Schmeichel [HS 91]. Let $S_{ij}=\{p_i, p_{i+1}, \cdots, p_j\}$ and H_{ij} be the convex hull of S_{ij}. ϵ_{ij} is defined to be 0 for $(j-i)<2$ and the definition of it for $(j-i)\geq2$ is as follows: Find two parallel lines l_1 and l_2 such that one of the lines, say l_1, passes through two consecutive points p_r and p_s on the hull, and the other line l_2 passes through a point p_k on the hull with $x(p_k)$ between $x(p_r)$ and $x(p_s)$. Then ϵ_{ij} is defined as half of the vertical (Chebyshev) distance between l_1 and l_2 when $(j-i)\geq2$, as shown in Fig. 2. It was shown in [HS 91] that l_1 and l_2 are unique.

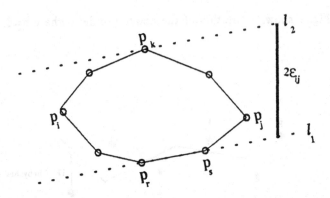

Fig. 2

In [HS 91], the following lemma was proved:

Lemma 1. For any t, the value of ϵ of a solution of the minimum Chebyshev ϵ polygonal approximation problem must be one element of Σ where Σ is equal to the set $\{\epsilon_{ij} \mid (j-i) \geq 1\}$.

For every ϵ_{ij}, in [HS 91], Hakimi and Schmeichel can find a polygonal approximation function with minimum number of turns such that the Chebyshev error of this function is not greater than ϵ_{ij} in $O(n)$ time. Therefore, the above theorem implies that we can solve the minimum Chebyshev ϵ polygonal approximation problem by finding the smallest ϵ_{ij} in Σ such that the number of turns of its corresponding polygonal approximation function is not larger than t. Moreover, it was shown in [HS 91] that if $\epsilon_{ij} < \epsilon_{pq}$ then the number of turns of the polygonal approximation function of ϵ_{ij} is greater than or equal to that of ϵ_{pq}. Based upon these properties, The algorithm proposed by Hakimi and Schmeichel [HS 91] is as follows: Find all elements of Σ, sort them and conduct a binary search to find the minimum value of ϵ_{ij} corresponding to a solution of the minimum Chebyshev ϵ polygonal approximation problem whose number of turns is less than or equal to t.

In this paper, we will show that there exists a common geometrical property among all elements in Σ and we can use the plane sweep strategy [PS 85] to solve the minimum Chebyshev ϵ polygonal approximation problem. The major difference between Hakimi and Schmeichel's algorithm and ours is the generation of the elements of Σ. In [HS 91], Hakimi and Schmeichel generate all elements in Σ to find the solution of the minimum Chebyshev ϵ polygonal approximation problem. In this paper, we can obtain it without generating all elements of Σ.

The following lemma is based upon the update procedure of [HS 91].

Lemma 2:   (a) For each i, $\epsilon_{i(j-1)} \leq \epsilon_{ij}$.

   (b) For each j, $\epsilon_{(i-1)j} \geq \epsilon_{ij}$.

Proof: (a) For this part of the proof, we can use the last step of the update procedure of finding ϵ_{ij} in [HS 91]. In this last step, suppose that $\epsilon_{i(j-1)}$ has already been found and some ϵ' will be found next. This parameter ϵ' is not necessarily ϵ_{ij}. In [HS 91], it was shown that the following relation holds: If $\epsilon' > \epsilon_{i(j-1)}$, then $\epsilon_{ij} = \epsilon'$; otherwise $\epsilon_{ij} = \epsilon_{i(j-1)}$. Thus, it is obvious that $\epsilon_{i(j-1)} \leq \epsilon_{ij}$.

(b) For this part, let us note that the update procedure mentioned above adds a point further to the right. If the updating procedure adds a point to the left, we can still find $\epsilon_{(i-1)j}$ from ϵ_{ij} because all of the arguments mentioned above still hold. Therefore, this part of the proof can be obtained by using similar reasoning.   Q.E.D.

Based upon Lemma 2, we can solve the minimum Chebyshev ϵ polygonal approximation problem according to the following procedure. Our algorithm uses two pointers, Front and Rear. If the Front points to a point p_j, then the Rear always points to a point p_i in which $i \leq j$, as shown in Fig. 3.

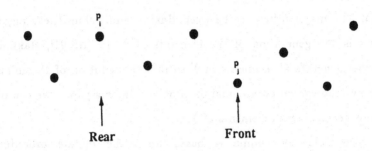

Fig. 3

Suppose that the Rear and the Front point to p_i and p_j respectively, we will find ϵ_{ij} immediately. Let t_{ij} denote the number of turns of the polygonal approximation function corresponding to ϵ_{ij}, then we have two cases. **Case 1:** $t_{ij} > t$. Then ϵ_{ij} must be smaller than our optimal Chebyshev error. Thus we should seek for a larger error. According to Lemma 2, $\epsilon_{ij} \leq \epsilon_{i(j+1)}$. Therefore we should move the Front to the right.

Case 2: $t_{ij} \leq t$. Then we should seek for an even smaller error. Again, according to Lemma 2, $\epsilon_{(i+1)j} \leq \epsilon_{ij}$. Therefore we should move the Rear to the right.

3 The Algorithm

The following algorithm returns a solution of the minimum Chebychev ϵ polygonal approximation function of n given points.

Algorithm Sweep;

Input: n points $p_1, p_2, \cdots, p_n$ with $x(p_1) < x(p_2) < \cdots < x(p_n)$ and an integer t.

Output: A polygonal approximation function f such that the number of turns of f is not larger than t and the Chebyshev error is minimized.

Step 1: Rear:=1; Front:=2;

Step 2: while ((Rear$\leq$n) **and** (Front$\leq$n)) **do** Step 2.1 to 2.3.

    Step 2.1:   Find ϵ_{ij} where i is equal to the Rear and j is equal to the Front.

    Step 2.2:   Find a polygonal approximation function f corresponding to ϵ_{ij}. Let the number of turns of f be t_{ij}.

    Step 2.3:   If $t_{ij} > t$, then Front:=Front+1;

                  else put ϵ_{ij} into the array A and Rear:=Rear+1;

Step 3: Return the polygonal approximation function corresponding to the minimum element in the array A.

Note that not every ϵ_{ij} is examined by Algorithm Sweep. In the following, we shall prove that at least one optimal ϵ_{ij} is stored in array A.

Lemma 3: In array A produced by Algorithm Sweep, there exists an ϵ_{ij} which corresponding to an optimal solution.

Proof: Let ϵ_{pq} be the error corresponding to the optimal solution. We have two cases.

Case 1: During the execution of Algorithm Sweep, the Front hits q before the Rear hits p. Suppose the Rear points to r when the Front hits q. Since r is at the left of p, according to Lemma 2, $\epsilon_{pq} \leq \epsilon_{rq}$. Let t_{rq} be the number of turns of the polygonal approximation function corresponding to ϵ_{rq}.

    Case 1.1: $\epsilon_{pq} < \epsilon_{rq}$. Based upon the fact of [HS 91], t_{rq} is no larger than t. Thus the Rear will be moved to the right and finally it will hit p. Then, t_{pq} will be smaller than or equal to t. Thus ϵ_{pq} will be stored into array A by Step 2.3.

    Case 1.2: $\epsilon_{pq} = \epsilon_{rq}$. Then t_{rq} must not be greater than t. Thus ϵ_{rq} will be stored into array A in which ϵ_{rq} is equal to ϵ_{pq}.

Case 2: During the execution of Algorithm Sweep, the Rear hits p before the Front hits q. Suppose the Front points to r when the Rear hits p. Since r is at the left of q, according to Lemma 2, $\epsilon_{pr} \leq \epsilon_{pq}$. Let t_{pr} be the number of turns of the polygonal approximation function corresponding to ϵ_{pr}.

Case 2.1: $\epsilon_{pr} < \epsilon_{pq}$. Then t_{pr} must be greater than t. Thus the Front will be moved to the right until it hits q. By then, t_{pq} will be smaller than or equal to t and ϵ_{pq} will be stored into array A..

Case 2.2: $\epsilon_{pr} = \epsilon_{pq}$. Then t_{pr} must not be greater than t. Thus ϵ_{pr} will be stored into array A in which ϵ_{pr} is equal to ϵ_{pq}.   Q.E.D.

Theorem 4: The minimum Chebyshev ϵ polygonal approximation problem can be solved in $O(n^2)$ time.

Proof: The correctness of Algorithm Sweep is due to Lemma 3. Now, let us analyze the time complexity of it. Step 2 executes at most 2×n steps. For Rear=1 and Front=2, it needs O(1) time to find the convex hull H_{12}, and ϵ_{12}. Later, for all other cases of Step 2.1, it takes O(*logn*) time to delete or insert a point from the existing convex hull to find a new convex hull [PS 85]. After a new convex hull H_{ij} is found, ϵ_{ij} can be found in O(n) time [HS 91]. Therefore, it needs $O(1)+O(2n-1)\times O(n)=O(n^2)$ time for executing Step 2.1. As for Step 2.2, it needs O(n) time to find the polygonal approximation function corresponding to ϵ_{ij} [HS 91]. Thus it takes $O(2n)O(n)=O(n^2)$ time to perform Step 2.2. The time for Step 2.3 is obviously O(n). Therefore the time complexity of Algorithm Sweep is $O(n^2)$.

    Q.E.D.

4 Concluding Remarks

In this paper, we present an algorithm for solving the minimum Chebyshev ϵ polygonal approximation problem in $O(n^2)$ time which is better than the previous best result.

Intuitively, we believe that it is still possible to have linear algorithms for this problem. It will be interesting to come up with such optimal algorithm.

References

[CC 92] W. S. Chan and F. Chin "Approximation of Polygonal Curves with Minimum Number of Line Segments" International Symposium on Algorithms, JAPAN, 1992.

[HS 91] S. L. Hakimi and E. F. Schmeichel "Fitting Polygonal Functions to a Set of Points in the Plane" Computer Vision, Graphics and Image Processing: GRAPHICAL MODELS AND IMAGE PROCESSING Vol. 53, No. 2, pp 132–136 (1991).

[PS 85] F. P. Preparata and M. Shamos, Computational Geometry: An Introduction [1985].

Optimal Rectilinear Steiner Tree for Extremal Point Sets*

Siu-Wing Cheng[1], Andrew Lim[2], Ching-Ting Wu[3]

[1] Department of Computer Science, The Hong Kong University of Science and Technology, Clear Water Bay, Hong Kong
[2] Information Technology Institute, National Computer Board, 71 Science Park Drive, Republic of Singapore
[3] 3M, St. Paul, Minnesota, 55144, USA

Abstract. We present an $O(n^3)$-time algorithm to construct an optimal rectilinear Steiner tree for an extremal point set of size n. Our result subsumes those given in [1, 4] and it partially improves on the result in [5].

1 Introduction

Although the general rectilinear Steiner tree problem is NP-complete, it becomes solvable in polynomial time when the point set is known to be *extremal*. A point set S is extremal if and only if all the points in S lie on the boundary of the rectilinear convex hull $RC(S)$ of S. The problem have been considered in [1, 3, 5]. The best algorithm is given in [5] which runs in $O(k^4 n)$ time, where k is the number of boundary edges of $RC(S)$. In the worst-case scenario, an extremal point set can have $\Theta(n)$ boundary edges, and the running time becomes $O(n^5)$. In this paper, we present an algorithm that runs in $O(n^3)$ time and it compares favorably with the result in [5] when $k = \Omega(\sqrt{n})$.

For abbreviation, we shall omit "rectilinear" in the rest of this paper. The terminologies about Steiner tree that we shall use are taken from [3, 5]. Given an extremal point set S, we call the elements of S *terminal points*. An *edge* connects two adjacent vertices.[4] An *interior edge* is an edge whose relative interior lies strictly within $RC(S)$. An *interior line* is a maximal sequence of adjacent and collinear interior edges. A *complete interior line* is an interior line whose endpoints lie on the boundary of $RC(S)$. A horizontal line and a vertical line intersect to form a degree-two *corner-vertex*, or a degree-three *T-vertex*, or a degree-four *cross-vertex* (can be viewed as the union of two T's). The two lines incident to a corner-vertex are called the *legs* of the corner. If both legs intersect the boundary, the corner is called a *complete interior corner*. The line that contains the two collinear edges incident to the T-vertex is called the *head*. The other line is called the *body*. With respect to the

---

* Part of the work was done while the authors were at University of Minnesota, Department of Computer Science. Work on this paper by the first author was partially supported by RGC grant DAG92/93.EG09.
[4] We still call an edge of $RC(S)$ a *boundary edge* though some exterior vertices may lie on it.

region enclosed by $RC(S)$, an *inner boundary corner* is a reflex angle on $RC(S)$ and an *outer boundary corner* is a convex angle on $RC(S)$. The inner boundary corners and the terminal points are called *nodes*. Consider the grid graph formed by drawing horizontal and vertical grid lines through each node on $RC(S)$. The intersection between two grid lines in the interior of $RC(S)$ is a *grid vertex*. Each intersection between a horizontal boundary edge and a vertical grid line or between a vertical boundary edge and a horizontal grid line is an *exterior vertex*. Note that a node must be an exterior vertex but not vice versa. Given exterior vertices c and d, we use $B(c, d)$ to denote the portion of the boundary of $RC(S)$ traversed from c to d in the counterclockwise order. Define $[c, d]$ to be the counterclockwisely ordered sequence of exterior vertices on $B(c, d)$. Define $S[c, d]$ to be the subsequence of $[c, d]$ that contains c, d, and all the terminal points in $[c, d]$.

Given a tree T, $L(T)$ denotes the total length of T. Throughout this paper, b_i denotes some exterior vertex on the boundary. In all drawings in this paper, dashed line represents the boundary of a rectilinear convex hull and solid line represents a line in a Steiner tree. Due to space limitation, some proofs and details are omitted which can be found in the full paper [2].

2 The overall strategy and the canonical form

Suppose we know that an optimal Steiner tree contains a vertical complete interior line ℓ which connects two exterior vertices, say b_r and b_t. Then to compute an optimal Steiner tree, it suffices to compute optimal Steiner trees connecting $S[b_r, b_t]$ with ℓ and $S[b_t, b_r]$ with ℓ and then put them together. We actually do not know the actual positions of b_r and b_t, so we repeat the above for all possible positions of b_r and b_t. The minimal tree obtained must then be optimal if the optimal Steiner tree does contain a vertical complete interior line. Since optimal Steiner tree may not contain a complete interior line, we need to handle other possible topologies similarly (there will be only a constant number of them to consider) and return the minimal tree obtained. Therefore, our strategy is to compute many subtrees and compose them to form an optimal Steiner tree according to its possible topologies. We shall develop efficient algorithms for computing a common substructure in order to implement the above strategy efficiently.

To limit the possible topologies of an optimal Steiner tree, the following tie-breaking rules are proposed in [5]. Let $\tau(S)$ be the set of all the optimal Steiner trees lying within $RC(S)$. First, select a subset $\tau_1(S) \subseteq \tau(S)$ of trees which maximizes the boundary length. Let $\tau_2(S) \subseteq \tau_1(S)$ be the set of trees in which the sum of degrees of nodes is maximized. Finally, let $\tau_3(S) \subseteq \tau_2(S)$ be the set of trees which are leftmost (in the sense described in [5]). It is proved in [5] that a tree in $\tau_3(S)$ possesses the following topological properties (the list is not exhaustive): (1) An interior corner bends towards the right. (2) The body of a T in the interior of $RC(S)$ intersects the boundary at a node. Furthermore, no edges are incident to the relative interior of the body of a T in the interior of $RC(S)$. (3) The legs of an interior corner hit the boundary, forming a complete interior corner. (4) The head of a T in the interior of $RC(S)$ is either a leg of a complete interior corner or a complete interior line.

By the edge sliding technique [5], it can also be proved that:

Theorem 1. *Let S be an extremal point set and $T \in \tau_3(S)$. For all interior line in T, at least one of its endpoint is a node. Moreover, if an interior line ℓ is the body of a T whose head lies on the boundary and the T-vertex is not a node, then ℓ hits the opposite boundary at a node and does not contain any Steiner point in its relative interior.* □

3   Common substructures

Let ℓ be an interior line that links an exterior vertex b_0 and another point v which is either a grid vertex in the interior of $RC(S)$ or an exterior vertex on $RC(S)$. Since we shall distinguish between b_0 and v, we view ℓ as a directed line (b_0, v). For each exterior vertex b_k such that $S[b_0, b_k] \cup \{v\}$ (resp. $S[b_k, b_0] \cup \{v\}$) is an extremal point set, we define $\tau(b_0, b_k, \ell)$ (resp. $\tau_c(b_0, b_k, \ell)$) to be the set of minimal Steiner trees that : (1) lie within $RC(S[b_0, b_k] \cup \{v\})$ (resp. $RC(S[b_k, b_0] \cup \{v\})$), (2) interconnect points in $S[b_0, b_k] \cup \{v\}$ (resp. $S[b_k, b_0] \cup \{v\}$), and (3) contain ℓ.

Define $CS(b_0, b_k, \ell)$ (resp. $CS_c(b_0, b_k, \ell)$) to be the set of minimal Steiner trees under the above three constraints and an additional *anchoring constraint*: any edge incident to v that does not lie on ℓ is perpendicular to ℓ and any edge that is perpendicularly incident to the relative interior of ℓ or v hits $B(b_0, b_k)$ (resp. $B(b_k, b_0)$) at an exterior vertex. $CS(\cdot)$'s and $CS_c(\cdot)$'s are the set of "common substructures".

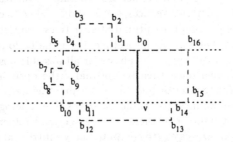

Fig. 1. $[b_{14}, b_{11}]$ is the open-span of (b_0, v). $[b_4, b_{11}] \cup [b_{14}, b_0]$ is the span of (b_0, v).

Draw an imaginary line through v perpendicular to ℓ to divide the plane into two strips. (Refer to Fig. 1.) The contiguous portion of the boundary that lies in the the half plane containing b_0 is called the *open-span* of ℓ. Draw one more imaginary line through b_0 perpendicular to ℓ. The portion of the boundary that touches both the imaginary line through v and the imaginary line through b_0 is called the *span* of ℓ.

Lemma 2. *If b_k lies in the open-span of ℓ, then $CS(b_0, b_k, \ell) \subseteq \tau(b_0, b_k, \ell)$ and $CS_c(b_0, b_k, \ell) \subseteq \tau_c(b_0, b_k, \ell)$.*

Proof. Let T be a tree in $\tau(b_0, b_k, \ell)$. (The argument for $\tau_c(b_0, b_k, \ell)$ is similar.) Among all the edges incident to ℓ and violating the anchoring constraint, let e be the one closest to v. e does not end at the corner-vertex of a corner; otherwise, that corner can be flipped to overlap with ℓ and decreases the total length. So, e must end at the relative interior of another line. Then e can be slid in a direction so that it will hit an exterior vertex. Thus, we can incrementally transform the tree until it satisfies the anchoring constraint. □

3.1 Algorithms

We want to compute an element $C(b_0, b_k, \ell)$ of $CS(b_0, b_k, \ell)$ (resp. $C_c(b_0, b_k, \ell)$ of $CS_c(b_0, b_k, \ell)$). For all $T \in CS(b_0, b_k, \ell)$, we call each edge in T that is incident to v and perpendicular to ℓ or perpendicularly incident to the relative interior of ℓ an *anchoring edge*. We call the endpoint of an anchoring edge not lying on ℓ an *anchor*. Sometimes, we refer to an anchor/anchoring edge according to their distances from b_0 on the axis parallel to ℓ (e.g., the farthest anchor, the farthest anchoring edge). For any $S[c, d]$, define $Tree(c, d)$ to be a tree in $\tau(S[c, d])$.

Lemma 3. *If $[b_0, b_q]$ are at nondecreasing distances from the supporting line of ℓ, then for each $b_k \in [b_0, b_q]$, $C(b_0, b_k, \ell)$ can be computed inductively in $O(1)$ time if $Tree(b_0, b_k)$ is available. Moreover, each $C(b_0, b_k, \ell)$ belongs to $\tau(b_0, b_k, \ell)$. A similar statement can be made for $C_c(b_0, b_k, \ell)$.*

Proof. Compute the following trees and return the minimal one: $Tree(b_0, b_{k+1}) \cup \{\ell\}$, $C(b_0, b_k, \ell) \cup B(b_k, b_{k+1})$, and $C(b_0, a_i, \ell) \cup \{\text{edge from } b_{k+1} \text{ to } \ell\}$, where a_i is the last point in $S[b_0, b_{k+1}] - \{b_{k+1}\}$. (Ignore the last candidate if b_{k+1} lies outside the open-span of ℓ.) The correctness follows from Lemma 2 and the assumption in the lemma. □

Lemma 4. *If $[b_0, b_q]$ lies in the open-span of ℓ, then for each $b_k \in [b_0, b_q]$, $C(b_0, b_k, \ell)$ can be computed inductively in linear time if $Tree(b_i, b_k)$ is available for all $[b_i, b_k] \subseteq [b_0, b_q]$. Moreover, each $C(b_0, b_k, \ell)$ belongs to $\tau(b_0, b_k, \ell)$. A similar statement can be made for $C_c(b_0, b_k, \ell)$.*

Proof. Consider $C(b_0, b_{k+1}, \ell)$, where $b_{k+1} \in [b_0, b_q]$. If b_{k+1} is an anchor, then $L(C(b_0, b_{k+1}, \ell)) = L(C(b_0, a_i, \ell) \cup Tree(a_{i+1}, b_{k+1}) \cup \{\text{edge from } b_{k+1} \text{ to } \ell\})$ for some neighbouring points a_i, a_{i+1} in $S[b_0, b_{k+1}]$. Else, $L(C(b_0, b_{k+1}, \ell)) = L(C(b_0, b_i, \ell) \cup Tree(b_i, b_{k+1}))$, where b_i is the farthest anchor. Since there are $O(|[b_0, b_{k+1}]|)$ possible positions for a_i, a_{i+1}, and b_i, we compute $O(|[b_0, b_{k+1}]|)$ candidates and return the minimal one. □

After we have computed $C(b_0, b_k, \ell)$ and $C_c(b_0, b_k, \ell)$ for some interior line $\ell = (b_0, v)$, we actually have also obtained some extra information almost for free as stated in the following lemma.

Lemma 5. *Let x be some point in the relative interior of ℓ and let ℓ' be (b_0, x). If b_k lies in the open-span of ℓ', then $C(b_0, b_k, \ell')$ can be set to be $(C(b_0, b_k, \ell) - \{\ell\}) \cup \{\ell'\}$. A similar statement can be made for $C_c(b_0, b_k, \ell')$.* □

Lemma 3, Lemma 4, and Lemma 5 are not sufficient to generate $C(\cdot)$'s and $C_c(\cdot)$'s in general. To this end, we need the following additional structures. Suppose that $\ell = (b_0, b_f)$ is a complete interior line linking two exterior vertices b_0 and b_f. For all $b_k \in [b_0, b_f]$ lying in the span of ℓ and for all b_j whose distance from b_0 on the axis parallel to ℓ is larger than that between b_k and b_0, define $PC(\ell, b_k, b_j)$ to be the minimal forest in the set $\{(C(b_0, b_i, \ell) - \{\ell\}) \cup Tree(b_i, b_j) : b_i \in [b_0, b_k] \text{ lying in the span of } \ell\}$. Now, for any b_j outside the open-span of ℓ', $C(b_0, b_j, \ell')$ can be obtained in $O(1)$ time as follows. If $C(b_0, b_j, \ell')$ does not contain an anchor, then we return $Tree(b_0, b_j) \cup \{\ell'\}$. Else, let b_i be its farthest anchor. By Lemma 5, we obtain $L(C(b_0, b_j, \ell')) = L((C(b_0, b_i, \ell) - \{\ell\}) \cup Tree(b_i, b_j) \cup \{\ell'\})$. So, $L(C(b_0, b_j, \ell')) = L(PC(b_0, b_q, b_j) \cup \{\ell'\})$, where b_q is the last exterior vertex in $[b_0, b_j]$ lying in the span of ℓ'. Similarly, we can define $PC_c(\cdot)$'s and obtain $C_c(\cdot)$'s in $O(1)$ time.

The above implies that we only need to compute $C(\cdot)$'s and $C_c(\cdot)$'s for ℓ equal to some complete interior line. Afterward, we can compute $PC(\cdot)$'s and $PC_c(\cdot)$'s which can then be used to generate the other more general $C(\cdot)$'s and $C_c(\cdot)$'s.

Given $\ell = (b_0, b_f)$ and b_j, $PC(\cdot)$'s can be computed efficiently as follows. Let b_p be the first point in $[b_0, b_f]$ that lies in the span of ℓ. We initialize $PC(\ell, b_p, b_j)$ to be $(C(b_0, b_p, \ell) - \{\ell\}) \cup Tree(b_p, b_j)$. Then for each subsequent $b_{k+1} \in [b_0, b_f]$ lying in the span of ℓ, we set $PC(\ell, b_{k+1}, b_j)$ to be the shorter tree between $PC(\ell, b_k, b_j)$ and $(C(b_0, b_{k+1}, \ell) - \{\ell\}) \cup Tree(b_{k+1}, b_j)$ in $O(1)$ time.

3.2 Two usages of $C(\cdot)$ and $C_c(\cdot)$

Let $\ell = (b_0, b_f)$ be a complete interior line. $C(\cdot)$'s can be used to compute a tree in $\tau(b_0, b_f, \ell)$.

Lemma 6. *There exists adjacent points $a_i, a_{i+1} \in S[b_0, b_f]$ such that $C(b_0, a_i, \ell) \cup Tree(a_{i+1}, b_f) \in \tau(b_0, b_f, \ell)$. Hence, if $C(b_0, a_i, \ell)$ and $Tree(a_{i+1}, b_f)$ are available for all adjacent points $a_i, a_{i+1} \in S[b_0, b_f]$, then a tree in $\tau(b_0, b_f, \ell)$ can be computed in $O(n)$ time.* □

Let b_0 and b_f be two exterior vertices and let $\ell_1 = (b_0, v)$ and $\ell_2 = (b_f, v)$ be two perpendicular interior lines that share a common endpoint.

Lemma 7. *There exists adjacent points $a_i, a_{i+1} \in S[b_0, b_f]$ such that $C(b_0, a_i, \ell_1) \cup C_c(b_f, a_{i+1}, \ell_2)$ is a minimal tree connecting $S[b_0, b_f]$ with ℓ_1 and ℓ_2, in which any edge in the interior of $RC(S[b_0, b_f])$ perpendicularly incident to ℓ_1 or ℓ_2 hits the boundary at some exterior vertex. Hence, if $C(b_0, a_i, \ell_1)$ and $C_c(b_f, a_{i+1}, \ell_2)$ are available for all neighbouring points $a_i, a_{i+1} \in S[b_0, b_f]$, then that minimal tree can be computed in $O(n)$ time.* □

4 Computing Steiner trees for three sides of $RC(S)$

We cut $RC(S)$ at the highest, lowest, leftmost, and rightmost exterior vertices to divide it into four rectilinear straight lines. Let B_1 be on the SE side, B_2 on the SW side, B_3 on the NW side, and B_4 on the NE side. For each $[b_r, b_t]$ in $B_1 \cup B_2 \cup B_3$, we want to compute the structures as listed below. Our computation proceeds in iterations $1, 2, 3, \ldots$. In iteration $i \geq 1$, all $[b_r, b_t]$ (not $S[b_r, b_t]$) of length i are considered.

- If b_r and b_t can be connected by a complete line in the interior of $RC(S)$, then we compute:

- $C(b_r, b_i, (b_r, b_t))$ for all $b_i \in [b_r, b_t]$ lying in the open-span of (b_r, b_t) and $C_c(b_t, b_i, (b_t, b_r)))$ for all $b_i \in [b_r, b_t]$ lying in the open-span of (b_t, b_r).
- $PC((b_r, b_t), b_k, b_j)$ for all $b_j \in [b_r, b_t]$ and for all $b_k \in [b_r, b_j]$ lying in the span of (b_r, b_t), and the symmetric $PC_c(\cdot)$'s.

– If b_r can be connected by a complete line in the interior of $RC(S)$ to some $b_s \in [b_r, b_t] - \{b_t\}$, then we compute $PC((b_r, b_s), b_k, b_t)$ for all $b_k \in [b_r, b_s]$ such that $PC((b_r, b_s), b_k, b_t)$ is defined. We also compute the symmetric $PC_c(\cdot)$'s

– Compute $Tree(b_r, b_t)$.

By previous results in Section 3.1, all the $C(\cdot)$'s, $C_c(\cdot)$'s, $PC(\cdot)$'s, and $PC_c(\cdot)$'s desired can be computed in $O(n^3)$ total time summed over all iterations. The remaining task is to compute $Tree(b_r, b_t)$. In the following, we introduce five classes of Steiner trees such that at least one class contains an optimal tree. Thus, we just need to compute a "minimal" tree for each class and return the shortest one. We shall need access to a lot of $C(\cdot)$ and $C_c(\cdot)$ structures which have usually been computed before or can be constructed in $O(1)$ time by previous results unless stated explicitly otherwise.

I. CLASS 1 : trees that are equal to $Tree(b_{r+1}, b_t) \cup B(b_r, b_{r+1})$ or $Tree(b_r, b_{t-1}) \cup B(b_{t-1}, b_t)$ which can easily be computed in $O(1)$ time.

II. CLASS 2 : trees that contain a complete line in the interior of $RC(S)$ that links b_r and some exterior vertex in $[b_r, b_t]$, or b_t and some exterior vertex in $[b_r, b_t]$. The tree can be divided into two subtrees by cutting along that complete line. One subtree can be computed in $O(n)$ time by Lemma 6, the other one can be computed in $O(n)$ time by Lemma 3. The total processing time is $O(n)$.

III. CLASS 3 : trees that contain a complete corner in the interior of $RC(S)$ that links b_r (lying on B_3) and some exterior vertex in $[b_r, b_t]$ lying on B_1 such that: (1) the corner bends to the right; (2) any interior edge, which is perpendicularly incident to the left side of the vertical leg or the lower side of the horizontal leg, hits the boundary at some exterior vertex.

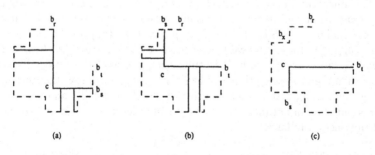

(a)　　　　(b)　　　　(c)

Fig. 2.

Refer to Fig. 2(a). Define $D(b_r, b_s, b_t)$ to be a Class 3 tree with the complete corner connecting b_r and b_s. $D(b_r, b_t, b_t)$ can be computed in $O(n)$ time by Lemma 6. For $b_s \neq b_t$, we imitate the proof of Lemma 3 and set $D(b_r, b_s, b_t)$ to be shorter tree between $D(b_r, b_s, b_{t-1}) \cup B(b_{t-1}, b_t)$ and $D(b_r, b_s, a_j) \cup$ {edge from b_t to the vertical leg}, where a_j is the last point in $S[b_s, b_t] - \{b_t\}$ (or just $D(b_r, b_s, b_{t-1}) \cup B(b_{t-1}, b_t)$

if b_t cannot be connected to (b_r, c) by an edge). So $D(b_r, b_s, b_t)$, for all $b_s \in [b_r, b_t]$, can be obtained in $O(n)$ time and we return the minimal one.

IV. CLASS 4 : trees that contain a complete corner in the interior of $RC(S)$ that links b_t (lying on B_1) and some exterior vertex in $[b_r, b_t]$ lying on B_3 such that: (1) the corner bends to the right; (2) any interior edge, which is perpendicularly incident to the left side of the vertical leg or the lower side of the horizontal leg, hits the boundary at some exterior vertex. (See Fig. 2(b).) Class 4 is symmetric to Class 3 and a minimal Class 4 tree can be computed similarly in $O(n)$ time.

V. CLASS 5 : trees that contain a complete corner in the interior of $RC(S)$ that bends to the right and links b_t (lying on B_1) and some exterior vertex in $[b_r, b_t]$ lying on $B_1 \cup B_2$. (See Fig. 2(c).) In the following, we give an algorithm such that if there exists a Class 5 tree $T \in \tau_3(S[b_r, b_t])$, then the tree returned is minimal in Class 5.

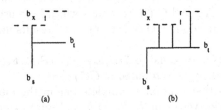

(a)   (b)

Fig. 3.

Refer to Fig. 2(c). Let c be the corner-vertex and let b_x be the exterior vertex closest to b_s with the same x-coordinate. If the solution should contain a line joining b_x to c (see Fig. 3(a)), then we can handle it as follows. Apply Lemma 6 to connect $S[b_x, b_s]$ to (b_x, b_s) and then imitate the proof of Lemma 3 to connect $S[b_r, b_x]$ with (b_t, c). Otherwise, the configuration looks like Fig. 3(b). We can cut at the c to divide the subtree further into one subtree T_1 connecting $S[b_r, a_i]$ with (b_t, c) and another subtree T_2 connecting $S[a_{i+1}, b_s]$ with (b_s, c), for some neighbouring points $a_i, a_{i+1} \in S[b_r, b_s]$. It can be proved that we can assume without loss of generality that a_{i+1} is the first terminal point in $S[b_x, b_s]$ or a_{i+1} is in $S[b_r, b_x] - \{b_x\}$. In the first case, T_2 is $C(b_s, a_{i+1}, (b_s, c))$ In the second case, T_2 is equal to $C(b_s, b_x, (b_s, c)) \cup B(a_{i+1}, b_x)$. In both cases, T_1 can be computed in $O(n)$ time using Lemma 3. If we actually perform the above, then the processing time will be $O(n^2)$. We can improve the running time to $O(n)$ by using an inductive strategy as for Class 3. The details can be found in [2].

It turns out that an optimal tree must exist in at least one class and so the main result of this section follows.

Theorem 8. $\tau(S[b_r, b_t])$ *contains a Class* k *tree, for some* k *equal to 1, 2, 3, or 4, or* $\tau_3(S[b_r, b_t])$ *contains a Class 5 tree. Thus, the total time needed to compute* Tree(b_r, b_t), *for all* b_r *and* b_t *lying on* $B_1 \cup B_2 \cup B_3$, *is* $O(n^3)$. □

The key ideas of the proof are as follows. First, it can be shown that if $\tau(S[b_r, b_t])$ does not contain a Class 1 tree or a Class 2 tree, then for all $T \in \tau_3(S[b_r, b_t])$, the path from b_r to b_t in T is actually an inner boundary corner of $RC(S[b_r, b_t])$. Thus, T is almost a Class 3 tree except that some edges which do not hit the boundary at some exterior vertices may be incident to the corner-vertex. Nevertheless, we can always slide these edges, if necessary, to hit some nodes so that the transformed tree will belong to a class. Since it takes $O(n)$ time to compute compute each $Tree(b_r, b_t)$, the total processing is $O(n^3)$ time.

5 Putting the trees together

In this section, we shall obtain the optimal Steiner tree for the extremal point set S. The process is divided into three major steps. Note that b_i may denote any exterior vertex on $RC(S)$ below (not necessarily on $B_1 \cup B_2 \cup B_3$). In step I, we apply Theorem 8 to compute $Tree(b_r, b_t)$, for all $b_r, b_t \in B_1 \cup B_2 \cup B_3$, in $O(n^3)$ time. By symmetry, we can also compute $Tree(b_r, b_t)$ for all $b_r, b_t \in B_2 \cup B_3 \cup B_4$ or $B_3 \cup B_4 \cup B_1$ or $B_4 \cup B_1 \cup B_2$. For every complete interior line (b_r, b_t), we shall also need access to $C(b_r, b_i, (b_r, b_t))$ (resp. $C_c(b_r, b_i, (b_r, b_t))$) for all $b_i \in [b_r, b_t]$ (resp. $b_i \in [b_t, b_r]$) lying in the open-span of (b_r, b_t). So in step II, we apply Lemma 4 to compute all these $C(\cdot)$'s and $C_c(\cdot)$'s in $O(n^3)$ time. Step III is the actual production of the optimal Steiner tree. We shall introduce six groups of trees in $\tau_3(S)$. Each tree in $\tau_3(S)$ must belong to least one group. We shall give an $O(n^3)$-time algorithm for each group such that if $\tau_3(S)$ contains a tree in that group, then the Steiner tree returned by the algorithm is optimal. Thus, we can invoke the six algorithms and return the minimal tree which is guaranteed to be optimal. Hence, the main result of this paper follows.

Theorem 9. *An optimal Steiner tree for an extremal point set of size n can be computed in $O(n^3)$ time.* □

In the following, we describe the six groups and the corresponding algorithms.

I. GROUP 1 : trees that do not contain any interior edge. A minimal tree in Group 1 can be obtained by taking $RC(S)$ with the removal of the longest portion of $RC(S)$ between two neighbouring terminal points that does not contain any terminal point. The time needed is $O(n)$.

II. GROUP 2 : trees that contain a complete interior corner whose legs end on B_1 and B_3 or B_2 and B_4. The tree can be divided along the legs of the complete corner to produce two subgraphs. Each of these subgraphs can be computed in $O(n)$ time by Lemma 7. The above is repeated for the $O(n^2)$ pairs of exterior vertices and so the total processing time is $O(n^3)$.

III. GROUP 3 : trees that contain a complete interior corner but are not in Group 2. (See Fig. 4.) Our strategy is to compute a tree for each of the configuration

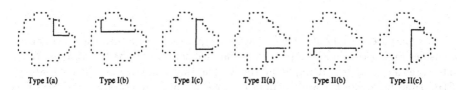

| Type I(a) | Type I(b) | Type I(c) | Type II(a) | Type II(b) | Type II(c) |

Fig. 4.

shown in Fig. 4 and then return the minimal tree among them as the answer. By rotating 90 degrees counterclockwisely, each Type II configuration becomes a Type I configuration. Hence, we shall focus on the Type I configurations. We shall need access to some $C(\cdot)$'s and $C_c(\cdot)$'s and so we first preprocess to compute some $PC(\cdot)$'s and $PC_c(\cdot)$'s.

For each vertical complete interior line $\ell = (b_r, b_s)$, where b_r is the upper endpoint, we compute in $O(n^2)$ time $PC(\ell, b_k, b_j)$ for all $b_k \in [b_r, b_s]$ lying in the span of ℓ and for all b_j lying on $B_1 \cup B_2 \cup B_3$ such that $PC(\ell, b_k, b_j)$ is defined. The symmetric $PC_c(\cdot)$'s can be computed similarly. Since there are at most $O(n)$ such complete interior lines, the total processing time is $O(n^3)$.

Consider Type I(a). (Type I(b) and (c) can be handled similarly.) Let b_r (resp. b_t) be the upper (resp. right) endpoint of the vertical (resp. horizontal) leg of the complete interior corner. The tree can be divided into two subtrees by cutting along the legs of the complete interior corner. The subtree connecting $S[b_t, b_r]$ with the complete interior corner can be computed in $O(n)$ time by Lemma 7. The other subtree has the same total length as $C(b_r, a_i, \ell_r) \cup C_c(b_t, a_{i+1}, \ell_t)$ for some neighbouring points $a_i, a_{i+1} \in S[b_r, b_t]$, where ℓ_r and ℓ_t are the vertical and horizontal legs of the complete interior corner, respectively. If $C(b_r, a_i, \ell_r)$ is not already available, then it can be obtained in $O(1)$ time using the $PC(\cdot)$'s precomputed. The same holds for $C_c(b_t, a_{i+1}, \ell_t)$. Since we have to iterate for all neighbouring points $a_i, a_{i+1} \in S[b_r, b_t]$, the above takes $O(n)$ time for a given pair of b_r and b_t. Hence, it takes a total of $O(n^3)$ time to iterate over all $O(n^2)$ pairs of b_r and b_t.

IV. Group 4 : trees that contain a complete interior line that ends on B_1 and B_3 or B_2 and B_4. The tree can be divided into two subtrees by cutting along the complete interior line. Each of the subtree can be computed in $O(n)$ time by Lemma 6. Since we have to iterate for the $O(n)$ positions for the complete interior line, the total processing time is $O(n^2)$.

V. Group 5 : trees that are not in Group 4 and contain a complete interior line ℓ such that ℓ ends on B_2 and B_3 and there is an edge perpendicularly incident to the right side and the relative interior of ℓ, or the tree has one of the three other symmetric configurations as shown in Fig. 5.

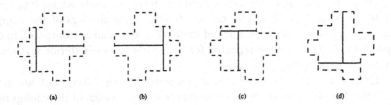

(a)   (b)   (c)   (d)

Fig. 5.

We compute a candidate tree for each configuration and return the minimal one. Consider Fig. 5(a). (Fig. 5(b)–(d) can be handled similarly.) The tree can be divided into three subtrees by cutting along the T. The three subtrees can be computed in $O(n)$ by Lemma 6 and Lemma 7. Since we need to iterate for $O(n^2)$ such possible T's, the total processing time is $O(n^3)$.

VI. GROUP 6 : trees that contain a complete interior line but are not in Group 2 or Group 3 or Group 4 or Group 5. Let T be a Group 6 tree in $\tau_3(S)$. By the definition

Fig. 6. $b_x \in [b_t, b_r]$ is a point closest to (b_r, b_t) among all those lying in $[b_t, b_r]$ and in the span of (b_t, b_r).

of Group 6, any complete interior line in T must end on B_2 and B_3, or B_3 and B_4, or B_4 and B_1, or B_1 and B_2. Without loss of generality, let the *longest* complete interior line in T be vertical and ends on B_3 and B_2 as shown in Fig. 6. The tree can be divided into two subtrees by cutting along the complete interior line. One subtree connects $S[b_r, b_t]$ with (b_r, b_t) and can be computed in $O(n)$ time by Lemma 6. Let $b_x \in [b_t, b_r]$ be a point closest to (b_t, b_r) among all those in $[b_t, b_r]$ lying in the span of (b_t, b_r). b_r, b_t, and b_x defines a rectangle (the dotted box shown in Fig. 6). It can be shown that no interior edge intersects the rectangle defined by b_r, b_t, and b_x. As a result, we should set the subtree connecting $S[b_t, b_r]$ with (b_r, b_t) to be the minimal tree among the following: $Tree(b_t, a_i) \cup Tree(a_{i+1}, b_x) \cup Tree(b_x, b_r) \cup \{(b_r, b_t)\}$ for all neighbouring points $a_i, a_{i+1} \in S[b_t, b_x]$, $Tree(b_t, a_i) \cup Tree(a_{i+1}, b_r) \cup \{(b_r, b_t)\}$ where a_i is the last terminal point in $S[b_t, b_x]$ and a_{i+1} is the first terminal point in $S[b_x, b_r] - \{b_x\}$, and $Tree(b_t, b_x) \cup Tree(b_x, a_i) \cup Tree(a_{i+1}, b_r) \cup \{(b_r, b_t)\}$ for all neighbouring points $a_i, a_{i+1} \in S[b_x, b_r]$. The above takes $O(n)$ time for a given (b_r, b_t). Since there are $O(n)$ such complete interior lines, the total processing time is $O(n^2)$.

References

1. M. Bern. Faster exact algorithms for Steiner trees in planar networks. *Networks*, 20:109–120, 1990.
2. S.W. Cheng, Andrew Lim, and C.T. Wu. Optimal rectilinear steiner tree for extremal point set. Technical Report Technical Report HKUST–CS93-6, HKUST, Department of Computer Science, April 1993.
3. J.P. Cohoon, D.S. Richards, and J.S. Salowe. An optimal steiner tree algorithm for a net whose terminals lie on the perimeter of a rectangle. *IEEE Transactions On Computer-Aided Design*, 9(4):398–407, 1990.
4. J.S. Provan. Convexity and the steiner tree problem. *Networks*, 18:55–72, 1988.
5. D.S. Richard and J.S. Salowe. A linear-time algorithm to construct a rectilinear Steiner minimal tree for k-extremal point sets. *Algorithmica*, 7:247–276, 1992.

Faster Approximation Algorithms
for the Rectilinear Steiner Tree Problem[*]

Ulrich Fößmeier[1], Michael Kaufmann[1], and Alexander Zelikovsky[2]

[1] Wilhelm-Schickard-Institut für Informatik, Universität Tübingen,
Sand 13, 72076 Tübingen, Germany
email: foessmei / mk @informatik.uni-tuebingen.de
[2] Institute of Mathematics, Academiei 5, 277028 Kishinev, Moldova,
email: 17azz@mathem.moldova.su

Abstract. The classical Steiner Tree Problem requires a shortest tree spanning a given vertex subset within a graph $G = (V, E)$. An important variant is the Steiner tree problem in rectilinear metric. Only recently two algorithms were found which achieve better approximations than the 'traditional' one with a factor of $3/2$. These algorithms with an approximation ratio of $11/8$ are quite slow and run in time $O(n^3)$ and $O(n^{5/2})$. A new simple implementation reduces the time to $O(n^{3/2})$. As our main result we present efficient parameterized algorithms which reach a performance ratio of $11/8 + \epsilon$ for any $\epsilon > 0$ in time $O(n \cdot \log^2 n)$, and a ratio of $11/8 + \log \log n / \log n$ in time $O(n \cdot \log^3 n)$.

Keywords: Algorithms, Steiner tree, Approximations

1 Introduction

Let $S = \{v_0, \ldots v_{n-1}\}$ be a set of points in the plane. A Steiner tree is a tree in the plane which contains the set S. The Steiner tree problem is to find the tree of minimal length. There are several versions of the problem, extensively described in the literature [4, 7, 8, 9, 15, 17].

We describe the three most important versions:
The *Euclidean Steiner problem* is the version where the edges are straight lines and the length of an edge from v_i to v_j is the Euclidean distance [3]. In the *rectilinear case*, the distance between two points is the sum of the differences of their x- and y-coordinates. This problem got recently new importance in the development of techniques for VLSI routing [12, 13]. The final variant we want to mention is the *Steiner problem on graphs* where the point set is a subset of the vertices, and the edges of the tree must be chosen from the set of edges of the graph. The edge lengths may be arbitrarily chosen. All three versions of the problem are known to be NP-hard since a long time [11, 7]. Therefore polynomial time algorithms for approximate solutions have been investigated.

The quality of an approximation is measured by its performance ratio: an upper bound on the ratio between achieved length and the optimal length.

For all three variants the well-known *MST-heuristic* for the Steiner tree problem approximates a Steiner minimal tree with a minimum length spanning tree MST of a complete graph G_S which has a vertex set S and edge lengths equal to shortest path lengths in the graph G. The lowest performance ratio of this heuristic equals 2 for the Steiner problem on graphs [10]. The fastest known implementation of the MST-heuristic runs in time $O(e + n \cdot \log n)$ [10], e being the number of edges and n the number of vertices. For many years, the problem of finding a better heuristic remained open. Surprisingly two better approximations were given recently [2, 19]. Their better performance ratios appear by a restriction of the general Steiner tree problem.

In this paper, we will consider the rectilinear case. It gained new attention in the research of VLSI-algorithms, many heuristics and special cases [9, 10, 12, 15] were devised. The MST-heuristic for the rectilinear case is due to Hwang [10]. He proved that the ratio between the length of the minimum spanning tree and the length of the minimal Steiner tree is $3/2$, and gave a $O(n \cdot \log n)$ time implementation for that approximation algorithm.
Analogously to the new approximation results in the variant mentioned above in [20] and [2] a ratio of $11/8$ was proved. The algorithm of Zelikovsky runs in time $O(n^3)$ and has been improved by the $O(n^{5/2})$ algorithm of Berman/Ramaiyer. The main results of this paper are:

---

[*] Parts of this work have been done at Max-Planck-Institut für Informatik, Saarbrücken and at the Fakultät für Mathematik und Informatik, Universität Passau, Passau

- An acceleration of the second algorithm to $O(n^{3/2})$.
- A new parameterized $O(r \cdot m \cdot n \cdot \log^2 n)$ algorithm with a performance ratio of $11/8 + \frac{1}{2m} + \frac{1}{8 \cdot 36^{r-2}}$ for any parameters $m, r > 0$.

First we introduce some notation: $ST(S)$ and $st(S)$ are a Steiner tree of S and its length, respectively, $SMT(S)$ is a Steiner minimal tree with length $smt(S)$. For a complete graph with a vertex set S, $M(S)$ denotes a minimum length spanning tree of S, and $m(S)$ denotes its length.

In general a Steiner tree for a set S may contain as vertices also other points of the plane than those in S. So $ST(S)$ contains the set S of *given vertices* and some additional vertices, the *steiner vertices*. $ST(S)$ is called a *full Steiner tree* if S coincides with the set of leaves of $ST(S)$. If $ST(S)$ is not full, then we can split it into edge-disjoint full Steiner subtrees. $ST(S)$ is called k−*restricted* if every full component has at most k given vertices. Let the shortest k−restricted Steiner tree for the set S, denoted by $SMT_k(S)$, have the length $smt_k(S)$. Notice, that $SMT_2(S) = M(S)$.

We will review in Section 2 the basic algorithm due to Berman/Ramaiyer [2]. In Section 3 we prove some facts that lead to a reduction of the necessary iteration steps. Then, we show how to precompute the data for each iteration efficiently, proving the time bound of $O(n^{3/2})$. Finally, in Section 5 the new fast algorithm is presented, which runs in $O(n \cdot \text{polylog}(n))$ time.

2   The Basic Algorithm

The idea is to start with a minimum spanning tree for S, then to add iteratively some new points which are the centers of triples τ of points in S and to construct new minimum spanning trees for the new point set. The triples are called *stars*. In the Steiner minimal tree of τ there is one additional Steiner point called *center*. At the end we get a 3-restricted Steiner tree for the point set S reaching the claimed approximation ratio. Berman/Ramaiyer did that also for general $k \geq 3$, considering not only triples but k-tupels. But the complexity of the algorithm increases very fast for higher k. Therefore we will restrict ourselves to $k = 3$.

We need some definitions for a triple τ of points in S:

For $u, v \in S$ $Bridge(u, v)$ is the longest edge on the path between u and v.
$Bridge(\tau) = \{Bridge(u, v); (u, v) \in \tau\}$

The criterion for the insertion of the center point of some star in Zelikovsky's algorithm is the $gain(\tau)$ which is defined to be the difference of the cost of $Bridge(\tau)$ and $smt(\tau)$. Zelikovsky proposed to add always the star with the maximal gain as long as there are stars with nonnegative gain. Thus the length of the actual minimum spanning tree MST will always be improved and the approximation ratio of $11/8$ could be proven.

The algorithm of [2] also operates with stars and their gains but it is more involved:

For $e = Bridge(u, v)$ with cost c, $alt(e) = (u, v)$ with cost $= (c - gain(\tau))$.
$Altbridge(\tau) = \{alt(u, v); (u, v) \in Bridge(\tau)\}$.

Intuitively, we simulate a $SMT(\tau) = M(\tau \cup c)$ where c is the center of τ by inserting two new edges between points of τ (namely the $Altbridge(\tau)$). Thus two cycles appear. $Bridge(\tau)$ denotes the set of edges which have to be removed from these cycles in order to get a new MST.

The algorithm of Berman/Ramaiyer consists of 3 phases: the evaluation phase, the selection phase and the construction phase. In the evaluation phase first $M(S)$ is computed. Then all stars τ are *treated* and the value $gain_M(\tau)$ is computed. M gives the actual edge set. If $gain_M \leq 0$ then τ is discarded, otherwise B_{old} is removed from M and B_{new} is inserted instead, with $B_{old} = Bridge_M(\tau)$, $B_{new} = Altbridge_M(\tau)$. Then store the triple (τ, B_{old}, B_{new}) in a stack. In the selection phase the stack is made empty and the actual τ is either *discarded* or *accepted* and then added to a list L if it might be included in the final ST. In the construction phase the stars stored in L are *realized* and so the output tree is built up from the entries of the list L. In more detail, in the selection phase (τ, B_{new}) is inserted in the list if $B_{new} \subseteq M$, i.e. the $Altbridge(\tau)$ were not used as bridges by another star at a later moment in the evaluation phase; if not, the set B_{new} is removed from E and the actual MST is recomputed.

The computations to be made in the course of the algorithm are the calculations of bridges, altbridges and gains of all stars as well as some operations which modify the actual set M and the corresponding MST. There are several methods which support manipulations on MSTs. Frederickson [6] gave a data structure for maintaining MST's in dynamic graphs. Insertions and deletions are allowed and also manipulations on the edge lengths. This method can also be used to compute (alt)bridges and gains. The update time per step is $O(\sqrt{e})$, where e is the number of edges. Another method [5] that only works on dynamic planar graphs takes only $O(\log n)$ time per operation.

Berman/Ramaiyer apply Frederickson's data structure and state that the number of edges to consider is only $O(n)$, namely the set of edges in the actual minimum spanning tree + the set of edges in $SMT(\tau)$. Thus each basic step in Berman/Ramaiyer's algorithm takes time $O(\sqrt{n})$. Trivially there are $\binom{n}{3} = \Theta(n^3)$ stars, so the algorithm needs $\Theta(n^{3.5})$ time. In both papers [20] and [2] it has been seen that only $O(n^2)$ stars have to be considered, improving the time bound to $\Theta(n^3)$ [20] or $\Theta(n^{5/2})$ [2].

In the next two sections we will show that only $O(n)$ stars have to be considered and how this set of stars is constructed efficiently.

3 $O(n)$ Stars are Enough

We shall prove

Lemma 1 *It is sufficient to consider only $O(n)$ stars and this set can be constructed in time $O(n \cdot \log^2 n)$.*

From this Lemma follows:

Theorem 1 *For any rectilinear Steiner tree problem an 11/8-approximation can be found in time $O(n^{3/2})$.*

For brevity, we assume that coordinates of all given points and subtractions between them are distinct.

First we need some definitions: A triple (s_0, s_1, s_2) is called a *star*, if $x_0 < x_1 < x_2$ and $(y_0 - y_1)(y_1 - y_2) < 0$ where $s_i = (x_i, y_i)$. There are four types of stars corresponding to the four possible orders of y_i. In the following, we consider the case of $y_1 < y_0 < y_2$ (similar arguments can be used for the other types of stars).

A *center* of a star is the point $c = (x_1, y_0)$ which is the additional point of $SMT(s_0, s_1, s_2)$. We use $z = (c; s_0, s_1, s_2)$ to denote such a star. The three given points defining the star z also define a rectangle R where they lie on the boundary. R is empty, if there is no point (x, y) such that $x_0 < x < x_2$ and $y_1 < y < y_2$.

Berman/Ramaiyer show that it is sufficient to consider a family of stars for which we know that there is a 3-restricted SMT with stars from this family. They also prove

Lemma 2 *There is an $SMT_3(S)$ using only stars defining empty rectangles.* $\Diamond$

A star $z = (c; s_0, s_1, s_2)$ is called a *tree star* if $M(S \cup c)$ contains the edges of the set $\bar{z} = \{(c, s_0), (c, s_1), (c, s_2)\}$.

Lemma 3 *There is an $SMT_3(S)$ using only tree stars defining empty rectangles.*

Proof. Let $T = SMT_3(S)$ and S_0, S_1, S_2 be given points of the three components of $T \backslash \bar{z}$. Then $\bar{z}$ should contain the three shortest edges which connect the center c with S_0, S_1, S_2, namely with s_0, s_1, s_2. Note, that the shortest edge from c to S, say (c, s_0), belongs to $M(S \cup c)$ and $\bar{z}$ simultaneously.

Let a and b be the two shortest edges between the sets S_0, S_1, S_2, say a connects S_0 with S_1 and b connects S_1 with S_2. Then the tree $T' = \{a, b\} \cup T \backslash \bar{z}$ is a 3-restricted $ST(S)$.

Suppose, that the edge $e_1 = (c, s_1)$ does not belong to $M(S \cup c)$. Then e_1 is the longest edge in the unique cycle $C \in M(S \cup c) \cup e_1$. Let $e'_1 = (c, s'_1)$ be the other edge in C with the end c. The longest edge in the path between s_1 and s'_1 in T' is not longer than in $M(S)$. So e_1 is the longest edge in the unique cycle of $T' \cup e_1 \cup e'_1$. If $s'_1 \in S_1$, then e_1 is not the shortest edge between s_1 and c. If $s'_1 \in S_0$ or $\in S_1$, then e_1 is longer than a or b and we can decrease the total length of T replacing e_1 with a or b respectively.

The case of $(c, s_2) \notin M(S \cup c)$ is similar. $\Diamond$

A necessary condition for a star to be a tree star is that the three points lie closer to the center than to any other points. In the following a star means a star with this condition and defining an empty rectangle.

A star is *positive* (*negative*) if $(y_2 - y_0) - (x_2 - x_1) > 0(< 0)$. This means that the top point lies above (below) the diagonal through the center of the star. Yao [18] introduced a graph associated with the set S: every point is connected with the nearest points in all eight angles defined by aces and bisectors. The Yao graph contains $M(S \cup c)$. This fact implies directly the next

Lemma 4 *There are at most two distinct stars with the same center, namely the positive and the negative with the shortest length.* ◇

For a center c of a star z, the nearest given point, s, should belong to z. In other words, c lies in the Voronoi region of s, i.e. the region of the plane containing points which are nearer to s than to any other given point. This point s is called the *root* of the star. A star is named *left*, *bottom* or *top*, if its root is the left, bottom or top point of a star, respectively.

Lemma 5 *For any point $s \in S$, there is at most one star with the top root s.*

Proof. Let $z = (c = (x_c, y_0); s_0 = (x_0, y_0), s_1, s)$ and $z' = (c' = (x_1', y_{c'}); s_0', s_1' = (x_1', y_1'), s)$ be two distinct top stars with a root $s = (x, y)$ and $y_{c'} > y_0$. We assumed above, that z and z' are empty, so $x_1' < x_0$ and $y_1' > y_0$. Note that $x_c - x_0 > y - y_0$, since c lies in the Voronoi region of s. Therefore,

$$|c', s| > x - x_1' > x_c - x_0 > y - y_0 > y_{c'} - y_1' > |c', s_1'|.$$

The last inequality means that c' does not lie in the Voronoi region of s. ◇

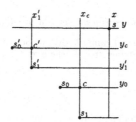

Lemma 6 *The number of left stars is at most $4n$.*

Proof. Let $s = (x, y)$ be a left root of two stars $z = (c = (x_1, y); s, s_1 = (x_1, y_1), s_2)$ and $z' = (c' = (x_1', y); s_1' = (x_1', y_1'), s_2')$ and $x_1 < x_1'$. Then $y_1 < y_1'$, since the both stars are empty, and $y - y_1' > x_1' - x$, since c' lies in the Voronoi region of s. Moreover, since the edge (c, s_1) belongs to $M(S \cup c)$, $y_1' - y_1 < x_1' - x_1$, and therefore,

$$y - y_1' > y_1' - y_1. \tag{1}$$

Claim 1 *There is no any other left star with the bottom point s_1.*

Proof. Indeed, let $(d = (x_1, y'); s' = (x', y'), s_1, s_3)$ be a such star. This star together with z and z' are empty, so $x' < x$ and

$$y_1' > y'. \tag{2}$$

Since c and d lie in the Voronoi regions of s and s' respectively,

$$|s, c| < |c, s_1|, \tag{3}$$

$$|s, s'| < |c, d| + |d, s'| < |c, d| + |d, s_1| = |c, s_1|. \tag{4}$$

Inequalities (1) and (2) imply

$$|s', s_1| = |s', d| + |d, s_1| < |s_1, d| + |c, d| = |c, s_1| \tag{5}$$

Inequalities (3)-(5) imply, that the edge (c, s_1) does not belong to $M(S \cup c)$. ◇

Consider a directed graph $H = (S, D)$ such that $(s, s') \in D$, if there is a left star with the left point s and the bottom point s'. To prove that $\#D \le 2n$ we will show that there is a vertex in H of degree at most 2.

Let s_1 be the point with the smallest abscissa. Then indegree of s_1 is $indeg(s_1) = 0$. Let $outdeg(s_1) > 2$ (otherwise s_1 will be the necessary vertex) and s_2 be the lowest bottom point for s_1. The claim implies that $indeg(s_1) = 1$, so $outdeg(s_2) > 1$ and $indeg(s_3) = 1$ for the lowest bottom point s_3, and so on. Finally, we will find a vertex of degree at most 2, since all points $s_1, s_2, s_3, \dots$ are distinct. Thus, there exist at most $2n$ distinct centers of the left stars and Lemma 6 follows from Lemma 4. ◇

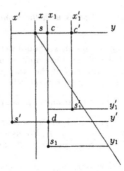

Similarly, the number of bottom stars is at most $4n$. Thus, for all possible four types of stars, Lemma 5 and Lemma 6 imply

Lemma 7 *The number of empty stars with the tree star condition is at most $36n$.* ◊

It follows directly from Lemma 3 and Lemma 7, that we only need a linear number of stars during the construction of a 3-restricted Steiner tree. In the next section we show how to determine these stars.

4 Computing the stars

We start by computing the Voronoi-diagram of the points in the plane. Let s_0, s_1 and s_2 denote the left, bottom and top point of a star. To find the corresponding star, we have to compute the center c. We only describe the type of stars which has also been described in Section 3.

4.1 Computing left stars

Since in the first case s_0 is the point closest to the center c, c has to lie in the Voronoi-region of s_0 and on the horizontal segment extending from s_0 to the border of its Voronoi-region. s_1 being the bottom point of the left star must lie closer to c than to any other point. Hence c must lie on a segment extending upwards from s_1 with a length being the distance from s_1 to its nearest neighbor. The proof of Lemma 6 showed that there are at most $4 \cdot n$ stars with this condition. An All-Nearest-Neighbors precomputation is necessary, but this can be obtained by an inspection of the planar graph being the dual of the Voronoi-diagram. Next we show how to compute the stars themselves:

We associate with each point s_0 its horizontal segment h and its vertical segment v. We have to check which segments intersect. The intersections of the segments are possible centers. The computation of the intersections can easily be done by a series of range queries. We store the vertical segments (x, y_1, y_2) in a range tree [14] and for every horizontal segment (x_l, x_r, y), we search for the vertical segment (x, y_b, y_a) such that $x_l < x < x_r$ and $y_a > y$ and $y_b < y$ maximal. Then we iterate the search with $h' = (x_l, x, y)$ until we cannot find any vertical segment with the query property. A standard implementation of the range tree supports the operations in time $O(\log^2 n)$ with a build-up time of $O(n \cdot \log^2 n)$. Since we perform at most $O(n)$ queries, the computation of the left stars stops after $O(n \cdot \log^2 n)$ steps.

4.2 Computing top stars

Since the center lies closest to the top root s_2, it has to lie in the Voronoi region of s_2. Analogously to Section 4.1 the center has to lie on the intersection of a horizontal and a vertical segment extending from the two other points s_0 and s_1 where the segments have the distance to the nearest neighbor as their length.

By contradiction to the tree star condition we can prove the following

Claim 2 *There is only one intersection in the Voronoi-region of s_2.*

The implementation is easy. We subsequently check all vertical segments. If we find the first intersection outside of the Voronoi-region of s_2, we can stop. Hence the running time is also $O(n \cdot \log^2 n)$ in this case.

5 A Fast Parameterized Version of the Algorithm

In this section we present a variant of the algorithm which saves the factor of $\sqrt{n}$ in the time bound. In time $O(n \cdot \text{polylog}(n))$ a performance ratio of $\frac{11}{8} + \varepsilon$ can be reached for any constant $\varepsilon > 0$. The 'O' in the running time of the algorithm is proportional to the parameters r and m, the performance ratio will be $\frac{11}{8} + \frac{1}{2m} + \frac{1}{8 \cdot 36^{r-2}}$. One can choose for example r = 3 and m = 8 (what are no big constants) to reach a performance ratio which difference to the $\frac{11}{8}$ is only about 0.06. The new algorithm combines the merits of the algorithms in [2] and [20]: While Zelikovsky always computes the actually best star, Berman/Ramaiyer allow the order of treating the stars in the evaluation phase to be arbitrary; this causes the need of the selection phase. Our idea is not to compute always the best star, but yet one laying in a range of $\frac{m}{m+1}$ of the optimal. So we save the selection phase (like Zelikovsky does) and are not forced to compute an optimal star $O(n)$ times.

Definition 1

a) g_i is the gain of star τ_i at the beginning of the repeat loop.
b) a_i is the gain of star τ_i at the time of its treatment (actual gain).
c) a_i' is the gain of star τ_i at the time of its treatment, where no artificial edges are used, i.e. bridges(τ_i) := edges of maximal length on the paths between the points of the star which are original, i.e. $\subseteq M(S)$.

In the algorithm always a_i' is computed, therefore a new edge (a B_{new} of an older star) will never be deleted and all stars once stored in the list ERG will be realized. We have to show that the cases where a star has a great a_i but small a_i' and is therefore discarded do not destroy the result.

Furthermore we only consider 'planar' stars, i.e. stars where the *altbridges* do not cross an existing edge. Here we also have to prove that non-planar stars would not achieve too much further gain, i.e. the performance ratio can be proved even without treating non-planar stars. The planarity of the structure is necessary to keep the frequent computing of the gains fast [5].

Algorithm

```
(1)  compute the stars τ_i (i = 1,...,n) and store them in a list L;
(2)  repeat r · m · log n times:
(3)      g_i := gain(τ_i) (i = 1,...,n);
(4)      sort L according to g_i (in decreasing order);
(5)      compute j : g_j ≥ g_1 (m/(m+1)), g_{j+1} < g_1 (m/(m+1))
(6)      for i := 1 to j do
(7)          a_i' := actual gain(τ_i) with bridges(τ_i) ⊆ E;
(8)          if a_i' ≥ g_1 (m/(m+1)) and τ_i is planar and B_old(τ_i) ⊆ E
(9)          then
(10)             store τ_i in a list ERG;
(11)             delete τ_i from L;
(12)             M := M\B_old(τ_i) ∪ B_new(τ_i);
(13)         endif
(14)     endfor
(15) endrepeat.
```

Note that if a star does not fulfil the condition $a_i' \geq g_1 \frac{m}{m+1}$ in line 8 it is not automatically refused; it remains in the list L and will be treated in subsequent rounds of the repeat loop.

In the next subsections we will prove

Theorem 2 *The algorithm computes in time $O(r \cdot m \cdot n \cdot \log^2 n)$ a steiner tree for a given set of points S, $|S| = n$. Its performance ratio is at most $\frac{11}{8} + \frac{1}{2m} + \frac{1}{8 \cdot 36^{r-2}}$.*

5.1 Performance Ratio

In this section we prove the performance ratio of the algorithm. We compare our accepted stars with the treated stars of the algorithm of Berman/Ramaiyer where we know the performance ratio to be $\frac{11}{8}$. Our aim is to reach any constant fraction of the gain they can realize.

Every star τ having positive gain at the end of the evaluation phase has (at least) one of the following characteristics:

a) $gain(\tau)$ (i.e. a_τ) $\leq g_1 \frac{m}{m+1}$, where g_1 is the gain of the best star at the beginning of the final round of the repeat loop.

b) τ is not planar.

c) $bridges(\tau) \not\subseteq E$, i.e. τ would delete an artificial edge.

For the following analysis let x be the sum of the gains of the treated stars in the algorithm of Berman/Ramaiyer.

Lemma 8 *After the* $m \cdot r \cdot \log n - th$ *execution of the repeat loop (line 2) the remaining gain of all planar stars* τ *for which the condition* $bridges(\tau) \subseteq E$ *holds sums to no more than* $\frac{x}{36^{r-2}}$.

Proof: First we consider the case $m = 1$. Let Max be the maximal gain of a star at the beginning of the first round. Then after i rounds every star has gain $\leq \frac{Max}{2^i}$ (in every round the 'better half' of the stars is executed), thus after $r \cdot \log n$ rounds $\leq \frac{Max}{n^r}$. Since there are at most $36n$ stars the whole gain left is $\leq \frac{36n Max}{n^r} = \frac{36 Max}{n^{r-1}}$, or (for $n \geq 36$) $\leq \frac{Max}{36^{r-2}} \leq \frac{x}{36^{r-2}}$.

In the case $m > 1$ we simulate one round for $m = 1$ by s rounds. How large should s be, such that after s rounds the gain of any remaining star is bounded through $\frac{Max}{2}$. After i rounds every star has gain $\leq Max \cdot \left(\frac{m}{m+1}\right)^i$. Setting $i = m$, the factor becomes $\left(\frac{m}{m+1}\right)^m$ which converges for large m to $\frac{1}{e} < \frac{1}{2}$. Therefore after m rounds the gain of every remaining star is at most $\frac{Max}{2}$. After $m \cdot r \cdot \log n$ rounds, the total gain of all non-treated stars is bounded to $\frac{x}{36^{r-2}}$. $\Diamond$

Lemma 9

$$\max_{\hat{S} \text{ independent set of stars}} \sum_{r \in \hat{S}} (a_\tau - a'_\tau) \leq \frac{2x}{m}$$

i.e. the sum of all gains we lose by refusing to delete artificial edges (B_{new} of older stars) is bounded to $\frac{2x}{m}$.

Proof: τ_{i_1} uses an edge $e^*_{i_1}$ as B_{old} and creates an edge e_{i_1} as B_{new}. Let S be the set of stars τ with positive gain and $e_{i_1} \in bridges(\tau)$. Let τ_{i_2} be the star for which $a_{\tau_{i_2}} - a'_{\tau_{i_2}}$ is maximal. $\Rightarrow$ At the time of the treatment of τ_{i_1} τ_{i_2} had a larger gain than τ_{i_1}, for τ_{i_2} would have had $e^*_{i_1}$ as bridge instead of e_{i_1}. So the gain we lose by the wrong choice of τ_{i_1} instead of τ_{i_2} is $\leq a_{i_2}$ $(gain(\tau_{i_2})$ at the time when τ_{i_2} was treated); for only *one* star $\in S$ could be realized using the edge e_{i_1} as bridge. The total loss is thus at most a_{i_2}.

τ_{i_2} had a gain of at least $a_{i_1} + a_{i_2}$ before the treatment of τ_{i_1} $(a_{i_1} = |e^*_{i_1}| - |e_{i_1}|)$. Thus $g_{i_2} \geq a_{i_1} + a_{i_2}$ or $g_{i_2} - a_{i_1} \geq a_{i_2}$.

From the acceptation of τ_{i_1} follows:

$$a_{i_1} \geq a'_{i_1} \geq \frac{m}{m+1} g_1 \geq \frac{m}{m+1} g_{i_1}.$$

$$\Rightarrow a_{i_2} \leq g_{i_2} - a_{i_1} \leq g_{i_1} - a_{i_1} \leq \frac{m+1}{m} a_{i_1} - a_{i_1} = \frac{a_{i_1}}{m}.$$

τ_{i_1} produces *two* new edges that could be used as B_{old} by other stars. Thus the loss resulting from the wrong choice of a_{i_1} can be bounded by $2\frac{a_{i_1}}{m}$.

$\Rightarrow$ The total loss resulting from the discard of stars with artificial edges $\leq \frac{2x}{m}$. $\Diamond$

Before we can complete our analysis we have to prove the following

Lemma 10 *[1] Let T be a 3-restricted SMT with two stars a,b intersecting each other. Then we can replace these stars with another star z without increasing the total length of the SMT.*

Proof: Let a, b be two intersecting stars of a 3-restricted SMT T. Note that these stars should be empty. So the height of one of them should be larger than the height of the other. This star, say a, and its induced rectangle will be called *vertical* and the other, say b, is *horizontal*. Then the two rectangles form a cross with the regions A, B, C, D, E. Note that some of the external regions (A, B, C, D) could have expansion 0, then the cross degenerates. We will look at this case later. The length of the stars a and b is equal to the half of the perimeter of the minimal rectangle which contains this cross, say p, plus the height of the horizontal rectangle, say h, plus the width of the vertical one, say w. Two of the external regions contain two points each, the other two regions contain one point each; and we know that the two regions containing two points do not lie at opposite sides of E. Without loss of generality regions A and D contain two points each (otherwise we have to turn the cross). We now have the situation that a_2 and a_3 lie in region D, b_2 and b_3 in region A, a_1 lies in region B and b_1 in C. Let P be the shortest path which connects the stars a and b in T. We distinguish two cases:

I) P connects points from A and D. Consider the star $z = \{a_1, b_1, a_2\}$ where a_2 has the larger y-coordinate than a_3. Then

$$d(z) + d(a_2, a_3) + d(b_2, b_3) \leq p + h + w = d(a) + d(b).$$

II) P connects e.g. B and A. Consider the star $z = \{b_1, a_2, b_2\}$ where a_2 has the larger y-coordinate among $\{a_2, a_3\}$ and b_2 has the larger x-coordinate among $\{b_2, b_3\}$. As in case I the cost of z is not higher than the cost of a and b together. The same holds for the star $z = \{a_1, b_1, a_2\}$.

Let us now consider the case where the stars a and b have a common end c such that one of the external regions, say D, has expansion 0, because c would be the only point in D; now c lies at the border of D and A. Then double the point c such that both A and D can be regarded as regions with one point (c_1 or c_2 respectively). Now we can apply the constructions above. $\Diamond$

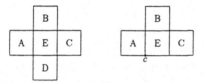

Corollary 1 *In the situation of Lemma 10 there is always a better star $\hat{\tau}_h$ that contains two points of the horizontal star, namely those that lie closely at the left and right of the intersecting vertical star. Analogously the existence of a better star $\hat{\tau}_v$ can be proven, containing the two corresponding points of the vertical star.*

Proof: Follows directly from the constructions in Lemma 10. $\Diamond$

Now we are able to prove

Lemma 11 *At the end of the evaluation phase all stars with positive gain that were refused because of non-planarity together have a gain of at most $\frac{2x}{m}$.*

Proof: Let τ_{i_2} be a star with positive actual gain that intersects an accepted star τ_{i_1}. Without loss of generality among all accepted stars producing an intersection with τ_{i_2} τ_{i_1} is the 'oldest', i.e. the earliest to be accepted. From Lemma 10 we know that there exists a star $\hat{\tau}$ which is not accepted and for which $a_{\hat{\tau}} \geq a_{\tau_{i_1}} + a_{\tau_{i_2}}$ holds. There can be four reasons why $\hat{\tau}$ was not accepted:

a) $\hat{\tau}$ was treated before τ_{i_1} but was not planar. From the construction of $\hat{\tau}$ follows that one of the stars τ_{i_1} or τ_{i_2} is not planar in this case, too. This star cannot be τ_{i_1}, because it is already accepted, therefore it has to be τ_{i_2}. But this means that τ_{i_2} intersects another star that was treated before $\hat{\tau}$ which contradicts the conditions of τ_{i_1}.

b) $\hat{\tau}$ was treated before τ_{i_1} but uses an artificial edge. $\hat{\tau}$ would be accepted using an artificial edge but refused when not using artificial edges. That means that the loss originating from the discard of $\hat{\tau}$ is already counted earlier (cf. Lemma 9), even if τ_{i_1} would be refused too.

c) $\hat{\tau}$ was treated before τ_{i_1} but its actual gain $a_{\hat{\tau}}$ was too small for an acceptation. In this case $a_{\tau_{i_1}}$ would also be too small and we have a contradiction to the acceptation of τ_{i_1}.

d) $\dot\tau$ was treated after τ_{i_1}, $\Rightarrow g_{\dot\tau} \le g_{\tau_{i_1}}$. Now we have

$$a_{\tau_{i_1}} + a_{\tau_{i_2}} \le a_{\dot\tau} \le g_{\dot\tau} \le g_{\tau_{i_1}} \le a_{\tau_{i_1}} \frac{m+1}{m}.$$

$$\Rightarrow a_{\tau_{i_2}} \le a_{\tau_{i_1}} \frac{m+1}{m} - a_{\tau_{i_1}} = a_{\tau_{i_1}} \frac{1}{m}.$$

So we lose very little.

The final question here is, how many stars we lose that are all intersecting the same star τ_{i_1}. For the notation see the figure below. Let $\hat S_l$ and $\hat S_r$ be the better stars instead of H and V_t, or H and V_{t+1}, each containing the points a and b, or b and c respectively (the existence of the stars $\hat S_l$ and $\hat S_r$ follows from Corollary 1). With similar arguments as in the case of one intersection holds: $\hat S_l$ and $\hat S_r$ were treated *after* H, therefore $g_{\hat S_l} \le g_H$ and $g_{\hat S_r} \le g_H$. $\hat S_l$ is better than H and V_t together and intersects V_{t-1}; $\Rightarrow$ there is a star $\hat S'_l$ being better than $\hat S_l$ and V_{t-1} together, therefore better than H, V_t and V_{t-1} together and $\hat S'_l$ intersects V_{t-2} and so on. The same construction is valid for the right side, i.e. the stars $V_{t+1}, \ldots, V_s$. Thus there are stars S^*_l and S^*_r with:

$$a_{S^*_l} + a_{S^*_r} \ge 2a_H + \sum_{i=1}^s a_{V_i}$$

And now, similar to the inequation above:

$$2a_H + \sum_{i=1}^s a_{V_i} \le a_{\hat S_l} + a_{\hat S_r} \le g_{\hat S_l} + g_{\hat S_r} \le 2g_H \le 2a_H \frac{m+1}{m}$$

$$\Rightarrow \sum_{i=1}^s a_{V_i} \le \frac{2a_H}{m}$$

Thus by discarding a family of non-planar stars we lose at most $\frac{2}{m}$ gain of a star already accepted. This completes the proof of Lemma 11. $\Diamond$

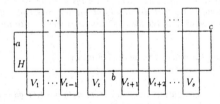

We conclude that the total gain g that could be realized is thus at least

$$\frac{smt_2(S) - smt_3(S)}{2} \left(1 - \left(\frac{1}{36^{r-2}} + \frac{2}{m} + \frac{2}{m}\right)\right).$$

This leads to a performance ratio of

$$\frac{3}{2} - \frac{\frac{3}{2} - \frac{5}{4}}{2} \left(1 - \frac{4}{m} - \frac{1}{36^{r-2}}\right) =$$

$$= \frac{11}{8} + \frac{1}{2m} + \frac{1}{8 \cdot 36^{r-2}}. \quad \Diamond$$

5.2 Running Time

Here we want to analyse the algorithm to prove the time bound of $O(r \cdot m \cdot n \cdot \log^2 n)$. As shown in Section 4 line 1 can be executed in time $O(n \cdot \log^2 n)$.

Lines 3,4 and 5 obviously need $O(n \cdot \log n)$ time for sorting $O(n)$ objects.

We cannot keep j small ($j \in \Theta(n)$); but since we ensure that the stars do not intersect (line 8) we can apply the methods described in [5], such that the recomputings of the gains can be done in time $O(\log n)$. Thus lines 7 - 12 can be executed in time $O(\log n)$. This is the reason why we ask the stars to be planar in line 8.

Totally, the algorithm needs $O(r \cdot m \cdot n \cdot \log^2 n)$ time which is $O(n \cdot \log^2 n)$ if we keep the parameters r and m constant. If we choose $r = \log \log n$ and $m = \frac{\log n}{\log \log n}$, we yield an $O(n \cdot \log^3 n)$ algorithm with a performance ratio of $\frac{11}{8} + O(\frac{\log \log n}{\log n})$ which is asymptotically equal to $\frac{11}{8}$.

6 Conclusion

In this paper refined versions of the new approximation algorithms for rectilinear Steiner trees of Zelikovsky and Berman/Ramaiyer were presented. Aiming for faster algorithms, not for better approximations, we improved the previous best time bound of $O(n^{5/2})$ considerably. For the 11/8-approximation we showed a running time of $n^{3/2}$. An alternative algorithm for a parameterized approximation gave an $11/8 + \varepsilon$-approximation for any $\varepsilon > 0$ within time $O(n \cdot \log^2 n)$, which is close to optimal. Allowing time $O(n \cdot \log^3 n)$, this algorithm provided an $11/8 + \log \log n / \log n$ approximation. To improve the running time and the approximation further, new methods seem to be necessary.

References

1. P. Berman. Personal communication. 1992.
2. P. Berman, V. Ramaiyer. *Improved approximations for the Steiner tree problem*. Proceedings, 3rd ACM-SIAM Symposium on Discrete Algorithms, 325-334, 1992.
3. M.W.Bern, R.L.Graham. *The shortest-network problem*. Sci.American,1, 66-71, 1989.
4. Ding-Zhu Du, Yanjun Zhang, Qing Feng. *On better heuristic for Euclidean Steiner minimum trees*. Proceedings, 32nd Foundations of Computer Science, 431-439, 1991.
5. D. Eppstein, G.F. Italiano, et al. *Maintenance of a Minimum Spanning Forest in a Dynamic Planar Graph*. Proceedings of 1st ACM-SIAM Symposium on Discrete Algorithms, 1-11, 1990.
6. G. Frederickson. *Data structures for On-Line Updating of Minimum Spanning Trees, with Applications*. SIAM J.Comp.,14, 781-789, 1985.
7. M.R.Garey, D.S.Johnson. *The rectilinear Steiner problem is NP-complete*. SIAM J.Appl.Math.,32, 826-834, 1977.
8. E.N.Gilbert, H.O.Pollak. *Steiner minimal trees*. SIAM J.Appl.Math.,16, 1-29, 1968.
9. M.Hanan. *On Steiner's Problem with Rectilinear Distance*. SIAM J.Appl.Math.,14, 255-265, 1966.
10. F.K.Hwang. *On Steiner minimal trees with rectilinear distance*. SIAM J.Appl.Math.,30, 104-114, 1976.
11. R. M. Karp. *Reducibility among combinatorial problems*. In Miller and Thatcher (eds.), Complexity of Computer Computations, Plenum Press, New York 85-103, 1972.
12. B.Korte, H.J.Prömel, A. Steger. *Steiner Trees in VLSI-Layout*. In Korte et al.: Paths, Flows and VLSI-Layout, Springer, 1990.
13. Th.Lengauer. *Combinatorial Algorithms for Integrated Circuit Layout*. John Wiley, 1990.
14. K. Mehlhorn. *Data Structures and Algorithms III: Multidimensional Data Structures and Comp. Geo.* Springer, 1985.
15. D.Richards. *Fast heuristic algorithms for rectilinear Steiner trees*. Algorithmica, 4, 191-207, 1989.
16. H. Takahashi, A. Matsuyama. *An approximate solution for the Steiner problem in graphs*. Math. Japonica, 24: 573-577, 1980.
17. Y.F.Wu, P.Widmayer, C.K.Wong. *A faster approximation algorithm for the Steiner problem in graphs*. Acta Informatica 23, 223-229, 1986.
18. A.C.Yao. *On constructing minimum spanning trees in $k-$dimensional space and related problems*. SIAM J. Comput., 11, 721 - 736, 1982.
19. A. Z. Zelikovsky. *An 11/6-approximation algorithm for the Steiner problem on graphs*. Algorithmica 9, 463-470, 1993.
20. A. Z. Zelikovsky. *The 11/8-approximation algorithm for the Steiner problem on Networks with rectilinear distance*. In Sets, Graphs and Numbers. Coll. Math. Soc. J. Bolyai 60, 733-745, 1992.

Springer-Verlag
and the Environment

We at Springer-Verlag firmly believe that an international science publisher has a special obligation to the environment, and our corporate policies consistently reflect this conviction.

We also expect our business partners – paper mills, printers, packaging manufacturers, etc. – to commit themselves to using environmentally friendly materials and production processes.

The paper in this book is made from low- or no-chlorine pulp and is acid free, in conformance with international standards for paper permanency.

Lecture Notes in Computer Science

For information about Vols. 1–685
please contact your bookseller or Springer-Verlag

Vol. 722: A. Miola (Ed.), Design and Implementation of Symbolic Computation Systems. Proceedings, 1993. XII, 384 pages. 1993.

Vol. 723: N. Aussenac, G. Boy, B. Gaines, M. Linster, J.-G. Ganascia, Y. Kodratoff (Eds.), Knowledge Acquisition for Knowledge-Based Systems. Proceedings, 1993. XIII, 446 pages. 1993. (Subseries LNAI).

Vol. 724: P. Cousot, M. Falaschi, G. Filè, A. Rauzy (Eds.), Static Analysis. Proceedings, 1993. IX, 283 pages. 1993.

Vol. 725: A. Schiper (Ed.), Distributed Algorithms. Proceedings, 1993. VIII, 325 pages. 1993.

Vol. 726: T. Lengauer (Ed.), Algorithms – ESA '93. Proceedings, 1993. IX, 419 pages. 1993

Vol. 727: M. Filgueiras, L. Damas (Eds.), Progress in Artificial Intelligence. Proceedings, 1993. X, 362 pages. 1993. (Subseries LNAI).

Vol. 728: P. Torasso (Ed.), Advances in Artificial Intelligence. Proceedings, 1993. XI, 336 pages. 1993. (Subseries LNAI).

Vol. 729: L. Donatiello, R. Nelson (Eds.), Performance Evaluation of Computer and Communication Systems. Proceedings, 1993. VIII, 675 pages. 1993.

Vol. 730: D. B. Lomet (Ed.), Foundations of Data Organization and Algorithms. Proceedings, 1993. XII, 412 pages. 1993.

Vol. 731: A. Schill (Ed.), DCE – The OSF Distributed Computing Environment. Proceedings, 1993. VIII, 285 pages. 1993.

Vol. 732: A. Bode, M. Dal Cin (Eds.), Parallel Computer Architectures. IX, 311 pages. 1993.

Vol. 733: Th. Grechenig, M. Tscheligi (Eds.), Human Computer Interaction. Proceedings, 1993. XIV, 450 pages. 1993.

Vol. 734: J. Volkert (Ed.), Parallel Computation. Proceedings, 1993. VIII, 248 pages. 1993.

Vol. 735: D. Bjørner, M. Broy, I. V. Pottosin (Eds.), Formal Methods in Programming and Their Applications. Proceedings, 1993. IX, 434 pages. 1993.

Vol. 736: R. L. Grossman, A. Nerode, A. P. Ravn, H. Rischel (Eds.), Hybrid Systems. VIII, 474 pages. 1993.

Vol. 737: J. Calmet, J. A. Campbell (Eds.), Artificial Intelligence and Symbolic Mathematical Computing. Proceedings, 1992. VIII, 305 pages. 1993.

Vol. 738: M. Weber, M. Simons, Ch. Lafontaine, The Generic Development Language Deva. XI, 246 pages. 1993.

Vol. 739: H. Imai, R. L. Rivest, T. Matsumoto (Eds.), Advances in Cryptology – ASIACRYPT '91. X, 499 pages. 1993.

Vol. 740: E. F. Brickell (Ed.), Advances in Cryptology – CRYPTO '92. Proceedings, 1992. X, 593 pages. 1993.

Vol. 741: B. Preneel, R. Govaerts, J. Vandewalle (Eds.), Computer Security and Industrial Cryptography. Proceedings, 1991. VIII, 275 pages. 1993.

Vol. 742: S. Nishio, A. Yonezawa (Eds.), Object Technologies for Advanced Software. Proceedings, 1993. X, 543 pages. 1993.

Vol. 743: S. Doshita, K. Furukawa, K. P. Jantke, T. Nishida (Eds.), Algorithmic Learning Theory. Proceedings, 1992. X, 260 pages. 1993. (Subseries LNAI)

Vol. 744: K. P. Jantke, T. Yokomori, S. Kobayashi, E. Tomita (Eds.), Algorithmic Learning Theory. Proceedings, 1993. XI, 423 pages. 1993. (Subseries LNAI)

Vol. 745: V. Roberto (Ed.), Intelligent Perceptual Systems. VIII, 378 pages. 1993. (Subseries LNAI)

Vol. 746: A. S. Tanguiane, Artificial Perception and Music Recognition. XV, 210 pages. 1993. (Subseries LNAI).

Vol. 747: M. Clarke, R. Kruse, S. Moral (Eds.), Symbolic and Quantitative Approaches to Reasoning and Uncertainty. Proceedings, 1993. X, 390 pages. 1993.

Vol. 748: R. H. Halstead Jr., T. Ito (Eds.), Parallel Symbolic Computing: Languages, Systems, and Applications. Proceedings, 1992. X, 419 pages. 1993.

Vol. 749: P. A. Fritzson (Ed.), Automated and Algorithmic Debugging. Proceedings, 1993. VIII, 369 pages. 1993.

Vol. 750: J. L. Díaz-Herrera (Ed.), Software Engineering Education. Proceedings, 1994. XII, 601 pages. 1994.

Vol. 751: B. Jähne, Spatio-Temporal Image Processing. XII, 208 pages. 1993.

Vol. 752: T. W. Finin, C. K. Nicholas, Y. Yesha (Eds.), Information and Knowledge Management. Proceedings, 1992. VII, 142 pages. 1993.

Vol. 753: L. J. Bass, J. Gornostaev, C. Unger (Eds.), Human-Computer Interaction. Proceedings, 1993. X, 388 pages. 1993.

Vol. 754: H. D. Pfeiffer, T. E. Nagle (Eds.), Conceptual Structures: Theory and Implementation. Proceedings, 1992. IX, 327 pages. 1993. (Subseries LNAI).

Vol. 755: B. Möller, H. Partsch, S. Schuman (Eds.), Formal Program Development. Proceedings. VII, 371 pages. 1993.

Vol. 756: J. Pieprzyk, B. Sadeghiyan, Design of Hashing Algorithms. XV, 194 pages. 1993.

Vol. 757: U. Banerjee, D. Gelernter, A. Nicolau, D. Padua (Eds.), Languages and Compilers for Parallel Computing. Proceedings, 1992. X, 576 pages. 1993.

Vol. 758: M. Teillaud, Towards Dynamic Randomized Algorithms in Computational Geometry. IX, 157 pages. 1993.

Vol. 759: N. R. Adam, B. K. Bhargava (Eds.), Advanced Database Systems. XV, 451 pages. 1993.

Vol. 760: S. Ceri, K. Tanaka, S. Tsur (Eds.), Deductive and Object-Oriented Databases. Proceedings, 1993. XII, 488 pages. 1993.

Vol. 761: R. K. Shyamasundar (Ed.), Foundations of Software Technology and Theoretical Computer Science. Proceedings, 1993. XIV, 456 pages. 1993.

Vol. 762: K. W. Ng, P. Raghavan, N. V. Balasubramanian, F. Y. L. Chin (Eds.), Algorithms and Computation. Proceedings, 1993. XIII, 542 pages. 1993.